# Welding Handbook

## Seventh Edition, Volume 4

# *Metals and Their Weldability*

# The Five Volumes of the Welding Handbook, Seventh Edition

1  Fundamentals of Welding

2  Welding Processes—

*Arc and Gas Welding and Cutting, Brazing, and Soldering*

3  Welding Processes—

*Resistance and Solid-State Welding and Other Joining Processes*

4  Metals and Their Weldability

5  Engineering, Costs, Quality, and Safety[a]

a. Scheduled to replace Sec. 5, 6th Ed., in 1984.

# Welding Handbook

## Seventh Edition, Volume 4

# Metals and Their Weldability

W. H. Kearns, Editor

AMERICAN WELDING SOCIETY
550 N.W. LeJeune Rd.
Miami, FL 33126

Library of Congress Number: 75-28707
International Standard Book Number: 0-87171-218-0

American Welding Society, 550 N.W. LeJeune Rd., Miami, FL 33126

The Welding Handbook is a collective effort of many volunteer technical specialists to provide information to assist with the design and application of welding and allied processes.

Reasonable care is taken in the compilation and publication of the Welding Handbook to insure authenticity of the contents. No representation or warranty is made as to the accuracy or reliability of this information.

The information contained in the Welding Handbook shall not be construed as a grant of any right of manufacture, sale, use, or reproduction in connection with any method, process, apparatus, product, composition or system, which is covered by patent, copyright, or trademark. Also, it shall not be construed as a defense against any liability for such infringement. No effort has been made to determine whether any information in the Handbook is covered by any patent, copyright or trademark, and no research has been conducted to determine whether an infringement would occur.

Printed in the United States of America

# Contents

# Welding Handbook Committee

# Preface

This is the fourth volume of the five volume series for the 7th Edition of the *Welding Handbook*. This volume updates and expands the information on the weldability, brazeability, and solderability of metals and their alloys. Information contained in Volumes 1, 2, and 3 supplements the material presented herein with respect to welding metallurgy, joining processes, testing methods, residual stresses, and distortion.

The order of presentation has changed somewhat from the corresponding volume of the 6th Edition. The information presented on many metals is more extensive. Chapters covering tool and die steels and commonly used, dissimilar metal combinations are now included. Carbon and low alloy steels, stainless steels, and high alloy steels (other than tool and die steels) are grouped in distinct chapters to show their similarities with respect to joining. The reactive metals, titanium, zirconium, hafnium, tantalum, and columbium, are grouped in a chapter for the same reason. Another chapter addresses the joining of nonstructural metals including beryllium, uranium, molybdenum, tungsten, lead, and the precious metals.

Some of the subjects covered in Section (Volume) 1 of the 6th Edition will be revised in Volume 5 of this Edition, when published. That volume should be retained until Volume 5 is available.

An index of major subjects precedes the volume index. It enables the reader to quickly locate information on a major subject in the current volumes of Welding Handbook, including Volumes (Sections) 1 and 5 of the 6th Edition.

This volume was a voluntary effort by the Welding Handbook Committee and the Chapter Committees. The Chapter Committee Members and the Handbook Committee Member responsible for a chapter are recognized on the title page of that chapter. Other individuals also contributed in a variety of ways, particularly in chapter reviews. All participants contributed generously of their time and talent, and the American Welding Society expresses herewith its appreciation to them and to their employers for supporting the work.

The Welding Handbook Committee expresses its appreciation to Richard French, Deborah Swain, Hallock Campbell and other AWS Staff Members for their assistance in the production of this volume.

The Welding Handbook Committee welcomes your comments on the Handbook. Please address them to the Editor, Welding Handbook, American Welding Society, P.O. Box 351040, Miami, FL 33135.

J. R. Condra, *Chairman*  
*Welding Handbook Committee*

W. H. Kearns, *Editor*  
*Welding Handbook*

# 1

# Carbon and Low Alloy Steels

## Chapter Committee

C.W. OTT, *Chairman*
    *U.S. Steel Corporation*
M.F. COUCH
    *Bethlehem Steel Corporation*
V.J. HUDACKO
    *The Babcock and Wilcox Company*

R.P. HURLEBAUS
    *The Budd Company*
F.A. WASSELL
    *Armco, Incorporated*
C.R. ZIMMERMAN
    *Alloy Rods Division*

## Welding Handbook Committee Member

L.J. PRIVOZNIK
    *Westinghouse Electric Corporation*

# 1

# Carbon and Low Alloy Steels

## WELDING CLASSIFICATION

From a weldability standpoint, carbon and low alloy steels can be divided into six general classes or groups according to composition, strength, heat treatment, or corrosion protection. There will be some overlap in these groups because some steels are used in more than one heat-treated condition. The groups are as follows:

(1) Carbon steels
(2) High-strength low alloy steels
(3) Quenched-and-tempered steels
(4) Heat-treatable low alloy steels
(5) Chromium-molybdenum steels
(6) Precoated steels

Carbon steels contain carbon up to about 1.00 percent, manganese up to about 1.65 percent, and silicon up to about 0.60 percent. Carbon steels are normally used in the as-rolled condition, although some may be in the annealed or normalized condition.

High-strength low alloy steels are designed to provide better mechanical properties than conventional carbon steels. They are generally classified according to mechanical properties rather than chemical compositions. Their yield strengths generally fall within the range of 42 to 70 ksi. These steels are usually welded in the as-rolled or normalized condition.

Quenched-and-tempered steels are a group of carbon and low alloy steels that are generally heat treated by the producer to provide yield strengths in the range of 50 to 150 ksi. In addition, they are designed to be welded in the heat treated condition. Normally, the weldments are not postweld heat treated except for a stress relieving operation to reduce welding or cold-forming stresses. Steels in this group include

carbon and alloy steels designed for welded construction.

Heat-treatable low alloy steels are, from a weldability standpoint, alloy steels that will not retain desirable properties after welding. They must be given a postweld heat treatment to obtain the best combination of properties offered by the steel. These steels generally have higher carbon contents than high-strength low alloy or quenched-and-tempered steels. Consequently, the steels are capable of higher mechanical properties but suffer from a lack of toughness.

Chromium-molybdenum steels are primarily used for service at elevated temperatures up to about 1300°F in applications such as power plants and petroleum refineries. They are welded in various heat-treated conditions: annealed, normalized and tempered, or quenched and tempered. Welded joints are often heat treated prior to use to improve ductility and toughness and reduce welding stresses.

Precoated steels have a thin coating of aluminum, zinc, or a zinc-rich primer to provide enhanced resistance to atmospheric corrosion or elevated temperature oxidation. Metal coatings may be applied by hot dipping, electrodeposition, or thermal spraying, depending upon the type. Zinc-rich primers are applied by roller coating or spray painting. Many applications involve sheet and strip, but numerous structural applications involve zinc-coated (galvanized) plate and structural sections. In all cases, the steel is coated prior to welding.

Steels in the six groups are available in various product forms including sheet, strip, plate, structural shapes, pipe, tubing, forgings, and castings. Regardless of the product form, the composition, mechanical properties, and condition of heat treatment must be known to establish satisfactory welding procedures. Although most steels are used in wrought form the same considerations apply to forgings and castings since weldability is primarily a function of composition, properties, and heat treatment. However, with large forgings and castings, consideration should be given to the effect of size or thickness with respect to heat input, cooling rate, and restraint. Also, the effects of residual elements in a casting, which may not be present in wrought steels, and localized variations in composition must also be taken into consideration.

# GENERAL CONSIDERATIONS

## HYDROGEN INDUCED CRACKING

One problem that may be encountered when welding many steels under certain conditions is hydrogen induced cracking.[1] Such cracking is known by various other names, such as underbead, cold, or delayed cracking. It generally occurs at some temperature below 200°F immediately upon cooling or after a period of several hours, the time depending upon the type of steel, the magnitude of the welding stresses, and the hydrogen content of the steel weld and heat-affected zones. Delayed cracking normally takes place after the weldment has cooled to ambient temperature. In any case, it is cold cracking as opposed to hot cracking, and is caused by dissolved hydrogen entrapped in small voids or dislocations in the weld metal or heat-affected zone. Sometimes, the weld metal may crack, although this seldom occurs when its yield strength is below about 90 ksi. Diffusion of hydrogen into the heat-affected zone from the weld metal during welding contributes to cracking in this zone. Microstructure of the steel is also a contributing factor.

### Contributing Conditions

Hydrogen-induced cracking in welded joints may be caused by stresses developed from transformation hardening or by excessive stresses imposed upon the joint. Such cracking is associated with the combined presence of three conditions, namely:

(1) The presence of hydrogen in the steel.

(2) A microstructure that is partly or wholly martensitic.

(3) A tensile stress at the sensitive location.

Hydrogen-induced cracking will not take place if one or more of these conditions is absent or at a low level.

### Hydrogen Sources

Hydrogen is generally introduced during welding. The source may be the filler metal, moisture in the electrode covering, welding flux, or shielding gas, or a contaminant on the filler or base metal. The filler wire or rod may be contaminated with dissolved hydrogen during processing.

Molten steel has a high solubility for atomic hydrogen that may be formed by the dissociation of water vapor or a hydrocarbon in the welding arc. The diffusion rate of atomic hydrogen in steel at or near its melting temperature is high. Therefore, the molten weld metal can rapidly pick up atomic hydrogen from the hot gas in the arc. Hydrogen atoms can diffuse rapidly from the weld metal into the heat-affected zone of the base metal.

During cooling and phase transformation, the absorbed hydrogen is rejected from the iron and tends to concentrate at microstructural dislocations and voids in the matrix. This tends to distort the iron and induce localized tensile stresses that may accumulate.

---

1. For additional information, refer to the *Welding Handbook*, Volume 1, 7th Edition, 1976: 122-42, and Linnert, G.E., *Welding Metallurgy*, Volume 2, 3rd Edition, 1967: 250-76, published by the American Welding Society.

## Microstructure

Cracking is most likely to be promoted by hydrogen when the steel has a martensitic microstructure. With this microstructure and a quantity of hydrogen present, a tensile stress much lower than the normal cohesive strength of the metal can initiate a crack. Possible sources of stress are phase transformation, thermal contraction, mechanical restraint, or applied loads. In general, the stress required to produce a crack in steel is progressively lower as the hydrogen content increases. The susceptibility of martensite to hydrogen-induced cracking is believed to be partly due to high local transformation stresses.

A bainitic microstructure in steel displays a markedly lower susceptibility to cracking from hydrogen. The local stresses are significantly lower in bainite even though it will have a hardness approaching that of any martensite in the microstructure.

A mixture of ferrite and high-carbon martensite or bainite is also quite susceptible to hydrogen embrittlement. This microstructure is produced during cooling from austenite at a slightly faster rate than the critical cooling rate for the steel. Therefore, any localized area of such a mixed microstructure is likely to display cracking. Because there is a range of cooling rates in a weld heat-affected zone, there is certain to be a narrow region in the zone where a crack-sensitive structure is produced.

Susceptibility to cracking can be reduced by minimizing the formation of martensite in the weld metal and heat-affected zone. This can be done by controlling the cooling rate of a weld. Cooling rate depends upon section thickness, preheat temperature, welding heat input, and postweld thermal treatment. With some steels, however, a change in welding procedures that reduces the amount of martensite in the microstructure may result in a detrimental change in certain mechanical properties of the welded joint.

## Stresses

Tensile stresses resulting from martensitic transformation can be intensified by additional stresses caused by (1) thermal contraction of the weld metal and heat-affected zone, (2) restraint on the welded joint imposed by the weldment design, or (3) the fabrication sequence. These stresses may be reduced by preheating to reduce the cooling rate, adjusting the welding procedure to reduce heat input, or redesigning the weldment or fabrication sequence to reduce restraint on the joint.

## Underbead Cracking

The greatest problem with hydrogen-induced cracking arises in the heat-affected zone of hardenable steel. Cracking in this particular location is called *underbead cracking* when it is present a short distance from the weld interface, as illustrated in Fig. 1.1. Sometimes, the crack will initiate in the heat-affected zone but follow some other path as it propagates through unaffected base metal.

As mentioned earlier, a mixed structure of ferrite in a matrix of martensite or bainite seems to display a high susceptibility to hydrogen embrittlement and cracking. A structure of this kind is likely to form in the heat-affected zone at a greater distance from the fusion line as the hardenability of the steel increases.

## Weld Metal Cracking

Weld metal normally presents fewer problems with hydrogen-induced cracking than the base metal. This is probably a result of the general use of a filler metal with a lower carbon content. However, hydrogen can embrittle the weld metal to a significant extent.

*Fig. 1.1—Underbead crack in a low alloy steel bead-on-plate weld (10% Nital etch) ( × 8)*

One form of hydrogen-induced cracking that occurs in weld metal appears as small bright spots on the fractured faces of broken specimens of weld metal. These spots are called *fisheyes.* The fisheye usually surrounds some discontinuity in the metal, such as a gas pocket or a non-metallic inclusion, which gives the appearance of a "pupil in an eye."

The conditions that lead to the formation of fisheyes in weld metal can be minimized by using dry low hydrogen electrodes or by heating the weldment for some period at a temperature in the range of 200° to 1300°F. Longer times are required with lower temperatures.

Microcracks may be observed in weld metal deposited with electrodes containing cellulose in the covering or low hydrogen electrodes with excessive moisture in the covering. These microcracks are generally oriented transverse to the axis of the weld. They should not be present in the weld metal deposited with dry low hydrogen electrodes.

## Avoiding Cracking

The tolerance of steels for hydrogen decreases as the carbon or alloy content is increased to achieve higher strength because the hardness of the martensite in the microstructure increases also. Welding stresses tend to increase at the same time. For example, some steels containing less than about 0.15 percent carbon, such as low carbon steel, may be welded with E6010 or E6011 covered electrodes. These electrodes are characteristically high in hydrogen because the coverings contain cellulose and 3 to 7 percent moisture. On the other hand, high-strength low alloy steels, such as HY-130,[2] must be welded with covered electrodes that contain no more than 0.1 percent moisture in the covering. Moisture or hydrogen limitations for covered electrodes vary between these two levels, depending upon the steel being welded.

Hydrogen-induced cracking can be controlled using (1) a welding process or an electrode that produces little or no hydrogen, (2) a combination of welding and thermal treatments that drive off the hydrogen or produce a microstructure that is insensitive to it, or (3) welding procedures that result in low welding stresses.

*Welding Process.* Available hydrogen during welding can be limited by using a "low hydrogen" process. Primary sources of hydrogen are cellulose, moisture, or both, in the electrode covering with shielded metal arc welding, moisture in submerged arc welding flux, and moisture in core ingredients of flux or metal cored electrodes. Other sources of hydrogen are adsorbed moisture, rust, or hydrocarbon contaminants (oil, grease, drawing lubricants, or paint) on the filler metal or the base metal. Shielding gas contaminated with humid air or moisture is another source of hydrogen.

The use of low hydrogen electrodes is recommended for shielded metal arc welding of crack-sensitive steels. However, the moisture content of the electrodes must be maintained within specification limits for welding a specific steel.

Electrodes are manufactured to be within acceptable moisture limits, consistent with the type of covering and strength of the weld metal. Low hydrogen electrodes are packaged in a container that provides the moisture protection necessary for the type of covering and the application.

Such electrodes can be maintained for many months in the containers when stored at room temperature with the relative humidity at 50 percent or less, or in electrode holding ovens for short times. However, the coverings may absorb excessive moisture if the containers are damaged or the electrodes are improperly stored.

The low hydrogen (EXX15 and EXX16) and low hydrogen, iron powder (EXX18, EXX28 and EXX48) electrodes are designed and developed to contain the minimum amount of moisture in their coverings. To maintain this low moisture level in the covering, hermetically sealed containers are mandatory for electrodes that deposit weld metal with a tensile strength of 80 ksi or higher. Such containers are optional for electrodes of lower strength classifications.[3] Electrodes that have been exposed to a humid atmosphere for an extended time may absorb excessive moisture. The moisture content of electrodes that have been exposed to the atmosphere should not exceed the limits given in the appropriate specification. If there is a possibility that the electrodes have picked up excessive moisture, they

---

2. Refer to Table 1.15 for the chemical composition of HY-130 alloy steel.

3. Refer to AWS A5.1-81, *Specification for Carbon Steel Covered Arc Welding Electrodes,* or AWS A5.5-81, *Specification for Low Alloy Steel Covered Arc Welding Electrodes.*

may be reconditioned by rebaking. Some electrodes require rebaking at a temperature as high as 800°F for approximately 2 hours. The proper temperature and time for reconditioning specific electrodes are determined by the time of exposure and the relative humidity and temperature conditions. The appropriate time and temperature for this reconditioning should be requested from the electrode manufacturer.

Current specifications for submerged arc, gas metal arc, and flux-cored arc welding electrodes do not specify limits on moisture or hydrogen content of the electrodes or flux. The welding code may place limits on them, and may specify the required conditions for baking submerged arc welding flux.[4] Flux-cored electrodes that have been contaminated with moisture or any other hydrogen-containing substance should not be used to weld steels that are sensitive to hydrogen-induced cracking.

*Thermal Treatments.* Preheating and postheating should be considered when there is danger of hydrogen-induced cracking in the welded joint.[5] Postheating should be done immediately after welding without intermediate cooling. The postheat temperature may be that used for preheating (200° to 600°F). The holding time at postheat temperature depends upon the joint thickness because the length of the path over which the hydrogen must diffuse to the surface is a controlling factor. Raising the temperature modestly will markedly reduce the time needed for removal of hydrogen. Heating at about 375°F for 24 hours is considered optimum for reducing the average hydrogen concentration to a safe level.

Since the solubility for hydrogen increases rapidly with rising temperature a postheat temperature must be a compromise between increased diffusion rate and greater solubility at elevated temperatures. Preheating and postheating temperatures also influence cracking indirectly because of their effect on cooling rate and resulting microstructure. Preheating can increase

---

4. AWS D1.1-81, *Structural Welding Code–Steel*, requires that flux be baked at 250°F min. for 1 hour if the packaging had been damaged. The top 1-inch thick layer of exposed flux in hoppers and wet flux must be discarded. These procedures should be followed for all applications.

5. When preheating, the metal must be heated to the appropriate temperature throughout the section thickness without intermediate cooling prior to or during welding, unless precautions are taken to avoid cracking.

the rate of hydrogen diffusion from the weld area, temper any martensite that forms, and decrease the cooling rate of the weld. The latter effect produces weld microstructures, other than martensite, that have lower susceptibility to cracking. For this reason, the engineer may specify a preheat temperature above 375°F when hydrogen solubility and diffusion rate are the only considerations.

Weldments of steels that are quenched and tempered to achieve desired properties require special treatment. They must be either welded with a low hydrogen process or postheated, as described previously, prior to the hardening heat treatment.

When a specific preheat temperature is not required, it is best to have the steel at a temperature above 32°F for welding to avoid cracking. If the steel temperature is below 32°F, it should be heated to at least 60°F before welding is started. Under humid conditions, the steel should be heated to a higher temperature to drive adsorbed moisture from its surfaces. This is especially true for steels that are sensitive to hydrogen-induced cracking.

# INTERRUPTION OF HEATING CYCLE

Where preheating, heating during welding, and postheating are employed in a welding procedure, there is sometimes a question as to whether the weldment should be allowed to cool to room temperature during or after welding but before final heat treatment. The effects of interrupting the heating cycle are both metallurgical and mechanical in nature. Metallurgical effects involve microstructural changes. The mechanical effects involve thermal contraction in the weldment that may produce localized distortion or high residual stresses. Accordingly, the greatest assurance of successful welding requires the use of continuous heating without interruption and postweld heat treatment immediately after completion of welding.

However, there may be operational or economical reasons for not carrying out a continuous heating procedure. Interrupted operations are necessary in some cases and quite common in many applications. It is difficult to make general rules for when interruptions are permissible because there are many factors that must be considered, some of which are discussed below.

## Ambient Air Temperature

With low ambient temperatures, it is good

practice to preheat prior to welding. Preheat temperature should be maintained during welding to avoid low toughness in the weld and heat-affected zones. This procedure will also reduce the possibility of hydrogen-induced cracking.

### Hydrogen-Forming Contaminants

Hydrogen can be introduced into the arc from water vapor and other materials in the coverings of certain types of welding electrodes, as discussed previously. If hydrogen is absorbed in the metal, it may produce high internal stresses and cracking when the weldment is cooled. Interrupting the heating cycle is safer when low hydrogen welding procedures are used. Even then, post heating the weld area for a few hours at approximately 400°F diffuses hydrogen away from the heat affected zone and permits cooling the weldment without risk of delayed cracking.

### Joint Restraint

Interruptions in heating are less desirable if a partially completed weld will be subjected to bending moments or to high restraint when cooled. All welding and postweld heat treatment should be completed before a weldment is exposed to any type of loading.

### Hardenability

Once welding has started, heating of steels with high hardenability should not be interrupted unless appropriate steps are taken to avoid cracking. These procedures are discussed later for the various types of hardenable steels.

### Base Metal Thickness

An increase in section thickness increases both the restraint on the weld and the cooling rate from welding temperatures of both the heat-affected zone and the weld metal. Accordingly, the weld area is subjected to high residual stresses.

For a given joint thickness, the greater the percentage of the weld completed, the more nearly the strength and rigidity of the joint approach those of a completed weld. Also, the unit stresses in the deposited weld metal will be lower. For this reason, an interruption of welding is generally not permitted until some minimum amount of weld metal or a given fraction of the joint thickness has been welded.

Wherever interruptions are permitted, the weldment must be cooled slowly and uniformly.

Welding should not resume until the weld area has been reheated uniformly to the specified preheat temperature.

## LAMELLAR TEARING

Groove welds, fillet welds, or combinations of these in corner or T-joints can result in high welding stresses in the base metal adjacent to the weld metal. High tensile stresses can develop perpendicular to the mid-plane of the steel plate (through-thickness direction) as well as parallel to it. The magnitude of these stresses depends upon the size of the weld, the welding procedures, and the restraint imposed on the joint by the weldment design.

When welding steel plate and structural shapes produced by conventional steel-making processes, any high internal tensile stresses developed perpendicular to the surface may cause tearing within the steel. The tears may or may not propagate to exposed surfaces. This tearing is usually associated with inclusions in the steel, and it progresses from one inclusion to another. On an etched cross section through the steel, lamellar tearing is characterized by a step-like or jagged crack with each step nearly parallel to the mid-plane of the plate. Figure 1.2 illustrates typical joint designs where lamellar tearing may take place, and its location with respect to the weld.

There is some evidence that sensitivity to lamellar tearing is increased by the presence of hydrogen in the steel. Various types of inclusions also increase the sensitivity of steel to lamellar tearing. Some manufacturers offer steel that is specially produced to reduce the sensitivity to this behavior. The following are some approaches to minimize lamellar tearing:

(1) Change the location and design of the welded joint to minimize through-thickness strains.

(2) Use a lower strength weld metal.

(3) Reduce the available hydrogen.

(4) Butter the surface of the plate with weld metal prior to making the weld.

(5) Use preheat and interpass temperatures of at least 200°F.

(6) Peen the weld beads.

(7) Use steel specially processed to have improved through-thickness properties.

The most reliable method of avoiding lamellar tearing is to use specially processed mill products. It is difficult to detect subsurface lamellar tearing that would be detrimental in critical applications.

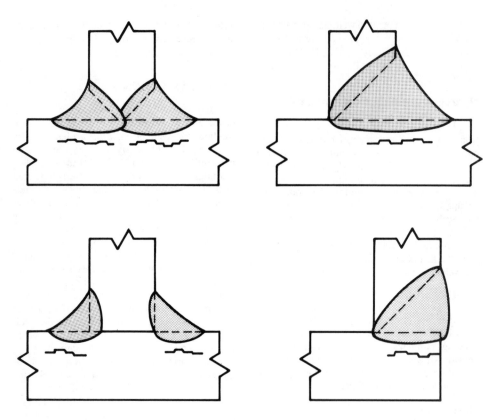

*Fig. 1.2—Weld joint designs in steel plate that are prone to lamellar tearing and the likely location of tears*

# CARBON STEELS

## DESCRIPTION

Carbon steels are alloys of iron and carbon in which carbon does not usually exceed 1 percent, manganese does not exceed 1.65 percent, and copper and silicon each do not exceed 0.60 percent. Other alloying elements are normally not present in more than residual amounts. The properties and weldability of these steels depend mainly on carbon content. Other elements have only a limited effect.

Carbon steel can be classified, according to various deoxidation practices, as rimmed, capped, semikilled, or killed steel. Deoxidation practice

and the steel making process will have an effect upon the characterisitics and properties of the steel. Variations in carbon have the greatest effect on mechanical properties with increasing carbon content leading to increased hardness and strength. Carbon steels are generally categorized according to their carbon content as indicated in Table 1.1.

AISI-SAE carbon steels are the 10XX, 11XX, 12XX, and 15XX groups. The 10XX group has a maximum of 1.0 percent manganese, but manganese in the 15XX group ranges from 1.00 to 1.65 percent. The 11XX resulfurized steels and the 12XX resulfurized and rephos-

**Table 1.1**
**Classification and weldability of carbon steels**

| Common name | Carbon content, percent | Typical hardness | Typical use | Weldability |
|---|---|---|---|---|
| Low carbon steel | 0.15 max | 60 HRB | Special plate and shapes, sheet, strip, welding electrodes | Excellent |
| Mild steel | 0.15-0.30 | 90 HRB | Structural shapes, plate, and bar | Good |
| Medium carbon steel | 0.30-0.50 | 25 HRC | Machine parts and tools | Fair (preheat and postheat normally required; low-hydrogen welding process recommended) |
| High carbon steel | 0.50-1.00 | 40 HRC | Springs, dies, railroad rail | Poor (low-hydrogen welding process, preheat, and postheat required) |

phorized steels are used for improved machinability. However, they are difficult to fusion weld because of their hot cracking tendencies.

ASTM Standards designate carbon steels on the basis of chemical or mechanical properties, or both, in conjunction with the product form and the intended application. Typical ASTM carbon steels used on construction, pressure vessels, and piping are listed in Table 1.2

## WELDABILITY CONSIDERATIONS

### Microstructure

Metal temperatures during welding may range from that of molten weld metal down to that of unaffected base metal. A continuous temperature gradient will exist between the two extremes. The microstructure and mechanical properties of a portion of the heated steel (heat-affected zone) will be changed as a result of welding. Such changes will depend upon the composition of the steel and the rate at which steel is heated and cooled.

With some steels, the thermal cycle may result in the formation of martensite in the weld metal and heat-affected zone.[6] The amount of

martensite formed and the hardness of the steel depend upon carbon content as well as the heating and cooling rates. Figure 1.3 gives the relationship between hardness and carbon content for steels that are 50 or 100 percent martensite after quenching. Martensitic transformation and the resulting high hardness can lead to cracking in the weld or heat-affected zone if the metal cannot yield to relieve welding stresses. The degree of hardening in the heat-affected zone is an important consideration determining the weldability of a carbon steel. Obviously, weldability generally decreases with increasing carbon or martensite in the weld metal or the heat-affected zone, or both.

### Carbon Equivalent

Although carbon is the most significant alloying element affecting weldability, the effects of other elements can be estimated by equating them to an equivalent amount of carbon. Thus, the effect of total alloy content can be expressed in terms of a carbon equivalency (CE). One empirical formula that may be used for judging the risk of underbead cracking in carbon steels is:

$$CE = \%C + \frac{\%Mn + \%Si}{4}$$

Figure 1.4 shows in a general way the relationships between carbon steel composition

6. Refer to the *Welding Handbook*, Volume 1, 7th Edition, 107-17, for additional information.

**Table 1.2**
**Composition and strength requirements of typical ASTM carbon steels**

| Application | ASTM Standard | Type or Grade | Typical composition limits, percent[a] | | | Tensile strength, ksi | Min. yield strength, ksi |
|---|---|---|---|---|---|---|---|
| | | | C | Mn | Si | | |
| *Structural steels* | | | | | | | |
| Welded buildings, bridges, and general structural purposes | A36 | | 0.29 | 0.80 – 1.20 | 0.15 – 0.40 | 58-80 | 36 |
| Welded buildings and general purposes | A529 | | 0.27 | 1.20 | – | 60-85 | 42 |
| General purpose sheet and strip | A570 | 30, 33, 36, 40, 45, 50 | 0.25 0.25 | 0.90 1.35 | – – | 49-55 60-65 | 30 45 |
| General purpose plate (improved toughness) | A573 | 58 | 0.23 | 0.60 – 0.90 | 0.10 – 0.35 | 58-71 | 32 |
| | | 65 | 0.26 | 0.85 – 1.20 | 0.15 – 0.40 | 65-77 | 35 |
| | | 70 | 0.28 | 0.85 – 1.20 | 0.15 – 0.40 | 70-90 | 42 |
| *Pressure vessel steels* | | | | | | | |
| Plate, low and intermediate tensile strength | A285 | A B C | 0.17 0.22 0.28 | 0.90 0.90 0.90 | – – – | 45-65 50-70 55-75 | 24 27 30 |
| Plate, manganese-silicon | A299 | | 0.30 | 0.90 – 1.40 | 0.15 – 0.40 | 75-95 | 40 |
| Plate, low temperature applications | A442 | 55 | 0.24 | 0.60 – 0.90 | 0.15 – 0.40 | 55-75 | 30 |
| | | 60 | 0.27 | 0.60 – 0.90 | 0.15 – 0.40 | 60-80 | 32 |
| Plate, intermediate and high temperature service | A515 | 55 | 0.28 | 0.90 | 0.15 – 0.40 | 55-75 | 30 |
| | | 60 | 0.31 | 0.90 | 0.15 – 0.40 | 60-80 | 32 |
| | | 65 | 0.33 | 0.90 | 0.15 – 0.40 | 65-85 | 35 |
| | | 70 | 0.35 | 1.20 | 0.15 – 0.40 | 70-90 | 38 |
| Plate, moderate and low temperature service | A516 | 55 | 0.26 | 0.60 – 1.20 | 0.15 – 0.40 | 55-75 | 30 |
| | | 60 | 0.27 | 0.85 – 1.20 | 0.15 – 0.40 | 60-80 | 32 |
| | | 65 | 0.29 | 0.85 – 1.20 | 0.15 – 0.40 | 65-85 | 35 |
| | | 70 | 0.31 | 0.85 – 1.20 | 0.15 – 0.40 | 70-90 | 38 |
| Plate, carbon-manganese-silicon heat-treated | A537 | 1[b] | 0.24 | 0.70 – 1.60 | 0.15 – 0.50 | 65-90 | 45 |
| | | 2[c] | | | | 75-100 | 55 |

**Table 1.2 (cont.)**
**Composition and strength requirements of typical ASTM carbon steels**

| Application | ASTM Standards | Type or Grade | Typical composition limits, percent[a] | | | Tensile strength, ksi | Min. yield strength, ksi |
|---|---|---|---|---|---|---|---|
| | | | C | Mn | Si | | |
| *Piping and tubing* | | | | | | | |
| Welded and seamless pipe, black and galvanized | A53 | A | 0.25 | 0.95 – 1.20 | – | 48 min. | 30 |
| | | B | 0.30 | 0.95 – 1.20 | – | 60 min. | 35 |
| Seamless pipe for high temperature service | A106 | A | 0.25 | 0.27 – 0.93 | 0.10 min. | 48 min. | 30 |
| | | B | 0.30 | 0.29 – 1.06 | 0.10 min. | 60 min. | 35 |
| | | C | 0.35 | 0.29 – 1.06 | 0.10 min. | 70 min. | 40 |
| Structural tubing | A501 | | 0.26 | – | – | 58 min. | 36 |
| *Cast steels* | | | | | | | |
| General use | A27 | 60-30 | 0.30 | 0.60 | 0.80 | 60 min. | 30 |
| Valves and fittings for high temperature service | A216 | WCA | 0.25 | 0.70 | 0.60 | 60-85 | 30 |
| | | WCB | 0.30 | 1.00 | 0.60 | 70-95 | 36 |
| | | WCC | 0.25 | 1.20 | 0.60 | 70-95 | 40 |
| Valves and fittings for low temperature service | A352 | LCA[c,d] | 0.25 | 0.70 | 0.60 | 60-85 | 30 |
| | | LCB[c,d] | 0.30 | 1.00 | 0.60 | 65-90 | 35 |
| | | LCC[c,d] | 0.25 | 1.20 | 0.60 | 70-95 | 40 |

a. Single values are maximum unless otherwise noted.
b. Normalized condition.
c. Quenched-and-tempered condition.
d. Normalized-and-tempered condition.

(CE) and hardness, underbead cracking sensitivity, or weldability based upon slow-bend capability of notched-weld-bead test bars.

Generally, steels with low CE values have excellent weldability, but a steel with only 0.20 percent C and 1.60 percent Mn will have a CE of 0.60, which has a relatively high crack sensitivity of about 56 percent. Generally, the susceptibility to underbead cracking from hydrogen increases when the CE exceeds 40.

## Nonmetallic Inclusions

Normal sulfur and phosphorus contents in carbon steels do not promote weld metal cracking. Large amounts of these elements are added to some steels to provide free-machining characteristics. These free-machining steels have relatively poor weldability because of hot tearing in the weld metal caused by low melting compounds of phosphorus and sulfur at the grain boundaries. The grains may be torn apart by thermal stresses during cooling. High sulfur content also promotes weld metal porosity.

Lead is also added to steel to improve machinability. It is nearly insoluble in steel and exists as distinct globules. The lead can melt during welding and volatilize into the weld fumes. Lead may, on occasion, cause porosity and embrittlement of steel. The major concern with lead, however, is its presence in the welding fumes and its toxicity. This requires special pre-

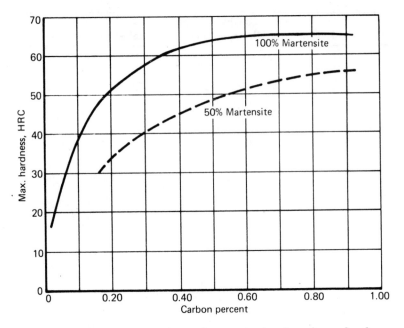

**Fig. 1.3—Relationship between carbon content and maximum hardness of steels with microstructures of 50 and 100 percent martensite**

cautions to assure good ventilation during welding. Normally, free machining steels should not be welded. If one of these steels must be welded, low hydrogen electrodes and low welding currents should be used to limit dilution, porosity, and cracking.

## Hydrogen

Carbon steels exhibit increasing susceptibility to hydrogen-induced cracking with increasing carbon content greater than about 0.15 percent. However, steels with 0.15 percent carbon or less are not immune to this problem, especially when thick sections are welded.

## Weld Cooling Rates

When arc welding thick sections, the weld metal and heat-affected zone can be hardened significantly as a result of quenching by the large mass of base metal. The cooling rate and the carbon equivalent of the steel are the controlling factors in determining the degree of hardening. The cooling rate depends primarily on the following factors: (1) the section thickness and joint geometry, (2) the base metal temperature before welding commences, and (3) the rate of heat input. Consequently, the use of higher welding current, slower welding speed (high heat input), or preheating of the base metal will reduce the cooling rate of the weld zone. Preheat should be maintained during the welding of successive beads. With higher carbon content or increased section thickness, a higher preheat and interpass temperature should be used to decrease the weld cooling rate and thus control the weld hardness and minimize the likelihood of cracking.[7]

In resistance spot welding of carbon steel sheets, the nugget may be hardened as a result of rapid cooling by the water-cooled copper alloy electrodes in contact with the sheet surfaces. Special electronic heat controls can provide a preheat or postweld heating cycle in the welding schedule to control the cooling rate and hardness of the nugget.

---

7. A comprehensive preheating guide for steels is *Suggested Arc Welding Procedures for Steels Meeting Standard Specifications,* by C. W. Ott and D. J. Snyder, Welding Research Council Bulletin No. 191, January 1974

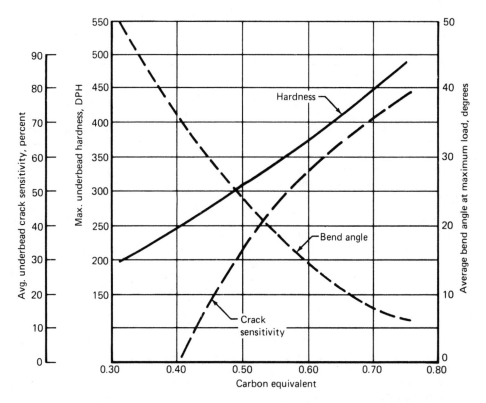

**Fig. 1.4 – Relationship between composition and underbead hardness, crack sensitivity, and notch-bead bend angle for 1-in. thick C-Mn steel plate welded with E6010 covered electrodes (CE = % C + % Mn/4 + % Si/4)**

## LOW CARBON STEEL

### Effect of Carbon Content

Steels with less than 0.15 percent carbon, known as *low carbon steels,* are easily joined by welding, brazing, or soldering. These steels have very low hardenability.

According to Fig. 1.3, a carbon steel containing 0.15 percent carbon is capable of being hardened to 30 to 40 HRC when cooled at a very high rate. However, the hardness of a weld heat-affected zone seldom reaches this level because the cooling rates are too low.

Rapidly cooled welded joints in steel containing carbon of about 0.10 percent and higher can develop cracks during severe cold forming operations because the weld area is harder than the unaffected base metal. When the welded joint

will be subjected to severe forming, the carbon content of the steel should be low.

Substantial hardness can be developed in resistance spot welds in thin steel sheet containing more than 0.08 percent carbon. Spot weld hardness is not a serious problem with these steels, except for some critical applications.

### Effect of Deoxidizing Practice

The deoxidation practice used in steel-making can influence weld metal soundness, particularly with autogenous welding. Rimmed and capped low carbon steels are not deoxidized. When these steels are remelted without the addition of deoxidizers, carbon and oxygen in the steel react to form carbon monoxide, which can be entrapped as porosity in weld metal. This is particularly true with high welding speeds where the gas does not have time to escape from the molten weld pool.

Porosity in welds in these steels can be minimized by adding a filler metal containing sufficient deoxidizers, such as aluminum, manganese, or silicon, to scavenge oxygen from the molten weld pool. With gas tungsten arc, plasma arc, or gas metal arc welding, E70S-2 filler metal should be used because it contains large amounts of deoxidizers.[8] The covering on shielded metal arc welding electrodes usually contains sufficient deoxidizers for welding rimmed or capped steels. Special aluminum paints are available that can be applied to the joint faces to deoxidize the weld metal during autogenous welding.

Submerged arc welding of rimmed or capped steel requires selection of an electrode and flux combination containing sufficient deoxidizers, such as silicon or manganese, to produce a sound weld, especially when high welding speeds are used.[9]

Weld metal soundness is not normally a problem with killed, low-carbon steels when good welding practices are used. A viscous, refractory slag sometimes forms on the surface of the molten weld pool and hampers metal flow. The slag results from deoxidation of the steel, particularly when shielding of the molten weld pool during welding is inadequate. The problem can be corrected by improving the welding practices, using a gas-shielded welding process, or adding a filler metal that contains sufficient manganese or silicon to make the slag more fluid.

## MILD STEEL

Carbon steels containing from about 0.15 to 0.30 percent carbon are commonly called mild steels. Underbead cracking or lack of toughness in the heat-affected zone is not usually encountered when welding mild steels containing no more than 0.20 percent carbon and 1 percent manganese. Such steels can be welded without preheat, postheat, or special welding procedures when the joint thickness is less than 1 in., and joint restraint is not severe.

As the carbon and manganese contents increase to about 0.30 and 1.40 percent respectively, weldability of the steels remains good, but the welds are susceptible to underbead cracking because of increased hardenability and strength. Welding with a low hydrogen process is recommended. Preheating and control of interpass temperature may be required, particularly when the joint thickness is greater than 1 in. or joint restraint is high. If hydrogen-induced cracking still is a problem with these procedures, hydrogen may be diffused from the joint using a postweld thermal treatment, as described previously.

Some mild steels are supplied in the normalized or quenched-and-tempered condition to provide good toughness or high strength properties. Tensile strengths may range from 65 to 100 ksi, depending upon the carbon and manganese content and the heat treatment.

The welding procedures for heat-treated mild steels are guided to a large extent by a need to have the weld metal, the heat-affected zone, and the unaffected base metal possess about the same toughness. Precautions should be taken to ensure that welding is done using low-hydrogen conditions. For example, the flux for submerged-arc welding should be dry.

Normal procedures are used with shielded metal arc, submerged arc, and gas metal arc welding because the cooling rates in the heat-affected zone are sufficiently rapid to reproduce a microstructure similar to that of the normalized or quenched steel. When carbon content is limited to about 0.20 percent, underbead cracking or lack of toughness in the heat-affected zone is not normally a problem. This is true even with low welding heat input that results in rapid cooling. In fact, high welding heat input or high preheat and interpass temperature results in a rather slow rate of cooling, which tends to increase the grain size and produce coarser pearlite in the heat-affected zone. These microstructural conditions are detrimental to good strength and toughness. If the welding process or procedure subjects the heat-affected zone to prolonged heating, high temperature, and slow cooling (electroslag welding, for example), the weldment will probably require a postweld heat treatment to restore good strength and toughness to this zone. When heat treatment of the weldment is uneconomical or impractical, the rate of cooling in the weld zone must be sufficiently rapid to produce a microstructure of adequate strength and toughness.

---

8. Refer to AWS A5.18-79, *Specification for Carbon Steel Filler Metals for Gas Shielded Arc Welding,* for additional information.

9. Refer to AWS A5.17-80, *Specification for Carbon Steel Electrodes and Fluxes for Submerged Arc Welding,* for further information.

In general, heat-treated mild steels are arc · welded without preheat, but a preheat should be used when the metal temperature is below about 50°F. In fact, a preheat of about 100°F or higher should be used if the plate thickness is over. 1 in. or the joint is highly restrained.

Dilution must be considered when selecting a filler metal to provide specified joint mechanical properties in a selected steel. Mechanical properties for weld metal given in filler metal trade literature are for undiluted metal. The properties of the weld metal in an actual fabrication may differ from the reported values because of dilution effects.

For heat-treated mild steels, low alloy steel filler metal may be required to meet mechanical property requirements. However, the weld metal strength should not greatly exceed the strength of the base metal. High-strength weld metal may require a softer heat-affected zone to undergo a relatively large amount of strain when the joint is subjected to deformation near room temperature. Under such conditions, fracture may occur prematurely in the heat-affected zone because of excessive localized strain.

In a butt joint, the filler metal should be selected to provide weld metal with essentially the same strength as the base metal. For fillet welds, a filler metal of lower strength is sometimes used to provide sufficient ductility to accommodate stress concentrations. However, a low strength filler metal should not be used indiscriminately as a remedy for cracking difficulties.

## MEDIUM CARBON STEEL

A pronounced change in the weldability of carbon steel takes place when the carbon content is in the 0.30 to 0.50 percent range. Steels containing about 0.30 percent carbon and relatively low manganese content have good weldability. As the carbon content of the steel is increased, the welding procedures must be designed to avoid the formation of large amounts of hard martensite in the heat-affected zone. If a steel containing about 0.50 percent carbon is welded with procedures commonly used for mild steel, the heat-affected zone is likely to be quite hard, low in toughness, and susceptible to cold cracking.

For most applications, medium carbon steel should be preheated prior to welding to control the cooling rate in the weld metal and heat-affected zone, and thus the formation of marten-

site. The appropriate preheat temperature depends upon the carbon equivalent of the steel, the joint thicknesses, and the welding procedure. In general, preheat temperature requirements increase with higher carbon equivalent (CE), greater joint thickness, or increased hydrogen in the arc. With a carbon equivalent in the 0.45 to 0.60 range, a preheat temperature in the 200° to 400°F range is recommended, depending upon the welding process and the joint thickness.[10] The interpass temperature should be the same as the preheat temperature.

A stress-relief heat treatment is recommended immediately after welding, especially with thick sections, high joint restraint, or service conditions involving impact or dynamic loading. If possible, the welded joint should be heated to stress-relief temperature without intermediate cooling to ambient temperature. This may be done by transferring the weldment to a preheated furnace. Slow cooling to room temperature following stress-relief is recommended to avoid introduction of thermal stresses.

When immediate stress relief is impractical, the welded joint should be maintained at or slightly above the specified preheat temperature for 2 to 3 hours per inch of joint thickness. This procedure promotes the diffusion of hydrogen from the weld zone and reduces the possibility of cracking during intermediate handling. However, it should not be considered a substitute for an appropriate stress-relief heat treatment.

Low hydrogen welding procedures are mandatory with these steels. Selection of filler metal for arc welding becomes more critical as the carbon content increases. Pickup of carbon from a steel containing 0.5 percent carbon by dilution will usually result in high weld metal hardness, susceptibility to cracking, and tendency for brittle failure. Dilution can be minimized by depositing small weld beads or using a welding procedure that provides shallow penetration. Low heat input is generally recommended for the first few layers in a multiple-pass weld to limit dilution. Higher heat input can be used to complete the joint. It is good practice to deposit the final weld bead, or beads, entirely on previously deposited weld metal without melting any base

_____

10. Refer to Welding Research Council Bulletin 191 for recommended preheat temperatures for specific steels.

metal. This practice has the effect of tempering the heat-affected zones of previously deposited weld beads, especially those in the base metal. Postweld heat treatment is still a requirement with this technique.

Medium carbon steels are used extensively in machinery and tools. Many of the steels are selected for wear resistance rather than high strength, and parts are frequently heat treated to meet desired properties. Welding may be performed before or after final heat treatment, and selection of filler metal and welding procedures must be considered in this context.

When welding is performed prior to heat treatment, special care must be exercised in choosing the filler metal if the weld metal properties must match the base metal properties after heat treatments. If welding is required on a previously heat-treated component, precautions are necessary to avoid cracking problems. A hard component may crack during welding if a suitable preheat is not used. The welding heat will soften the heat-affected zone, and reheat treatment of the weldment may be required to restore desired properties in the heat-affected zone.

## HIGH CARBON STEEL

The weldability of high carbon steels is poor because of their high hardenability and sensitivity to cracking in the weld metal and heat-affected zone. Low hydrogen welding procedures must be used for arc welding. Preheat and interpass temperatures of 400°F and higher are required to retard the formation of brittle, high carbon martensite in the weld.

A postweld stress relief is recommended, particularly for welded joints in thick sections. The stress relieving procedure described previously for medium carbon steels should be used.

Selection of an appropriate filler metal depends upon the carbon content of the steel, the weldment design, and service requirements. Steel filler metals are not normally produced with high carbon content. However, a low alloy steel filler metal may be suitable for many applications. Effects of dilution on the response of the weld metal to postweld heat treatment must be evaluated. Pickup of carbon in an alloy steel filler metal may significantly increase the hardenability of the weld metal. Consequently, the welding procedures should be designed to minimize dilution.

High carbon steels are commonly used for applications requiring high hardness or abrasion resistance, which is imparted by heat treatment. The steel should be welded in the annealed condition and then heat treated. Annealing is recommended prior to repair welding of broken parts. Successful welding requires development and testing of specific welding procedures for each application. Service requirements of the weldment must be considered when developing the welding and testing procedures.

## ARC WELDING

### Shielded Metal Arc Welding

Most carbon steels can be welded using covered electrodes and appropriate welding procedures including preheat when required. Covered electrodes are classified on the basis of the chemical composition and mechanical properties of undiluted weld metal, and also the type of covering. Carbon steel covered electrodes that produce weld metal with 60 or 70 ksi minimum tensile strength are the E60XX and E70XX types respectively.[11]

The E60XX classifications are suitable for welding low carbon and mild steels, provided the weld metal strength is adequate. They are not produced with low hydrogen coverings and should not be used for welding steels that are sensitive to hydrogen-induced cracking.

E70XX covered electrodes are produced with both regular and low hydrogen coverings. They should be used where higher strength welds or low hydrogen welding conditions, or both, are required. The low hydrogen types (E7015, E7016, E7018, E7028, and E7048) must be handled and stored under conditions that prevent moisture pickup in the coating.

Low alloy steel electrodes, Types E70XX-XX through E120XX-XX, are designed to produce weld metals with minimum tensile strengths of 70 to 120 ksi.[12] These electrodes are also produced with both regular and low hydrogen coatings. The carbon content of the deposited weld metal will be 0.15 percent or less, depending

---

11. Refer to AWS A5.1-81, *Specification for Carbon Steel Covered Arc Welding Electrodes.*

12. Refer to AWS A5.5-81, *Specification for Low Alloy Steel Covered Arc Welding Electrodes.*

upon the electrode type. Low hydrogen electrodes can be selected from this group to match the tensile strength of a medium or high carbon steel. The electrodes must be stored and used under conditions that prevent moisture pickup by the coating, as discussed previously.

## Gas Shielded Arc Welding

Carbon and low alloy steel bare electrodes and welding rods are available for use with the gas metal arc, gas tungsten arc, and plasma arc welding processes.[13] The electrodes are classified on the basis of chemical composition and mechanical properties of undiluted weld metal. Minimum tensile strengths range from 70 to 120 ksi. Selection of a bare electrode is made using the same considerations as for a covered electrode.

Shielding gases commonly used for gas metal arc welding of carbon steels are carbon dioxide ($CO_2$) and mixtures of argon with carbon dioxide and argon with oxygen. Selection of the gas depends primarily upon the electrode composition and the type of metal transfer: spray, globular, or short-circuiting. Generally, carbon dioxide shielding is suitable for low carbon and mild steels. A mixture of argon-oxygen or argon-$CO_2$ is suitable for all carbon steels, and is recommended with low alloy steel electrodes. Weld metal toughness is improved when one of these gas mixtures is used.

Argon is generally used for shielding with gas tungsten arc and plasma arc welding. However, helium-argon mixtures may be used to provide deeper joint penetration or permit faster travel speed with automatic welding.

Gas metal arc welds in rimmed or capped steels may be quite porous unless the electrode contains sufficient deoxidizers. An ER70S-2 electrode is a good selection. Other electrodes high in manganese and silicon, such as ER70S-6 or -7, may be satisfactory in many cases.

The low hydrogen characteristics of the three processes will be lacking if the filler metal, shielding gas, or both are contaminated. The filler metal may be contaminated with rust, moisture, oil, grease, drawing compounds, or other hydrogen-bearing materials. Therefore, proper cleaning, packaging, rust prevention, and storage are important to avoid hydrogen-induced cracking.

Shielding gases may be contaminated by moisture or hydrocarbons in the gas system or by aspiration of moist air into the system through leaks. Only welding grade shielding gases should be used, and the delivery system must be leak-tight.

## Flux Cored Arc Welding

Flux cored arc welding electrodes consist of a steel tube surrounding a core of fluxing ingredients and, sometimes, alloying additions. They are designed to deposit either carbon or a low alloy steel weld metal.[14] Some electrode classifications are self-shielding, as are covered electrodes, while other classifications use $CO_2$ or argon-$CO_2$ mixtures for shielding.

The operating characteristics of the electrodes vary with the core ingredients and the shielding gas, if used. In general, the gas-shielded electrodes give better notch toughness, particularly with argon-rich shielding mixtures and a basic slag (EXXT-5). However, some self-shielded electrodes can provide weld metal with adequate notch toughness for many low carbon or mild steel applications.

Carbon steel flux-cored electrodes are designed to produce undiluted weld metal with a minimum tensile strength of 60 or 70 ksi. These are suitable for welding low carbon and mild steels. Medium and high carbon steels can be welded with these electrodes if the weld metal will have adequate strength for the application. For weld strength requirements above 70 ksi, a suitable low alloy steel flux-cored electrode should be used. Low alloy steel flux-cored electrodes are available that deposit undiluted weld metals having tensile strengths ranging from 60 to 120 ksi.

As stated previously, flux-cored electrodes are generally considered to be low in hydrogen. However, contamination of the electrode or the shielding gas with moisture or hydrocarbon products can destroy this feature. Some electrodes may require baking to drive off absorbed moisture.

---

13. Refer to AWS A5.18-79, *Specification for Carbon Steel Filler Metals for Gas Shielded Arc Welding*, or AWS A5.28-79, *Specification for Low Alloy Steel Filler Metals for Gas Shielded Arc Welding*.

14. The electrodes are covered by AWS A5.20-79, *Specification for Carbon Steel Electrodes for Flux Cored Arc Welding*, and AWS A5.29-80, *Specification for Low Alloy Steel Electrodes for Flux Cored Arc Welding*.

## Submerged Arc Welding

Submerged arc welding is done with an electrode, either solid or metal-cored, and a granular flux. The flux performs the same functions as the flux covering on shielded metal arc welding electrodes. Solid wire electrodes, both carbon and low alloy steel, are classified according to chemical composition. Metal cored, low alloy steel electrodes together with an appropriate flux are classified on the basis of the chemical composition of the deposited metal.[15] Fluxes are classified on the basis of the chemical composition and mechanical properties of the weld metal that is deposited with a particular type of electrode. As with other electrodes, the various weld metals are classified using specific preheat and interpass temperatures and postweld heat treatments.

Weld metal properties are determined either in the as-welded condition or after a specified postweld heat treatment. Some weld metals may be used in either condition. Specific applications may require the use of welding and heat treating conditions that produce weld metal having mechanical properties different from those required by the filler metal specification. In such cases, expected weld metal mechanical properties should be established by appropriate tests.

Carbon steel electrode-flux combinations are designed to produce weld metal having 60 or 70 ksi minimum tensile strength. These combinations are recommended for welding low carbon and mild steels as well as medium carbon, high carbon, and low alloy steels when high weld joint strength is not required. Low alloy steel electrode-flux combinations are recommended for welding alloy steels of similar composition, as well as medium and high carbon steels when high strength joints are needed to meet service requirements. Response to postweld heat treatment must be considered.

As described previously, flux must be kept clean and dry to maintain low hydrogen welding conditions. Also the electrode must be clean and free of contaminants as discussed previously for the gas shielded arc welding processes.

## ELECTROSLAG AND ELECTROGAS WELDING

Electroslag and electrogas welding are used primarily for producing single pass groove welds in the vertical position. Electroslag welding uses a resistance-heated, molten slag pool to melt the electrode. This process is generally used for welding steel plates of 1.25 to 12 inches in thickness. Electrogas welding is a variation of gas metal arc or flux cored arc welding, depending upon the type of electrode. This process is best suited for welding plates over 0.5 inch in thickness.

These processes are used mostly to weld low carbon and mild steels, but medium carbon steels may also be welded. When the weldment requires postweld heat treatment, the electrode composition is generally selected to produce weld metal that will respond satisfactorily to the heat treatment. Dilution by the base metal must be considered during selection.

Electrodes and fluxes for electroslag welding are classified in a manner similar to that for submerged arc welding.[16] Solid or metal cored electrodes are used to produce weld metal having a minimum tensile strength of 60 ksi or higher, depending upon the specific electrode-flux combination and the welding conditions. Most solid electrodes are carbon-manganese steels while metal cored electrodes generally deposit low alloy steel weld metal.

Solid carbon steel electrodes for electrogas welding are essentially the same as those used for gas metal arc welding. Flux cored electrodes are designed specifically for use with this process to deposit low carbon or low alloy steel weld metal.[17] Welding may be done using either $CO_2$ or mixtures of argon with $CO_2$ or oxygen. One of the carbon steel flux-cored electrodes can be used without shielding gas. Minimum tensile strength requirements for deposited weld metal are the same as for electroslag welding.

## OXYACETYLENE WELDING

Most low carbon and mild steels can be

---

15. Refer to AWS A5.17, *Specification for Carbon Steel Electrodes and Fluxes for Submerged Arc Welding*, and AWS A5.23-80, *Specification for Low Alloy Steel Electrodes and Fluxes for Submerged Arc Welding*.

16. Refer to AWS A5.25-78, *Specification for Consumables Used for Electroslag Welding of Carbon and High-Strength Low Alloy Steels*.

17. Refer to AWS A5.26-78, *Specification for Consumables Used for Electrogas Welding of Carbon and High-Strength Low Alloy Steels*.

joined by oxyacetylene welding, but the process is much slower than arc welding. The slow heating characteristics of the process result in rather extensive heating of the steel. As a result, mechanical properties developed in the base metal by prior heat treatment or cold working may be impaired. On the other hand, the cooling rate in the weld zone will be comparatively slow.

Oxyacetylene welding of steel is done without flux. Therefore, the weld metal is not protected from the atmosphere by a slag or shielding gas cover. Welded joints are likely to contain discontinuities that are unacceptable for many applications and their mechanical properties may be too low for the intended service.

Carbon steels containing more than 0.35 percent carbon require special precautions with oxyacetylene welding. Preheating with the welding torch or another heat source is recommended to retard the cooling rate of the weld and heat-affected zone. In some cases, postweld heat treatment may be needed to refine the grain size of the weld zone and improve toughness.

Steel welding rods are available for oxyacetylene welding of carbon and low alloy steels.[18] They are classified on the basis of the minimum tensile strength of as-deposited weld metal. Type R45 welding rods should be used to deposit low-carbon steel weld metal for general applications. Type R60 welding rods are recommended for welding carbon steels having tensile strengths in the range of 50 to 65 ksi, and Type R65 rods for welding both carbon and low alloy steels with tensile strengths in the range of 65 to 75 ksi. Carbon and low-alloy steel welding rods designed for gas tungsten arc welding can also be used for oxyacetylene welding. If a suitable welding rod is not available for welding a specific steel, strips sheared from the base metal may be used as filler metal.

## RESISTANCE WELDING

Carbon steels can be joined by all of the resistance welding processes, i.e. spot, seam, projection, flash, upset, and high-frequency welding. The heating and cooling rates with these processes are very high compared to those in arc welding. Consequently, the hardenability and critical cooling rate of a steel must be considered when selecting the process and the welding procedures.

The weld and heat-affected zone are quenched by the watercooled copper alloy electrodes bearing on the workpieces. The severity of the quench depends upon the heat conduction path length, the electrode contact area, and the quenching time. It is more severe when spot, seam, or projection welding thin sheet and flash or upset welding of rod or wire with a small cross section. Retraction of the electrodes from the weldment as soon as practical will reduce the quenching rate. With some processes, such as seam and high-frequency welding, the weld may be quenched by the water spraying or flooding used to cool the electrodes.

If the hardness of the weld and heat-affected zone is excessive, postheat or tempering in the welding machine may reduce the hardness to an acceptable level. In some cases, heat treatment of the weldment in other equipment may be appropriate.

### Low Carbon Steel

Low carbon steel is readily resistance welded by all processes because it has low hardenability. Welds normally have adequate ductility for the application. Spot and seam welds in thin sheet can have relatively high hardness when the carbon content exceeds about 0.08 percent. If this is objectionable, sheet with a nominal carbon content of 0.08 percent or less should be specified.

Suggested schedules for spot, seam, and projection welding low carbon steel sheet are given in Tables 1.3 through 1.6.[19] Variation from these conditions will be necessary to account for the weldment design and the configuration of the welding machine. Suitable welding conditions for a specific application should be determined by appropriate testing.

Data for flash welding of carbon steel shapes are given in the chapter covering the process.[20] Recommended procedures for high-frequency and upset welding should be obtained from the equipment manufacturer.

18. Refer to AWS A5.2-80, *Specification for Iron and Steel Gas Welding Rods.*

19. Appropriate projection designs for steel are discussed in the *Welding Handbook,* Vol. 3, 7th Ed., 30-31.

20. Refer to the *Welding Handbook,* Vol. 3, 7th Ed., 58-76.

## Table 1.3
## Suggested schedules for spot welding low carbon steel sheet

| Thickness, in.[a] | Electrode face diam., in.[b] | Static electrode force, lbs | Weld time, cycles[c] | Approx. welding current, kA[d] | Approx. nugget diam., in. | Minimum pitch, in.[e] | Min. shear strength, lbs., when base metal tensile strength is | |
|---|---|---|---|---|---|---|---|---|
| | | | | | | | Below 70 ksi | Above 70 ksi |
| 0.010 | 0.13 | 200 | 4 | 4.0 | 0.10 | 0.25 | 130 | 180 |
| 0.021 | 0.19 | 300 | 6 | 6.5 | 0.13 | 0.37 | 320 | 440 |
| 0.031 | 0.19 | 400 | 8 | 8.0 | 0.16 | 0.50 | 570 | 800 |
| 0.040 | 0.25 | 500 | 10 | 9.5 | 0.19 | 0.75 | 920 | 1200 |
| 0.050 | 0.25 | 650 | 12 | 10.5 | 0.22 | 0.87 | 1350 | |
| 0.062 | 0.25 | 800 | 14 | 12.0 | 0.25 | 1.00 | 1850 | |
| 0.078 | 0.31 | 1100 | 17 | 14.0 | 0.29 | 1.25 | 2700 | |
| 0.094 | 0.31 | 1300 | 20 | 15.5 | 0.31 | 1.50 | 3450 | |
| 0.109 | 0.38 | 1600 | 23 | 17.5 | 0.32 | 1.62 | 4150 | |
| 0.125 | 0.38 | 1800 | 26 | 19.0 | 0.33 | 1.75 | 5000 | |

a. Thickness of thinnest outside sheet. Data applicable to a total metal thickness of four times the given thickness. Maximum ratio of two adjacent thicknesses is 3 to 1.
b. Flat-faced electrode of RWMA Group A, Class 2 copper alloy.
c. Single impulse, frequency of 60Hz.
d. Single-phase ac machine.
e. Minimum pitch between adjacent spot welds between two sheets without adjustment of the welding schedule for shunting.

## Table 1.4
## Suggested schedules for multiple-impulse spot welding low carbon steel sheet

| Thickness, in.[a] | Electrode face diam, in.[b] | Static electrode force, lbs | No. of impulses[c] | | | Approx. welding current, kA[d] | Approx. nugget diam., in. | Min. shear strength, lbs |
|---|---|---|---|---|---|---|---|---|
| | | | Single spot | Multiple spot spacing | | | | |
| | | | | 1-2 in. | 2-4 in. | | | |
| 0.125 | 0.44 | 1800 | 3 | 5 | 4 | 18.0 | 0.37 | 5,000 |
| 0.188 | 0.50 | 1950 | 6 | 20 | 14 | 19.5 | 0.56 | 10,000 |
| 0.25 | 0.56 | 2150 | 12 | 24 | 18 | 21.5 | 0.75 | 15,000 |
| 0.31 | 0.62 | 2400 | 15 | 30 | 23 | 24.0 | 0.87 | 20,000 |

a. Thinnest of two sheets. Maximum ratio of sheet thickness is 2 to 1.
b. Flat faced electrode of RWMA Group A, Class 2 copper alloy.
c. Heat time—20 cycles, cool time—5 cycles (60 Hz).
d. Single-phase ac machine.

## Mild and Medium Carbon Steels

When resistance welding these steels, control of the cooling rate from welding temperature or a subsequent tempering cycle is usually required to control the hardness and associated cracking in the weld zone. With mild steels, postheating with a lower current for several cycles (60 Hz) may be sufficient to avoid the formation of hard martensite in the weld. Spot welds in medium carbon steels can be quenched

to martensite and then tempered by resistance heating to soften the martensite and improve ductility. Suitable postheat or quench and temper cycles must be established by welding tests. Examples of quench and temper cycles when spot welding AISI 1020, 1035, and 1045 carbon steel sheet are given in Table 1.7. As expected, the shear strength of the spot welds increases as the carbon content of the steel increases.

Normally, mild and medium carbon steels can be flash welded without cracking if the cross

**Table 1.5**
**Suggested schedules for seam welding low carbon steel sheet**

| Thickness, in. [a] | Electrode face width, in. [b] | Static electrode force, lbs | Heat time, cycles[c] | Cool time, cycles[c] | Travel speed, in./min | Welds per in. | Approx. welding current, kA[d] |
|---|---|---|---|---|---|---|---|
| 0.010 | 0.19 | 400 | 2 | 1 | 80 | 15 | 8.0 |
| 0.021 | 0.19 | 550 | 2 | 2 | 75 | 12 | 11.0 |
| 0.031 | 0.25 | 700 | 3 | 2 | 72 | 10 | 13.0 |
| 0.040 | 0.25 | 900 | 3 | 3 | 67 | 9 | 15.0 |
| 0.050 | 0.31 | 1050 | 4 | 3 | 65 | 8 | 16.5 |
| 0.062 | 0.31 | 1200 | 4 | 4 | 63 | 7 | 17.5 |
| 0.078 | 0.38 | 1500 | 6 | 5 | 55 | 6 | 19.0 |
| 0.094 | 0.44 | 1700 | 7 | 6 | 50 | 5.5 | 20.0 |
| 0.109 | 0.50 | 1950 | 9 | 6 | 48 | 5 | 21.0 |
| 0.125 | 0.50 | 2200 | 11 | 7 | 45 | 4.5 | 22.0 |

a. Thickness of thinnest outside piece. Data applicable to a total metal thickness of 4 times the given thickness. Maximum ratio of two adjacent thicknesses is 3 to 1.
b. Flat-faced wheel electrode of RWMA Group A, Class 2 copper alloy.
c. 60 Hz
d. Single-phase ac machine.

section is at least 1 inch in diameter. Preheat or postheat is not required, but application of one or both can improve weld joint ductility. When flash welding small or thin cross sections, postheat in the welding machine can be used to control weld hardness and cold cracking. The welding machine should be equipped with appropriate current and time controls to program a postheat cycle. An alternative is to heat treat the weldment in a furnace to homogenize the metallurgical structure in the weld zone and provide the required mechanical properties.

## High Carbon Steel

High carbon steels are seldom joined by resistance welding processes because of their high hardenability and limited use for products where these processes are applicable. The procedures given for welding mild and medium carbon steels are applicable also to high carbon steels, with appropriate adjustments to accommodate the higher hardenability.

## ELECTRON AND LASER BEAM WELDING

Low carbon steels are readily welded by electron and laser beam welding. The rapid heating and cooling rates with these processes result in weld and heat-affected zones with relatively small grain size when compared to arc welds. Also the heat-affected zone is much narrower.

Fully killed steel is preferred for applications involving these welding processes. Welding of rimmed or capped steel requires the addition of deoxidizers to avoid porosity caused by violent gas-metal reactions in the molten weld pool. Highly deoxidized filler metal or preplaced aluminum on the joint faces, in the form of foil, paint, or thermal sprayed deposit, will reduce the amount of porosity in the joint. Semikilled steels are somewhat easier to weld than rimmed or capped steels with respect to porosity. Some deoxidation may be needed but to a lesser extent.

## FRICTION WELDING

Low carbon and mild steels can be readily joined by friction welding. Welding conditions are not critical from a metallurgical standpoint. Medium and high carbon steels can be friction welded but the welding conditions must be controlled within narrow ranges. The heating time for these steels should be relatively long to slow the cooling rate of the weld.

## BRAZING

### Processes and Equipment

Carbon steels can be joined by virtually all of the brazing processes. Torch, furnace, and

## Table 1.6
### Suggested schedules for projection welding low carbon steel sheet

| Thickness, t in.[d] | Projection Diam., in. | Projection Height, in. | Minimum pitch, in. | Minimum contact overlap, in. | Welding Schedule A for single projection[a] | | | | Welding Schedule B for 1-3 projections, per projection[b] | | | | Welding Schedule C for 3 or more projections, per projection[c] | | | |
|---|---|---|---|---|---|---|---|---|---|---|---|---|---|---|---|---|
| | | | | | Weld time, cycles[e] | Electrode force, lb | Welding current, A | Shear strength, lb | Weld time, cycles[e] | Electrode force, lb | Welding current, A | Shear strength, lb | Weld time, cycles[e] | Electrode force, lb | Weld current, A | Shear strength, lb |
| 0.022 | 0.090 | 0.025 | 0.38 | 0.25 | 3 | 150 | 4,400 | 370 | 6 | 150 | 3,850 | 325 | 6 | 80 | 2900 | 290 |
| 0.028 | 0.090 | 0.025 | 0.38 | 0.25 | 3 | 195 | 5,500 | 500 | 6 | 150 | 4,450 | 425 | 8 | 100 | 3300 | 340 |
| 0.034 | 0.110 | 0.035 | 0.50 | 0.38 | 3 | 240 | 6,600 | 700 | 6 | 150 | 5,100 | 525 | 11 | 125 | 3800 | 425 |
| 0.043 | 0.110 | 0.035 | 0.50 | 0.38 | 5 | 330 | 8,000 | 1060 | 10 | 210 | 6,000 | 875 | 15 | 160 | 4300 | 720 |
| 0.049 | 0.140 | 0.038 | 0.75 | 0.50 | 8 | 400 | 8,800 | 1300 | 16 | 270 | 6,500 | 1100 | 19 | 220 | 4600 | 875 |
| 0.061 | 0.150 | 0.042 | 0.75 | 0.50 | 10 | 550 | 10,300 | 1800 | 20 | 365 | 7,650 | 1575 | 25 | 330 | 5400 | 1225 |
| 0.077 | 0.180 | 0.048 | 0.88 | 0.50 | 14 | 800 | 11,850 | 2425 | 28 | 530 | 8,850 | 2150 | 34 | 470 | 6400 | 1750 |
| 0.092 | 0.210 | 0.050 | 1.06 | 0.62 | 16 | 1020 | 13,150 | 3250 | 32 | 680 | 9,750 | 2800 | 42 | 610 | 7200 | 2325 |
| 0.107 | 0.240 | 0.055 | 1.25 | 0.75 | 19 | 1250 | 14,100 | 3850 | 38 | 830 | 10,600 | 3450 | 50 | 740 | 8300 | 2900 |
| 0.123 | 0.270 | 0.058 | 1.50 | 0.81 | 22 | 1500 | 14,850 | 4800 | 45 | 1000 | 11,300 | 4200 | 60 | 900 | 9200 | 3600 |
| 0.135 | 0.300 | 0.062 | 1.63 | 0.88 | 24 | 1650 | 15,300 | 5500 | 48 | 1100 | 11,850 | 4850 | 66 | 1000 | 9900 | 4250 |

a. Schedule A is usable for welding more than one projection if current is decreased but excessive weld expulsion may result and power demand will be greater than with Schedules B or C.

b. Schedule B is usable for welding more than three projections but some weld expulsion may result and power demand will be greater than with Schedule C.

c. Schedule C is usable for welding less than three projections with welding current increased approximately 15% and possible objectionable final sheet separation.

d. For unequal sheet thickness ratios, T/t, up to 3 to 1: (1) The weld time, cycles, should be increased by a factor $f_t$, determined by the formula $f_t = 1.5(T/t) - 0.5$. (2) The welding current per projection should be increased by a factor, $f_c$, determined by the formula $f_c = 0.1 (T/t) + 0.9$.

e. 60 Hz

**Table 1.7**
**Typical spot welding schedules for medium carbon and low alloy steel sheet**

| AISI No. | Thick-ness, in.[a] | Elec-trode face diam, in.[b] | Elec-trode force, lbs | Weld time cycles[c] | Quench time, cycles[c] | Temper time cycles[c] | Approx. welding current, kA[d] | Temper current, percent[e] | Approx. nugget diam., in. | Mini-mum pitch, in.[f] | Min. shear strength, lbs |
|---|---|---|---|---|---|---|---|---|---|---|---|
| 1020 | 0.040 | 0.25 | 1475 | 6 | 17 | 6 | 16.0 | 90 | 0.23 | 1 | 1360 |
| 1035 | 0.040 | 0.25 | 1475 | 6 | 20 | 6 | 14.2 | 91 | 0.22 | 1 | 1560 |
| 1045 | 0.040 | 0.25 | 1475 | 6 | 24 | 6 | 13.8 | 88 | 0.21 | 1 | 2000 |
| 4130 | 0.040 | 0.25 | 1475 | 6 | 18 | 6 | 13.0 | 90 | 0.22 | 1 | 2120 |
| 4340 | 0.031 | 0.19 | 900 | 4 | 12 | 4 | 8.3 | 84 | 0.16 | 0.75 | 1080 |
|  | 0.062 | 0.31 | 2000 | 10 | 45 | 10 | 13.9 | 77 | 0.27 | 1.50 | 3840 |
|  | 0.125 | 0.63 | 5500 | 45 | 240 | 90 | 21.8 | 88 | 0.55 | 2.50 | 13,700 |
| 8630 | 0.031 | 0.19 | 800 | 4 | 12 | 4 | 8.7 | 88 | 0.16 | 0.75 | 1220 |
|  | 0.062 | 0.31 | 1800 | 10 | 36 | 10 | 12.8 | 83 | 0.27 | 1.50 | 4240 |
|  | 0.125 | 0.63 | 4500 | 45 | 210 | .90 | 21.8 | 84 | 0.55 | 2.50 | 13,200 |

a. Two equal thicknesses
b. Flat-faced electrode of RWMA Group A, Class 2 copper alloy
c. Cycle of 60 Hz
d. Single-phase ac machine
e. Percentage of welding current with phase-shift heat control
f. Minimum pitch between adjacent spot welds without adjustment of the welding schedule for shunting

induction heating methods are commonly used. Filler metals in the form of continuous wire or strip can be automatically applied using electromechanical wire feeders. Filler metal in powder form can be blended with flux and paste forming ingredients, and then automatically applied with pressurized dispensing equipment. For torch brazing, the equipment would include a standard oxyacetylene or similar torch. Brazing furnaces can be batch or conveyor type, with or without control of the atmosphere.

## Filler Metals

Carbon steels can be brazed with copper-, gold-, nickel-, and silver-base filler metals.[21] All BAg brazing filler metals are suitable with carbon steel. Those containing nickel usually have better wettability and are preferred for good joint strength.

The BCu filler metals are mainly used for furnace brazing where preplacement is needed. The RBCuZn filler metals in rod form are generally used with torch brazing but can be used

with furnace or induction brazing. The high brazing temperatures of these filler metals often permit simultaneous brazing and heat treating operations.

The BNi filler metals are used when their unique properties are needed for special applications. Brazing is normally done in a controlled atmosphere.

## Fluxes and Atmospheres

Flux or a proper atmosphere is generally required when brazing steels. Selection will depend upon brazing filler metal. AWS brazing flux Type 3A, 3B, or 4 is suitable for BAg filler metals. Type 5 flux is normally used with the RBCuZn filler metals. Fluxes and atmospheres may be used together. Flux can be used in either paste or powder form, or combined with the filler metal. In a face-fed operation, the hand-held filler metal can be coated with the appropriate flux. In atmosphere brazing, the filler metal is preplaced in or near the joint and the assembly charged into the brazing chamber. Brazing temperature and time must be controlled to ensure proper melting and flow of the filler metal into the joint.

---

21. Refer to AWS A5.8-81, *Specification for Brazing Filler Metals,* for additional information.

## Joint Clearance

Joint clearances in the range of 0.002 to 0.005 in. produce the best mechanical properties with most filler metals when using a mineral flux. Light press fits are preferred for BCu and other filler metals when furnace brazing with a protective atmosphere. Filler metals with relatively narrow melting ranges are required for close-fitting joints. Conversely, filler metals with wide melting ranges have good bridging characteristics when wide clearances are involved. Furnace dew point can be used to control the fluidity of BCu filler metal when brazing joints with large gaps.

## Metallurgical Considerations

The mechanical properties of the heated area in cold worked steel may be impaired as a result of annealing during brazing. With hot-rolled steel, brazing above the austenitizing temperature will alter its mechanical properties. The changes in properties may result from decarburization of the steel in some furnace atmospheres, a change in grain size, or both. Original grain size can be restored by subsequent heat treatment below the remelt temperature of the filler metal. Loss of carbon through decarburization is generally unimportant in low carbon steels. However, surface hardness of medium and high carbon steels may be substantially lowered.

The brazing of high carbon steel is best done prior to or during the hardening operation. Brazing after hardening will soften the steel.

The hardening temperature for carbon steels is normally in the 1400° to 1600°F range. If the brazing is done prior to the hardening operation, the filler metal must melt well above the hardening temperature so that the brazed joints will have sufficient strength during that operation. Copper filler metal (BCu) is frequently used for this purpose.

At times, the high temperature (2000°–2100°F) required for copper brazing adversely affects the metallurgical structure of the steel. In such cases, silver and copper-zinc filler metals with brazing temperatures in the 1700° to 1800°F range can be used.

When brazing and hardening operations are combined, a filler metal that solidifies above the austenitizing temperature of the steel is generally used. The brazement is cooled to below the filler metal solidus temperature, and then it is quenched to harden the steel. Particular attention must be given to the brazement design and the handling procedures because the brazed joint strength will be very low at the austenitizing temperature. The brazement design should place the joint in compression rather than tension during quenching.

## THERMAL CUTTING

Carbon steels are easily cut or gouged by oxyfuel gas, air-carbon arc, plasma arc, and other thermal cutting processes. Low carbon and mild steels are cut using standard procedures. However, with medium and high carbon steels, the heat-affected zone may be quench-hardened significantly. Preheating, postheating, or both may be necessary to control the hardness of the cut edge to avoid cracking or heat checking.

Preheat temperatures similar to those recommended for welding with a low hydrogen process are normally satisfactory for cutting. The temperature must be nearly uniform through the section when using oxyfuel gas cutting. If it is not, the oxidation reaction will proceed faster in the hotter zone and result in a rough cut surface.

The cut surface may be heat treated to relieve stresses, reduce hardness, or alter the metallurgical structure. Furnace or local torch heating may be suitable.

Air-carbon arc gouging may leave areas of carburized steel on the cut surface when improper cutting conditions are used. The high-carbon areas may result in excessive hardening of the steel during welding if they are not removed beforehand. It is good practice to grind to clean metal after manual air-carbon arc gouging to remove any carburized areas on the surface. This procedure is not normally required with mechanized gouging.

# HIGH-STRENGTH LOW ALLOY STEELS

## DESCRIPTION

A group of low alloy steels that are designed to provide better mechanical properties and, sometimes, greater resistance to atmospheric corrosion than conventional carbon steels are known as *high-strength low alloy (HSLA) steels*. They are not considered to be alloy steels in the normal sense because they are designed to meet specific mechanical properties rather than a chemical composition. The chemical composition of a specific HSLA steel may necessarily vary within limits for different product thicknesses to meet mechanical property requirements.

The principal difference between structural grades of carbon steel and HSLA steels is chemical composition. The alloying additions to HSLA steels strengthen the ferrite, promote hardenability, and helps to control grain size. Figure 1.5 shows typical microstructures of a mild steel and a HSLA steel. The HSLA steel has a smaller grain size and increased pearlite, which accounts for its higher strength.

The HSLA steels are commonly furnished in the as-rolled condition to meet required mechanical properties. They may also be furnished in a controlled-rolled, normalized, or precipitation-hardened condition to meet specific notch toughness requirements. The minimum yield strengths of the steels are in the 42 to 80 ksi range, and the tensile strengths in the 60 to 90 ksi range.

The weldability of most HSLA steels is similar to that of mild steel. The carbon equivalents are kept low for good weldability.[22]

## COMPOSITION AND PROPERTIES

Most HSLA steels that are used in the as-rolled or normalized condition are covered by ASTM Specifications while other HSLA steels are covered by SAE Recommended Practice J410c. These steels are designed to provide a combination of higher strength, better corrosion resistance, or improved notch toughness not available with mild steels, and yet possess good weldability. Chromium, columbium, copper, molybdenum, nickel, nitrogen, phosphorus, titanium, vanadium, and zirconium are used in various combinations.

Table 1.8 lists a number of ASTM Specifications covering structural quality HSLA steels together with their alloying elements, mill forms, and tensile strength ranges. Table 1.9 lists several ASTM HSLA steels for pressure vessel applications. High-strength low alloy steel pipe, tubing, and castings with similar mechanical prop-

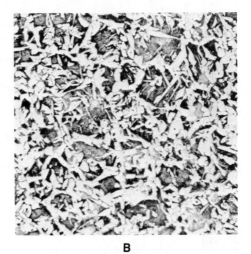

**A**

**B**

*Fig. 1.5—Typical microstructures of (A) mild steel and (B) a high-strength low-alloy steel ($\times 100$) (10% Nital etch)*

erties and chemical compositions are covered by other ASTM Specifications.

HSLA steels may be divided into two general groups, A and B. The steels in Group A are designed for high strength while those in Group B have both high strength and improved atmospheric corrosion resistance.

Group A steels contain vanadium, columbium, or both with nitrogen to improve the yield strength by grain refinement and precipitation hardening. The notch toughness, as measured by

---

22. One carbon equivalent formula used to judge the weldability of low alloy steels is $CE = \%C + \dfrac{\%Mn}{6} + \dfrac{\%Cr + \%Mo + \%V}{5} + \dfrac{\%Si + \%Ni + \%Cu}{15}$

## Table 1.8
## Typical ASTM high-strength low alloy structural steels

| ASTM Specification | Type or Grade | Composition, percent[a] | | | | | | | | | | Min tensile strength, ksi | Min yield strength, ksi |
|---|---|---|---|---|---|---|---|---|---|---|---|---|---|
| | | C | Mn | P | S | Si | Cr | Ni | Mo | V | Other | | |
| A242 | 1 | 0.15 | 1.00 | 0.15 | 0.05 | | | | | | 0.20 min Cu | 63-70 | 42-50 |
| A441 | | 0.22 | 0.85-1.25 | 0.04 | 0.05 | 0.30 | | | | | 0.20 min Cu; 0.02 min V | 60-70 | 40-50 |
| A572 | 42[b] | 0.21 | 1.35 | 0.04 | 0.05 | 0.30 | | | | | 0.20 min Cu | 60 | 42 |
| | 50[b] | 0.23 | 1.35 | 0.04 | 0.05 | 0.30 | | | | | 0.20 min Cu | 65 | 50 |
| | 60[b] | 0.26 | 1.35 | 0.04 | 0.05 | 0.30 | | | | | 0.20 min Cu | 75 | 60 |
| | 65[b] | 0.26 | 1.65 | 0.04 | 0.05 | 0.30 | | | | | 0.20 min Cu | 80 | 65 |
| A588 | A | 0.10-0.19 | 0.90-1.25 | 0.04 | 0.05 | 0.15-0.30 | 0.40-0.65 | | | 0.02-0.10 | 0.25-0.40 Cu | 63-70 | 42-50 |
| | B | 0.20 | 0.75-1.25 | 0.04 | 0.05 | 0.15-0.30 | 0.40-0.70 | 0.25-0.50 | | 0.01-0.10 | 0.20-0.40 Cu | | |
| | C | 0.15 | 0.80-1.35 | 0.04 | 0.05 | 0.15-0.30 | 0.30-0.50 | 0.25-0.50 | | 0.01-0.10 | 0.20-0.50 Cu | | |
| | D | 0.10-0.20 | 0.75-1.25 | 0.04 | 0.05 | 0.50-0.90 | 0.50-0.90 | | | | 0.30 Cu; 0.05-0.15 Zr; 0.04 Cb | | |
| | E | 0.15 | 1.20 | 0.04 | 0.05 | 0.15-0.30 | | 0.75-1.25 | 0.10-0.25 | 0.05 | 0.50-0.80 Cu | | |
| | F | 0.10-0.20 | 0.50-1.00 | 0.04 | 0.05 | 0.30 | 0.30 | 0.40-1.10 | 0.10-0.20 | 0.01-0.10 | 0.30-1.00 Cu | | |
| | G | 0.20 | 1.20 | 0.04 | 0.05 | 0.25-0.70 | 0.50-1.00 | 0.80 | 0.10 | | 0.30-0.50 Cu; 0.07 Ti | | |
| | H | 0.20 | 1.25 | 0.035 | 0.040 | 0.25-0.75 | 0.10-0.25 | 0.30-0.60 | 0.15 | 0.02-0.10 | 0.20-0.35 Cu; 0.005-0.030 Ti | | |
| | J | 0.20 | 0.60-1.00 | 0.04 | 0.05 | 0.30-0.50 | | 0.50-0.70 | | | 0.30 min Cu; 0.03-0.05 Ti | | |
| A633 | A | 0.18 | 1.00-1.35 | 0.04 | 0.05 | 0.15-0.50 | | | | | 0.05 Cb | 63-83 | 42 |
| | C | 0.20 | 1.15-1.50 | 0.04 | 0.05 | 0.15-0.50 | | | | | 0.01-0.05 Cb | 65-90 | 46-50 |
| | D | 0.20 | 0.70-1.60 | 0.04 | 0.05 | 0.15-0.50 | 0.25 | 0.25 | 0.08 | | 0.35 Cu | 65-90 | 46-50 |
| | E | 0.22 | 1.15-1.50 | 0.04 | 0.05 | 0.15-0.50 | | | | 0.04-0.11 | 0.01-0.03 N | 75-100 | 55-60 |
| A710 | A | 0.07 | 0.40-0.70 | 0.025 | 0.025 | 0.35 | 0.60-0.90 | 0.70-1.00 | 0.15-0.25 | | 1.00-1.30 Cu; 0.02 min Cb | 65-90 | 55-85 |
| | B | 0.06 | 0.40-0.65 | 0.025 | 0.025 | 0.20-0.35 | | 1.20-1.50 | | | 1.00-1.30 Cu; 0.02 min Cb | 88-90 | 75-85 |

a. Single values are maximum unless otherwise noted.
b. These grades may contain columbium, vanadium, or nitrogen.

## Table 1.9
## Typical ASTM high strength low alloy steels for pressure vessel plate

| ASTM Specifi-cation | Type or Grade | Heat composition, percent[a] | | | | | | | | | | Min tensile strength, ksi | Min yield strength, ksi |
|---|---|---|---|---|---|---|---|---|---|---|---|---|---|
| | | C | Mn | P | S | Si | Cr | Ni | Mo | V | Other | | |
| A202 | A | 0.17 | 1.05-1.40 | 0.035 | 0.040 | 0.60-0.90 | 0.35-0.60 | | | | | 75-95 | 45 |
| | B | 0.25 | 1.05-1.40 | 0.035 | 0.040 | 0.60-0.90 | 0.35-0.60 | | | | | 85-110 | 47 |
| A203 | A | 0.23 | 0.80 | 0.035 | 0.040 | 0.15-0.30 | | 2.10-2.50 | | | | 65-85 | 37 |
| | B | 0.25 | 0.80 | 0.035 | 0.040 | 0.15-0.30 | | 2.10-2.50 | | | | 70-90 | 40 |
| | D | 0.20 | 0.80 | 0.035 | 0.040 | 0.15-0.30 | | 3.25-3.75 | | | | 65-85 | 37 |
| | E | 0.23 | 0.80 | 0.035 | 0.040 | 0.15-0.30 | | 3.25-3.75 | | | | 70-90 | 40 |
| A204 | A | 0.25 | 0.90 | 0.035 | 0.040 | 0.15-0.30 | | | 0.45-0.60 | | | 65-85 | 37 |
| | B | 0.27 | 0.90 | 0.035 | 0.040 | 0.15-0.30 | | | 0.45-0.60 | | | 70-90 | 40 |
| | C | 0.28 | 0.90 | 0.035 | 0.040 | 0.15-0.30 | | | 0.45-0.60 | | | 75-95 | 43 |
| A225 | C | 0.25 | 1.60 | 0.035 | 0.040 | 0.15-0.40 | | 0.40-0.70 | | 0.13-0.18 | | 105-135 | 70 |
| | D | 0.20 | 1.70 | 0.035 | 0.040 | 0.10-0.50 | | 0.40-0.70 | | 0.10-0.18 | | 75-105 | 55-60 |
| A302 | A | 0.25 | 0.95-1.30 | 0.035 | 0.040 | 0.15-0.30 | | | 0.45-0.60 | | | 75-95 | 45 |
| | B | 0.25 | 1.15-1.50 | 0.035 | 0.040 | 0.15-0.30 | | | 0.45-0.60 | | | 80-100 | 50 |
| | C | 0.25 | 1.15-1.50 | 0.035 | 0.040 | 0.15-0.30 | | 0.40-0.70 | 0.45-0.60 | | | 80-100 | 50 |
| | D | 0.25 | 1.15-1.50 | 0.035 | 0.040 | 0.15-0.30 | | 0.70-1.00 | 0.45-0.60 | | | 80-100 | 50 |
| A353 | | 0.13 | 0.90 | 0.035 | 0.040 | 0.15-0.30 | | 8.50-9.50 | | | | 100-120 | 75 |
| A735 | | 0.06 | 1.20-2.20 | 0.04 | 0.025 | 0.40 | | | 0.23-0.47 | | 0.20-0.35 Cu; 0.03-0.09 Cb | 80-115 | 65-80 |
| A736 | | 0.07 | 0.40-0.70 | 0.025 | 0.025 | 0.35 | 0.60-0.90 | 0.70-1.00 | 0.15-0.25 | | 1.00-1.30 Cu; 0.02 min Cb | 72-105 | 55-75 |
| A737 | B | 0.20 | 1.15-1.50 | 0.035 | 0.030 | 0.15-0.50 | | | | | 0.05 Cb | 70-90 | 50 |
| | C | 0.22 | 1.15-1.50 | 0.035 | 0.030 | 0.15-0.50 | | | | 0.04-0.11 | 0.03 N | 80-100 | 60 |

a.  Single values are maximum unless otherwise noted.

the Charpy impact test, is improved substantially over that of carbon steel. The weldability of this group of HSLA steels is considered to be excellent.

Typical Group A steels are covered by ASTM Specifications A441, A572, and A633 for structural applications, and A225 and A737 for pressure vessels. Structural applications include pipelines, buildings, bridges, construction equipment and machinery, and railroad equipment.

Group B HSLA steels have higher yield strengths and better corrosion resistance than carbon steel. Alloying elements used to improve corrosion resistance include Cu, P, Ni, Cr, and Si in various combinations and amounts. The corrosion resistance of some of these steels is up to four times better than that of carbon steel.

In addition to improved corrosion resistance, Group B steels have improved notch toughness when compared to ordinary carbon-manganese steels, such as ASTM A36. They are well suited for many low temperature applications. Typical Group B steels are covered by ASTM Specifications A242 and A588.

As-rolled HSLA steels have relatively poor and variable notch toughness at low temperature. If notch toughness is important in the application, an adequate minimum level should be specified. Steels that are designed specifically for low temperature applications are made to special chemical compositions using killed, fine-grain practice. They are used in the normalized or quenched-and-tempered condition.

The steels specifically intended for low temperature applications are covered in ASTM Specifications A203 and A353. ASTM A353 steel contains 9 percent nickel, and is intended for service at temperatures as low as $-320°F$. (ASTM A553 steels are quenched-and-tempered versions of ASTM A353 steel.)

# WELDING

## Preheat

Successful welding of HSLA steels requires consideration of preheat and also control of hydrogen in the welding process. Preheat is beneficial in controlling the cooling rate of the weld to reduce or avoid martensitic transformation in the heat-affected zone. This, in turn, controls the hardness of this zone. Preheat also tends to reduce welding stresses somewhat.

Table 1.10 gives the recommended mini-

mum preheat and interpass temperatures for arc welding several HSLA structural steels. Similar information for other HSLA steels may be obtained from WRC Bulletin 191[23] or the steel producer. The preheat temperature must be high enough to prevent cracking, particularly with highly restrained joints. When welding two different HSLA steels together, the higher of the two preheat temperatures should be used. The base metal on each side of the joint must be preheated to temperature over a distance equal to the section thickness or 3 in., whichever is the larger.

## Welding Processes

High-strength low alloy steels can be welded by all of the commonly used arc welding processes. Shielded metal arc, gas metal arc, flux cored arc, and submerged arc welding are used for most applications. Low hydrogen practices should be employed with all processes.

Electroslag and electrogas welding may be used, but there are certain loading restrictions for bridge and other applications. Preheat is not normally required with these two processes because of their high heat input characteristics. Welding with one of these processes or with multiple-electrode submerged arc welding, requires careful evaluation where good notch toughness in the weld metal and heat-affected zone is an important requirement. However, heat input must be considered also with the other processes for good notch toughness.

The HSLA steels can also be joined by resistance spot, seam, projection, upset, and flash welding. When spot, seam, or projection welding, these steels can be welded with about the same current and time settings used for low carbon steel. However, higher electrode force may be needed because of the higher strength of the steel. Higher upsetting force may be required with flash or upset welding for the same reason, but the other welding variables should be similar to those used for low carbon steel. Preheat or postheat cycles may be helpful during resistance welding of some HSLA steels to avoid excessive hardening of the weld and heat-affected zone.

---

23. Ott, C. W. and Snyder, D. J. *Suggested Arc Welding Procedures for Steels Meeting Standard Specifications,* Bulletin 191, Welding Research Council, 1974 Jan.

**Table 1.10**
**Minimum preheat and interpass temperature for ASTM high-strength low alloy structural steels using low hydrogen welding procedures**

| ASTM steel | Thickness, in[a] | Minimum temperature, °F |
|---|---|---|
| A242<br>A441<br>A572, Gr 42, 50<br>A588<br>A633, Gr A, B, C, D | Up to 0.75<br>0.81 to 1.50<br>1.56 to 2.50<br>Over 2.50 | 32<br>50<br>150<br>225 |
| A572, Gr 60, 65<br>A633, Gr E | Up to 0.75<br>0.81 to 1.50<br>1.56 to 2.50<br>Over 2.50 | 50<br>150<br>225<br>300 |

a. Thickness of thicker section at the joint

## Filler Metals

The considerations governing the choice of welding electrodes or rods for arc welding as-rolled HSLA steels are often very similar to those for the structural carbon steels, except that it is generally necessary to use low hydrogen welding practices with these higher strength steels. Although it is possible to weld some of these steels with non-low hydrogen electrodes (EXX10-X, EXX11-X), most steel manufacturers recommend that low hydrogen electrodes always be used.

For normalized HSLA steels, which have good notch toughness, additional consideration must be given to the choice of arc welding electrodes and procedures that will provide adequate levels of notch toughness in both the weld metal and the heat-affected zone. Welding consumables capable of producing weld metals with various levels of notch toughness are available for use with the commonly used arc welding processes. In some cases, the weld metal composition can be similar to that of the steel. The specific welding procedures used, including preheat, heat input, and weld bead size, can have an effect on the notch toughness of both the weld metal and the heat-affected zone. If notch toughness of those areas is of concern, a welding procedure should be developed, tested, and then rigorously followed during fabrication to assure that satisfactory mechanical properties will be achieved in the welded joint.

Suggested electrodes for arc welding several HSLA steels for structural applications are given in Table 1.11. For ASTM A242 and A588 steels used in the bare, unpainted condition, welding electrodes should be specified that will provide weld metal having similar corrosion resistance and color match in the weathered condition. Welding consumables are available to provide various weld metal compositions to meet those requirements. Table 1.12 suggests specific welding consumables for these steels that will provide a suitable weld metal for such applications.

Table 1.13 gives the covered electrodes typically used to arc weld some HSLA pressure vessel steels. The electrodes shown generally match the composition of the respective steels. Similar consumables are available for other welding processes.

In the case of ASTM A353 steel, E310 or ER310 stainless steel electrodes can be used when weld metal with good toughness at cryogenic temperatures is required. If that choice is contemplated, consideration must be given to the difference in thermal expansion characteristics of the base metal and the austenitic stainless steel weld metal. The welded joint must be capable of withstanding the stresses developed at the weld interface if it is thermally cycled in service.

Low alloy steel electrodes are available for electroslag and eletrogas welding of HSLA steels.[24] A flux is used in combination with an

24. Refer to AWS A5.25-78, *Specification for Consumables Used for Electroslag Welding of Carbon and High-Strength Low Alloy Steels,* or AWS A5.26-78, *Specification for Consumables Used for Electrogas Welding of Carbon and High-Strength Low Alloy Steels.*

### Table 1.11
### Suggested consumables for welding high-strength low alloy steels for structures

| | Welding process | | | |
|---|---|---|---|---|
| ASTM steel | Shielded metal arc | Submerged arc | Gas metal arc | Flux covered arc |
| A242[a]<br>A441<br>A572, Grade 42<br>A588[a] (4 in. and under)<br>A633, Grades A, B,[a] C, D<br>(2.5 in. and under) | E7015<br>E7016<br>E7018<br>E7028 | F7XX-EXXX | ER70S-X | E7XT-1<br>E7XT-4, 5, 6, 7, or 8<br>E7XT-11<br>E7XT-G |
| A572 Grade 60, 65<br>A633 Grade E[a] | E8015-XX<br>E8016-XX<br>E8018-XX | F8XX-EXXX[b] | ER80S-XX[b] | E8XTX-XX[b] |

a. Another filler metal, such as E80XX-XX covered electrodes, special welding procedures, or both, may be required for good notch toughness or for atmospheric corrosion and weathering applications.
b. For bridges, the deposited weld metal must have a minimum Charpy V-notch impact strength of 20 ft·lbs at 0°F.

electrode with electroslag welding. The flux is classified on the basis of the mechanical properties of a weld deposit made with a particular electrode. With both processes, the weld metal is required to have a tensile strength, as-welded, in the range of 70 to 90 ksi and a minimum yield strength of 50 ksi.

When welding ASTM A242 or A588 steel for bare exposed service, the deposited weld metal is required to have a chemical composition similar to that obtained from one of the shielded metal arc welding electrodes listed in Table 1.12.

### Table 1.12
### Suggested welding consumables for exposed bare applications of ASTM A242 and A588 steels

| | Welding process | | |
|---|---|---|---|
| Shielded metal arc | Submerged arc | Gas metal arc | Flux cored arc[2] |
| AWS A5.5 | AWS A5.23 | AWS A5.28 | AWS A5.29 |
| E8016 or 18-G[1,2]<br>E8016 or 18-B1[2]<br>E8016 or 18-B2[2]<br>E8015 or 18-B2L[2] | F7XX-EXXX-W[2,3]<br><br>F7XX-EXXX-B2[2,3] | ER80S-G[2]<br>ER80S-B2[2,3]<br>ER80S-B2L[2,3] | E80T1-W<br>E81T1-B1<br>E8XTX-B2<br>E80T5-B2L |
| E8016 or 18-C1<br>E8016 or 18-C2<br>E8016 or 18-C3 | F7XX-EXXX-Ni1[3]<br>F7XX-EXXX-Ni2[3]<br>F7XX-EXXX-Ni3[3]<br>F7XX-EXXX-Ni4[3] | ER80S-Ni1[3]<br>ER80S-Ni2[3]<br>ER80S-Ni3[3] | E8XTX-Ni1<br>E8XTX-Ni2<br>E80T5-Ni3 |

1. Deposited weld metal shall have the following chemical composition percent: 0.12 C max, 0.50-1.30 Mn, 0.03 P max, 0.04 S max. 0.35-0.80 Si, 0.30-0.75 Cu, 0.40-0.80 Ni, 0.45-0.70 Cr.
2. Deposited weld metal shall have a minimum Charpy V-notch impact strength of 20 ft·lb (27.1J) at 0°F(− 18°C) for bridges.
3. The use of the same type of filler metal having next higher mechanical properties as listed in the AWS specifications is permitted.

**Table 1.13**
**Suggested electrodes for shielded metal arc welding high strength low-alloy steels for pressure vessels**

| ASTM steel | AWS Specification | Electrode Classification | Nominal weld metal composition, percent |
|---|---|---|---|
| A203 Gr A, B Gr C, D | A.5.5-81 | E80XX-C1 E80XX-C2 | 2.5Ni 3.5Ni |
| A204 Gr A, B Gr C | A5.5-81 | E70XX-A1 E80XX-B2[a] | 0.5Mo 1.25Cr-0.5Mo |
| A225 Gr C Gr D | A5.5-81 | E120XX-M E100XX-M | 1.8Mn-2.1Ni-0.7Cr-0.4Mo 1.2Mn-1.8Ni-0.3Cr-0.4Mo |
| A302 Gr A, B Gr C, D | A5.5-81 | E90XX-D1 E100XX-D2 | 1.5Mn-0.4Mo 1.75Mn-0.4Mo |
| A353 | A5.11-76 A5.4-78 | ENiCrFe-2[b] ENiCrMo-3 E310 | 70Ni-15Cr-12Fe 55Ni-22Cr-9Mo 25Cr-20Ni |

a. Do not use for low temperature applications.
b. Joint efficiency will be only about 95 percent with this electrode.

## Postweld Heat Treatment

Weldments in structures of HSLA steels are seldom postweld heat-treated. However, a heat treatment may be required if the weldment must maintain dimensions during machining, is exposed to an environment that can cause stress corrosion cracking, or requires softening to im-

**Table 1.14**
**Suggested postweld heat-treating temperatures for high-strength low alloy steels**

| ASTM Specification | Temperature range, °F[a] |
|---|---|
| A203 A225 A302 A441[b] A572[b] A588[b] A633[b] A737 | 1100-1250 |
| A204 | 1150-1350 |
| A353 | 1025-1085 |
| A710 | 1100-1200 |
| A735 A736 | 1000-1200 |

a. For boilers and pressure vessels, refer to the appropriate section of the ASME Boiler and Pressure Vessel Code for postweld heat-treating requirements.
b. Several propriety HSLA steels are produced to this Specification. Consult the producer for recommendations on a particular steel.

prove ductility. Suggested postweld heat-treating temperatures for several ASTM HSLA steels are given in Table 1.14. For proprietary steels, the recommendations of the producer should be followed. Some fabrication codes require that welded joints or the entire weldment be heat-treated for certain applications.

## BRAZING

High-strength low alloy steels can be brazed with the same brazing filler metals, processes, and procedures used for carbon steels. When selecting a brazing filler metal and a brazing process, the effects of the proposed brazing cycle on the mechanical properties of the steel must be considered. Exposure of a steel to a particular brazing cycle may alter its mechanical properties, particularly those of heat treated steel. In most cases, the brazing temperature should be below the lower transformation temperature of the steel.

## THERMAL CUTTING

High-strength low alloy steels can be thermally cut using conventional oxyfuel gas, air-carbon arc, plasma arc, and other thermal cutting processes. The procedures are the same as those used for carbon steels, described previously. Preheating temperatures suggested for arc welding in Table 1.10 are recommended for cutting operations to minimize cracking in the cut edge.

The cut surface should be clean and free of cracks for subsequent welding. Oxide or other residue on the surface and any cracks should be removed mechanically.

# QUENCHED-AND-TEMPERED STEELS

## DESCRIPTION

Quenched-and-tempered (Q&T) steels are furnished in the heat-treated condition with yield strengths ranging from 50 to 150 ksi depending on chemical composition, thickness, and heat treatment. The heat treatment for most of these steels consists of austenitizing, quenching, and tempering. A few are given a precipitation hardening (aging) treatment following hot rolling or a hardening treatment. Some Q & T steels fall within the structural carbon steel classification, some within the low alloy steel classification, and others within other alloy steel classifications of the American Iron and Steel Institute (AISI). Welded structures fabricated from these steels generally do not need further heat treatment except for a postweld heat treatment (stress relief) in special situations.

## COMPOSITION AND PROPERTIES

The Q&T steels combine high yield and tensile strengths with good notch toughness, ductility, corrosion resistance, or weldability. The various steels have different combinations of these characteristics based on their intended applications. The chemical compositions and strengths of typical Q&T steels are given in Table 1.15. Many of the steels are covered by ASTM Specifications. However, a few steels, such as HY-80, HY-100, and HY-130, are covered by Military Specifications. The steels listed are used primarily as plate. Some of these steels, as well as other similar steels, are produced as forgings or castings.

ASTM A514 and A517 plate and ASTM A592 forgings are used in a wide range of applications where good notch toughness is needed in a steel having a yield strength of 100 ksi. Typical applications are earth-moving equipment, pressure vessels, bridges, steel mill and mining equipment, TV towers, fans, and ships.

ASTM A533 Grade B steel is used in thick sections for nuclear pressure vessel construction. At its highest strength level (Class 3) it may be used for thin-wall or multiple-layer pressure vessels. ASTM A537 Class 2 carbon steel is intended for use in pressure vessels where high notch toughness and a minimum yield strength of 60 ksi are required.

ASTM A543 Type B, Classes 1 and 2 steels

are modifications of HY-80 and HY-100 steels. These steels are intended for nuclear reactor vessels and other applications requiring high yield strength and good toughness. HY-80 and HY-100 steels are used in construction of ships, submarines, and other marine equipment. HY-130 steel is similar to the other two HY steels, but it has a higher yield strength and better toughness.

ASTM A553 steels contain 8 or 9 percent nickel and are essentially quenched and tempered variations of ASTM A353 normalized and tempered 9 percent nickel steel. The welding considerations are the same for both types of steels regardless of heat treatment. Therefore, the information presented previously for ASTM A353 steel is applicable to A553 steel.

ASTM A678 Q&T carbon steel is somewhat similar to ASTM A537 Class 2 steel, but is intended for structural applications where good notch toughness and a yield strength of about 60 ksi are required.

The carbon content of Q&T steels generally does not exceed 0.22 percent for good weldability. The other alloying elements are carefully selected to provide the most economically heat-treated steel with the desired properties and acceptable weldability. The heat treatment used for the steels in Table 1.15 by the manufacturer and the resulting microstructure are listed in Table 1.16. Austenitizing and tempering temperatures for other Q&T steels may be obtained from the applicable specification or the producer.

## WELDING METALLURGY

Various Q&T steels respond differently to heat treating and welding, depending upon their chemical composition and thickness. In general, the steels are alloyed to favor the formation of tempered martensite and lower bainite structures (Table 1.16). Some of the metallurgical features of Q&T steels are discussed below using ASTM A514/A517 steels as an example. These ASTM steels are frequently specified for high-strength welded structures.

### Isothermal Transformation

An isothermal transformation diagram for a typical ASTM A514 or A517 steel is shown in Fig. 1.6. The transformation behavior of this steel has several significant features. One feature is the long time before transformation of austenite to pearlite begins in the temperature range of about 1100° to 1300°F. This long time precludes

formation of upper bainite, ferrite, and pearlite unless the cooling rate is very slow. The presence of these microstructures would have a detrimental effect on strength and toughness.

Another feature is that a moderately long time elapses before transformation to ferrite and upper bainite starts in the temperature range of about 950° to 1100°F. This permits the quenching of plates up to at least 2 inches thick with little or no transformation in this temperature range. Transformation to ferrite and upper bainite is best avoided because the remaining austenite, enriched in carbon, may be retained or transformed to high carbon martensite or bainite during cooling to room temperature. If considerable transformation occurs in the temperature range of 950° to 1100°F as a result of slow cooling, the final microstructure will consist of a heterogeneous mixture of ferrite, upper bainite, and retained austenite, together with high carbon martensite or bainite. Such a combination of soft and hard microconstituents, even when tempered, will not have a high level of toughness.

If the steel is cooled at a relatively rapid rate, transformation to lower bainite occurs in a relatively short period at temperatures below 900°F. Such a microstructure is essentially homogeneous and has good toughness, particularly after tempering.

Martensite begins to form at a relatively high temperature of 740°F. This characteristic is a significant factor in the freedom from quench-cracking displayed by this steel. It is also an important consideration in welding because cooling of martensite from such a relatively high temperature provides some degree of self-tempering.

### Hardenability

The specific chemical composition of a steel plays a major role in its hardenability. End-quench hardenability bands for typical heats of ASTM A514 or A517 Grades B and F steels are shown in Fig. 1.7. The lower alloy content of Grade B steel is reflected in lower hardenability as shown by the more rapid drop in hardness with distance from the quenched end of the test specimen. However, the hardenabilities of the two steels are sufficiently similar to produce nearly identical microstructures at a point slightly more than 0.5 in. from the quenched end. Such hardenability response would be expected to provide similar microstructures in plate thicknesses up to 1.25 in. when similarly quenched. Grade F steel may be used for plate thicknesses exceed-

## Table 1.15
### Typical quenched-and-tempered steels

| ASTM Specification or common designation | Grade, type, or class | Composition, percent | | | | | | | | | | Min strength, ksi | |
|---|---|---|---|---|---|---|---|---|---|---|---|---|---|
| | | C | Mn | P | S | Si | Cr | Ni | Mo | Cu | Others | Tensile | Yield |
| A514/A517 | A | 0.15-0.21 | 0.80-1.10 | 0.035 | 0.04 | 0.40-0.80 | 0.50-0.80 | | 0.18-0.28 | | Zr, 0.05-0.15; B, 0.0025 | 100-130/ 105-135 | 90/100 |
| | B | 0.12-0.21 | 0.70-1.00 | 0.035 | 0.04 | 0.20-0.35 | 0.40-0.65 | | 0.15-0.25 | | V, 0.03-0.08; Ti, 0.01-0.03; B, 0.0005-0.005 | | |
| | C | 0.10-0.20 | 1.10-1.50 | 0.035 | 0.04 | 0.15-0.30 | | | 0.20-0.30 | | B, 0.001-0.005 | | |
| | D | 0.13-0.20 | 0.40-0.70 | 0.035 | 0.04 | 0.20-0.35 | 0.85-1.20 | | 0.15-0.25 | 0.20-0.40 | Ti, 0.004-0.10[b]; B, 0.0015-0.005 | | |
| | E | 0.12-0.20 | 0.40-0.70 | 0.035 | 0.04 | 0.20-0.35 | 1.40-2.00 | | 0.40-0.60 | 0.20-0.40 | Ti, 0.04-0.10[b]; B, 0.0015-0.005 | | |
| | F | 0.10-0.20 | 0.60-1.00 | 0.035 | 0.04 | 0.15-0.35 | 0.40-0.65 | 0.70-1.00 | 0.40-0.60 | 0.15-0.50 | V, 0.03-0.08; B, 0.0005-0.006 | | |
| | G | 0.15-0.21 | 0.80-1.10 | 0.035 | 0.04 | 0.50-0.90 | 0.50-0.90 | | 0.40-0.60 | | Zr, 0.05-0.15; B, 0.0025 | | |
| | H | 0.12-0.21 | 0.95-1.30 | 0.035 | 0.04 | 0.20-0.35 | 0.40-0.65 | 0.30-0.70 | 0.20-0.30 | | V, 0.03-0.08; B, 0.0005-0.005 | | |
| | J | 0.12-0.21 | 0.45-0.70 | 0.035 | 0.04 | 0.20-0.35 | | | 0.50-0.65 | | B, 0.001-0.005 | | |
| | K | 0.10-0.20 | 1.10-1.50 | 0.035 | 0.04 | 0.15-0.30 | | | 0.45-0.55 | | B, 0.001-0.005 | | |
| | L | 0.13-0.20 | 0.40-0.70 | 0.035 | 0.04 | 0.20-0.35 | 1.15-1.65 | | 0.25-0.40 | 0.20-0.40 | Ti, 0.04-0.10[b]; B, 0.0015-0.005 | | |
| | M | 0.12-0.21 | 0.45-0.70 | 0.035 | 0.04 | 0.20-0.35 | | 1.20-1.50 | 0.45-0.60 | | B, 0.001-0.005 | | |
| | N | 0.15-0.21 | 0.80-1.10 | 0.035 | 0.04 | 0.40-0.90 | 0.50-0.80 | | 0.25 | | Zr, 0.05-0.15; B, 0.0005-0.0025 | | |
| | P | 0.12-0.21 | 0.45-0.70 | 0.035 | 0.04 | 0.20-0.35 | 0.85-1.20 | 1.20-1.50 | 0.45-0.60 | | B, 0.001-0.005 | | |
| | Q | 0.14-0.21 | 0.95-1.30 | 0.035 | 0.04 | 0.15-0.35 | 1.00-1.50 | 1.20-1.50 | 0.40-0.60 | | V, 0.03-0.08 | | |

a. When a single value is shown, it is a maximum limit.
b. Vanadium may be substituted for part or all of titanium content on a one for one basis.
c. Limiting values vary with the plate thickness.
d. When specified.

**Table 1.15 (cont.)**
**Typical quenched-and-tempered steels**

| ASTM Specification or common designation | Grade, type, or class | Composition, percent | | | | | | | | | | Min strength, ksi | |
|---|---|---|---|---|---|---|---|---|---|---|---|---|---|
| | | C | Mn | P | S | Si | Cr | Ni | Mo | Cu | Others | Tensile | Yield |
| A533 | A | 0.25 | 1.15-1.50 | 0.035 | 0.040 | 0.15-0.30 | | | 0.45-0.60 | | | 80-125 | 50-82.5 |
| | B | 0.25 | 1.15-1.50 | 0.035 | 0.040 | 0.15-0.30 | | 0.40-0.70 | 0.45-0.60 | | | | |
| | C | 0.25 | 1.15-1.50 | 0.035 | 0.040 | 0.15-0.30 | | 0.70-1.00 | 0.45-0.60 | | | | |
| | D | 0.25 | 1.15-1.50 | 0.035 | 0.040 | 0.15-0.30 | | 0.20-0.40 | 0.45-0.60 | | | | |
| A537 | 2 | 0.24 | 0.70-1.60 | 0.035 | 0.040 | 0.15-0.30 | 0.25 | 0.25 | 0.08 | 0.35 | | 70-100 | 46-60 |
| A543 | B | 0.23 | 0.40 | 0.020 | 0.020 | 0.20-0.40 | 1.50-2.00 | 2.60-4.00[c] | 0.45-0.60 | | V, 0.03 | 90-135 | 70-100 |
| | C | 0.23 | 0.40 | 0.020 | 0.020 | 0.20-0.40 | 1.20-1.50 | 2.25-3.50 | 0.45-0.60 | | V, 0.03 | | |
| A678 | A | 0.16 | 0.90-1.50 | 0.04 | 0.05 | 0.15-0.50 | 0.25 | 0.25 | 0.08 | 0.20[d] | | 70-90 | 50 |
| | B | 0.20 | 0.70-1.60 | 0.04 | 0.05 | 0.15-0.50 | 0.25 | 0.25 | 0.08 | 0.20[d] | | 80-100 | 60 |
| | C | 0.22 | 1.00-1.60 | 0.04 | 0.05 | 0.20-0.50 | 0.25 | 0.25 | 0.08 | 0.20[d] | | 85-115 | 65-75 |
| HY-80 | | 0.12-0.18 | 0.10-0.40 | 0.025 | 0.025 | 0.15-0.35 | 1.00-1.80 | 2.00-3.25 | 0.20-0.60 | 0.25 | V, 0.03; Ti, 0.02 | — | 80 |
| HY-100 | | 0.12-0.20 | 0.10-0.40 | 0.025 | 0.025 | 0.15-0.35 | 1.00-1.80 | 2.25-3.50 | 0.20-0.60 | 0.25 | V, 0.03; Ti, 0.02 | — | 100 |
| HY-130 | | 0.12 | 0.60-0.90 | 0.010 | 0.015 | 0.15-0.35 | 0.40-0.70 | 4.75-5.25 | 0.30-0.65 | | V, 0.05-0.10 | — | 130 |

a. When a single value is shown, it is a maximum limit.
b. Vanadium may be substituted for part or all of titanium content on a one for one basis.
c. Limiting values vary with the plate thickness.
d. When specified.

### Table 1.16
### Heat treatments for typical quenched-and-tempered steels

| Steel | Austenitizing temperature, °F | Quenching medium | Tempering temperature, °F | Microstructure |
|---|---|---|---|---|
| ASTM A514, A517 | 1650 | Water or oil | 1150 | Bainite and martensite |
| ASTM A533 Type B | 1550 | Water | 1100 | Bainite and martensite (thin plate) or ferrite and bainite (thick plate) |
| ASTM A537 Class 2 | 1650 | Water | 1100 | Ferrite and bainite, martensite, or all three |
| ASTM A543 Type B | 1650 | Water | 1100 | Bainite and martensite |
| ASTM A678 Grade C | 1650 | Water | 1100 | Ferrite and bainite, martensite, or all three |
| HY-80, HY-100 | 1650 | Water | 1200 | Bainite and martensite |
| HY-130 | 1500 | Water | 1000 | Bainite and martensite |

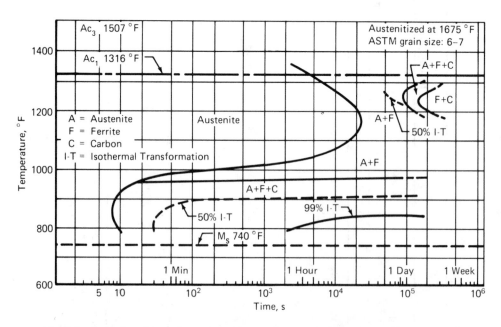

*Fig. 1.6—Isothermal transformation diagram for a typical ASTM A514 or A517 steel*

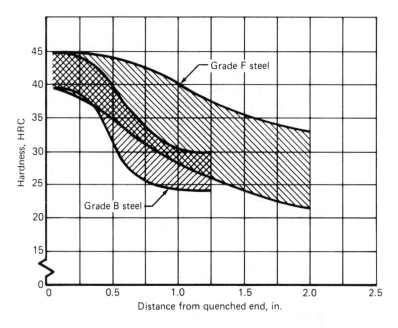

**Fig. 1.7—End quench hardenability bands for ASTM A514 or A517 Grades B and F steels**

ing 1.25 in. because of its better hardenability. The hardenability bands also give an indication of maximum hardness that may be expected in the heat-affected zone of a weld in the as-welded condition.

### Tempering Response

The effect of tempering on the hardness and strength of ASTM A514 or A517 Grades B and F steels is shown in Fig. 1.8. The change in slope of the hardness curves for both steels at about 950°F is a result of the secondary hardening effect of vanadium in these steels. Secondary hardening, from the precipitation of fine vanadium carbides, occurs at tempering temperatures up to about 1250°F, but is most pronounced in the range of 950° to 1150°F.

### JOINT DESIGN

Appropriate joint design, good workmanship, and adequate inspection are needed to take advantage of the high strength of quenched and tempered steels and to optimize the serviceability of weldments made from them. Abrupt changes in cross section or stress raisers in regions of high stress, which may be tolerable with low carbon and mild steels, are detrimental with steels having yield strengths of about 80 ksi or higher. Incorrectly located welds can cause, or contribute to, abrupt changes in cross section. Welded joints in Q&T steels should be carefully designed and properly located to provide adequate access for good welding practices and proper inspection. Groove welds are generally better than fillet welds for many applications because of the inherent stress concentrations usually associated with fillet welds. Reliable inspection can be performed on groove welds using radiographic or ultrasonic methods. Fillet welds are difficult to inspect by these methods.

Weld joint designs employing V- or U-grooves are preferred. Bevel- or J-grooves may be used but the welding procedures must ensure complete fusion with both groove faces. Complete joint penetration is desirable in most cases. Double welded joints should be used where prac-

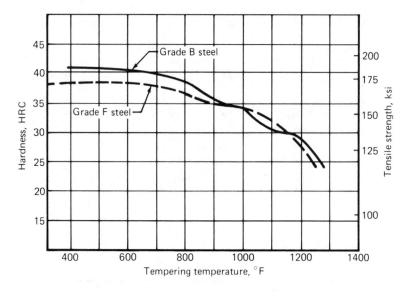

**Fig. 1.8—Relationship between hardness, tensile strength, and tempering temperature for quench-hardened ASTM A514 or A517 Grades B and F steels**

tical to help balance the welding stresses across the joints.

Stress raisers, such as those found most commonly at the toes of improperly contoured welds, are of great concern with high strength steels. As the yield strength increases, plastic deformation takes place in a smaller volume of metal in the immediate vicinity of the stress raiser. During that deformation, the plasticity of the weld metal or heat-affected zone may be exceeded and result in formation of a crack at the stress raiser.

Regardless of whether a groove or fillet weld is used to join a high strength steel, the weld metal should blend smoothly with the base metal at the toes and the root of the weld. Excessive weld reinforcement should be avoided.

Structural weldments are usually complex geometries that have abrupt changes in section that act as stress raisers. The stresses at these locations may be as high as the yield strength of the steel. Joint design and external geometry, combined with the range and direction of stress, control the fatigue life of a sound weld. Other factors such as steel composition, strength, toughness, microstructure, heat treatment, and minimum stress have only a secondary effect on

fatigue life. The toughness of the steel, as it relates to fracture toughness, determines the length to which a fatigue crack must propagate (critical crack size) before it progresses rapidly through the weldment. Use of a high-toughness steel would not be expected to extend significantly the fatigue life of a weldment because the extension of existing discontinuities to fatigue cracks of detectable length requires a large percentage of the loading cycles to failure. Therefore, a high strength steel offers no improvement over a carbon steel having a yield strength of about 36 ksi for weldments exposed to fatigue loading. Weld joint designs typically used for structural applications are categorized according to their expected fatigue performance in many codes and specifications.[25]

## PREHEAT

Preheat for welding Q&T steels must be used with caution because it reduces the cooling

---

25. AWS D1.1-82, *Structural Welding Code–Steel*, published by the American Welding Society, contains this type of information.

rate of the weld heat-affected zone. If the cooling rate is too slow, the reaustenitized zone adjacent to the weld metal can transform to ferrite with regions of high carbon martensite or to coarse bainite. Both microstructures lack high strength and good toughness. Next to this zone, a band of previously tempered steel may be overtempered with a decrease in strength.

The effect of two preheat temperatures on the toughness of simulated heat-affected zones in ASTM A514 or A517 steel plate, 0.5-in. thick, is illustrated by Fig. 1.9. The curves indicate the toughness of the grain-coarsened area in the heat-affected zone on a final pass or a multiple-pass weld or a single-pass weld. Such an area is considered to be the worst condition because it does not receive the benefits of tempering from additional weld passes. The toughness of the heat-affected zone of prior passes in multiple-pass welds is expected to be significantly better. Nevertheless, the higher preheat temperature (500°F) resulted in significantly decreased toughness when compared to that with the lower preheat temperature (200°F).

In some cases, a moderate preheat can provide insurance against cracking from several causes. Preheating is recommended when the joint to be welded is thick and highly restrained or when some other condition is present that may cause cracking.

Suggested minimum preheat and interpass temperatures for welding several ASTM Q&T steels are given in Table 1.17. The maximum temperature should not exceed the suggested minimum temperature by more than 150°F because the upper limit for acceptable welding heat input, described later, may be too restrictive. Table 1.18 gives the preheat temperature ranges for HY-80, HY-100, and HY-130 steels for military applications. If a preheat of less than 100°F is used, precautions are necessary to ensure that the surfaces to be welded are free of moisture.

## WELDING PROCESSES

Any of the commonly used welding processes, such as shielded metal arc, submerged arc, gas metal arc, flux cored arc, and gas tungsten arc welding can be used to join quenched and tempered steels having minimum yield strengths up to 150 ksi. However, the gas metal arc and gas tungsten arc welding processes are

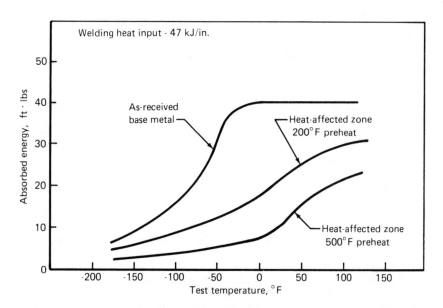

**Fig. 1.9—Effect of preheat temperature on the Charpy V-notch toughness of the heat-affected zone (2400°F) in 0.5-in. thick ASTM A514 or A517 steel**

**Table 1.17**
**Suggested minimum preheat and interpass temperatures for arc welding typical ASTM quenched-and-tempered steels**

| Thickness range, in. | Minimum preheat and interpass temperature, °F[a] | | | | |
|---|---|---|---|---|---|
| | A514/A517 | A533 | A537 | A543 | A678 |
| Up to 0.50 | 50 | 50 | 50 | 100 | 50 |
| 0.56 to 0.75 | 50 | 100 | 50 | 125 | 100 |
| 0.81 to 1.00 | 125 | 100 | 50 | 150 | 100 |
| 1.1 to 1.5 | 125 | 200 | 100 | 200 | 150 |
| 1.6 to 2.0 | 175 | 200 | 150 | 200 | 150 |
| 2.1 to 2.5 | 175 | 300 | 150 | 300 | 150 |
| over 2.5 | 225 | 300 | 225 | 300 | — |

a. With low hydrogen welding practices. Maximum temperature should not exceed the given value by more than 150°F.

**Table 1.18**
**Recommended preheat and interpass temperature ranges for HY-80, HY-100, and HY-130 steels**

| Thickness range, in. | Temperature, °F | |
|---|---|---|
| | HY-80, HY-100 | HY-130 |
| Up to 0.5 | 60-300 | 75-150 |
| 0.51 to 0.63 | 125-300 | 75-150 |
| 0.64 to 0.88 | 125-300 | 125-200 |
| 0.89 to 1.13 | 125-300 | 200-275 |
| 1.14 to 1.38 | 200-300 | 200-275 |
| Over 1.38 | 200-300 | 225-300 |

best for steels with minimum yield strengths over 100 ksi, such as HY-130, to achieve a satisfactory level of notch toughness in the weld metal.

The cooling rate of an arc weld is usually sufficiently rapid under normal conditions so that the mechanical properties of the heat-affected zone approach those of the unaffected base metal. A hardening heat treatment (quenching and tempering) after welding is usually not necessary.

Excessive heat input when welding a quenched and tempered steel can reduce the strength and toughness of the welded joint. Such reductions can occur in the heat-affected zone or the weld metal, or both. Welding with a process that uses an extremely high heat input with an accompanying slow cooling rate, such as electroslag, electrogas, or multiple-wire submerged arc welding, may require complete reheat treatment of the welded joint (quenching and tempering) to obtain acceptable mechanical properties.

Quenched-and-tempered steels can be joined by electron beam (EBW) and laser beam (LBW) welding. The heat inputs with these processes will be much lower than those with the arc welding processes. In laboratory tests, autogeneous welds in 0.5 in. thick HY-130 steel plate made by these processes were as strong as the base metal. The weld metal and heat-affected zone microstructures consisted of untempered martensite and bainite, both of which were harder than the unaffected base metal.

## HEAT INPUT

The cooling rates of the weld metal and

heat-affected zone are a function of welding heat input as well as the temperature of the adjacent base metal and the joint thickness. The relative effects of welding heat input and preheat temperature are the same. The higher the heat input, the slower is the cooling rate. For a given joint thickness, and preheat temperature, heat input during welding must not exceed a specific value to have an acceptable cooling rate in the heat-affected zone. If the preheat temperature is increased, the welding heat input must be decreased to maintain the same cooling rate.

The minimum cooling rate required to produce a microstructure with suitable mechanical properties will vary with the particular steel being welded. A cooling rate that is sufficiently fast for one type of steel may be too slow for another. In other words, a heat input suitable for welding a joint in one Q&T steel with a specific thickness and preheat may be too high for a similar joint in another Q&T steel with the same thickness and preheat.

Welding heat is dissipated more rapidly into a thick cross section of steel than into a thin one. For this reason, the maximum heat input can be increased as the joint thickness increases.

Prior to welding a particular Q&T steel, recommendations should be obtained from the steel producer concerning its weldability, hardenability, and preheat and heat input limitations. Maximum heat inputs for welding various thicknesses of ASTM A514 or A517 Grade F steel are given in Table 1.19; for HY-80, HY-100, and ASTM A678 Grade C steels in Table 1.20; and for HY-130 steel in Table 1.21.

The heat input limitations for most Q&T steels are based on toughness requirements because this property decreases more from excessive heat input than does strength. With a few Q&T steels, strength is degraded more severely

### Table 1.19
### Maximum welding heat input for butt joints in ASTM A514 or A517 Grade F steel

| Preheat temperature °F | Maximum heat input, kJ/in.[a] | | | | | | | |
|---|---|---|---|---|---|---|---|---|
| | Section thickness, in. | | | | | | | |
| | 0.19 | 0.25 | 0.5 | 0.75 | 1.0 | 1.25 | 1.5 | 2.0 |
| 70 | 27 | 36 | 70 | 121 | b | b | b | b |
| 200 | 21 | 29 | 56 | 99 | 173 | b | b | b |
| 300 | 17 | 24 | 47 | 82 | 126 | 175 | b | b |
| 400 | 13 | 19 | 40 | 65 | 93 | 127 | 165 | b |

a. Heat input limit for intermediate preheat temperatures and thicknesses may be obtained by interpolation. Heat input limit may be increased 25 percent for fillet welds.

b. No limit with respect to the heat-affected zone cooling rate with most arc welding processes. Welding processes of high heat input should not be used to weld these steels.

### Table 1.20
### Maximum welding heat inputs for joints in HY-80, HY-100, and ASTM A678 Grade C steels

| Section thickness, in. | Maximum heat input, kJ/in. |
|---|---|
| Up to 0.50 | 45 |
| Over 0.50 | 55 |

than is toughness, and the heat input limitations for those steels would be based on achieving adequate joint strength. Heat input limitations have not been established for every Q&T steel. However, excessively high heat inputs should be avoided as a matter of good welding practice. Heat input limitations for a particular Q&T steel may be a specification requirement, as in the case of HY-80, HY-100, and HY-130 steels used for fabrication of naval ships.

Welding heat input may be calculated using the formula

Heat input, J/in. = (60 AV)/s where:

A = welding current, A

V = arc voltage, V

s = arc travel speed, in./min.

Heat input limitations are applicable to each weld pass and are not considered cumulatively. They are also applicable only to single-arc welding processes. Multiple-arc processes with the arcs in tandem generally do not provide time for the first weld bead to cool sufficiently before the trailing arc passes over it and adds additional heat. Therefore, such processes should not be used for welding Q&T steels unless a detailed evaluation proves their suitability.

Welding heat input also has an effect on the mechanical properties of the weld metal. This effect becomes of concern for welding steels having minimum yield strengths greater than 100 ksi. The heat input limitation needed to assure adequate mechanical properties in the weld heat-affected zone also limits the size of the weld beads. Large weld beads characteristically have poor notch toughness. Good practice for welding Q&T steels is to deposit many small stringer beads. This technique produces weld metal with good notch toughness as a result of grain refining and tempering action of succeeding passes. Such practice is especially beneficial if the welded joint must undergo severe plastic deformation during subsequent fabrication.

## WELDING CONSUMABLES

Welding electrodes or electrode-flux combinations are available for arc welding most of the Q&T steels by the commonly used processes. Suggested consumables for typical Q&T steels are given in Table 1.22. Their selection is based on the tensile strength, composition, and notch toughness of the weld metal. Some entries cover two or more classifications that have similar characteristics. Classifications other than those listed may be suitable for some applications.

Selection of the consumables for a specific Q&T steel application should be based on a thorough evaluation of expected weld joint properties. In general, the welding consumables chosen should deposit weld metal having strength and toughness for the application.

A joint efficiency of 100 percent can be obtained in butt joints in these steels when proper welding consumables and procedures are used. However, consumables that deposit weld metal having a strength lower than that of the steel itself are often adequate for welds that are subject to relatively low stresses. Fillet welds stressed in longitudinal shear, welds carrying secondary stresses, and welds joining a lower strength steel to a Q&T steel are typical examples. Low strength weld metal may be also desirable for highly restrained welds to reduce susceptibility to toe cracking or lamellar tearing in the steel, especially with corner and T-joints.

Hydrogen is very detrimental when welding Q&T steels; a very small quantity may cause hydrogen-induced cracking. Since moisture is a source of hydrogen, the moisture content of covered electrodes must be kept below certain levels depending upon the steel being welded. It should not be allowed to exceed 0.4 weight percent when welding Q&T carbon steels having minimum yield strengths up to 70 ksi, such as ASTM A537 steels and A678 steels; and 0.1 weight percent with Q&T low alloy steels having minimum yield strengths of 100 ksi and above, such as HY-130 steel. If the permitted moisture content of the selected covered electrode type exceeds the appropriate limit for the steel, the electrodes must be reconditioned prior to use regardless of the type of packaging. The maximum time of electrode exposure to specific humidity and temperature conditions to keep the moisture content under the appropriate limit should be determined by tests[26] or obtained from the electrode manufacturer. If the data are not available, the maximum exposure time to ambient conditions should not exceed that listed in Table 1.23 for the electrode classification.

Proper packaging and handling procedures can minimize moisture pickup and the need for reconditioning of covered electrodes used for welding Q&T steels. Electrodes should be pack-

---

26. The test to determine coating moisture content is described in AWS A5.5-81, *Specification for Low Alloy Steel Covered Arc Welding Electrodes.*

**Table 1.21**
**Maximum welding heat inputs for joints in HY-130 steel**

| Section thickness, in.[a] | Maximum heat input, kJ/in. | |
|---|---|---|
| | Shielded metal arc welding | Gas metal arc and gas tungsten arc welding |
| 0.37-0.56 | 40 | 35 |
| 0.63-0.88 | 45 | 40 |
| 1.0-1.38 | 45 | 45 |
| 1.5-4.0 | 50 | 50 |

a. Welding of thin sections by shielded metal arc or gas metal arc process is not recommended. Gas tungsten arc welding is recommended for plates under 0.38-in. thick.

**Table 1.22**
**Suggested consumables for arc welding typical quenched-and-tempered steels.**

| ASTM Specification or common name | Grade or Type | Class | Welding process | | | |
|---|---|---|---|---|---|---|
| | | | SMAW[a] | GMAW,[b] GTAW [b] | SAW[c] | FCAW[d] |
| A514/A517 | All | | E1X01X-M | ER1X0S-1 | F1XXX-EXXX-MX | E1X0TX-KX |
| A533 | B | 1, 2 | E901X-D1 E901X-M | ER100S-1 | F9XX-EXXX-FX | E9XTX-NiX E9XTX-K2 |
| | | 3 | E1101X-M | ER110S-1 | F11XX-EXXX-M2 | E110TX-XX |
| A537 | | 2 | E801X-CX E901X-D1 E9018-M | ER80X-NiX | F8XX-EXXX-NiX F9XX-EXXX-NiX | E8XTX-NiX E9XTX-NiX |
| A543 | B | 1, 2 | E1101X-M | ER110S-1 ER120S-1 | F11XX-EXXX-FX | E11XTX-KX |
| A678 | C | | E901X-D1 E9018-M E1001X-D2 E10018-M | ER100S-1 | F9XX-EXXX-FX F10XX-EXXX-MX | E9XTX-NiX |
| HY-80, HY-100 | | | E1101X-M | ER110S-1 ER120S-1 | F11XX-EXXX-MX | E11XTX-KX |

a. Shielded metal arc welding electrodes with low hydrogen coverings in AWS Specification A5.5-81.
b. Electrodes for gas metal arc or gas tungsten arc welding in AWS Specification A5.28-79. Shielding gas for GMAW is argon plus 2 percent oxygen.
c. Submerged arc welding with electode-flux combinations classified in AWS Specification A5.23-80.
d. Flux cored arc welding electrodes classified in AWS Specification A5.29-80. Classification of EXXT1 and EXXT5 electrodes is done with $CO_2$ shielding gas. However, Ar-$CO_2$ gas mixtures may be used when recommended by the manufacturer to improve usability.

**Table 1.23**
**Maximum atmospheric exposure times for low hydrogen covered electrodes**

| Electrode classification | Maximum time, hours[a] |
|---|---|
| E70XX | 4 |
| E80XX | 2 |
| E90XX | 1 |
| E1XXXX | 0.5 |

a. Longer times are permissible when appropriate tests show that the moisture content of the electrodes will not exceed specification limits under job-site conditions.

aged by the manufacturer in hermetically sealed containers. If the container is damaged, the electrodes should be reconditioned prior to use.

Covered electrodes should be placed in a holding oven maintained at 250°F or a low-humidity storage cabinet immediately after opening the container or removing them from a reconditioning oven. Electrodes should be removed from the holding oven or storage cabinet as needed. If the electrodes are exposed to ambient conditions longer than the approved time, they should be reconditioned according to the manufacturer's recommendations and then returned to the holding oven or low-humidity cabinet for storage. When welding under very humid conditions, portable holding ovens should be used at the welding site and the electrodes removed individually, as needed, by the welder. Electrodes contaminated by contact with water, paint, oil, grease, or other hydrocarbons should be discarded.

Bare arc welding electrodes (including flux core types) and welding rods should be produced, packaged, and stored in a manner that will limit the hydrogen content to a low level. It has been suggested that the hydrogen content of filler wire or rod should not exceed 5 parts per million when welding Q&T steels with yield strengths below 100 ksi or 3 parts per million for higher strength steels.

Fluxes for submerged arc welding should be purchased in moisture-resistant packages and stored in a dry location. Packages should not be opened until needed. Immediately after opening, the flux should be placed in the dispensing system. Excess flux from opened packages should be stored in a holding oven operating at about 250°F. Flux from damaged packages should

either be discarded or dried in an oven at about 500°F for at least one hour. After drying, the flux should be used immediately or stored in a holding oven. Any flux that has been contaminated by foreign materials, such as water, oil, or grease, or that was fused during welding should be discarded.

Shielding gas used for gas metal arc, flux cored arc, or gas tungsten arc welding should have a low moisture content, as indicated by a dew point of −40°F or lower.

## WELDING TECHNIQUE

As mentioned previously, deposition of stringer beads without appreciable transverse oscillation of the electrode is preferred for welding Q&T steel. A weave bead technique generally requires a slower travel speed along the joint, with a corresponding increase in heat input. This, in turn, produces welds with lower strength and toughness. For vertical welding, slight weaving of the electrode for no more than two electrode diameters is usually satisfactory.

## POSTWELD HEAT TREATMENT

Service experience with structures and pressure vessels, and also full-scale laboratory tests, have shown that a postweld stress-relief heat treatment is generally not required to prevent brittle fracture in welded joints in most Q&T steels. Of the steels listed in Table 1.15, only welds in ASTM A533 and A537 steels are usually stress relieved when the section thickness is greater than a specified amount. A postweld stress-relief heat treatment with any of the Q&T steels should be used only after assurance that the benefits can be expected and possible harmful effects can be tolerated.

Stress relief is necessary for some applications when (1) the steel has inadequate notch toughness after cold forming or welding, (2) dimensional stability must be maintained during close tolerance machining, or (3) the weldment is susceptible to stress corrosion after cold forming or welding. These conditions are also generally applicable to most other steels.

The need for a postweld stress-relief heat treatment should be thoroughly investigated for each application because many alloy steels are designed for service in the as-welded condition. The mechanical properties of the welded steel or the weld metal itself may be adversely affected by a stress relief. The alloying elements that contribute most to the high strength and notch toughness of Q&T steel and weld metal usually are those that cause adverse effects when the weldment is postweld heat treated.

A postweld heat treatment in the temperature range of 950° to 1200°F may decrease the toughness of the weld metal and the heat-affected zone. The extent of such impairment depends on the chemical composition of the weld or base metal, the stress relieving temperature, and the time at temperature. The impairment is greater with slow cooling, as is commonly used with a stress-relief heat treatment.

Furthermore, when weldments of many high-strength alloy steels are heated above about 950°F, intergranular cracking may take place in the coarse-grained region of the weld heat-affected zone. The intergranular cracking occurs by stress rupture, usually in the early stage of the heat treatment. Susceptibility to cracking increases with increasing weld restraint and severity of stress concentrations. Chromium, molybdenum, and vanadium are the major contributors to this type of crack susceptibility, but other carbide-forming elements contribute to the problem. The precipitation of carbides during stress relaxation at elevated temperature alters the delicate balance between resistance to grain boundary sliding and resistance to deformation within the coarse grains in the heat-affected zone. The cracking is variously known as *stress-rupture cracking*, *stress-relief cracking*, and *reheat cracking*.

Some procedures that may be used singly or in combination to minimize stress-relief cracking in steels include (1) select a weld joint design, weld location, and sequence of assembly that will minimize restraint; (2) use a weld joint design and contour that will minimize stress concentrations; (3) provide a weld metal that has significantly lower strength than that of the heat-affected zone at the heat treating temperature; and (4) peen each layer of weld metal to reduce shrinkage stresses.

If postweld stress relief is required, the temperature must not exceed that used for tempering the steel. In fact, a temperature about 50°F lower than the tempering temperature is desirable to avoid lowering the strength of the steel.

## THERMAL CUTTING

Oxyfuel gas cutting of Q&T steels can be done readily either manually or mechanically. Machine cutting is preferred to obtain relatively smooth, uniform cut edges for welding. Generally, the cutting conditions are the same for carbon steels and Q&T steels. If scale on the steel surface causes erratic cutting action, the travel speed should be reduced about 10 to 15 percent, or a cutting tip of the next larger size should be used. Otherwise, the scale should be removed from the surface by grinding or blast cleaning. Stack cutting of thin plates should be avoided because the high heat input tends to overheat the plate closest to the cutting tip. Plasma arc cutting may also be used on Q&T steels.

Thermally cut edges of Q&T steels will be hardened by the cutting operation but will be relatively tough. These hard edges may be tempered to soften them (preferably in a furnace) to facilitate machining. The tempering temperature should be at least 50°F below the original tempering temperature of the steel to avoid softening.

Quenched and tempered steels can generally be thermally cut without preheat. However, the steel temperature should not be lower than 50°F during cutting. For sections over about 4-in. thick, preheat is required. For example, a preheat of 300°F is suggested by one steel manufacturer for ASTM A514 or A517 steels. For cutting the higher carbon steels, such as ASTM A533 and A543, the preheat temperature recommended for welding is suggested for cutting.

All slag or loose scale from thermal cutting should be removed before welding. Any area of excessive roughness from the cutting should be smoothed by grinding.

Air-carbon arc gouging may be used to remove welds, portions of welds, or base metal. Heat input must be controlled as with arc welding. Proper current, air pressure, and operating technique must be used to minimize the possibility of carbon deposits. The gouged surface should be ground to remove any carburized steel. Thermal gouging using an oxyfuel gas process should not be used on Q & T steels because of the excessive heat input and the slow cooling rate inherent with the process.

## BRAZING

Quenched-and-tempered steels can be brazed using filler metals, brazing methods, and procedures suitable for carbon steels. However, the brazing temperatures used with these filler metals generally exceed the tempering temperatures of Q & T steels. Consequently, the mechanical properties of the steel would be impaired by a brazing cycle. A quench-and-temper heat treatment of the brazement would be required to restore mechanical properties. Therefore, a heat-treatable alloy steel should be considered for a brazing application.

# HEAT TREATABLE LOW ALLOY STEELS

## DESCRIPTION

The high hardness of many low alloy steels precludes welding them in the hardened condition because of their cold-cracking tendencies. The carbon contents of these steels generally range from about 0.25 to 0.45 percent, compared to 0.10 to 0.25 percent for the quenched and tempered low alloy steels. They are normally welded in the annealed or overtempered condition, and then the entire weldment is heat treated to the desired strength or hardness. To differentiate them from other steels, they are called *heat treatable low alloy (HTLA) steels* in this discussion.

## COMPOSITIONS AND PROPERTIES

The compositions of several HTLA steels that are welded and then quenched and tempered are given in Table 1.24. Such steels can be heat treated to very high strength and hardness. Moreover, some HTLA steels have sufficient carbon and alloy content to give them high hardenability. High hardenability makes it necessary to properly preheat the steel for welding to avoid cold cracking and obtain a sound weld. Controlled preheat and interpass temperatures are necessary for the weld metal as well as for the heat-affected zone. Otherwise, the heat-affected zone and the weld metal would be very hard in the as-welded condition.

Welding at a preheat temperature below 400°F will require that the hydrogen content be kept extremely low to prevent cracking. Proper control of sulfur and phosphorus is also very important, especially in steels of such high hardenability. The maximum limit for phosphorus and sulfur is 0.025 percent for AISI alloy steels made by the basic electric furnace process, and 0.035 and 0.040 percent, respectively, for such steels made by the basic open hearth or basic oxygen furnace processes. Presence of these elements in excess of 0.020 percent increases the crack sensitivity of steel. Therefore, sulfur and phosphorus contents are considered very critical. Sulfur increases the sensitivity of the weld metal to hot cracking. Phosphorus reduces ductility and toughness and increases the sensitivity to cold cracking in both the base metal and the weld metal. Sulfur and phosphorus limits of less than 0.015 percent each should be specified for these steels when they are to be heat treated to very high strength levels of about 200,000 psi. Other restrictions on impurities include microcleanliness ratings that require the use of vacuum-melted base metals and filler metal.

The relatively high carbon and alloy content of HTLA steels compared to mild steels also tends to contribute to hot cracking in diluted weld metal by increasing the temperature range over which solidification takes place. The extended solidification temperature range increases segregation that, in turn, reduces high-temperature strength and ductility. These conditions in combination with the shrinkage of the metal can cause hot cracks in the weld metal in much the same manner that hot tears occur in steel castings. Hot cracks are encountered most frequently in craters in the first pass of a multiple-pass weld and in a concave fillet weld.

Studies of weld metal of 4340 steel composition indicate that its resistance to hot cracking can be improved by maintaining the sum of the sulfur and phosphorus contents below 0.025 percent. Improved cracking resistance can be obtained also by using a filler metal with a lower carbon and alloy content. These conditions can be applied to other low alloy steels.

The approximate relationship between tempering temperature and tensile strength for several HTLA steels is given in Table 1.25. The tempering temperatures shown are approximations because the specific temperature required in any given case depends on the following four factors:

**Table 1.24**
**Compositions of typical heat treatable low alloy steels**

| Designation, AISI-SAE or other | Composition, percent | | | | | | |
|---|---|---|---|---|---|---|---|
| | C | Mn | Si | Ni | Cr | Mo | V |
| 4027 | 0.25-0.30 | 0.70-0.90 | 0.15-0.35 | — | — | 0.20-0.30 | — |
| 4037 | 0.35-0.40 | 0.70-0.90 | 0.15-0.35 | — | — | 0.20-0.30 | — |
| 4130 | 0.28-0.33 | 0.40-0.60 | 0.15-0.35 | — | 0.80-1.10 | 0.15-0.25 | — |
| 4135 | 0.33-0.38 | 0.70-0.90 | 0.15-0.35 | — | 0.80-1.10 | 0.15-0.25 | — |
| 4140 | 0.38-0.43 | 0.75-1.00 | 0.15-0.35 | — | 0.80-1.10 | 0.15-0.25 | — |
| 4320 | 0.17-0.22 | 0.45-0.65 | 0.15-0.35 | 1.65-2.00 | 0.40-0.60 | 0.20-0.30 | — |
| 4340 | 0.38-0.43 | 0.60-0.80 | 0.15-0.35 | 1.65-2.00 | 0.70-0.90 | 0.20-0.30 | — |
| 5130 | 0.28-0.33 | 0.70-0.90 | 0.15-0.35 | — | 0.80-1.10 | — | — |
| 5140 | 0.38-0.43 | 0.70-0.90 | 0.15-0.35 | — | 0.70-0.90 | — | — |
| 8630 | 0.28-0.33 | 0.70-0.90 | 0.15-0.35 | 0.40-0.70 | 0.40-0.60 | 0.15-0.25 | — |
| 8640 | 0.38-0.43 | 0.75-1.00 | 0.15-0.35 | 0.40-0.70 | 0.40-0.60 | 0.15-0.25 | — |
| 8470 | 0.38-0.43 | 0.75-1.00 | 0.15-0.35 | 0.40-0.70 | 0.40-0.60 | 0.20-0.30 | — |
| AMS 6434 | 0.31-0.38 | 0.60-0.80 | 0.20-0.35 | 1.65-2.00 | 0.65-0.90 | 0.30-0.40 | 0.17-0.23 |
| 300M | 0.40-0.46 | 0.65-0.90 | 1.45-1.80 | 1.65-2.00 | 0.70-0.95 | 0.30-0.45 | 0.05 min |
| D-6a | 0.42-0.48 | 0.60-0.80 | 0.15-0.30 | 0.40-0.70 | 0.90-1.20 | 0.90-1.10 | 0.05-0.10 |

**Table 1.25**
**Approximate heat-treating conditions for several low alloy steels**

| SAE, AISI, or other designation | Austenitizing temperature, °F | Quenching Medium | Room temperature tensile strength, ksi | | | | | | | |
|---|---|---|---|---|---|---|---|---|---|---|
| | | | 100 | 120 to 140 | 140 to 160 | 160 to 180 | 180 to 200 | 200 to 220 | 250 | 300 |
| | | | Approximate tempering temperature, °F | | | | | | | |
| 4037 | 1525–1575 | Oil or water | 1200 | 1100 | 925 | | | | | |
| 4130 | 1550–1625 | Oil or water | 1250 | 1050 | 925 | 850 | 725 | | | |
| 4135 | 1550–1625 | Oil | | 1125 | 1025 | 900 | 800 | | | |
| 4140 | 1525–1600 | Oil | 1300 | 1175 | 1075 | 950 | 850 | 725 | | |
| 4340 | 1475–1550 | Oil | | | 1175 | 1050 | 925 | 850 | | |
| 8630 | 1550–1625 | Oil or water | 1225 | 1050 | 925 | 850 | 725 | | | |
| 8735 | 1525–1600 | Oil | | 1125 | 1025 | 800 | 785 | | | |
| 8740 | 1525–1600 | Oil | | 1175 | 1075 | 950 | 850 | 725 | | |
| D-6 | 1550–1650 | Air or oil | | | | | | 1000 | 650 | 450 |

(1) Thickness of the weldment

(2) Chemical composition of the particular heat of steel

(3) Prior processing of the steel

(4) Method of tempering

Adjustment of the tempering temperature to produce the desired properties in a particular weldment can be determined by preliminary tension or hardness tests. A low tempering temperature promotes high strength and hardness, but low ductility and toughness.

## WELDING METALLURGY

The combined carbon and alloy contents of the HTLA steels are sufficient to promote the formation of martensite from austenite when cooled rapidly to below the appropriate transformation temperature. The carbon content is sufficiently high to form hard martensite that may be brittle (see Fig. 1.3). During welding, a portion of the weld heat-affected zone will transform to austenite. If the weld metal and the austenitic heat-affected zone are cooled too fast, they will transform to martensite or a combination of martensite and bainite, as illustrated in Fig. 1.10 for AISI 4340 steel. The internal stresses that develop during cooling from both the contraction of the weld and heat-affected zone and the transformation of austenite may cause hard martensite to crack. Steels in this group are very sensitive to hydrogen-induced cracking. Therefore, the welding process and procedures should minimize the presence of hydrogen during welding as well as the formation of martensite.

The best approach to welding these steels is to preheat the joint area to 600°F or higher so that the cooling rate of the weld will be slow enough to form softer bainite in preference to hard martensite. A bainitic microstructure in the

weld and heat-affected zone will have sufficient toughness to permit intermediate handling between welding and postweld heat treatment.

In some applications, a practical preheat temperature may be too low for complete transformation to bainite. Then, the weld and heat-affected zone may contain some martensite and retained austenite. Appropriate postweld operations are necessary to transform the austenite to martensite or bainite, depending upon the intermediate processing prior to hardening.

## PREHEAT

The minimum preheat and interpass temperature required to prevent cracking with a given steel depends on

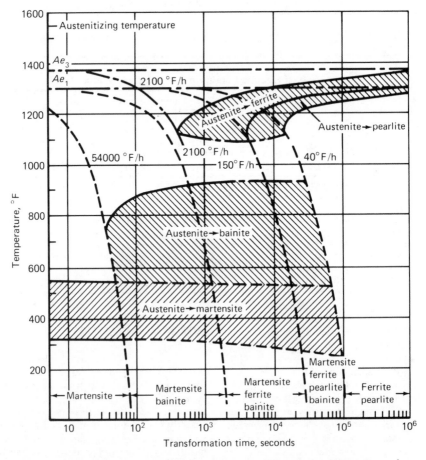

**Fig. 1.10—Continuous cooling transformation diagram for AISI 4340 steel**

(1) Its carbon and alloy content

(2) Condition of heat treatment

(3) Section thickness or amount of restraint on the joint

(4) Available hydrogen during welding

A change in process or procedures to reduce available hydrogen or a decrease in thicknesses or joint restraint may permit the use of a lower preheat temperature.

The ideal preheat temperature is about 50°F above the temperature at which martensite starts to form on cooling ($M_s$). Holding at this temperature for a time after welding will produce a bainitic structure in the weld and heat-affected zones. It will also permit dissolved hydrogen to escape from thin sections. The volumetric expansion that takes place during transformation will not produce localized peak stresses at this preheat temperature that lead to cracking.

However, the $M_s$ temperature of many of the HTLA steels is above 500°F. A preheat temperature of 550°F or higher will contribute to welder discomfort and promote the formation of a thin oxide layer on the joint faces. Such oxidation may cause unacceptable discontinuities in the weld. Consequently, a welding process that permits the use of a lower preheat temperature should be used when practical. Otherwise, the welding process and procedures must be designed to minimize the problems associated with a high preheat temperature. Some low alloy steels with high carbon content will require high preheat and interpass temperatures with low hydrogen welding processes. Recommended minimum preheat and interpass temperatures for several AISI low alloy steels are given in Table 1.26.

### Table 1.26
### Recommended minimum preheat and interpass temperatures for several AISI low alloy steels

| AISI steel | Thickness range, in. | Minimum preheat and interpass temperature ,°F[a] |
|---|---|---|
| 4027 | Up to 0.5 | 50 |
| | 0.6–1.0 | 150 |
| | 1.1–2.0 | 250 |
| 4037 | Up to 0.5 | 100 |
| | 0.6–1.0 | 200 |
| | 1.1–2.0 | 300 |
| 4130, 5140 | Up to 0.5 | 300 |
| | 0.6–1.0 | 400 |
| | 1.1–2.0 | 450 |
| 4135, 4140 | Up to 0.5 | 350 |
| | 0.6–1.0 | 450 |
| | 1.1–2.0 | 500 |
| 4320, 5130 | Up to 0.5 | 200 |
| | 0.6–1.0 | 300 |
| | 1.1–2.0 | 400 |
| 4340 | Up to 2.0 | 550 |
| 8630 | Up to 0.5 | 200 |
| | 0.6–1.0 | 250 |
| | 1.1–2.0 | 300 |
| 8640 | Up to 0.5 | 200 |
| | 0.6–1.0 | 300 |
| | 1.1–2.0 | 350 |
| 8740 | Up to 1.0 | 300 |
| | 1.1–2.0 | 400 |

a. Low hydrogen welding processes only

When a preheat temperature below the $M_s$ temperature of the steel is used, some austenite will immediately transform to hard martensite. The balance will remain unchanged until the temperature is increased above the $M_s$ or decreased to room temperature. In the latter case, transformation to martensite will increase the danger of cold cracking. Therefore, the interpass temperature must not fall below the preheat temperature, and certain postweld treatments must be performed before the weld is cooled to room temperature. These treatments are discussed later.

## ARC WELDING PROCESSES

### Shielded Metal Arc Welding

Covered electrodes that deposit matching

weld metal are available for welding some HTLA steels that are quenched and tempered after welding. These include AISI 4130 and 4340 steels. The electrode manufacturers should be consulted for their recommendations for specific applications, particularly for multiple pass welds where high joint strength is required and dilution is limited.

The effect of tempering temperature on the tensile strength and ductility of weld metals of 4130, 4140, and 4340 compositions is shown in Fig. 1.11. The responses of the weld metals to heat treatment are similar to those of the base metals. (see Table 1.25.)

Covered electrodes with comparable composition but lower carbon content may be suitable for applications where a joint strength somewhat lower than base metal strength is satisfactory. Joint strength will depend upon the following:

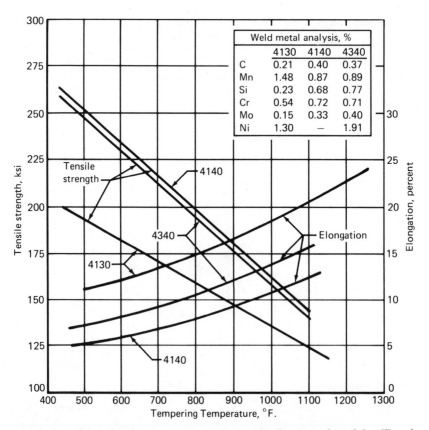

Fig. 1.11—Effect of tempering temperature on tensile strength and ductility of 4130, 4140, and 4340 all-weld-metal test specimens (oil quenched from 1575°F and tempered for four hours)

(1) Electrode selection
(2) Base metal thickness
(3) Joint design
(4) Welding procedure
(5) Amount and uniformity of dilution
(6) Response of the weld metal to post-weld heat treatment

The welding procedure and expected results should be evaluated by appropriate tests, including response to heat treatment. In any case, low hydrogen electrodes must be used, and they must be kept dry to avoid hydrogen-induced cracking.

## Gas Shielded Arc Welding

HTLA steels can be welded by gas tungsten arc (GTAW), gas metal arc (GMAW), and flux cored arc (FCAW) welding. Gas tungsten arc welding is capable of producing welds with the lowest hydrogen content, and is preferred for thin, highly stressed joints. Bare wire of a composition similar to the base metal may be used for filler metal with GTAW and GMAW. The carbon, phosphorus, and sulfur contents of the filler metal are generally low to reduce weld metal hot cracking tendencies and improve weld metal ductility. These processes provide good control of weld metal composition and cleanli-ness. Argon, helium, or mixtures of the two may be used for GTAW. Argon-oxygen or argon-$CO_2$ mixtures may be used with gas metal arc welding. Typical transverse tensile properties of GTAW and GMAW joints in typical HTLA steels after heat treatment are given in Table 1.27.

With flux cored arc welding, electrode selection may be limited. Welds made with this process are comparable to those made with low-hydrogen covered electrodes. Generally, $CO_2$ gas is used for shielding, but argon-$CO_2$ mixtures are sometimes used, especially with electrodes of 0.062 in. diameter and smaller.

## Submerged Arc Welding

Some low alloy steels that are quenched and tempered after welding may be welded by the submerged arc process. The most important part of a welding procedure is the selection of a filler metal and a flux that will produce weld metal with the desired tensile strength, ductility, and notch-toughness after heat treatment. The recommendations of the manufacturer should be followed in selecting the optimum combination of flux and filler metal. Fluxes that produce a neutral or a basic slag generally give better mechanical properties.

## Table 1.27
## Typical transverse tensile properties of arc welded joints in low alloy steels after quenching and tempering

| Steel designation | Thickness, in. | Welding process | Filler metal[a] | Tempering temperature, °F | Welded joint Tensile strength, ksi | Yield strength, ksi | Elongation in 2 in. percent | Approx. base metal tensile strength, ksi |
|---|---|---|---|---|---|---|---|---|
| 4130 | 0.25 | GMAW | (1) | 950 | 170 | 166 | 7 | 170 |
| 4140 | 0.50 | GMAW | 4140 | 900 | 189 | 178 | 8[b] | 190 |
| 4340 | 1.0 | GMAW | 4340 | 950 | 189.5 | 181.5 | 11 | 190 |
| 4340 | 1.0 | GMAW | (2) | 950 | 191.5 | 177.5 | 8 | 190 |
| 4335V | 0.25 | GTAW | 4340 | 400 | 255 | 222 | 9 | 259 |
| D6 | 0.093 | GTAW | (3) | 600 | 270 | 237 | 6[b] | 275 |
| D6 | 0.093 | GTAW | D6 | 600 | 258 | 218 | 6 | 265 |
| D6 | 0.50 | GTAW | (4) | 1000 | 224 | 207 | 7 | 230 |

a.

| Filler metal | Compostion, percent | | | | | | |
|---|---|---|---|---|---|---|---|
| | C | Mn | Si | Ni | Mo | Cr | V |
| (1) | 0.18 | 1.50 | 0.44 | 1.2 | 0.34 | 0.65 | — |
| (2) | 0.25 | 1.17 | 0.65 | 1.8 | 0.80 | 1.17 | 0.21 |
| (3) | 0.25 | 0.28 | 0.03 | — | 1.29 | 0.98 | 0.56 |
| (4) | 0.25 | 0.55 | 0.65 | — | 0.50 | 1.25 | 0.30 |

b. Elongation in 1 in. gage length

Submerged arc welding normally is used where a high deposition rate is used, but the weld metal usually has lower toughness than weld metal deposited by gas tungsten arc welding, particularly after heat treatment to a high strength level. The toughness obtained will be affected by the characteristics of the slag produced by the flux.

## Weld Backing

Various methods of weld backing are used to shield the back side of a joint, support the weld metal, and provide the proper root contour. Successful methods include the use of ceramic tape and copper bars with drilled holes for applying shielding gas to the underside of the joint. Some high strength weld metals have a tendency to crack when they contact a cold copper bar. In that case, a material with a lower thermal conductivity should be used for the backing bar. Copper is not recommended for backing when a relatively high preheat temperature is required because of its rapid absorption of heat and the likelihood of the weld metal fusing to it.

## Welding Procedure

Deposition of filler metal in the joint usually is done with a relatively low heat input. High heat input tends to produce a wide heat-affected zone with enlarged grain size, and to increase the likelihood of hot cracking in the weld metal and heat-affected zone. Automatic welding is preferred over manual welding for linear or simple circumferential joints. Automatic operation produces more uniform welds with fewer stops and starts, which frequently give rise to weld defects. All materials involved in the welding operation must be clean. This includes base metal, welding rod or electrodes, and fixtures. For example, a problem with porosity and nonmetallic inclusions in a welded joint in heat treatable alloy steel was traced to abrasive grit on ground faces of the joint preparation.

## FLASH WELDING

The procedure used for flash welding the HTLA steels is very much like that used for medium carbon steels. In some cases, the welding current is lower and the flashing time is longer to compensate for the higher electrical resistance of an HTLA steel. Welding procedures should be established for the particular steel and

qualified by suitable tests to ensure that the required mechanical properties will be obtained. Preheat and postheat (tempering) are sometimes recommended for flash welding air-hardening steels to avoid cracking.

## SPOT WELDING

HTLA steels can be spot welded using procedures similar to those for medium carbon steels. Application of a tempering cycle in the welding machine to soften the quench-hardened weld nugget is strongly recommended. Examples of spot welding schedules for AISI 4130, 4340, and 8630 low alloy steels are given in Table 1.7.

Overlapping spot welds can be made with wheel electrodes to produce a seam weld. Intermittent travel should be used with the work stationary during the welding and tempering cycles.

A welding schedule is qualified by mechanical testing, metallurgical examination, and hardness tests. The welds must be free of cracks and other defects before and after subsequent heat treatments.

## ELECTRON BEAM WELDING

The HTLA steels can be readily joined by electron beam welding. This process can produce deep, narrow welds with a narrow heat-affected zone and lower welding stresses than with arc welding. As a result, these steels can be electron beam welded at room temperature or with moderate preheat, depending upon the alloy composition and thickness.

Electron beam welds generally solidify with a columnar grain structure which tends to promote segregation and cracking in the center of the weld. With some low alloy steels, oscillation of the electron beam may be required to avoid these problems.

## POSTWELD HEAT TREATMENT

The heat treatment required immediately after fusion welding of these steels depends on the preheat and interpass temperature and any subsequent processing prior to hardening. When the preheat and interpass temperature is below the $M_s$ temperature of the steel, the weld must not be cooled to room temperature until after it is given a thermal treatment to avoid cracking.

When immediate stress relief after welding is not practical, the welded joint should be heated from the preheat temperature to 50° to 100°F above the $M_s$ temperature of the steel. The remaining austenite will transform to a reasonably ductile bainitic structure after about one hour at temperature. Then, the weldment can be cooled to room temperature without danger of cracking.

If the weldment is to be stress relieved in the 1100° to 1250°F range immediately after welding, it should first be cooled from the preheat temperature to a lower temperature where transformation of austenite to martensite will be essentially complete, as determined from the isothermal transformation diagram for the steel. Then the weldment should be immediately heated to the stress relieving temperature where the martensite in the welded joint will be tempered and softened. After holding it at temperature for the specified time, the weldment can be cooled to room temperature without danger of cracking.

The weld should be inspected for discontinuities and any defects repaired before final heat treatment. The repair welding procedures should be the same as those for the initial welding.

Finally, the weldment should be austeni- tized, quenched, and tempered to achieve the desired mechanical properties. The procedures should ensure that all of the austenite transforms to martensite before the weldment is tempered.

## BRAZING

Low alloy steels can be brazed using processes, procedures, and filler metals commonly used for carbon steels. When the steel is to be quenched and tempered to achieve desired properties, brazing and hardening operations can be combined. The solidus of the filler metal must be above the austenitizing temperature of the low alloy steel. The joint is made at normal brazing temperature. The brazement is then removed from the heat source to permit the filler metal to solidify. After it cools to the hardening temperature of the steel, the brazement is uniformly quenched in the appropriate cooling medium. Finally, the brazement is tempered to the desired hardness. Proper support must be provided at high tempering temperatures to avoid rupturing the brazed joint.

# CHROMIUM-MOLYBDENUM STEELS

## DESCRIPTION

Alloy steels of this family contain 0.5 to 9 percent chromium and 0.5 or 1 percent molybdenum. The carbon content is normally less than 0.20 percent for good weldability, but the alloys have high hardenability. The chromium provides improved oxidation and corrosion resistance and the molybdenum increases strength at elevated temperatures. They are normally supplied in the annealed or normalized-and-tempered condition.

Chromium-molybdenum (Cr-Mo) steels are widely used in the petroleum industry and in steam power generating for elevated temperature applications. They are used in various product forms to various ASTM Specifications, as shown in Table 1.28.

## COMPOSITION AND PROPERTIES

The nominal chemical compositions of the Cr-Mo steels are given in Table 1.29. Some alloys may contain small additions of columbium, titanium, or vanadium, or increased amounts of carbon or silicon for specific applications. Some castings or forging alloys may contain up to 0.35 percent carbon.

Oxidation resistance, elevated temperature strength, and resistance to sulfide corrosion all increase as the chromium or molybdenum content is increased. However, the corrosion rate in high temperature steam does not vary significantly when the chromium content is increased beyond 2.25 percent. Generally, the best elevated temperature strength is obtained with the 2-1/4 Cr-1Mo steel. However, recent studies on the higher chromium grades show very good promise for elevated temperature tensile and creep strength. These improvements appear to be related to judicious control of residual elements and heat treatment.

Chromium-molybdenum steels are air hardenable, and undergo the high- and low-temperature metallurgical transformations common to low alloy steels. The mechanical properties depend upon the condition of heat treatment. The tensile property requirements of the ASTM Specifications for these steels vary with the product form and the type of heat treatment. Table 1.30 summarizes these requirements. Because of the wide range of properties of some product forms, the user must be aware of the actual specification to which the product was produced (Table 1.28).

When these steels are cooled rapidly from above their upper-critical temperatures, hardness and strength increase with a reduction in ductility and toughness. Since the carbon content is usually below about 0.15 percent, they cannot be quenched to high hardness (Fig. 1.3). Therefore, ductility is greater at any given strength level than when a high carbon variation is used, as in some specifications for forgings and castings. Because of their hardenability, these steels may require further heat treatment to restore toughness, ductility, or other desired mechanical prop-

## Table 1.28
### ASTM Specifications for chromium-molybdenum steel product forms

| Type | Forgings | Tubes | Pipe | Castings | Plate |
|------|----------|-------|------|----------|-------|
| 1/2Cr-1/2Mo | A182-F2 | A213-T2 | A335-P2 A369-FP2 A426-CP2 | | A387-Gr 2 |
| 1Cr-1/2Mo | A182-F12 A336-F12 | A213-T12 | A335-P12 A369-FP12 A426-CP12 | | A387-Gr 12 |
| 1-1/4Cr-1/2Mo | A182-F11 A336-F11/F11A A541-Cl5 | A199-T11 A200-T11 A213-T11 | A335-P11 A369-FP11 A426-CP11 | A217-WC6 A356-Gr6 A389-C23 | A387-Gr 11 |
| 2Cr-1/2Mo | | A199-T3b A200-T3b A213-T3b | A369-FP3b | | |
| 2-1/4Cr-1Mo | A182-F22/F22a A336-F22/F22A A541-Cl6/6A | A199-T22 A200-T22 A213-T22 | A335-P22 A369-FP22 A426-CP22 | A217-WC9 A356-Gr10 A643-GrC | A387-Gr 22 A542 |
| 3Cr-1Mo | A182-F21 A336-F21/F21A | A199-T21 A200-T21 A213-T21 | A335-P21 A369-FP21 A426-CP21 | | A387-Gr 21 |
| 5Cr-1/2Mo | A182-F5/F5a A336-F5/F5A A473-501/502 | A199-T5 A200-T5 A213-T5 | A335-P5 A369-FP5 A426-CP5 | A217-C5 | A387-Gr 5 |
| 5Cr-1/2MoSi | | A213-T5b | A335-P5b A426-CP5b | | |
| 5Cr-1/2MoTi | | A213-T5c | A335-P5c | | |
| 7Cr-1/2Mo | A182-F7 A473-501A | A199-T7 A200-T7 A213-T7 | A335-P7 A369-FP7 A426-CP7 | | A387-Gr 7 |
| 9Cr-1Mo | A182-F9 A336-F9 A473-501B | A199-T9 A200-T9 A213-T9 | A335-P9 A369-FP9 A426-CP9 | A217-C12 | A387-Gr 9 |

**Table 1.29**
**Nominal chemical compositions chromium-molybdenum steels**

| Type | Composition, percent[a] | | | | | | |
|------|------|------|------|------|------|------|------|
| | C | Mn | S | P | Si | Cr | Mo |
| 1/2Cr-1/2Mo | 0.10-0.20 | 0.30-0.60 | 0.045 | 0.045 | 0.10-0.30 | 0.50-0.80 | 0.45-0.65 |
| 1Cr-1/2Mo | 0.15 | 0.30-0.60 | 0.045 | 0.045 | 0.50 | 0.80-1.25 | 0.45-0.65 |
| 1-1/4Cr-1/2Mo | 0.15 | 0.30-0.60 | 0.030 | 0.045 | 0.50-1.00 | 1.00-1.50 | 0.45-0.65 |
| 2Cr-1/2Mo | 0.15 | 0.30-0.60 | 0.030 | 0.030 | 0.50 | 1.65-2.35 | 0.45-0.65 |
| 2-1/4Cr-1Mo | 0.15 | 0.30-0.60 | 0.030 | 0.030 | 0.50 | 1.90-2.60 | 0.87-1.13 |
| 3Cr-1Mo | 0.15 | 0.30-0.60 | 0.030 | 0.030 | 0.50 | 2.65-3.35 | 0.80-1.06 |
| 5Cr-1/2Mo | 0.15 | 0.30-0.60 | 0.030 | 0.030 | 0.50 | 4.00-6.00 | 0.45-0.65 |
| 7Cr-1/2Mo | 0.15 | 0.30-0.60 | 0.030 | 0.030 | 0.50-1.00 | 6.00-8.00 | 0.45-0.65 |
| 9Cr-1Mo | 0.15 | 0.30-0.60 | 0.030 | 0.030 | 0.25-1.00 | 8.00-10.00 | 0.90-1.10 |

a. Single values are maximum.

**Table 1.30**
**Tensile property requirements for chromium-molybdenum steel products manufactured to ASTM Specifications**

| Product form | Tensile strength, ksi | Yield strength, ksi | Elongation, percent | Reduction of area, percent |
|------|------|------|------|------|
| Forgings | 60-90 | 30-65 | 20-22 | 45-50 |
| Tubing | 60[a] | 30[a] | 30[a] | — |
| Pipe | 60-90 | 30-60 | 18-20 | 35[b] |
| Castings | 70-90 | 40-60 | 18-20 | 35[b] |
| Plate | 60-115 | 30-100 | 13-18 | 40[b] |

a. Minimum
b. Minimum, but may not apply to all grades.

erties after being heated above their transformation temperatures, as in welding or hot forming operations.

## HEAT TREATMENT

The types of heat treatment normally applied to Cr-Mo steels are the same as those applied to other hardenable steels, namely: annealing, normalizing and tempering, quenching and tempering, or tempering only. Annealing, normalizing-and-tempering, and quenching-and-tempering treatments are used to obtain desired grain size or required mechanical properties for the application.

A Cr-Mo steel is annealed by heating it to a temperature in the range of 1550° to 1675°F and holding at temperature for one hour per inch of thickness. It is then cooled to 1000°F at a maximum rate of 50°F per hour, followed by furnace or air cooling to room temperature. This procedure produces a relatively soft, ferritic structure throughout the steel.

Normalizing-and-quenching treatments require heating the steel to a temperature in the range of 1550° to 1675°F, holding for one hour per inch of thickness, and then cooling rapidly to room temperature. Normalizing requires cooling in still air; quenching requires immersion or spray quenching in water. These treatments harden the steel which is usually tempered after hardening.

A Cr-Mo steel is tempered at a selected temperature below its lower critical temperature (approximately 1400°F) for an appropriate time. It is then cooled in still air or in the furnace.

Heat treatment of these steels is used to advantage when fabricating large heavy-walled pressure vessels. Such vessels are usually constructed from 2-1/4 Cr-1 Mo steel that has been normalized or quenched and tempered. These treatments result in a microstructure of bainite or low-carbon martensite which has good ductility. Regardless of the heat treatment given to the steel, proper control of the tempering treatment is required. The tempering temperature and time at temperature are both important.

## WELDING METALLURGY

The welding metallurgy of the chromium-molybdenum steels is similar to that of the other hardenable, low alloy steels discussed previously. These steels will harden when quenched from the austenitizing temperature, and are sensitive to hydrogen-induced cracking.

The welding procedures must contain the necessary safeguards to prevent cracking in the weld metal and the heat-affected zone. This includes quench cracking and hydrogen-induced cracking. Appropriate preheat and welding consumables must be used to avoid cracking. A postweld heat treatment is normally used to im-

prove the toughness of the weld metal and heat-affected zone. The carbon content of the base metal and weld metal, as well as dilution effects, must also be taken into account.

Low hydrogen welding processes and procedures must be used. The filler metal composition should be nearly the same as that of the base metal, except for carbon content. The carbon content of the filler metal is usually lower than that of the base metal. However, matching carbon content is required when the weldment is to be quenched and tempered or normalized and tempered and 100 percent joint efficiency is required. The effect of increasing carbon content on the tensile strength and ductility of 2-1/4 Cr-1Mo weld metal is shown in Fig. 1.12.

The Cr-Mo steels, as with other low alloy steels, are subject to creep (temper) embrittlement during longtime service at temperatures in the 700° to 1100°F range. The degree of embrittlement of the steel is influenced by chemical composition as well as heat treatment. There is evidence that nonequilibrium segregation of re-

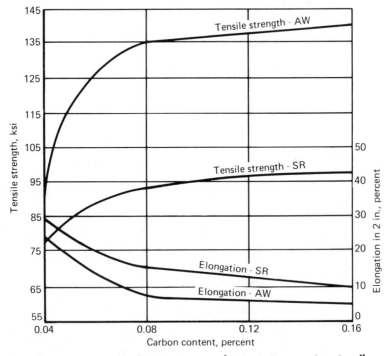

**Fig. 1.12—Effect of carbon content on the room temperature tensile strength and ductility of 2-1/4Cr-1Mo weld metal, as-welded (AW) and stress relieved (SR) at 1300°F for one hour**

sidual elements in prior austenite grain boundaries is a contributing factor to creep embrittlement of Cr-Mo steel weld metal. Since weld metal composition is affected by the welding consumables (including fluxes) and the welding procedures, the fabricator should consult the manufacturers of welding consumables for specially balanced compositions for welding Cr-Mo steels that will see long-time elevated temperature service. The welding consumables should be evaluated using the welding and heat-treating procedures specified for production.

## JOINT DESIGN

Joint designs used for welding carbon steel by a specific process are suitable for welding the Cr-Mo steels. The joint geometry should minimize any notch conditions that might contribute to stress concentration. Sharp corners and rapid changes in section size are to be avoided. Butt joints are preferred; fillet welds should only be used under conditions of low stress. Fit-up for single-welded joints should assure complete joint penetration without excessive melt-thru.

## PREHEAT

Preheat is required in most cases to prevent hardening and cracking. Recommended minimum preheat and interpass temperatures for various thicknesses of Cr-Mo steels are given in Table 1.31. They generally increase with the alloy content and the section thickness.

Lower preheat temperatures may be used if the welding heat input is relatively high or the available hydrogen is very low, as with gas tungsten arc welding. The preheat temperature should be increased if cracking is encountered, particularly if hydrogen is suspected as the cause. The temperature of the completed weld should be raised 100°F and held at temperature for a sufficient time to allow excessive hydrogen to diffuse to the surface and escape. The preheat temperature should be increased when the carbon content of the steel exceeds 0.15 percent.

The method of heating should be one that will provide a uniform temperature along the entire joint length before welding is started. Also the width of the heated area should be sufficient to ensure that the temperature is nearly uniform through the thickness.

Based on industrial experience, certain general suggestions can be made regarding the interruption of the heating cycle during the welding of chromium-molybdenum steels. These suggestions should not be applied indiscriminately but interpreted in the light of the general factors given previously and the specific job conditions.

In any case where an interrupted procedure is followed in the welding of a section thickness of less than 1 in., the weld deposit prior to inter-

### Table 1.31
### Recommended minimum preheat temperatures for welding chromium-molybdenum steels with low hydrogen covered electrodes

| Steel[a] | Thickness, in. | | |
|---|---|---|---|
| | Up to 0.5 | 0.5 to 1.0 | Over 1.0 |
| 1/2Cr-1/2Mo<br>1Cr-1/2Mo<br>1-1/4Cr-1/2M0 | 100°F | 200°F | 300°F |
| 2Cr-1/2Mo<br>2-1/4Cr-1Mo | 150 | 200 | 300 |
| 3Cr-1Mo | 250 | 300 | 400 |
| 5Cr-1/2Mo<br>7Cr-1/2Mo<br>9Cr-1Mo | 400 | 400 | 500 |

a. Maximum carbon content of 0.15 percent. For higher carbon content, the preheat temperature should be increased 100° to 200°F. Lower preheat temperatures may be used with gas tungsten arc welding.

ruption should be at least 33 percent of the thickness, but not less than two weld layers, whichever is greater. For very thick sections, 33 percent of the thickness may be excessive, but a minimum weld thickness should be specified.

The heating cycle may be safely interrupted with Cr-Mo steels containing less than 4 percent chromium in thickness under 1 in. provided the welding is done under low hydrogen conditions.

When the welding consumables are a potential source of hydrogen, such as covered electrodes, the temperature of the weld should be raised 100°F and held for 1 hour before cooling to room temperature to allow hydrogen to escape. This step is not necessary when the welding consumables are essentially free of hydrogen, as with gas tungsten arc welding and gas metal arc welding using a solid wire electrode. When the thickness exceeds 1 in., welding should be completed and the postweld heat treatment started without interruption. Alternatively, the weld joint can be given a short-time tempering treatment at 1200° to 1300°F before interruption of the heating cycle.

## WELDING PROCESSES

The Cr-Mo steels are readily joined by arc welding and brazing processes commonly used for carbon steel. They can also be joined by electroslag, electron beam, laser beam, friction, and resistance welding. Brazing procedures are similar to those given for other steels, but the thermal effects of the brazing cycle on the properties of Cr-Mo steel must be considered.

## ARC WELDING CONSUMABLES

The filler metal should have the same nominal composition as the base metal, except for carbon content. Normally, the carbon content is lower than that of the base metal. Suggested welding consumables (rods, electrodes, fluxes, and shielding gases) for arc welding the Cr-Mo steels to themselves are given in Table 1.32.

**Table 1.32**
**Suggested welding consumables for joining chromium-molybdenum steels**

| Steel | SMAW[a] | GTAW,[b] GMAW[b] | FCAW[c] | SAW[d] |
|---|---|---|---|---|
| 1/2Cr-1/2Mo | E80XX-B1 | ER80X-B2L | E81T1-B1 | F8XX-EXXX-F4 |
| 1Cr-1/2Mo, 1-1/4Cr-1/2Mo | E80XX-B2 or B2L | ER80X-B2 or B2L | E8XTX-B2 or B2X | F8XX-EXXX-B2 or B2H |
| 2Cr-1/2Mo, 2-1/4Cr-1Mo | E90XX-B3 or B3L | ER90X-B3 or B3L | E9XTX-B3 or B3X | F9XX-EXXX-B3 or B4 |
| 3Cr-1Mo | E90XX-B3 E502-XX | ER90X-B3 ER502 | E9XTXS-B3 E502T-1 or 2 | F9XX-EXXX-B3 |
| 5Cr-1Mo | E502-XX | ER502 | E502T-1 or 2 | F9XX-EXXX-B6 or B6H |
| 7Cr-1/2Mo | E7Cr-XX | ER502 ER505 | E502T-1 or 2 E505T-1 or 2 | |
| 9Cr-1Mo | E505-XX | ER505 | E505T-1 or 2 | |

a. Shielded metal arc welding electrodes from AWS Specification A5.4-78 or A5.5-81.
b. Welding rod or electrode from AWS Specification A5.9-77 or A5.28-79. Argon or argon plus 1 to 5 percent oxygen is used for shielding with GMAW.
c. Flux cored electrodes from AWS Specification A5.22-80 or A5.29-80 using $CO_2$ or argon − 2 percent oxygen shielding.
d. Electrode-flux combination to produce the desired weld metal composition and strength from AWS Specification A5.23-80.

Other consumables may be acceptable for specific applications.

When several grades of Cr-Mo steels are to be welded on one job, limiting the number of different filler metals used will simplify material control. Filler metal of the same or slightly higher alloy content can be used for welding several Cr-Mo steels. For example 1-1/4Cr-1/2Mo filler metal can be used for welding 1/2Cr-1/2Mo, 1Cr-1/2Mo, and 1-1/4Cr-1/2Mo steels. Similarly, 2-1/4Cr-1Mo filler metal can be used for 1-1/4Cr-1/2Mo, 2Cr-1/2Mo, 2-1/4Cr-1Mo, and 3Cr-1/2Mo steels.

In any case, each welded joint must possess the required properties for the intended service after postweld heat treatment. Where service requires corrosion or oxidation resistance, the characteristics of the filler metal and base metal should be matched as closely as possible.

Covered electrodes for welding the Cr-Mo steels are low hydrogen types. Procedures for storing and handling low hydrogen covered electrodes were described previously.

The most used electrode-flux combinations for submerged arc welding deposit 1-1/4Cr-1/2Mo and 2-1/4Cr-1Mo weld metals. In general, the weld metal is lower in carbon, manganese, and chromium and higher in silicon than the electrode. Usually, the electrode contains less than 0.50 percent silicon, but the weld metal may contain up to 0.80 percent because of flux reactions. Typical carbon content of weld metal is about 0.07 percent.

Filler metal should always be used when joining a Cr-Mo steel to carbon steel with the gas tungsten arc welding process. Deoxidized welding rod containing at least 0.50 percent silicon should be used on rimmed or semikilled carbon steel to avoid porosity in the weld.

Weld metal with good ductility can be obtained using a low carbon filler metal. The carbon content of the filler metal may be 0.05 percent or less. (ER80S-B2L is an example.) Postweld heat treatment often may be omitted for low carbon welds in thin sections of steels containing 0.5 to 2.25 percent Cr and 0.5 or 1 percent Mo.

Type 309 or Type 310 austenitic stainless steel filler metals are often employed for minor repair welding of chromium-molybdenum steels. They are sometimes preferred for those applications where the weldment cannot be given a postweld heat treatment. Stainless steel weld metal has a lower yield strength than Cr-Mo steels and excellent as-welded ductility. For these reasons, it easily yields and relieves the majority of the welding stresses.

However, an austenitic stainless steel filler metal is not satisfactory if the welded joint will be subjected to cyclic temperature service, or a service temperature where either carbon migration or sigma formation can take place. The difference in coefficients of thermal expansion of an austenitic weld metal and ferritic Cr-Mo steel results in internal stress at the weld interfaces during cycling temperature service. This condition can promote early failure, particularly when aided by carbon migration.

When joining dissimilar Cr-Mo steels, a filler metal with a composition similar to the lower alloy steel or to an intermediate composition is commonly used for butt joints. Normally, the weld metal need not be stronger or more resistant to creep or corrosion than the lower alloy base metal. An exception to this is the attachment of auxiliary parts where the weld metal becomes an integral part of the main structure. In this case, the filler metal should provide weld metal with mechanical and chemical properties equivalent to those of steel in the main structure.

Carbon migration occurs in a chromium-molybdenum steel when it is welded with a higher chromium filler metal and then subjected to high temperature service. The carbon diffuses from the lower chromium base metal to the higher chromium weld metal and forms chromium carbide adjacent to the weld interface. This occurs very slowly at about 1000°F but much faster at higher temperatures. There is considerable speculation about whether carbon migration weakens the heat-affected zone or embrittles the weld metal in a weldment. Many welds between 2Cr-1/2Mo and 5Cr-1/2Mo steels made with 2-1/4Cr-1Mo filler metal have been in service for many years without failure.

## RESISTANCE UPSET WELDING

The longitudinal seams in Cr-Mo steel pipe and tubing are sometimes made by resistance upset welding.[27] Strip is roll formed to shape and welded continuously in a tube mill. The square

27. Refer to the *Welding Handbook*, Vol. 3, 7th Ed., 78-82, 146-167 for additional information on the welding process.

edges are progressively resistance heated to welding temperature with alternating current of 60Hz or high-frequency current. The hot edges are forged together by pressure rolls. The upset is usually trimmed off while the seam is still hot. The high localized heat inputs and fast cooling rates typical of these welding operations result in an as-welded Cr-Mo steel microstructure of low ductility. A postweld heat treatment is required to restore ductility and toughness. Postweld heat treatments may be performed in the welding machine, if it is designed for this operation, or by other thermally applied treatments.

## ELECTROSLAG WELDING

The Cr-Mo steels can be joined by electroslag welding. The process is primarily used to weld sections that are 3 in. or more in thickness. A submerged arc welding electrode of the same nominal composition as the base metal is generally used for filler metal, together with an appropriate flux designed for electroslag welding.

One advantage of this process is that the large quantity of heat generated during welding preheats the base metal ahead of the molten weld pool. This large buildup of heat also provides, in effect, a stress relief. As a result, the weldment can be cooled to room temperature before it is given a postweld heat treatment. An annealing or normalizing heat treatment is necessary to refine the grain structure of the weld metal and heat-affected zone.

## POSTWELD HEAT TREATMENT

A welded joint in a Cr-Mo steel may be placed in service (1) in the as-welded condition, (2) after a stress-relief heat treatment (PWHT), or (3) after annealing or normalizing and tempering.

Some Cr-Mo steel weldments can be put in service as-welded when a high preheat is used and the section thickness is relatively thin. Steels containing up to 1.25 percent Cr and 0.5 percent Mo have sufficient ductility as-welded to meet the requirements of many codes, particularly when welded with a low carbon filler metal (0.05 percent C or lower). For example, welds in Cr-Mo steel pipe or tubing having a diameter of 4 in. or less, a wall thickness under 0.5 in., and a chromium content of 1.25 percent or smaller can be placed in service as-welded when a proper

preheat is used during welding.[28] Welds in Cr-Mo steels with high hardenability should be postweld heat-treated regardless of dimensions.

A stress-relief heat treatment is used to reduce welding stresses and also to increase the ductility and toughness of the weld metal and heat-affected zone. The welded joint or the entire weldment is heated to some temperature below the temperature at which the steel will begin transformation to austenite $(Ac_1)$. Table 1.33 gives recommended stress-relief temperature ranges for chromium-molybdenum steels.

The stress-relief temperature should not exceed the tempering temperature of normalized and tempered or quenched and tempered steel. Also, the temperature should be in the low end of the recommended range for high creep strength, and in the high end of the range for good resistance to corrosion or hydrogen embrittlement. Welded joints to be used in acid or caustic service should always be heat treated.

It may be necessary to anneal or normalize and temper a weldment that is fabricated with a high heat-input welding process, such as electroslag welding. One of these heat treatments is required to refine the grain size in the weld metal and heat-affected zone and thus improve toughness.

Whether the welded joint must be held at the preheat temperature until it is heat-treated depends upon the compositions of the base metal and weld metal, the welding procedures, the degree of restraint, and the steps taken to avoid the formation of martensite prior to cooling. Welds in Cr-Mo steels containing 1.25 percent Cr or less can be safely cooled to ambient temperature prior to heat treatment, provided the carbon content is in the normal range. When it is not practical to postweld heat-treat the higher Cr steels without first cooling to ambient temperature, the procedures given previously for HTLA steels should be followed. Such procedures minimize the likelihood of cracking until the weld can be heat-treated.

Postweld heat treatments tend to lower the strength of the base metal, particularly if it is in the normalized-and-tempered or quenched and tempered condition. The effect is more pronounced with long times at temperature. There-

---

28. Refer to AWS D10.8-78, *Recommended Practices for Welding of Chromium-Molybdenum Steel Piping and Tubing,* for further information.

**Table 1.33**
**Recommended stress-relief temperature ranges for chromium-molybdenum steels**

| Steel | Temperature range,°F[a] |
|---|---|
| 1/2Cr-1/2Mo | 1150-1300 |
| 1Cr-1/2Mo<br>1-1/4Cr-1/2Mo | 1150-1350 |
| 2-1/4Cr-1Mo<br>3Cr-1Mo | 1250-1375 |
| 5Cr-1/2Mo<br>7Cr-1/2Mo<br>9Cr-1Mo | 1250-1400 |

a. Temperature should not exceed the tempering temperature of the steel.

fore, the time should be about one hour per inch of thickness (0.5 hour minimum).

For welded joints in piping and tubing, the method of heating and the equipment used should provide uniform heating around the circumference. With local heating, as opposed to furnace heating, no matter how long the holding time, there is always a temperature gradient through the thickness. The thicker the section, the greater this gradient will be, and the higher will be the stresses produced by it. The amount of this gradient is proportional to the width of the heated band on the surface. Therefore, it is recommended that the width of the heated band, that is the width of the surface directly under the heat source, be at least five times the thickness.

When the minimum width of the heated band is five times the thickness, the through-temperature gradient will be less than the amount that produces stresses greater than the yield strength of the metal. This will eliminate the need for very slow heating rates and will prevent damage to the pipe even if the pipe is heated as quickly as the commercially available power sources permit. Heating rates in furnaces are usually slower than with localized heating. In either case, if it is possible to measure the gradients through the wall, the maximum temperature differential through the wall thickness should not exceed 150°F.

# PRECOATED STEELS

## DESCRIPTION

Steel sheet and other product forms are precoated with an oxidation- or corrosion-resistant material to extend the service life of the product. Coatings commonly used on steels are aluminum (aluminized steel), zinc (galvanized steel), and zinc-rich primers. When precoated steels are fusion welded, the effectiveness of the coating adjacent to the weld is significantly decreased by the welding heat. The coating metal melts and then may alloy with the steel, oxidize, or volatilize. This reduces the effectiveness of the coating adjacent to the weld. A reconditioning operation is required to restore corrosion resistance to the affected areas.

## ALUMINIZED STEEL

Two different aluminum coatings are used on steel sheet: commercially pure aluminum and aluminum-8 percent silicon alloy. Both coatings are applied to sheet by continuous hot dipping. The pure aluminum coating provides improved resistance to atmospheric corrosion near room temperature. The alloy coating is used to provide oxidation resistance up to about 1200°F, but its formability is not as good as the pure aluminum coating.

## Resistance Welding

Resistance spot welding of aluminized steel sheet is similar to that for bare low carbon steel sheet of the same thickness. The welding current and electrode force should be slightly higher and the weld time somewhat longer for aluminized steel. Truncated cone electrodes of RWMA Group A, Class 2 copper alloy are recommended for the best life between dressings.

Typical spot welding schedules for aluminized steel sheet are given in Table 1.34. They are intended as guides in setting up production schedules. Some variations may be necessary because of the part or welding machine configuration.

Aluminized steel sheet can be seam welded using schedules similar to those for bare low carbon steel sheet (Table 1.5), except that the welding current should be increased slightly. Similar electrode wheels should be suitable.

Aluminized steel tubing is produced in tube mills using the resistance upset welding process. Both 60 Hz and high-frequency currents are used for welding.

## Arc Welding

*Shielded Metal Arc Welding.* Welding of aluminized steel by the shielded metal arc process requires an electrode with a basic type covering, such as Type E7015. The formation of aluminum oxide has an adverse effect on the characteristics of the molten weld pool, and it may prevent proper wetting and shaping of the weld bead. E7016 electrodes and ac power may have some advantage, but erratic arc operation may be encountered.

*Gas Tungsten Arc Welding.* When aluminized steel is welded by the tungsten arc welding (GTAW) process without filler metal, almost all of the aluminum coating alloys with the steel weld metal and provides adequate deoxidation, even with rimmed or capped steel. The thinner the sheet, the higher is the aluminum content of the weld metal. When the aluminum content of the weld metal exceeds about 1 percent, the ductility of the metal is low, and formability may be a problem. High aluminum content could cause brittle weld failure in an application involving impact loading. A thick aluminum coating adds more aluminum to the weld metal and is more likely to give brittle welds than a thin coating.

Filler metal additions should be used with GTAW to control the aluminum content of the weld metal and provide acceptable weld ductility. Good shielding gas coverage is needed to avoid oxidation of the aluminum coating.

*Gas Metal Arc Welding.* Loss of ductility in

## Table 1.34
## Typical spot welding schedules for aluminized steel sheet

| Sheet thickness, in.[a] | Electrode face diam., in.[b] | Electrode force, lbs | Approx. welding current, kA[c] | Weld time, cycles[d] | Approx. nugget diam., in. | Minimum shear strength, lbs |
|---|---|---|---|---|---|---|
| 0.018 | 0.16 | 300 | 8.6 | 8 | 0.15 | 500 |
| 0.021 | 0.16 | 350 | 8.8 | 9 | 0.16 | 570 |
| 0.027 | 0.19 | 400 | 9.0 | 10 | 0.17 | 840 |
| 0.033 | 0.19 | 500 | 9.2 | 11 | 0.18 | 1080 |
| 0.039 | 0.25 | 600 | 11.4 | 14 | 0.21 | 1500 |
| 0.052 | 0.25 | 800 | 11.8 | 18 | 0.23 | 2000 |
| 0.063 | 0.25 | 1100 | 12.0 | 24 | 0.24 | 2200 |
| 0.078 | 0.31 | 1400 | 16.0 | 30 | 0.33 | 4300 |
| 0.093 | 0.38 | 1800 | 19.0 | 36 | 0.40 | 4500 |

a. Two equal thicknesses
b. Truncated cone eletrodes of RWMA Group A, Class 2 copper alloy
c. Single-phase, ac welding machine
d. 60 Hz

the aluminized weld metal is not a problem with gas metal arc welding. A deoxidized electrode is not required because the aluminum from the coating will deoxidize the weld metal. As with GTAW, adequate shielding gas coverage is required to avoid oxidation of the aluminum coating. With improper shielding, aluminum oxide can form a film on the surface of the molten weld pool and cause weld contour or soundness problems. Argon with a minimum addition of oxygen or $CO_2$ to stabilize the arc is recommended for shielding the weld pool.

## Soldering

Chemical or mechanical cleaning methods should be used to remove the oxide film on aluminum-coated steel before soldering. A dip in 5 percent trisodium phosphate solution followed by water rinsing and drying will assist in the preparation of the aluminum coating for soldering.

Heating of the aluminum-coated steel to soldering temperature must be rapid. Electric or ultrasonic soldering irons with sufficient heating capacity should be used.

Some aluminum-coated steels may be abrasion soldered without fluxes by heating the metal surface sufficiently and melting a small amount of solder on the hot surface. The aluminum under the solder pool is then abraded using a stick of solder, the tip of the soldering iron, or a specially designed brush that assists in displacing the oxide film.

Specially formulated fluxes are commercially available for soldering aluminum-coated steel sheets. These fluxes should be applied sparingly with a fine brush. Soldering should be performed quickly to avoid excessive oxidation of the aluminum surface and undesirable alloying of the aluminum coating with the steel.

Solders suitable for joining aluminum can be used to join aluminum-coated steel.[29] The solders are available in the form of sticks, flux cored wire, and paste.

## GALVANIZED STEEL

Galvanized steel has a zinc coating for corrosion protection. Most steel products are coated with zinc by dipping in a molten bath or by

---

29. Refer to Chapter 8 for information on the soldering of aluminum.

electroplating (electrogalvanizing). Hot-dipped galvanizing gives coating weights of about 0.60 to 1.25 oz/ft² or about $5 \times 10^{-4}$ to $1 \times 10^{-3}$ in. thick coating per side on steel sheet and strip. Electrogalvanized coatings are generally 0.10 to 0.25 oz/ft² or about $8 \times 10^{-5}$ to $2 \times 10^{-4}$ in. thick per side. However, thicker coatings can be produced by electroplating.

## Weld Cracking

Arc welded joints in galvanized steel made with carbon steel electrodes are subject to cracking. Cracking is caused by intergranular penetration of zinc into the weld metal and is sometimes called *zinc penetration cracking*. It occurs most often across the throat of a fillet weld, as shown in Fig. 1.13.

Steel is embrittled by molten zinc. Zinc has considerable solid solubility in iron, and the two metals form an intermetallic compound at the melting temperature of zinc. Molten zinc may attack carbon steel weld metal along the grain boundaries and form a brittle compound that fractures when a residual tensile stress of sufficient magnitude is present. Cracking is most prevalent when the coating is present at the root of the weld, particularly with fillet welds.

Whether cracking will occur in fillet welds depends upon several factors, including the following:

(1) Thickness of the galvanized coating
(2) Method of galvanizing
(3) Thickness of the galvanized steel
(4) Root opening of the joint
(5) Joint restraint
(6) Welding process
(7) Electrode classification

Cracking appears to be most severe with thick coatings applied by hot dipping, and least severe or nonexistent with a thin electrogalvanized coating. It tends to be less prevalent with shielded metal arc welding (SMAW) than with gas metal arc welding (GMAW). For example, intergranular penetration of zinc often occurs in fillet welds made by the GMAW process on 0.25 in. and thicker plate, but with SMAW, cracking usually does not occur with plate thicknesses less than 0.5 inch. The higher heat input and slower welding speed with shielded metal arc welding volatilizes away a greater amount of zinc ahead of the molten weld pool.

Welding consumables low in silicon are recommended. Weld metal containing about 0.2 percent silicon or less is usually free from zinc

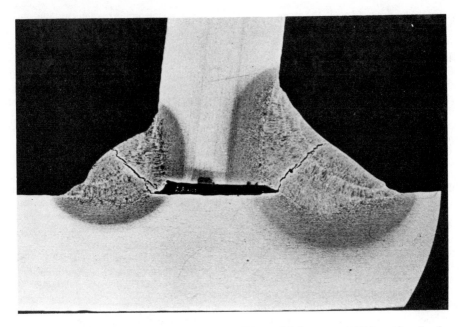

*Fig. 1.13—Cross section through cracked fillet welds between 0.25-in. galvanized steel plates made by gas metal arc welding with $CO_2$ shielding ( ×5)*

penetration cracking. Gas metal arc welds made with silicon deoxidized electrodes, such as ER70S-6, and welds made with a covered electrode with a basic coating, such as E7015 or E7016, are highly susceptible to cracking. On the other hand, GMAW welds made with low silicon electrodes, such as ER70S-3, or with rutile covered electrodes, such as E6012 or E6013, have low cracking susceptibility.

Cracking is less severe when the zinc fume can escape rapidly ahead of the weld pool. Therefore, a joint with a large root opening is best.

Cracking susceptibility of fillet welds can be minimized if the following steps are taken:

(1) Use a single- or double-bevel on the butting edge of the plate, either before or after galvanizing.

(2) Remove the coating from both faying surfaces by burning with an oxyfuel gas torch or by shot blasting.

(3) Maintain a suitable root opening in the joint, at least 0.06 inch.

(4) Run procedure qualification tests with the electrodes.

Both the weld metal silicon content and the penetration characteristics of the electrode are important. Electrode selection should not be based on silicon content alone. Until experience is obtained with a particular electrode type, or until electrodes are developed specifically for welding galvanized steel, qualification tests should be made to find a suitable electrode type for the application.

Batches of galvanized steel products produced under different conditions may behave differently with respect to welding. This should be kept in mind as qualification tests are made. Zinc coated products that are specially processed to provide a zinc-iron alloy coating with little free zinc give less weld cracking problems than steel coated with a high proportion of unalloyed zinc.

## Shielded Metal Arc Welding

The procedures and amperages for welding galvanized steels with covered electrodes are generally similar to those for welding uncoated steels. Electrode manipulation is modified to melt and vaporize the zinc coating from the steel ahead of the molten weld pool. The exposed steel and weld metal will have impaired corrosion resistance, and they may need to be coated to protect them in service.

Covered electrodes generally recommended for welding galvanized steel sheet and plate are E6012, E6013, and E7016. The first two have rutile coverings and the latter has a titania covering. Similar electrodes with iron powder additions, such as E7014 (rutile), E7018 (titania), or E7024 (titania), may be preferred for welding plate thicknesses because of their higher deposition rates. E6010 or E6011 electrodes may be suitable for some applications.

Electrodes with rutile covering and low silicon content (0.2 percent or lower) generally produce fillet welds free of cracks. Conversely, titania electrodes with high silicon content (0.4 percent or more) usually produce cracked welds because of the greater susceptibility of the weld metal to zinc penetration.

*Groove Welds.* The root opening for a groove weld in galvanized steel should be wider than for a similar weld in bare steel to obtain complete joint penetration. For example, a root opening of 0.09 in. is recommended with 0.125 in. thick galvanized sheet.

The welding arc should be repeatedly advanced onto the zinc coating ahead of the molten weld pool as welding progresses to melt and vaporize the coating from the steel. Removal of the coating will reduce weld metal porosity caused by entrapment of zinc vapor. The welding speed will be 80 to 90 percent of that with uncoated steel.

With multiple-pass welds in plate, a root opening of 0.13 to 0.19 in. is recommended. The arc should be manipulated during the first pass to vaporize the zinc coating. Subsequent passes may be made with conventional techniques if they are deposited largely on bare weld metal and only partially on galvanized joint faces. Otherwise, the special arc manipulation should be used for all passes. The need to burn off the zinc coating will reduce the welding speed. An alternative is to remove the zinc coating from the joint faces and adjacent surfaces prior to welding.

*Fillet Welds.* The heat of the arc is used to volatilize the zinc coating ahead of the molten weld pool, as with groove welds. This action should be incorporated in the electrode manipulation needed to produce an acceptable fillet weld geometry. The arc should be moved ahead about 0.3 in. to burn off the coating and then returned as welding progresses along the joint.

A frequent problem when welding in the horizontal and vertical positions is undercut. It may occur on the vertical member in the horizontal position and on both members in the vertical position. The condition is more prevalent with E6013 electrodes where the slag is very fluid and the weld face is concave. Electrodes that produce a faster freezing slag and a convex face have less tendency to produce this condition. Proper electrode manipulation is required for the first pass in any case, to vaporize the coating ahead of the weld pool.

Iron powder electrodes, such as E7014 or E7018, may be used where higher deposition rates are desirable. Electrode manipulation should be that normally used for bare steel, but the welding speed should be slower to allow for vaporization of the coating.

*Mechanical Properties.* Transverse tensile tests, all-weld-metal tensile tests, bend tests, and impact tests carried out over a range of temperatures indicate that mechanical properties of sound welds in galvanized steel made with either rutile- or basic-covered electrodes are equivalent to those of joints in uncoated steel.

## Gas Metal Arc Welding

Galvanized steel can be joined by gas metal arc welding (GMAW) using a steel electrode. The welding procedures are similar to those for bare low carbon steel. Travel speed should be slower on galvanized steel to allow the zinc coating to volatize ahead of the weld pool. The required reduction in speed depends upon the thickness of the zinc coating, the joint design, and the welding position. Root openings must be greater than for bare steel to obtain good joint penetration and to provide a path for vaporized zinc to escape. Slight weaving of the arc is required to control excessive melt-thru with groove welds. Typical conditions for welding butt, lap, and T-joints in galvanized steel sheet using short-circuiting transfer are given in Tables 1.35, 1.36, and 1.37.

*Mechanical Properties.* Transverse tensile specimens of groove welds with the reinforcement removed flush with the sheet surface usually fracture in the base metal, remote from the weld. Fillet welds in cruciform tensile test specimens made from galvanized steel have strengths equivalent to fillet welds on uncoated steel.

*Spatter.* With the short-circuit transfer and $CO_2$ shielding, spatter is greater when welding galvanized steel than when welding bare steel.

**Table 1.35**
**Typical conditions for gas metal arc welding square-groove butt joints in galvanized steel sheet with short-circuiting transfer and $CO_2$ shielding**

| Sheet thickness, in. | Root opening, in. | Welding position | Electrode feed rate,[a] in./min | Arc voltage, V | Welding current, A | Travel speed, in./min |
|---|---|---|---|---|---|---|
| 0.062 | 0 | Flat | 140-190 | 17-20 | 70-90 | 12-17 |
| | | Vertical-down | 195 | 17 | 90 | 14 |
| | | Horizontal | 120 | 18 | 100 | 20 |
| | | Overhead | 120-130 | 18-19 | 100-110 | — |
| 0.125 | 0.03-0.06 | Flat | | | | 13 |
| | 0.03-0.06 | Vertical-down | 170 | 20 | 135 | 18 |
| | 0.03-0.06 | Horizontal | | | | 16 |
| | 0.03 | Overhead | | | | 13 |

a. Electrode—ER70S-3
Electrode diam.—0.035 in.
Electrode extension—0.25-0.38 in.

**Table 1.36**
**Typical conditions for gas metal arc welding T-joints (fillet welds) in galvanized steel sheet with short-circuiting transfer and $CO_2$ shielding**

| Sheet thickness, in. | Welding position | Electrode feed rate,[a] in./min | Arc voltage, V | Welding current, A | Travel speed, in./min |
|---|---|---|---|---|---|
| 0.062 | Flat | 120-130 | 18 | 100-110 | — |
| | Vertical-down | 130-155 | 19 | 110-120 | — |
| | Overhead | 130 | 19-20 | 110 | 14 |
| | Horizontal | 140 | 20 | 120 | 12 |
| 0.125 | Flat | | | | 11 |
| | Vertical-down | 170 | 20 | 135 | 14 |
| | Overhead | | | | 10 |
| | Horizontal | | | | 12 |

a. Electrode—ER70S-3
Electrode diam.—0.035 in.
Electrode extension—0.25-0.38 in.

Spatter adherence can be minimized by spraying the work and the welding gun nozzle with an antispatter compound before welding. Any silicon-based anti-spatter compound remaining on the weldment can adversely affect paint adherence.

When welds are made in the flat position with no root opening, spatter particles tend to build up in the root of the joint and make the arc unstable. A root opening in the joint will reduce this problem.

*Weld Soundness.* Butt joints in galvanized steel welded with GMAW short-circuit transfer are normally free from discontinuities. Fillet welds in T-joints may contain variable amounts of porosity that tend to be more extensive in the second weld in the joint. Fillet welds made with

**Table 1.37**
**Typical conditions for gas metal arc welding lap joints (fillet welds) in galvanized steel sheet with short-circuiting transfer and CO₂ shielding**

| Sheet thickness, in. | Welding position | Electrode feed rate,[a] in./min | Arc voltage, V | Welding current, A | Travel speed, in./min |
|---|---|---|---|---|---|
| 0.062 | Horizontal | 120 | 18 | 100 | 12-16 |
|  | Vertical-down |  |  |  | 13-16 |
|  | Overhead |  |  |  | 10-12 |
| 0.125 | Horizontal | 160 | 19 | 130 | 9-10 |
|  | Vertical-down | 160 | 19 | 130 | 12 |
|  | Overhead | 140 | 19 | 120 | 8-9 |

a. Electrode—ER70S-3
   Electrode diam.—0.035 in.
   Electrode extension—0.25-0.38 in.

vertical-down welding procedures contain more porosity than similar welds made vertical-up or in other positions.

## Flux Cored Arc Welding

Galvanized steels may be arc welded with flux cored electrodes. The recommendations of the electrode manufacturer should be followed, and the welding procedures should be qualified by appropriate tests.

## Submerged Arc Welding

*Butt Joints.* Butt joints in galvanized steel plate with zinc-free joint faces can be submerged arc welded using the same procedures as for bare plate.

When the joint faces are galvanized, a 70-degree V-groove with a 0.25-in. wide root face should be used to permit the zinc vapor to escape. Square-groove joints with galvanized edges can be welded, but weld soundness will depend on the welding current, travel speed, and zinc coating thickness. A root opening of 0.06 to 0.13 in. will permit the zinc vapor to escape from the back side of an unbacked joint. The face of the weld is sometimes quite uneven, and slag removal is more difficult than with bare steel plate.

*T-Joints.* When making fillet welds on galvanized plate, the travel speed should be slower than for bare plate to obtain sound weld metal. Porosity is significantly reduced when the abutting plate edge is free of galvanize and a 0.06-in. root opening is used.

## Gas Tungsten Arc Welding

Gas tungsten arc welding of galvanized steel is not recommended unless the zinc coating is removed first. The zinc vapor may contaminate the tungsten electrode, which, in turn, will cause erratic arc operation, and poor weld quality. If the zinc coating is removed, the bare steel is welded using procedures suitable for low carbon steel.

## Oxyacetylene Welding

The heated zone with oxyacetylene welding is quite wide because of the low heat input of the process. As the heat spreads, the zinc coating vaporizes and leaves the steel exposed. The welding procedure is the same as that for bare low carbon steel. If necessary, the joint can be heated ahead of the weld pool by moving the flame forward for a brief time and then returning it to the weld pool.

## Braze Welding

Galvanized steel can be braze welded using a brass or bronze filler metal deposited by shielded metal arc, gas metal arc, gas tungsten arc, or oxyacetylene welding. Dilution of a copper alloy filler metal with steel is undesirable because of increased hardness and brittleness. Therefore, the heat input should be only that required for good wettability. Welding amperages for arc welding should be kept low to minimize melting of the steel. Short-circuiting transfer should be used with gas metal arc welding.

Filler metals commonly used with the arc welding processes are copper-silicon (silicon bronze), copper-tin (phosphor bronze), and copper-aluminum (aluminum bronze). They are available as covered electrodes and bare welding rods or electrodes.[30] Bare rods are used with the gas tungsten arc torch. Small diameter electrodes should be used to limit the heat input. Typical conditions for braze welding galvanized steel by gas metal arc welding are given in Table 1.38.

With gas tungsten arc welding, the filler rod is fed at a low angle into the joint while a fairly short arc is directed on the rod. A flux is not necessary. Sound welds can be made with this procedure.

Bare copper-zinc (brass) welding rods are commonly used for oxyacetylene braze welding although the filler metals mentioned previously may be used also.[31] The welding rod diameter should be about 1.5 times the sheet thickness, but not over 0.25 in.

A borax-boric acid braze welding flux is needed for good wetting. A generous application of flux to the faying surfaces and the welding rod as welding progresses, or the use of flux coated rods, will usually reduce the loss of galvanize coating.

The smallest oxyacetylene torch tip consistent with the thickness of galvanized steel should be used to limit flame spread. A neutral or slightly oxidizing flame with the lowest practical heat input should be used. After the work has reached a dull red heat, the flame is directed onto the welding rod as welding progresses along the joint. The flame should not be oscillated.

Joint strength of a properly made braze weld in galvanized steel will equal or exceed the strength of the steel. The joint corrosion resistance is excellent because the filler metal wets and covers the exposed steel surfaces.

Cross sections of braze welds in galvanized sheet made by oxyacetylene, gas tungsten arc and shielded metal arc welding are shown in Fig. 1.14. The top edges of the steel sheets melted and diluted with the filler metal in the arc welded joints, but this did not happen with oxyacetylene braze welding.

## Resistance Welding

*Spot Welding.* Spot welding on galvanized steel sheet and strip is widely practiced, but the zinc coating affects the welding conditions in two ways.

First, the zinc coating at the electrode-to-work interface is melted by resistance heating. The molten zinc readily alloys with and sticks to the copper electrode and builds up on the electrode face. Zinc buildup rapidly increases the electrode face area and lowers the current density in the weld zone. Consequently, the welding current must be progressively increased to maintain satisfactory weld strength.

Second, the presence of the soft, highly conductive zinc coating changes the electrical and thermal characteristics of the current path. Galvanized steel has a lower contact resistance between the faying surfaces than does uncoated low carbon steel. In addition, the zinc coating provides an excellent path for conduction of heat from the weld to the water-cooled electrodes. As a result, high welding current is needed to melt the zinc coating so it can be displaced from between the steel surfaces by the electrode force. Both the welding current and the weld time for spot welding galvanized steel should be increased by 25 to 50 percent over those recommended for bare steel sheet of the same thickness. The acceptable ranges of welding current and weld time are more restrictive for galvanized steel than for uncoated steel. On the other hand, variations in electrode force are more tolerable for galvanized steel because of its lower contact resistance.

Schedules for spot welding galvanized steel are listed in Table 1.39. These schedules should be considered as guides for good welding practice but they may require modification for specific applications. Standard truncated cone electrodes of RWMA Group A, Class 2 copper alloy with face diameters of four to five times the thickness of the sheet being welded are recommended. An adequate flow of internal cooling water to the

---

30. Refer to AWS A5.6-76. *Specification for Copper and Copper Alloy Covered Electrodes,* or AWS A5.7-77, *Specification for Copper and Copper Alloy Bare Welding Rods and Electrodes.*

31. Refer to AWS A5.8-81, *Specification for Brazing Filler Metal.*

**Table 1.38**
**Typical conditions for braze welding galvanized steel sheet with the gas metal arc short-circuiting transfer method**

| Joint type | Weld type | Sheet thickness, in. | Electrode[a] | | Arc voltage, V | Welding current,[b] A | Travel speed, in./min |
|---|---|---|---|---|---|---|---|
| | | | Diam., in. | Feed rate, in./min | | | |
| Butt | Square-groove | 0.062 | 0.030 | 255 | 18 | 100 | 18 |
| T | Fillet | 0.062 | 0.030 | 280 | 19 | 130 | 18 |
| | | 0.080 | 0.045 | 160 | 16 | 150 | 16 |
| Corner | Fillet on outside corner | 0.062 | 0.030 | 255 | 18 | 100 | 14 |
| | | 0.080 | 0.045 | 150 | 16 | 140 | 18 |
| Lap | Fillet | 0.062 | 0.030 | 260 | 18 | 110 | 18 |

a. ERCuAl-A2 electrode with argon shielding
b. Direct current, electrode positive

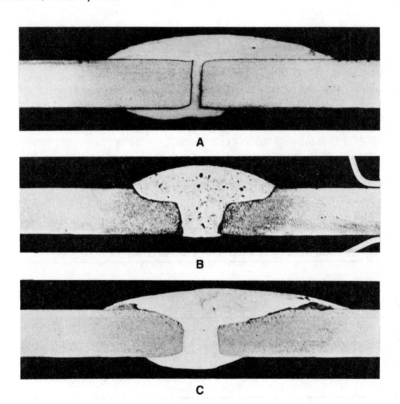

*Fig. 1.14—Cross sections of braze welds in 0.060-in. thick galvanized steel sheet: (A) oxyacetylene weld with Cu-Zn filler metal, (B) gas tungsten arc weld with Cu-Si filler metal, (C), shielded metal arc weld with Cu-Al covered electrode (×8)*

**Table 1.39**
**Typical spot welding schedules for galvanized low carbon steel sheet**

| Sheet thickness, in.[a] | Electrode face diam., in.[b] | Electrode force, lb | Approx. welding current, kA[c] | Weld time, cycles[d] | Approx. nugget size, in. | Minimum shear strength, lbs |
|---|---|---|---|---|---|---|
| 0.022 | 0.19 | 300 | 13.0 | 8 | 0.15 | 550 |
| 0.030 | 0.19 | 400 | 13.0 | 10 | 0.16 | 1000 |
| 0.036 | 0.25 | 500 | 13.5 | 12 | 0.19 | 1180 |
| 0.039 | 0.25 | 650 | 14.0 | 13 | 0.21 | 1400 |
| 0.052 | 0.25 | 725 | 14.5 | 18 | 0.22 | 1700 |
| 0.063 | 0.25 | 850 | 15.5 | 22 | 0.24 | 2500 |
| 0.078 | 0.31 | 1200 | 19.0 | 24 | 0.28 | 3200 |
| 0.093 | 0.38 | 1400 | 21.0 | 30 | 0.34 | 4200 |
| 0.108 | 0.38 | 1750 | 20.0 | 37 | 0.40 | 5900 |
| 0.123 | 0.38 | 2000 | 20.0 | 42 | 0.48 | 7200 |

a. Two equal thicknesses
b. Truncated cone eletrodes of RWMA Group A, Class 2 alloy
c. Single-phase ac machine
d. 60 Hz

electrode tip is necessary to help minimize zinc pickup.

*Seam Welding.* Typical schedules for pressure-tight seam welds in galvanized steel sheet are listed in Table 1.40. Higher welding speeds than those given will require higher welding currents. Excessive welding current tends to cause small transverse cracks in the weld that originate at the outer surfaces. Intermittent current seam welding is recommended for galvanized steel to avoid surface melting and the possibility of severe cracking normally experienced with continuous current seam welding.

Alloying between the zinc coating and copper alloy electrodes rapidly builds up a layer on the electrode faces, as with spot welding. Flood cooling of the weld area helps to minimize this action. A knurl drive system for electrode wheels tends to break up the alloy layer as it forms on the electrode face.

An alternative is to use narrow electrode wheels and high-pressure, profiled drive rollers

**Table 1.40**
**Typical seam welding schedules for galvanized low carbon steel sheet**

| Sheet thickness, in.[a] | Electrode face width, in.[b] | Electrode force, lb. | Approx. welding current, kA[c] | Heat time, cycles[d] | Cool time, cycles[d] | Welding speed, in./min | Spot weld per in. |
|---|---|---|---|---|---|---|---|
| 0.015 | 0.25 | 900 | 15.0 | 2 | 2 | 120 | 7.5 |
| 0.036 | 0.25 | 1100 | 18.0 | 4 | 2 | 60 | 10.0 |
| 0.039 | 0.25 | 1200 | 19.0 | 4 | 3 | 60 | 9.0 |
| 0.052 | 0.25 | 1350 | 20.0 | 5 | 1 | 90 | 7.0 |
| 0.063 | 0.31 | 1500 | 19.8 | 8 | 2 | 54 | 7.0 |
| 0.078 | 0.31 | 1850 | 23.0 | 10 | 7 | 30 | 7.0 |

a. Two equal thicknesses
b. Wheel electrodes, 8- to 10-in. diameter with flat faces and 30 deg bevel edges, of RWMA Group A, Class 2 copper alloy
c. Single-phase ac machine
d. 60 Hz

to clean and shape the wheels. Marked improvement in electrode life is possible with this arrangement.

The profiled drive rollers are knurled only on the angled sides that engage and drive the wheel electrodes. The root face of each roller is smooth and contoured to match the face of the mating wheel electrode. Each roller is hydraulically loaded with sufficient force to reshape the wheel face, as welding progresses. This type of drive provides uniform welding speed, independent of the electrode wheel diameter. It also maintains a relatively clean surface condition on the electrodes. The applied force of the roller breaks up the coating layer and removes it from the electrode face. Excessive profiling pressure, however, can cause a nonuniform surface condition on the electrode face. A spring-loaded scraping device is placed ahead of the profiling roller to remove heavy build-up of coating on the electrode.

The electrode wheel is 0.31- to 0.38-in. thick with a face radius of 0.20 to 0.24 inch. This gives a seam width of about $0.5t^{1/2}$ inch, where "t" is the sheet thickness. These dimensions are suitable for sheet thicknesses up to 0.047 inch. Electrode contact width is 0.06 to 0.10 in., depending upon wheel profile. The welds are narrow (0.06 to 0.14 in.) compared with those of conventional seam welding (0.12 to 0.24 in.). However, such welds possess strength levels in shear greater than the strength of the galvanized steel sheet, and are suitable for gas-tight applications. The narrowness of the weld makes it possible to reduce the power requirements of the welding machine. With the lower heat input, coating damage adjacent to the weld zone can be minimized. Typical seam welding schedules for this technique are given in Table 1.41.

*Projection Welding.* During projection welding of galvanized steel, the cold set-down of any one projection should not exceed 25 percent of its height. Appropriate tests must be made to establish a suitable electrode force so that this limit is not exceeded. Excessive set-down causes the interface area of the projection to be too large, and the projection may not weld. It may also cause a large variation in the amperages flowing through the several projections.

The welding machine should be equipped with a low-inertia head for fast followup as the projections collapse during heating. The pressure system should be capable of moving the head at about 13 inches per second.

For sheet thicknesses below 0.094 in., the projection diameter and height should be slightly larger than those used with bare low carbon steel sheet to obtain adequate shear strength. The contact resistance at the faying surfaces will be lower because of the deformation and subsequent melting of the soft zinc coating. Therefore, somewhat higher welding currents, lower electrode forces, and shorter weld times are recommended for projection welding galvanized steel.

Projection welds in sheet thicknesses less than 0.094 in. might not contain a nugget. Excessive welding current melts and expels the projection in preference to forming a molten nugget. However, useful shear strengths are obtainable when projections welds are made with a suitable welding schedule.

## Soldering

The chief problem in soldering galvanized steel is the difficult of wetting the zinc coating by the solder. A corrosive flux, consisting of zinc-, ammonium-, and stannous-chlorides with a small amount of muriatic acid, and a 60/40 or 50/50 tin-lead solder give satisfactory results. The flux is applied and allowed to stand on the zinc coating for a few minutes. The solder is then applied to the joint with a large soldering iron that is rubbed on the joint surface to expedite the wetting action of the solder. Flux residues must be thoroughly removed.

Electronic assemblies using galvanized sheet must first be precoated with solder and cleaned before joining. They are then soldered in the usual manner with a noncorrosive flux. Rosin core solders rapidly remove the light oxide film and promote solder flow. Their residues are noncorrosive and nonconductive.

## Protection of Welds

Arc welded joints on galvanized steel made with steel filler metal will rust in damp atmospheres, as will exposed areas of steel base metal, unless a protective coating is applied.

The exposed steel surfaces should be cleaned by wire brushing, preferably with a rotary brush. Heat damaged areas on the back side of welds should also be brushed. The welds and damaged areas should then be coated with a zinc-rich paint.

Alternatively, various proprietary zinc alloys are available in stick form. These can be melted onto the cleaned steel surface with an

**Table 1.41**
**Typical seam welding schedules for galvanized steel sheet with radius-faced electrode wheels**

| Sheet thickness, in. | Electrode force,[a] lbs | Welding current, kA, with heat and cool times of | |
|---|---|---|---|
| | | 40ms[b] | 20ms[c] |
| 0.016-0.024 | 500-675 | 10.2-11.1 | 13.0-14.1 |
| 0.024-0.031 | 675-775 | 11.1-12.0 | 13.7-14.8 |
| 0.031-0.039 | 725-900 | 11.8-12.8 | 14.2-15.4 |
| 0.039-0.047 | 900-1000 | 12.5-13.6 | 15.0-16.2 |

a. Copper alloy wheel electrodes with 0.2 in. face radius
b. Welding speed—80 in./min (9.4 welds/in.)
c. Welding speed—160 in./min (9.4 welds/in.)

oxyfuel gas torch. With gas shielded arc welding processes, the zinc alloy stick can be rubbed on the weld while it is still hot enough to melt the zinc. Flame sprayed zinc can also be applied to the exposed areas. With resistance spot and seam welds, a thin layer of zinc-iron alloy that usually remains on the electrode contact areas provides some corrosive protection.

## ZINC-RICH COATINGS

An alternative to galvanized coating on steel is the application of zinc-rich paints containing 85 to 97 percent metallic zinc. They are commonly referred to as *weld-through primers*.

### Resistance Spot Welding

Spot welding is facilitated if the zinc-rich coating is on the faying surface of the joint only, not on the surfaces in contact with the electrodes. The paint will contaminate the electrodes and necessitate frequent electrode dressing.

Although this coating is considered to have good electrical conductivity, the binder surrounding the zinc particles has sufficient electrical resistance to cause problems with spot welding. The open circuit voltage of the welding machine must be high enough to break down this dielectric binder. Arcing between the faying surfaces, a popping noise, and the generation of acrid fumes are characteristic of spot welding through zinc-rich paints.

When possible, it is best to weld through the paint before it has completely dried. This allows the welding pressure to displace the paint from between the faying surfaces at the weld location and provide a more conductive current path.

### Arc Welding

Arc welding of painted steel without first removing the paint is not generally recommended because the pigments can contaminate the weld metal. When zinc-rich paint is used on steels prior to arc welding, porosity may be a problem because of the gas-forming constituents in the paint. Paint may also cause cracking in welds in steels that are sensitive to hydrogen induced cracking. Tests must be made to determine if there are such problems when welding a particular steel.

## SAFE PRACTICES

Zinc fume is potentially a hazard to health. Specific precautions are given in ANSI Publication Z49.1, *Safety in Welding and Cutting.* It is essential to consult this Standard before starting to weld or thermally cut zinc coated metals.

Metal fume fever, more commonly called zinc chills, may follow exposure to zinc fume released during welding or cutting operations on zinc-coated or zinc-containing metals, including brazing rods. The chills are caused by colloidal zinc oxide penetrating to the lungs. Inhalation of iron oxide fume may also cause metal fume fever. Aluminum oxide fume is presently considered nontoxic, but inhalation should be minimized by appropriate ventilation practices.

This reaction to metal fume is an acute self-limiting condition without known complications, after effects, or chronic form. The illness begins a few hours after exposure or more frequently during the night, and may cause a sweet taste in the mouth, dryness of the throat, coughing, fatigue, yawning, weakness, head and body aches, vomiting, chills or a fever rarely exceeding 102°F. A second attack seldom occurs during repeated exposure unless there has been an interval of several days between exposures.

Threshold limit values (TLV) of iron and zinc oxides in breathing air, as published by the American Conference of Governmental Industrial Hygienists and OSHA, are approximately 5.0 mg/m³. These values are subject to review and may change.

Repeated exposure to moderate concentrations of iron or zinc oxide in air have not proved permanently harmful. However, exposure to concentrations high enough to cause discomfort to welders or cutting operators should be avoided.

When welding of materials containing zinc is to be done in a confined space, ventilation must be provided. If adequate ventilation cannot be provided, personnel who may be exposed to the fume must be equipped with hose masks or air-supplied respirators approved by the U.S Bureau of Mines. For gas metal arc welding, fume extractor nozzles are sometimes used in conjunction with the welding gun.

## Metric Conversion Factors

$1 \text{ ksi} = 6.89 \text{ MPa}$
$1 \text{ lb·f} = 4.45 \text{ N}$
$1 \text{ in.} = 25.4 \text{ mm}$
$1 \text{ in./min} = 0.423 \text{ mm/s}$
$1 \text{ oz/ft}^2 = 0.305 \text{ kg/m}^2$
$1 \text{ ft·lb} = 1.36 \text{ J}$
$1 \text{ kJ/in.} = 39.4 \text{ J/mm}$
$t_C = 0.56 (t_F - 32)$

# SUPPLEMENTARY READING LIST

*Brazing Manual,* 3rd Ed., Miami: American Welding Society, 1976.

Caplan, J. S., and Landerman, E., Preventing Hydrogen-Induced Cracking After Welding of Pressure Vessel Steels by the Use of Low Temperature Postweld Heat Treatments, New York: Welding Research Council, 1976, June; Bulletin 216.

Coe, F. R., Welding Steels Without Hydrogen Cracking, Cambridge, England: The Welding Institute, 1973.

Dorschu, K. E., Factors Affecting Weld Metal Properties in Carbon and Low Alloy Steel Pressure Vessel Steels, New York: Welding Research Council, 1977 Oct.; Bulletin 231.

Emmer, L. G., Clauser, C. D. and Low, J. R., Critical Literature Review of Embrittlement in 2-1/4Cr-1Mo Steel, New York: Welding Research Council, 1973 May; Bulletin 183.

Interpretive Report on Effect of Hydrogen in Pressure Vessel Steels, New York: Welding Research Council, 1969 Oct.; Bulletin 145.

Jubb, J. E. M., Lamellar Tearing, New York: Welding Research Council, 1971 Dec.; Bulletin 168.

Linnert, G. E., *Welding Metallurgy,* Vol. 1 (Fundamentals), 3rd Ed., Miami: American Welding Society, 1965.

Linnert, G. E., *Welding Metallurgy,* Vol. 2 (Technology), 3rd Ed., Miami: American Welding Society, 1967.

Meitzner, C. F., Stress Relief Cracking in Steel Weldments, New York: Welding Research Council, 1975 Nov.; Bulletin 211.

*Metals Handbook—Welding and Brazing,* Vol. 6, 8th Ed., Metals Park, OH: American Society for Metals, 1971.

Oldroyd, P. S., and Williams, N. T., Overcoming the inherent problems of coated steels, *Welding and Metal Fabrication,* 48(2): 97-103; 1980, Mar.

Orts, D., The do's and don't's of welding galva-nized and aluminum coated steels, *Welding Design and Fabrication,* 49(2): 62-67; 1976 Dec.

*Resistance Welding, Recommended Practices for,* AWS C1.1-66, Miami: American Welding Society, 1966.

*Resistance Welding Coated Low Carbon Steels, Recommended Practices for,* AWS C1.3-70, Miami: American Welding Society, 1970.

Stout, R. D., Hardness as an Index to the Weldability and Service Performance of Steel Weldments, New York: Welding Research Council, 1971 Dec.; Bulletin 168.

Stout, R. D., and Doty, W. D., *Weldability of Steels,* 3rd Ed., (Edited by Epstein, S., and Somers, R. E.) New York: Welding Research Council, 1978.

Shackleton, D. N., Welding HY-100 and HY-130 Steels, Cambridge, England: The Welding Institute, 1973 Sept.

The Toughness of Weld Heat-Affected Zones, Cambridge, England: The Welding Institute, 1975.

*Welding and Weldability of C-Mn Steels and C-Mn Microalloyed Steels, Guide to the,* IIS/IIW-382-71: International Institute of Welding, 1971 (Available from the American Welding Society).

Welding Coated Steels, Cambridge, England: The Welding Institute, 1978.

*Welding of Chromium-Molybdenum Steel Piping and Tubing, Recommended Guide to the,* IIS/IIW-382-71: International Institute of Welding, 1971 (Available from the American Welding Society).

*Welding of Zinc Coated Steels,* Miami: American Welding Society 1972.

William, N. T., Recent developments in resistance welding of zinc coated steels, *Welding and Metal Fabrication,* 45(5): 275-88; 1977 June.

# 2

# Stainless Steels

## Chapter Committee

S.D. REYNOLDS, *Chairman*
  *U.S. Nuclear Regulatory Commission*
K.E. DORSCHU
  *Consultant*
D.P. EDMONDS
  *Oak Ridge National Laboratories*
L. POOLE
  *Westinghouse Electric Corporation*

H.N. FARMER, JR.
  *Stoody Company*
D.W. HARVEY
  *U.S. Welding*
D.J. KOTECKI
  *Teledyne-McKay Company*

## Welding Handbook Committee Member

R.L. FROHLICH
  *Westinghouse Electric Corporation*

# 2

# Stainless Steels

## GENERAL DESCRIPTION

### DEFINITION

Stainless steels are those alloy steels that have a normal chromium content of at least 12 percent with or without other alloy additions. The stainlessness and corrosion resistance of these alloy steels are attributed to the presence of a passive oxide film on the surface. When exposed to conditions that remove the passive oxide film, stainless steels are subject to corrosive attack. The rate at which a stainless steel develops a passive film in the atmosphere depends on its chromium content. Polished stainless steels remain bright and tarnish-free under most atmospheric conditions. Exposure to elevated temperatures increases the thickness of the oxide film and tarnishes the metal.

Stainless steels are commonly divided into the following general groups:

(1) Chromium Martensitic
(2) Chromium Ferritic
(3) Austenitic
(4) Precipitation-hardening

The first three groups are characterized by the predominant metallurgical phase present when the stainless steel is placed in service. The fourth group contains those stainless steels that can be strengthened by an aging heat treatment.

The metallurgical characteristics of stainless steels are related to the effect of alloying additions on the allotropic transformation of austenite. The alloying additions essentially expand, constrict, or eliminate the gamma or austenite phase of the steel. Some of the resultant stainless steels are hardenable by heat treatment as a result of austenite-to-martensite transformation (martensitic types) or by precipitation of a second phase in low-carbon martensite (precipitation-hardening types). Other stainless steels are non-hardenable because (1) they cannot be transformed to austenite on heating (ferritic types) or (2) the austenite is stable down to room temperature and below (austenitic types). The major alloying elements that effect these characteristics are chromium, nickel, and carbon. Other alloying elements contribute to the metallurgical characteristics and corrosion resistance of stainless steels.

### CLASSIFICATION

Wrought stainless steels are classified and assigned designations by the American Iron and Steel Institute (AISI) according to chemical composition. The Cr-Ni-Mn austenitic stainless steels are the 2XX series, the Cr-Ni austenitic stainless steels, the 3XX series; and the Cr ferritic and martensitic stainless steels, the 4XX series. The precipitation-hardening grades are assigned designations based on their Cr and Ni contents. With the 2XX series, manganese (Mn) is substituted for part of the nickel (Ni).

In addition, there are a number of wrought stainless steels that are not classified by AISI. Some of these are designated by the same numbering system as the AISI types, and others are known by trade names. There are minor variations to many of the stainless steels. These variations include special carbon control for corrosion or high temperature applications; chemical stabilization with Al, Cb, or Ti; high S or Se for better machinability, and other alloy additions for special characteristics.

Corrosion resistant (stainless) steel castings

are standardized by the Alloy Casting Institute (ACI) Division of the Steel Founders Society of America. They are designated by a two-letter system, HX, or a letter-number system, CX-XXX, such as CF-8 or CF-12M. Many cast types are similar to counterparts in the AISI wrought stainless steel system.

Most stainless steels have been assigned numbers under the SAE-ASTM Unified Numbering System (UNS)[1]. Wrought and cast stainless steels are identified by the letters *S* and *J* respectively followed by five digits.

## PROPERTIES

The alloy system and the metallurgical phases present control the general physical and mechanical properties of stainless steel.[2] Typical physical properties of the four groups of wrought stainless steels are given in Table 2.1, together with those of carbon steel for comparison. Thermal expansion, thermal conductivity, and electrical resistivity have significant effects on the weldability of steel.

The relatively high coefficient of thermal expansion and the low thermal conductivity of austenitic stainless steel require better control of distortion during welding than is needed for the other types of stainless steels and carbon steel. The low thermal conductivities indicate that stainless steels generally require less welding heat than carbon steel. Also, stainless steels can be resistance spot and seam welded with lower welding current because their electrical resistivities are higher than that of carbon steel.

The austenitic stainless steels are essentially nonmagnetic (magnetic permeability of about 1.02), but cold working to improve strength properties increases magnetic permeability. Ferritic and martensitic stainless steels have magnetic permeabilities in the range of 600-1100, but those of the precipitation-hardenable types are below 100.

------

1. Refer to SAE HS 1086, ASTM-DS-56A, *Unified Numbering System for Metals and Alloys,* 2nd Ed., 1977.

2. The mechanical and physical properties, heat treatment, corrosion resistance, and fabrication of specific stainless steels are discussed in the *Metals Handbook,* Vol. 3, 9th Ed.; American Society for Metals, 1980: 1-124, 189-206, 269-313, 741-55.

## GENERAL WELDING CHARACTERISTICS

The metallurgical features of each group generally determine the weldability characteristics of the steels in that group. The weldability of martensitic stainless steels is affected greatly by hardenability that can result in cold cracking. Welded joints in ferritic stainless steel have low ductility as a result of grain coarsening that is related to the absence of allotropic (phase) transformation. The weldability of austenitic stainless steels is governed by their susceptibility to hot cracking, similar to other single-phase alloys with a face-centered cubic crystal structure. With the precipitation-hardening stainless steels, weldability is related to the mechanisms associated with the transformation (hardening) reactions.

Stainless steels can be joined by most welding processes, but with some restrictions. In general, those steels that contain aluminum or titanium, or both, can be arc welded only with the gas-shielded processes. The weld joint efficiency depends upon the ability of the welding process and procedures to produce nearly uniform mechanical properties in the weld metal, heat-affected zone, and base metal in the as-welded or postweld heat treated condition. These properties can vary considerably with ferritic, martensitic, and precipitation-hardening steels.

Weldability and various suitability-for-service conditions, including temperature, pressure, creep, impact, and corrosion environments, require careful evaluation because of the complex metallurgical aspects of stainless steels. When specific information is not available, the steel manufacturer should be consulted for technical data on the suitability-for-service of weldments.

## WELDING PROCESSES

The stainless steels are readily joined by arc, electron beam, laser beam, resistance, and friction welding processes. Gas metal arc, gas tungsten arc, flux cored arc, and shielded metal arc welding are commonly used. Plasma arc and submerged arc welding are also suitable joining methods.

Sound welds can be produced by submerged arc welding, but there are certain restrictions on the process. Generally, the composition of weld metal deposited by this process is more difficult to control than that produced with other arc welding processes because of the effect of arc voltage

**Table 2.1**
**Typical physical properties of wrought stainless steels in the annealed condition**

| Property | Units | Stainless steel | | | | Carbon Steel |
| --- | --- | --- | --- | --- | --- | --- |
| | | Cr-Ni austenitic | Cr ferritic | Cr martensitic | Precipitation-hardening | |
| Density | lb/in.³ | 0.28-0.29 | 0.28 | 0.28 | 0.28 | 0.28 |
| | Mg/m³ | 7.8-8.0 | 7.8 | 7.8 | 7.8 | 7.8 |
| Elastic modulus | 10⁶ psi | 28-29 | 29 | 29 | 29 | 29 |
| | GPa | 193-200 | 200 | 200 | 200 | 200 |
| Mean coef. of thermal expansion, 32° to 1000°F (0° to 538°C) | $10^{-6}$ in./in./°F | 9.4-10.7 | 6.2-6.7 | 6.4-6.7 | 6.6 | 6.5 |
| | $10^{-6}$ m/m/°C | 17.0-19.2 | 11.2-12.1 | 11.6-12.1 | 11.9 | 11.7 |
| Thermal conductivity, 212°F (100°C) | Btu/h·ft·°F | 10.8-12.8 | 15.0-15.8 | 16.6 | 12.6-13.1 | 34.7 |
| | W/(m·K) | 18.7-22.8 | 24.4-26.3 | 28.7 | 21.8-23.0 | 60 |
| Specific heat, 32°-212°F (0°-100°C) | Btu/(lb·°F) | 0.11-0.12 | 0.11-0.12 | 0.10-0.11 | 0.10-0.11 | 0.12 |
| | J/(kg·K) | 460-500 | 460-500 | 420-460 | 420-460 | 480 |
| Electrical resistivity | $10^{-8}$ Ω·m | 69-102 | 59-67 | 55-72 | 77-102 | 12 |
| Melting range | °F | 2550-2650 | 2700-2790 | 2700-2790 | 2560-2625 | 2800 |
| | °C | 1400-1450 | 1480-1530 | 1480-1530 | 1400-1440 | 1538 |

variations. For example, the silicon content might be high and result in hot cracking of the weld metal. Heat input is higher and solidification of the weld metal is slower than with other arc welding processes. These conditions can lead to large grain size weld metal that is usually low in toughness.

Submerged arc welding is not recommended when an austenitic stainless steel weld deposit must be fully austenitic or low in ferrite content. However, it is suitable when a ferrite content of over 4 percent is permissible in the weld metal.

Proprietary fluxes are available for submerged arc welding of stainless steels. Alloying elements, including chromium, nickel, molybdenum, and columbium, can be added to the weld metal by suitable fluxes.

Spot, seam, projection, flash, and high-frequency resistance welding processes are well suited for joining stainless steels. The higher electrical resistance and higher strength of stainless steel require lower welding current and higher electrode or upsetting force than carbon steel.

Oxyacetylene welding is not recommended except for emergency repairs where suitable arc welding equipment is not available. A neutral or slightly excess acetylene (reducing) flame is recommended. A welding flux with good solvent power for chromium oxide is essential. Careful flame control with minimum puddling and proper heat input is required.

## PROTECTION AGAINST OXIDATION

To be suitable for stainless steels, a welding process must protect the molten weld metal from the atmosphere during transfer across the arc and during solidification. With some welding processes, fluxing is required to remove chromium and other oxides from the surfaces to be joined and from the molten weld metal. Chromium oxide is an extremely refractory material that melts well above the melting temperature of the stainless steels. Fluorides are the most effective agents for removing chromium oxide during welding. Calcium and sodium fluorides are used in covered electrode coatings and submerged arc welding fluxes for stainless steels.

The fluoride residuals in slags on a weldment can be quite corrosive, and can attack the metal during service at elevated temperatures. Welding slag should be thoroughly removed by chipping or brushing before postweld heat treatment or elevated temperature service.

Flux is seldom used with the gas shielded welding processes because the shielding gas cover usually prevents excessive oxidation. Inert gas protection is frequently provided in flash welding to prevent the formation of chromium oxide that can be trapped at the interface during upsetting.

## PREWELD AND POSTWELD CLEANING

The relatively high chromium content of stainless steels promotes the formation of tenacious oxides that must be removed for good welding practices. Surface contaminants affect stainless steel welds to a greater degree than carbon and low alloy steel welds.

To obtain sound welds, the surfaces to be joined require cleaning prior to welding. The area to be cleaned should include the weld groove faces and the adjacent surfaces for at least 0.5 in. on each side of the groove. Cleaning of a wider band is recommended on thick plate. The surfaces of parts to be resistance spot or seam welded should be cleaned.

The degree of cleaning necessary depends on the weld quality requirements for the application and the welding process employed. Special care is required with gas shielded welding processes because of the absence of fluxing.

Carbon contamination can adversely affect the metallurgical characteristics, the corrosion resistance, or both, of stainless steel. Pickup of carbon from surface contaminants or embedded particles must be prevented.

The surfaces of areas to be welded must be completely cleansed of all hydrocarbon and other contaminants, such as cutting fluids, grease, oil, waxes, and primers, by suitable solvents. Light oxide films can be removed by pickling, but normally are removed by mechanical methods. Acceptable preweld cleaning techniques include the following:

(1) Stainless steel wire brushes that have not been used for any other purpose

(2) Blasting with clean sand or grit

(3) Machining and grinding using a suitable tool and chloride-free cutting fluid

(4) Pickling with 10 to 20 percent nitric acid solution

Thorough postweld cleaning is required to remove welding slag when present. Objectionable

**Table 2.2**
**Chemical composition of all-weld-metal deposited from stainless steel covered electrodes[a]**

| AWS Classification[c] | Chemical composition, weight percent[b] | | | | | | | | | | |
|---|---|---|---|---|---|---|---|---|---|---|---|
| | C | Cr | Ni | Mo | Cb plus Ta | Mn | Si | P | S | N | Cu |
| E209[d] | 0.06 | 20.5-24.0 | 9.5-12.0 | 1.5-3.0 | — | 4.0-7.0 | 0.90 | 0.03 | 0.03 | 0.10-0.30 | 0.75 |
| E219 | 0.06 | 19.0-21.5 | 5.5-7.0 | 0.75 | — | 8.0-10.0 | 1.00 | 0.03 | 0.03 | 0.10-0.30 | 0.75 |
| E240 | 0.06 | 17.0-19.0 | 4.0-6.0 | 0.75 | — | 10.5-13.5 | 1.00 | 0.03 | 0.03 | 0.10-0.20 | 0.75 |
| E307 | 0.04-0.14 | 18.0-21.5 | 9.0-10.7 | 0.5-1.5 | — | 3.3-4.75 | 0.90 | 0.04 | 0.03 | — | 0.75 |
| E308 | 0.08 | 18.0-21.0 | 9.0-11.0 | 0.75 | — | 0.5-2.5 | 0.90 | 0.04 | 0.03 | — | 0.75 |
| E308H | 0.04-0.08 | 18.0-21.0 | 9.0-11.0 | 0.75 | — | 0.5-2.5 | 0.90 | 0.04 | 0.03 | — | 0.75 |
| E308L | 0.04 | 18.0-21.0 | 9.0-11.0 | 0.75 | — | 0.5-2.5 | 0.90 | 0.04 | 0.03 | — | 0.75 |
| E308Mo | 0.08 | 18.0-21.0 | 9.0-12.0 | 2.03-3.0 | — | 0.5-2.5 | 0.90 | 0.04 | 0.03 | — | 0.75 |
| E308MoL | 0.04 | 18.0-21.0 | 9.0-12.0 | 2.0-3.0 | — | 0.5-2.5 | 0.90 | 0.04 | 0.03 | — | 0.75 |
| E309 | 0.15 | 22.0-25.0 | 12.0-14.0 | 0.75 | — | 0.5-2.5 | 0.90 | 0.04 | 0.03 | — | 0.75 |
| E309L | 0.04 | 22.0-25.0 | 12.0-14.0 | 0.75 | — | 0.5-2.5 | 0.90 | 0.04 | 0.03 | — | 0.75 |
| E309Cb | 0.12 | 22.0-25.0 | 12.0-14.0 | 0.75 | 0.70-1.00 | 0.5-2.5 | 0.90 | 0.04 | 0.03 | — | 0.75 |
| E309Mo | 0.12 | 22.0-25.0 | 12.0-14.0 | 2.0-3.0 | — | 0.5-2.5 | 0.90 | 0.04 | 0.03 | — | 0.75 |
| E310 | 0.08-0.20 | 25.0-28.0 | 20.0-22.5 | 0.75 | — | 1.0-2.5 | 0.75 | 0.03 | 0.03 | — | 0.75 |
| E310H | 0.35-0.45 | 25.0-28.0 | 20.0-22.5 | 0.75 | — | 1.0-2.5 | 0.75 | 0.03 | 0.03 | — | 0.75 |
| E310Cb | 0.12 | 25.0-28.0 | 20.0-22.0 | 0.75 | 0.70-1.00 | 1.0-2.5 | 0.75 | 0.03 | 0.03 | — | 0.75 |
| E310Mo | 0.12 | 25.0-28.0 | 20.0-22.0 | 2.0-3.0 | — | 1.0-2.5 | 0.75 | 0.03 | 0.03 | — | 0.75 |
| E312 | 0.15 | 28.0-32.0 | 8.0-10.5 | 0.75 | — | 0.5-2.5 | 0.90 | 0.04 | 0.03 | — | 0.75 |
| E316 | 0.08 | 17.0-20.0 | 11.0-14.0 | 2.0-3.0 | — | 0.5-2.5 | 0.90 | 0.04 | 0.03 | — | 0.75 |
| E316H | 0.04-0.08 | 17.0-20.0 | 11.0-14.0 | 2.0-3.0 | — | 0.5-2.5 | 0.90 | 0.04 | 0.03 | — | 0.75 |
| E316L | 0.04 | 17.0-20.0 | 11.0-14.0 | 2.0-3.0 | — | 0.5-2.5 | 0.90 | 0.04 | 0.03 | — | 0.75 |
| E317 | 0.08 | 18.0-21.0 | 12.0-14.0 | 3.0-4.0 | — | 0.5-2.5 | 0.90 | 0.04 | 0.03 | — | 0.75 |
| E317L | 0.04 | 18.0-21.0 | 12.0-14.0 | 3.0-4.0 | — | 0.5-2.5 | 0.90 | 0.04 | 0.03 | — | 0.75 |
| E318 | 0.08 | 17.0-20.0 | 11.0-14.0 | 2.0-2.5 | 6 × C, min to 1.00 max | 0.5-2.5 | 0.90 | 0.04 | 0.03 | — | 0.75 |
| E320 | 0.07 | 19.0-21.0 | 32.0-36.0 | 2.0-3.0 | 8 × C, min to 1.00 max | 0.5-2.5 | 0.60 | 0.04 | 0.03 | — | 3.0-4.0 |
| E320LR | 0.035 | 19.0-21.0 | 32.0-36.0 | 2.0-3.0 | 8 × C, min to 0.40 max | 1.50-2.50 | 0.30 | 0.020 | 0.015 | — | 3.0-4.0 |
| E330 | 0.18-0.25 | 14.0-17.0 | 33.0-37.0 | 0.75 | — | 1.0-2.5 | 0.90 | 0.04 | 0.03 | — | 0.75 |
| E330H | 0.35-0.45 | 14.0-17.0 | 33.0-37.0 | 0.75 | — | 1.0-2.5 | 0.90 | 0.04 | 0.03 | — | 0.75 |
| E347 | 0.08 | 18.0-21.0 | 9.0-11.0 | 0.75 | 8 × C, min to 1.00 max | 0.5-2.5 | 0.90 | 0.04 | 0.03 | — | 0.75 |

## Table 2.2 (continued)

| AWS Classification[b] | Chemical composition, weight percent[b] | | | | | | | | | | |
|---|---|---|---|---|---|---|---|---|---|---|---|
| | C | Cr | Ni | Mo | Cb plus Ta | Mn | Si | P | S | N | Cu |
| E349[e] | 0.13 | 18.0-21.0 | 8.0-10.0 | 0.35-0.65 | 0.75-1.2 | 0.5-2.5 | 0.90 | 0.04 | 0.03 | — | 0.75 |
| E410 | 0.12 | 11.0-13.5 | 0.60 | 0.75 | — | 1.0 | 0.90 | 0.04 | 0.03 | — | 0.75 |
| E410NiMo | 0.06 | 11.0-12.5 | 4.0-5.0 | 0.40-0.70 | — | 1.0 | 0.90 | 0.04 | 0.03 | — | 0.75 |
| E430 | 0.10 | 15.0-18.0 | 0.60 | 0.75 | — | 1.0 | 0.90 | 0.04 | 0.03 | — | 0.75 |
| E630 | 0.05 | 16.0-16.75 | 4.5-5.0 | 0.75 | 0.15-0.30 | 0.25-0.75 | 0.75 | 0.04 | 0.03 | — | 3.25-4.0 |
| E16-8-2 | 0.10 | 14.5-16.5 | 7.5-9.5 | 1.0-2.0 | — | 0.5-2.5 | 0.60 | 0.03 | 0.03 | — | 0.75 |

a. Refer to AWS A5.4-81, *Specification for Covered Corrosion Resisting Chromium and Chromium-Nickel Welding Electrodes.*

b. Single values shown are maximum percentages except where otherwise specified.

c. Suffix-15 electrodes are classified with direct current, electrode positive. Suffix-16 electrodes are classified with alternating current and direct current, electrode positive. Electrodes up to and including 5/32 in. (4.0mm) in size are usable in all positions. Electrodes 3/16 in. (4.8mm) and larger are usable only in the flat and horizontal-fillet positions.

d. Vanadium shall be 0.10 to 0.30 percent.

e. Titanium shall be 0.15 percent max and tungsten shall be from 1.25 to 1.75 percent.

Table 2.3
Chemical composition of bare, metal cored, and stranded stainless steel welding electrodes and rods[a]

| AWS Classification | Composition, percent[b] | | | | | | | | | | | |
|---|---|---|---|---|---|---|---|---|---|---|---|---|
| | C | Cr | Ni | Mo | Cb plus Ta | Mn | Si | P | S | N | Cu |
| ER209[c] | 0.05 | 20.5–24.0 | 9.5–12.0 | 1.5–3.0 | — | 4.0–7.0 | 0.90 | 0.03 | 0.03 | 0.10–0.30 | 0.75 |
| ER218 | 0.10 | 16.0–18.0 | 8.0–9.0 | 0.75 | — | 7.0–9.0 | 3.5–4.5 | 0.03 | 0.03 | 0.08–0.18 | 0.75 |
| ER219 | 0.05 | 19.0–21.5 | 5.5–7.0 | 0.75 | — | 8.0–10.0 | 1.00 | 0.03 | 0.03 | 0.10–0.30 | 0.75 |
| ER240 | 0.05 | 17.0–19.0 | 4.0–6.0 | 0.75 | — | 10.5–13.5 | 1.00 | 0.03 | 0.03 | 0.10–0.20 | 0.75 |
| ER307 | 0.04–0.14 | 19.5–22.0 | 8.0–10.7 | 0.5–1.5 | — | 3.3–4.75 | 0.30–0.65 | 0.03 | 0.03 | — | 0.75 |
| ER308[d] | 0.08 | 19.5–22.0 | 9.0–11.0 | 0.75 | — | 1.0–2.5 | 0.30–0.65 | 0.03 | 0.03 | — | 0.75 |
| ER308H | 0.04–0.08 | 19.5–22.0 | 9.0–11.0 | 0.75 | — | 1.0–2.5 | 0.30–0.65 | 0.03 | 0.03 | — | 0.75 |
| ER308L[d] | 0.03 | 19.5–22.0 | 9.0–11.0 | 0.75 | — | 1.0–2.5 | 0.30–0.65 | 0.03 | 0.03 | — | 0.75 |
| ER308Mo | 0.08 | 18.0–21.0 | 9.0–12.0 | 2.0–3.0 | — | 1.0–2.5 | 0.30–0.65 | 0.03 | 0.03 | — | 0.75 |
| Er308MoL | 0.04 | 18.0–21.0 | 9.0–12.0 | 2.0–3.0 | — | 1.0–2.5 | 0.30–0.65 | 0.03 | 0.03 | — | 0.75 |
| ER309[e] | 0.12 | 23.0–25.0 | 12.0–14.0 | 0.75 | — | 1.0–2.5 | 0.30–0.65 | 0.03 | 0.03 | — | 0.75 |
| ER309L | 0.03 | 23.0–25.0 | 12.0–14.0 | 0.75 | — | 1.0–2.5 | 0.30–0.65 | 0.03 | 0.03 | — | 0.75 |
| ER310 | 0.08–0.15 | 25.0–28.0 | 20.0–22.5 | 0.75 | — | 1.0–2.5 | 0.30–0.65 | 0.03 | 0.03 | — | 0.75 |
| ER312 | 0.15 | 28.0–32.0 | 8.0–10.5 | 0.75 | — | 1.0–2.5 | 0.30–0.65 | 0.03 | 0.03 | — | 0.75 |
| ER316[d] | 0.08 | 18.0–20.0 | 11.0–14.0 | 2.0–3.0 | — | 1.0–2.5 | 0.30–0.65 | 0.03 | 0.03 | — | 0.75 |
| ER316H | 0.04–0.08 | 18.0–20.0 | 11.0–14.0 | 2.0–3.0 | — | 1.0–2.5 | 0.30–0.65 | 0.03 | 0.03 | — | 0.75 |
| ER316L[d] | 0.03 | 18.0–20.0 | 11.0–14.0 | 2.0–3.0 | — | 1.0–2.5 | 0.30–0.65 | 0.03 | 0.03 | — | 0.75 |
| ER317 | 0.08 | 18.0–20.0 | 13.0–15.0 | 3.0–4.0 | — | 1.0–2.5 | 0.30–0.65 | 0.03 | 0.03 | — | 0.75 |
| ER317L | 0.03 | 18.5–20.5 | 13.0–15.0 | 3.0–4.0 | — | 1.0–2.5 | 0.30–0.65 | 0.03 | 0.03 | — | 0.75 |
| ER318 | 0.08 | 18.0–20.0 | 11.0–14.0 | 2.0–3.0 | 8 × C min to 1.0 max | 1.0–2.5 | 0.30–0.65 | 0.03 | 0.03 | — | 0.75 |
| ER320 | 0.07 | 19.0–21.0 | 32.0–36.0 | 2.0–3.0 | 8 × C min to 1.0 max | 2.5 | 0.60 | 0.03 | 0.03 | — | 3.0–4.0 |
| ER320LR[e] | 0.025 | 19.0–21.0 | 32.0–36.0 | 2.0–3.0 | 8 × C min to 0.40 max | 1.5–2.0 | 0.15 | 0.015 | 0.020 | — | 3.0–4.0 |
| ER321[f] | 0.08 | 18.5–20.5 | 9.0–10.5 | 0.75 | — | 1.0–2.5 | 0.30–0.65 | 0.03 | 0.03 | — | 0.75 |
| ER330 | 0.18–0.25 | 15.0–17.0 | 34.0–37.0 | 0.75 | — | 1.0–2.5 | 0.30–0.65 | 0.03 | 0.03 | — | 0.75 |
| ER347[d] | 0.08 | 19.0–21.5 | 9.0–11.0 | 0.75 | 10 × C min to 1.0 max | 1.0–2.5 | 0.30–0.65 | 0.03 | 0.03 | — | 0.75 |

## Table 2.3 (continued)

| AWS Classification | C | Cr | Ni | Mo | Cb plus Ta | Mn | Si | P | S | N | Cu |
|---|---|---|---|---|---|---|---|---|---|---|---|
| | | | | | | Composition, percent[b] | | | | | |
| ER349[g] | 0.07-0.13 | 19.0-21.5 | 8.0-9.5 | 0.35-0.65 | 1.0-1.4 | 1.0-2.5 | 0.30-0.65 | 0.03 | 0.03 | – | 0.75 |
| ER410 | 0.12 | 11.5-13.5 | 0.6 | 0.75 | – | 0.6 | 0.50 | 0.03 | 0.03 | – | 0.75 |
| ER410NiMo | 0.06 | 11.0-12.5 | 4.0-5.0 | 0.4-0.7 | – | 0.6 | 0.50 | 0.03 | 0.03 | – | 0.75 |
| ER420 | 0.25-0.40 | 12.0-14.0 | 0.6 | 0.75 | – | 0.6 | 0.50 | 0.03 | 0.03 | – | 0.75 |
| ER430 | 0.10 | 15.5-17.0 | 0.6 | 0.75 | – | 0.6 | 0.50 | 0.03 | 0.03 | – | 0.75 |
| ER630 | 0.05 | 16.0-16.75 | 4.5-5.0 | 0.75 | 0.15-0.30 | 0.25-0.75 | 0.75 | 0.04 | 0.03 | – | 3.25-4.00 |
| ER26-1 | 0.01 | 25.0-27.5 | h | 0.75-1.50 | – | 0.40 | 0.40 | 0.02 | 0.02 | 0.015 | 0.20[h] |
| ER16-8-2 | 0.10 | 14.5-16.5 | 7.5-9.5 | 1.0-2.0 | – | 1.0-2.5 | 0.30-0.65 | 0.03 | 0.03 | – | 0.75 |

a. Refer to AWS A5.9-81, *Specification for Corrosion Resisting Chromium and Chromium-Nickel Bare and Composite Metal Cored and Chromium-Nickel Bare and Composite Metal Cored and Stranded Welding Electrodes and Rods*.

b. Single values shown are maximum percentages except where otherwise specified.

c. Vanadium—0.10-0.30 percent

d. These grades are available in high silicon classifications that have the same chemical composition requirements as tabulated here with the exception that the silicon content is 0.65 to 1.00 percent. These high silicon classifications are designated by the addition 'Si' to the standard classification designations in the table. The fabricator should consider carefully the use of high silicon filler metals in highly restrained fully austenitic welds. A discussion of the problem is presented in the Appendix to the specification.

e. Carbon shall be reported to the nearest 0.01 percent except for the classification E320LR for which carbon shall be reported to the nearest 0.005 percent.

f. Titanium—9 × C min to 1.0 max

g. Titanium—0.10 to 0.30 percent; tungsten—1.25 to 1.75 percent

h. Nickel, max—0.5 minus the copper content, percent

## Table 2.4
### Chemical composition of all-weld-metal deposited from flux cored stainless steel electrodes[a]

| AWS Classification[b] | C | Cr | Ni | Mo | Cb plus Ta | Mn | Si | P | S | Fe | Cu |
|---|---|---|---|---|---|---|---|---|---|---|---|
| E307T-X | 0.13 | 18.0-20.5 | 9.0-10.5 | 0.5-1.5 | 0.5-1.5 | 3.3-4.75 | 1.0 | 0.04 | 0.03 | Rem | 0.5 |
| E308T-X | 0.08 | 18.0-21.0 | 9.0-11.0 | 0.5 | – | 0.5-2.5 | 1.0 | 0.04 | 0.03 | Rem | 0.5 |
| E308LT-X | d | 18.0-21.0 | 9.0-11.0 | 0.5 | – | 0.5-2.5 | 1.0 | 0.04 | 0.03 | Rem | 0.5 |
| E308MoT-X | 0.08 | 18.0-21.0 | 9.0-12.0 | 2.0-3.0 | – | 0.5-2.5 | 1.0 | 0.04 | 0.03 | Rem | 0.5 |
| E308MoLT-X | d | 18.0-21.0 | 9.0-12.0 | 2.0-3.0 | – | 0.5-2.5 | 1.0 | 0.04 | 0.03 | Rem | 0.5 |
| E309T-X | 0.10 | 22.0-25.0 | 12.0-14.0 | 0.5 | – | 0.5-2.5 | 1.0 | 0.04 | 0.03 | Rem | 0.5 |
| E309CbLT-X | d | 22.0-25.0 | 12.0-14.0 | 0.5 | 0.70-1.00 | 0.5-2.5 | 1.0 | 0.04 | 0.03 | Rem | 0.5 |
| E309LT-X | d | 22.0-25.0 | 12.0-14.0 | 0.5 | – | 0.5-2.5 | 1.0 | 0.04 | 0.03 | Rem | 0.5 |
| E310T-X | 0.20 | 25.0-28.0 | 20.0-22.5 | 0.5 | – | 1.0-2.5 | 1.0 | 0.03 | 0.03 | Rem | 0.5 |
| E312T-X | 0.15 | 28.0-32.0 | 8.0-10.5 | 0.5 | – | 0.5-2.5 | 1.0 | 0.04 | 0.03 | Rem | 0.5 |
| E316T-X | 0.08 | 17.0-20.0 | 11.0-14.0 | 2.0-3.0 | – | 0.5-2.5 | 1.0 | 0.04 | 0.03 | Rem | 0.5 |
| E316LT-X | d | 17.0-20.0 | 11.0-14.0 | 2.0-3.0 | – | 0.5-2.5 | 1.0 | 0.04 | 0.03 | Rem | 0.5 |
| E317LT-X | d | 18.0-21.0 | 12.0-14.0 | 3.0-4.0 | – | 0.5-2.5 | 1.0 | 0.04 | 0.03 | Rem | 0.5 |
| E347T-X | 0.08 | 18.0-21.0 | 9.0-11.0 | 0.5 | 8 × C min to 1.0 max | 0.5-2.5 | 1.0 | 0.04 | 0.03 | Rem | 0.5 |
| E409T-X[e] | 0.10 | 10.5-13.0 | 0.60 | 0.5 | – | 0.80 | 1.0 | 0.04 | 0.03 | Rem | 0.5 |
| E410T-X | 0.12 | 11.0-13.5 | 0.60 | 0.5 | – | 1.2 | 1.0 | 0.04 | 0.03 | Rem | 0.5 |
| E410NiMoT-X | 0.06 | 11.0-12.5 | 4.0-5.0 | 0.40-0.70 | – | 1.0 | 1.0 | 0.04 | 0.03 | Rem | 0.5 |
| E410NiTiT-X[e] | d | 11.0-12.0 | 3.6-4.5 | 0.05 | – | 0.70 | 0.50 | 0.03 | 0.03 | Rem | 0.5 |
| E430T-X | 0.10 | 15.0-18.0 | 0.60 | 0.5 | – | 1.2 | 1.0 | 0.04 | 0.03 | Rem | 0.5 |
| E307T-3 | 0.13 | 19.5-22.0 | 9.0-10.5 | 0.5-1.5 | – | 3.3-4.75 | 1.0 | 0.04 | 0.03 | Rem | 0.5 |
| E308T-3 | 0.08 | 19.5-22.0 | 9.0-11.0 | 0.5 | – | 0.5-2.5 | 1.0 | 0.04 | 0.03 | Rem | 0.5 |
| E308LT-3 | 0.03 | 19.5-22.0 | 9.0-11.0 | 0.5 | – | 0.5-2.5 | 1.0 | 0.04 | 0.03 | Rem | 0.5 |
| E308MoT-3 | 0.08 | 18.0-21.0 | 9.0-12.0 | 2.0-3.0 | – | 0.5-2.5 | 1.0 | 0.04 | 0.03 | Rem | 0.5 |
| E308MoLT-3 | 0.03 | 18.0-21.0 | 9.0-12.0 | 2.0-3.0 | – | 0.5-2.5 | 1.0 | 0.04 | 0.03 | Rem | 0.5 |
| E309T-3 | 0.10 | 23.0-25.5 | 12.0-14.0 | 0.5 | – | 0.5-2.5 | 1.0 | 0.04 | 0.03 | Rem | 0.5 |
| E309LT-3 | 0.03 | 23.0-25.5 | 12.0-14.0 | 0.5 | – | 0.5-2.5 | 1.0 | 0.03 | 0.03 | Rem | 0.5 |
| E309CbLT-3 | 0.03 | 23.0-25.5 | 12.0-14.0 | 0.5 | 0.70-1.00 | 0.5-2.5 | 1.0 | 0.04 | 0.03 | Rem | 0.5 |
| E310T-3 | 0.20 | 25.0-28.0 | 20.0-22.5 | 0.5 | – | 1.0-2.5 | 1.0 | 0.03 | 0.03 | Rem | 0.5 |
| E312T-3 | 0.15 | 28.0-32.0 | 8.0-10.5 | 0.5 | – | 0.5-2.5 | 1.0 | 0.04 | 0.03 | Rem | 0.5 |
| E316T-3 | 0.08 | 18.0-20.5 | 11.0-14.0 | 2.0-3.0 | – | 0.5-2.5 | 1.0 | 0.04 | 0.03 | Rem | 0.5 |

**Table 2.4(continued)**

| AWS Classification[b] | Composition, percent[c] | | | | | | | | | | |
|---|---|---|---|---|---|---|---|---|---|---|---|
| | C | Cr | Ni | Mo | Cb plus Ta | Mn | Si | P | S | Fe | Cu |
| E316LT-3 | 0.03 | 18.0-20.5 | 11.0-14.0 | 2.0-3.0 | — | 0.5-2.5 | 1.0 | 0.04 | 0.03 | Rem | 0.5 |
| E317LT-3 | 0.03 | 18.5-21.0 | 13.0-15.0 | 3.0-4.0 | — | 0.5-2.5 | 1.0 | 0.04 | 0.03 | Rem | 0.5 |
| E347T-3 | 0.08 | 19.0-21.5 | 9.0-11.0 | 0.5 | 8 × C min to 1.0 max | 0.5-2.5 | 1.0 | 0.04 | 0.03 | Rem | 0.5 |
| E409T-3[e] | 0.10 | 10.5-13.0 | 0.60 | 0.5 | — | 0.80 | 1.0 | 0.04 | 0.03 | Rem | 0.5 |
| E410T-3 | 0.12 | 11.0-13.5 | 0.60 | 0.5 | — | 1.0 | 1.0 | 0.04 | 0.03 | Rem | 0.5 |
| E410NiMoT-3 | 0.06 | 11.0-12.5 | 4.0-5.0 | 0.40-0.70 | — | 1.0 | 1.0 | 0.04 | 0.03 | Rem | 0.5 |
| E410NiTiT-3[e] | 0.04 | 11.0-12.0 | 3.6-4.5 | 0.5 | — | 0.70 | 0.5 | 0.03 | 0.03 | Rem | 0.5 |
| E430T-3 | 0.10 | 15.0-18.0 | 0.60 | 0.5 | — | 1.0 | 1.0 | 0.04 | 0.03 | Rem | 0.5 |

a. Refer to AWS A5.22-80, *Specification for Flux Cored Corrosion Resisting Chromium and Chromium-Nickel Electrodes*.
b. The letter 'X' as specifically presented in this table indicates a classification covering the shielding designation for both the '1' and '2' categories. See Table 2.5 for details on these suffixes.
c. Single values shown are maximum percentages.
d. The carbon content is 0.04% maximum when the suffix 'X' is 1, and 0.03% maximum when the suffix X is 2.
e. Titanium—10 × C min to 1.5% max

surface discoloration from welding is best removed by wire brushing or mechanical polishing.

## FILLER METALS

Covered electrodes and bare solid- and composite-cored electrodes are commercially available for arc welding stainless steels. Bare welding rods and wire are also produced for use with gas tungsten arc, plasma arc, and electron beam welding. A number of suitable filler metals are available for brazing stainless steel for room and elevated temperature service.

Specifications for welding and brazing filler metals commonly used for stainless steels are issued by the *American Welding Society* (AWS), and by the *Society of Automotive Engineers* (SAE) as Aerospace Materials Specifications (AMS). Military standards cover filler metals suitable for military products. There are also other filler metals designed for joining stainless steels that fall outside of specification requirements but are suitable for many applications.

The stainless steel electrodes and welding rods covered by AWS Specifications are given in Tables 2.2, 2.3, and 2.4 with the chemical composition requirements. In some cases, the chemical composition of the electrode, welding rod, or all-weld-metal deposit from a particular electrode varies slightly from the corresponding base metal with the same three-digit designation. Such adjustments in the filler metal chemistry are necessary to produce weld metal having the desired metallurgical microstructure that is free of cracks or other unacceptable discontinuities.

Covered electrodes for stainless steels are available with either lime or titania type coverings. Electrodes with lime type coverings, designated EXXX-15, are useable with direct current, electrode positive (dcep) only. Electrodes designed for ac or dcep are designated EXXX-16, and can have either type of covering with sufficient ionizing elements, such as potassium, to stabilize an ac arc.

Some foreign suppliers provide rutile or acid type (-13) covered stainless steel electrodes for pipe welding and root pass applications.

Greater penetration is obtained with a Type -15 covering, and smoother surface finish with Type -16 covering. While the choice between the two electrode types is not critical for many applications, Type -15 electrodes are usually preferred for welding narrow grooves and for welding in positions other than flat. Type -16 electrodes are commonly used for weld overlays. Type -15 coatings produce a lower oxygen potential in the weld pool, which results in metallurgically cleaner weld metal.

Stainless steel electrode coverings can absorb moisture that can cause porosity in the weld metal and also cracking with martensitic chromium steels. When removed from hermetically-sealed containers or baking ovens, stainless steel covered electrodes should be stored in holding ovens at about 200° to 250°F until withdrawn for welding. Electrodes that have been exposed to humid conditions should be baked prior to use, as recommended by the manufacturer.

Flux-cored stainless steel electrodes are designed for welding without external shielding and with external shielding of either carbon dioxide ($CO_2$) or argon plus 2 percent oxygen. The type of shielding is indicated by a suffix digit as listed in Table 2.5. Other suitable external shield-

**Table 2.5**
**External shielding medium recommended for flux-cored stainless steel electrodes**

| AWS designation | External shielding[a] |
|---|---|
| EXXXT-1 | $CO_2$ |
| EXXXT-2 | $Ar + 2\%O_2$ |
| EXXXT-3 | None |

a. Other external shielding gases recommended by the electrode manufacturer can be used.

ing gases recommended by the manufacturer can be used.

A wide variety of filler metals are commercially available for the brazing of stainless steel parts. The selection of a brazing filler metal for a stainless steel assembly is dependent on its end use because of the wide variation in their melting points, base metal interface reactions, cost, and resulting service properties. Commercial brazing fillet metals are available that have copper, silver, nickel, cobalt, platinum, palladium, manganese, and gold as the base or as added elements.[3]

---

3. Additional information can be found in the AWS *Brazing Manual,* 3rd Ed., 1976, and AWS A5.8-81, *Specification for Brazing Filler Metal.*

# MARTENSITIC STAINLESS STEELS

## COMPOSITION

Martensitic stainless steels are essentially Fe-Cr-C alloys with nominally 11.5 to 18 percent chromium that are capable of the allotropic austenite-to-martensite transformation under almost all cooling conditions. These steels can be annealed to provide a ferritic structure by austenitizing and slow cooling, but are normally used with a martensitic structure.

The chemical compositions of typical weldable martensitic stainless steels are given in Table 2.6. There are one or more variations to some of those listed to provide special properties. Molybdenum, vanadium, and tungsten are added to the steels to improve elevated temperature properties. The martenistic stainless steels can be grouped into the 12 percent chromium, low-carbon engineering grades and the higher chromium, high-carbon cutlery grades. Type 431 with a nominal 16 percent chromium is considered an engineering grade.

These steels are known for their moderate corrosion resistance, heat resistance up to about 1100°F, ability to develop a wide range of mechanical properties, and relatively low cost.

## METALLURGICAL CHARACTERISTICS

These steels are classed as stainless steels because they contain sufficient chromium to develop the characteristic passive oxide film that renders them resistant to oxidizing corrosion conditions. The effect of 12 percent chromium addition on the phase diagram of iron-carbon alloys is illustrated in the pseudo-binary diagram of Fig. 2.1. The chromium addition constricts the size of the gamma ($\gamma$) or austenite region, decreases the eutectoid composition to 0.35 percent carbon, decreases the maximum carbon solubility to 0.7 percent, and raises the eutectoid temperature. From a metallurgical standpoint, the martensitic stainless steels response to hardening and tempering is similar to hardenable carbon and low alloy steels.

The one significant difference is that martensitic stainless steels contain sufficient chromium to render them air hardening from temperatures above 1500°F for all but very thick sections. Maximum hardness is achieved by quenching from above 1750°F. Figure 2.2 shows the TTT diagram for Type 410 stainless steel,

illustrating the ease of quenching to martensite. Stainless steel with a chromium content above about 17 percent cannot be hardened because increasing the chromium continually decreases the size of the austenitic field, as shown in Fig. 2.3, and it eventually disappears.

The hardness of steel is related to carbon content up to about 0.60 percent. Higher carbon content does not appreciably increase the hardness, and excess carbon is observed as primary carbides that enhance abrasion resistance. Types 440A, B, and C steels are examples. Other alloying elements are sometimes added in proprietary alloys to stabilize the microstructure, retard the effect of tempering, improve tensile strength, ductility, and toughness, and improve hot strength.

Within the limits of chromium and carbon contents, this group of stainless steels transforms completely to austenite at about 1850°F. Rapid cooling from this temperature results in maximum martensite in the microstructure. When heated to temperatures between 1500° and 1750°F, transformation to austenite can be incomplete, and cooling from this temperature range results in a microstructure of ferrite and martensite.

These steels lack toughness in the as-hardened condition, and generally require tempering to provide adequate toughness. The tempering treatment also can be adjusted to provide a variety of strength levels.

Chromium content also influences the metallurgical behavior of a martensitic stainless steel during welding. A significant change takes place in a hardened steel as the chromium content increases from about 11 to 17 percent. With a carbon content of about 0.08 percent, a steel containing 12 percent chromium (Type 410) should have a fully martensitic structure in the heat-affected zone of a weld. If the chromium content of the steel is increased to about 15 percent, the ferrite stabilizing effect of chromium can inhibit complete transformation to austenite, and some untransformed ferrite can remain in the microstructure. Consequently, only a portion of a rapidly cooled heat-affected zone would be martensite; the remainder would be ferrite. The presence of soft ferrite in a martensitic structure decreases the hardness of the steel and reduces the likelihood of cracking.

The hardness of the heat-affected zone depends primarily upon the carbon content. It

**Table 2.6**
**Chemical compositions of typical martensitic stainless steels**

| Type | UNS Number | Composition, percent[a] | | | | | | | |
|------|------------|------|------|------|------|------|------|------|------|
| | | C | Mn | Si | Cr | Ni | P | S | Other |
| Wrought alloys | | | | | | | | | |
| 403 | S40300 | 0.15 | 1.00 | 0.50 | 11.5-13.0 | | 0.04 | 0.03 | |
| 410 | S41000 | 0.15 | 1.00 | 1.00 | 11.5-13.0 | | 0.04 | 0.03 | |
| 414 | S41400 | 0.15 | 1.00 | 1.00 | 11.5-13.5 | 1.25-2.50 | 0.04 | 0.03 | |
| 416 | S41600 | 0.15 | 1.25 | 1.00 | 12.0-14.0 | | 0.04 | 0.03 | |
| 420 | S42000 | 0.15 min | 1.00 | 1.00 | 12.0-14.0 | | 0.04 | 0.03 | |
| 422 | S42200 | 0.20-0.25 | 1.00 | 0.75 | 11.0-13.0 | 0.5-1.0 | 0.025 | 0.025 | 0.75-1.25 Mo; 0.75-1.25 W; 0.15-0.3 V |
| 431 | S43100 | 0.20 | 1.00 | 1.00 | 15.0-17.0 | 1.25-2.50 | 0.04 | 0.03 | |
| 440A | S44002 | 0.60-0.75 | 1.00 | 1.00 | 16.0-18.0 | | 0.04 | 0.03 | 0.75 Mo |
| 440B | S44003 | 0.75-0.95 | 1.00 | 1.00 | 16.0-18.0 | | 0.04 | 0.03 | 0.75 Mo |
| 440C | S44004 | 0.95-1.20 | 1.00 | 1.00 | 16.0-18.0 | | 0.04 | 0.03 | 0.75 Mo |
| Casting alloys | | | | | | | | | |
| CA-6NM | J91540 | 0.06 | 1.00 | 1.00 | 11.5-14.0 | 3.5-4.5 | 0.04 | 0.04 | 0.40-1.0 Mo |
| CA-15 | J91150 | 0.15 | 1.00 | 1.50 | 11.5-14.0 | 1.0 | 0.04 | 0.04 | 0.5 Mo |
| CA-40 | J91153 | 0.20-0.40 | 1.00 | 1.50 | 11.5-14.0 | 1.0 | 0.04 | 0.04 | 0.5 Mo |

a. Single values are maximum.

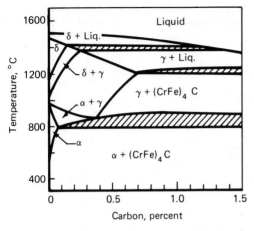

**Fig. 2.1—Pseudo-binary phase diagram of iron-carbon 12 percent chromium alloys**

can be controlled to some extent by the welding procedure. The steep thermal gradients that are accentuated by the low thermal conductivity combined with volumetric changes during phase transformation can cause high internal stresses. The stresses can be sufficiently high to cause cold cracking unless suitable precautions are taken during welding to minimize them.

## WELDABILITY

Martensitic stainless steels can be welded in the annealed, semihardened, hardened, stress-relieved, or tempered condition. The condition of heat treatment has minimal effect on the hard-enability of the weld heat-affected zone, and thus on the weldability.

The hardness of the heat-affected zone depends primarily upon the carbon content of the

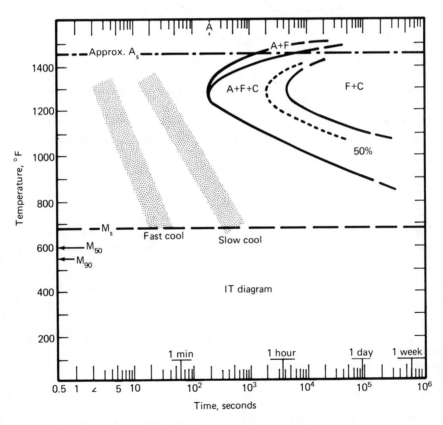

**Fig. 2.2—TTT diagram for Type 410 stainless steel**

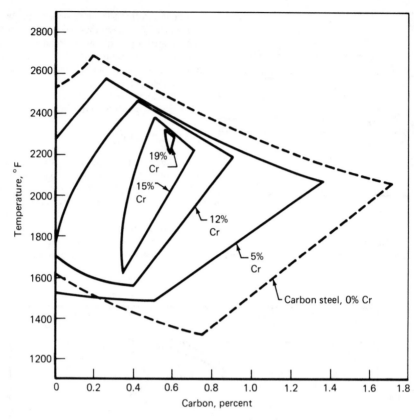

**Fig. 2.3—Effect of chromium content on the carbon range for austenite at elevated temperature**

steel, and can be controlled only to a degree by the welding procedure. As the hardness of the heat-affected zone increases, its susceptibility to cold cracking becomes greater and its toughness decreases.

Weldability is improved when an austenitic stainless steel filler metal is used to take advantage of the low yield strength and good ductility of the weld metal. The austenitic weld metal can yield during welding and minimize the strain imposed on the hardened heat-affected zone. In the case of hydrogen-bearing welding processes, such as shielded metal arc and submerged arc, austenitic weld metal provides an additional advantage in that it does not reject hydrogen to the base metal on cooling.

The ability to use martensitic stainless steel in the as-welded condition with a hard heat-affected zone is a function of the balance of mechanical properties in that zone, including hard-

ness and ductility. In general, welded joints in martensitic stainless steels should be given a post-weld heat treatment for optimun weldability.

Martensitic stainless steels are subject to hydrogen-induced cracking in the same manner as low alloy steels.[4] Appropriate precautions must be taken in welding process selection, storage and handling of the filler metal, cleanliness, and welding procedures to avoid hydrogen pickup during welding and associated cracking problems. In particular, covered electrodes must be low-hydrogen types, and they must be stored and handled in the same manner as low-hydrogen, low alloy steel covered electrodes.

---

4. A general discussion on hydrogen-induced cracking in low alloy steels and appropriate precautions to avoid cracking in welds is included in Chapter 1.

# ARC WELDING

## Filler Metals

Only Types 410, 410NiMo, and 420 martensitic stainless steel filler metals are available as standard grades (see Tables 2.2, 2.3, and 2.4). Other proprietary filler metals are commercially available. Type 410 filler metal is used to weld Types 403, 410, 414, and 420 martensitic stainless steels. Type 410NiMo filler metal is designed to weld Type CA-6NM castings or similar alloys.

When matching carbon content is desired, Type 420 stainless steel is welded with ER420 filler metal. This filler metal is sometimes used for surfacing of carbon steels to provide corrosion resistance.

Martensitic stainless steel weld metals, except for ER410NiMo, lack good toughness in the as-welded condition. A weldment should be given a suitable postweld heat treatment to provide acceptable weld toughness before it is placed in service.

An austenitic stainless steel filler metal, such as Type 308, 309, or 310, is often used to weld martensitic stainless steels to themselves or to other types of stainless steels to provide weld metal with good toughness, as-welded. Nonhardenable Ni-Cr and Ni-Cr-Fe filler metals can also be used, but at increased cost. The differences in the mechanical properties of the weld metal, heat-affected zone, and base metal as well as possible adverse effects from a post-weld heat treatment must be carefully evaluated before a dissimilar filler metal is used in production. Weldments made with these filler metals are normally placed in service in the as-welded condition.

## Preheating

Application of preheat and good interpass temperature control are the best means of avoiding cracking in welds in martensitic stainless steels. Preheating temperature, when used, is usually in the range of 400° to 600°F.

The martensitic transformation temperature ranges of these steels are sufficiently high that preheating at 600°F or below has minimal effect on the hardness of the heat-affected zone or weld metal (see Fig. 2.2). Preheating at a temperature in or above the martensitic transformation range is not recommended. Martensitic transformation during cooling to room temperature or following an immediate postweld tempering operation could contribute to weld cracking.

The carbon content of the steel is the most important factor in determining preheating re-

quirements. Joint thickness, filler metal, welding process, and degree of restraint are other considerations. Suggested preheat temperature, welding heat input, and postweld requirements based on carbon content are given in Table 2.7. Preheating is generally beneficial regardless of carbon content.

Whether interruption of preheat is acceptable also depends upon carbon content. When the carbon content is over 0.20 percent, it is best to maintain the interpass temperature during welding and to heat-treat the weldment before it is allowed to cool below 200°F.

## Postweld Heat Treatment

The functions of a postweld heat treatment are (1) to temper or anneal the weld metal and heat-affected zone to decrease hardness and improve toughness or strength, and (2) to decrease residual stresses associated with welding. Postweld heat treatments normally used for martensitic stainless steels are subcritical annealing and full annealing.

The necessity for a postweld heat treatment depends upon the composition of the steel, the filler metal, and the service requirements. Full annealing transforms a multiple-phase weld zone to a wholly ferritic structure. This annealing procedure requires proper control of the complete

---

**Table 2.7**

**Suggested preheat, welding heat input, and postweld requirements for martensitic stainless steels**

| Carbon content, percent | Approx. preheat temperature, °F[a] | Welding heat input | Postweld requirements |
|---|---|---|---|
| Less than 0.10 | 60 min. | Normal | Heat treatment optional |
| 0.10 to 0.20 | 400-500 | Normal | Cool slowly; heat treatment optional |
| 0.20 to 0.50 | 500-600 | Normal | Heat treatment required |
| Over 0.50 | 500-600 | High | Heat treatment required |

a. The ASME Boiler and Pressure Vessel Code recommends a minimum preheat of 400°F regardless of carbon content.

thermal cycle. It should not be used unless maximum softness is required because of the formation of coarse carbides in the microstructure that take longer to dissolve at the austenizing temperature. Typical postweld annealing temperatures are given in Table 2.8.[5]

If the carbon content of the steel is 0.20 percent or more, the weldment should be given a subcritical annealing heat treatment immediately upon completion of welding, as indicated in Table 2.7. It should be held at temperature for one hour per inch of weld thickness, with a minimum of one hour. The weldment can then be air cooled.

When the filler metal composition closely matches that of the base metal, including carbon content, the weldment can be quench hardened and tempered to produce uniform mechanical properties throughout the weldment.

## Welding Precautions

Types 416 and 416Se steels are free-machining grades that are weldable provided the welding process or filler metal does not supply hydrogen that can react with the sulfur or selenium in the base metal to produce porosity. The amount of sulfur and selenium that enters the weld metal by dilution must be held to a minimum, and the filler metal selection should provide weld metal that can tolerate these elements without hot cracking. ER312 (29Cr-9Ni) austenitic stainless steel filler metal is one choice.

If the actual carbon content of Type 431 steel approaches the permissible maximum of 0.20 percent, care is required in welding to avoid cracking. A hardened heat-affected zone can be kept from cracking by the use of a soaking preheat, maintenance of the preheat during welding, and slow cooling of the weld.

High-carbon Types 420 and 440 martensitic stainless steel weldments are usually heat treated to a high hardness immediately following welding. These steels require appropriate welding procedures to avoid cracking because of their relatively high carbon contents.

When welding casting alloy CA-6NM, 410NiMo filler metal is normally used. Preheat is not normally required but a subcritical anneal is recommended to improve mechanical properties.

5. Refer to the *Metals Handbook,* Vol. 4, *Heat Treating,* 9th Ed.: Metals Park, Ohio: American Society for Metals; 1981: 623-46.

### Table 2.8
### Annealing treatments for martensitic stainless steels

| Type | Subcritical annealing temperature range, °F[a] | Full annealing temperature range, °F[b] |
|---|---|---|
| 403, 410, 416 | 1200-1400 | 1525-1625 |
| 414 | 1200-1350 | Not recommended |
| 420 | 1250-1400 | 1525-1625 |
| 431 | 1150-1300 | Not recommended |
| 440A, 440B, 440C | 1250-1400 | 1550-1650 |
| CA-6NM | 1100-1150 | 1450-1500 |
| CA-15, CA-40 | 1150-1200 | 1550-1650 |

a. Air cool from temperature; lowest hardness is obtained by heating near the top of the range.
b. Furnace cool to 1100°F; weldment can then be air-cooled.

## RESISTANCE WELDING

### Spot Welding

Martensitic stainless steels can be spot welded in the annealed, hardened, or quenched-and-tempered conditions. Regardless of the initial base metal hardness and the welding schedule, the heat-affected zone adjacent to the weld nugget quenches to martensite. The hardness of the weld nugget and heat-affected zone mainly depends upon the carbon content of the steel, although it can be controlled somewhat with preheat, postheat, and tempering during the spot welding cycle. The likelihood of cracking in the heat-affected zone increases with the carbon content of the steel.

Satisfactory spot welds can often be obtained in the martensitic stainless steels containing 0.15 percent carbon or less (Types 403, 410, 414, and 416) without postweld heat treatment. Spot welded assemblies of steels containing higher carbon content (Types 420, 422, and 431) should be given a postweld heat treatment.

### Flash Welding

The martensitic stainless steels can be joined by flash welding. As with spot welding, a hard heat-affected zone is formed that can be softened somewhat by a tempering cycle in the welding machine. Alternatively, the weldment can be given a postweld heat treatment.

The high chromium content of martensitic

stainless steels requires precise control of flashing and upsetting during welding to avoid entrapment of oxides at the weld interface. Oxide inclusions in the weld are unacceptable. The use of a protective atmosphere, such as dry nitrogen or inert gas, will prevent oxidation during welding.

## OTHER WELDING PROCESSES

Martensitic stainless steels can be joined by other welding processes including electron beam welding, friction welding, and high frequency welding. The precautions required for arc and resistance welding also apply to these processes.

# FERRITIC STAINLESS STEELS

## COMPOSITION

Ferritic stainless steels are Fe-Cr-C alloys with sufficient chromium or chromium plus another ferrite stablizer, such as Al, Cb, Mo, and Ti, to inhibit the formation of austenite on heating. Therefore, they are nonhardenable by heat treatment. The chemical compositions of typical ferritic stainless steels are given in Table 2.9. Several wrought alloy compositions are also used for castings.

The first-generation ferritic stainless steels (Types 430, 442, and 446) contain only chromium as a ferrite stablizer and appreciable carbon. They are subject to intergranular corrosion after welding unless a postweld heat treatment is applied. They also exhibit low toughness.

The second-generation ferritic stainless steels (Types 405 and 409) are lower in chromium and carbon, but have powerful ferrite formers added to the melt. For example, aluminum is added in Type 405 and titanium in Type 409. Titanium and columbium react with carbon to form carbides, decreasing the amount of carbon in solid solution. These steels are largely ferritic, although welding or heat treating can result in a small amount of martensite. They are lower in cost, have useful corrosion resistance, and possess better fabrication characteristics than the first-generation ferritic stainless steels, but they have low toughness.

Recently, improvements in melting and refining procedures have resulted in economical production of large heats of ferritic stainless steels with very low carbon and nitrogen contents. These steels are the third generation, with lower interstitial content than those available previously. Types 444 (18Cr-2Mo) and 26-1 (26Cr-1Mo) are notable examples. When they are stabilized with powerful carbide forming elements, including Ti and Cb, these steels are generally not susceptible to intergranular corrosion after welding. In addition, they have improved toughness and good resistance both to pitting corrosion in chloride environments and to stress-corrosion cracking.

The coefficients of thermal expansion of ferritic stainless steel and mild steel are similar, but the thermal conductivity ratio is approximately one half (see Table 2.1).

## METALLURGICAL CHARACTERISTICS

The important metallurgical feature of ferritic stainless steels is that they contain sufficient chromium alone or in combination with stabilizing additions and getters of interstitial elements to essentially eliminate the formation of austenite at elevated temperature. Because austenite does not form and the ferrite is essentially stable at all temperatures up to melting, these steels cannot be hardened by quenching. As shown in Fig. 2.3, the minimum chromium addition necessary to prevent austenite formation in steel is a function of the carbon content. Nitrogen has an effect similar to carbon. Because these elements promote the formation of austenite on heating, their presence must be offset by additional chromium or other ferrite stabilizers, including aluminum, columbium, titanium, and molybdenum. Titanium and columbium can also form stable carbonitrides at elevated temperatures, which remove carbon and nitrogen from solution and lessen the tendency of the steel to transform to austenite.

Except for the types that have very low in-

**Table 2.9**
**Chemical compositions of typical ferritic stainless steels**

| Type | UNS number | Composition, percent[a] | | | | | | | |
|---|---|---|---|---|---|---|---|---|---|
| | | C | Mn | Si | Cr | Ni | P | S | Others |
| Wrought alloys | | | | | | | | | |
| 405 | S40500 | 0.08 | 1.00 | 1.00 | 11.5-14.5 | | 0.04 | 0.03 | 0.10-0.30 AL |
| 409 | S40900 | 0.08 | 1.00 | 1.00 | 10.5-11.75 | | 0.045 | 0.045 | Ti min—6 × %C |
| 429 | S42900 | 0.12 | 1.00 | 1.00 | 14.0-16.0 | | 0.04 | 0.03 | |
| 430 | S43000 | 0.12 | 1.00 | 1.00 | 16.0-18.0 | | 0.04 | 0.03 | |
| 434 | S43400 | 0.12 | 1.00 | 1.00 | 16.0-18.0 | | 0.04 | 0.03 | 0.75-1.25 Mo |
| 436 | S43600 | 0.12 | 1.00 | 1.00 | 16.0-18.0 | | 0.04 | 0.03 | 0.75-1.25 Mo; (Cb + Ta)min—5 × %C |
| 442 | S44200 | 0.20 | 1.00 | 1.00 | 18.0-23.0 | | 0.04 | 0.03 | |
| 444 | S44400 | 0.025 | 1.00 | 1.00 | 17.5-19.5 | 1.00 | 0.04 | 0.03 | 1.75-2.5 Mo; 0.035 N max; (Cb + Ta)min—0.2 + 4(%C + %N) |
| 446 | S44600 | 0.20 | 1.50 | 1.00 | 23.0-27.0 | | 0.04 | 0.03 | 0.25 N |
| 26-1 | S44626 | 0.06 | 0.75 | 0.75 | 25.0-27.0 | 0.50 | 0.04 | 0.020 | 0.75-1.50 Mo; 0.2-1.0 Ti; 0.04 N; 0.2 Cu |
| 29-4 | S44700 | 0.010 | 0.30 | 0.20 | 28.0-30.0 | 0.15 | 0.025 | 0.020 | 3.5-4.2 Mo., 0.020 Ni, 0.15 Cu |
| 29-4-2 | S44800 | 0.010 | 0.30 | 0.20 | 28.0-30.0 | 2.0-2.5 | 0.025 | 0.020 | 3.5-4.2 Mo; 0.020 Ni, 0.15 Cu |
| Casting alloys | | | | | | | | | |
| CB-30 | J91803 | 0.30 | 1.00 | 1.50 | 18.0-21.0 | 2.0 | 0.04 | 0.04 | |
| CC-50 | J92616 | 0.50 | 1.00 | 1.50 | 26.0-30.0 | 4.0 | 0.04 | 0.04 | |

a. Single values are maximum.

terstitial content or contain more than 21 percent Cr, formation of some austenite in ferritic stainless steel is possible at temperatures above about 1500°F. When air cooled, the austenite transforms to martensite and results in slightly hardened steel. Because only part of the structure transforms to austenite, the remaining soft ferrite easily accomodates much of the strain associated with martensite formation. The martensite forms primarily at ferrite grain boundaries, and can adversely affect the ductility of the steel. However, the likelihood of cracking is much less than with the martensitic stainless steels. Ferritic stainless steels that contain some martensite as a result of hot working or welding should be annealed at 1400° to 1500°F to obtain optimum formability and corrosion resistance. The annealing treatment causes the martensite to transform to ferrite and spheroidize carbides.

Although the ferritic grades that contain more than 21 percent chromium do not normally transform to austenite on heating, chromium carbides can form at the ferrite grain boundaries of those containing appreciable carbon, even when cooled rapidly from high temperatures. Depletion of chromium from the matrix makes the steel susceptible to intergranular corrosion.

An annealing treatment is needed to restore optimum corrosion resistance. An iron-chromium intermetallic, normally sigma phase, tends to form in these high chromium steels in the temperature range of 800° to 1400°F. The recommended annealing treatment is generally 1400° to 1500°F, followed by rapid cooling to prevent sigma formation. All ferritic stainless steels are susceptible to severe grain growth when heated above about 1700°F and, as a result, their toughness decreases. Toughness can be regained only by refining the grain size through cold working and annealing procedures. Fine-grained ferritic stainless steels that contain appreciable carbon also have low toughness. Typical Charpy V-notch impact transition temperatures can be above room temperature. In contrast, low-interstitial fine-grained types can have transition temperatures well below room temperature.

## WELDABILITY

Precautions during welding to prevent martensite formation are generally not necessary with the ferritic stainless steels because they cannot be hardened by quenching. Martensite for-

mation does not occur in many ferritic grades. Types 430, 434, 442, and 446 stainless steels, however, are exceptions because they have both high chromium and high carbon contents. They can have ductile-to-brittle fracture transition temperatures above room temperature. Welds in these steels are susceptible to cracking upon cooling when they are made under conditions of high restraint, such as heavy weldments or surfacing welds on carbon steel. A preheat of 300°F or higher can be used to minimize residual stresses that contribute to weld cracking.

The corrosion resistance of low-chromium, aluminum- or titanium-stabilized, ferritic stainless steels (Types 405 and 409) is generally not affected by the heat of welding. Consequently, these steels are often used in the as-welded condition. However, those steels that are higher in chromium and carbon (Types 430, 434, 442, and 446) tend to form chromium carbides at grain boundaries in the weld heat-affected zone; this makes the weld area susceptible to intergranular corrosion. These steels generally require annealing after welding to redissolve carbides and restore corrosion resistance.

The low-interstitial ferritic stainless steels (Types 444, 26-1, 29-4, 29-4-2) are not sensitized by grain-boundary carbide precipitation during welding. With some compositions, however, intermetallic compounds (sigma or chi phase) can form at grain boundaries during welding, and this can make the weld area susceptible to intergranular corrosion in certain environments. The steel manufacturers should be consulted for recommendations when selecting a low interstitial ferritic stainless steel for a new welding application

Weld heat-affected zones in ferritic stainless steel are subject to grain growth as a result of welding heat. The extent of grain growth depends upon the highest sustained temperature of exposure and the time at temperature. The result of grain growth is the loss of toughness in the heat-affected zone. This is especially serious with the low-interstitial ferritic types because they generally have rather good toughness prior to welding.

## ARC WELDING

### Filler Metals

Filler metals used for welding ferritic stainless steels to themselves and to other steels are generally of three types: (1) compositions approximately matching the base metals, (2) austenitic stainless steels, and (3) nickel alloys. The application of ferritic stainless steels for weldments is limited because of the lack of toughness in the heat-affected zone. This, in turn, limits the requirements for and availability of matching filler metal compositions. The difficulty of transferring aluminum and titanium across the arc further limits the availability of matching covered and flux cored electrodes.

Standard filler metals matching Types 409 and 430 stainless steels are readily available (see Tables 2.2, 2.3, and 2.4). Type 430 filler metal is available as solid and composite flux cored electrodes, covered electrodes, and welding rods. Type 409 filler metal is produced only as flux cored electrodes, both gas shielded and self-shielded variations.

Nonstandard electrodes that match other ferritic stainless steels are sometimes available, including Types 442 and 446 covered electrodes.

Bare solid wires of Types 434, 442, and 446 steels are seldom available because of their poor wire drawing characteristics and the limited demand. When matching base and weld metals are requried, one solution is the use of metal cored electrodes, if available. These compositions are often used for surfacing on mild or low alloy steel.

A modified Type 409 bare electrode that contains 4 percent nickel, known as AM363 alloy, is sometimes used to join Type 409 stainless steel by gas metal arc welding with argon-oxygen shielding gas. This electrode produces low-carbon martensitic weld metal with excellent resistance to cracking.

For the low-interstitial ferritic stainless steels, filler metal of matching composition is generally available as bare wire from the steel producer. The wire must be carefully processed and handled to avoid contamination that would raise the interstitials above specification limits.

Austenitic stainless steel or nickel alloy filler metals are often selected for joining ferritic stainless steels to themselves or to dissimilar metals. Austenitic stainless filler metals that are relatively high in delta ferrite, such as Types 309 and 312, are preferred for joining ferritic stainless steel to other types of stainless steel and to mild or low alloy steel. Nickel alloy filler metal, such as ENiCrFe-3 covered electrodes, can provide sound joints when welding ferritic stainless steels to themselves and to mild or low alloy steel, nickel alloys, and copper-nickel alloys.

The low-interstitial ferritic stainless steels can present special corrosion problems when they are welded with dissimilar filler metals. For certain applications, Type 444 stainless steel can be successfully joined to itself with Type 316L weld metal. In general, the producer of the steel should be consulted for recommendations on joining these steels with dissimilar filler metals.

## Preheating

When welding ferritic stainless steels, preheating has metallurgical effects different from the welding of martensitic stainless steels. Welded joints in ferritic stainless steels experience grain growth and loss of ductility when slowly cooled. Certain ferritic stainless steels have a tendency to form martensite in the grain bounderies. Preheating these steels helps to eliminate cracking in the weld heat-affected zone, and tends to limit welding stresses. The need for preheating is determined to a large extent by composition, desired mechanical properties, thickness, and conditions of restraint. Preheat, when employed, is normally in the 300° to 450°F range. Interpass temperature should be limited to the lowest practical level above the preheat termperature.

## Shielded Metal Arc Welding

Standard covered electrodes of matching analyses are available for welding all of the ferritic chromium steels, except for (1) the free-machining grades, (2) those containing titanium or aluminum, and (3) the low interstitial grades.

When a chromium stainless steel is welded with austenitic stainless steel electrodes, the electrodes should be higher in chromium than the base metal to allow for dilution. Type E310 electrodes (25Cr-20Ni) are commonly used but E309 electrodes (22Cr-12Ni) are a satisfactory substitute for welding most of the chromium stainless steels.

For the free-machining grades, E309-15 lime-type electrodes usually are preferred because they will produce about 10 percent ferrite in the diluted weld metal. For the titanium- or aluminum-bearing chromium stainless steels, electrode selection should be based on the service requirements of the weldment.

Chromium stainless steel electrodes require a short welding arc to prevent excessive chromium oxidation and nitrogen contamination. The short arc length also reduces the likelihood of porosity in the weld metal. Conversely, a long arc results in a loss of chromium, increases nitrogen contamination, and might cause porosity in the weld metal. The weave bead technique is not recommended, for the same reasons.

The low-interstitial ferritic stainless steel compositions are not suitable for shielded metal arc welding because of the inherent contamination from carbon, nitrogen, and oxygen. If the process must be used, austenitic stainless steel or nickel alloy covered electrodes are recommended. The manufacturer of the base metal should be consulted for recommendations.

## Gas Tungsten Arc Welding

Direct current, electrode negative is preferred for gas tungsten arc welding (GTAW) the chromium stainless steels. Alternating current welding is not recommended. Helium, argon, or mixtures of the two are used for shielding.

## Gas Metal Arc Welding

Gas metal arc welding (GMAW) is normally done with direct current, electrode positive. Argon-1 percent oxygen shielding is recommended with spray transfer and helium-argon-2.5 percent $CO_2$ mixture for short-circuiting transfer. The best shielding gas for an application depends upon the particular steel to be welded as well as the type of metal transfer.

Short-circuiting transfer requires a small diameter electrode with low arc voltage and welding current, which makes it well-suited for welding thin sections. An advantage of short-circuiting transfer is the relatively low heat input that tends to limit grain growth in the heat-affected zone, which is advantageous when welding the ferritic stainless steels. Unfortunately, low heat input can result in incomplete fusion and, as a result, the use of short-circuiting transfer is often restricted to noncritical applications. This type of transfer has an advantage when an austenitic stainless steel filler metal is used to weld a chromium stainless steel. Dilution rates as low as 10 percent can be obtained with a relatively low heat input.

With spray-type transfer and its high dilution rates, suitable austenitic stainless filler metals might not be available to weld all of the chromium stainless steels and the various dissimilar stainless steel combinations. Spray transfer is normally done using larger electrodes, and higher arc voltages and welding currents than those used for short-circuiting transfer. It gives better assurance against incomplete fusion and lack of penetration than short-circuiting transfer. However, welding is normally restricted to the flat or horizontal position.

Pulsed spray welding can be used with large diameter electrodes in all welding positions because of the better control of the molten weld pool. While the deposition rate is lower than with continuous spray transfer, the pulsed spray variation develops less total heat input, and minimizes undesirable grain growth in stainless steels.

## Submerged Arc Welding

Some of the chromium stainless steels can be welded with the submerged arc welding (SAW) process using single or multiple pass procedures. Consideration must be given to the recovery of chromium and molybdenum when selecting an appropriate flux for these steels because the efficiency of transferring these elements across the arc varies with the type of flux. Neutral fluxes require highly alloyed electrodes to compensate for losses across the arc. Bonded or agglomerated fluxes can be used to add alloying elements to the weld metal, which permits the use of lower alloy electrodes. The flux must not contribute excessive quantities of carbon, manganese, or silicon to the weld metal. Sulfur and phosphorus must also remain low because these elements promote embrittlement and cracking.

Heat input can vary widely with submerged arc welding, and this should be considered when selecting this process for joining a ferritic stainless steel. Austenitic stainless steel electrodes might be desirable for certain applications to avoid the problem with coarse-grained ferritic weld metal. In such cases, the effects of dilution must be considered when selecting the electrode-flux combination. Dilution in submerged arc welds can range from 30 to 50 percent or greater. Therefore, the electrode-flux combination should be selected to provide weld metal rich enough in chromium and nickel to compensate for the expected dilution.

Submerged arc welding is not suitable for joining low-interstitial ferritic stainless steels where corrosion resistance must be maintained.

## Postweld Heat Treatment

Postweld heat treatment of the ferritic stainless steels is conducted at subcritical temperatures to prevent further grain coarsening. Temperatures in the 1300° to 1550°F range are used. Care must be taken during heat treatment to minimize oxidation, and during cooling through the 1000° to 700°F range to avoid embrittlement and loss of toughness. Rapid cooling through this range is a metallurgical necessity with due consideration to distortion and residual stresses.

## RESISTANCE WELDING

### Spot Welding

General considerations regarding grain growth and lack of toughness in the ferritic stainless steels apply to spot welding. Spot welding of these steels is not recommended when weld ductility is the important criterion.

Specific recommendations for spot welding the chromium stainless steels have not been compiled, but welding schedules recommended for the austenitic stainless steels can be used as guides.

### Seam Welding

The same limitations given above for spot welding apply to seam welding of ferritic stainless steels. Welding schedules recommended for the austenitic stainless steels can be used as guides when developing suitable schedules for ferritic stainless steels.

### Flash Welding

Flash welding is used to join the ferritic stainless steels, provided the low ductility associated with welds in these steels can be tolerated. Standard flash welding techniques are used.

Steels containing as much as 16 percent chromium are flash welded to themselves and to other straight chromium steels, chromium-molybdenum steels, and plain carbon steel without difficulty. Higher chromium steels can also be welded with this process, but there will be a loss in ductility resulting from exposure to high temperatures during welding.

The use of inert gas atmosphere to protect the steel from oxidation during the flashing period improves the mechanical properties of flash welds. The inert atmosphere limits the amount of chromium oxide that forms, thus providing cleaner weld metal. A relatively long flashing time together with a large upset distance should be used to ensure that all oxides are expelled from the weld interface.

## OTHER WELDING PROCESSES

These steels are upset welded in production. They are amenable to joining by friction, electron beam, plasma arc, and laser beam welding.

# AUSTENITIC STAINLESS STEELS

## COMPOSITION

The austenitic stainless steels contain a combined total chromium, nickel, and manganese content of 24 percent or more with the chromium content generally above 16 percent. The chromium provides oxidation resistance and resistance to corrosion in certain media. The nickel and manganese stabilize the austenite phase sufficiently to retain most or all of it when the steel is cooled rapidly to room temperature. The amount of retained austenite depends upon the composition of the steel; the microstructure is either all austenite, or ferrite in a matrix of austenite.

### Wrought Alloys

The compositions of typical wrought austenitic stainless steels are given in Table 2.10. There are several variations to some of those listed. Nitrogen addition, denoted by the suffix $N$, increases the strength of the steel. The carbon content of some types, denoted by the suffix $H$, is controlled between specific levels for high temperature strength, while low carbon variations are denoted by the suffix $L$. Improved machineability is achieved by increasing the phosphorus or sulfur content or by the addition of selenium (Se); Types 303 and 303Se are examples. Increasing the silicon content improves heat resistance, Type 302B for example.

**Table 2.10**

**Compositions of typical wrought austenitic stainless steels**

| Type | UNS number | Composition, percent[a] | | | | | | | |
|------|-----------|------|------|------|------|------|------|------|------|
| | | C | Mn | Si | Cr | Ni[b] | P | S | Others |
| 201 | S20100 | 0.15 | 5.5-7.5 | 1.00 | 16.0-18.0 | 3.5-5.5 | 0.06 | 0.03 | 0.25 N |
| 202 | S20200 | 0.15 | 7.5-10.0 | 1.00 | 17.0-19.0 | 4.0-6.0 | 0.06 | 0.03 | 0.25 N |
| 301 | S30100 | 0.15 | 2.00 | 1.00 | 16.0-18.0 | 6.0-8.0 | 0.045 | 0.03 | |
| 302 | S30200 | 0.15 | 2.00 | 1.00 | 17.0-19.0 | 8.0-10.0 | 0.045 | 0.03 | |
| 302B | S30215 | 0.15 | 2.00 | 2.0-3.0 | 17.0-19.0 | 8.0-10.0 | 0.045 | 0.03 | |
| 303 | S30300 | 0.15 | 2.00 | 1.00 | 17.0-19.0 | 8.0-10.0 | 0.20 | 0.15 min | 0-0.6 Mo |
| 303Se | S30323 | 0.15 | 2.00 | 1.00 | 17.0-19.0 | 8.0-10.0 | 0.20 | 0.06 | 0.15 Se min |
| 304 | S30400 | 0.08 | 2.00 | 1.00 | 18.0-20.0 | 8.0-10.5 | 0.045 | 0.03 | |
| 304L | S30403 | 0.03 | 2.00 | 1.00 | 18.0-20.0 | 8.0-12.0 | 0.045 | 0.03 | |
| 305 | S30500 | 0.12 | 2.00 | 1.00 | 17.0-19.0 | 10.5-13.0 | 0.045 | 0.03 | |
| 308 | S30800 | 0.08 | 2.00 | 1.00 | 19.0-21.0 | 10.0-12.0 | 0.045 | 0.03 | |
| 309 | S30900 | 0.20 | 2.00 | 1.00 | 22.0-24.0 | 12.0-15.0 | 0.045 | 0.03 | |
| 309S | S30908 | 0.08 | 2.00 | 1.00 | 22.0-24.0 | 12.0-15.0 | 0.045 | 0.03 | |
| 310 | S31000 | 0.25 | 2.00 | 1.50 | 24.09-26.0 | 19.0-22.0 | 0.045 | 0.03 | |
| 310S | S31008 | 0.08 | 2.00 | 1.50 | 24.0-26.0 | 19.0-22.0 | 0.045 | 0.03 | |
| 314 | S31400 | 0.25 | 2.00 | 1.5-3.0 | 23.0-26.0 | 19.0-22.0 | 0.045 | 0.03 | |
| 316 | S31600 | 0.08 | 2.00 | 1.00 | 16.0-18.0 | 10.0-14.0 | 0.045 | 0.03 | 2.0-3.0 Mo |
| 316L | S31603 | 0.03 | 2.00 | 1.00 | 16.0-18.0 | 10.0-14.0 | 0.045 | 0.03 | 2.0-3.0 Mo |
| 317 | S31700 | 0.08 | 2.00 | 1.00 | 18.0-20.0 | 11.0-15.0 | 0.045 | 0.03 | 3.0-4.0 Mo |
| 317L | S31703 | 0.03 | 2.00 | 1.00 | 18.0-20.0 | 11.0-15.0 | 0.045 | 0.03 | 3.0-4.0 Mo |
| 321 | S32100 | 0.08 | 2.00 | 1.00 | 17.0-19.0 | 9.0-12.0 | 0.045 | 0.03 | 5×%C Ti min |
| 329 | S32900 | 0.10 | 2.00 | 1.00 | 25.0-30.0 | 3.0-6.0 | 0.045 | 0.03 | 1.0-2.0 Mo |
| 330 | N08330 | 0.08 | 2.00 | 0.75-1.5 | 17.0-20.0 | 34.0-37.0 | 0.04 | 0.03 | |
| 347 | S34700 | 0.08 | 2.00 | 1.00 | 17.0-19.0 | 9.0-13.0 | 0.045 | 0.03 | c |
| 348 | S34800 | 0.08 | 2.00 | 1.00 | 17.0-19.0 | 9.0-13.0 | 0.045 | 0.03 | 0.2 Cu, c, d |
| 384 | S38400 | 0.08 | 2.00 | 1.00 | 15.0-17.0 | 17.0-19.0 | 0.045 | 0.03 | |

a. Single values are maximum unless indicated otherwise.
b. Higher percentages are required for certain tube manufacturing processes.
c. (Cb+Ta) min—10×%C.
d. Ta—0.10% max.

## Cast Alloys

Typical cast austenitic stainless steels are listed in Table 2.11. Those designated by CX-XXX are generally similar to the corresponding wrought types, as noted. The HX steels are used under oxidizing or reducing conditions at elevated temperatures. The carbon contents of the HX steels are generally higher than those of the CX-XXX alloys for better strength at elevated temperatures, resulting from finely dispersed carbides in the austenitic matrix.

## Filler Metals

Standard filler metals for welding austenitic stainless steels are given in Tables 2.2, 2.3, and 2.4. Other filler metals are available for welding proprietary austenitic stainless steels. Standard consumable inserts of Types 308, 308L, 310, 312, 316, 316L, and 348 stainless steels are available.[6] They are used primarily for pipe welding.

## PROPERTIES

The nominal physical properties of austenitic stainless steels are listed in Table 2.1. These steels exhibit higher thermal expansion than the chromium stainless steels, which means that distortion during welding can be greater. Type 301 stainless steel is often considered to be magnetic. The remaining alloys are most often used in the solution annealed condition, and have low magnetic permeability.

Austenitic stainless steels have better ductility and toughness than carbon and low alloy steels because of the face-centered cubic crystal structure. Their notch toughness at cryogenic temperatures is excellent. They are stronger than carbon and low alloy steels at temperatures above 1000°F while maintaining good oxidation resistance. Type 316 stainless steel has the highest stress-rupture capability of the 300 series of alloys.

## METALLURGICAL CHARACTERISTICS

### Base Metals

Steels of the austenitic group are characteristically resistant to corrosion and oxidation, primarily as a result of the chromium. Chromium,

when added alone to steel in the range of 16 to 25 percent, forms a solid solution in alpha iron. The addition of one or more strong austenite-forming elements, usually nickel, to the iron-chromium alloy in sufficient proportions suppresses the ferritic structure so that the resulting structure is essentially austenitic. The resultant austenitic alloy is a solid solution of chromium, nickel, and carbon in gamma iron, which is nonmagnetic. The nickel addition also increases the high temperature strength and corrosion resistance of the steel.

The austenitic stainless steels contain other useful elements. Manganese and nitrogen augment the austenite stabilizing action of the nickel. Molybdenum, columbium, and titanium promote the formation of delta ferrite in the austenitic matrix and also form carbides in a manner similar to that of chromium. The ratio of austenite- and ferrite-forming elements is adjusted to provide suitable properties for the various applications of these steels. Table 2.12 summarizes the effects of the various alloying elements in austenitic stainless steels.

Silicon is a particularly important element in 310 and 314 wrought alloys and in the HK, HT, and HU casting alloys. As the silicon content is increased, the resistance of the steels to oxidation and carburization at elevated temperatures improves. Higher silicon content also improves the fluidity of the molten metal, and thus the casting properties. The carbon-to-silicon ratio is important in fully austenitic stainless steel weld metal. Figure 2.4 shows that the highest weld ductility can be obtained when the ratio of C-to-(Si + 0.8) is 1 to 6. The manufacturers of stainless steel welding electrodes and rods rigorously control the carbon content of the high-silicon types to produce the optimum combination of soundness and ductility.

### Weld Metal

The microstructures of austenitic stainless steel weld metals are quite different from those of similar wrought base metals. Alloys that are fully austenitic in the wrought form often contain small pools of ferrite in weld metal of equivalent composition. With some filler metals, such as Type 16-8-2, small amounts of martensite can form in the weld metal. Carbides might also be present in the weld metal, particularly with titanium or columbium stabilized filler metals, such as ER321 and ER347. The microstructure of as-deposited weld metal is dependent upon its com-

---

6. Refer to AWS A5.30-79, *Specification for Consumable Inserts,* for additional information.

## Table 2.11

### Compositions of typical cast austenitic stainless steels

| Alloy designation | UNS Number | Similar wrought type[b] | Composition, percent[a] | | | | | |
|---|---|---|---|---|---|---|---|---|
| | | | C | Si | Cr | Ni | Mo[c] | Other |
| CE-30 | J93423 | 312 | 0.30 | 2.00 | 26-30 | 8-11 | – | – |
| CF-3 | J92700 | 304L | 0.03 | 2.00 | 17-21 | 8-12 | – | – |
| CF-3M | J92800 | 316L | 0.03 | 1.50 | 17-21 | 9-13 | 2.0-3.0 | – |
| CF-8 | J92600 | 304 | 0.08 | 2.00 | 18-21 | 8-11 | – | – |
| CF-8C | J92710 | 347 | 0.08 | 2.00 | 18-21 | 9-12 | – | d |
| CF-8M | J92900 | 316 | 0.08 | 1.50 | 18-21 | 9-12 | 2.0-3.0 | – |
| CF-12M | – | 316 | 0.12 | 1.50 | 18-21 | 9-12 | 2.0-3.0 | – |
| CF-16F | J92701 | 303 | 0.16 | 2.00 | 18-21 | 9-12 | 1.5 | 0.20-0.35 Se |
| CF-20 | J92602 | 302 | 0.20 | 2.00 | 18-21 | 8-11 | – | – |
| CG-8M | – | 317 | 0.08 | 1.50 | 18-21 | 9-13 | 3.0-4.0 | – |
| CH-20 | J93402 | 309 | 0.20 | 2.00 | 22-26 | 12-15 | – | – |
| CK-20 | J94202 | 310 | 0.20 | 2.00 | 23-27 | 19-22 | – | – |
| CN-7M | J95150 | – | 0.07 | 1.50 | 18-22 | 27.5-30.5 | 2.0-3.0 | 3-4 Cu |
| HE | J93403 | – | 0.2-0.5 | 2.0 | 26-30 | 8-11 | – | – |
| HF | J92603 | 304 | 0.2-0.4 | 2.0 | 19-23 | 9-12 | – | – |
| HH | J93503 | 309 | 0.2-0.5 | 2.0 | 24-28 | 11-14 | – | 0.2 N |
| HI | J94003 | – | 0.2-0.5 | 2.0 | 26-30 | 14-18 | – | – |
| HK | J94224 | 310 | 0.2-0.6 | 2.0 | 24-28 | 18-22 | – | – |
| HL | J94604 | – | 0.2-0.6 | 2.0 | 28-32 | 18-22 | – | – |
| HN | J94213 | – | 0.2-0.5 | 2.0 | 19-23 | 23-27 | – | – |
| HP | – | – | 0.35-0.75 | 2.0 | 24-28 | 33-37 | – | – |
| HT | J94605 | 330 | 0.35-0.75 | 2.5 | 15-19 | 33-37 | – | – |
| HU | – | – | 0.35-0.75 | 2.5 | 17-21 | 37-41 | – | – |

a. Single values are maximum.
   Manganese — 1.50% max in CX-XX types; 2.0% max in HX types
   Phosporus — 0.04% max except for Cf-16F — 0.17% max
   Sulfur — 0.04% max
b. Compositions are not exactly the same.
c. Molybdenum in HX types is 0.5% max.
d. Cb — $8 \times$ %C (1.0% max), or Cb + Ta — $9 \times$ %C (1.1% max)

position and solidification rate. Figures 2.5 and 2.6 show typical austenitic weld metal microstructures with and without delta ferrite present, respectively. As-welded microstructures can be predicted from the chemical composition of the deposited weld metal using empirical constitution diagrams.

Typical constitution diagrams plot chromium equivalents on the abscissa and nickel equivalents on the ordinate. Figure 2.7 shows the Schaeffler diagram, which plots these equivalents from zero, and indicates the approximate microstructures for stainless steel welds. The DeLong constitution diagram, Fig. 2.8, allows for the significant influence that nitrogen has on the phase balance of austenitic stainless steels.

Percentage of ferrite in Schaeffler's work was determined using metallographic measurement methods. Magnetic measuring devices are more convenient, but these are secondary measurement systems. Therefore, an arbitrary standard or *ferrite number* (FN) has been established to designate the ferrite content. At low levels of ferrite, the ferrite number and the percentage of ferrite are identical; however, the two values diverge significantly when large amounts of ferrite are present in the stainless steel (see Fig. 2.8). The nickel equivalent is calculated using the weight percentages of the common austenite stabilizing elements (Ni, C, N, and Mn), and the chromium equivalent using the common ferrite stabilizers (Cr, Mo, Si, and Cb). The relative

**Table 2.12**
**Effects of alloying elements in austenitic stainless steels**

| Element | Types of steels | Effects |
|---|---|---|
| Carbon | All types | Strongly promotes the formation of austenite. Can form a carbide with chromium that can lead to intergranular corrosion. |
| Chromium | All types | Promotes formation of ferrite. Increases resistance to oxidation and corrosion. |
| Nickel | All types | Promotes formation of austenite. Increases high temperature strength, corrosion resistance, and ductility. |
| Nitrogen | XXXN | Very strong austenite former. Like carbon, nitrogen is thirty times as effective as nickel in forming austenite. Increases strength. |
| Columbium | 347 | Primarily added to combine with carbon to reduce susceptibility to intergranular corrosion. Acts as a grain refiner. Promotes the formation of ferrite. Improves creep strength. |
| Manganese | 2XX | Promotes the stability of austenite at or near room temperature but forms ferrite at high temperatures. Inhibits hot shortness by forming MnS. |
| Molybdenum | 316, 317 | Improves strength at high temperatures. Improves corrosion resistance to reducing media. Promotes the formation of ferrite. |
| Phosphorus, selenium, or sulfur | 303, 303Se | Increases machinability, but promotes hot cracking during welding. Lowers corrosion resistance slightly. |
| Silicon | 302B | Increases resistance to scaling and promotes the formation of ferrite. Small amounts are added to all grades for deoxidizing purposes. |
| Titanium | 321 | Primarily added to combine with carbon to reduce susceptibility to intergranular corrosion. Acts as a grain refiner. Promotes the formation of ferrite. |
| Copper | CN-7M | Generally added to stainless steels to increase corrosion resistance to certain environments. Decreases susceptibility to stress-corrosion cracking and provides age hardening effects. |

stabilizing strength of each element is indicated by the coefficients in the equations. For example, carbon and nitrogen are each about 30 times more effective in stabilizing austenite than nickel. Other less common elements in austenitic stainless steel weld metal can affect the microstructure. An example is titanium; small additions can increase the ferrite content of the deposit by several ferrite numbers.

## FERRITE

### Measurement

It is difficult to determine accurately how much ferrite is present in stainless steel weld metal. The ferrite content should be measured and specified as a ferrite number, which is not considered an absolute ferrite percentage. There is excellent agreement between laboratories

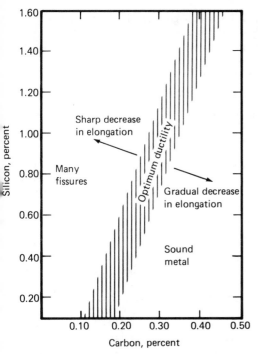

**Fig. 2.4—Effect of the combined carbon and silicon contents on the ductility of welds in austenitic stainless steels**

weld metal composition is sufficiently favorable toward the ferrite phase, residual islands of ferrite will persist in the microstructure. Weld metal less favorable to ferrite and those that first solidify as austenite remain fully austenitic down to room temperature.

Resistance of stainless steel weld metal to hot cracking increases when low-melting tramp elements, especially sulfur and phosphorus, are kept at low levels. The transformation of initial delta ferrite to austenite gives rise to a duplex structure soon after solidification has taken place. This transformation greatly increases the number of phase boundaries on which low-melting point constituents can be distributed, thereby further minimizing the likelihood of fissures in the weld metal.

Fissures might be observed in fully austenitic weld deposits that have been strained 20 percent, as in a bend test. Hot-short cracks caused by low-melting elements at grain boundaries can be found in heavily restrained welds. A delta ferrite level of at least 3FN will eliminate fissuring in weld metal deposited from austenitic filler metals E16-8-2, E308, E308L, E316, and E316L. Higher ferrite contents are recommended for other weld metals. For example, ferrite level of 4FN is required in E309, 5FN in E318, and 6FN in E347 weld metal to assure freedom from fissures.

when measuring ferrite using standard technique and ferrite numbers.

A standard procedure for calibrating magnetic instruments to measure the delta ferrite content of austenitic stainless steel weld metal is available from the American Welding Society.[7]

## Importance

Austenitic stainless steels can initially solidify as grains of delta ferrite or as grains of austenite. The scenario is controlled by the preponderance of ferrite-forming or austenite-forming elements in the weld metal.

As solidification continues, weld metal with initial ferrite grains begins to form austenite grains by a diffusion-controlled solid state transformation, usually in the grain centers. If the

---

7. AWS A4.2-74, *Standard Procedure for Calibrating Magnetic Instruments to Measure the Delta Ferrite Content of Austenitic Stainless Steel Weld Metal*

**Fig. 2.5—Microstructure of E308 austenitic stainless steel weld metal with pools of delta ferrite**

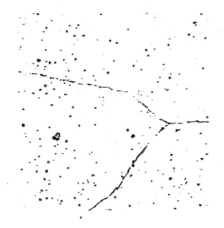

*Fig. 2.6—Microstructure of E310 austenitic stainless steel weld metal void of delta ferrite*

Manufacturers of stainless steel electrodes and welding rods have found empirically that ferrite-containing weld deposits have fewer fissures and hot-short cracks than deposits without ferrite. Under the FN system, electrodes and welding rods are now manufactured to deposit weld metal with specified amounts of ferrite.

Dependence on ferrite to assure sound weld deposits can lower the corrosion resistance of certain weld deposits in hot, oxidizing acids. The 18Cr-12Ni austenitic filler metals with molybdenum additions, such as Types E316, E317, and E318, might display poor corrosion resistance to certain media in the as-welded condition. The corrosion mechanism is usually localized attack on the ferrite. Preventive measures are to anneal the joint after welding or to specify a weld metal composition free from ferrite.

For most applications, ferrite-containing welds within the recommended FN range are desirable because they are less susceptible to fissuring and stress-corrosion cracking. The electrode manufacturer should be consulted for information about special applications.

Type 347 weld metal is usually formulated to contain a large amount of ferrite as a means of suppressing hot cracking. Ferrite is particularly helpful in this weld metal because columbium forms a low-melting eutectic phase that promotes cracking in a fully austenitic weld metal under a tensile stress.

Weld metal from E310Cb electrodes re-quires special consideration. It is not possible to obtain ferrite in the weld metal with this composition, and the weld metal can be especially crack-sensitive. Careful restriction on the Si, S, and P contents can reduce the fissuring tendency.

## Effects of Welding Conditions

The method of depositing austenitic stainless steel weld metal can alter its ferrite content. Ferrite content is modified by variations in weld metal cooling rates, arc length, and atmospheric contamination. Weld passes that have high dilution with the base metal can have altered ferrite content. Significant variations in ferrite content can exist from weld to weld, and within a weld from the root to face. For example, production pipe welds have shown that nearly 50 percent of all welds examined differed by at least 2 FN from procedure qualification values.

For applications where ferrite content of the weld metal is critical, welding procedures must be rigorously controlled. Nitrogen and chromium variations can significantly influence the ferrite content. Improper welding technique can result in nitrogen contamination because of excessive arc length, and the loss of chromium through oxidation. The result can be a weld with far less ferrite than specified.

## HIGH TEMPERATURE EXPOSURE

The thermal history of an austenitic stainless steel weldment can vary from the melting point associated with welding, through intermediate temperatures associated with postweld heat treatments, to elevated temperature service. These high temperature exposures produce varied effects on the microstructures and properties of the weldment. The actual effects depend upon the temperature, time at temperature, and cooling rate after exposuure.

## Carbide Precipitation and Intergranular Corrosion

Under certain conditions, austenitic stainless steel welds are subject to intergranular corrosion. When the unstabilized steels are heated or slowly cooled within the temperature range of 800° to 1600°F, known as the sensitizing temperature range, carbon is precipitated from solid solution mainly at the grain boundaries where it combines with chromium from the adjacent metal to form chromium-rich carbide ($M_{23}C_6$). The adjacent metal, impoverished in uncom-

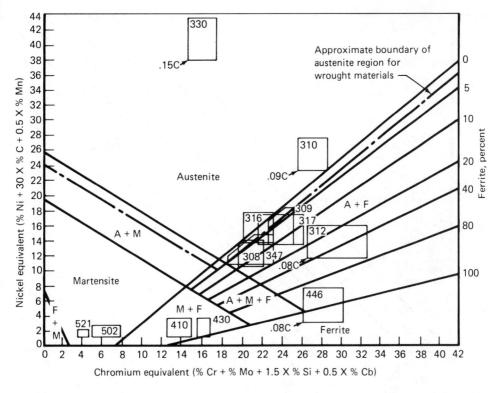

**Carbon**
Minimum value of 0.03 for all alloys except those otherwise indicated.

**Silicon**
Minimum value of 0.30.

**Nitrogen**
Nitrogen equals carbon in potency as austenite former. Nitrogen values vary from 0.05 for low chromium alloys, through 0.10 for the high chromium steels.

***Fig. 2.7—The Schaeffler diagram for estimating the microstructure of stainless steel weld metal***

bined chromium, exhibits reduced corrosion resistance. Under certain corrosive conditions, localized intergranular attack takes place. The severity of corrosion depends on the time and temperature of exposure, as well as the composition and prior heat treatment of the steel. Accepted methods for determining susceptibility to intergranular corrosion are described in ASTM Standard A262.[8]

8. ASTM A262-79, *Standard Recommended Practices for Detecting Susceptibility to Intergranular Attack in Stainless Steels*

During welding, the weld heat-affected zone experiences peak temperatures between the melting point and the temperature of the base metal, depending on the distance from the weld interface. A narrow band of metal will be heated for a sufficient time within the 800° to 1600°F temperature range to precipitate intergranular carbides. The amount of precipitation or degree of sensitization is approximately proportional to the carbon content of the austenitic stainless steel.

The 18-8 steels can retain a maximum of about 0.02 percent carbon in solid solution under all conditions. As the carbon content of the steel

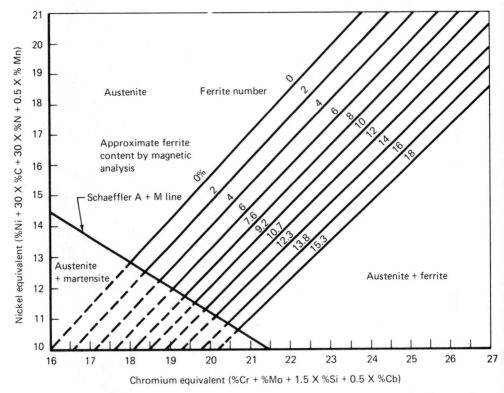

**Fig. 2.8—The DeLong constitution diagram for austenitic stainless steel weld metal**

increases, the amount of carbide precipitation increases. In many cases, a carbon content of up to about 0.08 percent is insufficient to be detrimental to corrosion resistance. At higher carbon levels, carbide precipitation increases rapidly. Therefore, the low-carbon stainless steels are preferred for service where the weldments must be used in the as-welded condition and be subjected to severely corrosive conditions. Two alternatives are (1) to use one of the stabilized grades, such as Type 321 or 347, or (2) anneal the weldment prior to service.

A typical service failure, shown in Fig. 2.9(A), illustrates the effect of carbide precipitation in conjunction with a severely corrosive medium. The section of pipe was from a piping system carrying hot, dilute nitric acid. The pipe was in service only two months before it failed. Apparently, welded Type 304 pipe without heat treatment requirements was used, and intergranular corrosion took place in the heat-affected zones of the longitudinal seam weld.

Figure 2.9(B) shows a cross section through the welded seam. The outside weld pass was made first. The attack took place about 0.06 to 0.12 in. from the weld. This indicates that the base metal from the weld to the point where corrosion started had reached temperatures of approximately 1600° to 2550°F. During welding, the corroded portion had been heated in the 800° to 1600°F range long enough to sensitize it to corrosion. Figure 2.10 shows the microstructure at the bottom of one of the corroded areas in the weld heat-affected zone. The thick, heavily etched grain boundaries are characteristic of carbide precipitation. As a result of the corrosion, some grains dropped out entirely during sectioning and polishing.

Weld metal containing a small percentage of delta ferrite is not susceptible to such severe sensitization as wrought material of similar composition. In weld metal, carbide precipitation occurs more uniformly throughout the structure at the interphase or substructure boundaries in

**A**

**B**

***Fig. 2.9–(A) Section from a 4-in. Type 304 stainless steel pipe with intergranular corrosion along the heat-affected zones of the welded seam (B) Cross section through the longitudinal seam of the pipe showing the extent of corrosion***

preference to the grain boundaries. Therefore, the degree of chromium depletion along the grain boundaries is lower than in the heat-affected zone. Consequently, weld metal is not generally considered to be so susceptible to intergranular corrosion as the base metal. Nevertheless, intergranular corrosion can occur in weld metal exposed to serverely corrosive environments.

## Stress Corrosion

Stress corrosion cracking of austenitic stainless steels can occur when the alloy is subjected simultaneously to a tensile stress and a specific corrodent. The cracking can be either intergranular or transgranular. Intergranular stress corrosion cracking can occur even though the

*Fig. 2.10—Microstructure of the sensitized Type 304 stainless steel at the corrosion face in the weld heat-affected zone*

alloy is insensitive to intergranular corrosion, described previously. Figure 2.11 shows an example of stress corrosion cracking in Type 316 stainless steel.

The ions of the halogen family (fluorine, chlorine, bromine, and iodine) are largely responsible for the stress corrosion cracking of austenitic stainless steels. Of the halides, the chloride ion causes the greatest number of failures. The common test for checking susceptibility to stress corrosion cracking is by immersing a stainless steel sample in boiling 41 percent magnesium-chloride solution.[9] Methods of controlling stress corrosion cracking include (1) a

stress-relieving or annealing heat treatment to reduce stress to a safe level, (2) substitution of more resistant high nickel- or straight-chromium alloys, or (3) removal of chlorides and oxygen from the ennvironment. Austenitic stainless steel is also subject to stress-corrosion cracking in high concentrations of hydroxyl ($OH^-$) ions or in lower concentration solutions where service conditions can promote chemical hideout. Intergranular stress corrosion cracking can occur in oxygenated boiling water environments and in pressurized water reactors in stagnant borated water environments.

## Phase Transformation

When austenitic stainless steel weldments are exposed to temperatures above 800°F, a variety of microstructural changes can occur. Car-

9. ASTM G36-73 (1979), *Recommended Practice for Performing Stress Corrosion Cracking Tests in a Boiling Magnesium Chloride Solution*

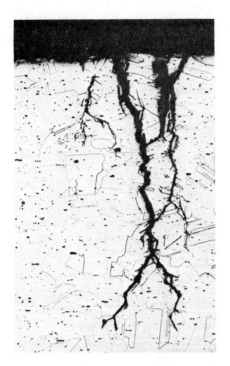

*Fig. 2.11—Stress-corrosion cracks in Type 316 stainless steel*

peratures from austenite. The phases and amount of each phase that forms depend upon the chemical composition, the ferrite content, and the time at temperature. Higher ferrite content tends to accelerate transformation and to lower the temperature where transformation will occur. Generally, sigma phase transformation occurs in the non-molybdenum-bearing stainless steels, and both sigma and chi phase transformations in the molybdenum-bearing steels. Nucleation of these phases occurs at the interphase (austenite-ferrite) boundaries, and growth is primarily toward the center of the ferrite. The presence of these intermetallics is generally considered to be detrimental to the mechanical properties, particularly ductility and impact resistance.

At temperatures above 1650°F, these carbides and intermetallics begin to go into solution in the austenite. The amount of dissolution is dependent upon the temperature and time at temperature. For example, considerable ductility can be restored to steel that has been embrittled by sigma phase by heating it at 1900°F for a short time. Complete conversion of sigma to austenite, however, usually requires heating to higher temperature, as high as 2250°F. If a high temperature is required, grain growth and any associated changes in mechanical properties should be anticipated in service.

## ARC WELDING

Austenitic stainless steels can be readily fabricated by arc welding. The commonly used processes are shielded metal arc, gas metal arc, gas tungsten arc, flux cored arc, plasma arc, and submerged arc welding. Other arc welding processes, such as electrogas and stud welding, are used, but only to a limited extent.

With the arc welding processes, the weld metal is protected from atmospheric oxidation by flux, slag, or inert gas. Protection must be sufficient to retain all essential alloying elements, and exclude all foreign matter affecting the corrosion resistance or properties of the welded joint. Carbon rapidly decreases the corrosion resistance and changes the properties of stainless steel. Therefore, its presence in flux, slag, or shielding gas must be controlled. The moisture content of electrode coverings, submerged arc welding fluxes, and shielding gases must be maintained at low levels because moisture can cause porosity in the weld metal.

bide precipitation can take place at the interphase (substructure) boundaries in preference to the grain boundaries, as discussed previously. Growth of these carbides depends upon the temperature and exposure time. During long times at temperature, these carbides can grow at the expense of the ferrite phase. When Type 308 shielded metal arc welds were aged at 1100°F for times up to 10,000 hours, carbide precipitation and growth were found to predominate over other transformations, for welds with 4FN and less in the as-welded structure. Depletion of chromium from the small ferrite islands, resulting from carbide growth, caused some ferrite to transform to austenite. Also, some of the ferrite in these welds transformed to sigma phase, but to a lesser extent.

When weldments are exposed to temperatures in the 900° to 1650°F range, various intermetallic phases can form. The phases form more readily from ferrite in the form of islands in weld metal and stringers in base metals, but they can also form after longer times and at higher tem-

When arc welding stainless steel less than 0.25 in. thick, distortion or warpage may be a serious problem. This is caused by the high expansion coefficient and the poor thermal conductivity of stainless steel (slow dissipation of welding heat). Rigid fixturing of the weldment can help control distortion of thin sheets during welding. Plates above 0.25 in. thickness might require special welding techniques, such as step-welding. The use of a chilled metal backing also helps by providing an excellent heat sink.

## Filler Metal Selection

In general, the deposited weld metal composition should nearly match the base metal composition when welding austenitic stainless steels to themselves. Recommended filler metals for use with the various arc welding processes are given in Table 2.13.

Other austenitic stainless steel filler metals can be used to weld a particular base metal, provided the weld metal will have suitable corrosion or mechanical properties, or both, for the required service. Generally, the alternate filler metal should be more highly alloyed than the base metal, but the ferrite content of the resultant weld metal should be considered for corrosion or high temperature service. Some filler metals might give too high a ferrite content, particularly the high chromium, low-nickel types, such as Type 312.

Consumable inserts of Types IN308, IN308L, IN310, IN312, IN316, IN316L, and IN348 are available for welding austenitic stainless steels.[10] Standard designs are shown in Fig. 2.12. They are used as preplaced filler metal in the root opening for the first weld pass, and are completely fused into the root of the joint. All configurations except the rectangular solid rings are available commercially in the form of coils and preformed rings.

Recommended filler metals for joining an austenitic stainless steel to a martensitic or ferritic stainless steel were given previously. One recommended procedure is to butter the chromium steel with a Type 309 filler metal and then fill the joint with a Type 308 filler metal. One of the nickel-chromium-iron filler metals might be suitable for some applications.

## Joint Designs

Typical dimensions for common groove-weld joint designs are given in Table 2.14. Some

10. Refer to AWS A5.30-79, *Specification for Consumable Inserts,* for additional information.

dimensions vary slightly to accommodate the welding process. When a backing strip is used, the root opening can be increased to permit good fusion with the strip, and the groove angle can be decreased to minimize filler metal requirements.

Backing rings and consumable inserts are often employed for better fit-up and easier root pass welding of stainless steel piping that cannot be back welded from the inside.

Nonremovable backing rings are undesirable for chemical and nuclear power piping because they encourage crevice corrosion, restrict the flow through the pipe, and interfere with mechanical and chemical cleaning processes. Elimination of backing makes the root pass difficult to deposit by shielded metal arc or gas metal arc welding in small diameter stainless steel pipe. Gas tungsten arc welding with a consumable insert is the preferred method for making the root pass. This technique makes it easier to weld stainless steel piping.

## Shielded Metal Arc Welding

Stainless steel covered electrodes use three types of core wire: mild steel, stainless steel, and metal cored. Alloying elements are sometimes added to the deposited metal by adding appropriate ingredients to the covering. Electrode design influences the appropriate amperage range for specific electrode sizes. Therefore, the amperage and voltage recommendations of the electrode manufacturer should be followed to obtain good deposition efficiency and to avoid overheating the electrode or weld metal. In general, the amperage ranges for austenitic stainless steel covered electrodes are 10 to 15 percent lower than for the same size of carbon steel covered electrodes.

Covered stainless steel electrodes have deposition efficiencies of about 75 percent at normal current and voltage. Stub losses may vary from 7 to 10 percent depending on shop practice and the electrode diameter.

Weaving of the electrode during welding is not recomended practice. For optimum quality, bead width should not exceed four times the core wire diameter. Individual layers of weld metal in a deep groove should not exceed a thickness of 0.125 in. to avoid discontinuities in the weld.

To eliminate possible slag entrapment and lack of fusion, individual weld beads should be thoroughly cleaned with a stainless steel chipping tool and then with a stainless steel wire

**Table 2.13**
**Recommended filler metals for welding wrought and cast austenitic stainless steels**

| Type of stainless steel | | Recommended filler metals | | |
|---|---|---|---|---|
| Wrought | Cast[a] | SMAW[b] | GMAW, GTAW, PAW, SAW[c] | FCAW[d] |
| 201<br>202 | – | E209<br>E219<br>E308 | ER209<br>ER219<br>ER308 | E308T-X |
| 301<br>302<br>304<br>305 | CF-20<br>CF-8 | E308 | ER308 | E308T-X |
| 304L | CF-3 | E308L<br>E347 | ER308L<br>ER347 | E308LT-X<br>E347T-X |
| 309 | CH-20 | E309 | ER309 | E309T-X |
| 309S | – | E309L<br>E309Cb | ER309L | E309LT-X<br>E309CbLT-X |
| 310<br>314 | CK-20 | E310 | ER310 | E310T-X |
| 310S | – | E310<br>E310Cb | ER310 | E310T-X |
| 316 | CF-8M | E316 | ER316 | E316T-X |
| 316L | CF-3M | E316L | ER316L | E316LT-X |
| 316H | CF-12M | E16-8-2<br>E316H | ER16-8-2<br>ER316H | E316T-X |
| 317 | – | E317 | ER317 | E317LT-X |
| 317L | – | E317L | ER317L | E317LT-X |
| 321 | – | E308L<br>E347 | ER321 | E308LT-X<br>E347T-X |
| 330 | HT | E330 | ER330 | – |
| 347 | CF-8C | E308L<br>E347 | ER347 | E308LT-X<br>E347T-X |
| 348 | – | E347 | ER347 | E347T-X |

a. Castings higher in carbon but otherwise of generally corresponding compositions are available in heat resisting grades. These castings carry the "H" designation (HF, HH, and HK, for instance). Electrodes best suited for welding these high-carbon versions are the standard electrodes recommended for the corresponding lower carbon corrosion-resistant castings shown above.

b. Covered electrodes for shielded metal arc welding (SMAW)

c. Bare welding rods and electrodes for gas metal arc (GMAW), gas tungsten arc (GTAW), plasma arc (PAW), and submerged arc welding (SAW).

d. Tubular eletrodes for flux cored arc welding (FCAW). The suffix -*x* can be -1, -2, or -3 (see Table 2.5).

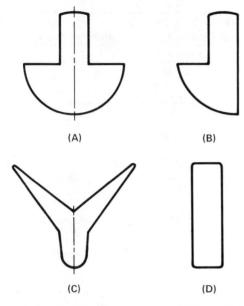

(A)          (B)

(C)          (D)

*Fig. 2.12—Standard consumable insert designs: (A) inverted T-shape, (B) J-shape, (C) Y-shape, (D) rectangular*

brush. The root of unbacked or double-V-groove welds must be thoroughly cleaned before welding the second side. This root surface might require grinding, chipping, or arc gouging to expose sound weld metal and ensure complete fusion by the second or back weld.

The covering employed for stainless steel electrodes and the resulting slag contain fluorides to flux chrome and nickel oxides from the weld metal. In some service environments, these fluorides become highly corrosive to the stainless steel, accelerate crevice and pitting corrosion, and lead to stress-corrosion cracking. All slag must be carefully removed from all stainless steel surfaces, including the root of the weld. If the weldment design will not permit access to the back side of the weld, shielded metal arc welding should not be used to make the root pass. Instead, gas metal arc, gas tungsten arc, or plasma arc welding should be used.

Occasionally, a weld is undercut along one or both sides of the joint. It is more prevalent when welding in the vertical and overhead positions where the molten weld pool is too large to control and flows to the lowest point as it solidifies. Undercutting can be caused by excessive welding current, too large an electrode, or improper weaving of the electrode.

As mentioned previously, moisture in the covering of the electrodes can result in porosity in the weld. The electrodes must be dry, and stored properly to avoid absorption of moisture from the atmosphere. Improper cleaning of the base metal can also result in porous welds.

## Gas Metal Arc Welding

When gas metal arc welding with adequate inert gas shielding, there is very little loss in alloying elements during transfer of filler metal across the arc. Reactive elements, like titanium, can be transferred across the arc. Therefore, Type ER321 stainless steel filler metal can be used with this process. Transfer efficiencies of over 95 percent are commonly reported with argon shielding. Standard sizes of wire electrodes range from 0.020 to over 0.125 in. diameter.

Using argon shielding gas and direct current, electrode positive, metal transfer from the electrode to the work is in minute droplets, or spray transfer, provided the current density is above a transition value and the arc voltage is in the 26V to 33V range. Below this minimum, the transfer will be in the form of large drops, or globular transfer, that results in excessive spatter and arc instability. The threshold current value for stable spray transfer is about 300 amperes for 0.062 in. electrode.

Metal transfer can also be accomplished by short-circuiting and pulsed spray transfer. These variations of GMAW operate at lower effective currents, and at an arc voltage typically between 18 and 24V, which can extend the application of GMAW to include 0.010 in. thick material. Welds can be made with low distortion because the heat input is lower than with spray transfer.

Shielding gases used to weld austenitic stainless steels include argon, argon-oxygen, argon-helium, and helium-argon-carbon dioxide. Argon-oxygen mixtures result in some oxidation in the weld pool that promotes better wetting action and improved arc stability than is obtained with pure argon. Argon-1 percent oxygen is very popular for welding with spray transfer. Helium-argon mixtures with 2 to 3 percent carbon dioxide additive are frequently used to weld stainless steels with short-circuiting transfer. Pure carbon dioxide causes a large loss of silicon and manganese, and can lead to an increase in carbon content in the low-carbon grades of stainless steel. Carbon absorption could impair the corrosion resistance of welds. Accordingly, $CO_2$ is not recommended for welding stainless steel.

Helium additions to argon widen the pene-

**Table 2.14**
**Typical groove weld joint designs for austenitic stainless steel**

| Joint design | Welding process[a] | Thickness, in. | Root opening, in. | Root face, in. | Groove angle, degrees |
|---|---|---|---|---|---|
| Square-groove, one pass | SMAW | 0.04-0.13 | 0-0.04 | – | – |
| | GTAW | 0.04-0.13 | 0-0.08 | – | – |
| | GMAW | 0.08-0.16 | 0-0.08 | – | – |
| Square-groove, two-pass | SMAW | 0.12-0.25 | 0-0.08 | – | – |
| | GTAW | 0.12-0.25 | 0-0.04 | – | – |
| | GMAW | 0.12-0.32 | 0-0.04 | – | – |
| | SAW | 0.15-0.32 | 0 | – | – |
| Single-V-groove[b] | SMAW | 0.12-0.50 | 0-0.08 | 0.06-0.12 | 60 |
| | GTAW | 0.15-0.25 | 0-0.01 | 0.06-0.08 | 90 |
| | GTAW | 0.25-0.63 | 9-0.02 | 0.04-0.06 | 70 |
| | GMAW | 0.15-0.50 | 0-0.08 | 0.06-0.12 | 60 |
| | SAW | 0.31-0.50 | 0-0.08 | 0.06-0.16 | 60 |
| Single-U-groove | GMAW | 0.5-0.75 | 0-0.08 | 0.08-0.12 | 15[c] |
| Double-V-groove[b] | SMAW | 0.5-1.25 | 0.04-0.13 | 0.06-0.16 | 60 |
| | SAW | | | | |
| | GMAW | 0.5-1.25 | 0-0.08 | 0.08-0.12 | 60 |
| Double-U-groove | SMAW | over 1.25 | 0.04-0.08 | 0.08-0.12 | 10-15[c] |

a. SMAW—Shielded metal arc welding
GTAW—gas tungsten arc welding
GMAW—gas metal arc welding
SAW—submerged arc welding

b. For welding in the horizontal position, the lower member should be beveled only 10 to 15 degrees, and the top member 45 to 55 degrees.

c. Groove radius—0.25 to 0.31 in.

tration profile of the weld bead. The penetration with argon is narrowly aligned with the electrode axis, and incomplete fusion is possible at the root of a tight butt joint if the torch is misaligned. A mixture of 50 Ar – 50 He provides a wider penetration profile, and virtually eliminates the problem.

When welding at speeds above 30 ipm, a trailing shield might be required to protect the weld metal.

## Flux Cored Arc Welding

Austenitic stainless steels are joined using flux cored electrodes described previously (see Table 2.4). They are designed for gas shielding with $CO_2$ or argon-2 percent oxygen and without gas shielding (see Table 2.5). Direct current, electrode positive is used for welding with these electrodes.

The chemical composition for each type of electrode allows adequate latitude for the manufacturer to control the Ferrite Number of the undiluted deposit. With the EXXXT-1 classifications using carbon dioxide shielding, there is some minor loss of oxidizable elements and some pickup of carbon. With the EXXXT-2 classifications using argon-oxygen shielding, there is some minor loss of oxidizable elements. With the EXXXT-3 classifications that are used without external shielding, there is some minor loss of oxidizable elements and a pickup of nitrogen, which may range from quite low to over 0.20 percent. Low welding currents coupled with long arc lengths (high arc voltages) should be avoided because they result in excessive nitrogen pickup and excessive loss in the ferrite content of the weld.

The E307T-X, E308T-X, E308LT-X,

E308MoLT-X, and E347T-X grades are normally ferrite controlled. When used with the recommended shielding gases and with reasonable and conventional welding currents and arc lengths, they produce weld metal with a typical ferrite level of 4 to 14 FN.

## Gas Tungsten Arc Welding

Gas tungsten arc welding (GTAW) is well suited for joining austenitic stainless steels in all welding positions and most thicknesses. However, it is more costly than the consumable electrode welding processes because deposition rates are lower.

Austenitic stainless steels are welded with this process using direct current, electrode negative. The power source should have a drooping volt-ampere characteristic or constant current output.

Argon, helium, and mixtures of the two gases are used for shielding the austenitic stainless steels with this process. Argon is generally preferred for manual welding because it is easy to start the arc, to control the molten weld pool, and to maintain good gas coverage over the weld. It is also preferred for welding of thin sheet because the heat input is lower than with helium for a given welding current. Helium can be used to advantage with thick sections for deeper penetration. Helium is also used for mechanized welding to take advantage of high welding speeds with less likelihood of undercut and a better bead shape. However, the arc is difficult to start in helium.

Inadequate gas shielding can result in nitrogen contamination of the weld metal. This, in turn, influences the ferrite content of certain weld metal compositions.

The process is well suited to weld the first pass in groove welds when the back side of the joint is inaccessible for a backing weld, and complete joint penetration is required. Although not always necessary, a consumable insert can be used, or the root edge can be flared as shown in Fig. 2.13, to provide filler metal for the first pass. Acceptable and unacceptable root surface contours are illustrated in Fig. 2.14. A common root surface defect caused by interruption of the welding arc is show in Fig. 2.15.

## Plasma Arc Welding

Plasma arc welding is readily applied for joining thin sections of austenitic stainless steels using the melt-in technique. The plasma arc has a characteristically long length, and is insensitive

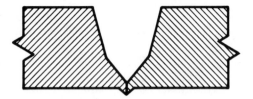

**Fig. 2.13—Groove weld joint design with a flared root edge**

**A**

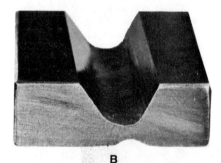

**B**

**Fig. 2.14—Root pass made using a consumable insert: (A) acceptable root surface, (B) unacceptable, concave root surface**

to variations in arc length. Tungsten contamination of the weld is unlikely because the tungsten electrode is contained within the welding torch. Square-groove joints can sometimes be welded in thicknesses from 0.090 to 0.250 in. without filler wire addition using the keyhole technique.

High welding speeds are attained on continuously formed stainless steel tubing, and high speeds and uniform underbeads on square-groove

**Fig. 2.15—Crater on the root surface of a stainless steel weld caused by abrupt interruption of the gas tungsten arc**

welds in plate. Circumferential pipe joints can be welded without backing rings or inserts, and with reduced filler metal requiremments.

## Submerged Arc Welding

Submerged arc welding (SAW) is commonly used to join most austenitic stainless steels to themselves, and in cladding or overlaying carbon and low alloy steel with austenitic stainless steel. Welded joints and weld cladding that are sound, serviceable, uniform in quality, and excellent in appearance can be made with this process, provided qualified materials and procedures are used.

Typical problems that are encountered with SAW are traceable to high heat input, large weld beads, and low cooling and solidification rates. These factors tend to increase both segregation of alloying elements and coarseness of microstructure. Large weld beads, for instance, can lead to increased crack sensitivity, particularly in weld metals that are fully austenitic. Cracking of this type appears to be related to impurities in the weld metal and to the solidification pattern of the bead.

Elements such as silicon, sulfur, and phosphorus are especially harmful when present in excess of specification limits. Carbon and manganese, on the other hand, are beneficial within limits. Silicon transfer from the flux into the weld metal can be appreciably higher in submerged

arc welds than it is for welds made by other processes. Quantities of silicon can be transferred to the weld metal as a result of high-temperature slag-metal reactions, and it either goes into solution or, more likely, forms inclusions in the weld metal. Sulfur and phosphorus also can be transferred from the flux and, through dilution, from the base metal. Many SAW fluxes are formulated to match specific heats of electrodes so that the weld metal meets a specified chemistry range.

*Welding Procedures.* Alternating current and direct current power supplies are commonly used for submerged arc welding, although direct current is preferred, particularly for welding thin material. Many joint designs and welding conditions used for carbon steel are useful as a first approximation for stainless steel, with the exception that the welding current for stainless steel should be about 20 percent lower than that normally used for a similar weld in carbon steel. The higher electrical resistivity and slightly lower melting temperature of austenitic stainless steel results in a deposition rate that is 20 to 30 percent higher than that of carbon steel, under otherwise identical conditions. Table 2.15 gives typical welding conditions for double-V-groove welds in stainless steel plate.

The electrical resistance of austenitic stainless steel makes the electrode extension beyond the contact tip more critical than with carbon steel. Resistance heating the electrode before it enters the arc can appreciably affect the deposition rate. Because of its higher resistivity and lower melting temperature, the melting rate of a stainless steel electrode is approximately 30 percent higher than that of a carbon steel electrode under the same welding conditions. This should be considered when selecting the electrode size for a particular application.

*Dilution.* Control of dilution can be the most important factor in submerged arc welding of austenitic stainless steel. Base metal dilution with this process can vary from less than 10 to as much as 75 percent, exceeding that of all other consumable electrode welding processes. It usually is desirable, and sometimes mandatory, to hold the weld metal composition to within narrow limits; control of dilution is necessary to accomplish this.

About 4 to 10 percent of delta ferrite is desired in the weld metal to control hot shortness. Excessive ferrite can lead to sigma embrittlement during elevated temperature exposure.

Base metal dilution below 40 percent is

**Table 2.15**
**Typical conditions for submerged arc welding double-V-groove joints in stainless steel plate**

| Plate thickness, in. | Root face, in. | First Weld[a] | | | | Second Weld[a] | | | |
|---|---|---|---|---|---|---|---|---|---|
| | | Electrode diam., in. | Welding current, A | Voltage, V | Travel speed in./min. | Electrode diam., in. | Welding current, A | Voltage, V | Travel speed in./min. |
| 0.375 | 0.25 | 0.188 | 525 | 30 | 20 | 0.188 | 575 | 32 | 24 |
| 0.50 | 0.25 | 0.188 | 700 | 35 | 18 | 0.188 | 900 | 33 | 18 |
| 0.625 | 0.25 | 0.188 | 700 | 33 | 16 | 0.250 | 900 | 35 | 12 |
| 0.75 | 0.25 | 0.250 | 700 | 33 | 15 | 0.250 | 950 | 35 | 12 |
| 0.875 | 0.31 | 0.250 | 715 | 33 | 15 | 0.250 | 1025 | 35 | 12 |

a. Groove angle–90 degrees

commonly required to produce sound welds in austenitic stainless steel. The effect of dilution is of special concern in single and double pass welds, and in the root pass of multiple-pass welds because of the normal compositional differences between the base metal and the filler metal. In these cases, small changes in penetration and bead shape, and hence in dilution, can produce significant changes in the composition and properties of the deposit.

*Electrodes.* Standard types of austenitic stainless steel electrodes used for submerged arc welding are listed in Table 2.3. With proper composition control, all types except fully austenitic Types ER310 and ER320 can be used effectively for submerged arc welding. Fully austenitic stainless steels tend to be hot short, and submerged arc welding of these is not recommended.

*Flux.* Fluxes used for submerged-arc welding of austenitic stainless steel are metallurgically neutral or basic in their effect on the weld metal. There are no standard specifications for stainless steel fluxes. Selection remains on a proprietary basis, and consultation with the manufacturer or supplier is recommended. Fluxes primarily used for submerged arc welding carbon steel are not suitable for welding stainless steel because of loss of chromium to the slag and pickup of manganese and silicon in weld metal from the flux.

Neutral, fused flux for welding stainless steel permits some oxidation of chromium by high temperature slag-metal reactions, but sound welds can be deposited. Basic bonded fluxes, developed specifically for stainless steels, are reinforced with alloying elements by the manu-

facturer to control the weld metal composition.

A variety of alloying elements can be added with bonded flux. Carbon, chromium, cobalt, columbium, copper, manganese, molybdenum, nickel, tungsten, and vanadium are the most common additions. Bonded SAW flux can absorb moisture when exposed to damp and humid conditions. Excessive moisture in the flux can cause porosity, worm holes, and surface imperfections in the weld bead. Fluxes that have been inadequately baked may behave similarly. Damp flux needs to be rebaked according to the manufacturers recommendations. Fluxes should be properly handled and stored to prevent contamination.

## Mechanical Properties

Austenitic stainless steels are often used because of their excellent strength and oxidation resistance at elevated temperatures. Table 2.16 lists stress-rupture strengths for several types at temperatures of 1000° to 1600°F. Although these data are for the base metals, weld metals of similar compositions will show similar performance, as indicated in Table 2.17. For long term service, up to 300,000 hour design for some applications, austenitic stainless steel welds tend to be weaker and lower in ductility than the corresponding base metals.

Exposure to elevated temperatures can have significant effects on properties of austenitic stainless steel weldments. The exact effect is dependent upon the chemical composition of the weld metal, the ferrite content, the temperature, and the time at temperature.

The presence of intermetallic phases in aus-

## Table 2.16
## Typical stress-rupture strengths of austenitic stainless steels

| | Stress, ksi | | | | | | | |
|---|---|---|---|---|---|---|---|---|
| | 1000 h failure | | | | 10,000 h failure | | | |
| | Test temperature, °F | | | | Test temperature, °F | | | |
| Stainless steel | 1000 | 1200 | 1400 | 1600 | 1000 | 1200 | 1400 | 1600 |
| 201 | | 21.0 | 7.2 | | 16.5 | | | |
| 304 | 35.0 | 14.4 | 6.0 | 2.7 | 27.0 | 9.7 | 3.9 | 1.9 |
| 309 | | 20.0 | 7.4 | 2.8 | | 15.0 | 4.3 | 1.7 |
| 310 | 32.0 | 13.7 | 5.1 | 2.4 | 24.0 | 8.5 | 3.0 | 1.3 |
| 316 | | 24.7 | 11.1 | 4.0 | | 18.2 | 6.9 | 1.8 |
| 321 | | 17.5 | 5.6 | | | 9.8 | 3.6 | |
| 347 | | 17.4 | 7.4 | 3.0 | | 11.2 | 2.5 | |

**Table 2.17**
**Typical stress-rupture properties of austenitic stainless steel weld metal**

| | Stress, ksi | | | | | |
|---|---|---|---|---|---|---|
| | 1000 h failure | | | 10,000 h failure | | |
| | Test temperature, °F | | | Test temperature, °F | | |
| Weld metal | 1000 | 1100 | 1200 | 1000 | 1100 | 1200 |
| E308 | 30 | 23 | 15 | 22 | 16 | 10 |
| E309 | | | 16 | | | |
| E310 | 31 | 21 | 12 | 21 | 14 | 8 |
| E316 | | | 17 | | | 10 |
| E347 | 48 | 37 | 26 | 40 | 29 | 19 |
| E16-8-2 | | | 25 | | | 19 |

tenitic stainless steel weld metal significantly decreases toughness. The presence of coarse carbides does not appear to have so significant an effect. The effects of long time exposure at elevated temperature have been evaluated. Welds made with Type E308 covered electrodes, with as-welded ferrite levels ranging from 2 to 15 FN, were aged for times up to 10,000 hours at 1100°F. After 10,000 hours, the ferrite in welds that originally had an FN of 15 had decreased to about 3 FN, and the 2FN welds had decreased to less than 0.5 FN. These reductions in measured FN resulted primarily from the transformation of ferrite to sigma in the higher ferrite welds and from carbide precipitation during ferrite to austenite transformation in the low ferrite welds. Charpy impact tests of aged specimens revealed that the impact strength dropped off rapidly with aging time for the higher ferrite welds because of increased sigma phase. However, for the lower ferrite welds, where sigma phase formation was not the primary transformation, impact strength decreased only slightly with increased aging time. These results are illustrated in Fig. 2.16. Exposure at lower temperatures does not have so significant an effect on impact properties.

Both intermetallic phases and coarse carbides in austenitic stainless steel weld metal tend to decrease ductility under creep conditions. Welds made with Type E308 covered electrodes having different ferrite contents have been creep-rupture tested for times up to 10,000 hours. For all ferrite levels, elongation decreased rapidly to values below 5 percent as the rupture lives approach 10,000 hours as shown in Fig. 2.17. Rupture failures in long-time tests were found to propagate along austenite-to-sigma phase boundaries in the higher ferrite welds, and along austenite-to-carbide boundaries in the lower ferrite welds. Therefore, both transformations appear to be equally detrimental to creep ductility. In general, creep-rupture life decreases as the ferrite content of the weld metal increases. However, when the inherent scatter in creep data is considered, these decreases may not always be significant.

The toughness of austenitic stainless steels at subzero temperatures is of considerable interest to producers of cryogenic equipment. Table 2.18 shows the impact strengths of weld metal deposited from several types of austenitic stainless steel covered electrodes. In the as-deposited condition, these weld metals usually contain some delta ferrite, which decreases low temperature toughness. If a postweld stress-relieving or a stabilizing heat treatment in the range of 1200 to 1700°F is given to such weld metal, the delta ferrite can transform to intermetallic phases or form carbides. These heat treatments can reduce the low-temperature toughness of the weld metal, as the data in the Table indicate.

## OXYACETYLENE WELDING

Oxyacetylene welding is infrequently selected for joining the austenitic stainless steels, and should only be used when appropriate arc welding equipment is not available. Applications are generally limited to sections less than about 0.13-in. thick.

For making butt joints in stainless steel sheet, the edges can be flanged so that the flange height is approximately equal to the thickness of the sheet. The joint can then be welded without

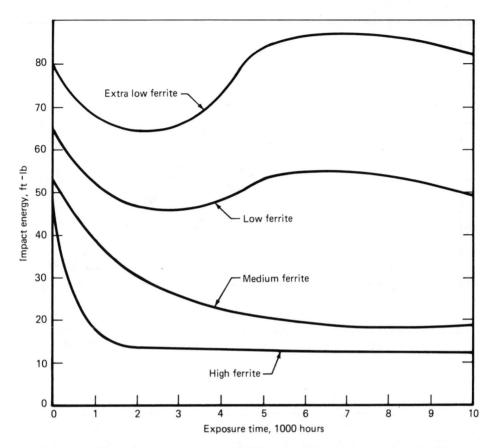

**Fig. 2.16—Effect of exposure time at 1100°F on the Charpy impact strength of Type E308 stainless steel weld metal with various original ferrite contents**

filler metal. Sheet thicknesses up to 0.063 in. sometimes can be joined with a square-groove weld. A V-groove weld is required for thicker sections.

The welding tip should be one or two sizes smaller than those used for carbon steel. Proprietary fluoride-type welding fluxes are used to flux away chromium oxides. Borax-type fluxes are not strong enough to reduce chromium oxides. A neutral flame should be used.

The weld should be made with one or more stringer beads. The puddling technique recommended for welding carbon steel, where the molten pool is stirred with the end of the filler rod, should never be used for welding stainless steels.

## RESISTANCE WELDING

Austenitic stainless steels can, in most cases, be welded by any of the resistance welding processes. High strength welded joints are easily produced using appropriate welding conditions for these steels. However, close control of welding variables must be exercised to maintain good joint corrosion resistance.

Welding currents used for spot and seam welding of austenitic stainless steels are generally lower than those used for mild steel because of their higher electrical resistance. Electrode pressures are consistently higher because of the high strength of these steels at elevated temperatures.

The lower heat conductivity of stainless steels produces a steeper thermal gradient than in mild steel welds. As a result, the quality of the weld is sensitive to variations in conditions, such as magnitude and duration of the welding current and welding pressure, and the electrode

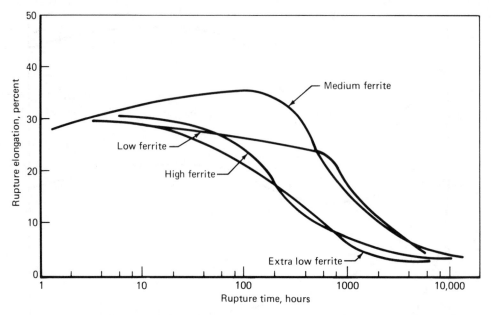

**Fig. 2.17—Relationship of rupture elongation and time-to-rupture at 1100°F for Type E308 stainless steel weld metal with various original ferrite contents**

shape. Control is made critical by the comparatively short welding times involved. However, the low magnetic permeability of austenitic stainless steels is an asset in spot or seam welding. It is unnecessary to increase the current in compensation for changes in reactance, as is usually done for welding large masses of carbon and low alloy steels.

## Spot Welding

Successful spot welding of austenitic stainless alloys has eight basic requirements, as follows:

(1) The welding current must remain the same for each weld; random transients cannot be tolerated. The welding current should be nearly sinusoidal. The current should be applied synchronously; that is, always on the same point of the voltage wave, preferably one that will not result in a transient current. The voltage variation at the transformer should not be more than 5 percent. The use of autotransformers between the welding transformer and the contactor should be avoided.

(2) Duration of the welding current should

be kept constant. Welding at time intervals of less than 5 cycles (1/12 sec) should not be attempted unless accuracy of better than 1/4 of a cycle (1/240 sec) can be assured. When using very short time intervals (as in welding thin gages), electronic timing equipment independent of mechanical current switching should be used. A timing variation of more than 1 cycle (1/60 sec) in the duration of the weld should not be permitted.

(3) The electrode force must be uniform. Variation in force will affect both the current density by changing the contact area and the electrical resistance of the weld. For this reason, the faying surfaces must be in contact prior to welding.

(4) The surfaces to be welded must be thoroughly clean and completely free of scale, oxide, dirt, grease, or drawing compound.

(5) An electrode pressure of two to three times that used for welding of mild steel should be used to obtain good results.

(6) RWMA Group A, Class 3 copper alloy electrodes are recommended, with Group A, Class 2 and Group B, Class 11 electrodes as

**Table 2.18**
**Impact strengths of austenitic stainless steel weld metal at room and low temperatures**

| Base metal | Filler metal | Heat treatment[a] | Ferrite in weld metal, percent[b] | Charpy impact strength, ft·lbs[c] Testing temperature, °F | | |
|---|---|---|---|---|---|---|
| | | | | 68 | −105 | −300 |
| 304 | E308-15 | AW | − | 32 | 23 | 19 |
| | | AN | − | 37 | 30 | 30 |
| 310 | E310-15 | AW | − | 36 | 30 | 24 |
| | | AN | − | 32 | 28 | 19 |
| 316 | E316-15 | AW | 0.5 | 31 | 26 | 18 |
| | | SR | − | 32 | 26 | 14 |
| | | ST | − | 28 | 21 | 13 |
| | | AN | − | 34 | 29 | 22 |
| 316 | E316-15 | AW | 8.0 | 32 | 28 | 19 |
| | | SR | − | 26 | 17 | 8 |
| | | ST | − | 11 | 7 | 3 |
| | | AN | − | 36 | 33 | 26 |
| 317 | E317-15 | AW | 2.0 | 22 | 17 | 11 |
| | | AN | − | 21 | 16 | 14 |
| 321 | E347-15 | AW | 3.5 | 32 | 26 | 21 |
| | | AN | − | 31 | 32 | 24 |
| 347 | E347-15 | AW | 3.5 | 30 | 25 | 19 |
| | | SR | − | 25 | 15 | 11 |
| | | ST | − | 21 | 17 | 14 |
| | | AN | − | 26 | 24 | 24 |
| 347 | E347-16 | AW | − | 27 | 21 | 19 |
| | | AN | − | 30 | 23 | 22 |

a. AW–as-welded
AN–annealed at 1950°F for 0.5 h, water-cooled
SR–Stress-relieved at 1200°F for 2 h
ST–Stabilized at 1550°F for 2 h
b. Ferrite content was determined with a ferrite magnetic gage.
c. Keyhole notch

alternates. To promote rapid cooling of the weld, the electrodes should be internally water cooled. External flood cooling is sometimes used.

(7) Proper electrode pressure should be maintained at all times. Insufficient pressure may cause blowholes inside the nugget, and in extreme cases, actual surface melting at the electrode.

(8) Because of the comparatively high resistance of austenitic stainless steels, the short-circuiting of current through the preceding welds can be significant, particularly when more than three metal thicknesses are welded together simultaneously and the spacing of spot welds is comparatively close. This effect should be determined and compensated for by a suitable increase in the welding current.

Table 2.19 tabulates recommended electrodes, machine settings, and spot weld spacing for resistance spot welding all standard grades of austenitic stainless steels. This table covers single impulse welds in thicknesses from 0.006 to 0.125 inch. Table 2.20 similarly covers multiple-impulse welds in stainless steel pieces from 0.156-to 0.250-inch thick.

With appropriate welding procedure technique, it is feasible to spot weld the unstabilized stainless steels with normal carbon contents without harmful carbide precipitation or loss of ductility. Spot welds in austenitic stainless steels are extremely tough and ductile. High joint efficiencies are obtained in spite of the rapid cooling inherent in spot welding.

For thinner gages, there is usually little dif-

## Table 2.19
### Suggested practices for spot welding stainless steels

| Thickness "T" of thinnest outside piece in.[a] | Electrodes[b] | | Net electrode force, lb | Weld time (single impulse), Cycles 60 Hz | Minimum shear strength, lb Ultimate tensile strength of metal | | | Welding current, A (approx) | | Diameter of fused zone, in. (approx) | Minimum weld spacing, in.[c] | Minimum contacting overlap, in. |
|---|---|---|---|---|---|---|---|---|---|---|---|---|
| | Diam., in. | Face diam., in | | | 70,000 to 90,000 psi | 90,000 to 150,000 psi | 150,000 psi and Higher | Tensile strength below 150,000 psi | Tensile strength 150,000 psi and higher | | | |
| 0.006 | 3/16 | 3/32 | 180 | 2 | 60 | 70 | 85 | 2000 | 2000 | 0.045 | 3/16 | 3/16 |
| 0.008 | 3/16 | 3/32 | 200 | 3 | 100 | 130 | 145 | 2000 | 2000 | 0.055 | 3/16 | 3/16 |
| 0.010 | 3/16 | 1/8 | 230 | 3 | 150 | 170 | 210 | 2000 | 2000 | 0.065 | 3/16 | 3/16 |
| 0.012 | 1/4 | 1/8 | 260 | 3 | 185 | 210 | 250 | 2100 | 2000 | 0.076 | 1/4 | 1/4 |
| 0.014 | 1/4 | 1/8 | 300 | 4 | 240 | 250 | 320 | 2500 | 2200 | 0.082 | 1/4 | 1/4 |
| 0.016 | 1/4 | 1/8 | 330 | 4 | 280 | 300 | 380 | 3000 | 2500 | 0.088 | 5/16 | 1/4 |
| 0.018 | 1/4 | 1/8 | 380 | 4 | 320 | 360 | 470 | 3500 | 2800 | 0.093 | 5/16 | 1/4 |
| 0.021 | 1/4 | 5/32 | 400 | 4 | 370 | 470 | 500 | 4000 | 3200 | 0.100 | 5/16 | 5/16 |
| 0.025 | 3/8 | 5/32 | 520 | 5 | 500 | 600 | 680 | 5000 | 4100 | 0.120 | 7/16 | 3/8 |
| 0.031 | 3/8 | 3/16 | 650 | 5 | 680 | 800 | 930 | 6000 | 4800 | 0.130 | 1/2 | 3/8 |
| 0.034 | 3/8 | 3/16 | 750 | 6 | 800 | 920 | 1100 | 7000 | 5500 | 0.150 | 9/16 | 7/16 |
| 0.040 | 3/8 | 3/16 | 900 | 6 | 1000 | 1270 | 1400 | 7800 | 6300 | 0.160 | 5/8 | 7/16 |
| 0.044 | 3/8 | 3/16 | 1000 | 8 | 1200 | 1450 | 1700 | 8700 | 7000 | 0.180 | 11/16 | 7/16 |
| 0.050 | 1/2 | 1/4 | 1200 | 8 | 1450 | 1700 | 2000 | 9500 | 7500 | 0.190 | 3/4 | 1/2 |
| 0.056 | 1/2 | 1/4 | 1350 | 10 | 1700 | 2000 | 2450 | 10,300 | 8300 | 0.210 | 7/8 | 9/16 |
| 0.062 | 1/2 | 1/4 | 1500 | 10 | 1950 | 2400 | 2900 | 11,000 | 9000 | 0.220 | 1 | 5/8 |
| 0.070 | 5/8 | 1/4 | 1700 | 12 | 2400 | 2800 | 3550 | 12,300 | 10,000 | 0.250 | 1 1/8 | 5/8 |
| 0.078 | 5/8 | 5/16 | 1900 | 14 | 2700 | 3400 | 4000 | 14,000 | 11,000 | 0.275 | 1 1/4 | 11/16 |
| 0.094 | 5/8 | 5/16 | 2400 | 16 | 3550 | 4200 | 5300 | 15,700 | 12,700 | 0.285 | 1 3/8 | 3/4 |
| 0.109 | 3/4 | 3/8 | 2800 | 18 | 4200 | 5000 | 6400 | 17,700 | 14,000 | 0.290 | 1 1/2 | 13/16 |
| 0.125 | 3/4 | 3/8 | 3300 | 20 | 5000 | 6000 | 7600 | 18,000 | 15,500 | 0.300 | 2 | 7/8 |

a. Types of steel—301, 302, 303, 304, 308, 309, 310, 316, 317, 321 and 347
Material should be free from scale, oxides, paint, grease and oil
Welding conditions determined by thickness of thinnest outside piece "T"
Data for total thickness of pile-up not exceeding 4 "T". Maximum ratio between two thicknesses 3 to 1
b. Truncated electrodes of RWMA Group A, Class 3 or Group B, Class II material. Electrodes with 3-in. spherical faces are also used.
c. Minimum weld spacing is that spacing for two pieces for which no special precautions need be taken to compensate for shunted current effect of adjacent welds. For three pieces increase spacing 30%

**Table 2.20**
**Multiple-impulse spot welding of stainless steel**

| Thickness "T" of thinnest outside piece, in.[a] | Electrodes[b] | | | No. of impulses[c] | Welding current, A (Approx.) Base metal tensile strength | | Minimum joint overlap, in. | Minimum weld spacing, in.[d] | Minimum diameter of nugget, in. | Minimum shear strength, lb Ultimate tensile strength of metal | |
|---|---|---|---|---|---|---|---|---|---|---|---|
| | Diam. in. | Face diam., in. | Net electrode force, lb | | Below 150,000 psi | 150,000 psi and higher | | | | 90,000 to 150,000 psi | 150,000 psi and higher |
| 0.156 | 1 | 1/2 | 4000 | 4 | 20700 | 17500 | 1 1/4 | 1 7/8 | 0.440 | 7600 | 10000 |
| 0.187 | 1 | 1/2 | 5000 | 5 | 21500 | 18500 | 1 1/2 | 2 | 0.500 | 9750 | 12300 |
| 0.203 | 1 | 5/8 | 5500 | 6 | 22000 | 19000 | 1 5/8 | 2 1/8 | 0.530 | 10600 | 13000 |
| 0.250 | 1 | 5/8 | 7000 | 7 | 22500 | 20000 | 1 3/4 | 2 3/8 | 0.600 | 13500 | 17000 |

a. Types of steel–301, 302, 303, 304, 308, 309, 310, 316, 317, 321, 347 and 349
Material should be free from scale, oxides, paint, grease and oil
Welding conditions determinede by thickness of thinnest outside piece "T"
Data for total thickness of pile-up not exceeding 4 "T" maximum ratio between two thicknesses 3 to 1
b. Truncated electrodes of RWMA Group A, Class 3 or Group B, Class 11 material. Electrodes with 3-in. spherical faces are also used.
c. Heat time–15 cycles (60HZ), Cool time–6 cycles(60HZ).
d. Minimum weld spacing is that spacing for two pieces which no special precautions need be taken to compensate for shunted current effect of adjacent welds. For three pieces increase spacing 30 percent

ficulty in keeping the weld time below the danger point of producing carbide precipitation. With thicker sections, however, the weld time and welding current must be kept within specific limits. When the thickness is greater than 0.12 in., higher carbon alloys might show carbide precipitation, which is usually undesirable in severely corrosive applications. If thick sections must be spot welded, either a low-carbon or a stabilized stainless steel should be used, depending upon the corrosive environment.

For thin sheet, a hold time of 0.5s is recommended to allow the water-cooled electrodes to quickly cool the nugget to below the carbide precipitation range. For heavier sheet, additional cooling time is needed. A hold time of about 1s is suggested for 0.125-in. thick sheet.

The appearance of spot welds made on austenitic chromium-nickel steels is usually very good. Maintaining the electrode force until the weld has cooled below the oxidizing temperature often will avoid discoloration. Where slight discoloration is present, it can easily be removed by cleaning with powdered pumice, scrubbing with alkali cleansers, or by brief immersion in a solution of 10 percent nitric acid plus 2 percent hydrofluoric acid or 50 percent hydrochloric acid at room temperature. Welds should not be exposed for long periods to these pickling solutions, or surfaces might be etched.

The nugget cross section can be determined readily by cutting through the center of a spot weld and, after sufficient polishing, etching it in a hot solution of 20 percent nitric acid plus 5 percent hydrofluoric acid. The weld nugget will be clearly outlined; the quality and approximate strength of the weld can be estimated by measuring its diameter and depth of fusion. This method can reveal the presence of any blowholes or cracks indicative of a poor weld.

The weld nugget should be free from blowholes and porosity, except for spot welds in 0.093-in. and thicker sheet and multiple thicknesses exceeding 0.188 in. that might contain a small amount of porosity unless extremely high welding pressures are used. Generally, a small amount of porosity can be tolerated in the nugget.

The coefficient of thermal expansion of austenitic stainless steel is considerably higher than that of carbon steel, and the thermal conductivity is much lower. Therefore, distortion and stress as a result of weld shrinkage are greater. When resistance welding austenitic stainless steel structures, suitable allowances must be made in the initial design to compensate for the effects of shrinkage on the completed structure.

In spot welding, the extent of weld shrinkage can be controlled by limiting the size of the individual spot welds. Generally, the safe upper limit of nugget diameter is four to five times the thickness of the thinnest sheet welded. A number of small diameter spot welds is preferred to a few of large diameter, even if the total area of weld metal is the same in both cases.

## Seam Welding

Austenitic stainless steels can be seam welded using the same types of machines used for carbon steels. As with spot welding, higher welding pressures and lower welding currents are used than for carbon steel. Precision control of the welding variables is required.

Distortion will be greater with seam welding than with spot welding because of the higher heat input. Abundant water cooling should be used to remove heat from the weld and to cool the electrode wheels. Maximum cooling is obtained if seam welding can be done under fast flowing water, but this is impractical in most applications. Copious water sprays on both sides of the seam directed at the point of welding are normally recommended.

Table 2.21 tabulates suggested electrodes and machine settings for resistance seam welding of all standard types of austenitic stainless steel sheet.

## Projection Welding

Almost all stainless steels are weldable by the projection welding process. Most of the work has been done with Types 309, 310, 316, 317, 321, 347 and 348 stainless steels, and operating data have been published, as shown in Tables 2.22 and 2.23. Other types not listed are also weldable.

The recommended type machine for projection welding is the press-type welder in which the movable electrode and welding head are moved in a straight line and guided by bearings or ways. Generally, press-type welding machines up to 300 kVA are operated by air, and between 300 and 500 kVA by either air or hydraulic power. Above 500 kVA, the machine will generally be operated by hydraulic power.

Fast follow-up is desirable when projection welding stainless steel. Therefore, air operation is recommended where possible because it is quicker in response.

The current required for projection welding

**Table 2.21**
**Welding schedules suggested for seam welding stainless steels**

| Thickness of thinnest outside piece, in.[a] | Electrode width, in., min[b] | Net electrode force, lb | Heat time, cycles (60Hz) | Cool time for maximum speed (pressure-tight), cycles (60Hz) | Maximum welding speed, in./min | Welds /in. | Welding current, A (approx) | Minimum joint overlap, in.[c] |
|---|---|---|---|---|---|---|---|---|
| 0.006 | ³⁄₁₆ | 300 | 2 | 1 | 60 | 20 | 4000 | ¼ |
| 0.008 | ³⁄₁₆ | 350 | 2 | 1 | 67 | 18 | 4600 | ¼ |
| 0.010 | ³⁄₁₆ | 400 | 3 | 2 | 45 | 16 | 5000 | ¼ |
| 0.012 | ¼ | 450 | 3 | 2 | 48 | 15 | 5600 | ⁵⁄₁₆ |
| 0.014 | ¼ | 500 | 3 | 2 | 51 | 14 | 6200 | ⁵⁄₁₆ |
| 0.016 | ¼ | 600 | 3 | 2 | 51 | 14 | 6700 | ⁵⁄₁₆ |
| 0.018 | ¼ | 650 | 3 | 2 | 55 | 13 | 7300 | ⁵⁄₁₆ |
| 0.021 | ¼ | 700 | 3 | 2 | 55 | 13 | 7900 | ³⁄₈ |
| 0.025 | ³⁄₈ | 850 | 3 | 3 | 50 | 12 | 9200 | ⁷⁄₁₆ |
| 0.031 | ³⁄₈ | 1000 | 3 | 3 | 50 | 12 | 10,600 | ⁷⁄₁₆ |
| 0.040 | ³⁄₈ | 1300 | 3 | 4 | 47 | 11 | 13,000 | ½ |
| 0.050 | ½ | 1600 | 4 | 4 | 45 | 10 | 14,200 | ⁵⁄₈ |
| 0.062 | ½ | 1850 | 4 | 5 | 40 | 10 | 15,100 | ⁵⁄₈ |
| 0.070 | ⁵⁄₈ | 2150 | 4 | 5 | 44 | 9 | 15,900 | ¹¹⁄₁₆ |
| 0.078 | ⁵⁄₈ | 2300 | 4 | 6 | 40 | 9 | 16,500 | ¹¹⁄₁₆ |
| 0.094 | ⁵⁄₈ | 2550 | 5 | 6 | 36 | 9 | 16,600 | ¾ |
| 0.109 | ¾ | 2950 | 5 | 7 | 38 | 8 | 16,800 | ¹³⁄₁₆ |
| 0.125 | ¾ | 3300 | 6 | 6 | 38 | 8 | 17,000 | ⅞ |

a.  Types of steel–301, 302, 303, 304, 308, 309, 310, 316, 317, 321 and 347
Material should be free from scale, oxides, paint, grease and oil.
Welding conditions determined by thickness of thinnest outside piece.
Data for total thickness of pile-up not exceeding 4 "T". Maximum ratio between thickness 3 to 1
b.  Electrode material RWMA Group A, Class 3 copper alloy. Face radius–3 in.
c.  For large assemblies, minimum joint overlap should be increased 30 percent.

is slightly less than that used for spot welding. It is suggested that as high a current as possible, without excessive splashing, be used in conjunction with proper electrode force. The weld time should be adjusted after the current and force are established.

If large, flat electrodes and high welding currents of short time duration are used, the welds will show less discoloration than with longer durations of lower currents. High welding current will cause quicker collapse of the projection, while the surface is protected by a large, flat, water-cooled electrode. The designs for projections are shown in Table 2.22.

## Cross Wire Welding

Cross wire welding is widely used in the fabrication of stainless steel wire into shelves, grates, baskets, guards, and many other wire products.[11]

Cut and formed wires are usually held in frames or jigs during the welding operation. Welds may be made singly, using conventional flatface tips or tips in which a V-groove has been cut in the face to provide additional contact area on the wire. For rapid production, large flat bar-type electrodes may be employed to make multiple welds. As many as 40 or more cross-wire joints may be welded at one time.

Satisfactory welds are made on stainless

11.  Additional information on cross wire welding is given in the *Welding Handbook*, Vol. 3, 7th Ed., 36-38.

### Table 2.22
### Projection welding design data

| Thickness of thinnest outside piece, in.[a] | Diameter of projection in.[b] | Height of projection in.[c] | Minimum shear strength, lb (Single projection weld) | | | Diameter of fused zone, in. min | Minimum joint overlap, in.[d] |
|---|---|---|---|---|---|---|---|
| | | | Tensile strength below 70,000 psi | Tensile strength 70,000 up to 150,000 psi | Tensile strength 150,000 psi and higher | | |
| 0.010 | 0.055 | 0.015 | 130 | 180 | 250 | 0.112 | 1/8 |
| 0.012 | 0.055 | 0.015 | 170 | 220 | 330 | 0.112 | 1/8 |
| 0.014 | 0.055 | 0.015 | 200 | 280 | 380 | 0.112 | 1/8 |
| 0.016 | 0.067 | 0.017 | 240 | 330 | 450 | 0.112 | 5/32 |
| 0.021 | 0.067 | 0.017 | 320 | 440 | 600 | 0.140 | 5/32 |
| 0.025 | 0.081 | 0.020 | 450 | 600 | 820 | 0.140 | 3/16 |
| 0.031 | 0.094 | 0.022 | 635 | 850 | 1100 | 0.169 | 7/32 |
| 0.034 | 0.094 | 0.022 | 790 | 1000 | 1300 | 0.169 | 7/32 |
| 0.044 | 0.119 | 0.028 | 920 | 1300 | 2000 | 0.169 | 9/32 |
| 0.050 | 0.119 | 0.028 | 1350 | 1700 | 2400 | 0.225 | 9/32 |
| 0.062 | 0.156 | 0.035 | 1950 | 2250 | 3400 | 0.225 | 3/8 |
| 0.070 | 0.156 | 0.035 | 2300 | 2800 | 4200 | 0.281 | 3/8 |
| 0.078 | 0.187 | 0.041 | 2700 | 3200 | 4800 | 0.281 | 7/16 |
| 0.094 | 0.218 | 0.048 | 3450 | 4000 | 6100 | 0.281 | 1/2 |
| 0.109 | 0.250 | 0.054 | 4150 | 5000 | 7000 | 0.338 | 5/8 |
| 0.125 | 0.281 | 0.060 | 4800 | 5700 | 8000 | 0.338 | 11/16 |

a.  Types of steel:
309, 310, 316, 317, 321 and 347 (Max carbon content 0.15%).
Material should be free from scale, oxides, paint, grease and oil.
Size of projection normally determined by thickness of thinner piece, and projection should be on thicker piece, where possible.
Data based on thickness of thinner sheet, and for two thicknesses only.
b.  Projection should be made on piece of higher conductivity when dissimilar metals are welded. For diameter of projection, a tolerance of ± 0.003 in. in material up to and including 0.050 in. in thickness and ± 0.007 in. in material over 0.050 in. in thickness may be allowed.
c.  For height of projection, a tolerance of ± 0.002 in. in material up to and including 0.050 in. in thickness and ± 0.005 in. in material over 0.050 in. in thickness may be allowed.
d.  Overlap does not include any radii from forming, etc. Weld should be located in center of overlap.

steel wire using short welding times, in the range of 1 to 5 cycles of 60 Hz current.

## Flash and Upset Welding

These processes are commonly used to join stainless steel wires and bars. Current requirements for stainless steel are about 15 percent less than for mild steel, and the upsetting pressures are higher, e.g., 13 to 25 ksi. The higher contact resistance and greater upsetting forces require the clamping pressures on stainless steel to be 40 to 50 percent higher than those employed on mild steel.

The flash welding process has several inherent advantages over upset welding for stainless steel. Sound joints may be made by either process by exerting sufficient pressure to close any voids at the weld interface and extrude all molten metal along with any slag or other impurities. In an upset weld, a very large upset must be formed to ensure a clean joint. In a flash weld, the oxides are easily forced out of the joint together with a small amount of plastic metal with a smaller upset. Flash welding is also faster and more reliable.

## OTHER WELDING PROCESSES

Other welding processes can also be used to join austenitic stainless steels. Electroslag

**Table 2.23**
**Manufacturing Process Date for Projection Welding Stainless Steels**

| Thickness of thinnest outside piece, in.[a] | Electrode face diameter, in.[b] | Net electrode force, lb | Weld time, cycles (60 per sec) | Hold time, cycles (60 per sec) | Approx. welding current, A |
|---|---|---|---|---|---|
| 0.014 | 1/8 | 300 | 7 | 15 | 4500 |
| 0.021 | 3/32 | 500 | 10 | 15 | 4750 |
| 0.031 | 3/16 | 700 | 15 | 15 | 5750 |
| 0.044 | 1/4 | 700 | 20 | 15 | 6000 |
| 0.062 | 5/16 | 1200 | 25 | 15 | 7500 |
| 0.078 | 3/8 | 1900 | 30 | 30 | 10,000 |
| 0.094 | 7/16 | 1900 | 30 | 30 | 10,000 |
| 0.109 | 1/2 | 2800 | 30 | 45 | 13,000 |
| 0.125 | 9/16 | 2800 | 30 | 45 | 14,000 |

a. Types of steel – 309, 310, 316, 317, 321 and 347 (max carbon content – 0.15%).
Material should be free from scale, oxides, paint, grease and oil.
Data based on thickness of thinner sheet, and for two thicknesses only. Maximum ratio between two thicknesses 3 to 1.
b. Electrode material, RWMA Group A, Class 2, or Group B, Class 12 alloy. Truncated electrodes with flat faces.

welding has been used to join heavy sections. Extensive grain coarsening occurs in the heat-affected zone of electroslag welds, and the grain size cannot be refined by heat treatment. In addition, the weld area is exposed to high temperatures for relatively long periods of time that can result in sensitization of unstabilized types. Despite this, electroslag welding has been used successfully to join coarse-grained components, such as austenitic stainless steel castings for service where grain coarsening and sensitization to corrosion are not of serious concern.

Electron beam welding and laser beam welding are sometimes used. These processes feature precise control, high welding speeds when joining thin sections, and welds with high depth-to-width ratios in thick sections. The heat-affected zones in these welds are usually much narrower than those of welds made by the conventional welding processes, and are not likely to be sensitized to corrosion.

Other suitable welding processes applicable to joining austenitic stainless steels to themselves and to dissimilar metals include friction welding, explosion welding, ultrasonic welding, diffusion welding, and high frequency welding. These processes are used less frequently than those discussed in detail here. Equipment manufacturers, or fabricators in the case of explosion welding, should be consulted for specific imformation regarding the suitability of the process for a specific application.

## POSTWELD HEAT TREATMENT

Most austenitic stainless steel weldments do not require postweld heat treatment. However, a heat treatment is sometimes used to improve corrosion resistance or to relieve stresses, or both. Corrosion problems associated with welds in austenitic stainless steels are often localized in the heat-affected zone. If the service environment is known to attack sensitized areas containing intergranular carbides or if maximum corrosion resistance is required, a postweld solution heat treatment at the annealing temperature should be used to resolution the carbides.

For solution annealing, the furnace must be capable of heating the entire weldment to the annealing temperature. Localized heating methods, such as high- or low-frequency induction or resistance heating, must be carefully analyzed because local heat treatment can easily produce an adjacent sensitized zone. Residual stresses of yield strength level can also develop if sharp thermal gradients are present.

The optimum solution annealing temperature depends on the type of stainless steel, and the soaking time is determined by the section thickness. Table 2.24 gives the recommended

**Table 2.24**
**Recommended solution annealing**
**temperatures for austenitic stainless steels**

| Type | Temperature °F |
|------|----------------|
| 201, 202, 301, 302, 303, 304, 304L, 305, 308 | 1850-2050 |
| 309, 309S, 316, | 1900-2050 |
| 316L, 317L | 1900-2025 |
| 317 | 1950-2050 |
| 321 | 1750-1950 |
| 347, 348 | 1800-1950 |

annealing temperatures for standard austenitic stainless steels. A short time at the annealing temperature is preferred to avoid grain growth. As a general rule, soaking time at temperature should be 3 minutes for each 0.1 in. of thickness. The weldment must be cooled rapidly and uniformly, at least through the temperature range of approximately 1650° to 800°F, to retain carbon in solid solution. Water quenching or spraying is necessary for thick sections, while air cooling is suitable for thin sections.

For many weldments, a solution heat treatment is not appropriate because of problems that might be encountered in performing the operation. The following difficulties associated with this heat treatment must be addressed:

(1) Large annealing furnaces, high-speed handling equipment, and adequate cooling facilities are needed.

(2) Application of the cooling medium must be uniform to prevent high residual stresses and distortion.

(3) Oxide scale will form on the heat-treated steel surfaces unless a protective atmosphere is used. It has to be removed by machining, grinding, sandblasting, or acid pickling. This operation could be detrimental if the weldment has accurately premachined surfaces, such as flange faces.

(4) Thin weldments will sag at the high temperature unless adequately supported.

(5) Adequate precautions must be taken to assure that repair welding is not done on a solution heat-treated weldment if the properties obtained by the heat treatment will be destroyed.

While thermal treatments below the annealing temperature can reduce residual stresses, they can also destroy several essential or desirable properties of austenitic stainless steels. Even short-time exposure to temperatures in the 800° to 1650°F range might sensitize and reduce corrosion resistance of the steel unless it is a low-carbon or stabilized type. There is some evidence that even low-carbon and stabilized types can suffer carbide precipitation when exposed to this temperature range for prolonged periods.

For austenitic stainless steel welds intended for service above 900°F, postweld stress-relieving is often performed in the range of 1450° to 1550°F. This heat treatment can significantly reduce residual stresses in welds, but various effects on properties have been reported. Any phase transformations that occur during the stress-relieving are not considered to be detrimental because the weldments will be exposed to high temperatures in service.

Some fabricators have stress-relieved stainless steel welds between 400° and 800°F, and claimed improved dimensional stability. It appears that such a low-temperature treatment might reduce and equalize high peak stresses, but complete stress-relieving is not possible in this temperature range. However, any heating operation that does not exceed 800°F is safe, and any reduction in residual stresses will reduce the susceptibility to certain corrosion mechanisms.

Postweld heat treatment of austenitic stainless steel can be complicated when the weldment is made partly from carbon or low alloy steel and partly from austenitic stainless steel. A carbon steel pressure vessel lined with stainless steel is an example. Many steels require specific minimum heat treating cycles without defined maximum conditions. The upper heat treating limits are possibly most damaging to the stainless steel components. Thus, it is essential that for any dissimilar steel fabrication, a heat-treating cycle suitable for both steels must be selected. Prime consideration must be given to the tempering of a low alloy steel while preserving the corrosion resistance and ductility of the stainless steel. Stress-relieving of dissimilar steel weldments is difficult since differences in coefficients of thermal expansion can re-establish high residual stresses during cooling, regardless of cooling rate or procedure.

# PRECIPITATION-HARDENING STAINLESS STEELS

## COMPOSITIONS, TYPES, AND PROPERTIES

Precipitation-hardening (PH) stainless steels have the ability to develop high strength with reasonably simple heat treatments. In addition, these steels have good corrosion and oxidation resistance without loss of ductility and fracture toughness associated with steels of comparable strength levels. Hardening of the steels is achieved by martensite formation, by precipitation-hardening, or by both mechanisms. The nominal compositions of typical precipitation-hardening stainless steels are listed in Table 2.25.

Precipitation-hardening is promoted by one or more alloying elements, such as copper, titanium, columbium, and aluminum. These additions are dissolved during solution annealing or austenite conditioning, and produce normally submicroscopic precipitates during an aging heat treatment that increases the hardness and strength of the matrix.

Precipitation-hardening stainless steels are grouped into three types depending on the structure and behavior of the steel when it is cooled from the appropriate solutioning (austenitizing) temperature. These types are martensitic, semiaustenitic, and austenitic.

The martensitic types are so named because their compositions are balanced to provide a martensitic structure after cooling from the solutioning temperature. Additional strength is obtained by aging.

Semiaustenitic types remain austenitic when initially cooled from the appropriate annealing temperature. Then, the steel is re-heated to condition the structure so that it transforms to martensite on cooling to room temperature. Subsequent aging produces additional strength.

Austenitic precipitation-hardening stainless steels remain austenitic after cooling from the appropriate solutioning temperature. Strengthening is obtained by aging the austenitic structure.

## Metallurgical Properties

*Martensitic Types.* Several martensitic types of PH stainless steels are available. Some types can be classified as moderate strength with tensile strengths of less than 200 ksi. Other types are considered high strength with tensile strengths exceeding 200 ksi. Typical precipitation-hardening steels of this type are shown in Table 2.25.

Metallurgically, these steels are similar. Most of them are solution heat-treated at about 1900°F, where the structure is predominantly austenite. Upon quenching, the austenite transforms to martensite, usually between 300° and 200°F. Some types of these steels, such as 17-4 PH and Stainless W, contain a few stringers of ferrite, in a martensitic matrix. Other types, such as 15-5 PH and Custom 450, are essentially ferrite free after quenching.

After quenching to martensite, the steels are strengthened by a precipitation mechanism during an aging heat treatment. Typical heat treatments for several common martensitic grades are shown in Table 2.26. The mechanical properties of the steel depend primarily on the aging temperature and time at temperature. A wide variation in properties is available to satisfy specific user requirements.

*Semiaustenitic Types.* The semiaustenitic types of PH stainless steel are the most complex from a metallurgical standpoint. In the solution treated or annealed condition, they are essentially austenitic, although they usually contain 5 to 20 percent delta ferrite. The ferrite remains untransformed through the subsequent heat treatments.

When the semiaustenitic PH steels are cooled rapidly from the annealing temperature, they remain austenitic because the martensitic transformation temperature is below room temperature. In the annealed condition, the steels are similar to Type 301 stainless steel with respect to formability. After fabrication, the steels can be hardened by three simple heat treatment steps: (1) austenite conditioning, (2) martensitic transformation, and (3) age hardening. Heat treatments for typical semiaustenitic PH stainless steels are given in Table 2.26.

In the first step, the steel is *conditioned* by heating it in the range of about 1350° to 1750°F and air cooled to precipitate carbon from solution in the form of chromium carbides. The removal of dissolved carbon raises the martensitic transformation temperature. Hence, cooling from the conditioning temperature to room temperature or below causes the steel to transform to martensite. Then the steel is aged to stress-relieve and temper the martensite for increased toughness, ductility, and corrosion resistance, and also to provide additional hardening by precipitation of an intermetallic phase.

**Table 2.25**
**Nominal compositions of typical precipitation-hardening stainless steels**

| Type | Designation[a] | UNS No. | Nominal composition, weight percent | | | | | | | |
|---|---|---|---|---|---|---|---|---|---|---|
| | | | C | Mn | Si | Cr | Ni | Mo | Al | Other elements |
| Martensitic (moderate strength) | 17-4 PH | S17400 | 0.04 | 0.30 | 0.60 | 16.0 | 4.2 | — | — | 3.4 CU, 0.25 Cb |
| | 15-5 PH | S15500 | 0.04 | 0.30 | 0.40 | 15.0 | 4.5 | — | — | 3.4 Cu, 0.25 Cb |
| | Custom 450 | S45000 | 0.03 | 0.25 | 0.25 | 15.0 | 6.0 | 0.8 | — | 1.5 Cu, 0.3 Cb |
| | Stainless W | S17600 | 0.06 | 0.50 | 0.50 | 16.75 | 6.25 | — | 0.2 | 0.8 Ti |
| Martensitic (high strength) | PH 13-8 Mo | S13800 | 0.04 | 0.03 | 0.03 | 12.7 | 8.2 | 2.2 | 1.1 | — |
| | Custom 455 | S45500 | 0.03 | 0.25 | 0.25 | 11.75 | 8.5 | — | — | 2.5 Cu, 1.2 Ti, 0.3 Cb |
| Semi-austenitic | 17-7 PH | S17700 | 0.07 | 0.50 | 0.30 | 17.0 | 7.1 | — | 1.2 | — |
| | Ph 15-7 Mo | S15700 | 0.07 | 0.50 | 0.30 | 15.2 | 7.1 | 2.2 | 1.2 | — |
| | PH 14-8 Mo | S14800 | 0.04 | 0.02 | 0.02 | 15.1 | 8.2 | 2.2 | 1.2 | — |
| | AM-350 | S35000 | 0.10 | 0.75 | 0.35 | 16.5 | 4.25 | 2.75 | — | — |
| | AM-355 | S35500 | 0.13 | 0.85 | 0.35 | 15.5 | 4.25 | 2.75 | — | — |
| Austenitic | A-286 | K66286 | 0.05 | 1.45 | 0.50 | 14.75 | 25.25 | 1.30 | 0.15 | 0.30 V, 2.15 Ti, 0.005 B |
| | 17-10 P | — | 0.10 | 0.60 | 0.50 | 17.0 | 11.0 | — | — | 0.30 P |
| | HNM | — | 0.30 | 3.50 | 0.50 | 18.50 | 9.50 | — | — | 0.25 P |

a. Some of these designations are registered trademarks.

**Table 2.26**
**Typical heat treatments for precipitation-hardening stainless steels**

| Type of steel | Austenite conditioning | | Aging treatment[b] | |
| | Temperature, °F | Quenching media[a] | Temperature, °F | Time, h |
| --- | --- | --- | --- | --- |
| **Martensitic** | | | | |
| 17-4 PH | 1900 | O, A | 900 or 925-1150 | 1 4 |
| 15-5 PH | 1900 | W | 900 or 1025-1150 | 1 4 |
| Custom 450 | 1900 | W | 900-1150 | 4 |
| Custom 455 | 1525 | W | 900-1050 | 4 |
| Stainless W | 1900 | A | 950-1050 | 0.5 |
| **Semiaustenitic** | | | | |
| 17-7 PH, PH 15-7 Mo | 1750 | A | (1)–90 (2) 950-1100 | 8 1 |
| | 1400 | A | 1050-1100 | 1.5 |
| PH 15-7 Mo | 1750 | A | (1)–90 (2) 950-1050 | 8 1 |
| AM350, AM355 | 1710 | A | (1)–100 (2) 850-1000 | 3 3 |
| **Austenitic** | | | | |
| A-286 | 1800 | O | 1325 | 16 |
| 17-10 P | 2050 | W | 1300 | 24 |
| HNM | 2050 | O, A | 1350 | 16 |

a. O–oil, A–air, W–water
b. Air cool

*Austenitic Types.* These alloys retain a stable austenitic structure after quenching from the solutioning temperature and even after substantial cold working. Following solution treatment, hardening is achieved by aging at about 1300°F for several hours. During aging, elements such as aluminum, titanium, and phosphorus form intermetallic compounds that significantly increase the strength of the steel. Typical heat treatments for three austenitic PH stainless steels are also shown in Table 2.26.

## Corrosion Resistance

As a group, the precipitation-hardening stainless steels have corrosion resistance comparable to the more common austenitic stainless steels. The corrosion resistance can vary with the steel composition and microstructure for any given set of service conditions and corrosive environment. For example, the martensitic PH stainless steels are generally superior in corrosion resistance to quenched-and-tempered conventional martensitic stainless steels and comparable to the conventional austenitic stainless steels. Under low-stress conditions, their corrosion resistance is greatest in the fully hardened condition, but it decreases slightly as the aging temperature is increased. The martensitic types exhibit good resistance to stress-corrosion cracking, particularly when aged at about 1025°F and higher.

Even when PH stainless steels are not stressed significantly, overaging can decrease their corrosion resistance in many environments. Other factors, such as increased ferrite levels in an austenitic matrix, or lower levels of chromium in the matrix might reduce the corrosion resistance, as is often the case with conventional stainless steels. Thus, the corrosion resistance of the base metal or a welded joint is highly dependent on the service conditions, the environment, the microstructure, and the composition of the PH stainless steel.

## High Temperature Properties

All precipitation-hardenable stainless steels tend to become embrittled after exposure for thousands of hours at temperatures above about 550°F. In general, long-time exosure in the 700° to 800°F range results in sharply increased tensile and yield strengths and reduced fracture toughness. These steels are generally limited to a maximum service temperature of about 600°F for long-time exposure.

For short-time service, many of the martensitic and semiaustenitic PH stainless steels have suitable mechanical properties for service temperatures up to 900°F. For the austenitic types, good corrosion resistance can be maintained up to about 1500°F. The choice of a steel for a particular maximum service temperature depends upon the service environment and the stress requirements.

## HEAT TREATMENT FOR WELDING

The sequence of fabrication and heat treatment operations chosen to achieve the properties required in a PH stainless steel weldment can be varied to place the steel in the preferred condition for welding. After welding, the maximum mechanical properties and corrosion resistance can be achieved by a solution heat treatment followed by aging. In some cases, solution heat treatment might not be desirable because of possible distortion, cracking, or other limitations. In many cases, only an aging treatment after welding is needed. In brazing, the brazing cycle may be used to accomplish part of the heat treatment.

For the martensitic types of PH steel, thin sections are usually arc welded in the annealed condition. Thick sections or highly restrained joints are often welded in the overaged condition.

Resistance welded joints are usually made in aged material. Heat from welding causes the heat-affected zone immediately adjacent to the weld to be austenitized and transformed to martensite on cooling. This zone is effectively soft because of the low carbon content of the steel, and underbead cracking is not a problem. The heating time is normally too short to cause hardening at greater distances from the weld. Suitable weld joint properties can be obtained by welding the steel in the hardened condition, and then giving the welded joint a postweld aging treatment.

Many semiaustenitic PH steels are arc welded in the solution-treated or the annealed condition, although they can be welded in the hardened condition. Resistance spot or seam welding is often performed on transformed or aged sheet metal. Regardless of the condition of the base metal before welding, the heat of welding causes a narrow heat-affected zone to transform to and remain austenitic after cooling. Cracking in the heat-affected zone is not a problem, but an austenite conditioning treatment followed by martensitic transformation and aging must be used after welding if maximum mechanical properties are desired in the weldment.

The austenitic precipitation-hardened stainless steels are difficult to weld, and some types are considered unweldable because of cracking problems. Welding is usually performed on the steel in the solution-treated condition. However, the heat-affected zone is susceptible to hot cracking. Although the heat-affected zones are expected to transform to austenite on cooling, grain-boundary eutectics can form and cause cracking. Welding with minimum restraint and low heat input helps to prevent cracking, but the practical difficulties of welding increase with the thickness of the joint being welded.

## WELDABILITY

Service requirements, the geometry of the weldment, and the conditions under which the welds must be made affect the weldability of a PH stainless steel. As stated before, a complete cycle of heat treatment, including solution annealing, is desirable after welding, but this is not always practical because of size or geometry. If the weldment cannot receive a complete heat treatment, the components should be solution annealed prior to welding. The weldment is then aged prior to service. The recommendations of the steel producer should be followed.

The PH stainless steels do not require preheat. The martensitic and semiaustenitic types are not subject to cracking. However, the austenitic types suffer from heat-affected zone hot cracking that makes them very difficult to weld. The PH stainless steels are less ductile and more notch sensitive than the conventional austenitic stainless steels. Therefore, sharp stress concentrations should be avoided in the weldment design and at welded joints.

The weldable types of PH stainless steels can be joined by any of the standard welding processes used to join the conventional austenitic stainless steels. Gas tungsten arc, gas metal arc, and plasma arc welding are well suited. Shielded metal arc welding can be used but high joint strengths might not be obtainable with some of the steels because matching electrodes are not available. These steels can also be joined by resistance welding processes.

## FILLER METALS

If maximum strength properties and corrosion resistance are required in the weldment, a filler metal similar in composition to the base metal should be used. The weldment is then given the appropriate strengthening heat treatment. However, this procedure is not recommended with the austenitic PH stainless steels because of the cracking problems. Nickel-base filler metals are normally used for welding these steels, although conventional austenitic stainless steel filler metals can be used.

If high-strength weld metal is not required because the applied stresses are low, one of the conventional austenitic stainless steel or nickel-base filler metals can be used. Such filler metals can also be used to join the PH stainless steels to other steels. Suggested filler metals for welding several PH stainless steels are given in Table 2.27. For those steels not listed, the filler metal composition should be the same as the base metal, unless high-strength welds are not required.

## ARC WELDING

### Shielded Metal Arc Welding

Martensitic and semiaustenitic PH stainless steels, such as 17-4 PH, 17-7 PH, AM350, and AM355, can be welded with covered electrodes because they do not contain aluminum or titanium. Normally, electrodes similar in composition to the base metal are used, and welds can be made in all positions. Type AMS 5827B (17-4 PH) electrodes can be used to weld 17-7 PH steel,

and reasonable heat-treatment response can be obtained if the weld deposit is highly diluted with base metal.

The electrodes must be dry, and stored and handled in the manner described previously for other stainless steel covered electrodes. Welding conditions suitable for conventional stainless steels are generally applicable for joining the PH types. A short arc length should be used to minimize oxidation and loss of chromium.

Generally, the welds will develop suitable strength and hardness after postweld heat treatment. Where high strength welds are not required, standard austenitic stainless steel electrodes such as E308 or E309 can be used.

### Gas Tungsten Arc Welding

Manual and automatic gas tungsten arc welding (GTAW) processes are frequently used for joining PH stainless steels in thicknesses up to about 0.25 inch. Although thicker sections can be welded by GTAW, other processes, such as gas metal arc welding, are faster and more economical.

Generally, direct current, electrode negative from a power source with a drooping volt-ampere characteristic is used for welding. Alternating current is sometimes used to weld those steels containing aluminum to take advantage of the arc cleaning action.

Argon is usually used for shielding during manual welding. Helium or helium-argon mixtures are sometimes used for automatic welding to take advantage of higher travel speeds afforded by the hotter helium arc. Conditions used for joining the precipitation hardening steels are similar to those used for joining the conventional austenitic stainless steels. Good fixturing and assembly procedures are required. The back side or root of the joint must also be protected from the atmosphere by inert gas shielding during welding.

### Gas Metal Arc Welding

Gas metal arc welding with spray transfer is used to join sections thicker than about 0.25 inch because deposition rates are higher and welding can be done at higher speeds than with the GTAW process. Welding procedures are similar to those for conventional austenitic stainless steels.

The shielding gas is generally argon with 1- to 2-percent oxygen added for arc stability. Mixtures of argon and helium are employed if a hotter arc is desired. A small oxygen addition

**Table 2.27**
**Suggested filler metals for welding precipitation-hardening stainless steels**

| Designation | UNS No. | Covered electrodes | Bare welding wire | Dissimilar PH stainless steels |
|---|---|---|---|---|
| **Martensitic Types** | | | | |
| 17-4 PH and 15-5 PH | S 17400 S 15500 | AMS 5827B (17-4 PH) or E 308 | AMS 5826 (17-4 PH) or ER 308 | E or ER309, E or ER309Cb |
| Stainless W | S 17600 | E 308 or ENiMo-3[a] | AMS 5805C (A-286) or ERNiMo-3[b] | E or ERNiMo-3, E or ER 309 |
| **Semiaustenitic Types** | | | | |
| 17-7 PH | S 17700 | AMS 5827B (17-4 PH), E308, or E309 | AMS 5824A (17-7 PH) | E or ER 310, ENiCrFe-2, or ERNiCr-3 |
| PH 15-7 Mo | S 15700 | E308 or E309 | AMS 5812C (PH 15-7 Mo) | E or ER309, E or ER310 |
| AM350 | S 35000 | AMS 5775A (AM350) | AMS 5774B (AM 350) | E or ER308, E or ER309 |
| AM355 | S 35500 | AMS 5781A (AM355) | AMS 5780A (AM355) | E or ER308, E or ER309 |
| **Austenitic Type** | | | | |
| A-286 | K 66286 | E309 or E310 | ERNiCrFe-6 or ERNiMo-3 | E or ER309, E or ER310 |

a. See AWS A5.11-76, *Specification for Nickel amd Nickel Alloy Covered Welding Electrodes.*
b. See AWS A5.14-76, *Specification for Nickel and Nickel Alloy Bare Welding Rods and Electrodes.*

can be added to provide a stable arc, but some aluminum or titanium can be lost from certain filler metals during transfer across the arc as a result of oxidation. Response of the weld metal to heat treatment might be less because of this action.

For flat position welding, spray transfer is usually preferred. For other welding positions, short-circuiting or pulsed-spray transfer can be employed using argon or an argon-helium mixture with a small addition of oxygen or carbon dioxide.

## Submerged Arc Welding

Submerged arc welding (SAW) can be employed to join thick sections, usually thicker than 0.5 inch. If full strength welds are not required,

a conventional austenitic stainless steel electrode, such as ER308, ER310, and ER316, can be used with conventional stainless steel fluxes. To provide a proper weld metal composition for response to postweld heat treatment, special fluxes must be used. If special fluxes are not used, the weld metal will probably not respond to heat treatment. This is particularly true for aluminum-bearing electrodes where aluminum is lost through metal-slag reactions. The stainless flux manufacturers should be consulted for recommendations.

Welding conditions must be rigorously controlled. Voltage, current, and travel speed variations will influence the amount of flux melted and the resulting weld metal composition. Most welding is done using direct current, electrode positive. Alternating current is sometimes used for moderate penetration and good arc stability. The electrode and flux supplier should be consulted for suggested welding procedures.

### Postweld Heat Treatment

As discussed previously, the appropriate postweld heat treatment must be determined during material selection, design of the weldment, and establishment of welding procedures. Filler metal selection also must be considered. In most cases, a postweld aging heat treatment is recommended as a minimum to improve weld joint strength and reduce welding stresses. The appropriate postweld heat treatment is the one recommended to harden the base metal.

## RESISTANCE WELDING

Resistance spot welding has been used to join several PH stainless steels including 17-7

PH, A-286, PH 15-7 Mo, AM350, and AM355. Alloys such as 17-4 PH, 17-10 P, and HNM are not widely used for applications requiring spot welding. Welding equipment and schedules are usually based on those used to spot weld conventional austenitic stainless steels. Because PH stainless steels are generally welded in the aged condition, higher welding pressures than those used for softer austenitic stainless steels might be necessary. Weld time should be as short as possible.

Resistance seam welding has also been used to join PH stainless steels, particularly 17-7 PH. Again, electrode force should be increased above that required for seam welding conventional austenitic stainless steel. Otherwise, the welding schedules are similar.

Flash welding has been used successfully. The upset pressure must be higher than that required for conventional steels. A postweld heat treatment might be appropriate for applications requiring maximum joint strength.

# BRAZING AND SOLDERING

The tenacious oxide film that imparts corrosion resistance to stainless steels is more difficult to remove by flux or reducing atmospheres than the oxide film that forms on carbon steel. Therefore, stainless steels require more stringent preparation and process control for brazing and soldering to assure good wettability and a strong bond with the filler metal.

## BRAZING

The brazing of stainless steels, especially the austenitic types, is performed as a routine operation in industry.[12] A generally greater degree of process control is required for brazing stainless steels than for carbon steels. This is necessary because carbon steel and stainless steel

___

12. Refer to the *Welding Handbook*, Vol. 2, 7th Ed., 370-438, and the AWS *Brazing Manual* for additional information on brazing.

are chemically different, and the service environments for stainless steel are generally more demanding than those for carbon steel. Successful brazing of stainless steel components depends upon proper recognition of the characteristics of the various types of stainless steel, and rigid adherence to certain process controls required by these characteristics.

### Base Metals

*Martensitic stainless steels* require a brazing cycle compatible with the heat treatment planned for the completed brazement. This is usually done by selecting a brazing filler metal that permits austenitization of the base metal at the brazing temperature. Martensitic stainless steels are austenitized in the temperature range of 1700° to 1950°F, depending upon the particular steel and the tempering temperature. To secure proper transformation of the austenite to martensite, rather than austenite, rapid cooling from the brazing temperature is needed. The remelt temperature of the brazed joint must be well above the tempering temperature of the steel to avoid failure during the tempering cycle. When the brazement does not require a hardening heat treatment, these steels can be brazed with any suitable filler metals.

*Ferritic stainless steels* can be brazed over a wide range of temperatures because they are not hardenable by heat treatment. These stainless steels are subject to grain growth when exposed to high temperatures for long times. This fact should be considered when selecting the brazing filler metal process and procedures for a particular application of a ferritic stainless steel.

*Austenitic stainless steel* assemblies are commonly fabricated by both torch and furnace brazing. A major precaution in brazing these steels is that the nonstabilized types, such as Types 302 and 304, are subject to sensitization and consequent loss of corrosion resistance. The extent of sensitization in these steels is governed by the time of exposure to temperatures between 800° and 1600°F.

Sensitization can be minimized by using a very short brazing thermal cycle. With short-time brazing cycles, the nonstabilized types can be brazed without serious loss of corrosion resistance. Types 321, 347 and 348 stabilized stainless steels can be brazed without danger of impairing their corrosion resistance. The low-carbon grades, such as Type 304L, are relatively insensitive to carbide precipitation, and can be brazed with essentially the same filler metals and heating cycles used for the stabilized types.

*Precipitation-hardening stainless steels* require a brazing thermal cycle that is compatible with the heat treatment used to harden a specific steel. These heat treatments vary rather widely, and specific brazing procedures are required for a particular steel. Suppliers of these alloys and of brazing filler metals should be consulted for recommended brazing procedures.

## Filler Metals

A wide variety of filler metals are commercially available for brazing stainless steel assemblies. Some are standard types and others are proprietary. Selection of a filler metal for a stainless steel brazement depends upon the intended service. Commercial brazing filler metals have copper, silver, nickel, cobalt, platinum, palladium, manganese, or gold as the base or added as alloying elements. A convenient means of grouping these filler metals for brazing stainless steels is based on service temperature.

For service temperatures up to about 700°F, the silver base (BAg-X) filler metals are used extensively. The copper (BCu-X) types are suitable for service temperatures up to about 800°F. Filler metals suitable for 800° to 1000°F service are the copper-manganese-nickel alloys, typically 53Cu-37Mn-10Ni. For applications above 1000°F, nickel-base filler metals or other types containing gold, with additions of copper, nickel, and palladium are used. Brazements for service above 800°F are generally furnace brazed in a high-purity reducing or inert atmosphere, or in a vacuum.

Of the copper-base filler metals, only the BCu-X classifications are recommended for brazing stainless steels. High-purity reducing atmospheres of low moisture content are generally used when brazing with these filler metals. The RBCuZn-X and BCuP-X filler metal types are not recommended for brazing stainless steels.

Silver-base filler metals generally used to braze stainless steels are listed in Table 2.28. Type BAg-2 is particularly useful when the joint clearance cannot be closely controlled. However, this filler metal should not be used with long heating times because it tends to liquate. Where improved corrosion resistance is needed, Types BAg-3 and BAg-4 are recommended because of their nickel content. Types BAg-5 and BAg-6 are general purpose filler metals with high brazing temperatures for applications where cadmium is prohibited. Type BAg-7 is used for fabricating food-handling equipment because it is white in color and free of cadmium. The silver-copper eutectic, Type BAg-8, is often used for vacuum and atmosphere brazing where freedom from volatile cadmium and zinc is required. This composition with an addition of 0.25 percent lithium (BAg-8a) has better wettability on stainless steels, particularly the PH types. Type BAg-19 is basically sterling silver with lithium to promote self-fluxing behavior, and is used extensively for furnace brazing. Also, its brazing temperature is relatively high.

Typical standard nickel- and gold-base filler metals for stainless steels are listed in Table 2.29. Other nonstandard or proprietary alloys are also available commercially. Nickel-base filler metals are used primarily where extreme heat and corrosion resistance are required, as in jet and rocket engines, chemical processing equipment, and nuclear reactor components. These filler metals are normally supplied as powders, but can be obtained as sintered or cast rods, preforms, and plastic-bonded sheet, wire, and tape. Filler met-

Table 2.28
Silver-base brazing filler metals for stainless steel

| AWS Classification | Nominal composition, percent | | | | | | | | Melting temperature, °F | Brazing temperature range, °F | Color | Characteristics |
|---|---|---|---|---|---|---|---|---|---|---|---|---|
| | Ag | Cu | Zn | Cd | Ni | Sn | Li | | | | | |
| BAg1 | 45 | 15 | 16 | 24 | — | — | — | 1145 | 1145-1400 | Whitish yellow | Free-flowing |
| BAg1a | 50 | 15.5 | 16.5 | 18 | — | — | — | 1175 | 1175-1400 | Whitish yellow | Free-flowing |
| BAg2 | 35 | 26 | 21 | 18 | — | — | — | 1295 | 1295-1550 | Light yellow | Good for nonuniform clearance |
| BAg3 | 50 | 15.5 | 15.5 | 16 | 3.0 | — | — | 1270 | 1270-1500 | Whitish yellow | Retards corrosion at joint |
| BAg4 | 40 | 30 | 28 | — | 2.0 | — | — | 1435 | 1435-1650 | Light yellow | Flows better than BAg3 |
| BAg5 | 45 | 30 | 25 | — | — | — | — | 1370 | 1370-1550 | Light yellow | Not free-flowing, cadmium-free, useful in food industry |
| BAg6 | 50 | 34 | 16 | — | — | — | — | 1425 | 1425-1600 | Ligh yellow | Similar to BAg5 |
| BAg7 | 56 | 22 | 17 | — | — | 5 | — | 1205 | 1205-1400 | White | Good color match |
| BAg8 | 72 | bal. | — | — | — | — | — | 1435 | 1435-1650 | White | Wetting is slow |
| BAg8a | 72 | bal. | — | — | — | — | 0.25 | 1410 | 1410-1600 | White | For furnace brazing PH stainless steels |
| BAg13 | 54 | bal. | 5 | — | 1 | — | — | 1575 | 1575-1775 | White | Useful to 700°F |
| BAg18 | 60 | bal. | — | — | — | 10 | — | 1325 | 1325-1550 | White | Wets well for brazing PH stainless steels |
| BAg19 | 92.5 | bal. | — | — | — | — | 0.25 | 1635 | 1610-1800 | White | Good for furnace brazing |
| BAg21 | 63 | 28.5 | — | — | 2.5 | 6 | — | 1475 | 1475-1650 | — | Immune to crevice corrosion |

## Table 2.29
## Typical nickel- and gold-base filler metals for brazing stainless steel[a]

### Nickel-base

| AWS Classification | Cr | B | Si | Fe | C | P | S | Al | Ti | Mn | Cu | Zr | Ni | Other elements total | Solidus, °F | Liquidus, °F | Brazing temp. range, °F |
|---|---|---|---|---|---|---|---|---|---|---|---|---|---|---|---|---|---|
| BNi-1 | 13.0-15.0 | 2.75-3.50 | 4.0-5.0 | 4.0-5.0 | 0.6-0.9 | 0.02 | 0.02 | 0.05 | 0.05 | – | – | 0.05 | Bal | 0.50 | 1790 | 1900 | 1950-2200 |
| BNi-1a | 13.0-15.0 | 2.75-3.50 | 4.0-5.0 | 4.0-5.0 | 0.06 | 0.02 | 0.02 | 0.05 | 0.05 | – | – | 0.05 | Bal | 0.50 | 1790 | 1970 | 1970-2200 |
| BNi-2 | 6.0-8.0 | 2.75-3.50 | 4.0-5.0 | 2.5-3.5 | 0.06 | 0.02 | 0.02 | 0.05 | 0.05 | – | – | 0.05 | Bal | 0.50 | 1780 | 1830 | 1850-2150 |
| BNi-3 | – | 2.75-3.50 | 4.0-5.0 | 0.5 | 0.06 | 0.02 | 0.02 | 0.05 | 0.05 | – | – | 0.05 | Bal | 0.50 | 1800 | 1900 | 1850-2150 |
| BNi-4 | – | 1.5-2.2 | 3.0-4.0 | 1.5 | 0.06 | 0.02 | 0.02 | 0.05 | 0.05 | – | – | 0.05 | Bal | 0.50 | 1800 | 1950 | 1850-2150 |
| BNi-5 | 18.5-19.5 | 0.03 | 9.75-10.50 | – | 0.10 | 0.02 | 0.02 | 0.05 | 0.05 | – | – | 0.05 | Bal | 0.50 | 1975 | 2075 | 2100-2200 |
| BNi-6 | – | – | – | – | 0.10 | 10.0-12.0 | 0.02 | 0.05 | 0.05 | – | – | 0.05 | Bal | 0.50 | 1610 | 1610 | 1700-2000 |
| BNi-7 | 13.0-15.0 | 0.01 | 0.10 | 0.2 | 0.08 | 9.7-10.5 | 0.02 | 0.05 | 0.05 | 0.04 | – | 0.05 | Bal | 0.50 | 1630 | 1630 | 1700-2000 |
| BNi-8 | – | – | 6.0-8.0 | – | 0.10 | 0.02 | 0.02 | 0.05 | 0.05 | 21.5-24.5 | 4.0-5.0 | 0.05 | Bal | 0.50 | 1800 | 1850 | 1850-2000 |

### Gold-base

| AWS Classification | Au | Cu | Pd | Ni | Other elements, total | Solidus, °F | Liquidus, °F | Brazing temp. range, °F |
|---|---|---|---|---|---|---|---|---|
| BAu-1 | 37.0-38.0 | Bal | – | – | 0.15 | 1815 | 1860 | 1860-2000 |
| BAu-2 | 79.5-80.5 | Bal | – | – | 0.15 | 1635 | 1635 | 1635-1850 |
| BAu-3 | 34.5-35.5 | Bal | – | 2.5-3.5 | 0.15 | 1785 | 1885 | 1885-1995 |
| BAu-4 | 81.5-82.5 | – | – | Bal | 0.15 | 1740 | 1740 | 1740-1840 |
| BAu-5 | 29.5-30.5 | – | 33.5-34.5 | 35.5-36.5 | 0.15 | 2075 | 2130 | 2130-2250 |

a. Single values are maximum.

als BNi-1, -2, -3, and-4 tend to erode thin sheet as a result of interaction with the base metal. Time at brazing temperature should be minimized when these filler metals are used.

The boron-free filler metals BNi-5, -6, and -7 are used for nuclear reactor components, where boron cannot be tolerated because of its absorption of neutrons. The BNi-5 filler metal has the highest melting range of those listed.

Filler metals based on gold, platinum, and palladium, such as gold-nickel, gold-nickel-chromium, copper-platinum, silver-palladium-manganese, and palladium-nickel-chromium can be used for brazing heat- and corrosion-resistant assemblies if the high cost of such alloys can be justified by special service requirements.

## Interactions During Brazing

Interface corrosion of brazed joints in Type 430 stainless steel is a problem with some silver-base metals. The corrosion is apparently caused by electrochemical action, even in the presence of ordinary tap water. For most stainless steels, a small percentage of nickel in the filler metal will prevent interface corrosion. For Type 430 brazements, nickel-bearing BAg-21 filler metal that is free of zinc is preferred. Silver-base filler metals containing both nickel and zinc are not completely effective in reducing the corrosion.

Austenitic stainless steels are subject to intergranular penetration and stress-corrosion cracking by molten brazing filler metal, especially the silver-base types. The more highly alloyed stainless steels, such as Type 310, appear to be most susceptible. The interaction takes place during brazing when the base metal is under residual stress, or stress from applied loading. The components should be in the annealed condition prior to brazing, and properly assembled and supported to avoid stress during the brazing cycle. Joint location in a brazement should be chosen to avoid stresses from differential thermal expansion.

Many of the brazing alloys intended for high-temperature service, such as Types BNi-1 and BNi-3, exhibit a tendency toward very rapid interalloying with the base metal during brazing. This generally occurs either as intergranular penetration by certain elements in the filler metal, such as boron, silicon, or carbon, into the base metal or as dissolution of the base metal by the filler metal. Interalloying can embrittle or severely erode the base metal, or both. In many cases, embrittlement is not considered to be par-ticularly harmful because the brazing filler metal often has low ductility, and the joint is not expected to withstand severe deformation in service. If the sections being brazed are relatively thin, such as honeycomb structures, penetration and erosion can be very damaging because it deteriorates a large percentage of the base metal thickness. In such instances, the aggressive elements can easily penetrate completely through the base metal.

Sometimes, interalloying can be controlled by brazing at the minimum temperature and time at temperature, and also by restricting the amount of brazing filler metal used in the joint. Gold- and palladium-base filler metals do not normally cause this problem.

## Joint Design

Braze joint designs used for stainless steel brazements are similar to those used with other metals. They are basically lap and butt joints with variations to improve stress distribution. Recommended joint clearance at brazing temperature depends upon the brazing filler metal being used, ranging from 0.001 to 0.005 inch.

For wide joint clearances, similar brazing filler metals or a filler metal and a compatible metal, with different melting temperatures, can be combined to produce a sluggish-flowing filler metal that will stay in a wide clearance, and yet wet the stainless steel satisfactorily. However, the joint strength will be lower with a wide joint clearance.

## Processes and Equipment

Stainless steels can be brazed with any of the various brazing processes. A large volume of controlled-atmosphere brazing is performed on stainless steels, and the success of this type of brazing can be attributed to reliable atmosphere and vacuum furnaces. The primary requirements are that the furnaces have good temperature control and be capable of fast heating and cooling rates. All gases used in atmosphere furnaces must be of high purity.

Commercial vacuum brazing equipment operates at pressures varying from $10^{-1}$ to $10^{-5}$ torr. The vacuum requirements will depend upon (1) the type of stainless steel, (2) the filler metal being used, (3) the area of the brazing interfaces, (4) the degree to which gases are expelled by the base metals during the brazing cycle, and (5) the leak rate of the facility. Partial pressures employing dry atmosphere are used in specific conditions.

## Precleaning

Stainless steels require more stringent precleaning than do carbon steels because the tenacious oxide films on stainless steels are difficult to remove by fluxes or reducing atmospheres. Precleaning of stainless steels for brazing should include a degreasing operation to remove any grease or oil films. The joint surfaces to be brazed should also be cleaned mechanically or with an acid pickling solution, as described previously. Wire brushing is not recommended. Care should be taken after cleaning to prevent dirt, oil, or fingerprints from soiling the cleaned surfaces. The best practice is to braze immediately after cleaning. When this is not possible, the cleaned part should be sealed in polytheylene bags to exclude moisture and other contaminents until the part can be brazed.

## Fluxes and Atmospheres

Assemblies of many stainless steels can be furnace brazed in atmospheres of dry hydrogen, argon, helium, or dissociated ammonia without the aid of flux. However, fluxes are sometimes relied upon for the brazing of those stainless steels that contain titanium or aluminum, or both, unless they are preplated with nickel. Some fluxes require postbraze cleaning, and their use increases the probability of weak joints due to entrapment of residue.

The chromium oxides that impart corrosion resistance to stainless steels are more difficult to remove than the oxides formed on carbon steels. Atmospheres having dew points of -40°F or lower are necessary to guarantee water vapor contents below the levels at which metal constituents react with the water vapor to form the metal oxide and hydrogen at brazing temperatures above 1800°F. For stainless steels, it often is necessary to rely on special high-activity fluxes to reduce surface oxides, particularly at low brazing temperatures.

Vacuum atmospheres promote the dissociation of base metal oxides at brazing temperatures. A vacuum effectively protects all metals from reoxidation during brazing. The vacuum level required depends upon many variables: type of equipment, heating rate, base metal and brazing filler metal outgassing of the part and fixtures, and furnace leak rate. Manufacturers of vacuum equipment, base metals, and filler metals should be consulted for specific recommendations.

A precaution to be considered before using dissociated ammonia atmospheres is that certain stainless steels at some brazing temperatures can be inadvertently nitrided. Nitriding produces a hard surface that can be beneficial or detrimental, depending on the service requirements of the components. Nitriding can be detected by the increased surface hardness of the metal and by metallographic examination.

Oxides of aluminum and titanium cannot be reduced at ordinary brazing temperatures in atmosphere furnaces. If these elements are present in the stainless steel in small amounts, satisfactory brazes can be obtained using high-purity gas atmospheres. When these elements are present in quantities exceeding 1 to 2 percent, the surfaces to be brazed should be cleaned and nickel-plated in lieu of using fluxes or vacuum atmospheres. Nickel plating will not only prevent the formation of detrimental oxides, but it can effectively limit embrittlement and erosion of the base metal. The thickness of the nickel plating should be in the range of 0.0002 to 0.002 in. because brazed joint strength decreases as the thickness of plating increases, and failure can occur in the plated layer.

Brazing flux for stainless steel depends upon the type of stainless steel, the brazing filler metal composition, the brazing process, and heating rate. A flux containing fluorides is generally required to remove the tenacious surface oxides. An AWS Type 3A, 3B, or 4 flux is commonly recommended for brazing stainless steel.

## Postbraze Operations

Flux or stop-off residue removal is usually necessary in the brazing of stainless steels. No postbraze operations are required on nonhardenable stainless steel assemblies that were furnace brazed in reducing or inert atmospheres, provided no flux was used and removal of stop-off, if used, is not necessary.

Stop-off materials can be removed simultaneously with flux residue. However, if flux was not used, stop-off can be removed by mechanical or chemical methods. Recommendations of the manufacturer should be followed.

Depending on the flux and the brazing process used, removal of the flux residue can be accomplished by water rinsing followed by chemical or mechanical cleaning. If abrasive cleaning is used for removal of flux or oxide film adjacent to the braze, sand or other nonmetallic grits should be used. Metallic shot other than stainless steel should be avoided as the particles from the shot can become embedded in the stainless steel surface, and cause rusting or pitting corrosion in service.

Heat treatment after brazing might be required on brazements made of martensitic and precipitation-hardening stainless steels. After cleaning, the brazement can be aged or tempered to develop desired base metal properties.

# SOLDERING

As with brazing, the surface oxide film on stainless steels makes these steels relatively difficult to solder. However, all stainless steels can be joined by soldering when proper techniques are employed.[13]

## Solders

Commercially available solders can be used to join stainless steels. In general, the higher the tin content of the solder, the better is the wetting and flow on stainless steel. It is generally recommended that tin contents be at least 50 percent to provide good bond strength.

Because stainless steels are used in a wide variety of applications and subjected to environments of various degrees of corrosiveness, the solder must be chosen for compatibility with both the environment and stainless steel. Tin and high-tin alloys provide a good color match with stainless steel, and do not darken as noticeably in service as do high lead content solders. Solders must be carefully selected and joints properly designed to minimize mechanical loading of the solder if moderately elevated temperatures are expected in service. Solders are quite weak at even moderately elevated temperatures, and are subject to creep if directly loaded.

If articles of stainless steel are fabricated for food or beverage processing, the solders must not contain cadmium or lead.

## Surface Preparation

Standard shop practices suffice for preparing stainless steels for soldering. Appropriate procedures include vapor, solvent, or caustic degreasing; acid pickling; grit or shot blasting; wire brushing or abrading with stainless steel wool or emery cloth. The chosen method should be appropriate to the type of foreign material to be removed. Shot or wire brushes, if used, should be stainless steel to avoid rust spots. If surfaces are highly polished, it is best to roughen them slightly with emery cloth or other suitable means before cleaning and soldering.

Soldering should, if possible, be done immediately after cleaning. If not, the parts should be precoated with solder or tin immediately after cleaning.

Many joint designs have recessed and hidden surfaces that make post-solder cleaning to remove flux residues difficult. Furthermore, these recessed or blind areas present problems in soldering because they cannot be visually, inspected to verify that solder has flowed into these areas to complete the joint. Therefore, it is often desirable to precoat or tin the specific joint areas with solder before assembling the pieces for final soldering. Corrosive flux must be removed at this time from the tinned parts. Alternatively, a suitable electroplated coating can be applied. Final soldering of the joint is then accomplished with a rosin type flux, the residues of which are innocuous and cause no serious corrosion problem.

## Heating Methods

Stainless steel assemblies can be heated for soldering by all commonly used production methods. Because stainless steels have low thermal conductivities, the rate of travel along the joint should be one that allows the entire joint area to reach soldering temperature so that the solder will flow into all areas to be joined. Increasing the rate of travel by using higher heat input is not recommended because there is danger of destroying the flux and excessive oxidation of the solder and base metal. In general, soldering temperatures of approximately 50° to 150°F over the melting point of the solder are desired.

Austenitic stainless steel has a high coefficient of thermal expansion, which might cause buckling and warpage during soldering. Jigs and fixtures should be used to maintain proper alignment and fit-up. On long seams, it is helpful to solder the joint at intervals before completing the joint. If warpage becomes a serious problem, it is often helpful to solder short lengths of the seam at alternate positions along the joint so that the heat is spread more uniformly over the joint length.

## Flux

Fluxes suitable for soldering stainless steel are corrosive and care must be exercised in their

---

13. Refer to the AWS *Soldering Manual* for additional information.

use to prevent damage to eyes, skin, and clothing. Orthophosphoric acid and hydrochloric acid fluxes are satisfactory, as are aqueous solutions of zinc chloride with other compounds. If molybdenum, titanium, columbium, or aluminum are present, the flux should contain some hydrofluoric acid. There are also commercial fluxes that are satisfactory. Rosin fluxes are not satisfactory for soldering stainless steel unless the parts are first precoated with solder or plating.

Flux-cored solders containing acid flux are also useful, but it will probably be necessary to supplement the core flux by the addition of other flux, externally applied.

### Postsoldering Treatment

Residues of fluxes, except rosin types, on stainless steel are hygroscopic, and in the presence of moisture are corrosive to stainless steels. Similarly, flux fume generated during soldering can condense on colder parts of the assembly and leave a residue that is corrosive in the presence of moisture. Therefore, it is imperative that these residues be thoroughly removed after soldering, preferably immediately thereafter. All traces of flux residue must be removed by neutralizing and thorough rinsing with water. Rosin flux residues are noncorrosive and need not be removed, except for appearance.

If desired, excess solder can be removed from the joint area with a stainless steel scraping tool. The tool should be softer than the base metal to avoid scratching. Water spots or other minor surface discolorations can be removed by scrubbing with a powdered cleanser or buffing with metal polish.

# THERMAL CUTTING

## OXYFUEL GAS CUTTING

The absence of alloying elements in carbon steel permits the oxidation reaction to proceed rapidly. As the quantity and number of alloying elements in steel increase, the oxidation rate decreases and cutting becomes more difficult. Oxidation of the iron in a low alloy steel liberates a considerable amount of heat, and iron oxides produced have melting points near the melting point of iron. However, the oxides of many alloying elements in stainless steels, such as aluminum and chromium, have melting points much higher than those of iron oxides. These high melting oxides, which are refractory in nature, can shield the metal in the kerf so that fresh iron is not continuously exposed to the cutting oxygen stream. The speed of cutting decreases as the amount of refractory oxide-forming elements in the iron increases. Variations of OFC must be used to cut stainless steels.[14]

---

14. Additional information on arc and oxygen cutting is available in the *Welding Handbook,* Vol. 2, 7th Ed., 459-516.

Several methods for oxygen cutting of stainless steels are:

(1) Waster plate
(2) Wire feed
(3) Powder cutting
(4) Flux cutting

When the above methods are used, the quality of the cut surface is somewhat impaired. Scale can adhere to the cut faces. Carbon or iron pickup, or both, usually occurs on the cut surfaces of stainless steel. This might affect the corrosion resistance and magnetic properties of the metal. If these properties are important, approximately $\frac{1}{8}$ in. of metal should be machined from the cut edges.

### Waster Plate

One method of cutting oxidation resistant steels is to clamp a low carbon steel "waster" plate on the upper surface of the material to be cut. The cut is started in the low carbon steel plate. The heat liberated by oxidation of the low carbon steel provides additional heat at the cutting face of the stainless steel to sustain the oxidation reaction. The iron oxide from the low carbon steel helps to wash away the refractory oxides from stainless steel. The thickness of the waster plate must be in proportion to the thickness of the stainless steel being cut. Several undesirable features of this method are the cost of the waster plate, the additional setup time, the slow cutting speeds, and the rough quality of the cut.

## Wire Feed

With the appropriate equipment, a small diameter, low carbon steel wire is fed continuously into the cutting torch preheat flame, ahead of the cut. The end of the steel wire should melt rapidly into the surface of the stainless steel plate. The effect of the wire addition on the cutting action is the same as that of the waster plate. The deposition rate of the low carbon steel wire must be adequate to maintain the oxygen cutting action. It should be determined by trial cuts. The thickness of the alloy plate and cutting speed are also factors that must be considered in the process. A motor-driven wire feeder and wire guide, mounted on the cutting torch, are needed as accessory equipment.

## Metal Powder Cutting

The metal powder cutting process (POC) is a technique for supplementing an OFC torch with a stream of iron-rich powdered material. The powdered material accelerates and propagates the oxidation reaction, and also the melting and spalling action of hard-to-cut materials. The powder is directed into the kerf through either the cutting tip or through single or multiple jets external to the tip.

Some of the powders react chemically with the refractory oxides produced in the kerf and increase their fluidity. The resultant molten slags are easily washed out of the reaction zone by the oxygen jet. Fresh metal surfaces are continuously exposed to the oxygen jet. Iron powder and mixtures of metallic powders, such as iron and aluminum, are used.

Cutting of oxidation resistant steels by the powder method can be done at approximately the same speeds as oxygen cutting of carbon steel of equivalent thickness. The cutting oxygen flow must be slightly higher with the powder process.

## Flux Cutting

This process is primarily intended for cutting stainless steels. The flux is designed to react with oxides of alloying elements, such as chromium and nickel, to produce compounds with melting points near those of iron oxides. A spe-cial apparatus is required to introduce the flux into the kerf. With a flux addition, stainless steel can be cut at a uniform linear speed without torch oscillation. Cutting speeds approaching those for equivalent thicknesses of carbon steel can be attained. The tip sizes will be larger, and the cutting oxygen flow will be somewhat greater than for carbon steels.

## PLASMA ARC CUTTING

Plasma arc cutting employs an extremely high-temperature, high-velocity constricted arc between the cutting torch and the piece to be cut. Unlike oxygen cutting, which depends on a chemical reaction, the plasma arc cutting process depends on thermal and mechanical action. The plasma arc melts a localized portion of the workpiece, and the molten particles are then removed by a high-velocity gas jet. Good-quality, high-speed cuts can be made with this process, and slabs up to 7 inches thick have been severed.

Because of its high cutting speed, plasma arc cutting has little effect on the metallurgical or physical properties of the workpiece, in comparison with other thermal cutting processes. The depth of the heat-affected zone depends on the type and thickness of the metal and the cutting speed. For example, in 1-in. thick Type 304 plate, the heat-affected zone is only 0.005- to 0.007-in. deep. There is virtually no time for chromium carbide precipitation because the cut face passes through the critical temperature range very rapidly. Measurements of the magnetic properties of Type 304 stainless steel made on uncut base metal and on plasma arc cut samples indicate that magnetic permeability is unaffected.

## AIR CARBON ARC CUTTING

### Principles of Operation

In air carbon arc cutting (AAC), a high-current arc is established between a carbon-graphite electrode and the metal workpiece to be melted. A compressed air jet is continuously directed at the point of melting to eject the molten metal away. Metal removal is continuous as the carbon arc is advanced in the cut. The process is used for severing and gouging. Gouging is sometimes used for weld groove preparation, back gouging, and the removal of defective weld metal.

Both ac and dc power are used with the appropriate electrodes. The electrode tip is heated to a high temperature by the arc current, but it does not melt. The electrode is consumed during cutting, as carbon is lost by oxidation or sublimation at the tip.

Because metal removal is primarily by melting rather than oxidation, the process can be used to cut stainless steel.

## Metallurgical Effects

To avoid difficulties with carburized metal, users of the air carbon arc cutting process must be aware of the metallurgical changes that occur during gouging and cutting. When the carbon electrode is positive, the current flow carries ionized carbon atoms from the electrode to the melted metal. The free carbon particles are rapidly absorbed by the melted metal. Because of this absorption, it is important that all molten (carburized) steel be removed from the kerf, preferably by the air blast.

When the air carbon arc cutting process is used under improper conditions, the carburized molten steel left behind in the kerf or groove can usually be recognized by its dull gray-black color. This contrasts with the bright blue color in a properly made groove. Inadequate air flow can leave small pools of carburized steel in the bottom of the groove. Irregular electrode travel, particularly in a manual operation, will produce ripples in the groove wall that tend to trap carburized steel. An improper electrode angle might cause small beads of carburized steel to remain along the edge of the groove.

The effect of carburized steel that remains in the kerf or groove through a subsequent welding operation depends on many factors including the amount present, the welding process to be employed, the type of base metal, and the weld quality required. Although it seems likely that the filler metal deposited during welding would assimilate small pools or beads of carburized steel on the kerf, experience shows that traces of high carbon steel (containing approximately one percent of carbon) can remain along the weld interface.

High carbon content is detrimental to the metallurgical structure, mechanical properties, and corrosion resistance of stainless steels. Therefore, any carburized metal must be completely removed from the cut surface prior to welding, heat treatment, or service. Although carburized metal on the kerf or groove surface can be removed by grinding, it is more efficient to conduct air-carbon arc gouging and cutting operations properly within prescribed conditions, and completely avoid the retention of undesirable metal.

# SAFE PRACTICES

Compounds of chromium, including hexavalent chromium, and of nickel may be found in fume from welding processes. The specific compounds and concentrations will vary with (1) the compositions of the base metal and filler metals and (2) the welding processes. Immediate effects of overexposure to welding fume that contain chromium and nickel are similar to the effects produced by other metals in fume. The fume can cause symptoms such as nausea, headaches, and dizziness. Some persons may develop a sensitivity to chromium or nickel, which can result in dermatitis or skin rash.

Fume and gases should not be inhaled, and the face should be kept out of the fume. Sufficient ventilation, an exhaust at the arc or flame, or both, should be used to keep fume and gases from the breathing zone and local area. In some cases, natural air movement will provide sufficient ventilation. Where ventilation is questionable, air sampling should be used to determine whether corrective measures are necessary.

Specific precautionary measures that should be used to protect against exposure to fume and gases have been incorporated in current OSHA Standards.[15]

---

15. See ANSI Publication Z49.1, *Safety in Welding and Cutting,* available from The American Welding Society.

# Metric Conversion Factors

1 in. = 25.4 mm
1 in./min = 0.423 mm/s
1 psi = 6.89 kPa
1 ksi = 6.89 MPa
1 lb•f = 9.8 N
1 ft•lb = 1.36 J
1 torr = 0.13 kPa
$t_C = 0.56 (t_F - 32)$

# SUPPLEMENTARY READING LIST

*Brazing Manual,* 3rd Ed., Miami: American Welding Society, 1976.

Castro, R. J., and deCadenet, J. J., *Welding Met allurgy of Stainless and Heat Resisting Steels,* Cambridge University Press, 1975.

Cole, N. C., *Corrosion Resistance of Brazed Joints,* New York: Welding Research Council, Bulletin 247, 1979 April.

Epsy, R. H., Weldability of nitrogen-strengthened stainless steels, *Welding Journal,* 61(5): 149s-56s; 1982 May.

Gooch, T. G., and Honeycombe, J., Welding variables and microfissuring in austenitic stainless steel weld metal, *Welding Journal,* 59(8): 233s-41s: 1980 Aug.

Harkins, F. G., *Welding of Age-Hardenable Stainless Steels,* New York: Welding Research Council, Bulletin 103, 1965 Feb.

Hauser, D., and Van Echo, J. A., Effects of ferrite content on austenitic stainless steel welds, *Welding Journal,* 61(2): 37s-44s; 1982 Feb.

Johnson, E. W., and Hudak, S. J., *Hydrogen Embrittlement of Austenitic Stainless Steel Weld Metal with Special Consideration Given to the Effects of Sigma Phase,* New York: Welding Research Council, Bulletin 240, 1978 Aug.

Kah, D. W., and Dickinson, D. W., Weldability of ferritic stainless steels, *Welding Journal,* 60(8): 135s-42s; 1981 Aug.

Kaltenhauser, R., Weldability of precipitation-hardening stainless steels, *Metals Engineering Quarterly,* 9(1): 44-57; 1969 Jan.

Lippold, J. C., and Savage, W. F., Solidification of austenitic stainless steel weldments, Part 1 – A proposed mechanism, *Welding Journal,* 58(12): 362s-74s; 1979 Dec.; Part 2 – The effect of alloy composition on ferrite morphology, *ibid.* 59(2): 48s-58s; 1980 Feb.

*Metals Handbook,* Vol. 3, 9th Ed., Metals Park, OH: American Society for Metals, 1-124, 189-206; 1980.

Ogawa, T., and Tsunetomi, E., Hot cracking susceptibility of austenitic stainless steels, *Welding Journal,* 61(3): 82s-93s; 1982 March.

Ogawa, T., et. al., The weldability of nitrogen containing austenitic stainless steel: Part 1 – Chloride pitting corrosion resistance, *Welding Journal,* 61(5): 139s-48s; 1982 May.

Sawhill, J. M., and Bond, A. P., Ductility and toughness of stainless steel welds, *Welding Journal,* 55(2): 33s-41s; 1976 Feb.

Schwartz, M. M., *Source Book on Brazing and Brazing Technology,* Metals Park, OH: American Society for Metals; 1980.

*Soldering Manual,* 2nd Ed., Miami: American Welding Society, 1978.

Szumachow, E. R., and Reid, H. F., Cryogenic toughness of SMA austenitic stainless steel weld metals, Part 1 – Role of ferrite, *Welding Journal,* 57(11): 325s-33s; 1978 Nov.; Part 2 – Role of nitrogen, *ibid.,* 58(2): 34s-44s; 1979 Feb.

Vagi, J. J., Evans, R. M., and Martin, D. C., *Welding of Precipitation-Hardening Stainless Steels,* NASA Technical Memorandum TM X-53582, 1967 Feb. 28.

Wegrzyn, J., and Klimpel, A., The effect of alloying elements on the sigma phase formation in 18-8 weld metals, *Welding Journal,* 60(8): 146s-54s; 1981 Aug.

*Welding of Stainless Steels and Other Joining Methods,* Wash, DC: American Iron and Steel Inst.; 1979 Apr.

# 3

# Tool and Die Steels

## Chapter Committee

E.G. SIGNES, *Chairman*
*Bethlehem Steel Corporation*
F.P. BERNIER
*Welding Equipment and Supply*
*Company*
W.H. KEARNS
*American Welding Society*
W. WOLLERING
*Ladish Company*

W.L. LUTES
*Welding Consultant*
R.M. NUGENT
*Cameron Iron Works*
G.M. WALBERG
*Allstate Welding Products*

## Welding Handbook Committee Member

E.G. SIGNES
*Bethlehem Steel Corporation*

# 3

# Tool and Die Steels

## GENERAL DESCRIPTION

Carbon and alloy steels for metal forming and cutting tools are designed to provide specific properties for various applications. These steels are produced by processes that give clean, homogenous material with close control of chemical composition. Precise production requirements and quality control of these steels are justified because the manufacture of complicated cutting and forming tools and the down time associated with premature tool failure are very costly. these factors should be taken into account when welding on expensive tools. Such operations should be carefully planned, and proper welding procedures should be developed to produce the required properties in the weld deposit.

## CLASSIFICATION

Tool steels are classified by the American Iron and Steel Institute (AISI) and the Society of Automotive Engineers (SAE) into seven major groups.[1] These groups, listed in Table 3.1, generally reflect the normal hardening medium (water, oil, or air) or the general applications of the alloys. The groups are subdivided into types that are based upon chemical composition, as shown in Table 3.2.

In general, the maximum hardness, wear

resistance, and dimensional stability increase with hardenability. For example, the water-hardening steels are low and the air-hardening steels are high in hardenability. The weldability of tool steels is ranked in reverse order. Air-hardening steels with high hardenability require the greatest care during welding.

## WATER-HARDENING GROUP

These are essentially plain carbon steels, although some of the high carbon types have small amounts of chromium and vanadium added to improve toughness and wear resistance. The carbon content varies between 0.60 and 1.40 percent. In general, plain carbon tool steels are less expensive than alloy tool steels. With proper heat treatment, they will have a hard martensitic surface with a tough core. These steels must be water quenched for high hardness and are therefore subject to considerable distortion. They have the best machinability ratings of all the tool steels, and are the best with respect to decarburization. However, their resistance to elevated temperatures is poor.

## SHOCK-RESISTING GROUP

These steels are used for applications where toughness and the ability to withstand repeated shock are paramount. They are comparatively low in carbon content, varying between 0.40 and 0.65 percent. The principal alloying elements in these steels are silicon, chromium, tungsten, and sometimes molybdenum. Silicon strengthens the ferrite, while chromium increases hardenability and contributes slightly to wear resistance. Molybdenum aids in increasing hardenability, while

1. Additional information is available in the following publications:
    (a) ANSI/ASTM A600, *Standard Specification for High-Speed Tool Steel*
    (b) ANSI/ASTM A681, *Standard Specification for Alloy Tool Steels*
    (c) ANSI/ASTM A686, *Standard Specification for Carbon Tool Steels*

**Table 3.1**
**Major tool steel groups**

| Group | Letter symbol | Type |
|---|---|---|
| Water-hardening | W | Plain carbon |
| Shock-resisting | S | Medium carbon, low alloy |
| Cold-work | O | Oil-hardening |
|  | A | Medium alloy, air-hardening |
|  | D | High carbon, high chromium |
| Hot-work | H | Chromium (H1-H19) |
|  |  | Tungsten (H20-H39) |
|  |  | Molybdenum (H40-H59) |
| High-speed | T | Tungsten |
|  | M | Molybdenum |
| Mold | P | Low carbon |
| Special purpose | L | Low alloy |

**Table 3.2**
**Compositions of typical tool steels**

| Type | Nominal composition, % | | | | | | | | |
|---|---|---|---|---|---|---|---|---|---|
|  | C | Mn | Si | Cr | Ni | V | W | Mo | Co |
| **Water-hardening** | | | | | | | | | |
| W1 | 0.60/1.40[a] | | | | | | | | |
| W2 | 0.60/1.40[a] | ... | ... | ... | ... | 0.25 | | | |
| W5 | 1.10 | ... | ... | 0.50 | | | | | |
| **Shock-resisting** | | | | | | | | | |
| S1 | 0.50 | ... | ... | 1.50 | ... | ... | 2.50 | | |
| S2 | 0.50 | ... | 1.00 | ... | ... | ... | ... | 0.50 | |
| S5 | 0.55 | 0.80 | 2.00 | ... | ... | ... | ... | 0.40 | |
| S7 | 0.50 | ... | ... | 3.25 | ... | ... | ... | 1.40 | |
| **Cold-work** | | | | | | | | | |
| **Oil-hardening** | | | | | | | | | |
| O1 | 0.90 | 1.00 | ... | 0.50 | ... | ... | 0.50 | | |
| O2 | 0.90 | 1.60 | | | | | | | |
| O6[b] | 1.45 | ... | · 1.00 | ... | ... | ... | ... | 0.25 | |
| O7 | 1.20 | ... | ... | 0.75 | ... | ... | 1.75 | | |
| **Medium alloy air-hardening** | | | | | | | | | |
| A2 | 1.00 | ... | ... | 5.00 | ... | ... | ... | 1.00 | |
| A3 | 1.25 | ... | ... | 5.00 | ... | 1.00 | ... | 1.00 | |
| A4 | 1.00 | 2.00 | ... | 1.00 | ... | ... | ... | 1.00 | |
| A6 | 0.70 | 2.00 | ... | 1.00 | ... | ... | ... | 1.25 | |
| A7 | 2.25 | ... | ... | 5.25 | ... | 4.75 | 1.00 | 1.00 | |
| A8 | 0.55 | ... | ... | 5.00 | ... | ... | 1.25 | 1.25 | |
| A9 | 0.50 | ... | ... | 5.00 | 1.50 | 1.00 | ... | 1.40 | |
| A10[b] | 1.35 | 1.80 | 1.25 | ... | 1.80 | ... | ... | 1.50 | |
| **High carbon, high chromium** | | | | | | | | | |
| D2 | 1.50 | ... | ... | 12.00 | ... | 1.00 | ... | 1.00 | |
| D3 | 2.25 | ... | ... | 12.00 | | | | | |
| D4 | 2.25 | ... | ... | 12.00 | ... | ... | ... | 1.00 | |
| D5 | 1.50 | ... | ... | 12.00 | ... | ... | ... | 1.00 | 3.00 |
| D7 | 2.35 | ... | ... | 12.00 | ... | 4.00 | ... | 1.00 | |

# Table 3.2 (cont.)
## Compositions of typical tool steels

| Type | Nominal composition, % | | | | | | | | |
|------|------|------|------|------|------|------|------|------|------|
|      | C | Mn | Si | Cr | Ni | V | W | Mo | Co |

**Hot-work**

Chromium

| | | | | | | | | | |
|------|------|------|------|------|------|------|------|------|------|
| H10 | 0.40 | ... | ... | 3.25 | ... | 0.40 | ... | 2.50 | |
| H11 | 0.35 | ... | ... | 5.00 | ... | 0.40 | ... | 1.50 | |
| H12 | 0.35 | ... | ... | 5.00 | ... | 0.40 | 1.50 | 1.50 | |
| H13 | 0.35 | ... | ... | 5.00 | ... | 1.00 | ... | 1.50 | |
| H14 | 0.40 | ... | ... | 5.00 | ... | ... | 5.00 | | |
| H19 | 0.40 | ... | ... | 4.25 | ... | 2.00 | 4.25 | ... | 4.25 |

Tungsten

| | | | | | | | | | |
|------|------|------|------|------|------|------|------|------|------|
| H21 | 0.35 | ... | ... | 3.50 | ... | ... | 9.00 | | |
| H22 | 0.35 | ... | ... | 2.00 | ... | ... | 11.00 | | |
| H23 | 0.30 | ... | ... | 12.00 | ... | 1.00 | 12.00 | | |
| H24 | 0.45 | ... | ... | 3.00 | ... | ... | 15.00 | | |
| H25 | 0.25 | ... | ... | 4.00 | ... | ... | 15.00 | | |
| H26 | 0.50 | ... | ... | 4.00 | ... | 1.00 | 18.00 | | |

Molybdenum

| | | | | | | | | | |
|------|------|------|------|------|------|------|------|------|------|
| H42 | 0.60 | ... | ... | 4.00 | ... | 2.00 | 6.00 | 5.00 | |

**High-speed**

Tungsten

| | | | | | | | | | |
|------|------|------|------|------|------|------|------|------|------|
| T1 | 0.75[a] | ... | ... | 4.00 | ... | 1.00 | 18.00 | | |
| T2 | 0.80 | ... | ... | 4.00 | ... | 2.00 | 18.00 | | |
| T4 | 0.75 | ... | ... | 4.00 | ... | 1.00 | 18.00 | ... | 5.00 |
| T5 | 0.80 | ... | ... | 4.00 | ... | 2.00 | 18.00 | ... | 8.00 |
| T6 | 0.80 | ... | ... | 4.50 | ... | 1.50 | 20.00 | ... | 12.00 |
| T8 | 0.75 | ... | ... | 4.00 | ... | 2.00 | 14.00 | ... | 5.00 |
| T15 | 1.50 | ... | ... | 4.00 | ... | 5.00 | 12.00 | ... | 5.00 |

Molybdenum

| | | | | | | | | | |
|------|------|------|------|------|------|------|------|------|------|
| M1 | 0.85[a] | ... | ... | 4.00 | ... | 1.00 | 1.50 | 8.00 | |
| M2 | 0.85/1.00[a] | ... | ... | 4.00 | ... | 2.00 | 6.00 | 5.00 | |
| M3 | 1.05/1.20 | ... | ... | 4.00 | ... | 2.40/3.00 | 6.00 | 5.00 | |
| M4 | 1.30 | ... | ... | 4.00 | ... | 4.00 | 5.50 | 4.50 | |
| M6 | 0.80 | ... | ... | 4.00 | ... | 1.50 | 4.00 | 5.00 | 12.00 |
| M7 | 1.00 | ... | ... | 4.00 | ... | 2.00 | 1.75 | 8.75 | |
| M10 | 0.85/1.00[a] | ... | ... | 4.00 | ... | 2.00 | ... | 8.00 | |
| M30 | 0.80 | ... | ... | 4.00 | ... | 1.25 | 2.00 | 8.00 | 5.00 |
| M34 | 0.90 | ... | ... | 4.00 | ... | 2.00 | 2.00 | 8.00 | 8.00 |
| M36 | 0.80 | ... | ... | 4.00 | ... | 2.00 | 6.00 | 5.00 | 8.00 |
| M41 | 1.10 | ... | ... | 4.25 | ... | 2.00 | 6.75 | 3.75 | 5.00 |
| M42 | 1.10 | ... | ... | 3.75 | ... | 1.15 | 1.50 | 9.50 | 8.00 |
| M43 | 1.20 | ... | ... | 3.75 | ... | 1.60 | 2.75 | 8.00 | 8.25 |
| M44 | 1.50 | ... | ... | 4.25 | ... | 2.25 | 5.25 | 6.25 | 12.00 |
| M46 | 1.25 | ... | ... | 4.00 | ... | 3.20 | 2.00 | 8.25 | 8.25 |
| M47 | 1.10 | ... | ... | 3.75 | ... | 1.25 | 1.50 | 9.50 | 5.00 |

**Special purpose low alloy**

| | | | | | | | | | |
|------|------|------|------|------|------|------|------|------|------|
| L2 | 0.50/1.10[a] | ... | ... | 1.00 | ... | 0.20 | | | |
| L6 | 0.70 | ... | ... | 0.75 | 1.50 | ... | ... | 0.25 | |

**Table 3.2 (cont.)**
**Compositions of typical tool steels**

| Type | Nominal composition, % | | | | | | | | | |
| | C | Mn | Si | Cr | Ni | V | W | Mo | Co | Al |
|---|---|---|---|---|---|---|---|---|---|---|
| | **Mold steels** | | | | | | | | | |
| P2 | 0.07 | ... | ... | 2.00 | 0.50 | ... | ... | 0.20 | | |
| P3 | 0.10 | ... | ... | 0.60 | 1.25 | | | | | |
| P4 | 0.07 | ... | ... | 5.00 | ... | ... | ... | 0.75 | | |
| P5 | 0.10 | ... | ... | 2.25 | | | | | | |
| P6 | 0.10 | ... | ... | 1.50 | 3.50 | | | | | |
| P20 | 0.35 | ... | ... | 1.70 | ... | ... | ... | 0.40 | | |
| P21 | 0.20 | ... | ... | ... | 4.00 | ... | ... | ... | ... | 1.20 |

a. Various carbon contents are available.
b. Contains free graphite in the microstructure to improve machinability.

tungsten imparts some red hardness to these steels. Most of these steels are oil-hardening, although some must be water quenched to develop full hardness.

The high silicon content tends to accelerate decarburization and suitable precautions should be taken during heat treatment to minimize this. The steels are considered fair in regard to red hardness, wear resistance, and machinability. The hardness is usually kept below 60 HRC. They are used in the manufacture of forming tools, punches, chisels, pneumatic tools, shear blades, and other applications where both high resistance to impact loading and moderate wear resistance are needed.

## COLD-WORK GROUP

This group of tool steels is considered to be the most important one because the majority of tool applications can be served by one or more of these steels.

The oil-hardening low alloy types contain manganese and smaller amounts of chromium and tungsten. They have very good nondeforming properties and are less likely to distort or crack during heat treatment than are the water-hardening steels. These steels are relatively inexpensive, and their high carbon content produces adequate wear resistance for short-run applications at or near room temperature. They have good machinability and good resistance to decarburization; toughness is fair; red hardness is almost as poor as the plain carbon tool steels. Typical applications are thread taps, solid threading dies, forming tools, and expansion reamers.

The air-hardening medium alloy types with about 1 percent carbon contain up to 2 percent manganese, up to 5 percent chromium, and 1 percent molybdenum. The increased alloy content, particularly manganese and molybdenum, confers marked air-hardening properties and increased hardenability to these types. This group has excellent nondeforming properties, good wear resistance, and also fair toughness, red hardness, and resistance to decarburization. These steels are used for blanking, forming, trimming, and thread-rolling dies.

The high carbon, high chromium types contain up to 2.25 percent carbon and 12 percent chromium. They may also contain molybdenum, vanadium, and cobalt. The combination of high carbon and high chromium gives excellent wear resistance and nondeforming properties. They have good abrasion resistance. The low dimensional change in hardening makes these steels popular for blanking and piercing dies; drawing dies for wire, bars, and tubes; thread-rolling dies; and master gages. These alloys are not considered for cutting edges because of susceptibility to edge brittleness. They are also limited to working below 900° F.

## HOT-WORK GROUP

In many applications, a tool is exposed to high service temperatures because the material is being hot-worked, as in hot forging and extruding, die casting, and plastic molding. Tool steels developed for these applications have good resistance to softening at elevated temperatures. This property is known as good red hardness.

The alloying elements that promote red hardness are chromium, molybdenum, and tungsten. However, the sum of these elements must be at least 5 percent to avoid excessive softening at elevated temperatures. This group of tool steels is divided into three types.

## Chromium Type

These steels contain at least 3.25 percent chromium and smaller amounts of vanadium, tungsten, and molybdenum. They have good red hardness because of their medium chromium content, together with the three other strong carbide-forming elements. The low carbon and relatively low total alloy content promote toughness at hardnesses of 40 to 55 HRC.

Higher tungsten and molybdenum contents will increase red hardness but reduce the toughness slightly. These steels are extremely deep hardening, and may be air-hardened to full hardness in sections up to 12 in. thick. The air-hardening qualities and balanced alloy content are responsible for low distortion during hardening. These steels are especially adapted to hot die work of all kinds, particularly extrusion dies, diecasting dies, forging dies, mandrels, and hot shears.

## Tungsten Type

These steels contain 9 to 18 percent tungsten and 3 to 12 percent chromium. The high alloy content increases their resistance to softening at high temperatures when compared with the chromium type steels. However, it also makes them more susceptible to brittleness at hardnesses of 45 to 55 HRC. They can be air hardened for low distortion, or quenched in oil or hot salt to minimize scaling. These steels have many of the characteristics of the high-speed tool steels but have better toughness. They can be used for high-temperature applications, such as mandrels and extrusion dies for brass, nickel alloys, and steel.

## Molybdenum Type

The H42 steel is similar to the tungsten hot-work steels in characteristics and uses. It resembles in composition the various types of molybdenum high-speed tool steels, but it has a lower carbon content and greater toughness. Its principal advantages over the tungsten hot-work steels are lower initial cost and greater resistance to heat cracking or checking. As with all high molybdenum steels, it requires care during heat treatment to avoid decarburization.

## HIGH-SPEED GROUP

These steels are highly alloyed and usually contain large amounts of tungsten or molybdenum along with chromium, vanadium, and sometimes cobalt. The carbon content varies between 0.75 and 1.2 percent, although some types contain as much as 1.5 percent.

The major application of high-speed steels is for cutting tools, but they are also used for making extrusion dies, burnishing tools, and blanking punches and dies.

Compositions of high-speed steels are designed to provide excellent red hardness and reasonably good shock resistance. They have good nondeforming properties and may be quenched in oil, air, or molten salts. They are rated as deep-hardening, have good wear resistance, fair machinability, and fair to poor resistance to decarburization.

The high-speed steels are subdivided into two groups: molybdenum type (Type M) and tungsten type (Type T). From the standpoint of fabrication and tool performance, there is little difference between the molybdenum and tungsten grades. The important properties are about the same.

When high red hardness is required, those steels containing cobalt are recommended. The high cobalt steels require careful protection against decarburization during heat treatment, and they are less ductile than the others. A steel with high vanadium content is desirable when the material being cut is highly abrasive.

The presence of many wear-resistant carbides in a hard heat-resistant matrix makes these steels suitable for cutting tool applications. The tungsten and molybdenum high-speed steels are used in a wide variety of cutting tools, such as tool bits, drills, reamers, broaches, taps, milling cutters, hobs, saws, and woodworking tools.

## MOLD STEELS

These low carbon steels contain chromium and nickel as the principal alloying elements, with molybdenum and vanadium as minor additives. Most of them are carburizing grades produced to tool steel quality. They are generally characterized by very low hardness in the annealed condition and good resistance to work hardening. The steels are generally carburized and hardened to 58 to 64 HRC for wear resistance. They generally have poor red hardness and are used almost entirely to manufacture casting dies and for molds for injection or compression molding of plastics.

## SPECIAL-PURPOSE STEELS

These steels do not fall into the usual categories and, therefore, are designated as special-purpose types. They were developed to handle the requirements of certain applications and are more expensive for many applications than the standard AISI alloy steels.

They contain chromium as the principal alloying element in combination with vanadium, molybdenum, and nickel. The high chromium content not only promotes wear resistance by the formation of hard complex iron-chromium carbides but, together with molybdenum, also increases hardenability. Nickel increases toughness, while vanadium serves to refine the grain size. These steels are oil-hardening and thus only fair in resisting dimensional change. Typical uses are various machine tool applications where high wear resistance with good toughness is required, as in bearings, rollers, clutch plates, cams, collets, and wrenches. The high carbon types are used for arbors, dies, drills, taps, knurls, and gages.

# METALLURGY

Tool steels generally contain at least 0.6 percent carbon to provide a martensitic hardness capability of at least 60 HRC, as shown in Fig. 3.1. Carbon in excess of the eutectoid composition will be present in the steels as undissolved carbides in a martensitic structure. These hard carbides increase the wear resistance of the steel. Some types of tool steels contain less carbon to provide toughness or shock resistance.

The hardenability of tool steels follows the same general rules governing other alloy steels.[2] The water-hardening plain carbon steels obviously have low hardenability. The alloy tool steels generally have sufficient alloying elements to permit quenching at a slower rate in oil or air.

The effect of additions of several alloying elements on the eutectoid temperature ($A_1$) of steel is shown in Fig. 3.2(A). Chromium, molybdenum, and tungsten, which are commonly found in tool steels, increase the eutectoid temperature of the steel by stabilizing the ferrite phase. Vanadium additions have a similar effect; the curve for vanadium would fall between those for molybdenum and titanium in Fig. 3.2 (A). At the same time, the carbon content of the eutectoid composition decreases with increasing alloy addition, as shown in Fig. 3.2(B). This takes place because these elements have a greater affinity for carbon than does iron. As a result, higher temperatures are required to dissolve the carbides into austenite with tool steels.

The higher the carbon content of the austenite, the more sluggish is its transformation to other microstructures. Therefore, high carbon austenite may be retained down to room temperature with fast quenching rates. Subsequent cooling to below room temperature will transform most of the retained austenite to martensite. At the same time, the steel will expand and produce or increase the residual stresses.

In general, all of the alloying elements in tool steels except cobalt decrease the austenite-

2. Additional information on welding metallurgy may be found in the *Welding Handbook,* Vol. 1, 7th Ed., 100-142.

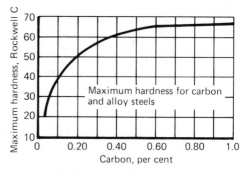

**Fig. 3.1 – Influence of carbon content on the maximum hardness of as-quenched steel**

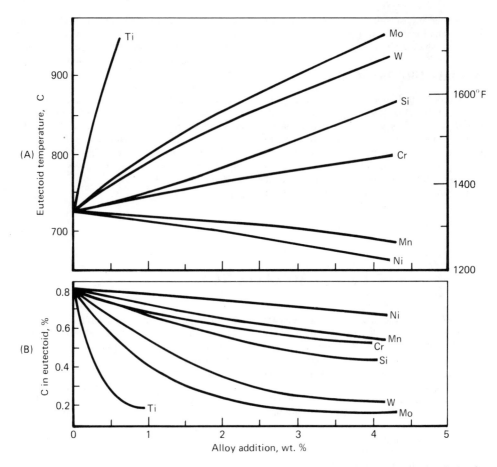

**Fig. 3.2 – Effect of some alloying elements on (A) the eutectoid temperature and (B) the eutectoid carbon content of alloy steels**

to-martensite transformation temperature. Cobalt, commonly found in high-speed tool steels, raises this temperature.

In most cases, the welding of tool steels involves the repair of tools or dies that have been hardened and machined to final shape. Due to the composition of these steels, they are highly susceptible to hydrogen-assisted cracking in the heat-affected zone when martensite is present. To avoid this cracking, care must be taken in selecting the welding electrode and welding procedures. High heat input with slow cooling rates will produce a weld with less susceptibility to cracking, but it may also produce an unacceptable soft region. Specific suggestions for repair procedure are given later in this chapter.

## HEAT TREATMENT

Tool steels usually are received from the supplier in the annealed condition. If practical, tools and dies should be welded in this condition because the steel will have the best ductility. Previously hardened tools should be annealed prior to welding if feasible. The welded tool should then be heat treated to provide the desired properties. Suggested annealing and hardening heat treatments for several tool steels are given in Table 3.3. However, the appropriate heat treat-

**Table 3.3**
**Typical welding and heat treating procedures for representative tool steels**

| Type | Filler metal type | Annealed base metal | | | | | | Hardened base metal | |
|---|---|---|---|---|---|---|---|---|---|
| | | Temperature, °F | | | | Quenching media[b] | HRC | Preheat & postheat[a] °F | HRC[c] |
| | | Preheat & postheat[a] | Annealing | Austenitizing | Tempering | | | | |
| W1,W2 | Water-hardening | 250-450 | 1360-1450 | 1400-1550 | 350-650 | B,W | 50-64 | 250-450 | 56-62 |
| S1 | Hot-work | 300-500 | 1450-1500 | 1650-1750 | 400-1200 | O | 40-58 | 300-500 | 52-56 |
| S5 | Hot-work | 300-500 | 1425-1475 | 1600-1700 | 350-800 | O | 50-60 | 300-500 | 52-56 |
| S7 | Hot-work | 300-500 | 1500-1550 | 1700-1750 | 400-1150[d] | A,O | 45-57 | 300-500 | 52-56 |
| O1 | Oil-hardening | 300-400 | 1400-1450 | 1450-1500 | 350-500 | O | 57-62 | 300-400 | 56-62 |
| O6 | Oil-hardening | 300-400 | 1410-1450 | 1450-1500 | 350-600 | O | 58-63 | 300-400 | 56-62 |
| A2 | Air-hardening | 300-500 | 1550-1600 | 1700-1800 | 350-1000[d] | A | 57-62 | 300-400 | 56-58 |
| A4 | Air-hardening | 300-500 | 1360-1400 | 1500-1600 | 350-800[d] | A | 54-62 | 300-400 | 60-62 |
| D2 | Air-hardening | 700-900 | 1600-1650 | 1800-1875 | 400-1000[d] | A | 54-61 | 700-900 | 58-60 |
| H11, H12, H13 | Hot-work | 900-1200 | 1550-1600 | 1825-1900 | 1000-1200[d] | A | 38-56 | 700-1000 | 46-54 |
| M1 | High-speed | 950-1100 | 1500-1600 | 2150-2225 | 1000-1100[d] | A,O,S | 60-65 | 950-1050 | 60-63 |
| M2 | High-speed | 950-1100 | 1600-1650 | 2175-2250 | 1000-1100[d] | A,O,S | 60-65 | 950-1050 | 60-63 |
| M10 | High-speed | 950-1100 | 1500-1600 | 2150-2225 | 1000-1100[d] | A,O,S | 60-65 | 950-1050 | 60-63 |
| T1,T2,T4 | High-speed | 950-1100 | 1600-1650 | 2300-2375 | 1000-1100[d] | A,O,S | 60-66 | 950-1050 | 61-64 |

a. Preheat and postheat temperatures for welding
b. A - air cool, B - brine quench, O - oil quench, S - salt bath quench, W - water quench
c. Hardness after postheat
d. Double temper

ment for a specific tool steel should be obtained from the manufacturer. When a hardening heat treatment is required after welding, the weld metal must also respond favorably to the treatment. This must be considered when selecting a proper filler metal for the job.

Where hardened tools must be welded, appropriate procedures must be followed to minimize cracking. Almost all weld repairs to tools fall into this category. Suitable preheat and postweld heat treatments should be used to provide the best properties under the circumstances. Such heat treatments may include stress relieving or tempering.[3] In general, the part temperature should not exceed the original tempering temperature, although sometimes this may not be the case.

## FULL ANNEALING

Annealing of a tool steel is accomplished by (1) heating it slowly to a temperature slightly above its transformation range, (2) holding at that temperature long enough for the heat to penetrate through the entire piece, and then (3) cooling it slowly at a controlled rate to room temperature. The cooling rate must be adjusted to suit the tool size to minimize thermal stresses. Small tools can be cooled at faster rates than large tools. A typical maximum cooling rate of a high-speed steel is 50° F per hour down to about 1000° F. Further cooling usually can be done in still air at a faster rate.

The annealing equipment must provide means to prevent carburization or decarburization. Atmosphere furnaces or salt baths may be used for heating. Pack annealing may also be used in some cases.

## STRESS RELIEVING

Stress relieving may be used to reduce internal stresses caused by welding or machining. It is done by heating the tool at some temperature below the transformation range of the steel. When the tool is in the hardened condition, it may be stress relieved by retempering it. Stress relief of a tool that was welded in the hardened condition should not be done above the temper-

---

3. Information on heat treating of tool steels is given in the *Metals Handbook*, Vol. 2, 8th Ed., Metals Park, Ohio: American Society for Metals, 1964:221-42.

ing temperature of the steel. Such a treatment would alter the hardness and toughness of the entire tool, whereas the welding only softens the heat-affected zone. Heating and cooling rates for stress relieving should be similar to those used for annealing.

## AUSTENITIZING

This operation is accomplished by slowly heating the steel to a temperature above its transformation range and holding at that temperature long enough for resolutioning of the carbides. Small tools may be heated more quickly than large ones. High alloy steels are normally heated very slowly to a temperature just below the transformation range of the steel. They are then heated quickly into the austenitizing temperature range, which may be several hundred degrees higher.

Steels will scale heavily in an oxidizing atmosphere, depending upon the temperature and time. Decarburization will also occur in the austenitizing temperature range. Tools and dies should be heated in a suitable protective atmosphere or vacuum to avoid these problems. Excessive soaking time at a high temperature may cause grain growth. The tool must be properly supported during austenitizing to prevent sagging and distortion.

## QUENCHING

Tool steels are quenched in water, brine, oil, polymers, salt, or air depending on alloy composition and section thickness. The quenching medium must cool the workpiece at a sufficient rate to obtain full hardness. However, an excessive cooling rate should be avoided because of the danger of cracking the tool.

Air-hardening tool steels may be hot quenched to between 1000° and 1200° F. The workpiece should only remain in the quenching medium until the tool temperature has stabilized. Then, the tool can be air cooled or oil quenched to about 150° F before tempering. If the holding time is too long, the austenite will start to transform.

Water-hardening steels tend to distort and change size during quenching. Internal stresses developed during water quenching may easily crack tool components having sharp corners and abrupt section changes that constitute stress raisers. For this reason, shallow hardening is often done using a fine water spray. Where submersion is required, a brine bath will cause less distortion than a water bath.

## TEMPERING

Tempering should be done immediately after quenching to relieve stresses, prevent cracking, and toughen the alloy. In most cases, the tool is not cooled to room temperature between quenching and tempering. The tool should be removed from the quenching medium while still at 150 to 200° F and tempered immediately.

Within the recommended tempering range for the specific steel, a higher temperature gives greater toughness at some sacrifice in hardness. Tempering at the low end of the range gives maximum hardness and wear resistance but lower toughness. Two or more tempering cycles with cooling to room temperature between them is recommended to produce an optimum metallurgical structure. Welding on hardened but untempered tool steel is not recommended, as it will probably crack the piece.

# ARC WELDING

## APPLICATIONS

Tool steels may be welded for one or more of the following purposes:

(1) Assembly of components to form a tool or die

(2) Fabrication of a composite tool by surfacing techniques

(3) Alteration of a tool or die to accommodate part design changes

(4) Repair of worn areas by weld build up

(5) Repair of cracks or other damaged areas

Assembly of components by welding permits the use of less expensive steels for those components that do not require the hardness or wear resistance of tool steels for satisfactory performance. Tough steel parts may be used to support tool steel components or weld deposits. Salvaging of tools or dies to produce modified parts or restore the tool to original dimensions is a significant application of arc welding.

## PROCESS SELECTION

All of the common arc welding processes may be used to fabricate or repair tools and dies. Shielded metal arc, gas metal arc, flux cored arc and plasma arc welding are particularly useful for repairing damaged or worn areas or for fabricating composite tools. Gas tungsten arc welding can be used for repairing small surface defects or for welding of thin sections. Submerged arc and flux cored arc welding may be used to build up large areas, such as steel mill rolls and large die blocks. The availability of a suitable electrode or rod must also be considered during process selection. Shielded metal arc welding with special large electrodes up to ¾ in. diameter may be used to deposit large volumes of weld metal buildup, as for die blocks, or to fill large cavities formed during defect removal.

## FILLER METALS

The important factors that should be considered in selecting a suitable filler metal for an application are as follows:

(1) The composition of the base metal

(2) The heat-treated condition of the base metal (annealed or hardened)

(3) The service requirements of the deposited weld metal

(4) Postweld heat treatment

When welding annealed tool steel, an approximate match of filler and base metal compositions is recommended. Then, the weld metal and base metal will respond similarly during heat treatment.

When welding hardened tool steel, the filler metal should produce a deposit with the desired properties as-welded or, if applicable, in the stress relieved or tempered condition without a hardening heat treatment. In this case, the filler metal composition may be significantly different from that of the base metal.

Electrodes and rods are available commercially for welding most types of tool steels. A matching filler metal may not always be available, but one that will produce a satisfactory weld deposit can usually be found. Some suitable filler metals may be included in specifications for low alloy steel or surfacing electrodes and rods.[4] However, most tool steel filler metals are proprietary and are not covered by applicable standards.

Filler metals may be divided into three main categories. The first category will produce weld deposits corresponding to the basic tool steels: water-, oil-, and air-hardening, hot-work, and high-speed. Applications of these filler metals to some tool steel types are given in Table 3.3. The deposits will be hard in the as-welded condition regardless of the base metal composition. The second category of the filler metals produces low alloy steel deposits that have moderate hardness and toughness at room temperature after peening. Some may respond to heat treatment. The third category includes the stainless steel, nickel, nickel-copper, and copper-nickel filler metals used for buildup of worn tools. These tools may be covered with a hard, wear-resistant weld deposit.

Shielded metal arc welding electrodes for tool steels are designed in the same manner as standard carbon and low alloy steel electrodes. The coverings contain specific ingredients to stabilize the arc, flux the molten metal, and form a protective slag over the weld bead. Alloying elements may also be added to the weld metal by incorporating them in the covering. Flux cored electrodes have similar ingredients incorporated in a core material within a steel tube.

The hardness of the deposited metal from a tool steel welding electrode will vary according to the following factors:

(1) Preweld conditions, such as the preheating temperature employed

(2) The welding sequence

(3) Dilution of the filler metal with the base metal, which can influence the weld metal hardness, as-welded or after heat treatment

(4) The cooling rate

(5) The heat treatment after welding

---

4. Additional information may be found in the following AWS Specifications:

(a) A.5.5-81, *Specification for Low Alloy Steel Coverd Arc Welding Electrodes*

(b) A5.13-80, *Specification for Solid Surfacing Welding Rods and Electrodes*

(c) A 5.21-80, *Specification for Composite Surfacing Welding Rods and Electrodes*

(d) A5.23-80, *Specification for Low Alloy Steel Electrodes and Fluxes for Submerged Arc Welding*

(e) A5.28-79, *Specification for Low Alloy Steel Filler Metals for Gas Shielded Arc Welding*

## SURFACE PREPARATION

Welding of a tool steel should not be attempted unless the surfaces are clean and dry. In cleaning such surfaces, care must be taken to avoid any tool marks that will act later as stress raisers and weaken the tool. Notch sensitivity is especially acute in the tool steels of high hardenability. All cracks should be removed from the surfaces to be welded. All surfaces of the workpiece should be smooth.

If the tool has been annealed, metal can be removed by grinding or thermal cutting. If hardened, grinding or air carbon arc gouging should be used. Glazing or discoloration of a hardened steel during grinding indicates damage to it. Preheating before grinding will lessen the possibility of damage. With oxygen cutting, preheat is indispensible, as is stress relieving before the cut area cools below about 200°F.

## PREHEATING

Tool steels should always be preheated for welding regardless of composition or condition. A suitable preheat temperature depends upon the composition, thickness, and sometimes the condition of the steel. Typical preheat temperature ranges for several types of tool steels are given in Table 3.3. These temperatures should not be confused with those used for preheating a tool prior to austenitizing during hardening of a tool.

When preheating a hardened tool steel, the temperature should not exceed the tempering temperature used previously. Heating to a higher temperature will overtemper and soften the tool. If the tempering temperature is unknown or the section is thick, the selected preheat temperature should be at the low end of the range recommended for the particular tool steel.

With annealed tool steel or thin sections, the component should be preheated at the upper end of the range recommended for that steel. The preheat temperature should be maintained between passes during welding.

## REPAIR PROCEDURES

The importance of the proper operation sequence and welding procedures cannot be overemphasized. Successful repair depends upon performing the appropriate operations properly and in the correct sequence. The main steps in a repair operation are as follows:

(1) Determine the type of steel to be re-

paired and its heat-treated condition.

(2) Select a welding electrode that will provide deposited metal with the desired properties as-welded or after heat treatment.

(3) Perform any suitable heat treatment needed to condition the steel for welding.

(4) Prepare the surface to be welded.

(5) Preheat the tool to the proper temperature.

(6) Deposit the filler metal using proper welding procedures.

(7) Postweld heat-treat the tool to relieve welding stresses or to produce desired properties.

The sequence of operations for repair welding of a tool depends upon its prior heat treatment and the location of the repair. When the tool is in the annealed condition, the suggested sequence is as follows:

(1) Preheat.

(2) Weld with an appropriate filler metal.

(3) Stress relieve.

(4) Machine to near dimensions.

(5) Harden and temper.

(6) Grind to final dimensions.

When the tool is in the hardened condition, the recommended repair sequence is as follows:

(1) Preheat.

(2) Weld with an appropriate filler metal.

(3) Temper.

(4) Grind to dimensions.

In come cases, the repair may be in a noncritical area and a nonhardening filler metal can be used. If the repair area will be subjected to wear, it may be partially filled with a nonhardening filler metal and then completed with a tool steel filler metal. The tool steel layer should be thick enough to accommodate some dilution with the nonhardening weld metal, and have properties similar to the hardened base metal after tempering only.

A common practice is to use a minimum tool steel filler metal thickness of 0.13 inch. The nonhardening buildup layer should have sufficient strength to support the surfacing layer under working conditions. With some hardfacing filler metals, there is a safe deposit thickness above which it will crack during welding.

When the situation permits, the tool should be repair welded in the hardened and tempered condition. This will avoid the need for annealing and hardening heat treatments, and possible costly correction of dimensional changes that these heat treatments may cause. If the base metal was previously hardened but not yet tempered, it cannot be safely welded. Annealing or tem-

pering will be necessary prior to welding. When repairing tool steel in the annealed condition, the filler metal used must produce a weld deposit that will respond to heat treatment, when that is required.

## Base Metal

The type of tool steel to be repaired and its condition of heat treatment are best determined from the original specifications for the tool. If these are not available, a chemical analysis must be made to determine the carbon and alloy content of the steel. The heat-treated condition can be determined using a hardness test. If the steel is hard, it has obviously been quenched and tempered. If it is soft, it is in the annealed condition.

## Welding Process

The shielded metal arc welding process is most versatile for repair welding small areas. Suitable covered electrodes are available for use with most tool steels. Large areas may be repaired more economically with the welding processes that use a continuous wire electrode.

Manual tungsten arc welding (GTAW) is generally used for repair of only small tools. Automatic GTAW, using hot or cold wire feed, is more economical for larger repairs because the smoother surface of the weld metal requires less machining.

Tool steels should not be repaired by oxyacetylene welding. The slowness of this process will introduce too much heat into the base metal. This may cause distortion, excessive softening of hardened metal, embrittlement of annealed metal, or cracking.

## Preparation for Welding

When making partial repairs of cutting edges or working surfaces, the damaged areas should be ground to a uniform depth that will provide a deposit thickness with the required hardness and wear characteristics. Dilution of the base metal into the weld metal must also be considered. A groove depth of 0.13 in. is a common practice. An example is shown in Fig. 3.3. When repairing the entire cutting edge of a tool or die, the edge to be welded should be grooved approximately 45 degress for a sufficient depth. This preparation is shown in Fig. 3.4.

When preparing for repairs of a cast drawing or forming die, the edges or areas should be prepared uniformly to provide deposits of sufficient thickness. It is recommended that studs to reinforce the repair welds be used on cast iron

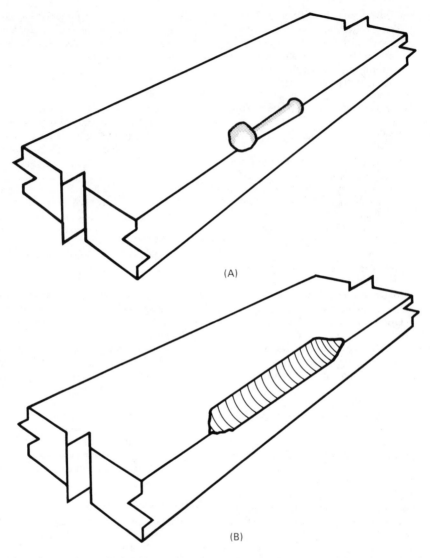

***Fig. 3.3 – Preparation for partial repair: (A) damaged edge, (B) grooved for welding***

that has been saturated with oil or contaminated with sand.

Damaged areas on forging die blocks should be chipped, ground, or machined as uniformly as possible to a finished depth of about 0.19 inch. In any case, all surface defects should be completely removed.

## Welding Procedures

Welding electrodes are available for most classifications of tool steels. The size of the elec-

trode to be used depends upon the width and depth of the damaged area. In general, a 3/32 in. covered electrode will repair a damaged area about 0.09 in. wide and 0.09 in. deep. The same rule applies to the other electrode sizes. Most tool steel covered electrodes are available in diameters of 1/16, 3/32, 1/8, and 3/16 inch. Large covered electrodes, up to 3/4 in. diameter, are available for repair of large areas in molds and dies. The electrode size should be the smallest one that will do the job, especially on sharp

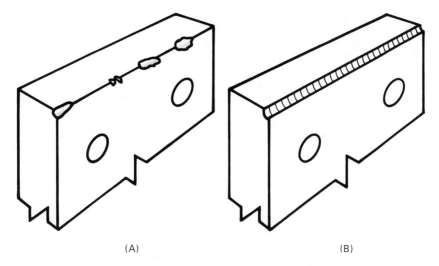

***Fig. 3.4 – Preparation for full repair: (A) damaged edge, (B) grooved for welding***

cutting edges, to keep the heat input to a minimum.

Direct current, electrode positive (reverse polarity) is recommended for most tool steel electrodes. Recommended welding current ranges are usually supplied with the welding electrodes.

When making partial repairs of cutting edges or working surfaces, the electrode should deposit metal that will respond satisfactorily to the heat treatment normally given to the base metal. In making full repairs to cutting edges or working surfaces, the electrode should deposit metal that will have characteristics best suited for the work that the tool will perform. Factors such as resistance to heat, abrasion, and shock and also the thickness of metal to be cut or formed should be considered during electrode selection. For repairing forging dies or the facing of cast iron or carbon steel drawing and forming dies, welding electrodes recommended for the specific application should be used.

When welding on one die unit with two or three dissimilar tool steel electrodes, the electrodes should be used in sequence according to the required preheat temperature. The first electrode should require the highest preheat temperature, and last electrode the lowest preheat temperature. This procedure prevents overtempering of previous weld deposits with a high preheat. The temperatures should, of course, be compatible with the base metal.

The temperature of the base metal should be kept as uniform as possible during the welding operation to ensure uniform hardness of the weld metal. It should not be permitted to exceed the tempering temperature of the tool steel being welded.

Each weld bead should be peened before the metal cools to below about 700° F. This might require occasional interruption of welding when the repair is extensive. Peening is a mechanical means of working the weld metal and relieving welding stresses. Ball peen hammers are generally used, but small pneumatic hammers are efficient for large repairs. The benefits of peening weld metal are well recognized, and the method is commonly practiced.

Peening must be correctly done or more harm than benefit may result. The type of peening hammer, the amount of peening, and the effects of cold working the weld metal should be considered. Too little peening will not properly relieve stresses. On the other hand, severe peening may cause cracking or other harmful effects especially under improper conditions.

When repair welding cutting edges, the blade should be positioned so that the weld metal will flow or roll over the edges. Arc striking on an adjacent plate will avoid starting marks.

Welding should be done in the flat position with the axis inclined slightly. The direction of welding should be uphill, if possible. Gravity will cause the molten weld metal to flow downhill and build up evenly. The slag will also flow back and keep the crater clean. Stringer beads, rather than weave beads, should be used.

Travel speed should be adjusted to produce an even deposit and to assure uniform fusion of the weld metal with the base metal. The weld should be cleaned by frequent chipping and brushing.

A number of small passes should be used to fill the groove. The bead size of final passes should be adjusted so that the repair will be as close as possible to final size to minimize the grinding operation. When breaking the arc, the arc length should be gradually decreased, and then the electrode rapidly moved back over the hot weld metal. This will avoid deep craters and the melting of adjacent sharp edges.

When repairing sections of cutting edges, the welding technique should avoid craters and melting of the edges at the extreme ends of the repair. Welding should first progress in one direction to within a short distance of the other end of the prepared groove as shown in Fig. 3.5(A). Then it should progress in the opposite direction and overlap the first bead as illustrated in Fig. 3.5(B). When repairing deeply damaged cutting edges or drawing and forming surfaces, welding should start at the bottom of the preparation and gradually fill it up. A slightly higher amperage should be used for the first and second passes than for the succeeding passes.

A thick buildup of weld metal is often required to restore a severely worn tool or die to original dimensions. Deposition of a thick layer of hard weld metal may cause it to crack. To minimize this possibility, the initial buildup should be done with a ductile, high-strength, nonhardenable filler metal, such as a low alloy steel. This layer is then covered with a weld metal that will provide the desired wear resistance and toughness as-welded. This procedure should not be used if the welded tool will be hardened and tempered because the deposited metal would likely crack during heat treatment. Intermittent welding should be used when repairing the entire cutting or forming edge of a draw ring, a female extrusion die, or a similar circular part, to ensure even distribution of heat.

During the welding operation, warpage or distortion is counteracted by preheating and peening. The use of shims and clamps is also helpful. For example, before welding the entire edge, a long shear blade can be clamped with a reverse curvature to allow for the shrinkage stress developed in the weld metal. If the curvature is proper, the welded blade will be straight when released.

## Postweld Heat Treatment

After welding, the repaired tool should be cooled to near 150° F and then reheated to the recommended tempering temperature. Typical tempering temperatures are given in Table 3.3.

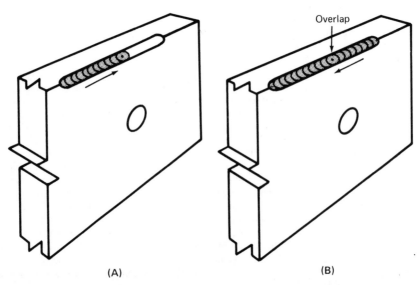

(A)                                    (B)

*Fig. 3.5 – Technique to avoid craters or edge damage during repair welding*

Tools that are repaired in the annealed condition should be hardened and tempered according to the recommendations of the material supplier. Table 3.3 may be used as a guide.

As a general rule, tools that require only partial repair of working areas should be tempered at the recommended temperature for the tool steel. When an entire cutting edge or working area has been repaired, the tool should be tempered at the temperature recommended for the tool steel filler metal.

## COMPOSITE CONSTRUCTION

For some applications, a tool can be compositely fabricated using a carbon or a low alloy steel base, and then building up the cutting edge or the working area with a tool steel electrode having the desired characteristics. An example of a composite tool is shown in Fig. 3.6.

This type of construction can provide a working surface of desired characteristics and a tough, shock-resistant core. Advantages include the elimination of a hardening heat treatment and the fact that machining or drilling can be a final operation. Many tools and dies are susceptible to breakage because of their inherent hardness. A resilient core with composite fabrication reduces the likelihood of this.

Sometimes it is impractical or impossible to construct a die of one type of tool steel that can satisfactorily perform several functions, such as trim, form, and restrike. This can be accomplished by building up the cutting or working areas of a composite die with weld metals having the desired characteristics to perform each operation.

To improve the life of an existing tool steel unit, cutting edges or working surfaces can be resurfaced with a more appropriate tool steel for the application. However, the procedures must be designed with the prior heat-treatment of the unit in mind.

### Base Metal

The base steel for a composite tool or die

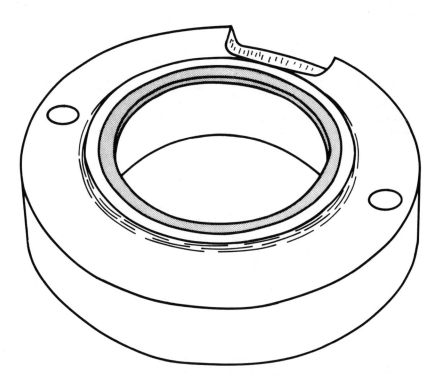

*Fig. 3.6 – A compositely fabricated female trimmer with an AISI 1035 steel base and a tool steel cutting edge*

must possess the required mechanical properties for the application. These may include strength, toughness, hardness, or a combination of these, at the design operating temperature. Toughness is important for impact loading. Hardness may be important for proper support of the tool steel weld metal under high compression loading.

Carbon and low alloy steels may be suitable for most applications. When the tool will operate at elevated temperature, an alloy steel with acceptable properties at temperature must be used for the base. Any heat treatment needed to provide the desired strength and hardness of a low alloy base should be done prior to welding.

### Electrode Selection

The electrode should be selected to provide a weld deposit having the characteristics best suited for the type of work that the tool will have to do. Factors, such as resistance to heat, abrasion, shock, and also the thickness of metal to be cut or formed, should be considered.

The size of the electrode depends upon the extent of welding and the type of preparation selected. Dilution of the base metal must also be considered in the selection.

### Groove Design

Many types of preparation may be used for cutting or forming edges in composite construction. Four•common preparations are the angle type, the 90° shelf type, straight shelf type, and the buildup type. These are illustrated in Fig. 3.7. A radius groove is also used. With the exception of the buildup type, the design should provide a thickness that will minimize the effects of dilution and provide the required strength after the edge has been machined to size. A· minimum thickness of 0.19 in. is common practice. Two or more weld passes should be used to fill the groove to minimize the dilution of the filler metal with the base metal.

On units that have large areas to be filled, the base metal surface should be prepared so that the finished weld deposit will have adequate thickness (0.19 in. is common practice). When converting existing tool steel units into composite ones, the angle type of preparation should be used. The edge should be machined back far enough to allow for a sufficient thickness of finished weld metal.

Composite fabrications should be made oversize to allow for distortion during welding and finish machining. On ring type or circular units, at least 0.125 in. is generally provided for grinding or machining to finish dimensions.

For the composite construction of cast iron drawing and forming dies, edges or areas to be faced are generally prepared uniformly so that finished deposits are at least 0.125 in. thick. Nickel, copper-nickel, or nickel-copper filler metal can be used as a buttering layer for the tool steel weld metal. Provisions should be made in the preparation to allow for the buttering passes.

### Preheat

A preheat in the rage of 200° to 400° F is recommended when using mild, medium carbon, or high carbon steel as a base. Small units can be heated to the low end of the preheat range, and large units to the high end. When using steels of the nickel-molybdenum or nickel-chromium type, a preheat of 600° to 700°F is recommended.

### Welding Procedures

The welding procedures for composite fabrication should be similar to those used in making repairs to existing units.

### Postweld Heat Treatment

A composite unit with welded tool steel edges should always be tempered after welding. The tempering temperature should be that recommended by the electrode manufacturer, but should not exceed the one specified for the base metal. This is true even when converting tool steel units to composite fabrication.

When nonhardenable low alloy steel electrodes are used to provide desirable wear characteristics at certain locations, a stress relief heat treatment is recommended.

## HIGH-SPEED STEEL CUTTING TOOLS

The most practical way to repair high-speed cutting tools .is by the gas tungsten arc welding (GTAW) process. Shielded metal arc welding is not suitable because the high welding heat may cause checking and cracking of the tools.

### Repair

Repairs can be made by the GTAW process to teeth and flutes on milling cutters, broaches, drills, ball and end mills, reamers, taps, and other cutting tools. In addition, tool life can be prolonged by building up worn areas. The as-welded deposits of high-speed steel welding rods are easily ground to original dimensions.

### Fabrication

All types of cutoff, turning, grooving, and

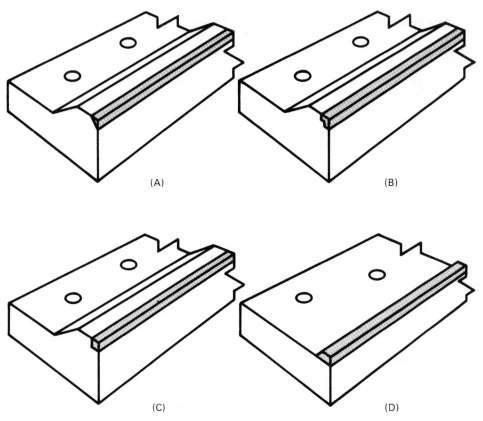

***Fig. 3.7 – Types of preparation for cutting or forming edges: (A) angle, (B) 90-degree shelf, (C) straight shelf, (D) buildup***

special shaped forming tools as well as lathe centers and centerless grinder blades can be compositely fabricated by using a low or medium carbon steel or SAE 1335 alloy steel as a base. Tool bases, to which the high-speed steel filler metals are applied, are machined in the same manner for carbide inserts. The appropriate location is built up with an air-hardening, high-speed steel filler metal. The welded tool can be annealed to facilitate machining. The machined tool can then be heat-treated.

### Welding Rods

There are two AWS classifications for high-speed steel welding rods, namely RFe5-A and RFe5-B.[5] The rods may be solid wire or com-

posite constructions. They are suitable for applications where high hardness is required at service temperatures up to 1100° F and good wear resistance and toughness are also required. These are high carbon, chromium-molybdenum-tungsten-vanadium alloy steels. There are also other proprietary high-speed steel welding rods available.

Deposits of RFe5 filler metals are well suited for metal-to-metal wear, especially at elevated temperatures. They have low coefficients of friction. Compressive strengths are good.

If the weld metal requires machining, it can be annealed in the 1550° to 1650° F range. After machining, the weld metal can be hardened by preheating to the 1300° to 1500° F range, austenitizing at 2200° to 2250° F, air or oil quenching and then double tempering at 1025° F for two hours. Appropriate procedures must be used to prevent decarburization during heat treatment.

5. Refer to *AWS A5.13-80, Specification for Solid Surfacing Welding Rods and Electrodes,* and *AWS A5.21-80, Specification for Composite Surfacing Welding Rods and Electrodes.*

## Welding Procedures

The surface to be welded must be free of cracked or unsound metal. Cutting edges to be repaired should be prepared by grinding using an angle type of preparation (see Fig. 3.7).

Recommended preheat, interpass, and post-weld temperatures are given in Table 3.3. In any case, the tempering temperature of a hardened base metal should not be exceeded. It is important that the interpass temperature be maintained during welding.

Welding should be done with the GTAW process in the flat position using DCEN(SP) power and argon shielding.[6] Welding current should be limited to that necessary for good fusion. Heat input into the base metal should be kept to a minimum by using a backhand welding technique. The arc should impinge upon the end of the welding rod and the molten weld pool. If more than one pass is used, each pass should be peened before depositing the succeeding one.

## Postheat

After welding is completed, the tool shoiuld be air cooled to about 200° F. It should then be double tempered at the temperature recommended for the filler metal.

# FLASH AND FRICTION WELDING

High-speed tool steel bodies can be flash or friction welded to carbon or alloy steel shanks in the manufacture of drills, reamers, and similar tools. This technique will provide a tough, ductile shank at lower material cost than will a one-piece unit.

Tool steels can be welded by both processes before and after heat treatment. Changes in hardness are restricted to a relatively narrow heat-affected zone. Rapid cooling will reharden the metal at the weld interface, including the flash. Preheating can be done at the beginning of the heating phase to reduce the cooling rate in the heat-affected zone. This will produce a more ductile joint and minimize the likelihood of cracking. These processes can be adapted to high production and can be automated.

# BRAZING

## BASE METALS

For brazing, it is convenient to group tool steels in two broad classifications: carbon and high-speed. The carbon tool steels depend primarily on the high carbon content (0.60-1.40%) for hardness. Except for thin sections, these steels must be quenched rapidly during heat treatment to achieve optimum properties after tempering. Alloying elements may be added to the carbon steels to impart special properties, such as reduced distortion during heat treatment, greater wear resistance and toughness, or better high-temperature properties. Such steels are the alloy tool steels.

High-speed steels are classified separately because their properties depend upon relatively high percentages of alloying elements including tungsten, molybdenum, chromium, and vanadium. These steels, as well as some alloy tool steels, require high austenitizing temperatures. This characteristic must be considered when developing a brazing procedure.

## FILLER METALS

The choice of brazing filler metal depends upon the properties of the tool steel to be brazed

---

6. The effects of gas tungsten arc welding variables on surfacing are given in Table 14.3, *Welding Handbook*, Vol. 2, 7th Ed., 539.

and the heat treatment required to develop its optimum properties. Most of the brazing filler metals of the silver, copper, and copper-zinc classifications may be used.[7] The best filler metal should be determined for each specific application.

## JOINT DESIGN

Lap or sleeve type joints are generally used when brazing tool steels. With silver or copper-zinc types of filler metals, joint clearances at the brazing temperature should be 0.002 to 0.005 inch. The filler metal should be preplaced so that it is protected from direct contact with the heat source. Where brazing and hardening operations are combined, the joint should be designed so that the brazing filler metal is under compression during quenching. A tensile stress on the joint at temperatures near the melting point of the filler metal may crack the joint.

## EQUIPMENT

Torch, furnace, and induction heating are the methods most commonly used for brazing tool steels. Available equipment is frequently the main factor when deciding the heating method to be used.

## SURFACE PREPARATION

The base metal surfaces must be clean and free from oil, oxide, or other foreign material to secure proper wetting and flow of the filler metal. A machined or roughened surface is always preferable to a smoothly ground or polished surface. Paste flux and brazing filler metal do not wet and flow well on extremely smooth surfaces.

## FLUXES AND ATMOSPHERES

In general, AWS brazing flux types 3A and 3B are used for brazing tool steels.[8] Some modification of the flux may be required for a particular tool steel and brazing temperature.

A controlled atmosphere may be used to

prevent oxidation during heating and to avoid the necessity of a postbraze cleaning operation. If a controlled atmosphere is used, steps must be taken to prevent carburization or decarburization of the tool steel.

## TECHNIQUES

The brazing of carbon tool steel is best done prior to or at the same time as the hardening operation. The hardening temperature for carbon steel is normally in the range of 1400° to 1500° F. If the brazing is done prior to the hardening operation, the filler metal must solidify well above this temperature range. Then, the assembly can be handled without joint failure when reheated to the hardening temperature. A copper filler metal is frequently used for this purpose. However, the high brazing temperature required by copper filler metal may adversely affect the structure of some steels. Silver and copper-zinc filler metals are available that can be brazed at temperatures in the range of 1700° to 1800° F.

When the brazing and hardening operations are combined, a filler metal that solidifies just above the austenitizing temperature is generally used. Particular attention must be given to the joint design and the handling of the assembly because the joint strength will be very low at the austenitizing temperature of the steel. Any stress developed during quenching should put the brazed joint in compression rather than in tension.

The brazing of the alloy tool steels depends upon knowledge of the particular steel involved. The alloy tool steels have a wide range of compositions and, therefore, wide differences in behavior during heat treatment or heating for brazing. The tool steel should be studied carefully to determine its proper heat-treating cycle, the quenching medium, the best brazing filler metal, and the proper technique for combining the heat-treating and the brazing operations to achieve maximum service.

Quenching may produce steep temperature gradients in a brazement, and the differential expansions and contractions in it may rupture the brazed joint. Initially, the austenitic steel contracts with falling temperature. When the transformation to martensite takes place, the steel expands. Finally, the martensitic steel contracts as the temperature continues to fall. These changes do not take place uniformly because cooling begins from the surface. If the assembly

---

7. Refer to *AWS A5.8-81, Specification for Brazing Filler Metals,* for the various AWS brazing alloy classifications, their compositions, and brazing ranges.
8. Brazing fluxes and atmospheres are discussed in the *Welding Handbook,* Vol. 2, 7th Ed., 397-407.

can be properly supported during quenching, a filler metal that solidifies well below the austenitizing temperature may be used.

High-speed tool steels require hardening treatments at temperatures above the usual silver brazing temperatures. It is, therefore, common practice to harden the steels prior to brazing, and then braze during or after the second tempering treatment. Tempering is usually done in the range of 1000° to 1200° F. Brazing filler metals such as BAg-1 or BAg-1a can be used at temperatures above 1150° F if short brazing cycles are used. Hardened high-speed tool steel may be brazed in this manner without overtempering.

Broken high-speed tools may be repaired by brazing. This may avoid delays in production while awaiting a replacement tool. Broaches, circular saws, and milling cutters are examples of tools that can be salvaged by brazing.

## SAFE PRACTICES

During the welding and heat treatment of tool steels, proper safe practices must be followed to prevent injury to personnel and damage to plant and equipment. The equipment must be installed properly and be equipped with appropriate safety devices for the particular operation.

Personnel must be equipped with appropriate eye, ear, and body protection to avoid burns from the arc, spatter, hot metal, or quenching media. Adequate ventilation must be provided to remove harmful fumes and gases from the breathing zone of persons working in the area.

The requirements of ANSI Standard Z49.1, *Safety in Welding and Cutting,* as well as appropriate federal, state, and local codes should be followed when welding and heat treating tool steels.

## Metric Conversion Factors

$$t_C = 0.56(t_F - 32)$$
$$1 \text{ in.} = 25.4 \text{ mm}$$

## SUPPLEMENTARY READING LIST

*Metals Handbook: Heat Treating, Cleaning and Finishing,* Vol. 2, 8th Ed., Metals Park, OH: Amer. Soc. for Metals, 1964.

*Metals Handbook: Properties and Selection: Stainless Steels, Tool Materials and Special Purpose Metals,* Vol. 3, 9th Ed., Metals Park, OH: Amer. Soc. for Metals, 1980.

Roberts, G. A., and Cary, R. A., *Tool Steels,* 4th Ed., Metals Park, OH: Amer. Soc. for Metals, 1980.

*Steel Products Manual – Tool Steels,* Washington, DC: Amer. Iron and Steel Inst., 1978 Mar.

Wilson, R., *Metallurgy and Heat Treatment of Tool Steels,* New York: McGraw-Hill, 1975.

# 4

# High Alloy Steels

## Chapter Committee

B.A. GRAVILLE *Chairman*
*Welding Institute of Canada*
D.W. DICKINSON
*Republic Steel Corporation*

D.W. HARVEY
*US Welding*
W.L. LUTES
*Welding Consultant*

## Welding Handbook Committee Member

J.R. HANNAHS
*Midwest Testing Laboratories, Incorporated*

# 4

# High Alloy Steels

## GENERAL CONSIDERATIONS

This chapter covers the weldability of alloy steels that contain more than about 5 percent total alloy additions except stainless steels, chromium-molybdenum steels, 9 percent nickel steels (ASTM A353 and A553), and tool steels. These excepted steels are discussed in other chapters. Four general classes of high alloy steels discussed here are (1) the nickel-cobalt family, (2) 5 percent chromium-molybdenum-vanadium steels, (3) maraging steels, and (4) austenitic manganese steels. These steels are designed to provide tensile and yield strengths significantly higher than those of common low alloy steels.

The tensile strength of plain carbon steels may be increased by simply raising the carbon content. However, this approach has limitations because plain high carbon steels, per se, are usually shallow hardening, and are not easily welded. Their range of mechanical properties is usually severely limited. Toughness and ductility are inadequate for many structural applications. Nevertheless, a number of high-strength carbon steels are used in the quenched and tempered condition. Notch sensitivity of carbon steels is greatly improved by addition of manganese. However, low carbon content and addition of strong carbide-forming elements give superior combinations of strength and toughness.

The toughness and weldability of alloy steels are related to the amount and type of non-metallic inclusions in them. Reduction of non-metallic inclusions by special melting techniques improves the mechanical properties of high alloy steels at high strength levels. Manipulation of the transformation characteristics of alloy steels also contributes to improved strength and toughness. This is achieved either by controlled processing of billet, plate, rod, bar, and other mill forms during hot working or by subsequent heat treatment. Combinations of alloying, refining, processing, and heat treatment provide a number of premium quality, low-residual, high-strength alloy steels that exhibit good fracture toughness and through-thickness properties.

The metallurgical principles involved in high alloy steels designed for high strength include the formation of fine carbides or other intermetallic compounds, dispersed in a fine-grained, solution-strengthened martensitic matrix as well as the absence of brittle grain-boundary films and non-metallic inclusions. A certain amount of carbon is always present in alloy steels in amounts up to 0.45 percent. Carbon invariably forms carbides with chromium, molybdenum, vanadium, or columbium when they are present singly or together in steels.

High-strength alloy steels generally are susceptible to embrittlement when certain impurity elements exceed some level. Such elements tend to segregate during solidification. Fracture commonly takes place along the prior austenitic grain boundaries where the impurities tend to concentrate. Temper embrittlement, for instance, is known to occur when certain steels are held or slowly cooled through the 1050° to 700°F range during or after tempering.

Special deoxidation practices and refinement techniques can substantially reduce the levels of undesired residual elements in steel. Sulfur and phosphorus contents too often are less than 0.01 percent.

Vacuum induction melting, argon rinsing, vacuum degassing, electroslag and vacuum remelting, rare earth treatment, and other special steel refining techniques are used to reduce inclusions and impurities to very low levels. These practices provide reasonable toughness and ductility at high strength levels and also improve the properties in the weld heat-affected zone.

Many alloy steels can be heat treated to tensile strengths exceeding 200 ksi with moderate ductility. However, the normal criterion for high strength steels that are designed for structural applications is good fracture toughness properties at low temperature. A second requirement is that the steels have nearly uniform isotropic mechanical properties. The number of commercial alloy steels that meet such criteria are relatively few and proprietary. The steels discussed here are representative.

# ULTRA HIGH STRENGTH STEELS

## DEFINITION AND GENERAL CHARACTERISTICS

The so-called ultra high strength steels are a group of alloy steels designed for structural applications having at least 180 ksi yield strength or 200 ksi tensile strength combined with good fracture toughness. They generally are used for critical applications requiring high strength-to-weight ratios as well as excellent reliability and consistant response to heat treatment. Maraging steels, which are also capable of high strength levels, are discussed in the following section.

The ultra high-strength steels are capable of high strength and good toughness as a result of alloying additions that restrict the decomposition of austenite by diffusion and, at the same time, promote the formation of tough martensite on quenching. Subsequent tempering may lead to secondary hardening that further increases the strength of these steels.

Welded joints in these steels generally do not require heat treatment except for a stress relief in certain applications. With specific alloys, the weldment may be quenched and tempered to obtain desired properties.

Two general types of ultra high-strength steels are the nickel-cobalt family and the 5 percent chromium-molybdenum-vanadium family. The latter steels are also known as hot-work die steels, and are discussed in Chapter 3 for tool and die applications. Chemical compositions of typical ultra high strength steels, other than the maraging group, are given in Table 4.1.

## NICKEL-COBALT ALLOY STEELS

### General Description

Several ultra high-strength steels contain 8 to 10 percent nickel, 4 to 14 percent cobalt, and small additions of chromium and molybdenum. The nickel provides good toughness, improves strength by solid solution, and promotes hardenability. Cobalt is added to maintain a high austenite-to-martensite transformation ($M_s$) temperature and thus limit the amount of retained austenite in the steel after quenching. It also contributes to solid solution strengthening. With low cobalt addition, refrigeration is required to transform retained austenite to martensite.

Chromium and molybdenum, two carbide forming elements, are added to promote hardenability and strength, but they do contribute to embrittlement. Additions of these elements are kept low to maintain good toughness and produce a fairly uniform response to tempering without pronounced secondary hardening. Silicon and other residual elements are kept low for optimum toughness.

The $M_s$ temperatures of these steels are sufficiently high to permit self-tempering of the martensite as the steel cools through the transformation range to room temperature. With relatively low carbon content, a strong, tough martensite is formed as-quenched. This behavior contributes to good weldability. As the carbon content is increased, weldability and toughness of the steel decrease.

**Table 4.1**
**Chemical compositions of typical ultra high-strength alloy steels**

| Designation | Composition, percent[a] | | | | | | | | | | | |
|---|---|---|---|---|---|---|---|---|---|---|---|---|
| | C | Mn | P | S | Si | Ni | Cr | Mo | V | Co | Al | Ti |
| AF 1410 | 0.16[b] | 0.05 | 0.01 | 0.01 | 0.1 | 10[b] | 2[b] | 1[b] | 0.1[b] | 14[b] | 0.015[b] | 0.01 |
| H-11 | 0.33-0.43 | 0.20-0.50 | 0.030 | 0.030 | 0.80-1.20 | 0.30 | 4.75-5.50 | 1.10-1.60 | 0.30-0.60 | | | |
| H-11 Modified | 0.37-0.43 | 0.20-0.40 | 0.030 | 0.030 | 0.80-1.00 | | 4.75-5.25 | 1.20-1.40 | 0.04-0.06 | | | |
| H-13 | 0.32-0.45 | 0.20-0.50 | 0.030 | 0.030 | 0.80-1.20 | | 4.75-5.50 | 1.10-1.75 | 0.08-1.20 | | | |
| HP9-4-20[c] | 0.16-0.23 | 0.20-0.40 | 0.010 | 0.010 | 0.10 | 8.50-9.50 | 0.65-0.85 | 0.90-1.10 | 0.06-0.12 | 4.25-4.75 | | |
| HP9-4-30 | 0.29-0.34 | 0.10-0.35 | 0.010 | 0.010 | 0.10 | 7.25-8.00 | 0.90-1.10 | 0.90-1.10 | 0.06-0.12 | 4.25-4.75 | | |
| HY-180 | 0.09-0.13 | 0.05-0.25 | 0.010 | 0.006 | 0.15 | 9.5-10.5 | 1.8-2.2 | 0.09-1.10 | | 7.5-8.5 | 0.025 | 0.02 |

a. Single values are maximum unless otherwise noted.
b. Nominal values.
c. Refer to ASTM Specification A605-72 (1977).

## Heat Treatment and Properties

*HY-180 Steel.* A tentative isothermal diagram for HY-180 steel is shown in Fig. 4.1. The steel is hardened by quenching from above the $Ae_1$ temperature which also increases its strength and toughness. Further strengthening occurs during aging with the precipitation of a fine dispersion of secondary hardening alloy carbides in a heavily dislocated martensitic matrix. The effect of aging temperature on the strength and toughness of HY-180 steel is shown in Fig. 4.2. Aging at 950°F provides maximum tensile and yield strengths, but these properties decrease with longer aging times.

Suggested heat treatments for HY-180 steel are given in Table 4.2. Typical mechanical properties of HY-180 steel plate in the quenched and aged condition are given in Table 4.3.

*HP 9-4-XX Steels.* Two premium grade ultra high-strength alloy steels are HP9-4-20 and HP9-4-30. Both steels are designed for applications requiring optimum combinations of high yield strength and toughness with good weldability. HP9-4-20 steel is capable of developing a

yield strength of 180 ksi in combination with a Charpy V-notch impact strength of 50 ft·lbs at room temperature in plate thicknesses up to 4 inches. HP9-4-30 steels can be heat treated to a yield strength of 220 ksi. HP9-4-20 steel has better weldability, temper resistance, and toughness than HP9-4-30 steel.

Figure 4.3 shows the isothermal transformation diagram for HP9-4-20 alloy steel. The diagram for HP9-4-30 steel would be very similar to this. Heat treatments for HP9-4-20 and HP9-4-30 steels are shown in Table 4.4.

Refrigeration to below −100°F after quenching from the austenitizing temperature will transform retained austenite to martensite. During aging, the response will be uniform and secondary hardening will not be significant. Figure 4.4 shows the microstructures of HP9-4-20 steel plate as-rolled and after quenching and tempering heat treatment.

Typical mechanical properties for HP9-4-20 and HP9-4-30 alloy steels after quenching and tempering are given in Table 4.5. The higher carbon content of HP9-4-30 accounts for its higher strength properties and somewhat lower

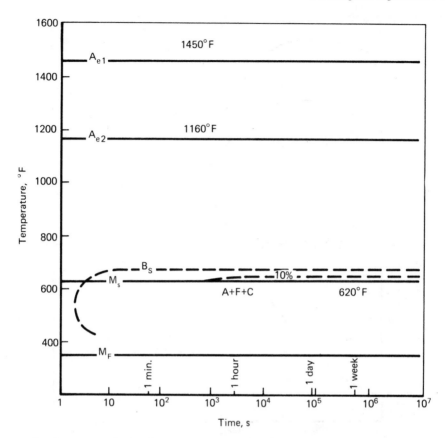

**Fig. 4.1—Tentative isothermal transformation diagram for HY-180 alloy steel**

ductility and toughness.

*AF1410 Steel.* This nickel-cobalt alloy steel has good hot and cold-working characteristics and is readily machined. Generally, this alloy is available as bar stock or as billet stock for forging. This alloy steel also uses the cobalt-nickel effect to prevent transformation of austenite on quenching to normal equilibrium eutectoidal phases. It is strengthened during tempering by carbide precipitation, not by the formation of intermetallic compounds. The hysteresis of the reverse transformation of martensite to austenite enables tempering to readily take place within a supersaturated martensitic matrix. An understanding of the metallurgical response of Ni-Co alloy steels to a welding thermal cycle aids in establishing successful welding procedures for them.

## Applications

Nickel-cobalt alloy steels are designed for structures where the metal will be loaded to near its yield strength to meet requirements of structural efficiency and economy. Such applications include helicopter components, armor, high-performance ships and shallow-draft vessels, rocket and missile cases, air frame and under-carriage components, aerospace and hydrospace structures, pressure vessels, and other heavy section structures.

HP9-4-20 steel has been used for pressure vessels, structural components in aircraft, rotor shafts for metal forming equipment, and high-strength, shock-absorbing automotive parts. HP9-4-30 steel, a forging grade alloy, has been used for a variety of aircraft forgings such as engine mounts and landing gear.

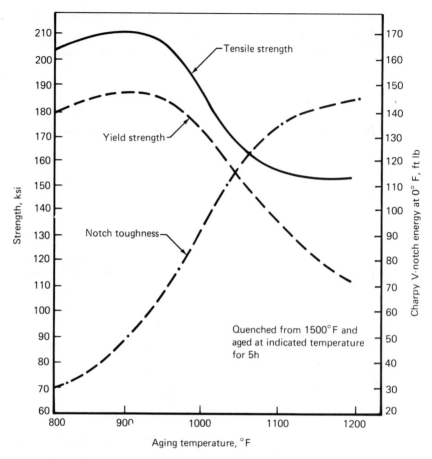

**Fig. 4.2 – Effect of aging temperature on the mechanical properties of HY-180 alloy steel plate, 0.5 in. thick**

AF1410 steel combines ultra high strength with superior fracture properties and stress corrosion resistance. The low carbon content of this alloy enables the material to be fabricated easily. The alloy has been designed to compete with titanium alloys at lower cost and superior strength-to-weight ratio. AF1410 steel may be used for aerospace applications, armor plate, gears, dies, fasteners, and bearings.

Suggested guidelines for designing welded ultra high-strength steel structures are as follows:

(1) Avoid fillet welded joints, such as T-joints, by use of forgings and butt joints.

(2) Place welded joints in low stress locations with good accessibility for welding and inspection.

(3) Weld with automatic or mechanized equipment to ensure consistent welding conditions.

(4) Use weld groove designs that minimize the amount of filler metal required to fill them.

(5) Use premium quality, clean filler metal.

(6) Provide good supervision for the welding operation.

## Welding Considerations

For acceptable welds in high-strength alloy steels, the mechanical properties of the weld metal must be equal to those of the base metal. The combination of strength and toughness can

**Table 4.2**
**Heat treatments for HY-180 steel**

| Heat treatment | Procedure |
|---|---|
| Normalizing | Hold at 1650°F ± 25°F for 1 h/in. of thickness (3 h maximum), air cool. |
| Hardening | Hold at 1500°F ± 25°F for 1 h/in. of thickness (3 h maximum), water quench. |
| Tempering | Hold at 950°F for 5 h for thicknesses through 1 in., and for 10 h for greater thicknesses. |
| Softening | Hold at 1150°F for 16 h (decreases hardness to 31 HRC) |

**Table 4.3**
**Typical mechanical properties of 2 inch HY-180 steel plate in quenched and tempered condition**

| | |
|---|---|
| Tensile strength, ksi | 200 |
| Yield strength, ksi | 185 |
| Elongation in 2 in., percent | 17 |
| Reduction of area, percent | 71 |
| Charpy V-notch impact strength at 0°F, ft·lbs | 90 |

**Table 4.4**
**Heat treatments for HP9-4-20 and HP9-4-30 alloy steels**

| Heat treatment | Procedure |
|---|---|
| Normalizing | Hold at 1600° to 1700°F for 1 h/in. of thickness, air cool. |
| Hardening | Hold at 1525° to 1575°F for 1 h/in. of thickness; water or oil quench; refrigerate to −125° to −150°F for 1 h; warm to room temperature. |
| Tempering | Hold at 1000° to 1075°F for 2 h, air cool; then repeat operation. |
| Softening | Hold at 1150°F for 24 h, air cool. |
| Stress relieving | Hold at 1000°F for 24 h, air cool. |

only be achieved with weld metal containing very low impurity levels. Oxygen, hydrogen, nitrogen, carbon, sulfur, and phosphorus must be controlled to specified levels. Welded joints must be free of defects including lack of fusion, microcracks, coarse grain structure, embrittling phases at grain boundaries, and non-metallic inclusions of sufficient size or number to initiate cracks or reduce the effective section thickness.

Ultra high-strength alloy steels can only be successfully welded under controlled conditions. Exceptional cleanliness of the filler metal, the joint area, and the weld metal itself are essential. Welding heat input must be controlled to a low, consistent level to take advantage of grain refinement in prior weld beads and to minimize structural degradation of the heat-affected zone. High energy welding processes can be used with some sacrifice in mechanical properties provided the weld metal *cleanness* level can be kept exceptionally low. An essential monitor of weld metal cleanness is the total oxygen content. For HY-180 alloy steel weldments, for example, oxygen content of less than 50 ppm is necessary.

Multiple pass welding of alloy steels of the Ni-Co family presents many interesting metallurgical features. Notably, the reheating cycle of succeeding weld passes gives rise to cyclic phase transformations. Originally-formed, self-tempered martensite in reheated zones may revert to austenite. Such austenite inherits an ultrafine dispersion of carbide particles. High stresses developed in the fast-cooling, restrained weld metal generate a high density of lattice defects. This strained austenite portion of weld metal retransforms to martensite which retains much of the strain and second-phase particles. This behavior results in a fine-grained martensite. Assuming that this sequence of events takes place, it explains why welded joints filled with small weld beads have better mechanical properties than joints made with large weld beads. Low energy welding processes provide interpass grain refinement of the type described. Thus, gas tungsten arc welding with cold wire feed is generally recommended for welding these steels in preference to gas metal arc welding or gas tungsten arc welding with hot wire feed.

Nickel-cobalt high-strength alloy steels are often welded in the heat-treated condition. Heat treatment after welding is not normally required, except for stress-relief in special situations, because low heat input welding with associated rapid cooling results in minimum deterioration

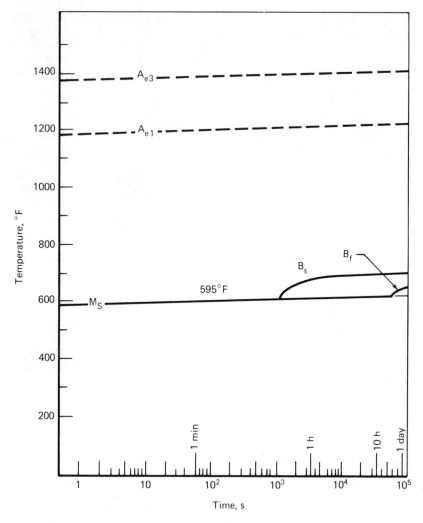

**Fig. 4.3—Isothermal transformation diagram for HP9-4-20 alloy steel**

of properties in the heat-affected zone.

Successful fabrication of these steels is dependent upon the use of low-hydrogen welding techniques and matching filler metals with low impurity levels. The filler metal generally must be manufactured from selected melting stock and double refined. An example of the cleanliness needed in filler metal is the AF1410 alloy steel wire composition given in Table 4.6. Steel of this quality is produced by vacuum induction melting followed by vacuum arc remelting.

Shielding gases should be essentially free of oxygen, nitrogen, and hydrogen. Dry, pure argon, helium or mixtures of both can be used. Filler rod or wire must be absolutely clean; no surface debris or lubricant can be permitted. Likewise, all joints must be thoroughly cleaned of all traces of cutting fluid, lubricants, oils and greases, scale, oxide films, and other contaminants.

Heat-affected zone and weld metal properties are maintained by controlling heat input and interpass temperature, and sometimes by a postweld heat treatment. With premium-quality high-strength alloys, only the gas tungsten arc

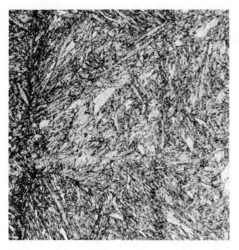 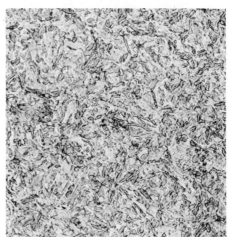

*Fig. 4.4–Microstructures of as-rolled (left) and heat treated (right) 1-inch HP9-4-20 alloy steel plate (×500)*

and plasma arc welding processes can provide clean weld metal with suitable mechanical properties. Gas metal arc welding is not recommended because oxygen levels of weld metals deposited by this process seldom fall below 150 ppm.

Electron beam welding can be used where production conditions permit. A postweld heat treatment may be required with some steels because of the absence of tempering with single pass welds.

### Welding HP9-4-20 Steel

The weldability of HP9-4-20 alloy steel is considered excellent provided the welding process and procedures prevent weld metal contamination and the heat input is not excessive. Weld metal having properties equivalent to heat treated base metal can be obtained with the gas tungsten arc welding process using the conditions listed in Table 4.7. Machine welding with cold wire feed is recommended. Preheat or postweld heat treatment is not required.

**Table 4.5
Typical mechanical properties of quenched and tempered (1025°F) HP9-4-20 and HP9-4-30 alloy steels at room temperature**

|  | HP9-4-20 | HP9-4-30 |
|---|---|---|
| Tensile strength, ksi | 205 | 220-240 |
| Yield strength, ksi | 185 | 190-200 |
| Elongation, percent | 17[a] | 12-16[b] |
| Reduction in area | 65 | 35-50 |
| Charpy V-notch impact strength, ft·lb | 60 | 18-25 |
| Fracture toughness, ksi·in.$^{1/2}$ | — | 90-105 |

a. In 1 in. gage length
b. In a 4D gage length

**Table 4.6
Composition of AF1410 alloy steel wire produced by double vacuum melting**

| Element | Weight percent |
|---|---|
| C | 0.15 |
| Mn | less than 0.05 |
| Si | less than 0.01 |
| S | 0.005 |
| P | 0.001 |
| Ni | 9.82 |
| Cr | 1.90 |
| Mo | 1.00 |
| Co | 13.76 |
| Al | 0.025 |
| V | less than 0.01 |
| Ti | less than 0.01 |
| H | 1 ppm[a] |
| O | 52 ppm |
| N | 4 ppm |

a. Parts per million

**Table 4.7**
**Typical condition for gas tungsten arc welding HP 9-4-20 and AF1410 alloy steel plate**

| | HP 9-4-20 | AF 1410 |
|---|---|---|
| Filler metal | HP 9-4-20 | AF1410 |
| Filler wire diam., in. | 0.062 | 0.062 |
| Filler wire feed, in./min. | 20-30 | 20-22 |
| Shielding gas | Ar | 75% He - 25% Ar |
| Welding current, A (dc) | 300-350 | 160-200 |
| Arc voltage, V | 10-12 | 13 |
| Travel speed, in./min. | 5-10 | 4 |
| Heat input, kJ/in. | 20-24 | 33-34 |
| Max. interpass temperature,°F | 200 | 160 |

Typical weld metal mechanical properties are shown by the test data in Table 4.8. The range of weld metal impact strengths that may be expected from various heats of HP9-4-20 alloy filler metal is shown in Fig. 4.5.

Weldments may be stress relieved at 1000°F

without significantly changing the mechanical properties of the weld metal. Toughness and yield strength of the weld metal tend to increase while the tensile strength decreases.

## Welding AF1410 Steel

As with other Ni-Co alloy steels, AF 1410 steel is best joined by gas tungsten arc or plasma arc welding using low heat input and multiple passes. Matching filler metal is recommended. (See Table 4.6.) Typical conditions for mechanized gas tungsten arc welding are given in Table 4.7. Mechanical properties of weld metal deposited in 0.625 in. thick AF1410 plate using the conditions listed in Table 4.7 are given in Table 4.9. Aging the weld metal at 900°F for 2 hours appears to improve yield strength and toughness.

## CHROMIUM-MOLYBDENUM-VANADIUM STEELS

### General Description

The 5 percent chromium-molybdenum-vanadium (5CrMoV) ultra high-strength steels are

**Table 4.8**
**Mechanical properties of HP9-4-20 alloy steel weld metal as-deposited by gas tungsten arc welding**

| Base plate thickness, in. | Tensile strength, ksi | Yield strength, ksi | Elongation, percent | Reduction of area, percent | Impact strength, ft·lb[a] |
|---|---|---|---|---|---|
| 1.0 | 207 | 186 | 17 | 59 | 63 at 70°F 50 at −80°F |
| 1.0 | 211 | 203 | 20 | 65 | 64 at 70°F 55 at −80°F |
| 2.0 | 209 | 200 | 18.5 | 60 | 59 at 70°F 57 at −80°F |

a. Charpy V-notch impact test

**Table 4.9**
**Mechanical properties of AF 1410 weld metal deposited by gas tungsten arc welding**

| | Weld metal | | Base plate |
|---|---|---|---|
| | As-welded | Aged[a] | |
| Tensile strength, ksi | 223 | 223 | 230 |
| Yield strength, ksi | 202 | 211 | 210 |
| Elongation in 1 in., percent | 16 | 16 | — |
| Reduction of area, percent | 57 | 63 | — |
| Charpy V-notch impact strength at 0°F, ft·lb | 41 | 44 | 35 min. |

a. Aged at 900°F for 2 h and water quenched.

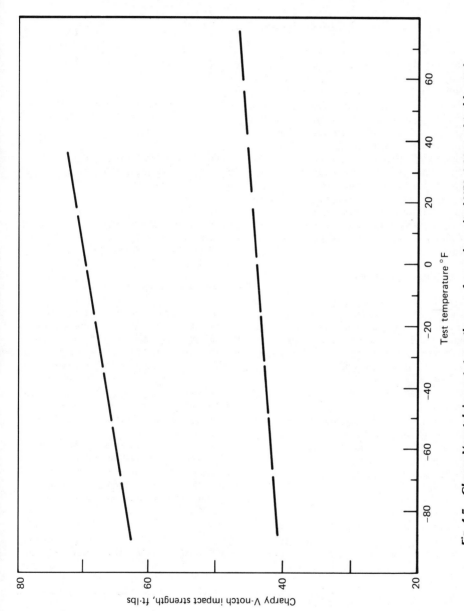

*Fig. 4.5—Charpy V-notch impact strength range for as-deposited HP9-4-20 steel weld metal*

tool steels that are suitable for structural applications.[1] They are Types H11, H11 Modified, and H13 alloy steels; the compositions are shown in Table 4.1. Type H11 Modified steel is the same as H11 steel except that its carbon content is maintained in the high end of the carbon range for H11 steel.

Type H13 steel is similar to Type H11 steel except for higher vanadium content. The increased vanadium increases the dispersion of vanadium carbides for good wear resistance. The rather wide carbon range for Type H13 steel permits a significant variation in mechanical properties with a given heat treatment. This behavior must be considered for applications requiring very high strength properties.

All three steels are secondary hardening, and develop optimum properties when tempered above 950°F. The tempering behaviours of H11 Mod. and H13 steels are shown in Fig. 4.6. These steels are air hardening, and therefore require care during welding and heat treatment to avoid cracking.

## Mechanical Properties

Typical mechanical properties of H11 Mod. and H13 steels after tempering at several temperatures are given in Table 4.10. Good notch toughness and high strength are available after an appropriate heat treatment. The change in hardness with temperature for H11 Mod. steel after tempering at three temperatures is shown in Fig. 4.7. Type H13 steel would behave in a similar manner.

Type H11 Mod. steel has good ductility and notch toughness in combination with high strength at temperatures up to about 1100°F. Hardness and tensile strength decrease rapidly at higher temperatures. It is weldable, but subject to hydrogen embrittlement.

The transverse properties and the weldability of these alloy steels are improved by vacuum arc remelting to remove impurities. Filler metal must also be low in impurities and be free of surface contaminants and defects.

## Applications

Typical applications of H11 Mod. steel are aircraft landing gear and frame components,

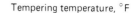

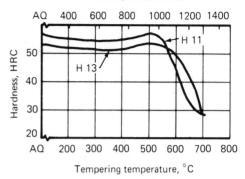

**Fig. 4.6—Variation of hardness with tempering temperature for H11 Mod. and H13 alloy steels**

parts for steam and gas turbines, high strength fasteners, springs, and hot work dies. H13 steel is used for dies, punches, shafts, beams, torsion bars, and ratchets.

## Preheat and Postheat

The 5CrMoV alloy steels are air-hardening and form hard martensite when quenched from the hardening temperature (1825°F). Martensite will form in the weld metal and weld heat-affected zones unless appropriate preheat and postheat procedures are followed to avoid it. As with all hardenable alloy steels, a martensitic heat-affected zone is very susceptible to underbead cracking, particularly when hydrogen is present during welding.[2]

To avoid weld cracking, 5 CrMoV alloy steels should be preheated for welding. The ideal preheat temperature is one that is above the temperature ($M_s$) at which martensite begins to form on cooling. This temperature is about 550°F for H11 steel and about 650°F for H13 steel. When the weld is held at or above this temperature, it remains austenitic, but it could transform to bainite after a period. If the weld is cooled rapidly prior to bainitic transformation, it transforms to hard martensite and may possibly crack.

High preheat and interpass temperatures

---

1. Refer to Chapter 3 for additional information.

2. Refer to Chapter 1 for information on the welding of hardenable low alloy steels and the role of hydrogen.

**Table 4.10**
**Typical longitudinal mechanical properties of H11 Mod. and H13 alloy steels**

| Tempering temperature, °F | Tensile strength, ksi | Yield strength, ksi | Elongation, % | Reduction in area, % | Charpy V-notch impact energy, ft·lb | Hardness, HRC |
|---|---|---|---|---|---|---|
| | | | H11 Mod.[a] | | | |
| 950 | 308 | 248 | 5.9[b] | 29.5 | 10.0 | 56.5 |
| 1000 | 291 | 243 | 9.6[b] | 30.6 | 15.5 | 56.0 |
| 1050 | 269 | 227 | 11.0[b] | 34.5 | 19.5 | 52.0 |
| 1100 | 223 | 192 | 13.1[b] | 39.3 | 23.0 | 45.0 |
| 1200 | 154 | 124 | 14.1[b] | 41.2 | 29.5 | 33.0 |
| 1300 | 136 | 101 | 16.4[b] | 42.2 | 66.8 | 29.0 |
| | | | H13[c] | | | |
| 980 | 284 | 228 | 13.0[d] | 46.2 | 12 | 52 |
| 1030 | 266 | 222 | 13.1[d] | 50.1 | 18 | 50 |
| 1065 | 251 | 213 | 13.5[d] | 52.4 | 20 | 48 |
| 1100 | 229 | 198 | 14.4[d] | 53.7 | 21 | 46 |
| 1120 | 217 | 187 | 15.4[d] | 54.0 | 22 | 44 |

a. Air cooled from 1850°F, double tempered 2 + 2 h at indicated temperature
b. In 2 inches
c. Round bars, oil quenched from 1850°F, and double tempered 2 + 2 h at indicated temperature
d. In 4D

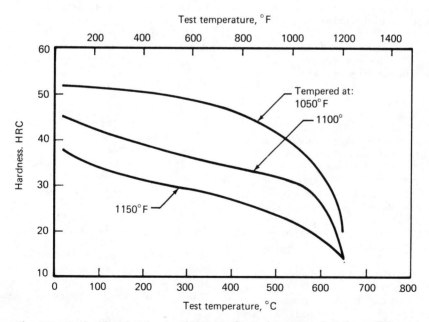

**Fig. 4.7—Typical hot hardness of H11 Mod. steel tempered at three different temperatures**

**Table 4.11**
**Transverse tensile strength of gas tungsten arc welds in 0.078 in. thick H-11 alloy steel using low alloy steel filler metals**

| Filler metal[a] | Tensile strength, ksi[b] | Yield strength, ksi | Elongation in 2 in., percent[c] |
|---|---|---|---|
| A | 242 | 222 | 2.4 |
| B | 270 | 232 | 2.7 |
| C | 274 | 228 | 4.1 |
| D | 266 | 227 | 3.6 |
| E | 259 | 232 | 3.2 |

a. Filler metal compositions:

| Filler metal | Composition, percent | | | | | | | | | |
|---|---|---|---|---|---|---|---|---|---|---|
| | C | Mn | P | S | Si | Cu | Ni | Cr | V | Mo |
| A | 0.05 | 1.17 | 0.01 | 0.02 | 0.41 | 0.47 | 0.01 | 0.01 | 0.01 | 0.06 |
| B | 0.16 | 1.50 | 0.01 | 0.02 | 0.55 | 0.30 | 0.01 | 0.01 | 0.01 | 0.16 |
| C | 0.21 | 0.55 | 0.01 | 0.01 | 0.51 | 0.87 | 0.01 | 1.07 | 0.05 | 0.35 |
| D | 0.16 | 1.13 | 0.01 | 0.03 | 0.40 | 0.43 | 0.01 | 0.01 | 0.01 | 0.03 |
| E | 0.27 | 0.40 | 0.01 | 0.01 | 0.29 | 0.10 | 0.01 | 0.80 | 0.01 | 0.19 |

b. Air quenched from 1900°-1950°F, and double tempered 2 + 2 h at 1000°F
c. Elongation tends to localize at the welded joint

can cause problems during welding from oxidation of the joint faces. The oxide may lead to occasional porosity or lack of fusion in the weld. Preheating to a lower temperature will avoid this problem, but some austenite in the weld will transform to martensite. If the weld is cooled rapidly to room temperature, the remaining austenite will transform to martensite. The volumetric change accompanying this transformation will produce local high stresses, and possibly result in microcracking in the weld and heat-affected zones.

The problem can be avoided by raising the temperature of the weld from the preheat level to about 100°F above the $M_s$ temperature and holding for about 1 hour. This procedure will transform the remaining austenite to a reasonably ductile bainitic structure. Cracking would not be expected after the weld is cooled to room temperature. Further heat treatment is required to produce a desirable microstructure in the welded joint.

An alternative is to cool the weld from the preheat temperature to about 150°F to transform the retained austenite to martensite. The weld must then be immediately heated to an appropriate higher temperature to relieve welding stresses and temper the martensite The welded joint

should not be stress relieved without the cooling step. Any weld repairs should be made prior to final heat treatment of the weldment.

A preheat temperature of about 450°F can be used without excessive oxidation or operator discomfort. When gas tungsten arc welding thin sections, a preheat of 300°F may be suitable.

## Welding Considerations

The 5 CrMoV steels can be welded using the processes and procedures suitable for low-alloy hardenable steels.[3] They are normally welded in the annealed condition.

When the weldment is to be used for a high strength, critical application requiring good notch toughness, the cleanness and welding requirements described previously for nickel-cobalt alloy steels must be applied with these steels. Gas tungsten arc welding is recommended.

When nonconsumable weld backing is required to provide desired root reinforcement, the backing must not quench harden the weld metal. Quench hardening may cause cracks in the root of the weld. Metal backing should have relatively

---

3. Refer to Chapter 1 for additional information.

low thermal conductivity, and it should be heated to the specified preheat temperature before welding commences. Copper backing is not recommended. A refractory backing material may be used, provided it is not a source of moisture or other contamination.

The filler metal may have the same composition as the base metal or a lower alloy content, particularly the carbon level. Low-alloy steel filler metal will produce weld metal with better ductility and toughness at some sacrifice in strength. Weld metal composition will depend on dilution as affected by joint design and welding procedures. Mechanical properties of gas tungsten arc welded joints made with five low alloy steel filler metals are given in Table 4.11. The properties are somewhat lower than those of H11 steel given a similar heat treatment, but they may be adequate for may applications.

### Postweld Finishing

The face and root surfaces of the welded joint should be machined flush with the base metal to remove stress concentrations. It may be necessary to provide additional thickness when designing the weld joint to allow for this machining.

# MARAGING STEELS

## GENERAL DESCRIPTION

Maraging steels are a group of iron-nickel alloys that are strengthened by precipitation of one or more intermetallic compounds in a matrix of essentially carbon-free martensite. In addition to the nickel, these steels generally contain either cobalt or chromium, molybdenum, titanium, and aluminum.

The nominal compositions of six commercial maraging steels are given in Table 4.12. For optimum properties, the carbon and impurity elements in maraging steels are deliberately kept very low. The maximum amount for each element is specified as follows:

Carbon—0.03 percent
Manganese and silicon—0.10 percent
Phosphorus and sulfur—0.010 percent

The 18 Ni (350) grade may have different molybdenum and titanium contents, depending on the supplier. Maraging steel ASTM A590, sometimes designated as 12-5-3, is designed for nuclear applications, but its strength is limited because of the absence of cobalt.

## METALLURGY

Maraging steels are characterized by high strength and excellent toughness. The strength is obtained as a result of age hardening of low-carbon martensite. The principal substitutional alloying elements are nickel, molybdenum, and cobalt or chromium. The iron-nickel equilibrium diagram, Fig. 4.8, shows a wide two phase region of $\alpha$ and $\gamma$. In practice, alloys containing about 18 percent nickel transform from austenite to body-centered cubic martensite during air cooling through a relatively narrow temperature range, as shown in Fig. 4.9. Transformation back to austenite on reheating occurs through a higher temperature range. This behaviour permits aging of the martensite at a temperature of 900°F for several hours without transformation to austenite. Extended heating at this temperature will, however, result in transformation back to austenite.

After solution annealing for typically one hour at 1500°F, the steel will transform to mar-

**Table 4.12**
**Nominal compositions of commercial maraging steels**

| Grade[a] | Nominal composition, percent | | | | | |
|---|---|---|---|---|---|---|
| | Ni | Co | Cr | Mo | Ti | Al |
| ASTM A538 | | | | | | |
| Gr A (200) | 18 | 8 | — | 4 | 0.2 | 0.1 |
| Gr B (250) | 18 | 8 | — | 5 | 0.4 | 0.1 |
| Gr C (300) | 18 | 9 | — | 5 | 0.7 | 0.1 |
| 18Ni (350) | 18 | 12 | — | 4 | 1.3 | 0.1 |
| Cast 18Ni | 17 | 10 | — | 5 | 0.3 | 0.1 |
| ASTM A590 | 12 | — | 5 | 3 | 0.3 | 0.4 |

a. Carbon—0.03 percent max

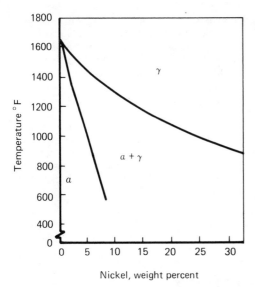

**Fig. 4.8—Iron-nickel equilibrium diagram**

tensite during air cooling. The annealing temperature must be high enough to cause precipitates to go into solution and to remove residual stresses. The cooling rate is not critical and, therefore, martensite formation is independent of section size. The martensite is relatively soft because of the low carbon content of the steel. Austenite-to-martensite transformation takes place at fairly low temperatures, about 310°F for 18 percent nickel maraging steel. Typical hardness in the solution annealed condition is about 30HRC (350DPH), which permits machining.

Subsequent aging at 900°F for 3 to 12 hours causes an increase in hardness and strength. The dimensional change during aging is very small, a decrease of 0.04 to 0.08 percent, which allows components to be machined in the solution annealed condition and then aged.

Age hardening appears to result from precipitation of an alloy phase having the structure of $Fe_2Mo$. Additional hardening also results from the precipitation of particles of $Ni_3Ti$.

The role of cobalt is more complex. It does not appear to produce precipitates but increases the precipitation of $Ni_3Mo$ by limiting molybdenum solubility in the martensitic matrix. Cobalt also raises the martensite start temperature which permits a martensitic structure to be obtained on cooling to room temperature.

With extended aging times, there is evidence of precipitation of $Fe_2Mo$. Ultimately, reversion to austenite will occur. This causes soft-

ening of the steel. Although overaging is usually avoided, some components may be slightly overaged deliberately to provide desired properties.

## PHYSICAL PROPERTIES

The physical properties of maraging steel of concern in welding are compared to those of mild steel and austenitic stainless steel in Table 4.13. Differences in properties between maraging and mild steels are not significant, except for thermal conductivity. Heat loss by conduction during welding would be lower with maraging steel.

The significant difference in coefficients of thermal expansion between maraging steel and austenitic stainless steel may cause some problems when welding them together.

Maraging steels are magnetic, therefore, arc blow may be encountered during arc welding operations, particularly if the metal becomes magnetized.

## MECHANICAL PROPERTIES

Typical mechanical properties of the common grades of maraging steel are given in Table 4.14. One of the notable features of these steels is the excellent combination of toughness and high strength. The Charpy V-notch (CVN) and fracture toughness properties are more than twice those of conventional quenched and tempered high-strength steels. Toughness is, to a great extent, dependent on the purity of the steel.

The maximum service temperature for maraging steels is about 750°F. Above this temperature, long term strength drops rapidly due to overaging.

## APPLICATIONS

Maraging steels are used for aircraft and aerospace components where both high strength and weldability are important considerations. They are also used for tooling applications because of the low dimensional change during aging. The tool can be machined to size while the steel is in the annealed condition and then aged to increase strength and hardness.

## WELDING METALLURGY

### Heat-Affected Zone

The weld heat-affected zone can be divided

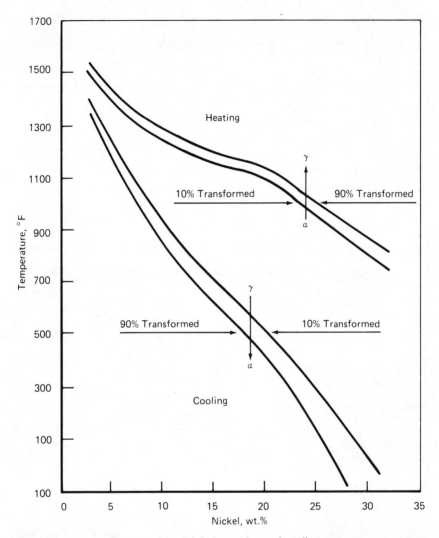

**Fig. 4.9—Iron-nickel transformation diagram**

into Regions A, B, and C, as shown in Fig. 4.10. Region A, next to the weld interface, becomes fully austenitic during welding and transforms to a coarse martensite on cooling. This region will be relatively soft in the as-welded condition, but it will harden during aging. Next to this is Region B, a narrow zone usually dark etching, which is reheated in the range of 1100° to 1350°F. It is martensite with fine, reverted austenite. Region C is martensitic and will have been aged to various extents at temperatures up to 1100°F. For practical purposes, the properties of this region are unchanged by welding.

The hardness across the heat-affected zone of a typical weld in aged maraging steel, as-welded and after re-aging, is shown in Fig. 4.11. The as-welded hardness of Region A, a coarse low-carbon martensite, decreased to approximately 330 DPH. After re-aging, the hardness increased to approximately that of the aged base metal.

The narrow Region B is a fine dispersion of austenite in martensite. Apparently under conditions of very rapid heating, the austenite can form by a shear mechanism. With a slow heating rate austenite formation is diffusion controlled

**Table 4.13**
**Physical properties of maraging steel, mild steel, and austenitic stainless steel**

| Property | Units | Maraging steel | Mild steel | Stainless steel |
|---|---|---|---|---|
| Density | Mg/m³ | 8.0 | 7.85 | 8.0 |
| Coefficient of thermal expansion | μm/(m·K) | 10.1 (24°-284°C) | 12.8 (20°-300°C) | 17.8 (0°-315°C) |
| Thermal conductivity (20°C) | W/(m·k) | 19.7 | 52 | 15 |
| Electrical resistivity | μΩ·m | 0.36 to 0.7 | 0.17 | 0.72 |
| Melting temperature | °C | 1430-1450 | 1520 | 1400-1450 |

**Table 4.14**
**Typical mechanical properties of aged commercial maraging steels**

| Grade[a] | Tensile strength, ksi | Yield strength, ksi | Elongation in 2 in., percent | Reduction in area, percent | Fracture toughness, ksi·in.$^{1/2}$ | Charpy V-notch impact strength, ft·lb (RT) |
|---|---|---|---|---|---|---|
| ASTM A538 | | | | | | |
| Gr A (200) | 218 | 203 | 10 | 60 | 140-220 | 35 |
| Gr B (250) | 260 | 247 | 8 | 55 | 110 | 20 |
| Gr C (300) | 297 | 290 | 7 | 40 | 73 | 15 |
| 18Ni (350) | 355 | 348 | 6 | 25 | 32-45 | 8 |
| Cast 18Ni | 225 | 240 | 8 | 35 | 95 | 50 |
| ASTM A590 | 190 | 180 | 14 | 60 | — | 10 |

a.  Solution treated at 1500°F, aged at 900°F

C  B    A                    A    B  C

*Fig. 4.10 – Three regions in the heat-affected zone of a weld in maraging steel*
*(X4)(Reduced to 79%)*

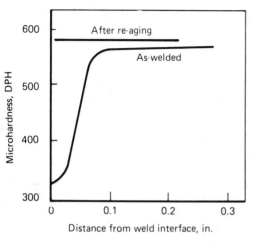

**Fig. 4.11 – Microhardness of a weld heat-affected zone in aged 18Ni(250) maraging steel as-welded and after re-aging**

(conventional austenite reversion). The amount of austenite formed will increase with increasing heat input.

The austenite in Region B will not harden during subsequent aging and, therefore, the zone will remain softer than the unaffected base metal. This may not be of practical significance if the zone is narrow and, in most cases, the joint strength will be controlled by the weld metal properties. With high welding heat input, welded joints may fail in Region B if a relatively large amount of austenite is formed. Therefore, control of input to restrict the width and microstructure of this region is recommended.

## Weld Metal

The weld metal and base metal have the same basic structure when a matching filler metal is used. Both structures are low carbon martensite which hardens during aging. The weld metal structure, however, is more complex. Both single and multiple pass welds have small, white patches of austenite in the martensitic matrix after aging. These austenitic patches form during reheating by subsequent weld passes or during the aging heat treatment. Local segregation of alloying elements allows stable austenite to form at relatively low temperatures. Martensitic weld metal containing small patches of austenite is shown in Fig. 4.12.

Aged weld metal containing patches of austenite will generally have a lower strength than

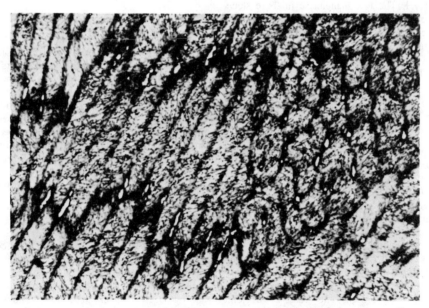

**Fig. 4.12 – Microstructure of aged maraging steel weld metal showing small white patches of austenite in martensite (X500)**

aged base metal. However, joint efficiencies exceeding 90 percent can be obtained in most applications.

## ARC WELDING

### Cracking

The low-carbon martensite formed in a weld heat-affected zone on cooling has a low susceptibility to hydrogen induced cracking. This is partly because the region close to the weld interface is relatively soft in the as-welded condition. Maraging steels are less sensitive to hydrogen embrittlement than are heat-treated low alloy steels at the same strength levels.

The danger of cold cracking is also lessened because of the residual welding stress pattern. The weld metal is under compressive longitudinal stress rather than a tensile stress, as shown in Fig. 4.13, because the martensitic transformation occurs at relatively low temperature. The stress changes from tension to compression when the weld metal expands during the phase transformation. At the same time, the stress in the heat-affected zone changes from compression to tension.

Hot cracking can be a potential problem in welding maraging steels and is related to impurity level. The low manganese in these steels makes them particularly sensitive to sulfur embrittlement. Hot cracking can occur with sulfur content as low as 0.005 percent when the joint fit-up is poor. With good fit-up, sulfur levels up to 0.010 percent can be tolerated. An example of a hot crack in a fillet weld is shown in Fig. 4.14.

Examination of the fracture surfaces of hot cracks in welds has revealed titanium sulfide inclusions that can liquate during welding and form fissures as the metal cools. The higher strength maraging steels that contain larger amounts of titanium are more susceptible to hot cracking and require greater care when welding. (See Table 4.12.) In view of the low austenite-to-martensite transformation temperature and the sensitivity to hot cracking, maraging steels should be welded without preheat and the interpass temperature be restricted to 250°F.

Control of heat input during welding is also necessary to avoid hot cracking and inferior mechanical properties. The minimum practical heat input should be used. The effect of the joint geometry on heat input requirements should be considered. Joint design, weld backing, and clamping fixtures should be chosen to minimize

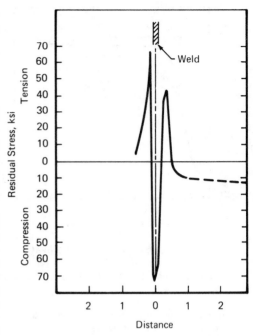

**Fig. 4.13 – The residual stress pattern in a maraging steel weld**

welding heat input requirements.

With gas metal arc welding, hot cracking in the weld metal of fillet welds is related to the depth-to-width ratio of the weld bead, as well as the alloy composition, the root opening, and the degree of joint restraint. In general, a high depth-to-width ratio (over 0.6) or a wide root opening increases the tendency to hot crack, as does increased joint restraint (increase in section thicknesses). Welding should be done using conditions that limit the depth of fusion yet provide complete fusion with the joint faces and previously deposited weld beads.

### Filler Metal Selection

The bare filler metal should have a composition similar to that of the base metal and should be made using vacuum melting techniques to obtain low levels of oxygen, nitrogen, and hydrogen. Oxygen and nitrogen should be below 50 ppm and hydrogen below 5 ppm. Control of the filler metal composition is critical. In particular, carbon and sulfur must be kept as low

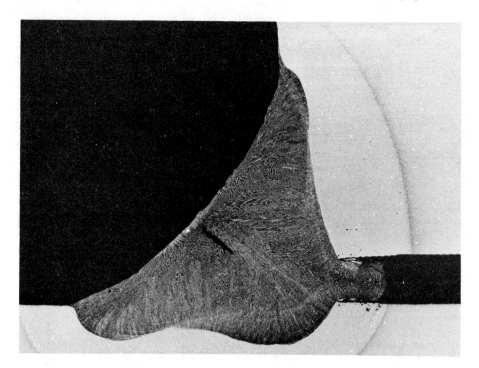

**Fig. 4.14—Hot crack in maraging steel weld metal (×7)**

as possible. Carbon can form brittle inclusions and tie up some solid solution strengthing alloy additions. Sulfur forms low melting sulfide inclusions. Excessive silicon can lower impact properties and increase the sensitivity to cracking. Titanium content must be controlled within limits. A low level of titanium results in low strength and a tendency to porosity; a high level can lead to hot cracking and increased tendency to reform austenite in the weld metal.

### Cleanness

To achieve good toughness at the high strength levels, impurities in maraging steel weld metal must be maintained at very low levels. Filler metal and welding procedures must be designed to ensure that impurities in welds are minimized. Prior to welding, joint surfaces should be cleaned with clean lint-free rags and suitable solvent. Welding wire feed rolls and guides should be clean to avoid contamination of the filler metal. Shielding gases must be dry and pure. Gas equipment and lines should be clean, free of leaks, and purged of air and moisture prior to use.

Cleanness of the filler wire is extremely important. Vacuum annealing and ultrasonic cleaning are recommended. To avoid contamination, cleaned wire should be stored in dry, inert gas filled containers.

Each bead of multiple pass welds should be cleaned of any surface contamination before depositing the next bead. A clean stainless steel wire brush or motor-driven metal cutting tools are recommended. A dark surface discoloration would indicate poor shielding of the weld zone.

### Joint Design

Typical joint designs for arc welding the maraging steels are shown in Fig. 4.15. The joint dimensions, groove angle, or root opening may need to be adjusted for specific welding processes and applications. The root of double welded joints should be mechancially gouged to sound metal before welding the second side.

### Gas Tungsten Arc Welding

The most widely used welding processes for maraging steels are the inert gas processes

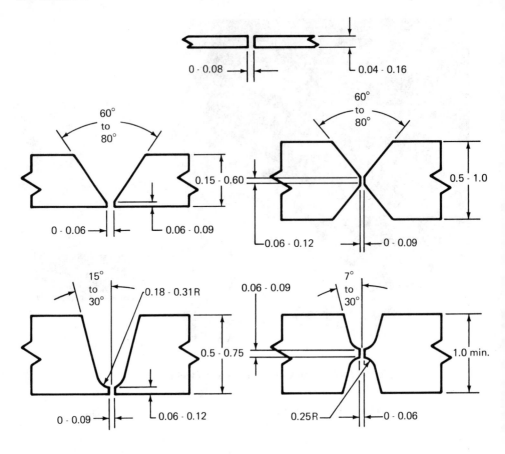

Dimensions are in inches.

***Fig. 4.15—Typical weld joint designs for maraging steels***

and, in particular, gas tungsten arc welding. This process allows good control of heat input and protection of the weld from oxidation. Relatively conventional joint designs are used (see Fig. 4.15). Typical welding procedures are shown in Table 4.15. These may be used as guides in establishing suitable welding procedures for specific applications. Gas tungsten arc welded joints have better toughness than those made by gas metal arc welding.

## Gas Metal Arc Welding

As described previously, the properties of maraging steels are dependent on the impurity levels. Carbon and oxygen levels should be kept low. The need to minimize contamination during welding presents a problem with gas metal arc welding of maraging steels.

For good arc stability, the use of oxidizing shielding gases is common when gas metal arc welding carbon and low alloy steels. With these steels, slight pick-up of carbon or oxidation from carbon dioxide or oxygen additions to the shielding gas is not considered detrimental. With maraging steel, however, carbon pick-up could result in embrittlement due to the formation of titanium carbides at the grain boundaries. Oxygen pick-up may result in the formation of titanium oxide and a reduction in fracture toughness, the extent of which depends on the oxide distribution.

Significant amounts of oxygen could be drawn into the arc atmosphere during gas metal arc welding. Therefore, oxygen should not be

**Table 4.15**
**Typical procedures for gas tungsten arc welding maraging steel**

| Joint thickness, in. | Joint design | Weld pass | Arc volts | Welding current, A[a] | Travel speed, in./min | Filler wire feed, in./min[b] | Shielding gas | Remarks |
|---|---|---|---|---|---|---|---|---|
| 0.080 | Square groove | 1 | 8 | 150 | 7 | 10 | Ar | Cold wire feed |
| 0.080 | Single-V-groove, 0.04 in. root face, 90° groove angle | 1<br>2 | 8 | 120<br>150 | 11 | None<br>10 | Ar | Cold wire feed |
| 0.5 | Single-U-groove, 0.09 in. groove radius, 0.06 in. root face, 30° groove angle | 1<br>2<br>3-5<br>6-12 | 9-10 | 100<br>150<br>200<br>225 | 8 | 10<br>10<br>20<br>30 | 80%He-20%Ar | Cold wire feed |
| 0.6 | Single-U-groove, 0.16 in. groove radius, 0.06 root face, 60 ° groove angle | 1<br>2-6 | 11-11.5 | 265<br>400 | 9<br>14 | 106<br>163 | 75%He-25%Ar | Hot wire feed[c] |
| 0.75 | Double-U-groove 0.09 in. groove radius, 0.06 in. root face, 60° groove angle | 1-10 | 10 | 340-400 | 12-15 | 60-70 | Ar | Cold wire feed |
| 1.0 | Double-V-groove, 0.06 in. root face, 60° groove angle | 1-30 | 10-12 | 210-230 | 4-6 | 20-24 | Ar | Cold wire feed |

a. Direct current, electrode negative
b. Filler wire diameter: 0.062 in.
c. Hot wire power: 5.5-6V, 135-170A, ac

deliberately added to the shielding gas unless it is necessary to stabilize the arc. Then, it should be limited to a 0.5 percent addition.

Typical procedures for gas metal arc welding of maraging steels are given in Table 4.16. Spray transfer is generally used with this process but short-circuiting transfer or pulsed power is also suitable.

## Shielded Metal Arc Welding

Although covered electrodes for welding maraging steels have been produced, they are not widely used. The composition of the core wire is generally similar to the base metal but the titanium content is increased to account for losses across the arc. The coatings are generally basic lime cryolite or lime titania with low levels of silica. Weld joint strengths matching those of the base metal have been achieved with the lower strength maraging steels, but weld metal toughness was lower.

## Submerged Arc Welding

Maraging steels have been submerged arc welded using an electrode composition similar to that of the base metal but with higher titanium content to deoxidize the weld metal. Basic fluxes containing no silica appear to be suitable. Conventional fluxes are not adequate because the titanium recovery is low, and the welds are crack sensitive and brittle. A typical composition of a basic flux for maraging steels is as follows:

## Table 4.16
## Typical conditions for gas metal arc welding maraging steels

| Joint thickness, in. | Joint design | Type of transfer | Shielding gas | Electrode Diam., in. | Feed, in./min | No. of passes | Arc volts | Welding current, A[a] | Travel speed, in./min |
|---|---|---|---|---|---|---|---|---|---|
| 0.50 | Single-V-groove, 60°-80° groove angle, 0.06 in. root face, | Spray | Ar | 0.062 | 200-220 | 3-5 | 30-34 | 290-310 | 10 |
| | 0.06 in. root opening | Short circuiting | He | 0.035 | 325 | 18 | 25 | 125 | 6-8 |
| 0.75 | Single-U-groove, 45° groove angle, 0.09 in. root face, 0.07 in. root opening | Spray | Ar + 2%$O_2$ | 0.062 | 240 | 10[b] | 28 | 350-375 | 15-20 |
| 1.0 | Double-V-groove, 60°-80° groove angle, 0.06 in. root face, | Spray | Ar | 0.062 | 200-220 | 8 | 30-34 | 290-310 | 10 |
| | 0.09 in. root opening | Pulsed spray | Ar + 0.3%$O_2$ | 0.045 | 180 | 24 | 70 peak 20 bkgd | 140 avg | 6 |

a. Direct current electrode positive.
b. First two passes made by gas tungsten arc welding using the same filler metal and Ar shielding. Arc volts: 8-10, welding current (dcen): 160-175A, travel speed: 4-6 in./min.

Aluminum oxide–37 percent
Calcium carbonate–28 percent
Calcium fluoride–15 percent
Magnesium oxide–14 percent
Ferro titanium–6 percent

Double V-groove and double U-groove joint designs with 0.25 and 0.31 in. root faces respectively have been used for submerged arc welding 0.75 to 1.25 in. plate. Typical welding conditions would be 500 to 600A dc, electrode positive, 27 to 30 arc volts, 10 to 12 in./min travel speed using a 0.125 in. diameter electrode.

The strength of submerged arc welds generally matches the base metal in the lower strength grades, but the weld metal ductility and toughness are considerably lower than those achieved with the inert gas shielded processes. The low weld metal toughness is presumably a result of contamination. Submerged arc weld metal generally has a higher inclusion content than metal deposited by an inert gas shielded welding process.

Attempts to produce multiple pass welds with the submerged arc process have generally been unsuccessful because of cracking in the underlying bead. The cracks are apparently associated with the high energy input of the process, but their exact cause is uncertain. The general problems associated with submerged arc welding severely limit its use with maraging steels.

## Mechanical Properties

Typical mechanical properties of gas tungsten arc and gas metal arc welds in maraging steel plate are shown in Table 4.17. Weld metal strength is generally below that of the base metal, particularly with the higher strength grades of maraging steel as a result of small patches of austenite in multiple pass welds. Joint efficiencies based on yield strength of about 95 percent can be achieved with appropriate welding procedures. The toughness of gas tungsten arc welds

**Table 4.17**
**Typical mechanical properties of arc welds in maraging steel plate at room temperature**

| ASTM Grade[a] | Process variation | Filler metal | TS, ksi[b] | YS, ksi | Elong., percent in 1 in. | RA, percent | Approx. joint efficiency, percent[c] | Toughness | |
|---|---|---|---|---|---|---|---|---|---|
| | | | | | | | | CVN, ft·lb | $K_{IC}$, ksi·in.$^{1/2}$ |
| | | | *Gas tungsten arc welds* | | | | | | |
| A 358 | | | | | | | | | |
| Gr A (200) | Cold wire | 18Ni (200) | 207-215 | 199-214 | 10-13 | 53-60 | 90-100 | 35-37 | 90-130 |
| | Hot wire | 18Ni (200) | 197-211 | 186-203 | 9-13 | 43-58 | 90-95 | | 132-148 |
| Gr B (250) | Cold wire | 18Ni (250) | 227-243 | 220-237 | 10-13 | 44-60 | 90-95 | 19-23 | 62-63 |
| | High-current | 18Ni (250) | 244 | 243 | 12-16 | 21-28 | 97 | | 70-80 |
| Gr C (300) | Cold wire | 18Ni (300) | 243 | 202 | 8 | 40 | 75 | | 59 |
| A 590 (12-5-3) | Cold wire | 12-5-3 | 182-192 | 176-179 | 14-15 | 58-65 | 95 | 45-79 | |
| | Hot wire | 12-5-3 | 170-172 | 168-169 | 12 | 55-57 | 90 | 42-59 | |
| | Cold wire | 17-2-3 | 183-187 | 174-179 | 14-15 | 55-63 | 95 | 47-54 | |
| | | | *Gas metal arc welds* | | | | | | |
| A 358 | | | | | | | | | |
| Gr A (200) | Spray | 18Ni (200) | 205-220 | 197-217 | 6-11 | 34-56 | 90-100 | 17-22 | 76 |
| | Pulsed spray | 18Ni (200) | 208 | 203 | 5 | — | 100 | 24 | |
| Gr B (250) | Spray | 18Ni (250) | 235-252 | 220-247 | 1.5-7 | 7-38 | 83-95 | 7-10 | 70-80 |
| | Short-circuiting | 18Ni (250) | 245 | 228 | 4 | 14 | 95 | 12 | 65-75 |
| Gr C (300) | Spray | 18Ni (300) | 245 | 232 | 3 | 13 | 85 | | |
| A 590 (12-5-3) | Spray | 12-5-3 | 182-191 | 176-187 | 10-12 | 47-58 | 90-95 | 34-50 | 54 |
| | Pulsed spray | 12-5-3 | 188 | 180 | 9 | 32 | 95 | 30 | |

a. Plate 0.4–1.0 in. thick.
b. Welds aged at 900°F for 3 to 10 hours. Transverse weld tension tests.
c. Based on yield strength.

**Table 4.18**
**Typical mechanical properties of electron beam welds in maraging steel plate**

| ASTM Grade | Thickness, in. | Processing sequence[a] | No. of passes | Welding conditions | | | Mechanical properties[b] | | | |
|---|---|---|---|---|---|---|---|---|---|---|
| | | | | Voltage, kV | Current, mA | Travel speed, in./min | TS, ksi | YS, ksi | Elong., percent[c] | RA, percent |
| A 358 Gr A (200) | 1.0 | S-A-W | 1 | 50 | 400 | 40 | 151.0 | 147.0 | 7 | 31 |
| | | S-A-W-A | 1 | 50 | 400 | 40 | 198.0 | 196.0 | 4 | 13 |
| | | S-A-W | 2 | 150 | 13 | 10 | 168.0 | 160.0 | 10 | 32 |
| | | S-A-W-A | 2 | 150 | 13 | 10 | 217.0 | 212.0 | 7 | 14 |
| Gr B (250) | 0.1 | S-A-W | 1 | 30 | 65 | 40 | 165.8 | 165.8 | 2.5 | 21.2 |
| | | S-A-W-A | 1 | 30 | 65 | 40 | 274.4 | 270.8 | 4.1 | 12.7 |
| | 0.3 | S-A-W | 1 | 150 | 20 | 60 | 242.5 | 242.5 | 4 | — |
| | | S-A-W-A | 1 | 150 | 20 | 60 | 265.0 | 262.0 | 4 | 30 |
| | 0.5 | S-A-W | 1 | 150 | 17 | 17 | 189.0 | 181.0 | 8 | 30 |
| | | S-A-W-A | 1 | 150 | 17 | 17 | 261.0 | 250.0 | 14 | 25 |
| | 1.0 | S-A-W-A | 1 | 50 | 320 | 40 | 260.2 | 256.1 | 4 | 27.8 |
| A 590 (12-5-3) | 1.0 | S-A-W | 1 | 50 | 400 | 40 | 147.0 | 130.0 | 13 | 61 |
| | | S-A-W-A | 1 | 50 | 400 | 40 | 186.0 | 181.0 | 9 | 33 |

a. S – Solution annealed, A – aged, W – welded.
b. All test specimens failed in the weld.
c. Elongation is not a true indication of weld ductility because an electron beam weld is a small portion of the gage length.

can match that of the base metal, but the toughness of gas metal arc welds tends to be somewhat lower.

## ELECTRON BEAM WELDING

Electron beam welding in vacuum is particularly suited for joining maraging steels because of the clean conditions during welding. In addition, the low heat input, narrow heat-affected zone, and small distortion are advantageous. Table 4.18 shows typical mechanical properties of electron beam welds in plate. In general, the weld metal toughness is below that of the base metal and, in some cases, below that achieved with inert gas shielded welding.

# AUSTENITIC MANGANESE STEEL

## GENERAL DESCRIPTION

Austenitic manganese steel, sometimes called *Hadfield* manganese steel, is an extremely tough, non-magnetic alloy in which the hardening transformation that occurs in low alloy steel is suppressed by the high manganese content. It is characterized by high strength, good ductility, and rapid work-hardening characteristics as well as good wear resistance.

The manganese content generally ranges from 11 to 14 percent and the carbon from 0.7 to 1.4 percent. Chromium, molybdenum, nickel, vanadium, copper, titanium, and bismuth are sometimes added singly or in combinations to provide special properties.

This steel is available as castings, hot-rolled billets, bars, plates, and wire. Both castings and rolled products are normally supplied in the quenched condition. Wire is usually available in the annealed condition.

## METALLURGY

Manganese in steel has a strong affinity for oxygen, sulfur, and carbon. Therefore, it acts as a deoxidizer, reduces hot cracking, and contributes to hardening. Free manganese will form a substitutional solid solution with iron and make the transformation of austenite very sluggish during cooling. Large manganese additions to steel suppress the transformation of austenite so effectively that the austenite is stable down to room temperature with moderately fast cooling.

To ensure an austenitic structure at room temperature, normal practice is to water-quench austenitic manganese steel from an elevated temperature where it is fully austenitic and the carbon is in solution. The appropriate austenitizing (solutioning) temperature depends upon the carbon content as well as the manganese content, as indicated in Fig. 4.16. The temperature required to dissolve all the carbides increases with carbon content. The desired microstructure after austenitizing and quenching is shown in Fig. 4.17.

For maximum toughness, the austenitizing temperature must be sufficiently high to insure complete dissolution of carbides. A temperature in the range of 1800° to 1950°F is commonly used, although higher temperatures are necessary for the high carbon grades. Time at temperature is not critical. Equilibrium is probably established within 20 to 30 minutes at temperatures above 1850°F, but a time allowance is needed for the center of heavy sections to reach the austenitizing temperature.

If an adequate austenitizing temperature is not reached and not all the carbides are dissolved, the steel may contain chains and clusters of carbides together with intermittent grain boundary carbide envelopes. Carbide precipitation will occur when the cooling rate from the austenitizing temperature is too slow, and this is the primary cause of impaired properties. High carbon manganese steels are especially sensitive to carbide formation.

Carbides may also appear in the grain boundaries of large sections when quenching is not effective because of the low thermal conductivity of the steel. With a slow cooling rate, typical of the center of heavy sections, a carbide layer may form around each grain as shown in Fig. 4.18. Then, the mechanical properties of the steel will be lower than normal. For this reason,

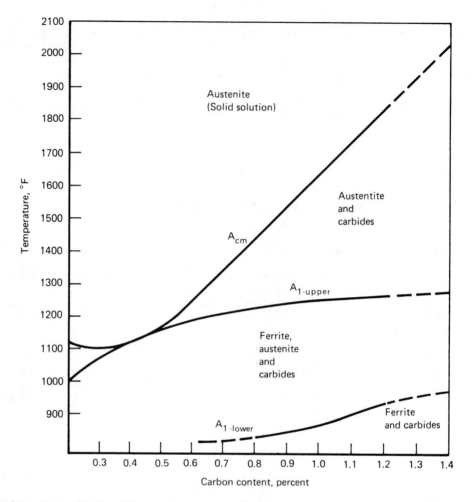

**Fig. 4.16 – Equilibrium diagram showing the effect of carbon on the microstructure of 12.5 percent manganese steel**

the maximum thickness for commercial castings of a standard composition is 6 inches.

The as-cast structure of these steels is austenite grains surrounded by carbides with patches of pearlite, depending on the cooling rate and the composition. An as-cast structure, shown in Fig. 4.19, is very brittle. Large castings are, therefore, crack sensitive until they are heat-treated.

The relatively high austenitizing temperature and the high carbon content of these steels combine to cause marked surface decarburization and some loss of manganese unless precautions are taken to prevent it. A decarburized skin,

which may be partially martensitic, is usually weaker than the underlying metal. Tensile deformation in service sometimes produces many cracks in the weaker skin that terminate in the very tough, austenitic structure beneath it. Service performance is not seriously affected unless critical fatigue conditions or very thin sections are involved.

A martensitic skin is also magnetic. For applications requiring non-magnetic properties (below 1.3 μ permeability), the skin must be removed by grinding or pickling.

Isothermal transformation can occur in

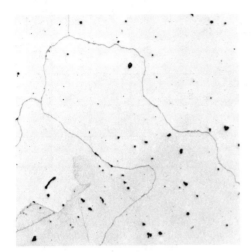

**Fig. 4.17–Microstructure of properly quenched austenitic maganese steel ( × 100)**

**Fig. 4.19–Microstructure of as-cast austenitic manganese steel ( × 100)**

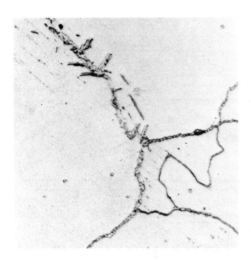

**Fig. 4.18–Microstructure of austenitic manganese steel improperly quenched from austenitizing temperature ( × 100)**

quench annealed steel if it is exposed to an elevated temperature below the $Ac_1$ temperature where carbides precipitate from the austenite and pearlite can form. The carbides precipitate at temperatures up to about 1000° or 1100°F, and pearlite forms from about 1000° to 1400°F. Transformation generally starts along the grain bound-

aries or the crystallographic planes, as shown in Fig. 4.20. Composition has a significant influence on the microstructure obtained but, in every case, such transformation results in a complete loss of ductility and a reduction in strength.

The effect of elevated temperature exposure on the ductility of austenitic manganese steel is shown in Fig. 4.21. Figure 4.22 shows the relationship between alloy composition, tensile strength, and elongation with reheat temperature. Steels with higher carbon or lower manganese content tend to embrittle at lower temperatures. Hence, alloy segregation during solidification can have a significant effect during reheating that accompanies welding.

To avoid embrittlement from reheating, manganese steel should not be tempered or stress relieved after quench annealing or welding. In general, this type of steel should not be heated above 600°F, except for very short times during welding, unless it will be given a quench anneal later.

## ALLOYS

The standard compositions for austenitic manganese steel castings are given in Table 4.19. Most commercial castings have carbon contents in the 1.05 to 1.35 percent range and manganese in the 11.0 to 14.0 percent range. The carbon content is normally kept close to the midpoint of

Fig. 4.20–Microstructure of austenitic
manganese steel after quench annealing and
reheating at 1000°F for 2 hours (×250)

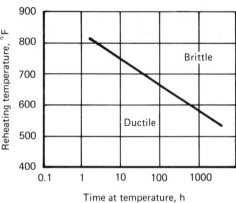

Fig. 4.21 – Effect of reheating temperature
and time on the ductility of austenitic
manganese steel (1.2%C-13%Mn-0.5%Si)
after solutioning at 2000°F for 2 hours and
water quenching (Based on metallographic
examination)

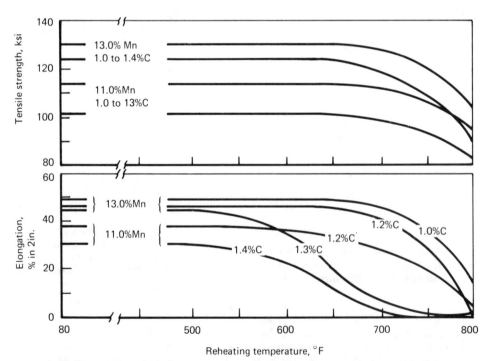

Fig. 4.22 – Relationship between composition, reheat temperature, strength, and ductility
of austenitic manganese steel (1 in. diam. bars solutioned, quenched, and reheated for 48
hours at temperature)

**Table 4.19**
**Standard composition ranges for austenitic manganese steel castings**

| ASTM A128, Grade | Composition, percent[a] | | | | |
|---|---|---|---|---|---|
| | C | Mn | Cr | Mo | Ni |
| A. | 1.05-1.35 | 11.0 min | — | — | — |
| B-1 | 0.9-1.05 | 11.5-14.0 | — | — | — |
| B-2 | 1.05-1.2 | 11.5-14.0 | — | — | — |
| B-3 | 1.12-1.28 | 11.5-14.0 | — | — | — |
| B-4 | 1.2-1.35 | 11.5-14.0 | — | — | — |
| C | 1.05-1.35 | 11.5-14.0 | 1.5-2.5 | — | — |
| D | 0.7-1.3 | 11.5-14.0 | — | — | 3.0-4.0 |
| E-1 | 0.7-1.3 | 11.5-14.0 | — | 0.9-1.2 | — |
| E-2 | 1.05-1.45 | 11.5-14.0 | — | 1.8-2.1 | — |
| F | 1.05-1.35 | 6.0-8.0 | — | 0.9-1.2 | — |

a. Si - 1.00%
   P - 0.07%
   Fe - remainder

the range, and the manganese between 12 and 13 percent. A manganese content near the lower limit of the range results in low tensile strength, and near the upper limit it provides no advantage.

## Carbon

Carbon has only a slight but distinct effect on yield strength. As the carbon content decreases, the yield strength also decreases. Its effect on other tensile properties is small compared to that of other variables such as grain size. The optimum carbon content is about 1.15 percent. Higher carbon levels may cause problems during heat treatment or foundry handling. Low carbon content helps to avoid the embrittling effect of carbide precipitation during cooling. For this reason, filler metals with low carbon content and other modifications are produced for use where normal solutioning and quenching heat treatment is impractical.

## Manganese

Manganese content has almost no effect on the yield strength of this steel but, within the range of 11.0 to 14.0 percent the tensile strength and ductility are maximum. As the manganese content is decreased, the tensile strength and ductility of the steel are lowered. At about 8 percent manganese, these properties are only about 50 percent of maximum. Below this level, the steel tends to become air hardening and a

variety of microstructures may form during quenching. An austenitic structure can be maintained by certain other alloying elements, such as nickel. Chromium, or molybdenum, is added to enhance specific properties.

A minimum manganese content of 11.5 percent is recommended for most applications. The maximum content is arbitrary, depending more on production costs than properties. During solidification, segregation within the microstructure can result in manganese variations of up to 2 percent above or below the nominal.

## Silicon

Silicon is added to the steel chiefly for steelmaking purpose. While silicon content does not generally exceed one percent, it may be increased up to 2 percent to provide a moderate increase in yield strength and improved resistance to plastic flow under repeated impact. Above about 2.2 percent silicon, a sharp reduction in strength and ductility takes place, and higher contents are undesirable for normal applications.

## Phosphorus and Sulfur

Phosphide and sulfide inclusions in wrought austenitic manganese steels may contribute to directional properties. They are relatively harmless in castings, and sulfur seldom influences the properties of austenitic manganese steel. The scavenging effect of the manganese results in the formation of manganese sulfide in the form of innocuous rounded inclusions. Although ASTM Specification A128 allows a maximum of 0.07 percent phosphorus, problems with cracking during fabrication and repair welding will be fewer if the phosphorus content of cast or wrought products does not exceed 0.04 percent. This element tends to cause weld hot cracking, and the phosphorus content of welding electrodes should not exceed 0.030 percent. Low phosphorus may be a problem with solid wire electrodes.

## Other Alloying Additions

Molybdenum or chromium is added to the steel as an alloying element to raise the yield strength. Nickel addition has little effect on yield strength, but it stabilizes the austenite and may slightly increase ductility. Bismuth additions can improve machineability. Titanium in the steel can lower the amount of carbon dissolved in the austenite by forming stable carbides.

## Wrought Compositions

The compositions of wrought products are similar to those of castings. Low carbon grades containing molybdenum and nickel are in wide use. The microstructure of as-rolled manganese steels is similar to the as-cast structure. High-carbon alloys may undergo substantial transformation of austenite during working, usually to bainite. Grain boundary carbides may also form. As with most metals, the grain size is reduced by the working operation.

## PHYSICAL PROPERTIES

The physical properties of austenitic manganese steel of concern in welding are given in Table 4.20. The thermal and electrical characteristics of this steel are similar to those of other austenitic steels. The dimensional change expected during heating is about 1.5 times that of mild steels (see Table 4.13). Carbide precipitation and transformation to pearlite will influence the physical properties in the range of 700° to 1400°F. Thermal conductivity is about 25 percent that of mild steel at room temperature, and this contributes to heat build up during welding.

The austenite in austenitic manganese steel is virtually non-magnetic, with a permeability at H-24 of about 1.03 or less. This permits the use of austenitic manganese steel where a strong and tough non-magnetic metal is required.

## MECHANICAL PROPERTIES

### Tensile Properties

Work hardening of an austenitic manganese steel develops approximately the same range of tensile properties that are produced in other steels by heat treatment. In a standard tensile test, there is little or no marked reduction of area by necking with austenitic manganese steel. As extension under load occurs, the strength of separate, favorably-oriented grains increases. As these grains become stronger, other grains are stretched and hardened progressively. Deformation is practically uniform along the reduced section.

Typical mechanical properties of various types of austenitic manganese steels are given in Table 4.21. The outstanding toughness of austenitic manganese steel is shown by the stress-strain curves in Fig. 4.23. The relatively low yield strength, a characteristic of austenitic manganese

**Table 4.20**
**Physical properties of austenitic manganese steel**

| Property | Units | Value |
|---|---|---|
| Melting temperature | °C (°F) | 1396 (2545) |
| Density | Mg/m³ | 7.92 |
| Specific heat (RT) | kJ/(kg·K) | 0.50 |
| Thermal conductivity (RT) | W/(m·K) | 13.4 |
| Electrical resistivity (RT) | μ Ω·m | 0.68 |
| Coefficient of thermal expansion, 0°-300°C | μm/(m·K) | 20.7 |

steel, is significant. It is a factor when deformation in service is not acceptable. If deformation is not critical, the low yield strength may be considered as temporary because it will increase rapidly with deformation.

The mechanical properties of austenitic manganese steel between −50° and 400°F are excellent for many applications. However, standard grades are not recommended for wear applications above 500°F because of structural instability. At high temperatures, these steels may lack the strength and ductility necessary to withstand high welding stresses. They are not oxidation resistant, and their creep properties are poor in comparison with the Cr-Ni austenitic stainless steels.

Ductility tends to increase with temperature until transformation of austenite begins. A hot short range starts between 1500°F and 1600°F and it may extend to the melting temperature. This behavior must be considered when establishing welding procedures to avoid hot cracking.

### Impact Properties

The impact properties of austenitic manganese steel are high, as evidenced by service experience and Charpy V-notch impact tests. The impact strengths of ASTM A128, Grade B-2 cast steel are about 90 to 110 ft·lbs at 75°F and about 45 to 65 ft·lbs at −100°F. The addition of nickel to the steel (Grade D) improves the low temperature impact strength to about 75 to 90 ft·lbs at −100°F.

As-rolled C-Mn-Ni steel bar has an impact strength of about 200 ft·lbs at room temperature and 100 ft·lbs at −100°F. The impact strength at −100°F is increased to about 170 ft·lbs by quench annealing.

**Table 4.21**

**Typical mechanical properties of austenitic manganese steel**

| Type | Chemical composition, percent | | | | | | Diam. or thickness, in. | Condition[a] | TS, ksi | YS, ksi | Elong, percent | RA, percent | Hardness, HB |
| | C | Mn | Si | Cr | Ni | Mo | | | | | | | |
|---|---|---|---|---|---|---|---|---|---|---|---|---|---|
| C-Mn | 1.0-1.4 | 11-14 | 0.2-1.0 | — | — | — | 1 D | C, QA | 100-145 | 50-57 | 30-65 | 30-45 | 185-210 |
| | 1.11 | 12.7 | 0.54 | — | — | — | 1 D | C | 65 | 52 | 4 | | |
| | | 11-14 | 0.2-0.6 | — | — | — | 4 t | C, QA | 90 | 52 | 25 | 35 | 170-200 |
| | | | | | | | 8 t | C, QA | 66 | 47 | 18 | 25 | |
| | 1.1-1.4 | | | | | | 1 D | R, QA | 131-158 | 43-67 | 40-63 | 35-50 | |
| C-Mn-Cr | 1.1-1.25 | 12.5-13.5 | 0.5 | 1.8-2.1 | — | — | 1 D | C, QA | 96-147 | 58-68 | 27-59 | 26-38 | 205-215 |
| | | | | | | | 4 t | C, QA | 82 | 53 | 31 | 29 | |
| | | | | | | | 6 t | C, QA | 81 | 56 | 20 | 19 | |
| C-Mn-Ni | 0.6-0.9 | 12.4-14.3 | 0.5-0.9 | — | 3.4-3.6 | — | 1 D | C, QA | 90-132 | 42-49 | 40-88 | | 150-180 |
| | 0.8-0.9 | 13.9-15.1 | 0.9-1.3 | — | 2.8-4.0 | — | — | R, QA | 134-146 | 46-56 | 74-87 | 45 | 180 |
| C-Mn-Mo | 0.75-1.0 | 12.1-14.1 | 0.4-0.6 | — | — | 1 | 1 D | C, QA | 106-137 | 50-59 | 37-67 | 30-39 | 179-207 |
| | | | | | | | 8 t | C, QA | 80-133 | 42-55 | 27-61 | 26-60 | |
| | 1.15 | 12.8-14.3 | 0.5 | — | — | 1 | 1 D | C, QA | 120-144 | 56-74 | 45-53 | 31-37 | 202-207 |
| | | | | | | | 8 t | C, QA | 76-77 | 50-56 | 16-33 | 12-29 | |
| | 0.72 | 13 | | | | 1 | 1 D | R, QA | 145-147 | 53-54 | 60-72 | 43-49 | 187 |

a. C – cast
R – rolled
QA – quenched annealed

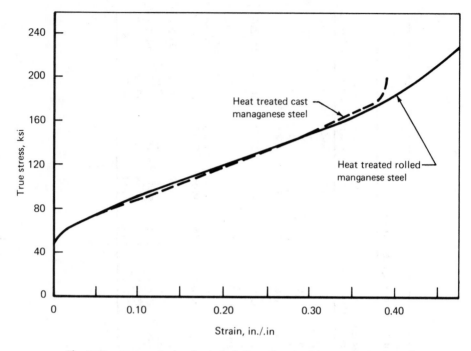

**Fig. 4.23—Stress-strain characteristics of austenitic manganese steel**

At all service temperatures experienced in railway track, mining, and construction service, cast manganese steel has outstanding toughness. This is a valuable asset for service at sub-zero temperatures.

Standard austenitic manganese steel is apparently immune to hydrogen induced embrittlement. Considerable hydrogen has been extracted from standard austenitic manganese steel, suggesting that up to perhaps 6.3cc per 100 grams can be accommodated without adverse effects.

### Work Hardening

Manganese steels have a very high coefficient of work hardening, exceeding that of austenitic stainless steel. The maximum hardness obtainable is about 550 HB. The work hardening properties make the manganese steels particularly suitable for applications where repeated impact loading occurs.

Work hardening is attributable frequently to formation of martensite, which is magnetic. Where magnetism is noted, the composition of that region is outside the standard analysis range. This often occurs in a thin skin on heat treated forgings or castings.

Since work hardening is produced by deformation, dimensional allowance must be made for the required deformation. The hard layer will vary in depth, depending on the processing, although it is always rather shallow. Applications involving both wear and impact continually harden the surface as hardened metal is worn away.

The majority of manganese steel applications do not lend themselves to work hardening. Therefore, other methods such as alloying, heat treatment, and age hardening are employed to increase hardness. Resistance to flow during cold working is increased as the yield strength is raised. Vanadium, chromium, silicon, molybdenum, and carbon additions are all effective in raising yield strength, although the first two elements may reduce ductility at the same time.

## APPLICATIONS

### Castings

Austenitic manganese steel castings are widely used in crushing, earthmoving, and material handling equipment, railroad track, and special applications where non-magnetic properties are important. Although austenitic manganese steel has relatively low yield and tensile strengths in the as-cast condition, it has good wear resistance under impact and abrasion. Gyratory and cone crusher parts such as concaves, bowl liners and mantles, jaw crushers, and hammer mill parts are examples of severe impact and abrasion service. Weights of these parts vary from 100 to 30,000 pounds. The steels are also used for railway frogs, switches, and crossings where impact is severe. The non-magnetic properties make them useful for parts for electromagnets, induction furnaces, and other electrical equipment.

## Wrought Products

Large sections of work-hardened manganese steel, ranging in thickness from 0.125 to 0.5 in., are widely used as replaceable wear shoes. They are used by railroads for applications such as pedestals and journals. The shoes provide for vertical motion of wheels on locomotives and cars while protected by replaceable manganese steel liners. The primary advantage of work-hardened austenitic manganese steel in this application is its ability to resist metal-to-metal wear without lubrication.

# WELDING

## Welding Processes

As discussed previously, reheating of austenitic manganese steel may cause carbide precipitation and some transformation of austenite which significantly reduces ductility. Therefore, this steel should be welded using a process that results in minimal heat input. Arc welding with a consumable electrode is generally used for joining and surfacing operations.

Oxyacetylene and gas tungsten arc welding are not recommended because of the characteristic heat build up in the parts during welding.

## Welding Electrodes

Austenitic manganese steel arc welding electrodes are available as bare rod or wire, and in the form of covered electrodes (rods) for shielded metal arc welding. The rod or wire may be solid, or tubular filled with alloying and fluxing ingredients. Typical arc welding electrode compositions are shown in Table 4.22. Other

proprietary electrodes are also available. They are suitable for build-up of worn parts as well as joining austenitic manganese steel components.

Some solid electrodes may contain over 0.030 percent phosphorus. These electrodes are not recommended for welding structures but they may be suitable for some build-up applications.

Solid austenitic manganese electrodes are sometimes used without external shielding (open arc). The manganese acts as a deoxidizer in the molten weld pool. A short arc length should be used to minimize the time that the molten metal droplets are exposed to the atmosphere. This procedure is not recommended for structural fabrication because of the potential for porosity and slag entrapment in the weld metal. A suitable shielding gas should be used with solid wire electrodes that are low in phosphorus for structural welding.

Bare tubular wire electrodes may be designed for self-shielding operation. Tubular rods may have a suitable covering for shielded metal arc welding. Such covered electrodes should be stored and used according to the manufacturers recommendations. Weld joints with excellent mechanical properties and cracking resistance can be produced using tubular electrodes because the phosphorus content is low. Generally, the comparatively light coverings used on covered tubular electrodes are designed to permit (1) welding with both ac and dc power, (2) ease of handling, and (3) out-of-position welding. Weld metal compositions are similar to those produced by corresponding bare wire electrodes, but cracking tendencies are lower.

Composite electrodes with a phosphorous content below 0.030 percent are recommended for fabrication and repair welding. In many applications, austenitic manganese steel can be directly joined to carbon or low alloy steel using low-phosphorus austenitic manganese steel electrodes. The welding procedures should minimize dilution with the carbon or low alloy steel.

In addition to austenitic manganese steel electrodes, several other electrode compositions are available for welding austenitic manganese steel. The corrosion resistant manganese-chromium and chromium-nickel-manganese compositions are among the most commonly used electrodes. They offer significant advantages for certain applications—namely, higher abrasion resistance, ability to work harden rapidly under impact, and high tolerance to dilution when joining austenitic manganese steel to carbon or low alloy steel. However, high chromium deposits

Table 4.22
Typical electrodes for arc welding austenitic manganese steels

| Type | AWS Class.[a] | Composition, percent[b] | | | | | | | | | Remarks |
|---|---|---|---|---|---|---|---|---|---|---|---|
| | | C | Mn | Ni | Cr | Mo | V | Si | P | Fe | |
| Mn-Ni | EFeMn-A | 0.5-0.9 | 11-16 | 3-6 | 0.5[c] | — | — | 1.3[c] | 0.03[c] | Remainder | |
| Mn-Ni-Cr | d | 0.6-0.85 | 14-17 | 2-4 | 2.5-4.0 | — | — | 0.2-0.7 | 0.02 | Remainder | |
| Mn-Mo | EFeMn-B | 0.5-0.9 | 11-16 | — | 0.5[c] | 0.6-1.4 | — | 0.3-1.3 | 0.03[c] | Remainder | |
| Mn-Cr | d | 0.3-0.6 | 14-15 | 1.0 | 14-15 | 0.3-1.7 | 0-0.6 | 0.2-0.5 | | Remainder | Corrosion resistant |
| Cr-Ni-Mn | d | 0.5 | 4.5 | 10 | 20 | 1.4 | — | 0.6 | | Remainder | Corrosion resistant |

a. Covered electrode requirements are covered in AWS A5.13-80, *Specification for Solid Surfacing Welding Rods and Electrodes*. Bare and covered composite electrodes are described in AWS A5.21-80, *Specification for Composite Surfacing Welding Rods and Electrodes*.
b. Composition given is for deposited weld metal with covered electrodes.
c. Maximum values.
d. Proprietary electrodes.

can not be cut by ordinary oxygen cutting methods. Austenitic chromium-nickel corrosion resistant steel electrodes may also be used to weld austenitic manganese steels to carbon steels or as a buttering layer on the carbon steel.

## Welding Procedures

Prior to welding, any work-hardened metal, cracks, or other defective areas should be removed by grinding or air carbon arc gouging. Work-hardened metal is more susceptible to cracking than is as-quenched material. As with welding, the steel should not be overheated during metal removal.

Preheat should not be used. The interpass temperature of the metal next to the weld should not exceed about 600°F after a one minute period to avoid excessive heating and embrittlement of the heat-affected zone. It should be possible to touch the base metal at a distance of 6 inches from the weld at all times without burning the skin.

The welding heat input should be the minimum required to obtain complete fusion and joint penetration. A short arc length and direct current, electrode positive should be used to deposit essentially stringer beads. Intermittent welding may be used to limit heat build up. The weld bead should be peened while very hot to upset it and relieve welding stresses.

Multiple pass welds are preferred to single pass welds because succeeding passes will temper any martensite that may form in the preceeding weld beads. However, heat input and interpass temperature restrictions must be adhered to.

Semi-automatic or automatic arc welding may be done using solid or tubular electrodes. These methods are advantageous for shop operations. Heat input into the base metal is less than with shielded metal arc welding because smaller electrodes and higher travel speeds can be used. A semi-automatic welding operation is shown in Fig. 4.24.

## Mechanical Properties

Typical mechanical properties of austenitic manganese steel weld metal are given in Table 4.23. The Mn-Cr electrode produced weld metal with significantly higher yield strength than the other electrodes. This can be attributed to its high chromium content. The excellent notch toughness of Mn-Ni weld metal is illustrated in Table 4.24.

## Repair Welding

Build up of worn areas is the most common repair application. The deposited metal is usually of the same type as the base metal, although hard surfacing of areas subject to extreme wear may appreciably extend service life.

In rebuilding worn areas, it must be determined if the surfaces have work hardened in service. Work-hardened surfaces should be removed before resurfacing to prevent possible heat-affected zone cracking and subsequent spalling of the weld metal. The hardness of the exposed surface should be checked to ensure that all work-hardened metal has been removed. The filler metal should match the base metal.

Cracked or broken castings can be repaired by welding. The crack should be removed by grinding or carbon arc gouging, and a generous radius left at the end of the groove. Welding should progress from the end to the start of the groove to minimize welding stresses. Skip welding can minimize heat buildup, and each bead can be peened while hot. Austenitic manganese steel or austenitic stainless steel electrodes, such as Type 308, 309, 310, or 312 may be used for the repair of cracks.

Foundry repairs of casting defects by welding are made after solution heat treatment and quenching because as-cast castings are too brittle, and would likely crack after welding. The faces of a cavity to be filled must have a bevel of at least 15 degrees to provide ready access for welding.

## Dissimilar Steels

When welding austenitic manganese steel to carbon or low alloy steel, manganese steel electrodes may be used with proper procedures. Best results are obtained when the filler metal contains more than 14 percent manganese, less than 0.03 percent phosphorus, and dilution with the carbon or low alloy steel is 25 percent or less. Otherwise cracking may occur. Carbon or low alloy steel filler metal should not be used.

Austenitic Mn-Cr electrodes are preferred when the weld will be subject to wear and high yield strength is needed. Type 308, 309, 310 or 312 stainless steel filler metal can be used when the welded joint will be exposed to little or no wear and moderate stresses. The weld interface may be hard and brittle if the dilution is excessive because of the high carbon in the austenitic manganese steel.

*Fig. 4.24 – Installing austenitic manganese steel wear bars onto crawler track links by semi-automatic arc welding*

**Table 4.23**
**Typical mechanical properties of austenitic manganese steel weld metal**

| Welding process[a] | Electrode type | Yield strength, ksi | Tensile strength, ksi | Elongation, % | Reduction of area, % | Brinell Hardness |
|---|---|---|---|---|---|---|
| SMAW | Mn-Ni | 64.1 | 121.3 | 47.0 | 37.6 | 207 |
| SMAW | Mn-Ni-Cr | 75.6 | 119.8 | 42.0 | 33.2 | 223 |
| AW | | 78.7 | 122.3 | 37.0 | 31.6 | 235 |
| SAW | | 79.4 | 120.6 | 38.0 | 34.0 | 207 |
| SMAW | Mn-Mo | 67.9 | 119.8 | 32.0 | 33.1 | 214 |
| SMAW | Mn-Cr | 120.0 | 146.0 | 30.0 | — | 194 |

a.  SMAW – Shielded metal arc welding
    AW – Semi-automatic without auxiliary shielding
    SAW – Submerged arc welding

**Table 4.24**
**Charpy V-notch impact strength of austenitic manganese-nickel steel weld metal at room and low temperatures.**

| Test temperature, °F | Impact strength, ft·lbs |
|---|---|
| 75 | 118 |
| 0 | 96 |
| −75 | 80 |
| −150 | 55 |

Figure 4.25 shows a weld interface between cast austenitic manganese steel and Type E308 chromium-nickel stainless steel weld metal deposited by shielded metal arc welding. The weld metal in Fig. 4.26 was deposited by gas metal arc welding using a steel electrode containing 16.5 percent manganese 16.6 percent chromium. The dilution at the interface is apparent in both microstructures. In contrast, Fig. 4.27 shows an interface between austenitic manganese base metal and austenitic Mn-Mo weld metal deposited by shielded metal arc welding. In this case, dilution has no significance.

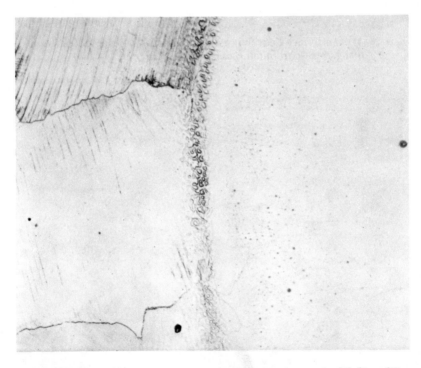

*Fig. 4.25–Weld interface between cast austenitic manganese steel (left) and Type E308 chromium-nickel stainless steel weld metal (right) (×250)*

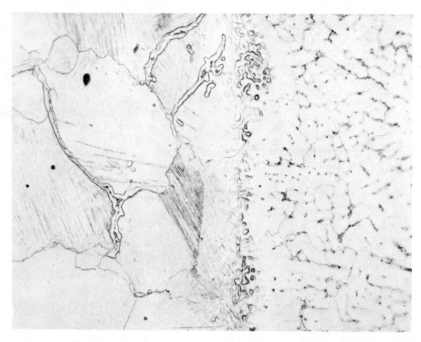

*Fig. 4.26–Weld interface between cast austenitic manganese steel (left) and manganese-chromium stainless steel weld (right) (×250)*

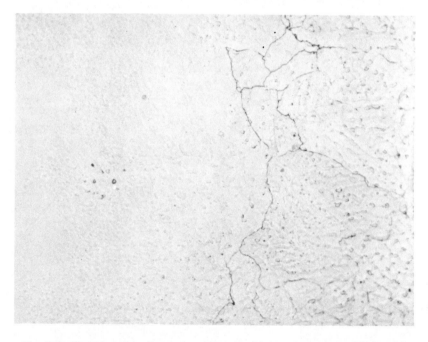

*Fig. 4.27–Weld interface between cast austenitic manganese steel (left) and austenitic manganese-molybdenum steel weld metal (right) (×250)*

# Metric Conversion Factors

$°C = 0.56 (°F - 32)$
1 ksi = 6.89 MPa
1 ft·lbf = 1.36 J
$1 \text{ ksi·in.}^{1/2} = 1.1 \text{ MN·m}^{-3/2}$
1 in./min = 0.42 mm/s
$1 \text{ lb/in.}^3 = 27.7 \text{ Mg/m}^3$
1 Btu /(ft·h·°F) = 1.73 W/(m·K)
$1 \text{ in.} \times 10^{-6} / (\text{in.·°F}) = 0.56\mu\text{m}/(\text{m·K})$
1 Btu/(lb·°F) = 4.19 kJ/(kg·K)

# SUPPLEMENTARY READING LIST

Albom, M. J., and Titherington, C.C., Investigation of weldability of ultra-high-strength steels, *Welding Journal*, 39(9): 385s-91s; 1960, Sept.

Avery, H. S., and Chapin, J. H., Austenitic manganese steel welding electrodes, *Welding Journal*, 33(5): 459-79; 1954 May.

Bailey, N., Weldability and toughness of maraging steel, *Metal Construction*, 3(1): 1-5; 1971 Jan.

Bailey, N. Solidification cracking in MIG fillet welds, *Metal Construction*, 6(5): 143-47; 1974 May.

Bailey, N., and Roberts, C., Maraging steel for structural welding, *Welding Journal*, 57(1): 15-28; 1978 Jan.

Blauel, J. G., Smith, H. R., and Schulze, G., Fracture toughness study of a Grade 300 maraging steel weld joint, *Welding Journal*, 53(5): 211s-18s; 1974 May.

Corrigan, D. A., Gas shielded arc welding of 18% Ni maraging steel, *Welding Journal*, 43(7): 289s-94s; 1964 July.

Duffey, F. D. and Sutar, W., Submerged arc welding of 18% Ni maraging steel, *Welding Journal*, 44(6): 251s-63s; 1965 June.

Kenyon, N., Effect of austenite on the toughness of maraging steel welds, *Welding Journal*, 47(5): 193s-98s; 1968 May.

Lang, F. H., and Kenyon, N., *Welding of Maraging Steels*, New York: Welding Research Council Bulletin 159; 1971 Feb.

Lutes, W. L. and Reid, H. F., Comparative properties of electrodes for arc welding austenitic manganese steel, *Welding Journal*, 35(8): 766-83; 1956 Aug.

*Metals Handbook*, Vol. 1, *Properties and Selection: Irons and Steels*, 1978, Vol. 3. *Properties and Selection: Stainless Steels, Tool Materials, and Special Purpose Metals*, 1980 Vol. 4, *Heat Treating*, 1981, 9th Ed., Metals Park, OH: American Society for Metals.

Pepe, J. J. and Savage, W. F., The weld heat-affected zone of the 18Ni maraging steels, *Welding Journal*, 49(12): 545s-53s; 1970 Dec.

# 5

# Cast Irons

## Chapter Committee

R.A. BISHEL, *Chairman*
*Huntington Alloys, Incorporated*

J.G. BIELENBERG
*Welding Consultant*

R.J. DYBAS
*General Electric Company*

K.L. JOHNSON
*Advanced Robotics Corporation*

A.B. MALIZIO
*U.S. Pipe and Foundry Company*

## Welding Handbook Committee Member

W.L. WILCOX
*Scott Paper Company*

# 5

# Cast Irons

## GENERAL DESCRIPTION

Cast iron is a common term for a series of ferrous alloys that normally contain more than 2 percent carbon and 1 to 3 percent silicon, as well as phosphorus and sulfur. Alloy cast irons may also contain one or more other elements deliberately added to provide desired properties such as strength, hardness, hardenability, or corrosion resistance. Common alloying additions are chromium, copper, molybdenum, and nickel.

In general, cast irons have lower melting ranges than those of steels, are quite fluid when molten, and undergo moderate shrinkage during solidification and cooling. The toughness and ductility of cast iron are lower than those of steel; these low properties limit the applications of cast iron.

The mechanical properties of an iron casting depend upon the type of microstructure as well as the form and distribution of the microstructural constituents. One microstructural constituent that has a significant effect is the free graphite (carbon). The amount, size, and shape of the graphite particles affect the strength and ductility of a cast iron. Consequently, cast irons may be classified by the characteristics of the graphite as it appears in a polished section.[1]

The microstructure of the matrix surrounding the graphite particles also influences the mechanical properties of a casting. The matrix is basically steel that may be ferritic, pearlitic, austenitic, or martensitic. The specific matrix in a casting will depend upon the chemical composition, cooling rate, and heat treatment of the casting.

There are four basic types of cast iron: gray iron, malleable iron, ductile or nodular iron, and white iron. Compacted graphite (GC) cast iron, a fifth type not yet standardized, has a microstructure intermediate to those of gray and ductile cast irons. Its mechanical properties fall within the range of 35 to 60 ksi yield strength (0.1 percent offset), 45 to 75 ksi tensile strength, and 1 to 6 percent elongation.

Cast irons contain more carbon than high carbon steels. They also have appreciable amounts of silicon that influence the structural form of the carbon in a casting. As a consequence, the metastable iron-iron carbide system coexists with the stable iron-graphite system in an iron casting. This duality of carbon form produces a metallic matrix in a casting that has the microstructure and attributes of steel. Additionally, the uncombined graphitic carbon is distributed throughout the matrix in a variety of geometric shapes including tiny particles, various flake-like forms, and spheroids. The size, shape, and distribution of the graphite are incorporated in casting specifications because they influence both the mechanical and the physical properties of a casting.

The matrix structure and graphite character in a particular casting are achieved by careful balance of its chemical composition, degree of inoculation, rate of solidification, and control of the cooling rate. Solidification and cooling rates are affected by the section thickness and the rate of heat transfer to the mold. Castings are heat treated to achieve properties that cannot be obtained in the as-cast condition. Annealing, austenitizing and quenching, and tempering heat treatments are used to achieve the desired microstructure in the metallic matrix.

---

1. See ANSI/ASTM A 247-§7 (1978), *Standard Method for Evaluating the Microstructure of Graphite in Iron Castings.*

## WHITE CAST IRON

White cast iron is formed when the carbon does not precipitate as graphite during solidification but remains in combination with iron, chromium, or molybdenum as carbides. This iron is hard and brittle, and has a white crystalline fracture appearance. White cast iron is normally considered unweldable because of the absence of adequate ductility to accommodate thermal stresses in the base metal.

## GRAY CAST IRON

Gray cast irons are iron-carbon-silicon alloys that contain uncombined carbon in the form of graphite flakes, as shown in Fig. 5.1. This type of cast iron is named from the gray appearance of a fractured surface.

Gray iron castings for general engineering and automotive applications are divided into ten classes or grades based on minimum tensile strength, as listed in Table 5.1. Hardness, matrix microstructure, chemical composition, pressure tightness, and radiographic soundness in various combinations are sometimes specified to meet service requirements.

The tensile strength, hardness, and microstructure of a gray iron casting are influenced by several factors including chemical composition, design, the characteristics of the mold, and cooling rate during and after solidification. Castings with similar properties can be produced using different raw material mixes or various melting and casting practices, or both.

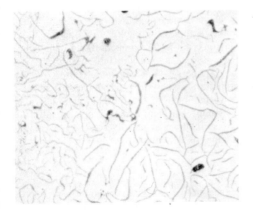

**Fig. 5.1 – Graphite form in gray cast iron**
*( × 250)*

Copper, chromium, molybdenum, and nickel are frequently added to gray cast iron to control matrix microstructure and graphite formation. They also can improve the corrosion resistance of the cast iron to certain media.

Castings are heat-treated to obtain desired mechanical properties. Gray iron, like steel, can be hardened by quenching from a suitable temperature (1600°F), but it will be brittle. Tempering after quenching is necessary to improve toughness and decrease hardness.

Gray cast iron lacks ductility because of the flake form of the graphite. Since the graphite has virtually no strength, the numerous flakes create internal notches in the steel matrix. Fracture can initiate easily at these sites and progress rapidly across the section without plastic strain.

## MALLEABLE CAST IRON

Malleable cast iron is produced by heat treating a white iron casting of suitable composition. The initial formation of white cast iron is promoted by (1) low carbon and low silicon contents, (2) the presence of carbide forming elements such as chromium, molybdenum, and vanadium, and (3) rapid solidification and cooling. The white iron casting is then heated in a controlled atmosphere furnace to a temperature above the eutectoid temperature (usually 1700°F) and held for several hours. This treatment permits the dissolved carbon in the austenite to precipitate as irregular nodules of graphite, known as *temper carbon*.

As such a casting cools slowly in a furnace through the eutectoid transformation range, the remaining excess carbon in the austenite precipitates onto the existing temper carbon nodules. The resulting graphite form is shown in Fig. 5.2. This treatment produces a ferritic matrix. A pearlitic or martensitic matrix can be obtained by alloying or by air or liquid quenching from the austenitizing temperature. These two matrix structures are normally tempered after quenching. The minimum mechanical properties of standard malleable irons are given in Table 5.2.

Malleable irons have some ductility because the temper carbon is in the form of nodules rather than flakes, as found in gray cast iron. The effect of these nodules on ductility is less severe than that of the flake graphite. The strength and ductility of malleable irons are related to the metallurgical structure of the matrix as well as the distribution of the free graphite. If the cast iron is reconverted to a metastable state by quenching

## Table 5.1
## Mechanical properties of standard gray cast irons

| Class or Grade | Minimum tensile strength, ksi[a] | Hardness range, HB[b] | Matrix microstructure |
|---|---|---|---|
| G1800[c] | 18 | 187 max | Ferritic-pearlitic |
| 20[d] | 20 | | |
| G2500 | 25 | 170 to 229 | Pearlitic-ferritic |
| 25 | 25 | | |
| G3000 | 30 | 187 to 241 | Pearlitic |
| 30 | 30 | | |
| G3500 | 35 | 207 to 255 | Pearlitic |
| 35 | 35 | | |
| G4000 | 40 | 217 to 269 | Pearlitic |
| 40 | 40 | | |
| 45 | 45 | | |
| 50 | 50 | | |
| 55 | 55 | | |
| 60 | 60 | | |

a. Minimum tensile strength of cast test bars based on the thickness of the controlling section of a casting. Refer to the appropriate specification.
b. Brinell hardness using a 10mm ball and 3000 kg load.
c. GXX00 grades refer to ANSI/ASTM A 159-77, *Specification for Automotive Gray Iron Castings.*
d. Two digit designations refer to ANSI/ASTM A48-76, *Standard Specification for Gray Iron Castings.*

from a high temperature, it must be malleabilized again to regain ductility.

## DUCTILE CAST IRON

Ductile cast iron and gray cast iron are similar with respect to carbon and silicon contents, and in terms of general foundry practice for the

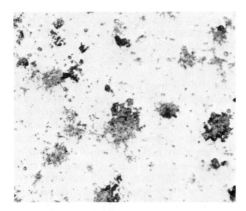

*Fig. 5.2–Graphite form in malleable cast iron (×100)*

production of castings. However, they differ in the geometric form assumed by the free graphite. In ductile cast iron, the graphite is caused to nucleate throughout the metal matrix in the form of spheroids, as shown in Fig. 5.3. As with malleable cast iron, the agglomeration of the free graphite into nodular form avoids the internal sharp notches associated with flake graphite. This accounts for the higher strength and ductility of ductile cast iron when compared to gray cast iron of similar composition. The production of ductile iron castings does not require the long-time heat treatment required to produce malleable iron castings. The properties of standard ductile cast irons are given in Table 5.3.

Spheroidization or nodulization of the graphite is achieved by introducing magnesium or cerium into a low-sulfur melt, preferably one containing not more than 0.02 percent sulfur. A low sulfur level is usually achieved by adding calcium oxide, calcium carbide, or sodium carbonate to the molten iron.

Magnesium can be introduced in elemental form if it is diluted or otherwise shielded to prevent instantaneous vaporization. Various proprietary alloys of magnesium combined with one or more of the elements nickel, iron, silicon, and calcium are frequently used. A residual concen-

### Table 5.2
### Mechanical properties of standard malleable iron castings

| Class or grade | Microstructure | Minimum tensile strength, ksi | Minimum yield strength, ksi | Minimum elongation in 2 in., percent | Hardness range, HB[a] | Heat treatment[b] |
|---|---|---|---|---|---|---|
| | | ASTM A47-77 | | | | |
| 32510 | Ferritic | 50 | 32.5 | 10 | | |
| 35018 | Ferritic | 53 | 35 | 18 | | |
| | | ASTM A220-76 | | | | |
| 40010 | Pearlitic | 60 | 40 | 10 | 149 to 197 | |
| 45006 | Pearlitic | 65 | 45 | 6 | 156 to 207 | |
| 50005 | Pearlitic | 70 | 50 | 5 | 179 to 229 | |
| 60004 | Pearlitic | 80 | 60 | 4 | 197 to 241 | |
| | | ASTM A602-70[c] | | | | |
| M4504 | Ferritic-pearlitic | 65 | 45 | 4 | 163 to 217 | AQ&T |
| M5003 | Ferritic-pearlitic | 75 | 50 | 3 | 187 to 241 | AQ&T |
| M5503 | Martensitic | 75 | 55 | 3 | 187 to 241 | LQ&T |

a. Brinell hardness using a 10 mm ball and 3000 kg load.
b. AQ&T–air quench and temper; LQ&T–liquid quench and temper.
c. Automotive castings.

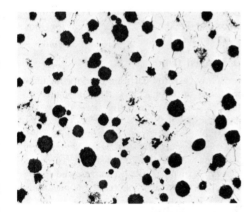

*Fig. 5.3–Graphite form in ductile cast iron*
*( × 100)*

tration of about 0.035 percent magnesium is necessary to produce full nodulization. If magnesium is lost from the melted zone of the base metal during welding of a ductile iron casting, the shape of the graphite along the weld interface can change. A welding process and procedure

that minimize heat input and time in the molten state will minimize graphite degradation.

## COMPACTED GRAPHITE CAST IRON

Compacted graphite (CG) cast iron is produced in a manner similar to ductile cast iron, but careful control of the magnesium addition to the melt is required. The graphite in this cast iron, shown in Fig. 5.4, forms in irregular geometric shapes that have a lesser surface-to-volume ratio than the flakes in gray cast iron; but it does not assume the spheroidal shape found in ductile cast iron. This intermediate form of graphite imparts a combination of desirable qualities to a casting. The mechanical properties of the three types of cast iron are compared in Table 5.4. One method for attaining the desired graphite form is to inoculate a low-sulfur melt with a magnesium ferro-silicon alloy that contains several additional elements to enhance the performance and reliability of the CG cast iron.

During solidification, CG cast iron shrinks less than ductile cast iron. It also has good thermal shock resistance. High-strength CG cast iron differs from gray cast iron in that smaller alloy

**Table 5.3**
**Mechanical properties of typical ductile cast irons**

| Grade[a] | Microstructure | Minimum tensile strength, ksi | Minimum yield strength, ksi | Minimum elongation in 2 in., percent | Heat treatment[b] |
|---|---|---|---|---|---|
| 60-40-18 | Ferritic | 60 | 40 | 18 | A |
| 65-45-12 | Ferritic-pearlitic | 65 | 45 | 12 | |
| 80-55-06 | Ferritic-pearlitic | 80 | 55 | 6 | |
| 100-70-03 | Pearlitic | 100 | 70 | 3 | AQ&T, LQ&T |
| 120-90-02 | Martensitic | 120 | 90 | 2 | LQ&T |

a. Refer to ANSI/ASTM A536, Standard Specification for Ductile Iron Castings.
b. A – ferritizing anneal; AQ – air quench; LQ – liquid quench; T – temper.

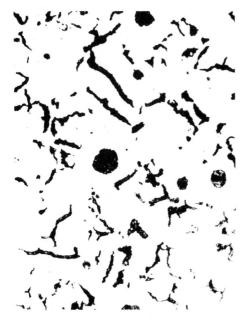

*Fig. 5.4–Graphite form in compacted graphite cast iron*

additions are needed to produce equivalent strength properties. The welding precautions described for ductile cast iron should also be followed with CG cast iron to minimize alteration of the graphite form.

## APPLICATIONS

Each type of cast iron is used for a number of applications, many of which are covered by standards of the American National Standards Institute (ANSI), the American Society for Testing and Materials (ASTM), and the Society of Automotive Engineers. A standard sometimes contains special requirements for a cast iron to meet specific service conditions. The appropriate specification should be consulted for those requirements.

### Gray Cast Iron

Gray cast iron parts are used for many applications in the machinery and automotive industries. Typical automotive parts covered by specifications are brake drums, clutch plates, and cam shafts. Furnace parts, ingot and glass molds, and melting pots that operate at elevated temperatures are made of gray cast iron, as are various types of pipe, valves, flanges, and fittings for both pressure and nonpressure applications.

### Ductile Cast Iron

Some culvert, sewer, and pressure pipe, as well as fittings, valves, and pumps are cast of ductile iron to specifications. The advantages of these products are their relatively good toughness and weldability when compared to similar components of gray cast iron. Ductile cast iron piping systems are commonly fabricated by welding.

### Malleable Cast Iron

Malleable cast iron is used for flanges, pipe

**Table 5.4**
**Comparison of the mechanical properties of gray, compacted graphite, and ductile cast irons**

| Type | Approx. carbon equivalent[a] | Tensile strength, ksi | Yield strength, ksi[b] | Elongation in 2 in., percent |
|---|---|---|---|---|
| Gray–Class 25 | 4.4 | 25 | — | <1 |
| Gray–Class 45 | 3.6 | 45 | — | <1 |
| Compacted graphite | 4.2 | 45-75 | 35-60 | 1-6 |
| Ductile | 4.2 | 60-100 | 40-80 | 6-22 |

a.  $CE = \%TC + 0.33\,(\%P + \%Si)$, where TC is total carbon
b.  0.1% offset

fittings, and valve parts. A number of automotive parts of malleable cast iron include steering components, compressor crank shafts and hubs, transmission and differential parts, connecting rods, and universal joints. These products are covered by appropriate specifications.

# WELDING CONSIDERATIONS

## UTILIZATION

Welding or braze welding of cast iron is generally used for (1) the repair of foundry defects, (2) the fabrication of components, or (3) the salvage of parts that break or wear out in service. All types of cast irons except white iron are considered to be weldable but to a lesser degree than carbon steels. The best results are achieved with low-strength cast irons. Good joint efficiency is possible with established welding procedures for cast irons with tensile strengths up to and including 70 ksi. At higher strength levels, matching strength in a welded joint is increasingly difficult to achieve.

## WELDABILITY

All cast irons have a common problem affecting their weldability, namely too much carbon. The manufacturing process, that is, casting and sometimes heat treating, is capable of producing microstructures that possess useful mechanical properties. However, a welding thermal cycle ordinarily does not produce a desirable microstructure. The iron immediately adjacent to the weld metal is heated to an excessive temperature, and the cooling rate of the entire heat-affected zone is too rapid. Carbides tend to form in the base metal immediately adjacent to the weld metal, and the remainder of the heat-affected zone tends to form high-carbon martensite. Both of these microstructures are very brittle and are subject to cracking, either spontaneously or during service. The degree of brittleness and propensity to cracking depend to some extent upon the type of cast iron, the condition of heat treatment, and the welding procedure.

Fusion welding, because of its localized nature, produces thermal stresses in the weld area. A base metal must be capable of some local plastic deformation to accommodate these welding stresses or cracking will likely result. Ductile and malleable cast irons having a ferritic matrix are better suited to absorb welding stresses than is gray cast iron.

As with steels, phosphorus in cast iron has an adverse effect on weldability; it should be held to less than 0.10 percent. Repair welding of high-phosphorus castings requires special techniques.

## Effects of Graphite Form

The composition and microstructure of a cast iron affect the amount of carbon in the heat-affected zone that is dissolved during welding. This, in turn, influences the brittleness and crack susceptibility of the welded joint. To minimize the formation of massive carbides and high-carbon martensite, it is most helpful to have the carbon present as spheroids that have a low surface-to-volume ratio. The smaller the surface area of the graphite in contact with an austenitic matrix, the lesser is the amount of carbon in the microstructure at room temperature. Graphite flakes in gray cast iron display the greatest tendency to dissolve in austenite because of their relatively large surface area. However, graphite in any form dissolves slowly and often remains in the weld metal. In general, melting during fusion welding is a reversal of the casting solidification process, and those areas last to solidify during casting are the first ones to melt during welding.

Cast iron typically includes the following:

3.5 percent carbon
2.5 percent silicon
0.5 percent manganese
0.04 percent phosphorus
0.06 percent sulfur

The addition of 0.07 percent of magnesium to a cast iron of such composition promotes the formation of nodular graphite. Increasing the manganese or sulfur content decreases graphitization, but higher silicon promotes it. High phosphorus content contributes to embrittlement of cast iron.

## Effects of Filler Metal

The composition and mechanical properties of the filler metal are very important considerations in the welding of cast iron.[2] As each weld bead is made, filler metal and base metal, or previously deposited weld metal, are melted together. The change in composition of the filler metal as a result of mixing with base metal or previously deposited metal is called dilution.[3]

Dilution should be minimized by welding with the lowest heat input consistent with sound weld metal. The high carbon, silicon, phosphorus, and sulfur levels in cast iron can have an adverse effect on the mechanical properties and soundness of the weld metal. For example, when cast iron is welded with carbon steel filler metal, the carbon content of the deposited metal will usually increase as a result of dilution. The weld metal can be hard with low ductility and difficult to machine unless a suitable postweld heat treatment is applied. It might also be crack sensitive. Inability of the weld metal to yield and relieve welding stresses can result in cracking in the adjacent cast iron heat-affected zone.

The mechanical properties of the weld metal employed on cast iron can play a major part in the success of the operation. When the yield strength of the weld metal is low, the stresses imposed on the cast iron during cooling are relatively low. This reduces the likelihood of cracking. During service, a soft weld metal can creep and relieve stresses in the cast iron. Nickel or nickel alloy weld metal is very effective in this respect, and considerable use is made of nickel and nickel alloy filler metals for arc welding cast iron. Another advantage of these types of weld metal is the ease with which they can be machined in the as-welded condition.

## PREHEATING

The formation of a hard and brittle heat-affected zone can lead to cracking during cooling or in service. Low heat input with arc welding limits the width of the heat-affected zone but a band of hard brittle iron can still form adjacent to the weld metal. Heat-affected zone hardness can be limited by preheating in combination with slow cooling after welding. Preheating provides lower cooling rates in both the weld metal and the heat-affected zone than occurs without it. A low cooling rate during and after austenitic transformation reduces the amount of martensite formed and thus the hardness.

The preheat temperature and extent of preheating depend upon the type of cast iron being welded, the mass of the casting, the welding process, and the type of filler metal. Recommended preheat and interpass temperature ranges are given in Table 5.5. In general, ferritic ductile and ferritic malleable cast irons can be arc welded with lower preheat than can the pearlitic types because ferritic cast irons have better ductility.

All cast irons generally need to be preheated when oxyacetylene welding to reduce the heat input requirements. High preheat is needed when using a cast iron filler metal because the weld

---

2. Refer to AWS A5.15-82, *Specification for Welding Rods and Covered Electrodes for Cast Iron.*

3. Dilution is discussed in more detail in Chapter 12.

**Table 5.5**
**Recommended preheat and interpass temperature range for welding cast irons**

| Type of cast iron | Matrix microstructure | Temperature range, °F[a] | |
|---|---|---|---|
| | | Arc welding | Oxyacetylene welding |
| Gray | — | 70 to 600 | 800 to 1200 |
| Malleable | Ferritic | 70 to 300 | 800 to 1200 |
| Malleable | Pearlitic | 70 to 600 | 800 to 1200 |
| Ductile | Ferritic | 70 to 300 | 400 to 1200 |
| Ductile | Pearlitic | 70 to 600 | 400 to 1200 |

a. A suitable preheat temperature for a particular application depends upon the welding process, filler metal, mass of the casting, welding process, and postweld heat treatment.

metal has low ductility near room temperature. A filler metal that deposits relatively low strength, ductile weld metal, such as ENi-CI or ENi-CI-A electrodes, can be used with the base metal at or slightly above room temperature. The weld metal can readily yield during cooling and relieve welding stresses that might otherwise cause cracking in the weldment.

Cracking from unequal expansion can take place during the preheating of complex castings or when the preheating is confined to a small area of a large casting. Local preheating should be gradual. Preheating of either a large section of the casting where the welding is to be done or the entire casting in a uniform manner is recommended. In any case, the preheating temperature should be maintained during welding, and welding should be completed before the casting is cooled slowly to room temperature.

Where possible, the preheating pattern should be designed to place the welded joint in compression after cooling to room temperature. The part configuration should be thoroughly analyzed before preheating to minimize the introduction of thermal stresses from this operation that might cause subsequent failure.

To ensure proper control, preheat and interpass temperatures should be monitored. Contact pyrometers, temperature-sensitive crayons, or thermocouples are means of monitoring temperature.

## SURFACE PREPARATION

All casting skin and foreign material should be removed from the joint surfaces and adjacent areas. Castings that have been in service are often impregnated with oil or grease that can be removed by solvents or steam cleaning. Where

possible, the casting should be heated uniformly at about 700°F for 30 minutes, or for a shorter time at 1000°F, using an oxyfuel gas torch or circulating air furnace. An alternative is to heat the weld area by depositing the first weld pass and then removing it by a suitable means because the weld metal is generally very porous. This welding and removal operation is repeated until the weld metal is sound. Then, the weld is completed in a normal fashion. Castings that have been impregnated with a plastic or glass sealer should not be repair welded in any case because the sealer will produce gross porosity or inhibit bonding.

Brazing or soldering filler metals will not wet exposed graphitic carbon in cast iron and will result in a poor metallurgical bond. Where wetting difficulties are encountered, several cleaning methods can be used. One method is a proprietary electrochemical cleaning operation that produces a surface essentially free of graphite, sand, silicon, oxides, and other contaminants. The process uses a catalyzed molten salt bath operating at 850° to 950°F in a steel tank. Direct current is passed through the bath with the work as one electrode and the steel tank as the other. The direction of current flow is occasionally reversed to produce alternate reducing and oxidizing effects. A water rinse completes the surface treatment.

Abrasive blasting with steel shot may be a suitable method for preparing the surfaces of ductile and malleable cast irons but not for gray cast irons. Searing the surface with an oxidizing flame or heating the casting to about 1650°F in a strongly decarburizing atmosphere may be suitable for some applications.

Before any cleaning procedure is used in production, wetting tests should be conducted using the proposed filler metal and joining procedures. The filler metal should be applied to a cleaned, flat surface and then examined visually for good wetting. If the surface is not uniformly wetted, it has not been sufficiently cleaned.

## POSTWELD HEAT TREATMENT

Thermal stress relieving is desirable for fully restrained welds, welds intended for use in severe service, and welds subject to close machining tolerances. Normally, stress relieving is done immediately after welding by increasing the temperature of the entire casting into the 1100° to 1150°F range. The casting is held at temperature for about one hour per inch of thickness. The cooling rate should not exceed 50°F per hour until the casting has cooled down to about 700°F. Some reduction in hardness can be achieved by this stress relieving treatment but it will be slight because the carbides are stable at 1100°F.

Heat treatment at 1650°F followed by slow cooling to 1000°F or lower results in maximum softening and stress relief. To obtain optimum ductility, the weldment should be given this heat treatment immediately after welding. Reduction in strength must always be considered when selecting a stress-relieving temperature.

For the best results with ductile cast iron, the welded casting should be immediately placed into a hot furnace (1100° to 1200°F) and the temperature raised to 1650°F. The casting should be held at temperature for two to four hours. It is then cooled to 1300°F, held there for five hours, and then cooled down to 1100°F in the furnace. The casting may then be cooled to room temperature in the furnace or in still air.

If a casting is welded with a high (1200°F) preheat and interpass temperature, it should be slowly cooled by covering it with an insulating blanket, vermiculite, or hot sand. Malleable iron castings may be reheat-treated after welding.

## FILLER METAL SELECTION

Several factors must be considered when choosing a filler metal for welding cast iron. The important ones are as follows:

(1) The type of cast iron
(2) Mechanical properties desired in the welded joint
(3) Tolerance of the filler metal to dilution by the base metal
(4) Ability of the weld metal to yield and relieve welding stresses
(5) Machineability of the weld zone
(6) Color matching
(7) Applicable welding process
(8) Cost

Filler metals are available for welding cast iron by the shielded metal arc, flux cored arc, gas metal arc, and oxyacetylene welding processes. In addition, filler metals are available for braze welding. In special cases, cast irons may be welded using gas tungsten arc welding and a suitable filler rod. Typical filler metals are given in Table 5.6. Other suitable proprietary filler metals are also available.

# ARC WELDING

## JOINT DESIGN

Joint designs commonly used for arc welding of carbon steels are generally suitable for welding cast iron sections together. The root opening should be wide enough to permit good fusion into the root faces and into a backing plate, if used. When practical, thick sections should be welded from both sides using either a double-V- or double-U-groove preparation.

When repairing cracked castings, a hole about 0.13 inch or larger in diameter should be drilled at the end of each crack to prevent further propagation. Then, sufficient cast iron should be removed to eliminate the crack and provide room to properly manipulate a welding electrode or torch during repair welding.

## SHIELDED METAL ARC WELDING

Cast irons can be arc welded with nickel, nickel alloy, mild steel, and copper alloy covered electrodes. Bare gray cast iron electrodes are used for special applications. Filler metal selection depends upon the type of cast iron to be joined and the application. In any case, dilution with the cast iron should be kept to a minimum. Covered electrodes for cast iron welding are listed in Table 5.6. Figure 5.5 shows the repair welding of a casting with a covered electrode.

## Nickel Alloy Electrodes

The nickel alloy electrodes, such as ENi-CI and ENiFe-CI, are specifically designed for the welding of cast irons. Deposited metal from each type of electrode has a carbon content well above the solubility limit. The excess carbon is rejected as graphite during solidification of the weld metal. This reaction causes an increase in volume that tends to minimize weld metal shrinkage dur-

ing solidification. This, in turn, reduces residual stresses in the weld metal and the cast iron heat-affected zone.

The nickel (ENi-CI) electrode produces a softer weld metal than the nickel-iron (ENiFe-CI) electrode. This is important where machinability of the weld metal is a factor. The ENiFe-CI weld metal has good machinability also, particularly with multiple-pass welds. In general, this electrode is a better choice because of the higher strength and ductility of the weld metal. Also, this weld metal has a greater tolerance for phosphorus from the cast iron and better resistance to hot cracking. Castings that offer a small

## Table 5.6
## Filler metals for welding or braze welding cast iron

| Description | Form | Applicable process[a] | AWS Specification | AWS Classification |
|---|---|---|---|---|
| | | Cast iron | | |
| Gray iron | Welding rod | OAW | A5.15 | RCI |
| Gray iron | Bare electrode | BMAW | – | – |
| Alloy gray iron | Welding rod | OAW | A5.15 | RCI-A |
| Ductile iron | Welding rod | OAW | A5.15 | RCI-B |
| | | Steel | | |
| Carbon steel | Covered electrode | SMAW | A5.15 | ESt |
| Carbon steel | Covered electrode | SMAW | A5.1 | E7018 |
| Carbon steel | Bare electrode | GMAW | A5.18 | E70S-2 |
| | | Nickel alloys | | |
| 93% Ni | Bare electrode | GMAW | A5.14 | ERNi-1 |
| | | | | ENi-CI |
| 95% Ni | Covered electrode | SMAW | A5.15 | ENi-CI-A |
| | | | | ENiFe-CI |
| 53 Ni-45 FE | Covered electrode | SMAW | A5.15 | ENiFe-CI-A |
| 53 Ni-45 Fe | Flux cored electrode | FCAW | – | – |
| 55 Ni-40 Cu-4 Fe | Covered electrode | SMAW | A5.15 | ENiCu-A |
| 65 Ni-30 Cu-4 Fe | Covered electrode | SMAW | A5.15 | ENiCu-B |
| | | Copper alloys | | |
| Low fuming brass | Welding rod | OAW | A5.27 | RCuZn-B |
| Low fuming brass | Welding rod | OAW | A5.27 | RCuZn-C |
| Nickel brass | Welding rod | OAW | A5.27 | RBCuZn-D |
| Copper-tin | Covered electrode | SMAW | A5.6 | ECuSn-A |
| Copper-tin | Bare electrode | GMAW | A5.7 | ERCuSn-A |
| Copper-aluminum | Covered electrode | SMAW | A5.6 | ECuAl-A2 |
| Copper-aluminum | Bare electrode | GMAW | A5.7 | ERCuAl-A2 |

a. OAW–oxyacetylene welding
   BMAW–bare metal arc welding
   SMAW–shielded metal arc welding
   GMAW–gas metal arc welding

*Fig. 5.5–Repairing a pump casting by shielded metal arc welding*

amount of restraint can be welded with either of the nickel alloy electrodes. When welding must be done under high restraint, the ENiFe-CI electrode is preferred. This electrode is also suitable for welding cast iron to mild steel, stainless steels, and nickel alloys.

For V-groove welds, a 60- to 80-degree groove angle is suitable. For thick sections, a U-groove with 20 to 25 degree groove angle and 0.19 to 0.25 in. root radius should be used.

Welding current for a particular electrode size should be within the range recommended by the manufacturer but as low as possible, consistent with smooth operation, desired bead contour, and good fusion. When used in other than the flat and horizontal positions, the welding current should be reduced about 25 percent for vertical welding and about 15 percent for overhead welding.

The electrode should be manipulated so that the bead width is no greater than three times the nominal diameter of the electrode. If a larger cavity must be filled, the sides should be buttered with weld metal first, and then the cavity gradually filled toward the center of the repaired area. With large castings, a backstep sequence provides a more even thermal distribution.

Use of preheat is not always necessary, but it is often used. It is especially helpful in overcoming the differential mass effect encountered when welding thick to thin sections. Preheat is also beneficial for applications requiring leak tightness because it reduces the likelihood of cracking in the weld or heat-affected zone.

Peening of the hot weld bead helps to reduce welding stresses and maintain dimensions. It should be done with repeated moderate blows using a round-nose tool and sufficient force to deform the weld metal but without rupturing it. A post-weld heat treatment is sometimes used to improve the machinability of the weld heat-affected zone.

The nickel-copper-iron covered electrodes (ENiCu-A and ENiCu-B) are used to weld cast iron in a manner similar to that for nickel (ENi-CI) and nickel-iron (ENiFe-CI) electrodes. The rejection of carbon as graphite in the weld metal is also similar. However, nickel-copper-iron weld metal is more susceptible to cracking from dilution by cast iron. Therefore, dilution must necessarily be limited using appropriate welding techniques.

## Mild Steel Electrodes

Mild steel electrodes, such as E7018 and

ESt types, are used primarily to repair small cosmetic casting defects where color match is desirable and machining is not of major concern. Dilution raises the carbon content of the steel weld metal and that, in turn, increases hardenability. Therefore, the welding procedure should be designed to minimize both the dilution and the cooling rate to keep the hardness of the weld as low as possible.

Mild steel electrodes can be acceptable for joining ferritic ductile or malleable iron castings to mild steel components. However, the steel weld metal shrinks more than the cast iron and causes stresses to develop at the weld interface. The stresses can be severe enough to cause cracking in the cast iron heat-affected zone. Use of steel electrodes should be restricted to applications where the joint is not loaded in tension or in bending. In any case, a suitable preheat should be used, and the welding procedures should be thoroughly evaluated prior to production use.

### Gray Iron Electrodes

Bare metal arc welding with gray cast iron electrodes has limited use for the repair welding of gray iron castings. The process is faster than oxyacetylene welding, and a machineable weld can be produced with special procedures.

The casting must be preheated to prevent cracking. During welding, the joint faces should be thoroughly melted and fused with the deposited metal. The arc should not be broken abruptly, but the arc length should be gradually increased and held for a short time to permit the crater to solidify slowly and thus avoid cracking.

For machineable deposits, the base metal should be preheated as for oxyfuel gas welding. Welding is done in the flat position with a large electrode and high amperage to produce a large, fluid weld pool. Carbon dams may be used to support the weld pool at the root and edges of the repair. Undercut is a problem with this technique, and undercut areas can be filled using oxyacetylene welding procedures. The cooling rate to room temperature should be controlled to avoid the formation of martensite in the weld zone.

### Copper Alloy Electrodes

Cast iron can be arc (braze) welded with copper alloy electrodes.[4] Copper alloy weld metal has appreciable amounts of alpha phase which is soft and ductile when hot. Yielding of the soft weld metal, during cooling, limits welding stresses and reduces the likelihood of cracking. A large part of the contraction strain takes place before the cooling weld metal reaches 500°F, and it is relieved by harmless plastic stretching of the weld metal. The strength of the weld metal increases rapidly as the temperature drops to ambient with little change in ductility.

Two types of copper alloy electrodes are listed in Table 5.6. The strength of copper-aluminum weld metal will be about twice that of copper-tin weld metal. Two advantages of arc welding over oxyacetylene welding are higher welding speed and lower distortion. Also, the heat-affected zone will be narrower with less likelihood of cracking.

The welding groove face area should be large to provide a large brazed area and adequate joint strength. A V-groove with a 90- to 120-degree included angle is recommended. The joint faces should be cleaned, as described previously, to remove any graphite that can inhibit wetting.

A preheat suitable for the type of cast iron should be used. The copper alloy filler metal is deposited in stringer beads using a heat input just sufficient to obtain good wetting of the cast iron joint faces but with minimum dilution. Where possible, the arc should be directed at a previously deposited bead, and never at a corner or edge on the cast iron. The joint faces can be buttered with filler metal prior to welding the joint. The weldment should be slowly cooled to room temperature.

## GAS METAL ARC WELDING

Gas metal arc welding with short-circuiting transfer is suitable for joining ductile iron. Because of the relatively low heat input with this process, the hard portion of the heat-affected zone is usually confined to a thin layer next to the weld metal. As a result, the strength and ductility of the welded joint are about the same as those of the base metal.

Bare electrodes, similar to covered electrodes for shielded metal arc welding, are recommended[5] (see Table 5.6). Typical electrodes

---

4. See *Welding Handbook,* Vol. 2, 7th Ed., Ch. 11 for information on braze welding.

5. See Specifications AWS A5.14-76 and AWS A5.18-79 for information on nickel alloy and carbon steel bare electrodes respectively.

are of nickel (ENi-1) and carbon steel (E70S-2). The ERNi-1 nickel electrode is low in carbon. Therefore, deposited metal from this electrode contracts more on cooling than deposited metal from the ENi-CI covered electrode. A shielding gas appropriate for the electrode should be used.

## FLUX CORED ARC WELDING

Flux cored arc welding can be used with cast irons. A flux cored electrode is available that produces weld metal with a composition and microstructure similar to that deposited by ENiFe-CI covered electrodes. The principle of graphite precipitation and the associated volume increase is the same for both electrodes.

The flux cored electrode is a self-shielding design, but it may be used with carbon dioxide ($CO_2$) shielding or as a submerged arc welding electrode. The electrode producer should be consulted for recommendations on usage and procedures for specific applications.

## BUTTERING

A suitable procedure for arc welding cast iron uses a buttering technique that provides good weld joint ductility without a postweld heat treatment. The object of the buttering is to place the heat-affected zone of the welded joint in the buttering layer rather than in the cast iron. A layer of weld metal about 0.3-in. thick is deposited on the joint faces, and then the components are annealed immediately.

The filler metal employed for buttering can be, but is not necessarily, the same one that is subsequently used for joining the cast pieces together. However, it must have properties commensurate with the base metal after heat treatment and welding. This approach has been successfully demonstrated by shielded metal arc welding with nickel-iron (ENiFe-CI) and mild steel (E7016) covered electrodes.

# OXYACETYLENE WELDING

Oxyacetylene welding procedures require large amounts of heat input during both the preheating and the welding operations. The extensive heating is a limiting factor in the application of this process to finished or semifinished castings where distortion or dimensional stability can be a problem. On the other hand, the resulting slow cooling rate lessens the tendency for brittleness in the weld heat-affected zone. Only limited success has been achieved in welding malleable cast irons by this process.

## JOINT DESIGN

The joint design is normally a double-V-groove with a 90-degree groove angle. If welding can be done only from one side, the groove angle for a single-V-groove may be increased to 120 degrees. The joint groove must have sufficient width to permit good fusion of the root faces and uninterrupted torch manipulation. If a deep hole is being repaired, the groove should be elongated to avoid torch backfire.

## FILLER METALS

For gray iron castings, gray iron welding rods of AWS Type RCI or RCI-A are used. The RCI-A rods contain small amounts of molybdenum and nickel, and the deposited metal is of higher strength than that of RCI rods. The melting temperature is slightly higher, the molten metal is more fluid, and welding can be done faster.

Type RCI-B rods are designed for welding ductile cast irons but can also be used with gray cast irons. These rods contain a small amount of cerium that tends to agglomerate the graphite in the weld metal during solidification and produce ductile weld deposits.

Other elements are added to the filler metals to provide desirable properties. Phosphorus additions improve the fluidity of the molten weld metal, while low phosphorus reduces the fluidity for surfacing operations. Small additions of chromium and vanadium improve strength. Cast iron weld deposits have excellent color match with the base metals.

## FLUX

A flux is required during oxyacetylene welding to increase the fluidity of the fusible iron silicate slag that forms in the molten weld metal.

This slag is difficult to remove when a flux is not used. Prefluxed welding rods are commercially available, but the more common practice is to dip the heated end of a bare welding rod into the flux and transfer it as required to the weld. Excessive flux can result in harmful slag entrapment.

Fluxes for welding gray cast iron contain borates or boric acid, soda ash, and other compounds such as iron oxide, sodium chloride, and ammonium sulfate. Most fluxes are of proprietary compositions, but a typical flux contains equal parts of boric acid and soda ash, 2 percent ammonium sulfate, and 15 percent powdered iron. Fluxes for welding ductile cast irons are similar to those used for gray cast irons except for ingredients that produce a lower melting slag.

## WELDING PROCEDURE

A preheat temperature between 1100° and 1200°F is normally employed to compensate for the low heat input of the process. Thin sections can be welded with a lower preheat temperature because the heat loss into the base metal will be less than with thick sections.

Welding is performed with a neutral or slightly reducing flame using welding tips of medium or high flame velocity. The backhand welding technique is recommended. The tip of the welding rod should be kept in the molten weld pool to minimize slag inclusions and porosity in the weld metal.

The weld metal should be deposited in layers less than 1/8-in. thick. For good fusion with the joint faces, the molten pool must be kept small and the torch must be directed on the base metal. Moving the welding rod about in the molten weld pool helps to float the slag to the surface.

For ductile cast irons, melting of the base metal should be restricted to that required for good fusion. An interpass temperature of between 1100° and 1200°F should be maintained. If the temperature falls below this range, porosity may result from gas evolved by the molten base metal. Immediately after completion of welding, the casting should be insulated to provide slow cooling. Stress relieving or annealing of the casting is recommended.

# BRAZE WELDING

Braze welding with an oxyacetylene torch can be used to make field repairs to castings, but it is not normally used to repair new castings because of poor color match.[6] A typical operation is show in Fig. 5.6. Copper alloy filler metals commonly used for braze welding are listed in Table 5.6. Joint strengths equivalent to fusion welds are possible with gray cast iron but not with ductile or malleable cast iron. With ductile cast iron, for example, a joint efficiency of only about 80 percent can be expected with RBCuZn-D filler metal and even less with the other RBCuZn filler metals. Corrosion resistance of the joints is generally poor.

V-groove joint designs similar to those used for arc or oxyacetylene welding are suitable. The joint faces and adjacent surfaces should be prepared as described for brazing.

The copper-zinc welding rods are used with a neutral or slightly oxidizing flame. To prevent

excessive oxidation, the molten weld pool should be kept covered with a thin oxide film. Preheating with the welding torch is satisfactory for small castings. Large castings require preheating to a temperature in the range of 600° to 750°F, or higher, prior to welding. Generally, a suitable flux is applied to the welding rod by warming one end with the flame and then dipping it into the powdered flux. The flux-coated rod is applied to the joint during welding. Flux-coated welding rods are commercially available. With thick sections, flux is sometimes applied to the preheated joint faces before welding begins.

The behavior of the filler metal on the cast iron indicates when the proper temperature has been reached. If the cast iron is too cold, the filler metal will not wet and spread over the surface. If the temperature of the iron is too high, the filler metal will ball up on the surface of the cast iron. The fact that braze welding is done at a temperature of several hundred degrees below the melting point of cast iron is an advantage with respect to dilution.

---

6. Refer to the *Welding Handbook*, Vol. 2, Ch. 11, for additional information on braze welding.

*Fig. 5.6–Braze welding a cast iron flange to a cast iron pipe with an oxyacetylene torch*

Welding is best done in the flat position. As welding progresses, the welding groove faces must be tinned with filler metal ahead of the weld to insure good wetting and fusion. When tinning is properly done, the filler metal penetrates the grain boundaries of the iron at the surface. After the groove faces are tinned, the weld metal is built up in layers until the joint is filled. The casting is then covered with insulation and allowed to cool slowly to room temperature.

# BRAZING

## FILLER METALS AND FLUXES

With proper surface preparation, any filler metal suitable for brazing carbon steel can be used for cast iron. The best suited ones are the lower melting silver brazing filler metals. Those containing nickel, such as Types BAg-3 and BAg-4, have good wettability on clean cast iron and, therefore, produce higher strength joints. Copper and copper-zinc filler metals can be used, but their high brazing temperatures must be considered. Filler metals containing phosphorus (BCuP) are not suitable for joining cast iron because the formation of brittle iron phosphide embrittles the brazed joint. Types 3A and 3B brazing fluxes are used with BAg brazing filler metals.

## BRAZING PROCESSES

Any brazing processes suitable for steel are applicable to cast irons. The choice of process depends upon the metals being joined, the brazing filler metal used, the design of the joint, and the relative masses of the parts. Those processes that use automatic temperature control are most desirable because the cast iron must not be overheated during the brazing operation.

## JOINT CLEARANCE

Joint clearance should be determined for a specific application after considering such factors as the joint design, the thermal expansion coefficients of the metals being joined, the method of heating, and the filler metal. The recommended joint clearance for BAg filler metals is 0.002 to 0.005 in. The proper clearance is easily maintained with small components, but some sacrifice must usually be made with large parts.

## BRAZING PROCEDURES

In general, the handling of cast iron parts during brazing is the same as for other base metals, assuming adequate preparation of the faying surfaces as described previously. Because cast irons expand significantly with temperature and conduct heat quite poorly, the heating and cooling cycles for brazing should be designed to minimize thermal stresses. The casting should be cooled slowly from the brazing temperature.

## POSTBRAZE OPERATIONS

Postbraze operations include the removal of excess flux. Warm water is usually adequate for dissolving the flux normally used with silver base filler metals. Inspection operations are then made on the cleaned joints.

# OTHER JOINING PROCESSES

## SOLDERING

Soldering is used to a limited extent for repair of small surface defects in iron castings. Such repairs can provide both water- and air-tight seals. Torch heating is usually employed with a slightly reducing flame. Soldering temperatures in the range of 450°-500°F are normal. The faying surfaces must be prepared as described previously. If tinning difficulties are experienced, a solder relatively high in tin (50 percent or more) might help the operation. A typical solder used to repair gray cast iron has the composition 35Sn-30Pb-35Zn.

Fluxes similar to those for soldering steel are normally used to assist wetting for a well-tinned surface. The joint should be cooled slowly to prevent tearing of the solder filler metal.

## THERMIT WELDING

Iron castings can be repaired by Thermit welding, especially large structures such as machine bases or frames.[7] The shrinkage of weld metal produced by the alumino thermic reaction is significantly greater than that of gray or ductile cast iron. For this reason, difficulties can arise when the process is used for the repair of cracks that are longer than eight times the section thickness or that do not extend through the section.

The process is essentially the same as that used for the welding of steel. Cast iron Thermit mixture normally consists of aluminum powder and iron oxide to which 3 percent ferrosilicon and 20 percent mild steel punchings are added. Because of the insulation and slow cooling provided by the sand mold surrounding the joint, the weld metal is generally somewhat harder and tougher than ordinary gray cast iron but is machinable. Stress relieving is usually not necessary unless a condition of high restraint is encountered.

---

7. Refer to the *Welding Handbook,* Vol. 3, Ch. 13, for more information on Thermit welding.

# CUTTING

## OXYFUEL GAS CUTTING

The high carbon content of cast iron hinders oxyfuel gas cutting operations.[8] The large amount of graphite and iron carbide in the metal interferes with the oxidation of the iron matrix. Poor fluidity of the molten metal and slag prevents clearing the kerf. High quality production cuts, typical with steels, cannot be obtained. However, cast iron can be severed using special techniques, and it is advisable to preheat the casting to the temperature recommended for oxyacetylene welding.

In general, a reducing preheat flame is used

to cause burning in the kerf to maintain preheat. After the starting edge is preheated through the full thickness, the cutting torch is directed backward at 45 degrees to the work surface to start the cut. After cutting is underway, the torch angle with the work surface is gradually increased to about 75 degrees or more. During cutting, the cutting torch is oscillated in a semicircular pattern transverse to the direction of travel. Cast irons are sometimes cut using a waster plate, metal powder, or chemical flux cutting techniques.

## ARC CUTTING

Cast irons can be cut by the air carbon arc and plasma arc cutting processes. The equipment manufacturer should be consulted for specific information and recommendations.

---

8. See the *Welding Handbook,* Vol. 2, Ch. 13, for additional information on oxyfuel gas and arc cutting methods.

## Metric Conversion Factors

1 ksi = 6.89 MPa
1 in. = 25.4 mm
$t_C = 0.56 (t_F - 32)$

# SUPPLEMENTARY READING LIST

Bishel, R. A., Flux-cored electrode for cast iron welding, *Welding Journal,* 52(6): 372-81; 1973 June.

Bishel, R. A., and Conaway, H. R., Flux-cored arc welding for high-quality joints in ductile iron, *Modern Casting,* 67(1); 1977 Jan.

Burgess, C. O., *Welding, Joining and Cutting of Gray Iron,* Cleveland: Gray Iron Founders Society, Inc., 1951.

Conway, H. R., Cored wire breakthrough speeds cast iron joining, *Metal Progress,* 106(6); 1975 Nov.

Cookson, C., Maintenance and repair welding of castings, *Welding of Castings,* Cambridge, England: The Welding Institute, 1977.

Electron-beam welding of pearlitic malleable iron, *Modern Casting,* 50(1); 1966 July.

Gregory, E. N., and Jones, S. B., Welding cast irons, *Welding of Castings,* Cambridge, England: The Welding Institute, 1977.

Hogaboom, A. G., Welding of gray cast iron, *Welding Journal,* 56(2): 17-22; 1977 Feb.

Klimek, J., and Morrison, A. V., Gray cast iron welding, *Welding Journal,* 56(3): 29-33; 1977 Mar.

Kotecki, D. J., and Braton, N. R., Preheat effects on gas metal arc welded ductile cast iron, *Welding Journal,* 48(4): 161-66; 1969 Apr.

Maintenance welding of cast iron, *Canadian Welding and Fabrication,* 65(4); 1974 Apr.

Short-arc welding of spheroidal-graphite iron in the SKF Katrincholm Works, Sweden, *Foundry Trade Journal,* 1968 May 23.

*Test Program on Welding Iron Castings,* AWS D11.1-65, Miami: American Welding Society, 1965.

*The Oxy-Acetylene Handbook,* 3rd ed., New York: Union Carbide Corporation, Linde Div., 1976.

Walton, C. F., *Gray and Ductile Iron Castings Handbook,* Rocky River, OH: Iron Castings Society, 1971.

*Welding Ductile Iron – Current Practices and Applications,* Mountainside, N.J.: The Welding Research Committee, Ductile Iron Society, 1977.

Welding gray iron with mild steel electrodes, *Foundry,* 96(1); 1968 Jan.

# 6

# Nickel and Cobalt Alloys

## Chapter Committee

W. YENISCAVICH, *Chairman*
   *Westinghouse Electric Corporation*
J.R. HOLLERAN
   *Pittsburgh Testing Laboratory*
A.C. LINGENFELTER
   *Huntington Alloys, Incorporated*

S.J. MATTHEWS
   *Cabot Corporation*
H.F. MERRICK
   *International Nickel Company*
R.K. WILSON
   *International Nickel Company*

## Welding Handbook Committee Member

A.F. MANZ
   *Linde Division*
   *Union Carbide Corporation*

# 6

# Nickel and Cobalt Alloys

## METAL PROPERTIES

Nickel has a face-centered cubic crystal structure up to its melting point. In this respect, nickel and copper are similar. Cobalt, however, undergoes a transition from a close-packed hexagonal crystal structure to a face-centered cubic structure above approximately 750° F. As with iron, the addition of nickel to cobalt stabilizes the face-centered cubic crystal structure to below room temperature. Most complex cobalt alloys are designed to retain this cubic structure to take advantage of its inherent ductility.

Typical physical and mechanical properties of nickel and cobalt are given in Table 6.1. Nickel and some of its alloys are magnetic at room temperature. Unalloyed cobalt is magnetic but its alloys are not.

Commercially pure nickel is weldable by most common welding processes. Typical applications are food processing equipment, caustic handling equipment, chemical shipping drums, and electrical and electronic parts. There are relatively few applications for pure cobalt, and none for welded structures.

### Table 6.1
### Physical and mechanical properties of cobalt and nickel

| Property | Units | Cobalt | Nickel |
|---|---|---|---|
| Density | lb /in.$^3$ | 0.322 | 0.321 |
| Melting point | °F (°C) | 2723 (1495) | 2647 (1453) |
| Coef. of thermal expansion (68° F) | in./(in.•°F) | $7.7 \times 10^{-6}$ | $7.4 \times 10^{-6}$ |
| Thermal conductivity | (cal/cm·s·°C) | 0.165 | 0.22 |
| Electrical resistivity | mΩ·cm | 5.3 | 6.8 |
| Modulus of elasticity in tension | psi | $30.6 \times 10^6$ | $29.6 \times 10^6$ |
| Tensile strength, annealed | ksi | 37[a] | 67 |
| Yield strength, 0.2% offset | ksi | 28-41[b] | 21.5 |
| Elongation in 2 in. | percent | 0-8 | 47 |

a. Compressive strength is about 117 ksi.
b. Compressive yield strength is about 56 ksi.

# NICKEL ALLOYS

Nickel is alloyed with a number of other metals to impart specific properties. These may include improved mechanical properties as well as corrosion or oxidation resistance at room and elevated temperatures. Alloying significantly decreases thermal and electrical conductivities. Typical nickel alloys and their compositions are given in Table 6.2. These alloys are representative of the large number of available alloys, some of which are referred to as superalloys. The list is not intended to be all-inclusive or to promote the alloys listed. Similar alloys with other designations may be available.

Nickel can be strengthened by solid-solution alloying and by dispersion strengthening with a metal oxide. Some nickel alloys may be further strengthened by a precipitation-hardening heat treatment or by dispersion strengthening.

The type of strengthening is a convenient means of classifying nickel alloys. In practice, some of the alloys classified as solid-solution types may contain minor amounts of elements that contribute to precipitation hardening. Their presence may cause some strengthening during heat treatment or service. Consequently, the classification of such alloys is somewhat arbitrary.

## SOLID-SOLUTION ALLOYS

All nickel alloys are strengthened by solid solution. Additions of aluminum, chromium, cobalt, copper, iron, molybdenum, titanium, tungsten, and vanadium to nickel contribute to solid-solution strengthening. Aluminum, chromium, molybdenum, and tungsten contribute strongly while the others have a lesser effect. Molybdenum and tungsten provide improved strength at elevated temperatures.

### Nickel-Copper Alloys

Nickel and copper form a continuous series of solid solutions with a face-centered cubic crystal structure. Commercial alloys contain from about 30 to 45 percent copper. They are tough and ductile. Except for free-machining (high sulfur) alloys, they are readily joined by welding, brazing, and soldering with proper precautions.

### Nickel-Chromium Alloys

Alloys of this family are used primarily for applications involving high temperatures, oxi-

dation, and corrosion. Some alloys are designed for thermocouples or for electrical resistance applications. Other alloys are designed for structural applications at elevated temperatures.

Some of the alloys contain iron, molybdenum, tungsten, cobalt, and copper in various combinations to enhance specific properties. These include improved corrosion resistance and high temperature strength.

In general, nickel-chromium alloys can be welded by processes and procedures that adequately protect the weld zone from oxidation. They may be brazed using special techniques to promote wetting of the base metal.

### Nickel-Iron-Chromium Alloys

These alloys contain about 20 to 45 percent nickel, 13 to 22 percent chromium, and the remainder iron. They are generally used for corrosion- or oxidation-resistant applications that can be fabricated by welding.

### Nickel-Molybdenum Alloys

These are nickel alloys that contain from 16 to 28 percent molybdenum and lesser amounts of chromium and iron. The alloys are used primarily for their corrosion resistance. They are not normally used for elevated temperature service. The nickel-molybdenum alloys are in general readily weldable.

### Nickel-Chromium-Molybdenum Alloys

These alloys are designed primarily for corrosion resistance at room temperature as well as resistance to oxidizing and reducing atmospheres at elevated temperatures. They are not particularly strong at elevated temperatures and, therefore, are used for low stress applications. All have good weldability.

## PRECIPITATION-HARDENABLE ALLOYS

These alloys are strengthened by controlled precipitation of a second phase, known as gamma prime, from a supersaturated solid solution. Precipitation occurs upon reheating a solution-treated and quenched alloy to an appropriate temperature for a specific time. Some cast alloys will age directly as the solidified casting cools in the mold.

The most important phase from a strength-

**Table 6.2**
**Nominal chemical composition of typical nickel alloys**

| UNS No. | Common designation[a] | Ni[b] | C | Cr | Mo | Fe | Co | Cu | Al | Ti | Cb[c] | Mn | Si | W | B | Other |
|---|---|---|---|---|---|---|---|---|---|---|---|---|---|---|---|---|
| | | | | | | | Commercially pure nickels | | | | | | | | | |
| N02200 | Nickel 200 | 99.5 | 0.08 | | | 0.2 | | 0.1 | | | | 0.2 | 0.2 | | | |
| N02201 | Nickel 201 | 99.5 | 0.01 | | | 0.2 | | 0.1 | | | | 0.2 | 0.2 | | | |
| N02205 | Nickel 205 | 99.5 | 0.08 | | | 0.1 | | 0.08 | | 0.03 | | 0.2 | 0.08 | | | 0.05Mg |
| | | | | | | | Solid solution types | | | | | | | | | |
| N04400 | Monel 400 | 66.5 | 0.2 | | | 1.2 | | 31.5 | | | | 1 | 0.2 | | | |
| N04404 | Monel 404 | 54.5 | 0.08 | | | 0.2 | | 44 | 0.03 | | | 0.05 | 0.05 | | | |
| N04405 | Monel R-405 | 66.5 | 0.2 | | | 1.2 | | 31.5 | | | | 0.1 | 0.02 | | | |
| N06001 | Hastelloy F | 47 | 0.05 | 22 | 6.5 | 17 | 2.5 | | | | 2 | 1 | 1 | 1 | | |
| N06002 | Hastelloy X | 47 | 0.10 | 22 | 9 | 18 | 1.5 | | | | | 1.5 | 1 | 0.6 | | |
| N06003 | Nichrome V | 76 | 0.1 | 20 | | 1 | | | | | | 2 | 1 | | | |
| N06004 | Nichrome | 57 | 0.1 | 16 | | 25 | | | | | | 2 | 1 | | | |
| N06007 | Hastelloy G | 44 | 0.1 | 22 | 6.5 | 20 | 2.5 | 2 | | | 2 | 1.5 | 1 | 1 | | |
| N06102 | IN 102 | 68 | 0.06 | 15 | 3 | 7 | | | 0.4 | 0.6 | 3 | | | 3 | | 0.03Zr, 0.02Mg |
| N06333 | RA 333 | 45 | 0.05 | 25 | 3 | 18 | 3 | | | | 1 | 1.5 | 1.2 | 3 | 0.005 | |
| N06600 | Inconel 600 | 76 | 0.08 | 15.5 | | 8 | | 0.2 | | | | 0.5 | 0.2 | | | |
| N06601 | Inconel 601 | 60.5 | 0.05 | 23 | | 14 | | | 1.4 | | | 0.5 | 0.2 | | | |
| N06625 | Inconel 625 | 61 | 0.05 | 21.5 | 9 | 2.5 | | | 0.2 | 0.2 | 3.6 | 0.2 | 0.2 | | | |
| N08020 | Carpenter 20Cb3 | 36 | 0.04 | 20 | 2.5 | 36 | | 3.5 | | | 0.5 | 1 | 0.5 | | | |
| N08800 | Incoloy 800 | 32.5 | 0.05 | 21 | | 46 | | | 0.4 | 0.4 | | 0.8 | 0.5 | | | |
| N08825 | Incoloy 825 | 42 | 0.03 | 21.5 | 3 | 30 | | 2.25 | 0.1 | 0.9 | | 0.5 | 0.25 | | | 0.02Mg |
| N10001 | Hastelloy B | 61 | 0.05 | 1 | 28 | 5 | 2.5 | | | | | 1 | 1 | | | |
| N10002 | Hastelloy C | 54 | 0.08 | 15.5 | 16 | 5 | 2.5 | | | | | 1 | 1 | 4 | | |
| NA | Hastelloy D | 82 | 0.10 | | | 1 | 1.5 | 3 | | | | 1 | 9 | | | |
| N10003 | Hastelloy N | 70 | 0.06 | 7 | 16.5 | 5 | | | | | | 0.8 | 0.5 | | | |
| N10004 | Hastelloy W | 60 | 0.12 | 5 | 24.5 | 5.5 | 2.5 | | | | | 1 | 1 | | | |

## Table 6.2 (continued)
### Nominal chemical composition of typical nickel alloys

| UNS No. | Common designation[a] | Ni[b] | C | Cr | Mo | Fe | Co | Cu | Al | Ti | Cb[c] | Mn | Si | W | B | Other |
|---|---|---|---|---|---|---|---|---|---|---|---|---|---|---|---|---|
| | | | | | | | | | | **Composition, percent** | | | | | | |
| | | | | | | | **Precipitation hardenable types** | | | | | | | | | |
| NA | Duranickel 301 | 96.5 | 0.15 | | | 0.3 | | 0.13 | 4.4 | 0.6 | | 0.25 | 0.5 | | | |
| N05500 | Monel K-500 | 66.5 | 0.10 | | | 1 | | 29.5 | 2.7 | 0.6 | | 0.08 | 0.2 | | | |
| N07001 | Waspaloy | 58 | 0.08 | 19.5 | 4 | | 13.5 | | 1.3 | 3 | | | | | 0.006 | 0.06Zr |
| N07041 | René 41 | 55 | 0.10 | 19 | 10 | 1 | 10 | | 1.5 | 3 | | | | | 0.005 | |
| N07080 | Nimonic 80A | 76 | 0.06 | 19.5 | | | | | 1.6 | 2.4 | | 0.05 | 0.1 | | 0.006 | 0.06Zr |
| N07090 | Nimonic 90 | 59 | 0.07 | 19.5 | | | 16.5 | | 1.5 | 2.5 | | 0.3 | 0.3 | | 0.003 | 0.06Zr |
| N07252 | M 252 | 55 | 0.15 | 20 | 10 | | 10 | | 1 | 2.6 | | 0.3 | 0.3 | | 0.005 | |
| N07500 | Udimet 500 | 54 | 0.08 | 18 | 4 | | 18.5 | | 2.9 | 2.9 | | 0.5 | 0.5 | | 0.006 | 0.05Zr |
| N07713 | Alloy 713C[d] | 74 | 0.12 | 12.5 | 4 | | | | 6 | 0.8 | 2 | 0.5 | 0.5 | | 0.012 | 0.10Zr |
| N07718 | Inconel 718 | 52.5 | 0.04 | 19 | 3 | 18.5 | | | 0.5 | 0.9 | 5.1 | 0.2 | 0.2 | | | |
| N07750 | Inconel X750 | 73 | 0.04 | 15.5 | | 7 | | | 0.7 | 2.5 | 1 | 0.5 | 0.2 | | | |
| N09706 | Inconel 706 | 41.5 | 0.03 | 16 | | 40 | | | 0.2 | 1.8 | 2.9 | 0.2 | 0.2 | | | |
| N09901 | Alloy 901 | 42.5 | 0.05 | 12.5 | | 36 | 6 | | 0.2 | 2.8 | | 0.1 | 0.1 | | 0.015 | |
| N09902 | Ni-Span-C 901 | 42.2 | 0.03 | 5.3 | | 48.5 | | | 0.6 | 2.6 | | 0.4 | 0.5 | | | |
| N13100 | IN 100[d] | 60 | 0.18 | 10 | 3 | | 15 | | 5.5 | 4.7 | | | | | 0.014 | 0.06Zr, 1.0V |
| | | | | | | | **Dispersion strengthened types** | | | | | | | | | |
| NA | TD Nickel | 98 | | | | | | | | | | | | | | 2 ThO$_2$ |
| NA | TD Ni Cr | 78 | | 20 | | | | | | | | | | | | 2 ThO$_2$ |

a. Several of these are registered tradenames. Alloys of similar compositions may be known by other common designations or tradenames.
b. Includes small amount of cobalt if cobalt content is not specified.
c. Includes tantalum also.
d. Casting alloys.

ening standpoint is the ordered face-centered cubic gamma prime that is based upon the compound $Ni_3Al$. This phase has a rather high solubility for titanium and columbium. Consequently, its composition will vary with the alloy composition and the temperature of formation. Aluminum has the greatest hardening potential, but this is moderated by titanium and columbium. The latter has the greatest effect on decreasing the aging rate.

These types of alloys are normally welded in the solution-treated condition. During welding, some portion of the heat-affected zone is heated into the aging temperature range. As the weld metal solidifies, the aging metal becomes subjected to welding stresses. Under certain postweld combinations of temperature and stress, the weld heat-affected zone may crack. This is known as *strain-age cracking*. Alloys high in aluminum are the most sensitive to this type of cracking. The problem is much less severe in those alloys where columbium has been substituted for a significant portion of the aluminum. Columbium retards the aging reaction. Consequently, the weld heat-affected zone can remain sufficiently ductile and yield during heat treatment to relieve high welding stresses without rupture. The relative weldability of several precipitation-hardenable alloys is indicated in Fig. 6.1.

## Nickel-Copper Alloys

The principal alloy in this group contains 66 percent nickel, 30 percent copper, 2.7 percent aluminum and 0.6 percent titanium. The recommended heat-treating procedures should be followed to avoid strain-age cracking when welding this alloy. The corrosion resistance of this alloy is similar to the solid-solution nickel-copper alloy of similar composition.

## Nickel-Chromium Alloys

The nickel-chromium alloys are strengthened by the addition of aluminum and titanium, and sometimes columbium. Chromium content ranges from about 13 to 20 percent for good high-temperature oxidation resistance.

The strength of these alloys after heat treatment is related to the combined aluminum, titanium, and columbium content. The higher this content, the higher is the strength of the alloy. Alloys that contain relatively large amounts of aluminum and titanium are considered unweldable because of their strain-age cracking tendencies. Carefully applied preweld and postweld heat-treating sequences can be used to reduce the strain-age cracking tendencies of these alloys. One of the principal advantages of columbium additions for strengthening is the improved weldability of such alloys compared to those alloys containing only aluminum and titanium. This is due to the sluggish formation of the columbium precipitate compared to the more rapidly forming aluminum precipitate.

Molybdenum and cobalt are often added to improve high-temperature strength. Their effect on weldability is minor. The principal areas of application for these alloys are gas turbine components, aircraft parts, and spacecraft.

## Nickel-Iron-Chromium Alloys

These alloys nominally contain 40 to 45 percent nickel, 13 to 15 percent chromium, 30 to 40 percent iron and small amounts of aluminum and titanium. Their weldability is similar to that of the nickel-chromium alloys. However, most applications involve forgings that require little welding. The same precautions necessary to avoid strain-age cracking with other aluminum-titanium-hardened nickel alloys apply to these alloys as well.

## DISPERSION-STRENGTHENED NICKEL

Nickel and nickel-chromium alloys can be strengthened by the uniform dispersion of very fine refractory oxide $(ThO_2)$ particles throughout the matrix. This is done using powder metallurgy techniques. When these metals are fusion welded, the oxide particles will agglomerate during solidification. This will destroy the original strengthening mechanism afforded by dispersion within the matrix. The weld metal will be significantly weaker than the base metal. The high strength of these metals can be retained by joining them with processes that do not involve melting of the base metal.

## CAST NICKEL ALLOYS

Many nickel alloys can be used in cast as well as wrought forms. Several of them are included in Table 6.2. Some alloys are designed specifically for casting. Table 6.3 gives the composition of several ASTM casting alloys. Casting alloys are strengthened by both solid-solution and precipitation hardening. Precipitation-hardening alloys high in aluminum content, such as Alloy 713C, will harden during slow cooling in

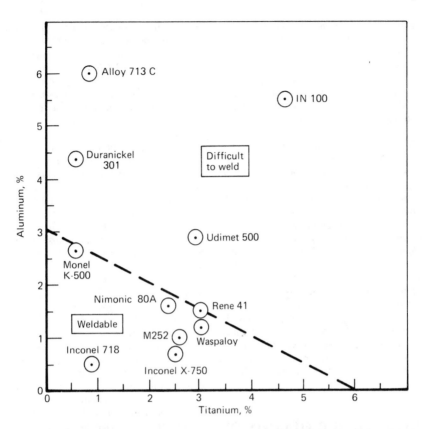

**Fig. 6.1—*Estimated weldability of some precipitation hardenable nickel alloys based upon aluminum and titanium content***

### Table 6.3
### Nominal chemical composition of standard ASTM nickel casting alloys

| Alloy grade | Composition, percent | | | | | | | | | |
|---|---|---|---|---|---|---|---|---|---|---|
| | Ni | C | Cr | Mo | Fe | Co | Cu | Mn[a] | Si[a] | W |
| | ASTM A 297-79[a] | | | | | | | | | |
| HW | 60 | 0.5 | 12 | | 25 | | | 2.0 | 2.5 | |
| HX | 66 | 0.5 | 17 | | 15 | | | 2.0 | 2.5 | |
| | ASTM A 494-79 | | | | | | | | | |
| CY-40 | 72 | 0.4[a] | 16 | | 11[a] | | | 1.5 | 3.0 | |
| CW-12M-1 | 55 | 0.12[a] | 16.5 | 17 | 6 | | | 1.0 | 1.0 | 4.5 |
| CW-12M-2 | 60 | 0.07[a] | 18.5 | 18.5 | 3[a] | | | 1.0 | 1.0 | |
| CZ-100 | 95 | 1.0[a] | | | 3[a] | | 1.25[a] | 1.5 | 2.0 | |
| M-35-1 | 68 | 0.35[a] | | | 3.5[a] | | 30 | 1.5 | 1.25 | |
| N-12M-1 | 65 | 0.12[a] | | 28 | 5 | | | 1.0 | 1.0 | |

a. Maximum

the mold and are essentially unweldable by fusion processes. However, defects or service damage in some of these alloys may be repaired by welding.

Many cast nickel alloys contain significant amounts of silicon to improve fluidity and castability. Most of these cast alloys are weldable by conventional means, but as the silicon content increases so does the weld cracking sensitivity. Cracking can often be avoided using welding techniques that minimize base metal dilution. The nickel casting alloy that contains 10 pecent silicon and 3 percent copper (Hastelloy D) is considered unweldable by arc welding methods, but it may be welded with the oxyacetylene process. Nickel alloys containing about 30 percent copper are considered unweldable when the silicon content is over about 2 percent because of their sensitivity to cracking.

Defective castings of weldable alloys may be repaired by suitable welding procedures. Generally, a filler metal of the same composition as the base metal is used. In some applications, the casting may be welded to a wrought product, such as a cast valve body to wrought pipe. In such cases, the filler metal must be compatible with both base metals and suitable for the intended service.

## EFFECTS OF MINOR ELEMENTS ON WELDABILITY

The presence of very small quantities of some elements can have a profound effect on the weldability of nickel alloys. The presence of sulfur frequently is related to hot cracking because it forms a low melting eutectic with nickel that will segregate to the grain boundaries of the weld metal. Manganese and magnesium are frequently added to combine with sulfur and prevent the formation of nickel sulfide.

Calcium and cerium are used as deoxidizers and also as malleabilizers interacting with sulfur. Small additions of aluminum and titanium also serve as deoxidizers. All of these elements tend to contribute to the formation of oxide films, islands, and slag spots, which form on the weld surface. In multipass welding, such tenacious slag films should be removed between passes to avoid discontinuities in the weld metal.

Phosphorus also forms a low melting eutectic with nickel that segregates to the grain boundaries. This contributes to hot cracking. Sulfur, phosphorus, and similar impurity elements tend to have an additive effect, and the total of all of these elements should be kept low.

Boron and zirconium are frequently added to nickel alloys to improve their hot malleability and to enhance stress-rupture life. However, they also tend to segregate at the grain boundaries and increase the tendency for cracking in the fusion and heat-affected zones in the base metal. The tendency for cracking is also increased if the base metal has a grain size coarser than ASTM No. 5. The effect of boron and zirconium tends to be additive.

Carbon is an interstitial strengthening element in nickel. During welding, the carbon in the heat-affected zone is dissolved at elevated temperature. When nickel is used in the 600° F range, the carbon will reprecipitate as graphite at the grain boundaries. This reduces the ductility of the heat-affected zone. This is not a problem with low carbon nickel or alloys that contain strong carbide-forming elements such as chromium, columbium, and titanium.

Some nickel-chromium and nickel-chromium-iron alloys, like the austenitic stainless steels, exhibit carbide precipitation (sensitization) in the weld heat-affected zone. Sensitization can make the alloys susceptible to intergranular corrosion. Those alloys stabilized with titanium and columbium are not sensitized by welding. An alternate approach is to use an extra low carbon version of the selected alloy.

Silicon causes hot-short cracking in nickel alloys. The severity of cracking varies with the alloy composition and the joining process, but it is especially severe in the high nickel-chromium alloys. Filler metals containing columbium are often used for welding castings with high silicon content to prevent hot cracking of the weld metal.

Lead will cause hot-shortness in nickel alloy weld metal. However, it is seldom found in high quality base and filler metals.

# COBALT ALLOYS

Cobalt alloys of commercial importance find their widest application in corrosion and high-temperature service. The compositions of typical cobalt alloys are given in Table 6.4.

The alloys commonly fabricated by welding generally contain two or more of the following elements: nickel, chromium, tungsten, and molybdenum. All of these elements, with the exception of nickel, form a second phase with cobalt in the amounts normally present in commercial alloys. However, this does not adversely affect the weldability of the alloys.

Other alloying elements including columbium, titanium, manganese, and silicon are not detrimental to welding if properly controlled within specified limits. Residual elements that are insoluble in cobalt or undergo eutectic reactions with cobalt are detrimental and may initiate weld hot cracking. These include sulfur, lead, phosphorus, and bismuth.

Most cobalt alloys can be welded by one or more of the commonly used welding processes. It is important to employ welding techniques that minimize joint restraint and energy input. The alloys should be welded and fabricated in the solution heat-treated condition to provide the desired cobalt-rich, single-phase solid solution. In general, the filler metals used in welding cobalt alloys are similar to specific base metal compositions. Since the alloys are susceptible to embrittlement by low melting point residual elements, cleanliness in the area to be welded is a prerequisite to weld soundness.

## MAJOR ALLOYING ELEMENTS

Tungsten, chromium, nickel, and molybdenum are added to cobalt alloys to provide solid-solution strengthening. Nickel is the only major alloying element that forms a series of solid solutions with cobalt. This element probably exerts the least effect on the weldability of cobalt. All of the other elements, when added to cobalt in the amounts present in commercial alloys, form two-phase alloys. However, they all form single-phase solid solutions at solution heat-treatment temperatures of commercial alloys. In addition, all of these elements, when alloyed with cobalt, produce a narrow liquidus-solidus range favorable to weldability. In general, the major alloy additions are not detrimental to the welding behavior of cobalt.

The loss in ductility experienced by some cobalt alloys in the 1200 to 1800° F range is characteristic of high alloy compositions. This behavior is believed to be associated with the precipitation of intermetallic compounds, which adversely affects crack resistance. This can be avoided by maintaining minimum restraint and energy input during welding.

## MINOR ALLOYING ELEMENTS

Columbium, titanium, vanadium, boron, and zirconium generally are added to improve the high-temperature properties of certain cobalt alloys. These elements should be controlled within specified limits to avoid possible cracking difficulties in the weld and heat-affected zones. Other elements normally present in small amounts in cobalt alloys are carbon, silicon, iron, and manganese. The latter two, like nickel, form a continuous series of solid solutions with cobalt and are not considered detrimental to welding. Carbon and silicon, which are used for deoxidation purposes, may be harmful if present in excessive amounts.

## WROUGHT ALLOYS

Wrought cobalt alloys are used mostly in sheet form. They combine high strength with good oxidation resistance at temperatures up to about 1800° F. They are readily weldable by conventional methods. Elgiloy and MP 35N alloys are used principally for springs and fasteners. Their high strength is achieved by cold working.

## CAST ALLOYS

Cast alloys are used for gas turbine and other applications where good oxidation and sulfidation resistance as well as strength are needed at high temperatures. Strengthening may be achieved by the precipitation of carbides, such as $M_{23}C_6$ and $M_6C$, through the use of solution and precipitation heat treatments. The precipitation rate is quite sluggish and does not inhibit weldability.

## Table 6.4
## Nominal chemical compositions of typical cobalt alloys

| UNS No. | Common designation[a] | C | Ni | Cr | Co | W | Ta | Mo | Al | Ti | Fe | Mn | Si | B | Zr | Other |
|---|---|---|---|---|---|---|---|---|---|---|---|---|---|---|---|---|
| | | | | | | | Wrought alloys | | | | | | | | | |
| R30605 | L-605 | 0.1 | 10 | 20 | Bal | 15 | | | | | 3[b] | 1.5 | 0.5 | | | – |
| R30188 | Haynes Alloy No. 188 | 0.1 | 22 | 22 | Bal | 14 | | | | | 3[b] | 1.2 | 0.4 | | | 0.08 La |
| | Elgiloy | 0.15 | 15 | 20 | 40 | | | 7 | | | Bal | 2 | | | | |
| R30035 | MP 35N | 0.05 | 35 | 20 | 35 | | | 10 | | | | | | | | |
| R30816 | S-816 | 0.38 | 20 | 20 | Bal | 4 | | 4 | | | 4 | 1.2 | 0.4 | | | 4 Cb |
| | | | | | | | Cast alloys | | | | | | | | | |
| | Haynes Alloy No. 21 | 0.25 | 2.5 | 27 | Bal | | | 5.5 | | | 2[b] | | | | | |
| | X-40 | 0.5 | 10.5 | 25.5 | Bal | 7.5 | | | | | – | 0.75 | 0.75 | | | |
| | FSX-414 | 0.25 | 10 | 29 | Bal | 7.5 | | | | | 1 | | | 0.01 | | |
| | WI 52 | 0.45 | | 21 | Bal | 11 | | | | | 2 | 0.25 | 0.25 | | | 2 Cb |
| | MAR-M 302 | 0.85 | | 21.5 | Bal | 10 | 9 | | | | 2 | | | 0.005 | 0.2 | |
| | MAR-M 509 | 0.6 | 10 | 23.5 | Bal | 7 | 3.5 | | | 2 | | | | | 0.5 | |
| | Haynes Alloy No. 1002 | 0.6 | 16 | 22 | Bal | 7 | 3.8 | | 0.3 | 0.2 | 1.5 | 0.7 | 0.4 | | | |

a. Several of these are registered tradenames. Alloys of similar composition may be known by other designations or tradenames.
b. Maximum.

# SURFACE PREPARATION

Cleanness is the single most important requirement for successful joining of nickel and cobalt alloys. Welding, brazing, soldering, and any postheating must be performed only on base metal that is clean and completely free of foreign material. Both nickel and cobalt alloys are susceptible to embrittlement by sulfur, phosphorus, and low melting point metals such as lead, zinc, and tin. Lead hammers, solder, and grinding wheels or belts loaded with these metals are frequent sources of contamination. Detrimental elements are often present in oils, paints, marking crayons, cutting fluids, and shop dust. Cracking from sulfur embrittlement of nickel sheet is evident in the heat-affected zone on one side of the weld in Fig. 6.2. That sheet was improperly cleaned with a soiled rag, while the other sheet was properly cleaned.

Cobalt alloys are highly susceptible to cracking at high temperatures when they are contaminated with copper. Molten copper will initiate liquid metal stress cracking in the base metal. For example, a minute amount of copper, inadvertently transferred onto the surface of a sheet from a backing bar, can cause severe cracking if the copper melts during welding. This may be prevented by either plating the copper backing with nickel or chromium to prevent accidental copper transfer or by using a stainless steel backing. Likewise, copper or brass wire brushes may contaminate the metal and should not be used for cleaning cobalt alloys.

Oxides should be thoroughly removed from the surfaces to be welded because they can inhibit wetting and fusion of the base metal with the weld metal. Their presence can also cause subsurface inclusions and poor bead contour. They may be removed by grinding, abrasive blasting, machining, or pickling.[1]

_____

1. Cleaning of nickel and cobalt alloys is discussed in the *Metals Handbook,* Vol. 2, 8th ed., Metals Park, OH: American Society of Metals, 1964; 607-10, 661-63.

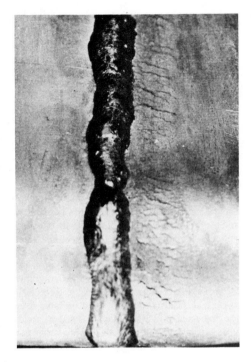

*Fig. 6.2—Cracking from sulphur contamination in the heat-affected zone of a weld in nickel sheet*

Wire brushes used for cleaning should be made of austenitic stainless steel. However, they will not remove tenacious surface oxides from welds. Those oxides must be removed by grinding with an aluminum oxide or silicon carbide wheel or a carbide burr. Any cleaning tools, including wire brushes and carbide burrs, should be clean and free of other metals that may be transferred to the base metal.

# ARC WELDING

The welding of nickel alloys is similar to the welding of austenitic stainless steels. Likewise, cobalt alloys stabilized with nickel have an austenitic structure and behave metallurgically similarly to nickel alloys. Unless otherwise noted, all welding techniques and procedures described for solid-solution strengthened nickel alloys apply also to cobalt alloys.

## APPLICABLE PROCESSES

Nickel and cobalt alloys can be joined by all of the processes commonly used for steel and other metals.[2] However, not all of the processes are applicable to every alloy because of metallurgical characteristics or the unavailability of suitable welding materials. In general, the precipitation-hardenable alloys require closer control of the welding process variables because of the possibility of aging and the formation of refractory oxides during welding. Arc welding processes that are broadly applicable to typical nickel and cobalt alloys are listed in Table 6.5. The indicated processes may be used to weld other alloys with compositions that are very similar to those alloys listed in the table.

## FILLER METALS AND FLUXES

### Covered Electrodes

Covered electrodes for shielded metal arc welding should have compositions similar to the base metals for which they are intended. They normally have additions of deoxidizing elements, such as titanium, manganese, and columbium. These elements also prevent weld metal cracking. Nickel alloy electrodes are divided into five families, namely, Ni, Ni-Cu, Ni-Cr-Fe, Ni-Mo, and Ni-Cr-Mo alloys. Each family contains one or more electrode classifications based upon the chemical composition of undiluted weld metal.[3]

The electrodes are generally designed for use with dc electrode positive (reverse polarity) power. Each electrode type and size has an optimum current range for good arc characteristics that is usually specified by the manufacturer. The welding conditions for a specific application can be developed by trial welds using the same electrode, base metal, thickness, and welding position to be used for production. Excessive welding current can cause a number of problems including an unstable arc, excessive spatter, electrode overheating, and spalling of the coating.

Nickel covered electrodes are used to join commercially pure nickel in wrought and cast forms to themselves and to steel. Sizes of 0.125 in. and under are suitable for welding in all positions. Other sizes are restricted to the flat and horizontal positions of welding.

Nickel-copper electrodes are used for welding this family of alloys to themselves and to steel. Sizes up to and including 0.125 in. may be used in all positions. Larger sizes are restricted to the flat position.

Nickel-chromium-iron electrodes are designed for welding the same family of alloys as well as dissimilar metal joints involving carbon steel, stainless steel, nickel, and nickel-base alloys. The restrictions on welding positions with respect to electrode size are the same as for nickel electrodes. Nickel-chromium-molybdenum electrodes are used for welding alloys of similar composition to themselves and to steel.

Nickel-molybdenum electrodes are designed for welding nickel-molybdenum alloys to themselves and to other nickel, cobalt, and iron base metals. They are normally used in the flat position only.

Several electrode classifications are used for surfacing of steel components for specific corrosion-resistant applications.[4]

Covered electrodes should be kept in the original sealed containers and stored in dry areas until used. Prior to use, electrodes that have been exposed to moisture should be baked in an oven according to the manufacturer's recommendations.

### Bare Rods and Electrodes

In general, nickel alloy bare rods and electrodes are divided into the same families as are covered electrodes.[5] However, there are more

---

2. Welding processes are discussed in the *Welding Handbook*, Vol. 2 and 3, 7th ed.

3. See AWS A5.11, *Specification for Nickel and Nickel Alloy Covered Electrodes*, published by the American Welding Society.

4. For information on surfacing, refer to the *Welding Handbook*, Vol. 2, 7th ed.; 518-62.

5. See AWS A5.14, *Specification for Nickel and Nickel Alloy Bare Welding Rods and Electrodes*, published by the American Welding Society.

**Table 6.5**
**Arc welding processes applicable to some nickel and cobalt alloys**

| UNS No. | Common designation[a] | SMAW | GTAW, PAW | GMAW | SAW |
|---------|----------------------|------|-----------|------|-----|
| | | | Process[b] | | |
| Commercially pure nickel | | | | | |
| N02200 | Nickel 200 | X | X · | X | X |
| N02201 | Nickel 201 | X | X | X | X |
| Solid-solution nickel alloys | | | | | |
| N04400 | Monel 400 | X | X | X | X |
| N04404 | Monel 404 | X | X | X | X |
| N04405 | Monel R-405 | X | X | X | |
| N06002 | Hastelloy X | X | X | X | |
| N06003 | Nichrome V | X | X | | |
| N06004 | Nichrome | X | X | | |
| N06007 | Hastelloy G | X | X | X | |
| N06333 | RA 333 | | X | | |
| N06600 | Inconel 600 | X | X | X | X |
| N06601 | Inconel 601 | X | X | X | X |
| N06625 | Inconel 625 | X | X | X | X |
| N08020 | Carpenter 20Cb3 | X | X | X | |
| N08800 | Incoloy 800 | X | X | X | X |
| N08825 | Incoloy 825 | X | X | X | |
| N10001 | Hastelloy B | X | X | X | |
| N10002 | Hastelloy C | X | X | X | |
| N10003 | Hastelloy N | ·X | X | | |
| Precipitation-hardenable nickel alloys | | | | | |
| N05500 | Monel K-500 | X | X | | |
| N07001 | Waspaloy | | X | | |
| N07041 | René 41 | | X | | |
| N07080 | Nimonic 80A | | X | | |
| N07090 | Nimonic 90 | | X | | |
| N07252 | M 252 | | X | | |
| N07500 | Udimet 500 | X | X | | |
| N07718 | Inconel 718 | | X | X | |
| N07750 | Inconel X-750 | | X | | |
| N09706 | Inconel 706 | | X | | |
| N09901 | Alloy 901 | | X | | |
| Cobalt alloys | | | | | |
| R30188 | Haynes Alloy No. 188 | | X | X | |
| R30605 | L-605 | X | X | X | |

a.   Several of these are registered tradenames. Alloys of similar composition may be known by other common designations or tradenames.

b.   SMAW - Shielded metal arc welding
GTAW - Gas tungsten arc welding
PAW - Plasma arc welding
GMAW - Gas metal arc welding
SAW - Submerged arc welding

classifications in most families of bare filler metals. They are designed to weld base metals of similar composition to themselves by the gas tungsten arc, gas metal arc, plasma arc, and submerged arc processes. As noted in Table 6.5, these processes are not applicable to all alloys. Some filler metals can be used to join dissimilar metal combinations and for surfacing applications on steel.

Cobalt alloy filler metals are available for fusion welding wrought and cast components. They are generally produced to an applicable Aeronautical Materials Specification.[6] Three applicable specifications are AMS 5789 (X-40), AMS 5796 (L-605), and AMS 5801 (Haynes Alloy No. 188).

The composition of the filler metal for gas tungsten arc and gas metal arc welding should, in general, be similar to that of the base metal. Nickel filler metals frequently contain additions, such as titanium, manganese, and columbium, to control porosity and hot cracking.

Precipitation-hardenable weld metal will normally respond to the aging treatment used for the base metal. However, the response is usually less and the weld joint strength generally will be slightly lower than that of the base metal after aging.

Precipitation-hardenable alloys may be welded with dissimilar filler metals to minimize processing difficulties. In some cases, the mechanical properties of the welded joints will be significantly lower than those of the base metals. This behavior must be factored into the weldment design.

The proper electrode diameter for gas metal arc welding will depend upon the base metal thickness and the type of metal transfer. For conventional spray, pulsed spray, and globular transfer, electrode diameters of from 0.035 to 0.093 in. are used. With short-circuiting transfer, diameters of 0.045 in. or smaller are usually required.

## Fluxes

Fluxes are available for submerged arc welding many nickel alloys. The electrodes used with these fluxes are the same as those used for gas metal arc welding. The flux composition must be suited to both the filler metal and the base metal being welded. An improper flux can cause slag adherence, weld cracking, inclusions,

poor weld bead contour, and undesirable changes in weld metal composition. Fluxes used for carbon and stainless steels are not suitable.

The resulting slag cover will be self-lifting when used with the proper welding conditions. Unfused flux can be collected and reused, provided it is kept dry and uncontaminated by other fluxes or foreign materials. Flux that has absorbed moisture should be dried according to the manufacturer's instructions. When recycled flux is used, it should always be mixed with at least 50 percent virgin flux.

## JOINT DESIGNS

Suggested designs for butt joints in nickel and cobalt alloys are shown in Fig. 6.3. Molten nickel and cobalt alloy weld metals do not flow and wet the base metal so readily as do carbon and stainless steel weld metals. The groove angle must be large enough to permit proper manipulation of the filler metal and deposition of stringer weld beads.

With the gas metal arc process, the high amperage used with a small diameter electrode produces a high level of arc force. Such an arc is not easily deflected from a straight line. Consequently, the joint design must permit the arc to be directed at all areas to be fused. U-groove joints should have a 30° bevel angle. This will permit proper manipulation of the arc to obtain good fusion with the groove faces.

Several factors influence joint design. The age-hardenable alloys are sluggish when molten, and full penetration is difficult to achieve in some joints. It is also difficult to avoid lack of fusion in deep groove welds. Another important factor is the lack of ductility in the welded joint at or near the aging temperature of the alloy. Fillet, flare, and edge welds should not be exposed to temperatures above 1000° F, because of the built-in notch at the root of the joint. If these types of welds must be used, the filler metal must be tolerant of notches at elevated temperatures to avoid cracking of the weld metal during heat treatment.

Precipitation-hardening alloys are occasionally welded with solution-strengthened filler metals to minimize processing difficulties. As a result, the mechanical properties of the welded joint are lower than those of the base metal. This is particularly true in heavy sections where dilution is low. To compensate for this, it is sometimes necessary to provide weld joint reinforcement or to modify the component design. The

---

6. Published by the Society of Automotive Engineers.

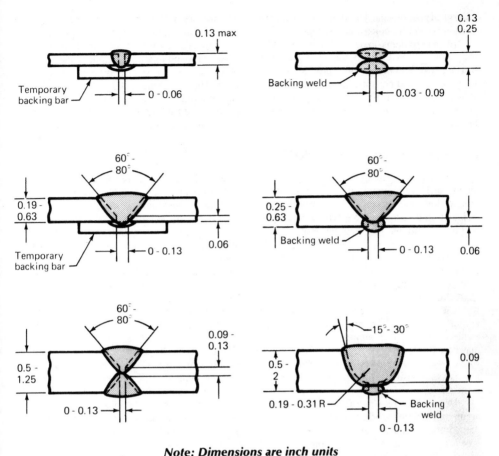

**Note: Dimensions are inch units**

*Fig. 6.3—Suggested designs for arc welded butt joints in nickel and cobalt alloys*

most common design adjustment is to locate the welded joints in regions of known low stress.

## PREHEAT AND INTERPASS TEMPERATURES

Preheat is not required nor recommended for welding nickel or cobalt alloys. However, the area to be welded should be at about 60° F or above to avoid moisture condensate that could produce weld metal porosity.

The interpass temperature should be low to help minimize total heat input. A maximum temperature of 200° F is recommended for some corrosion resistant alloys. Cooling methods used to reduce interpass temperature should not intro-duce contaminants that will cause weld discontinuities. Examples are traces of oil from compressed air or mineral deposits from a water spray.

## HEAT INPUT LIMITATIONS

High heat input during welding may produce undesirable changes in nickel and cobalt alloys. Some degree of annealing and grain growth can take place in the heat-affected zone. The heat input of the welding process and the interpass temperature will determine the extent of these changes.

High heat input may result in constitutional liquidation, carbide precipitation, or other metallurgical phenomena. These, in turn, may cause cracking or loss of corrosion resistance, or both.

When problems do occur, either the welding technique should be modified to decrease the heat input or another welding process of lower heat input should be substituted. Stringer beads should be used to fill the joint. A convex bead shape will help to minimize hot cracking tendencies.

## GAS TUNGSTEN ARC WELDING

Gas tungsten arc welding is widely used for joining nickel and cobalt alloys, especially for thin sections and for applications where a flux residue would be undesirable. The process is the best one for joining the precipitation-hardenable alloys.

Argon, helium, or a mixture of the two is normally used for shielding. The arc characteristics and heat pattern are affected by the choice of shielding gas. The choice should be based on welding trials for the particular production operation. Argon is normally used for manual welding. Hydrogen in amounts up to 5 percent may be added to argon to improve cleanness with single-pass welds. However, hydrogen addition may cause porosity in multiple-pass welds with some alloys.

Helium has shown some advantages over argon for machine welding thin sections without the addition of filler metal. Higher welding speeds are possible. Porosity-free joints in nickel-copper alloys are more readily obtained with helium, and porosity can be reduced in welds in commercially pure nickel. With dc electrode negative (straight polarity) power, welding speeds can be increased as much as 40 percent over those with argon shielding and the same welding current. However, arc initiation is more difficult with helium.

Welding grade shielding gas must be used, and an effective gas cover must always be maintained over the molten weld pool using an appropriate gas nozzle. Exposure of the molten metal to air can cause defects in the weld. Condensation in water-cooled torches can contaminate the shielding gas. This condition can be avoided by circulating lukewarm water through the torch.

During welding, the root of the joint must be shielded from the atmosphere to prevent oxidation of the weld and base metals. This can be done by purging the area with the same gas used for welding. The gas may be introduced through a temporary backing bar or cup, or contained by internal dams in the case of tubing or pipe.

Arc stability is best when the tungsten electrode is ground to a point. Cone angles of 30 to 60 degrees with a small flat apex are generally used. The tip geometry, however, should be designed for the particular welding conditions, and it can range from sharp to flat. Electrode extension below the gas nozzle should be a minimum, consistent with good visibility and torch manipulation.

Direct current electrode negative (straight polarity) power is recommended for both manual and automatic welding. Alternating current can be used for automatic welding if the arc length can be closely controlled. Superimposed high frequency power is required with ac for arc stabilization. It is also useful with dc power to initiate the arc. Touch starting can cause tungsten contamination of the weld metal.

The welding torch should be positioned with the electrode in a nearly vertical position. If the electrode is more than 35 degrees from the vertical, air may be drawn into the shielding gas and cause porosity in the weld metal with some nickel alloys. The shortest possible arc length must be maintained to ensure sound welds. When welding without filler metal, the arc length should not exceed 0.05 in., and preferably 0.02 to 0.03 inch. When filler metal is added, the arc length should be adjusted to permit proper manipulation of the filler rod and molten weld pool. Porosity in a nickel-copper alloy resulting from excessive arc length is illustrated in Fig. 6.4. The molten weld pool should not be agitated by arc manipulation. Otherwise, deoxidizing elements may be lost. The hot end of the filler metal must be kept under the protective gas shield to avoid oxidation and subsequent weld metal contamination.

Weld deposits should contain at least 50 percent filler metal for most applications. The filler rod melting rate and the other welding conditions should be adjusted to obtain the proper dilution.

Square groove welds can be made in sections up to about 0.10 in. thick in a single pass. The proper arc length must be used to ensure complete joint penetration. When automatic welding methods are employed, travel speed should be adjusted to avoid a teardrop-shaped weld pool. This weld pool shape is more prone to centerline cracking during solidification than is an elliptically shaped one. Welding speed also

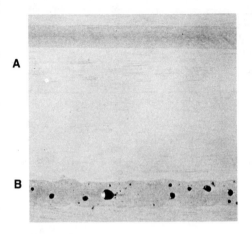

**Fig. 6.4—Effect of arc length on soundness of gas tungsten arc welds in a nickel-copper alloy: (A) proper arc length, (B) excessive arc length (porous weld)**

has an effect on porosity in the weld metal with some alloys. In general, porosity will be a minimum within some range of welding speed.

All of the solid-solution nickel and cobalt alloys, except high silicon casting alloys, are readily welded by this process. Commercially pure nickel and the nickel-copper alloys require special care to prevent porosity in welds made without filler metal. Filler metals normally contain deoxidizing elements to counteract porosity. Therefore, filler metal additions are generally recommended when welding these alloys. Nickel-chromium and nickel-iron-chromium alloys are less susceptible to weld metal porosity. Filler metals for these alloys contain alloying additions to overcome hot cracking tendencies.

The gas tungsten arc process is widely used for precipitation-hardenable alloys because it provides excellent protection against oxidation and loss of the hardening elements. A special precaution that must be observed, especially for those metals high in aluminum or titanium, is the removal of refractory oxides that form on the weld bead surface. If the oxides are not removed, they can cause lack of fusion with subsequent weld passes.

## PLASMA ARC WELDING

Nickel and cobalt alloys can be joined with the plasma arc welding process. The constricted arc permits deeper penetration than that obtainable with the gas tungsten arc, but the welding procedures with both processes are similar. Square-groove welds can be made in sections up to about 0.30 in. thick with a single pass when the keyhole technique is used. Thin sections can be welded with the melt-in technique as with gas tungsten arc welding. Sections over 0.3 in. thick can be welded using one of the groove weld joint designs. The first pass can be made with the keyhole technique and the succeeding passes with the melt-in technique. The root face should be about 0.18 in. wide, compared to 0.06 in. for gas tungsten arc welding.

Special techniques are required for keyhole welding of section thicknesses of 0.13 in. and greater. Upslope of the orifice gas flow and the welding current is required to initiate the keyhole; downslope of these variables is needed to fill the keyhole cavity at the end of the joint.

Argon or argon-hydrogen mixtures are normally recommended for the orifice and shielding gases. Hydrogen addition to argon increases the arc energy for keyhole welding and high-speed autogenous welding. Additions up to 15 percent may be used but only with care because hydrogen can contribute to porosity in the weld metal. Therefore, the gas mixture for a specific application should be determined by appropriate tests.

Typical conditions used for automatic plasma arc welding of four nickel alloys with the keyhole technique are given in Table 6.6. No filler metal was added. These conditions are not necessarily optimum for the particular alloy and thickness. Other conditions may also produce acceptable welds, and they should be evaluated by appropriate tests prior to production to ensure product reliability.

## GAS METAL ARC WELDING

This process can be used to weld all of the solution-strengthened nickel alloys except high silicon casting alloys. Many precipitation-hardenable alloys may also be welded, but electrodes of matching composition are seldom used. Such electrodes do not have adequate cracking resistance, particularly with multiple-pass welding.

Spray, pulsed spray, globular, or short-circuiting filler metal transfer may be used. Direct current electrode positive (reverse polarity) should be used with an electrode diameter of 0.035, 0.045, or 0.062 inch. The specific size depends

**Table 6.6**
**Typical conditions for autogenous plasma arc welding of nickel alloys with the keyhole technique**

| Alloy | Thickness, in. | Orifice gas flow, ft³/hª | Welding current, A | Arc voltage, V | Travel speed, in./min |
|---|---|---|---|---|---|
| Nickel | 0.125 | 10 | 160 | 31.0 | 20 |
| | 0.235 | 10 | 245 | 31.5 | 14 |
| | 0.287 | 10 | 240 | 31.5 | 10 |
| 67 Ni - 32 Cu | 0.250 | 12.5 | 210 | 31.0 | 14 |
| 76 Ni - 16 Cr - 8 Fe | 0.195 | 12.5 | 155 | 31.0 | 17 |
| | 0.260 | 12.5 | 210 | 31.0 | 17 |
| 46 Fe - 33 Ni - 21 Cr | 0.125 | 10 | 115 | 30.0 | 18 |
| | 0.187 | 10 | 185 | 27.0 | 16 |
| | 0.230 | 12.5 | 185 | 32.0 | 17 |

a. Orifice diam: 0.136 in.
Orifice and shielding gas: Argon-5%$H_2$
Backing gas: argon

upon the metal transfer mode and base metal thickness. Constant-potential dc power sources are normally used, but constant-current dc units may be acceptable in special cases.

Argon or argon-helium mixtures are normally used for shielding. Argon is normally recommended with spray or pulsed spray metal transfer. Addition of 15 to 20 percent helium to argon will increase bead width and decrease penetration. Argon-helium mixtures are recommended for short-circuiting transfer to take advantage of the higher arc energy due to the helium. The gas nozzle size and gas flow rate should provide adequate shielding of the weld area against oxidation. They will vary with the type of joint and position of welding.

As with gas tungsten arc welding, the root of the joint must be adequately shielded to prevent oxidation. A backing strip may be used for welding nickel or solution-strengthened alloys that are not crack sensitive.

The welding gun should be positioned nearly perpendicular to the axis of the weld, consistent with good visibility of the arc and good shielding. The arc length (voltage) should be adjusted to produce minimum spatter. A suitable length is usually about ¼ inch with spray transfer. Typical conditions for gas metal arc welding of nickel and nickel alloys are given in Table 6.7. These data should serve as a guide in developing appropriate welding procedures.

## SHIELDED METAL ARC WELDING

This process is used primarily for welding commercially pure nickel and solution-strengthened nickel alloys. The precipitation-hardenable alloys are seldom welded with this process. The alloying elements that contribute to precipitation hardening are difficult to transfer across the welding arc. Structures that are fabricated from these alloys are generally better welded by one of the gas shielded processes. If precipitation-hardenable alloys are welded by this process, the surface oxides that form on the weld bead must be removed before the next pass is deposited. Grit blasting or grinding is normally used for this operation.

In general, the minimum metal thickness that should be welded by this process is 0.06 inch. Nickel alloy covered electrodes are available for all-position welding of most alloy families. Some cobalt alloy covered electrodes are difficult to use out-of-position, and should be limited to the flat and horizontal positions. The recommendations of the electrode manufacturer should be followed. Vertical welds can best be accomplished using the gas tungsten arc process.

Decomposition of the electrode covering during welding provides shielding gas as well as a slag cover on the weld bead to isolate it from the atmosphere during cooling. However, protection afforded by the covering is not so good as that provided by the gas shielded processes.

Welding in the flat position should be done with the backhand technique using a drag (travel) angle of about 20 degrees. This technique will help with the control of the molten slag to avoid slag entrapment in the weld metal. Welding in the vertical and overhead positions should be

**Table 6.7**
**Typical conditions for gas metal arc welding of nickel alloys**

| Base metal | | Electrode | | | | | Arc voltage | | Welding |
| UNS. No. | Com. des.[a] | AWS Class.[b] | Diam., in. | Melting rate, in./min. | Shielding gas | Welding position | Av. | Peak | current, A |
|---|---|---|---|---|---|---|---|---|---|
| | | | Spray transfer | | | | | | |
| N02200 | Nickel 200 | ERNi-1 | 0.062 | 205 | Ar | Flat | 29-31 | NA[c] | 375 |
| N04400 | Monel 400 | ERNiCu-7 | 0.062 | 200 | Ar | Flat | 28-31 | NA | 290 |
| N06600 | Inconel 600 | ERNiCr-3 | 0.062 | 200 | Ar | Flat | 28-30 | NA | 265 |
| | | | Pulsed spray transfer | | | | | | |
| N02200 | Nickel 200 | ERNi-1 | 0.045 | 160 | Ar or Ar-He | Vertical | 21-22 | 46 | 150 |
| N04400 | Monel 400 | ERNiCu-7 | 0.045 | 140 | Ar or Ar-He | Vertical | 21-22 | 40 | 110 |
| N06600 | Inconel 600 | ERNiCr-3 | 0.045 | 140 | Ar or Ar-He | Vertical | 20-22 | 44 | 90-120 |
| | | | Short-circuiting transfer | | | | | | |
| N02200 | Nickel 200 | ERNi-1 | 0.035 | 360 | Ar-He | Vertical | 20-21 | NA | 160 |
| N04400 | Monel 400 | ERNiCu-7 | 0.035 | 275-290 | Ar-He | Vertical | 16-18 | NA | 130-135 |
| N06600 | Inconel 600 | ERNiCr-3 | 0.035 | 270-290 | Ar-He | Vertical | 16-18 | NA | 120-130 |
| N06007 | Hastelloy G | ERNiCrMo-1 | 0.062 | – | Ar-He | Flat | 25 | NA | 160 |
| N06455 | Hastelloy C-4 | ERNiCrMo-7 | 0.062 | – | Ar-He | Flat | 25 | NA | 180 |
| N10665 | Hastelloy B-2 | ERNiMo-7 | 0.062 | 185 | Ar-He | Flat | 25 | NA | 175 |

a. Some may be tradenames.
b. See AWS A5.14, Specification for Nickel Alloy Bare Welding Rods and Electrodes.
c. Not applicable.

done with a shorter arc length and lower current than for the flat position. For vertical welding, the electrode should be held normal to the weld axis.

With square and V-groove welds, the electrode should be positioned nearly perpendicular to the plane of the work in the transverse direction. On the other hand, the electrode should be held at a work angle of about 30 degrees from the vertical when welding U-groove joint designs. This is done to obtain good fusion with the joint faces. The work angle should be 40 to 45 degrees for fillet welding.

There should be no pronounced spatter during welding. Excessive spatter may be caused by excessive arc length or welding current. Arc blow may be a problem when welding high nickel alloys that are magnetic.

Nickel alloy weld metal does not flow readily, and it must be placed where required by manipulation of the electrode. Slight weaving of the electrode may be helpful, but it should not exceed three times the electrode diameter. Wide weld beads can result in slag entrapment, a large molten weld pool, an undesirable flat or concave bead surface, and disruption of the protective gaseous shield around the arc. Poor shielding may result in contamination of the weld metal. The use of high current to promote fluidity of the weld metal may cause overheating of the electrode, loss of arc stability, breakdown of the covering, and porosity in the weld metal.

The flux must be removed from each weld bead with a stainless steel wire brush or chisel hammer to avoid trapping slag in the weld metal.

The general procedures described are suitable for all alloys with some slight modifications to suit the characteristics of individual alloys. Commercially pure nickel, for example, is less fluid than the solid-solution alloys, and requires more direct placement of the filler metal in the joint.

Stringer beads are recommended for the nickel-molybdenum and nickel-chromium-molybdenum alloys. If a weave technique is used, it should not be greater than 1.5 times the electrode diameter. Welding in positions other than the flat is not recommended for these alloys. Porosity in the initial weld metal can be a problem with Ni-Mo covered electrodes. This can be minimized by starting on a tab. The best practice is to grind all arc starts and stops to expose sound metal.

## SUBMERGED ARC WELDING

Filler metals and fluxes are available for welding several nickel alloys, but cobalt alloys are seldom joined by this process (see Table 6.5). Submerged arc welding of low carbon nickel alloys may not be appropriate in some cases because the weld metal may pick up carbon and silicon from the flux. This will lower the corrosion resistance of the weld metal. In most cases, the corrosion resistance of submerged arc weld metal is generally lower than that of weld metal deposited with one of the inert gas arc welding processes. Submerged arc welding of nickel alloys can be done with direct current, electrode positive or negative. Electrode positive will produce a flat bead with deep penetration, and is generally used for welding. Electrode negative will provide a slightly higher deposition rate and shallow penetration and is generally used for surfacing.

Typical conditions for submerged arc welding of three nickel alloys are given in Table 6.8. Suitable fluxes and filler metals are available for nearly all of the nickel, nickel-copper, nickel-chromium, and nickel-iron-chromium solution-strengthened alloys. Those alloys that contain significant amounts of aluminum or titanium generally are not welded with this process because of the poor recovery of titanium and aluminum with the available fluxes. The process is also not generally recommended for joining thick sections of nickel-molybdenum alloys because the high heat input and slow cooling rate of the weld results in low weld ductility.

## POSTWELD HEAT TREATMENT

Postweld heat treatments are usually not needed to restore the corrosion resistance of nickel, nickel-copper, nickel-chromium, or nickel-iron-chromium alloys. However, some alloys may require a stress relief heat treatment for specific corrosion resistant applications.

The nickel-chromium and nickel-iron-chromium alloys, like some austenitic stainless steels, can exhibit carbide precipitation in the weld heat-affected zone. Such sensitization does not result in accelerated attack in most environments. Some alloys are stabilized by additions of titanium or columbium.

Welding of nickel-molybdenum and nickel-silicon alloys can influence the corrosion resist-

**Table 6.8**
**Typical conditions for submerged arc welding nickel alloys**

| Base metal | Nickel | 76Ni-16Cr-8Fe | 67Ni-32Cu |
|---|---|---|---|
| Electrode, AWS Class.[a] | ERNi-1 | ERNiCr-3 | ERNiCu-7 |
| Electrode diam., in. | 0.062 | 0.062 | 0.062 |
| Welding power | DCEP (RP) | DCEP (RP) | DCEP (RP) |
| Welding current, A | 250 | 250 | 260-280 |
| Voltage setting, V | 28-30 | 32-34 | 32-34 |
| Travel speed, in./min | 10-12 | 9-10 | 9-10 |
| Max. interpass temp., °F | 350 | 350 | 350 |

a. See AWS A5.14, *Specification for Nickel and Nickel Alloy Bare Welding Rods and Electrodes.* Appropriate proprietary fluxes are available for use with these electrodes.

ance of the metal in the heat-affected zones. Generally, the weldments should be solution annealed to restore corrosion resistance. The producers of the alloys should be consulted for the recommended heat treatments. Low carbon and low silicon alloys do not require a postweld heat treatment for corrosion resistance.

Cobalt alloys do not normally require postweld heat treatment. If a postweld heat treatment is desired to relieve stresses, a full solution anneal is recommended.

## STRAIN-AGE CRACKING

Most of the precipitation-hardenable nickel alloys are subject to strain-age cracking. Alloys containing columbium have a greater resistance to cracking because of the slow hardening response of the columbium precipitate compared to the aluminum-titanium precipitate. The rapid decrease in ductility of certain alloys at the aging temperature does not permit plastic flow to occur readily. High residual stresses may cause cracking. It may occur during aging, annealing, or stress relieving unless precautions are taken to prevent high residual stress. Although it is most commonly associated with welding, strain-age cracking can result from stresses induced by other processing operations such as machining and forming.

There is evidence that oxidation during heat treatment in air increases the incidence of cracking. Heat treating in an inert or reducing atmosphere will avoid this problem.

With a few exceptions, the susceptibility of nickel alloys to postweld strain-age cracking is closely related to the combined aluminum and titanium content. The higher the aluminum and titanium content, the greater is the possibility of cracking. Several techniques may be used to either control the magnitude of the residual stresses or provide a condition more tolerant of the stresses. Such techniques include welding the parts in an unrestrained condition, application of a suitable preweld heat treatment, welding the joint with an appropriate sequence, and annealing at an intermediate stage during welding.

Precipitation-hardenable nickel alloys develop their high strength properties from a series of heat treatments. The recommended postweld heat treatment is to solution anneal followed by precipitation hardening. Precipitation hardening an as-welded joint increases the tendency for strain-age cracking. During hardening, an overall volume contraction occurs. Welding stresses coupled with this contraction greatly increase the likelihood of strain-age cracking. Solution annealing after welding will relieve high welding stresses and decrease the possibility of strain-age cracking. However, it is important to heat the weldment through the hardening temperature range rapidly to avoid precipitation reactions.

One preweld heat treatment sometimes used to avoid strain-age cracking is to overage the components. Such a treatment is designed to precipitate the age-hardening constituent in a massive form. This reduces the hardness and yield strength of the alloy as well as the welding stresses that can be developed. Cracking of the weldment during solution annealing is then less likely to occur because further aging cannot take place during heating. Alloys containing relatively large amounts of aluminum and titanium, such as Udimet 500, have been successfully welded using this overaging technique.

Welding of precipitation hardened components should be avoided. Welding in this condition will result in resolutioning and overaging in the heat-affected zone. A postweld heat treatment is then necessary to restore the properties to this zone.

## MECHANICAL PROPERTIES

The yield and tensile strengths of welded nickel and cobalt alloys are, in general, equivalent to those of annealed base metal, but the ductility is slightly lower. Exceptions occur when a lower strength filler metal is used or when the base metal is strengthened by a combined thermal-mechanical treatment. The metal producer should be consulted for mechanical property data of weldments in a particular alloy.

Typical mechanical properties of all-weld-metal deposits of several nickel alloy filler metals are given in Table 6.9. Some nickel filler metals are not metallurgically stable during long-time exposure in certain temperature ranges. For example, the ERNiCrMo-3 and ENiCrMo-3 weld metals lose ductility after long-time exposure in the 1200° to 1500° F range due to precipitation of second phases in the microstructure, including $Ni_3Cb$. Carbide precipitation may take place in

some alloys during exposure at elevated temperatures.

### Table 6.9
### Typical mechanical properties of nickel alloy filler metal deposits

| AWS Classification[a] | Tensile strength, ksi | Yield strength, 0.2% offset, ksi | Elongation, % |
|---|---|---|---|
| ENi-1 | 72 | 42 | 35 |
| ERNi-1 | 75 | 39 | 45 |
| ENiCu-7 | 73 | 45 | 60 |
| ERNiCu-7 | 84 | 50 | 40 |
| ENiCrFe-3 | 92 | 52 | 43 |
| ERNiCr-3 | 95 | 55 | 50 |
| ENiCrFe-4 | 105 | 62 | 32 |
| ENiCrMo-3 | 115 | 72 | 32 |
| ERNiCrMo-3 | 115 | 70 | 43 |
| ERNiCrMo-4 | 106 | 70 | 46 |
| ERNiMo-3 | 123 | 80 | 46 |

a. Refer to AWS Specifications A5.11 and A5.14 for the chemical requirements and recommended applications for these electrodes. Those listed are not all inclusive.

# OXYACETYLENE WELDING

The oxyacetylene flame produces a sufficiently high temperature for welding commercially pure nickel and some solution-strengthened nickel alloys. However, this process should only be used in those situations where suitable arc welding equipment is not available. Welds can be made in all positions with practice.

## FLAME ADJUSTMENT

The torch tip should be large enough to provide a low velocity, soft flame. A high velocity, harsh flame is undesirable. Usually the tip should be the same size or one size larger than the one recommended for the same thickness of steel.

The welding torch should be adjusted with excess acetylene to produce a slightly reducing

flame. When chromium-bearing alloys are welded, the flame should not be excessively reducing because the weld metal might absorb carbon.

During welding, the molten weld pool should be kept quiet with the inner cone of the flame just touching its surface. Puddling of the molten metal should be avoided. This action can result in loss of the deoxidizing elements or exposure of the molten weld metal to the surrounding atmosphere. As a result, the weld metal may be porous.

The hot end of the filler rod should be kept within the flame envelope to minimize oxidation.

## FLUXES

Flux is required for welding nickel-copper, nickel-chromium, and nickel-iron-chromium alloys. Commercially pure nickel can be welded without flux.

The following mixture can be used for solution-strengthened nickel-copper alloys.[7]

| | |
|---|---|
| Barium fluoride | 60 percent |
| Calcium fluoride | 16 percent |
| Barium chloride | 15 percent |
| Gum arabic | 5 percent |
| Sodium fluoride | 4 percent |

Nickel-chromium and nickel-iron-chromium alloys can be fluxed with a mixture of one part sodium fluoride and two parts calcium fluoride with 3 percent hematite (red iron oxide) and a suitable wetting agent added to the mixture.

Precipitation-hardenable nickel-copper alloys can be fluxed with a water slurry of one part lithium fluoride and two parts of either of the two previously given flux compositions.

Nickel-silicon casting alloy (Hastelloy D) may be fluxed with a mixture of 65 percent cast iron welding flux and 35 percent boric acid. Proprietary fluxes are also available for use with this alloy.

The flux is mixed with water to produce a thin slurry. It should be applied to both sides of the joint and to the filler rod, and then allowed to dry before welding is started.

Borax must not be used as a flux when welding nickel alloys. It can form a brittle, low melting eutectic in the weld which is undesirable.

The flux residue must be removed from the joint· for high temperature service. Molten flux will corrode the base metal after an extended period. Unfused flux may be washed off with hot water. Fused flux is not soluble in water and must be removed mechanically by grit blasting or grinding.[8]

## NICKEL ALLOYS

The oxyacetylene welding process is not recommended for joining the low carbon nickel and nickel alloys, the nickel-molybdenum alloys, and the nickel-chromium-molybdenum alloys. These materials can readily pick up carbon from the flame, and this will reduce their corrosion resistance and high-temperature properties.

Oxyacetylene welding is the only recommended joining process for the Ni-10Si-3Cu-2Fe alloy (Hastelloy D). The welding procedure is similar to that used for cast iron. The filler rod should be the same composition as the base metal. A U-groove with a 45-degree bevel angle should be used for sections over 0.5 in. thick.

With the exception of the nickel-copper alloy, the precipitation-hardenable alloys should not normally be welded by the oxyacetylene process. The hardening elements are easily oxidized and fluxed away during welding.

## COBALT ALLOYS

The wrought cobalt alloys are usually low in carbon. As with low carbon nickel alloys, these alloys can pick up carbon during welding. This will alter the properties of the alloys. However, cobalt hardfacing alloys are high in carbon, and they are often deposited with the oxyacetylene process.

# RESISTANCE WELDING

Nickel and cobalt alloys are readily welded to themselves and to other metallurgically compatible metals by spot, seam, projection, and flash welding. The electrical resistivities of the alloys range from about 7 microhm•cm for commercially pure nickel to 138 microhm•cm for a resistance-heating nickel alloy. The welding current requirements are lower with the high-resistance alloys but the force requirements increase because of their high strengths at elevated temperatures.

---

7. Fluxes containing fluorides will give off fumes when heated that may irritate the eyes, nose, and throat. They must be used with adequate ventilation. Respiratory equipment should be used when welding in confined spaces.

---

8. Flux particles generated during cleaning must be collected and trapped by a suitable exhaust ventilation system.

For good electrical contact, the surfaces of the parts must be clean. All oxide, oil, grease, and other foreign matter must be removed by acceptable cleaning methods. Chemical pickling is the best method of oxide removal.

## SPOT WELDING

Nickel and cobalt alloys are spot welded in much the same manner as other metals. In many respects, these alloys are easy to spot weld. The configurations involved and the relatively short welding time tend to preclude any contamination from the atmosphere. As a result, auxiliary shielding is not normally needed during resistance spot welding. The thermal and electrical conductivities and the mechanical properties of the alloys vary depending upon their composition and condition. Conditions for spot welding, therefore, are adjusted to account for the material properties. Usually, several combinations of welding variables can produce similar and acceptable results.

### Equipment

Nickel and cobalt alloys can be welded successfully on almost all types of conventional spot welding equipment. The equipment must provide accurate control of welding current, weld time, and electrode force. Each of these settings may vary within a range without appreciably affecting weld quality. It is, however, desirable to have sufficient control over them to obtain reproducible results after the optimum settings are obtained for a given application.

Upslope controls will help prevent expulsion. Dual electrode force systems are sometimes used to provide a high forging force when welding high-strength, high-temperature alloys. No significant changes in welding characteristics or static weld properties can be attributed to the use of any specific type of spot welding equipment.

For most applications, the restricted dome electrode design shown in Fig. 6.5 is preferred. Truncated electrodes or ones with 5 or 8 in. radius faces are sometimes used for metal thicknesses in the range of 0.06 to 0.13 inch. Larger weld nuggets and correspondingly higher shear strengths can be obtained with these electrode shapes.

RWMA Group A, Class 1, 2, and 3 electrode alloys are recommended. Class 1 alloys are best for low resistivity alloys and for thin sheets to minimize sticking tendencies. Class 3 electrodes are recommended for high-temperature

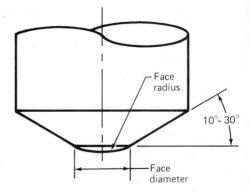

*Fig. 6.5—Restricted dome spot welding electrode design*

alloys to minimize mushrooming with high electrode forces and relatively long weld times.

### Welding Conditions

Spot welding conditions are primarily controlled by the total thickness of the assembly being welded and, to a rather large degree, by the welding machine being used. Similar welding conditions may be suitable for making welds in the same total thickness where the number of layers differs significantly. However, for any given thickness or total pile-up, various combinations of welding current, time, and electrode force may produce similar welds. Other variables such as electrode size and shape are important in controlling such characteristics as metal expulsion, sheet indentation, and sheet separation. Upslope, downslope, and forging force may be used for some high-strength, high-temperature alloys to control heating rate and soundness of the weld nugget.

Alloys that have properties similar to steel behave in a like manner. The high nickel and cobalt alloys generally are harder and stronger than low carbon steel, particularly at elevated temperatures. Higher electrode forces are therefore required during spot welding. The time of current flow should be as short as possible, but sufficiently long to gradually build up the welding heat. Welding current should be set somewhat above the value that produces a weak weld, but below the setting that causes weld metal expulsion. Current upslope is an asset in this respect.

Suitable conditions for spot welding annealed nickel alloys are given in Tables 6.10 and 6.11. A forging force is sometimes applied near

**Table 6.10**

**Conditions for spot welding annealed nickel alloys with single-phase machines**

| Thickness,[a] in. | Electrode[b] Face Radius, in. | Diam., in. | Force, lb | Weld time, cycles (60Hz) | Welding current, kA | Nugget diam., in. | Min. shear strength, lb |
|---|---|---|---|---|---|---|---|
| | | | | Nickel | | | |
| 0.021 | 3 | 0.16 | 370 | 4 | 7.8 | 0.12 | 350 |
| 0.031 | 3 | 0.19 | 900 | 4 | 15.4 | 0.18 | 760 |
| 0.063 | 3 | 0.25 | 1720 | 6 | 21.6 | 0.25 | 2400 |
| 0.094 | 3 | 0.31 | 2300 | 12 | 26.4 | 0.31 | 3600 |
| 0.125 | 3 | 0.38 | 3300 | 20 | 31.0 | 0.37 | 5600 |
| | | | | 67Ni-32Cu alloy[c] | | | |
| 0.021 | 3 | 0.19 | 300 | 12 | 6.2 | 0.13 | 450 |
| 0.031 | 3 | 0.19 | 700 | 12 | 10.5 | 0.17 | 845 |
| 0.063 | 3 | 0.31 | 2700 | 12 | 15.3 | 0.31 | 2060 |
| 0.093 | 3 | 0.38 | 2760 | 20 | 22.6 | 0.37 | 3880 |
| 0.125 | 3 | 0.50 | 5000 | 30 | 30.0 | 0.47 | 5850 |
| | | | | 76Ni-16Cr-8 Fe alloy[d] | | | |
| 0.021 | 3 | 0.16 | 300 | 12 | 4.0 | 0.12 | 545 |
| 0.031 | 3 | 0.19 | 700 | 12 | 6.7 | 0.18 | 920 |
| 0.063 | 3 | 0.31 | 2070 | 12 | 12.0 | 0.31 | 2750 |
| 0.093 | 3 | 0.38 | 3870 | 20 | 15.0 | 0.37 | 4400 |
| 0.125 | 3 | 0.44 | 5270 | 30 | 20.1 | 0.44 | 6400 |
| | | | | 73Ni-16Cr-7 Fe-3Ti alloy[e] | | | |
| 0.010 | 6 | 0.16 | 300 | 2 | 7.3 | 0.11 | |
| 0.015 | 6 | 0.16 | 400 | 4 | 7.4 | 0.11 | |
| 0.021 | 6 | 0.19 | 750 | 6 | 7.5 | 0.14 | |
| 0.031 | 6 | 0.22 | 1750 | 8 | 9.9 | 0.17 | |
| 0.062 | 10 | 0.31 | 4400 | 14 | 16.4 | 0.29 | |

a. Two equal thicknesses
b. Restricted dome electrode design
c. Commonly known as Monel
d. Commonly known as Inconel 600
e. Commonly known as Inconel X-750

**Table 6.11**

**Conditions for spot welding annealed nickel alloys with three-phase frequency converter machine**

| Thickness,[a] in. | Electrode[b] Face Radius, in. | Diam., in. | Force, lb | Time, cycles (60Hz) Weld | Pulse | Interpulse | Welding current, kA | Nugget diam., in. | Min. shear strength, lb |
|---|---|---|---|---|---|---|---|---|---|
| | | | | **67Ni-32Cu alloy[c]** | | | | | |
| 0.018 | 3 | 0.19 | 400 | 13 | 6 | 1 | 4.3 | 0.17 | 400 |
| 0.030 | 5 | 0.25 | 800 | 13 | 6 | 1 | 8.5 | 0.18 | 900 |
| 0.043 | 5 | 0.25 | 1600 | 17 | 8 | 1 | 11.5 | 0.26 | 1750 |
| 0.062 | 7 | 0.31 | 2200 | 21 | 10 | 1 | 14.5 | 0.32 | 2050 |
| 0.093 | 9 | 0.44 | 3800 | 39 | 9 | 1 | 22.5 | 0.40 | 5400 |
| 0.125 | 12 | 0.50 | 5000 | 65 | 10 | 1 | 31.0 | 0.48 | 7000 |
| | | | | **73Ni-16Cr-7 Fe-3Ti alloy[d]** | | | | | |
| 0.025 | 3 | 0.22 | 2000 | 9 | 8 | 1 | 6.0 | 0.16 | 900 |
| 0.031 | 5 | 0.25 | 2200 | 10 | 9 | 1 | 6.8 | 0.18 | 1150 |
| 0.043 | 5 | 0.25 | 2700 | 23 | 5 | 1 | 8.1 | 0.20 | 1800 |
| 0.062 | 8 | 0.31 | 3500 | 35 | 8 | 1 | 11.4 | 0.25 | 3300 |
| 0.093 | 8 | 0.44 | 5000 | 53 | 8 | 1 | 15.0 | 0.37 | 5700 |

a. Maximum thickness of multiple layers should not exceed four times this thickness. Maximum ratio for unequal thicknesses is 3 to 1.
b. Restricted dome electrode design.
c. Commonly known as Monel.
d. Commonly known as Inconel X-750.

the end of weld time to consolidate the weld nugget. For some alloys, forging force may be applied during a postheating impulse.

Precipitation-hardenable alloys are best welded in the solution annealed condition with settings similar to those used for similar solid-solution alloys. However, high electrode force and low welding current must be used to compensate for their high-temperature strength and high electrical resistance. The weld nuggets will be about the same size as those in the solid-solution alloys, but the shear strengths will be higher in the as-welded condition. Subsequent precipitation hardening will increase the shear strength by about 50 percent. Cracking will generally occur when these alloys are welded in the hardened condition. Postweld solution annealing followed by precipitation hardening is recommended to avoid strain-age cracking.

Some cracking may occur during spot welding of some precipitation-hardenable alloys if insufficient electrode force is used. If higher electrode force does not overcome cracking, increasing the weld time or lowering the welding current will help. Welding machines with low-inertia heads and current slope control are preferred.

## SEAM WELDING

This process is normally used to join sheet thicknesses ranging from 0.002 to 0.125 inch. Wheel electrodes of RWMA Group A, Class 1 or 2 alloy are used. Individual overlapping spots are created by coordinating the welding time and wheel rotation. The wheel electrodes can be rotated continuously or intermittently. Continuous seam welding imposes limitations on the weld cycle variations that can be used. For example, a forging force cannot be applied during continuous seam welding, but it can be used with intermittent motion. High-strength alloys, such as Hastelloy X and Inconel X-750, usually are welded with forging force and intermittent drive.

Suggested conditions for seam welding two nickel alloys with continuous motion are given in Table 6.12. Table 6.13 gives conditions for welding a different nickel alloy using intermittent motion and forging force. The force must be high to consolidate the weld nugget to prevent cracking and porosity.

## PROJECTION WELDING

This process requires die-formed projec-

tions, similar to those used for steel, in one or both parts. The parts are also usually die formed to shape in high-production operations. Nickel and cobalt alloys are seldom projection welded because production requirements are normally low.

Conditions for projection welding of nickel and cobalt alloys are influenced by the thickness and shape of the parts to be welded. Generally, higher electrode forces and longer weld times than those used for steels are required because of the higher strength of nickel and cobalt alloys at elevated temperature.

## FLASH WELDING

The nickel and cobalt alloys can be flash welded to themselves and to dissimilar metals. In two respects, flash welding is well adapted to the high-strength, heat-treatable alloys. First, molten metal is not retained in the joint. Second, the hot metal at the joint is upset. This upsetting operation may improve the ductility of the heat-affected zone.

The machine capacity required to weld nickel and nickel alloys does not differ greatly from that required for steel. This is especially true for transformer capacity. The upset force needed for making flash welds in nickel alloys is higher than that required for steel.

Joint designs for flash welds are similar to those used for other metals. Flat, sheared, or saw-cut edges and pinch-cut rod or wire ends are satisfactory for welding. The edges of thick sections are sometimes beveled slightly. The overall shortening of the parts due to metal lost during welding should be taken into account so that finished parts will be the proper length.

The flash welding conditions that are of greatest importance are flashing current, speed, and time, plus upset pressure and distance. With proper control of these variables, the molten metal at the interface will be forced out of joint, and the metal at the interface will be at the proper temperature for welding.

Generally, high flashing speeds and short flashing times are used to minimize weld contamination. Parabolic flashing is more desirable than linear flashing because maximum joint efficiency can be obtained with a minimum of metal loss.

Flash welding conditions vary with the machine size and the application. Table 6.14 gives typical conditions for flash welding 0.25 and 0.375 in. diameter rod of several nickel alloys.

**Table 6.12**
**Conditions for seam welding annealed nickel alloys with single-phase machines**

| Thickness,[a] in. | Electrode face | | Electrode force, lb | Time, Cycles (60Hz) | | Welding current, kA | Welding speed, in./min | Width of nugget, in. |
|---|---|---|---|---|---|---|---|---|
| | Width, in. | Radius, in. | | Heat | Cool | | | |
| 67Ni-32Cu alloy[b] | | | | | | | | |
| 0.010 | 0.16 | 3 | 200 | 1 | 3 | 5.3 | 75 | 0.09 |
| 0.015 | 0.16 | 6 | 300 | 1 | 3 | 7.6 | 75 | 0.10 |
| 0.021 | 0.19 | 6 | 500 | 2 | 6 | 8.7 | 38 | 0.15 |
| 0.031 | 0.19 | 6 | 700 | 4 | 12 | 10.0 | 19 | 0.15 |
| 0.062 | 0.38 | 6 | 2500 | 8 | 12 | 19.0 | 20 | 0.17 |
| 73Ni-16Cr-7 Fe-3Ti alloy[c] | | | | | | | | |
| 0.010 | 0.13 | 3 | 400 | 1 | 3 | 3.6 | 45 | 0.11 |
| 0.015 | 0.13 | 3 | 700 | 2 | 4 | 3.9 | 36 | 0.12 |
| 0.021 | 0.16 | 3 | 1400 | 3 | 6 | 8.0 | 30 | 0.14 |
| 0.031 | 0.19 | 3 | 2300 | 4 | 8 | 8.5 | 30 | 0.17 |
| 0.062 | 0.19 | 6 | 4000 | 8 | 16 | 10.3 | 12 | 0.18 |

a. Maximum thickness of multiple layers should not exceed four times this thickness. Maximum ratio for unequal thicknesses is 3 to 1.
b. Commonly known as Monel.
c. Commonly known as Inconel X-750.

**Table 6.13**
**Seam welding of 47Ni-22Cr-18Fe-9Mo alloy[a]**

| Thickness,[b] in. | Electrode[c] face width, in. | Electrode force, lb | | Weld Times, cycles (60Hz) | | | Forge time, cycles (60Hz) | Welding current, kA | Spots per in. |
|---|---|---|---|---|---|---|---|---|---|
| | | Weld | Forge | Total | Heat | Cool | | | |
| 0.030 | 0.19 | 1500 | - | 10 | 10 | - | - | 20.5 | 14 |
| 0.063 | 0.31 | 2000 | 4000 | 94 | 10 | 2 | 15 | 21.5 | 10 |
| 0.094 | 0.38 | 4500 | 4500 | 46 | 10 | 2 | 25 | 33.0 | 8 |

a. Commonly known as Hastelloy X
b. Two equal thicknesses
c. RWMA Class III copper alloy wheel, 12 in. diam., flat face, 15° double bevel

**Table 6.14**
**Typical conditions for flash welding nickel alloy rods**

| Metal | Diam.,[a] in. | Flash-off, in. | Flashing time, s | Upset current time, s | Upset, in. | Energy input, W·h | Joint efficiency, % |
|---|---|---|---|---|---|---|---|
| Nickel | 0.25 | 0.442 | 2.5 | 1.5 | 0.125 | 2.15 | 89 |
| Nickel | 0.375 | 0.442 | 2.5 | 2.5 | 0.145 | 4.87 | 98 |
| 67Ni-32Cu[b] | 0.25 | 0.442 | 2.5 | 1.5 | 0.125 | 1.93 | 97 |
| 67Ni-32Cu | 0.375 | 0.442 | 2.5 | 2.5 | 0.145 | 5.55 | 95 |
| 66Ni-30Cu-3Al[c] | 0.25 | 0.442 | 2.5 | 1.5 | 0.125 | 2.02 | 94 |
| 66Ni-30Cu-3Al | 0.375 | 0.442 | 2.5 | 2.5 | 0.145 | 4.79 | 100 |
| 76Ni-16Cr-8Fe[d] | 0.25 | 0.442 | 2.5 | 1.5 | 0.125 | 2.15 | 92 |
| 76Ni-16Cr-8 Fe | 0.375 | 0.442 | 2.5 | 2.5 | 0.145 | 5.19 | 96 |

a. Ends tapered with 110° included angle
b. Commonly known as Monel
c. Commonly known as Monel K-500
d. Commonly known as Inconel 600

Welding current is determined by the machine transformer tap setting. Since these alloys have higher strengths at elevated temperatures than steel, higher force is required to upset and extrude all the molten metal from the joint. The time of current flow during upsetting is critical.

If the time is too long, the joint may be overheated and oxidized. If the time is too short, the plasticity of the metal will not permit sufficient upsetting to force the molten metal from the joint. A properly made flash weld will not contain any cast metal.

# ELECTRON BEAM WELDING

All nickel and cobalt alloys that can be successfully joined by conventional arc welding processes can also be electron beam welded. Because of its lower heat input, this process may be suitable for joining some alloys that are considered difficult to arc weld. In general, the joint efficiencies of electron beam welds will equal or exceed those of gas tungsten arc welds.

Welding in vacuum provides excellent protection against atmospheric contamination. Porosity may be a problem when welding some alloys at high welding speeds because dissolved gases do not have time to escape to the surface of the molten weld pool. Oscillation of the beam to slightly agitate the weld pool may help the gases to escape and thus reduce porosity.

Hot cracking in the heat-affected zone may be a problem, particularly when welding thick sections of some alloys. The use of a cosmetic (second) pass to provide a good face contour may aggravate the problem because of the high restraint imposed by the first weld pass. In this case, cracking will likely take place in the heat-affected zone of the cosmetic pass.

An example of this type of cracking is shown in Fig. 6.6. The high degree of restraint imposed by the thick base metal on the narrow heat-affected zone can produce a high tensile stress during cooling. Microfissuring can readily take place, particularly if the alloy tends to be hot short. An obvious solution to this problem is to use a welding procedure that does not require a cosmetic pass. Filler metal addition and a subsequent finishing operation might be appropriate.

As with arc welding, the base metal should be in the annealed condition for welding. With precipitation-hardening alloys, the weldment should be solution treated and aged for optimum strength properties. However, distortion of the weldment must be considered when specifying a heat treatment.

Welding conditions used for electron beam welding depend upon the base metal compostion and thickness as well as the type of welding equipment. For a given thickness of an alloy, various combinations of accelerating voltage, beam current, and travel speed may produce satisfactory welds.

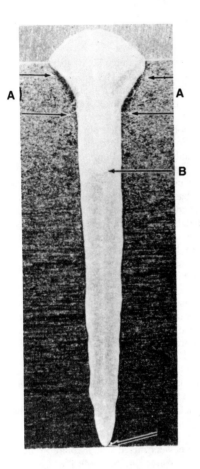 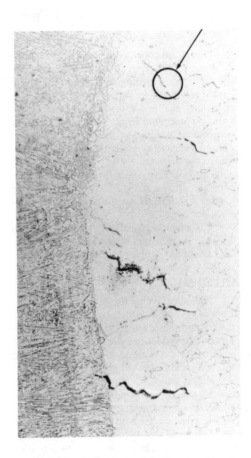

*Fig. 6.6—Left - Photomacrograph (3×) of an electron beam weld in 1.75 in. thick Ni-19Cr-19Fe-5Cb-3Mo alloy with a cosmetic pass penetrating to Point B. Right - Photomicrograph (50×) of Area A showing microfissures in the heat-affected zone at the weld interface. (Reduced 27%)*

## LASER BEAM WELDING

This process is being evaluated for welding various nickel alloys. Welding is normally done in the open, and a gas shield is recommended to protect the weld area from oxidation. Argon or helium is suitable.

Several nickel alloys have been welded with various laser beam systems in thickness ranging from 0.01 to 0.38 inch. The welded joint cross sections are similar to those produced by an electron beam. The metallurgical behavior of laser beam welds should be similar to that of electron beam welds. Several nickel alloys in sheet thicknesses of 0.039 and 0.079 in. have been successfully welded using a $CO_2$ laser beam with an output of up to 2.0 kW. Welding speeds of 80 and 15 in./min., respectively, were used.

# BRAZING

Nickel and cobalt alloys can be joined readily by most of the conventional brazing processes and by diffusion brazing techniques. The severe service conditions to which these alloys are generally subjected is one reason that all phases of a brazing procedure require careful control. The chemical composition of the alloy and its physical and mechanical properties should be considered so that the assembly may be fabricated without encountering avoidable problems. Two important factors are the effects of sulfur and low melting point metals and the possibility of stress cracking.

In general, cobalt alloys can be brazed with filler metals, procedures, and equipment that are similar to those used for nickel and nickel alloys.

## FILLER METALS

The selection of the brazing filler metal depends upon the service conditions of the assembly and the brazing process. It is also important to consider heat treatments that may be required for the brazement to be certain that it will withstand the temperatures involved.

Nickel and cobalt alloys can be conventionally brazed with a number of silver, copper, nickel, cobalt, and gold filler metals.[9] Diffusion brazing requires specially designed alloys that are metallurgically compatible with the base metal. A diffusion heat treatment is used to alloy the brazing filler metal and the base metal together.

### Nickel

These filler metals generally possess excellent oxidation and corrosion resistance as well as useful strengths at elevated temperatures. Some of them can be used to about 1800° F for continuous service and to 2200° F for short time service. They are generally designed for conventional brazing applications, but some alloys are suitable for diffusion brazing.

The filler metals are basically nickel or nickel-chromium, with additions of silicon, boron, manganese, or combinations of these, to depress the melting range below that of nickel- and cobalt-base metals. The brazing temperature ranges are usually between 1850 and 2200° F.

Filler metals that contain significant amounts of phosphorus to depress their melting ranges should not be used to braze nickel alloys. Brittle nickel phosphides can form at the filler metal-base metal interface. Boron-containing filler metals should not be used for brazing thin sections because of their erosive action and alloying with the base metal. Intergranular penetration of the base metal can also take place with some filler metals. In general, these filler metals are brittle, and this places some limitations on the design of brazed joints subject to bending or impact loading.

With diffusion brazing, either a brazing filler metal or a second metal that will react with the base metal at some elevated temperature to form a brazing alloy may be used. In the latter case, the alloy must be a eutectic that melts at a temperature compatible with the base metal. The brazement is heated for a sufficient period for the filler metal to diffuse with the base metal and thus raise the joint remelt temperature to nearly that of the base metal. This treatment produces a joint that is essentially free of a distinct filler metal layer.

Filler metals are commonly applied in the form of a powder mixed with a volatile vehicle, such as an acrylic resin, for application by brush, spray, or extrusion. Precautions must be taken in using the vehicle so that no harmful residue will remain. When properly used, the vehicle should completely disappear during the heating cycle. Filler metal is also available in tape and foil forms that permit the application of a controlled quantity of braze filler metal to a joint. Adjustments for density must be made when applying tape filler metals.

### Cobalt

These filler metals are generally used for their high-temperature properties and compatibility with cobalt alloys. For optimum results, brazing should be performed in a high-purity reducing or inert atmosphere. Special high-temperature fluxes are available. The brazing technique requires a degree of skill. By using diffusion brazing procedures, the brazements can be used at temperatures up to 1900° F with excursions to 2100° F.

---

9. AWS A5.8, *Specification for Brazing Filler Metals,* classifies many of these filler metals. There are also many other suitable filler metals available commercially.

## Silver

Silver brazing filler metals may be used to join nickel and cobalt alloys to themselves and to many other metals and alloys. With proper design and brazing technique, brazed joints can develop the full strength of the base metal at room temperature. However, nickel alloys are subject to stress corrosion cracking when exposed to molten silver filler metals. The base metal should be stress-free during the brazing operation.

The low melting filler metals, BAg-1, BAg-1a, and BAg-2, are commonly used, but for many corrosive environments filler metals containing at least 50 percent silver are preferred. The BAg-7 filler metal is useful where stress cracking in nickel alloys might be a problem. The maximum service temperature recommended for silver filler metals is about 400° F.

## Copper

Most high nickel alloys are capable of being brazed with BCu filler metal using the same equipment as used for steels. Minor changes in the brazing procedures may be necessary because of characteristics of some nickel alloys. Cobalt alloys should not be brazed with copper filler metals.

Copper filler metal will alloy more rapidly with nickel alloys than with steel. The molten copper will not flow far before it has picked up enough nickel to raise its liquidus and reduce its fluidity. Therefore, the filler metal should be placed as close to the joint as possible, and the assembly should be heated rapidly to brazing temperature. A slightly rough or lightly etched surface improves the capillary flow of copper filler metal. Maximum operating temperature with copper filler metal is about 400° F for continuous service and 900° F for short time exposure.

Copper phosphorus filler metals are not usually used with nickel-base alloys because of the formation of brittle nickel phosphide.

## Gold

Gold base filler metals are used for brazing nickel and cobalt alloys where good joint ductility and resistance to oxidation or corrosion are required. Because of their low rate of interaction with the base metal, they are commonly used on thin sections, usually with induction, furnace, or resistance heating in a reducing atmosphere or in vacuum without flux. For certain applications,

a borax-boric acid flux may be used. These filler metals are generally suited for continuous service up to 800° F.

## FLUXES AND ATMOSPHERES

Fluxes, gas atmosphere, and vacuum are used to avoid harmful reactions during brazing. Under some conditions they may also reduce surface oxides that are present. In no case can a brazing flux or atmosphere negate the need for thorough cleaning of parts prior to brazing.

### Fluxes

Brazing fluxes are usually mixtures of fluorides and borates that melt below the melting temperature of the filler metal. Standard fluxes are used on most solution-strengthened nickel and cobalt alloys. Other fluxes are available for alloys containing aluminum and titanium. Special fluxes may be needed with filler metals that require high brazing temperatures. Some fluxes are designed to have a long life during extended heating cycles.

The choice of flux and the technique for its use greatly influence the time required to make a braze and the quality of the joint. There is no single universal flux that is best for all brazing applications. Since there are many variables including base metal, filler metal, method of brazing, time required to braze, and joint design, there are many useful formulations of flux. Each flux is compounded somewhat differently, and each has its optimum performance region. Recommendations of the supplier should be sought, and trial runs are recommended when experience is lacking. For successful use, a flux must be chemically compatible with the base metals and filler metals involved and active throughout the brazing temperature range.

After brazing, a flux removal operation may be necessary. Flux removal is particularly important on brazements that are to operate at elevated temperatures or in corrosive environments.

### Controlled Atmospheres

Controlled atmospheres and vacuum environments are employed to prevent the formation of oxides during brazing and, in many instances, to remove the oxide films present on the metals so that the filler metal can wet and flow. The reactions resulting from the use of gas atmospheres and vacuum atmospheres are diverse. Certain conditions, however, apply to both.

Controlled atmospheres are most commonly used in furnace or retort brazing operations. However, they may also be used with induction and resistance brazing.

The general techniques of atmosphere brazing can involve the following:

(1) Gaseous atmospheres alone

(2) Gaseous atmospheres together with solid or liquid fluxes preplaced at the interfaces

(3) High vacuum

(4) Combinations of vacuum and gas atmospheres

Controlled atmospheres have the following advantages where additional solid or liquid fluxes are not required:

(1) The entire part is maintained in a clean, unoxidized condition throughout the brazing cycle. Parts may, therefore, frequently be machined to finished size prior to brazing.

(2) Usually, postbraze cleaning is not necessary.

(3) Intricate sealed parts from which fluxes cannot be removed, such as electronic tubes, can be brazed satisfactorily.

(4) Large surface areas can be brazed integrally and continuously without danger of entrapped flux pockets at the interfaces.

The most common method used for brazing nickel and cobalt alloys is a controlled atmosphere furnace. The furnace must be properly designed to maintain environment purity. There are three types of reducing atmospheres suitable for brazing, namely, combusted fuel gas, dissociated ammonia, and pure, dry hydrogen. An atmosphere of combusted fuel gas, containing not more than 20 grains of sulfur per 100 ft³, is satisfactory for nickel and nickel-copper alloys free of aluminum. Dissociated ammonia can be used with the same metals and also for nickel-chromium-iron alloys if the dew point is –80° F or lower. The hydrogen in these atmospheres will reduce most metal oxides on workpieces to be brazed provided the dew point is sufficiently low. Alloys containing chromium must be brazed at temperatures above 1500° F in dry hydrogen (–80° F dew point). Therefore, the brazing filler metal must have a liquidus above this temperature.

Oxides of aluminum and titanium cannot be reduced by hydrogen at ordinary brazing temperatures. If these elements are present in small amounts, satisfactory brazing can be done in gas atmospheres. When these elements are present in quantities exceeding 1 or 2 percent, the metal surface can be plated with a thin layer of pure metal that is easily cleaned by hydrogen, or a flux can be used in conjunction with the hydrogen. Copper plate is suitable with copper or silver filler metals. Nickel plate is generally used in conjunction with nickel- or cobalt-base filler metals.

Brazing can also be done in a pure inert gas or a vacuum environment. The mechanism of oxide removal in vacuum is not clear. It is likely that some oxides dissociate at elevated temperatures by diffusion of oxygen onto the base metal.

## STRESS CRACKING

Many nickel alloys in contact with molten brazing filler metal have a tendency to crack when in a highly stressed condition. Alloys with high annealing temperatures are subject to this stress cracking phenomenon, particularly those that are precipitation hardenable. The cracking occurs almost instantaneously during the brazing operation and is usually readily visible. The molten brazing filler metal will flow into the cracks and completely fill them.

The action is similar to stress corrosion cracking, and the molten filler metal is considered as the corrosive medium. Sufficient stress to cause cracking can be produced either by cold working prior to brazing or by an applied stress from mechanical or thermal sources during the brazing operation.

When stress cracking is encountered, its cause can usually be determined from a critical analysis of the brazing procedure. The usual remedy is to remove the source of stress. Stress cracking can be eliminated by one or more of the following actions:

(1) Use annealed material rather than cold worked stock.

(2) Anneal cold formed parts prior to brazing.

(3) Correct or remove the source of stresses from externally applied loads, such as improper fit of parts, high fixturing forces, or unsupported parts.

(4) Redesign the parts or revise the joint design.

(5) Heat at a slower rate.

(6) Select a brazing filler metal that is less likely to cause this type of damage.

The precipitation-hardenable nickel alloys are very susceptible to stress corrosion cracking. These alloys should be brazed in the annealed or solution heat-treated condition with a relatively high melting filler metal.

## POSTBRAZING THERMAL TREATMENT

It is frequently desirable to give brazed assemblies a postbrazing thermal treatment to improve mechanical properties. The treatment usually consists of heating to some intermediate temperature for a period of time followed by a controlled rate of cooling.

When a thermal treatment is performed subsequent to brazing, the brazed joint must have sufficient strength at temperature to withstand the necessary handling. Postbrazing thermal treatments may be a source of residual stresses in brazed joints. These stresses may cause microfissuring that may lead to premature failure in service.

# SOLDERING

## CONSIDERATIONS

Soldering can be employed to join nickel and high nickel alloys either to themselves or to any other solderable metal. Alloys containing chromium, aluminum, or titanium are more difficult to solder than are the other alloys. In designing a solder joint in a particular nickel alloy, any special characteristics of the base metal should be taken into consideration.

Many times, nickel alloys are used for a given application because of their resistance to corrosive attack. When this is a factor, the corrosion resistance of the solder must also be considered. In some cases, the joint should be located where the solder will not be exposed to the corrosive environment. The high tin solders, such as 95% tin-5% antimony, may produce a better color match, if that is important. However, the solder may eventually oxidize in a manner different from the base metal, and the joint may then be noticeable.

If soldering is to be done on a precipitation-hardenable alloy, it should be done after heat treatment. The temperatures involved in soldering will not soften precipitation-hardened parts.

Nickel alloys are subject to embrittlement at high temperatures when in contact with lead and many other low melting metals. This embrittlement will not take place at normal soldering temperatures. Overheating should be avoided. If welding, brazing, or other heating operations are to be done on an assembly, these operations must be done before soldering.

## SOLDERS

Any of the common types of solders may be used to join nickel alloys. A relatively high tin solder, such as the 60 percent tin – 40 percent lead or the 50 percent tin – 50 percent lead composition, is best for good wettability.

## SURFACE PREPARATION

Nickel and nickel alloys heated in the presence of sulfur become embrittled. These alloys should be clean and free from sulfur-bearing materials such as grease, paint, crayon, and lubricants before heating.

Joints with long laps and joints that will be inaccessible for cleaning after soldering should be precoated with solder prior to assembly. Precoating is generally done with the same alloy to be employed for soldering. The parts may be dipped in the molten solder or the surface may be heated, fluxed, and the solder flowed on. Excess solder may be removed by wiping or brushing the joint. Nickel alloys may also be precoated by tin plating or hot dipping.

## PROCESSES

Nickel alloys can be joined by any of the common soldering processes. Some minor differences in procedure may be required because of the low thermal conductivities of these alloys.

## FLUXES

Generally, rosin fluxes are not active enough to be used on the nickel alloys. A chloride flux is suitable for soldering nickel or the nickel-copper alloys. Fluxes containing hydrochloric acid are required for the chromium-containing alloys.

Many of the proprietary fluxes used for soldering stainless steel are satisfactory for use on nickel alloys.

## JOINT TYPES

The strength of soldered joints is low when compared to high nickel alloys, which have relatively high strength. Therefore, the strength of the joint should not depend on the solder alone. Lock seaming, riveting, spot welding, bolting, or other means should be employed to carry the structural load. The solder should be used only to seal the joint.

## POST TREATMENT

Because corrosive fluxes are required for soldering nickel alloys, the residue must be thoroughly removed after soldering. Residue from zinc chloride flux can be removed by first washing with a hot water bath containing 2 percent concentrated hydrochloric acid. Then the assembly should be rinsed with hot water containing some sodium carbonate, followed by a clean water rinse.

# THERMAL CUTTING

Nickel and cobalt alloys cannot be cut by conventional oxyfuel gas cutting techniques. Plasma arc cutting and air carbon arc gouging are normally used with nickel alloys.

## PLASMA ARC CUTTING

This process cuts rapidly and can produce good cut surfaces within a certain thickness range. Generally, low travel speeds with thick sections produce a rough surface.

A nitrogen-hydrogen mixture for the orifice gas is recommended for thickness up to 5 in. by

some equipment manufacturers. An argon-hydrogen mixture gives better cut surfaces for thicker sections. The recommendations of the equipment manufacturer should be consulted concerning orifice and shielding gas selections for the particular alloy. Conditions that may be acceptable for some nickel alloys are shown in Table 6.15.

## AIR CARBON ARC

This process may be used for gouging of nickel alloys to prepare a weld groove or to re-

## Table 6.15
## Suggested conditions for plasma arc cutting nickel alloys

| Alloy | Thickness, in. | Gas flow, ft³/h | Orifice gas | Power, kW | Speed, in./min |
|---|---|---|---|---|---|
| Commercially pure nickel | 1.5 | 220 | 85%$N_2$-15%$H_2$ | 95 | 26 |
| | 3 | 260 | 85%$N_2$-15%$H_2$ | 138 | 6 |
| | 6 | 150 | 65%Ar-35%$H_2$ | 104 | 5 |
| 67Ni-32Cu[a] | 2 | 270 | 85%$N_2$-15%$H_2$ | 155 | 35 |
| | 3 | 260 | 85%$N_2$-15%$H_2$ | 134 | 5 |
| 76Ni-16Cr-8Fe[b] | 1.75 | 220 | 85%$N_2$-15%$H_2$ | 95 | 25 |
| | 3 | 260 | 85%$N_2$-15%$H_2$ | 135 | 5 |
| | 6 | 150 | 65%Ar-35%$H_2$ | 103 | 5 |
| 73Ni-16Cr-7Fe-3Ti[c] | 2.5 | 270 | 85%$N_2$-15%$H_2$ | 148 | 20 |
| 46Fe-33Ni-21Cr[d] | 1.5 | 220 | 85%$N_2$-15%$H_2$ | 92 | 20 |
| | 3 | 260 | 85%$N_2$-15%$H_2$ | 163 | 5 |
| | 6 | 130 | 65%Ar-35%$H_2$ | 98 | 5 |

a.  Commonly known as Monel
b.  Commonly known as Inconel 600
c.  Commonly known as Inconel X750
d.  Commonly known as Incoloy 800

move a weld root or defective area. It may be used for cutting thin sections. Type AC carbon electrodes with ac or DCEN(SP) power are recommended.

The cutting conditions must be adjusted to ensure complete removal of all melted material from the cut surface. The molten metal may be carburized by carbon from the electrode. Its presence during a subsequent welding operation may cause undesirable metallurgical reactions in the weld metal.

## LASER BEAM

Nickel and cobalt alloys can be cut at high speeds with high-power laser beams. The power density of the focused beam permits cutting with a very narrow kerf, but there is a maximum thickness that can be cut economically with the process. The laser beam melts the metal, and a gas jet then blows the molten metal from the kerf. Clean cuts can be made with an inert gas jet.

# SAFE PRACTICES

Compounds of chromium, including hexavalent chromium, and of nickel may be found in fume from welding processes. The specific compounds and concentrations will vary with (1) the compositions of the base and filler metals, and (2) the welding processes. Immediate effects of overexposure to welding fumes containing chromium and nickel are similar to the effects produced by fumes of other metals. The fumes can cause symptoms such as nausea, headaches, and dizziness. Some persons may develop a sensitivity to chromium or nickel that can result in dermatitis or skin rash.

The fumes and gases should not be inhaled, and the face should be kept out of the fumes. Sufficient ventilation or exhaust at the arc, or both, should be used to keep fumes and gases from the welder's breathing zone and the general area. In some cases, natural air movement will provide enough ventilation. Where ventilation may be questionable, air sampling should be used to determine if corrective measures should be applied.

## Metric Conversion Factors

$t_C = 0.56 (t_F - 32)$
1 in. = 25.4 mm
1 in./min = 0.42 mm/s
1 cal/(cm•s• °C) = 418 W/(m•°K)
1μ in./(in.•°F) = 0.56μm/(m•°K)
1 ft³/h = 0.472 L/min
1 psi = 6.89 kPa
1 ksi = 6.89 MPa
1 lbf = 4.45 N
1 lb/in.³ = 2.77 × 10⁴ kg/m³

# SUPPLEMENTARY READING LIST

Amato, I., et al. 1972. Spreading and aggressive effects by nickel-base brazing filler metals on the Alloy 718. *Welding Journal* 15(7): 341s-45s; July.

Christensen, J. and Rørbo, K. 1974. Nickel brazing below 1025°C of untreated Inconel 718. *Welding Journal* 53(10): 460s-64s; Oct.

Coffee, D. L., et al. 1972. A welding defect related to the aluminum content in a nickel-base alloy. *Welding Journal* 51(1): 29s-30s; Jan.

Dix, A. W. and Savage, W. F. 1973. Short time aging characteristics of Inconel X-750. *Welding Journal* 52(3): 135s-39s; March.

Dix, A. W. and Savage, W. F. 1971. Factors influencing strain-age cracking in Inconel X-750. *Welding Journal* 50(6): 247s-52s; June.

Duvall, D. S. et al. 1974. TLP bonding: a new method for joining heat resistant alloys. *Welding Journal* 53(4): 203-14; April.

Eng, R. D., et al. 1977. Nickel-base brazing filler metals for aircraft gas turbine applications. *Welding Journal* 56(10): 15-21; Oct.

Kenyon, N., et al. 1975. Electroslag welding of high nickel alloys. *Welding Jounal* 54(7): 235s-39s; July.

Matthews, S. J. 1979. Simulated heat-affected zone studies of Hastelloy B-2. *Welding Jour-nal* 58(3): 91s-95s; Mar.

Mayor, R. A. 1976. Selected mechanical properties of Inconel 718 and 706 weldments. *Welding Journal* 55(9): 269s-75s; Sept.

Moore, T. J. 1974. Solid-state and fusion resistance spot welding of TD-NiCr sheet. *Welding Journal* 53(1): 37s-48s; Jan.

Prager, M. and Shira, C. S. 1968. *Welding of Precipitation-Hardening Nickel-Base Alloys.* New York: Welding Research Council, Bulletin 128; Feb.

Savage, W. F., et al. 1977. Effect of minor elements on hot cracking tendencies of Inconel 600. *Welding Journal* 56(8): 245s-53s; Aug.

Savage, W. F., et al. 1977. Effect of minor elements on fusion zone dimensions of Inconel 600. *Welding Journal* 56(4): 126s-32s; April.

Weiss, B. Z., et al. 1979. Static and dynamic crack toughness of brazed joints of Inconel 718 nickel-base alloy. *Welding Journal* 58(10): 287s-95s; Oct.

Welding Research Council. 1970. *Recent Studies of Cracking During Postweld Heat Treatment of Nickel-Base Alloys.* New York: Bulletin 150; May.

Yoshimura, H. and Winterton, K. 1972. Solidification mode of weld metal in Inconel 718. *Welding Journal* 51(3): 132s-38s; March.

# 7

# Copper Alloys

## Chapter Committee

M. PRAGER, *Chairman*
*Consultant*
C.W. DRALLE
*Ampco Metals*
C.J. GAFFOGLIO
*Copper Development Association*
K.G. WOLD
*Aqua-Chem, Incorporated*

S. GOLDSTEIN
*Kawecki-Berylco Industries, Incorporated*
P.W. HOLSBERG
*Department of the Navy*
W.V. WATERBURY
*Revere Copper and Brass, Incorporated*

## Welding Handbook Committee Member

I.G. BETZ
*Department of the Army*

# 7

# Copper Alloys

## PROPERTIES

Copper has a face centered cubic crystal structure, as do most of its alloys. This accounts for their good formability and malleability. Copper has a density of 0.32 lb/in.³, about three times that of aluminum. Its electrical and thermal conductivities are slightly lower than those of silver, but about 1.5 times those of aluminum. However, alloying significantly lowers these conductivities.

Copper is highly resistant to oxidation, fresh and salt water, ammonia-free alkaline solutions, and many organic chemicals. However, copper reacts readily with sulfur and its compounds to produce copper sulfide. Copper and its alloys are widely used for electrical conductors, water tubing, valves and fittings, heat exchangers, and chemical equipment.

## ALLOYS

Fourteen common alloying elements are added to copper, mostly within the limits of solid solubility. Most commercial alloys are solid solutions (single phase) and show no allotropic or crystallographic changes on heating and cooling. Some of those alloys are precipitation hardenable in the same manner as aluminum. Other alloys are composed of two phases; some of these alloys are heat treatable.

The principal alloying elements in copper alloys are aluminum, nickel, silicon, tin, and zinc. Small additions of other elements are added to the major alloy families to:
(1) Improve mechanical or corrosion properties
(2) Provide response to strengthening heat treatments
(3) Deoxidize the melt
(4) Improve machineability

### CLASSIFICATION

Copper and copper alloys are classified into eight major groups as follows:
(1) Coppers
(2) High copper alloys
(3) Brasses (Cu-Zn)
(4) Bronzes (Cu-Sn)
(5) Copper-nickels (Cu-Ni)
(6) Copper-nickel-zinc alloys (nickel silvers)
(7) Leaded coppers
(8) Special alloys

These groups are further divided into families consisting of wrought or cast alloys, as shown in Table 7.1. There are a number of specific alloys within each family. Some coppers and copper alloys have commonly used names, such as oxygen-free copper, beryllium copper, Muntz metal, and low fuming bronze.

270

**Table 7.1**
**Classification of copper and copper alloys**

| Family | Description | Range of UNS Numbers[a] |
|---|---|---|
| | Wrought alloys[b] | |
| Copper | Copper–99.3 percent minimum | C10100-C15735 |
| High copper alloys | Copper–96 to 99.2 percent | C16200-C19600 |
| Brasses | Copper-zinc alloys | C20500-C28200 |
| Leaded brasses | Copper-zinc-lead alloys | C31400-C38600 |
| Tin brasses | Copper-zinc-tin alloys | C40500-C48500 |
| Phosphor bronzes | Copper-tin alloys | C50100-C52400 |
| Leaded phosphor bronzes | Copper-tin-lead-alloys | C53200-C54800 |
| Aluminum bronzes | Copper-aluminum alloys | C60600-C64400 |
| Silicon bronzes | Copper-silicon alloys | C64700-C66100 |
| Miscellaneous brasses | Copper-zinc alloys | C66400-C69910 |
| Copper-nickels | Nickel–3 to 30 percent | C70100-C72500 |
| Nickel silvers | Copper-nickel-zinc alloys | C73200-C79900 |
| | Cast alloys[c] | |
| Coppers | Copper–99.3 percent minimum | C80100-C81100 |
| High copper alloys | Copper–94 to 99.2 percent | C81300-C82800 |
| Red brasses | | C83300-C83800 |
| Semi-red brasses | Copper-tin-zinc and copper-tin-zinc-lead alloys | C84200-C84800 |
| Yellow brasses | | C85200-C85800 |
| Manganese bronze | Copper-zinc-iron alloys | C86100-C86800 |
| Silicon bronzes and silicon brasses | Copper-zinc-silicon alloys | C87200-C87900 |
| Tin bronzes | Copper-tin-alloys | C90200-C91700 |
| Leaded tin bronzes | Copper-tin-lead alloys | C92200-C94500 |
| Nickel-tin bronzes | Copper-tin-nickel alloys | C94700-C94900 |
| Aluminum bronzes | Copper-aluminum-iron and copper-aluminum-iron-nickel alloys | C95200-C95800 |
| Copper-nickels | Copper-nickel-iron alloys | C96200-C96600 |
| Nickel silvers | Copper-nickel-zinc alloys | C97300-C97800 |
| Leaded coppers | Copper-lead alloys | C98200-C98800 |
| Special alloys | | C99300-C99750 |

a. Refer to SAE HS 1068a/ASTM DS-65A, *Unified Numbering System for Metals and Alloys*, 2nd Ed., Warrendale, PA: Society of Automotive Engineers, 1977 Sept. These numbers are expansions of the original three-digit numbering system for copper alloys.
b. For composition and properties, refer to *Standards Handbook, Part 2–Alloy Data, Wrought Copper and Copper Alloy Mill Products*, 2nd Ed., New York: Copper Development Assoc., Inc., 1968.
c. For composition and properties, refer to *Standards Handbook, Part 7–Data/Specifications, Cast Copper and Copper Alloy Products*, New York: Copper Development Assoc., Inc., 1970.

Physical properties of copper alloys important to welding, brazing, and soldering include melting temperature range, coefficient of thermal expansion, and thermal and electrical conductivities. These properties for widely used alloys are given in Table 7.2. The electrical and thermal conductivities significantly affect the heat input required for welding. They decrease rapidly when copper is alloyed with other metals.

## METALLURGY

Many copper alloys are solid solutions (single phase) while other alloys have two or more microstructural phases. Some multiple phase alloys can be hardened by precipitation of intermetallic compounds or by quenching from above a critical temperature (transformation hardening). In the latter case, hardening results from a

## Table 7.2
## Physical properties of typical wrought copper alloys

| Alloy | UNS No. | Melting range, °F | Coef of thrm exp, per °F (68° to 572°F) | Thermal conductivity, Btu/(ft·h·°F) | Electrical conductivity, percent, IACS |
|---|---|---|---|---|---|
| Oxygen free copper | C10200 | 1948-1991 | $9.8 \times 10^{-6}$ | 226 | 101 |
| Beryllium copper | C17200 | 1590-1800 | 9.9 | 62-75 | 22 |
| Commercial bronze, 90% | C22000 | 1870-1910 | 10.2 | 109 | 44 |
| Red brass, 85% | C23000 | 1810-1880 | 10.4 | 92 | 37 |
| Cartridge brass, 70% | C26000 | 1680-1750 | 11.1 | 70 | 28 |
| Phosphor bronze, 5%A | C51000 | 1750-1920 | 9.9 | 40 | 15 |
| Phosphor bronze, 10%D | C52400 | 1550-1830 | 10.2 | 29 | 11 |
| Aluminum bronze, D | C61400 | 1905-1915 | 9.0 | 39 | 14 |
| High silicon bronze, A | C65500 | 1780-1880 | 10.0 | 21 | 7 |
| Manganese bronze, A | C67500 | 1590-1630 | 11.8 | 61 | 24 |
| Copper-nickel, 10% | C70600 | 2010-2100 | 9.5 | 26 | 9 |
| Copper-nickel, 30% | C71500 | 2140-2260 | 9.0 | 17 | 4.6 |
| Nickel silver, 65-15 | C75200 | 1960-2030 | 9.0 | 19 | 6 |

martensitic type transformation. Some useful generalizations may be applied to copper alloys, based upon their common physical metallurgy.

First, solid-solution alloys are easily cold worked, although the rate of work hardening increases with increasing alloy content. However, their hot working characteristics deteriorate with increasing alloy content. Two-phase alloys harden rapidly during cold working, but they usually have better hot working and welding characteristics than do solid solutions of the same alloy system. Ductility usually decreases and yield strength increases as the proportion of second phase in the alloy increases.

Small additions of some elements often improve the corrosion resistance of copper alloys. Examples are iron, silicon, tin, arsenic, and antimony. Lead, selenium, and tellurium improve machinability, but they adversely affect hot working and weldability. Boron, phosphorus, silicon, and lithium are added as deoxidizers; silver and cadmium increase the resistance of copper to softening at soldering temperatures. Cadmium, cobalt, zirconium, chromium, and beryllium additions can produce high strength alloys as a result of (1) precipitation of a second phase during heat treatment, (2) a combination of cold working and precipitation hardening, or (3) rapid work hardening.

## MAJOR ALLOYING ELEMENTS

### Aluminum

The solubility of aluminum in copper is approximately 7.8 percent at elevated temperatures. The aluminum bronzes contain up to 15 percent aluminum as well as minor additions of iron, nickel, tin, and manganese. Alloys with less than 8 percent aluminum are single phase, with or without iron additions. When the aluminum content is between 9 and 15 percent, the system is two-phase and capable of either a martensitic or a eutectoid type of transformation. A refractory oxide forms on the surface of the metal, and it must be removed for soldering, brazing, and welding.

### Arsenic

Arsenic additions to copper alloys do not cause problems in welding unless the alloys also contain nickel. Together with nickel, it is detrimental. Arsenic is added to inhibit dezincification corrosion of brasses in salt water.

### Beryllium

The solubility of beryllium is approximately 2 percent at 1600°F and only 0.3 percent at room temperature. It easily forms a supersaturated solution with copper, from which it will precipitate in an age-hardening reaction.

## Boron

Boron in copper acts as a strengthener and deoxidizer. Boron deoxidized coppers are weldable with matching filler metals, and other coppers are weldable with boron-containing fillers.

## Cadmium

The solubility of cadmium in copper is approximately 0.5 percent at room temperature. The presence of cadmium in copper up to 1.25 percent causes no serious difficulty in fusion welding. It evaporates from copper rather easily at the welding temperature. A small amount of cadmium oxide may form in the molten metal but it can be fluxed without difficulty.

## Chromium

The solubility of chromium in copper is approximately 0.55 percent at 1900°F and less than 0.05 percent at room temperature. The phase that forms during precipitation hardening is almost pure chromium. Like aluminum and beryllium, chromium can form a refractory oxide on the molten weld pool, and this makes oxyacetylene welding difficult unless special fluxes are used. Arc welding should be done using a protective atmosphere over the molten weld pool.

## Iron

Solubility of iron is approximately 3 percent at 1900°F and less than 0.1 percent at room temperature. Iron is added to aluminum bronze, manganese bronze, and copper-nickel alloys to increase strength by solid solution and precipitation hardening. It also acts as a grain refiner. It has little effect on weldability when used within the alloy specification limits.

## Lead

Lead is the most harmful element with respect to the weldability of copper alloys. It is almost completely insoluble in copper at room temperature, and is added only to improve machineability or bearing properties. Leaded copper alloys are hot-short and will crack during fusion welding.

## Manganese

Manganese is highly soluble in copper. It is used in proportions of 0.05 to 1.25 percent in commercial copper alloys, including manganese bronze, deoxidized copper and copper-silicon alloys. Manganese additions are not detrimental to the weldability of copper alloys.

## Nickel

Copper and nickel have complete solid solubility in all proportions. The copper-nickel alloys are easily welded by most methods but are subject to embrittlement and hot cracking by many residual elements. With copper-nickel alloys, there must be sufficient deoxidizer or desulfurizer in the filler metal to provide a residual amount in the solidified weld metal. Manganese is most often used for this purpose. Zinc serves this function in the nickel silvers.

## Phosphorus

Phosphorus is beneficial to certain coppers and copper alloys as a strengthener and deoxidizer. It is soluble in copper up to 1.7 percent at the eutectic temperature of 1200°F, and approximately 0.4 percent at room temperature. When added to brass, phosphorus inhibits dezincification. It does not adversely affect weldability in the amounts normally present in copper alloys.

## Silicon

The solubility of silicon in copper is 5.3 percent at 1500°F and 3.6 percent at room temperature. Silicon is used both as a deoxidizer and as a major alloying element to improve strength, malleability, and ductility. Copper-silicon alloys have good weldability, but they tend to be hot-short at elevated temperatures. In welding, the cooling rate through this hot-short temperature range should be fast to prevent cracking.

Silicon oxide forms on copper-silicon alloys at temperatures as low as 400°F. It will interfere with brazing and soldering operations unless a suitable flux is applied prior to heating. When the oxide film forms on the molten weld pool during fusion welding, it protects the molten metal from further oxidation.

## Tin

The solubility of tin in copper increases rapidly with temperature. At 1450°F, the solubility is 13.5 percent; at room temperature, it is probably less than 1 percent. Alloys containing less than 2 percent tin may be single phase when cooled rapidly.

In welding operations, the tin oxidizes preferentially to the copper when exposed to the atmosphere. The strength of the weld may be reduced because of oxide entrapment in the weld metal. Tin oxide is, however, easily fluxed. Copper-tin alloys tend to be hot-short and to crack during fusion welding.

## Zinc

Zinc is soluble in copper up to 32.5 percent at 1700°F and 37 percent at room temperature. Zinc is the most important alloying element used commercially with copper. A characteristic of all copper-zinc alloys is the relative ease with which zinc will evaporate from the molten metal with very slight superheat.

In oxyacetylene welding, very little zinc oxide appears on the molten weld metal unless a strongly oxidizing flame is used. Then, an oxide film forms on the molten weld metal and suppresses evaporation of zinc, provided the weld metal is not overheated. Brass oxyacetylene welding rods contain sufficient zinc (38 to 41 percent) to develop in the weld metal a considerable proportion of the hard, strong beta phase. This beta phase is soft and ductile at elevated temperatures. Hence, cracking is not a problem.

Zinc does not cause an abrupt decrease in electrical and thermal conductivities of copper, as do many other alloying elements. The brasses, therefore, require considerable heat when being welded, especially those with zinc contents of 5 to 15 percent.

## MINOR ALLOYING ELEMENTS

Calcium, magnesium, lithium, sodium, or combinations of these may be added to copper alloys as deoxidizers. However, little if any remains in the copper alloy. They are rarely encountered in welding metallurgy. Antimony is used to some extent to raise the annealing temperature of copper, but in amounts so small as to have little influence on weldability. However, antimony, arsenic, phosphorus, bismuth, selenium, sulfur, and tellurium may cause hot cracking in copper alloys containing nickel.

Carbon is practically insoluble in copper alloys unless large amounts of iron, manganese, or other carbide formers are present. Carbon embrittles copper alloys by precipitating in the grain boundaries as graphite.

## WELDABILITY FACTORS

The high electrical and thermal conductivities of copper and certain high copper alloys have a marked effect on weldability. Copper, aluminum bronze, commercial bronze (90 percent), and red brass (85 percent) have high thermal conductivities. Welding heat is quickly conducted into the base metal. This may cause lack of fusion in weldments. Preheating of these alloys will reduce welding heat requirements for good fusion.

Many copper alloys are hardened by cold working. The application of heat in any quantity will soften them. After welding, the heat-affected zone will be softer and weaker than the adjacent base metal. This zone may also tend to hot crack with some cold worked alloys.

The most important precipitation hardening reactions in copper alloys are obtained with beryllium, chromium, boron, nickel-silicon, and zirconium. These alloy additions produce most of the alloys that are classified as precipitation hardenable. Care must be taken when welding the precipitation hardenable copper alloys to avoid oxidation and incomplete fusion. Wherever possible, the components should be welded in the annealed condition, and then the weldment given a precipitation hardening heat treatment.

Copper alloys with wide liquidus-to-solidus temperature ranges, such as copper-tin and copper-nickel, are susceptible to hot cracking at solidification temperatures. Low melting interdendritic liquid solidifies at a lower temperature than the bulk dendrite. Shrinkage stresses produce interdendritic separation during solidification. Hot cracking can be minimized by (1) reducing restraint during welding, (2) preheating to retard the cooling rate and reduce the magnitude of the welding stresses, and (3) reducing the size of the root opening and increasing the size of the root pass.

Certain elements such as zinc, cadmium, and phosphorus have low boiling points. Vaporization of these elements during welding will result in porosity. When welding copper alloys containing these elements, porosity can be minimized by fast weld speeds and use of filler metal low in these elements.

Surface oxides can cause a serious problem during welding, brazing, and soldering. The oxides on aluminum bronze, beryllium copper, chromium copper, and silicon bronze are difficult to remove. The surfaces to be joined must be clean, and special fluxing or shielding methods must be used to prevent the film from reforming during the joining operation.

# COPPER AND HIGH COPPER ALLOYS

## Oxygen-Bearing Coppers

The oxygen-bearing coppers include the fire-refined and the electrolytic tough-pitch grades. Fire-refined copper contains varying amounts of impurities including antimony, arsenic, bismuth, and lead. These impurities and the residual oxygen may cause porosity and other discontinuities during welding and brazing of this copper. Electrolytic tough-pitch copper contains less impurities and, as a result, it has more uniform mechanical properties. The residual oxygen content is about the same as that of fire-refined copper.

A copper-cuprous oxide eutectic is distributed as globules throughout wrought forms of these coppers. However, it has no serious effects on mechanical properties or electrical conductivity. However, it does make the copper susceptible to embrittlement when it is heated in the presence of hydrogen. Hydrogen can rapidly diffuse into the hot metal and reduce the oxides to form steam at the grain boundaries of the copper. This, in turn, will rupture the metal.

When oxygen-bearing coppers are heated to high temperatures, the copper oxide concentrates in the grain boundaries and causes major reductions in strength and ductility. Therefore, fusion welding of these coppers for structural applications is not recommended. Embrittlement will be less severe with a rapid solid-state welding process such as friction welding or with appropriate brazing procedures and filler metals.

## Oxygen-Free Coppers

Phosphorus-deoxidized copper has 0.004 to 0.04 percent residual phosphorus. Oxygen-free copper is produced by melting and casting under atmospheres that prevent oxygen contamination. Since no deoxidizing agent is introduced into this type of copper, it can absorb oxygen from the atmosphere during long heating periods at very high temperatures. Absorbed oxygen can cause problems during subsequent welding or brazing of the copper.

The oxygen-free coppers have mechanical properties similar to those of the oxygen-bearing coppers, but their microstructures are more uniform. They have excellent ductility and resistance to fatigue loading, and can be joined readily by welding, brazing, and soldering. Silver is sometimes added to oxygen-free copper to increase the elevated temperature strength properties with no change in electrical conductivity. This has no effect on the joining characteristic of copper.

## Free-Machining Coppers

Copper has very low solid solubility for lead, tellurium, and selenium. When added to copper, lead will disperse throughout the matrix as fine, discrete particles. Tellurium and sulfur will form hard stringers in the matrix. These inclusions reduce the ductility of copper but enhance its machinability. Free-machining coppers are difficult to fusion weld without cracking because of the inclusions in the metal. However, these coppers can be joined by brazing and soldering.

## Precipitation-Hardenable Coppers

When alloyed with small amounts of beryllium, chromium, or zirconium, the mechanical properties of copper can be significantly increased by a precipitation hardening heat treatment. The coppers are first solution heat treated to a soft condition by quenching from a high temperature. They are then precipitation hardened by aging at a moderate temperature. Sometimes, solution heat-treated alloys are cold worked prior to artificial aging. During aging, a second phase precipitates within the matrix and inhibits plastic deformation. Exposure of precipitation hardened alloys to high welding or brazing temperatures will soften them with a resultant decrease in mechanical properties in the weld zone.

# COPPER-ZINC ALLOYS (BRASSES)

Copper alloys in which zinc is the major alloying element are generally called brasses. However, some alloys are known by other names, such as commercial bronze, Muntz metal, manganese bronze, and low fuming bronze. Other elements are occasionally added to brasses to enhance particular mechanical or corrosion characteristics. Special brasses are identified by the addition of another element; two examples are aluminum and tin brasses. Additions of manganese, tin, iron, silicon, nickel, lead, or aluminum, either singly or collectively, rarely exceed 4 percent.

The addition of zinc to copper decreases the melting temperature, the density, the electrical and thermal conductivities, and the modulus of elasticity. It increases the strength, hardness, ductility, and coefficient of thermal expansion. The hot working properties decrease with increasing zinc content up to about 20 percent.

The color of brass also changes with zinc content from reddish brown through a tan gold to a light brown gold. This will affect the selection of filler metals when joint appearance is important.

Most brasses are single-phase, solid-solution alloys with good room temperature ductility. Brasses containing 40 percent or more of zinc are composed of two microstructural phases called alpha and beta. The latter phase improves the hot working characteristics of brass, but has little effect on electrical and thermal conductivities.

For joining considerations, brasses may be divided into three groups: (1) low zinc brasses with a maximum zinc content of 20 percent, (2) high zinc brasses containing more than 20 percent zinc, and (3) leaded brasses. The fusion welding characteristics of all alloys in one group are quite similar. The leaded brasses are considered unweldable by fusion but they can be brazed and soldered.

## COPPER-TIN ALLOYS (PHOSPHOR BRONZES)

Alloys of coppr and tin are known commercially as phosphor bronzes because 0.03 to 0.04 percent phosphorus is added during casting as a deoxidizing agent. Commercial phosphor bronzes contain from 1 to 10 percent tin and 0.03 percent residual phosphorus. When homogenized, they are single phase alloys. Phosphor bronzes have relatively high tensile strength, depending upon the tin content and the amount of cold work.

Phosphor bronzes are tough, hard, and fatigue resistant, particularly in the cold-worked condition. The copper-tin phase diagram suggests precipitation of an intermetallic compound at room temperature. However, this is not normally observed in the microstructure. Electrical and thermal conductivities decrease with increasing tin content. In a stressed condition, the phosphor bronzes are subject to hot cracking. Therefore, they should be stress relieved or annealed prior to welding.

## COPPER-ALUMINUM ALLOYS (ALUMINUM BRONZES)

Copper-aluminum alloys, called aluminum bronzes, contain from 3 to 15 percent aluminum,

with or without varying amounts of iron, nickel, manganese, and silicon. There are two types of aluminum bronzes based on metallurgical structure and response to heat treatment. The first includes the alpha or single-phase alloys that cannot be hardened by heat treatment. The second type includes the two-phase, alpha-beta alloys. Both types have low electrical and thermal conductivities which enhance weldability.

The alpha aluminum bronzes are readily weldable without preheating. When the aluminum content is below approximately 8.5 percent, the alloys have a tendency to be hot-short, and cracking may occur in the heat-affected zone of highly stressed weldments.

Generally, aluminum bronzes containing from 9.5 to 11.5 percent aluminum have both alpha and beta phases in their microstructures. These two-phase alloys can be strengthened by quenching to produce a martensitic type structure and then tempering in much the same manner as a hardenable steel. Structures found after heat treatment are analogous in many respects to those found in steels. Hardening is generally done by quenching from 1550° to 1850°F in water or oil, followed by tempering in the range of 800° to 1200°F. The specific heat treatment depends upon the composition of the alloy and the desired mechanical properties.

The two-phase alloys have very high tensile strengths compared to other copper alloys. As the aluminum content of these alloys increases, their ductility decreases and their hardness increases. They have a wider plastic range than the alpha alloys, which contributes to their good weldability.

The aluminum bronzes resist oxidation and scaling at elevated temperatures due to the formation of aluminum oxide on the surface. The oxide must be removed prior to welding, brazing, or soldering.

## COPPER-SILICON ALLOYS (SILICON BRONZES)

Copper-silicon alloys, known as silicon bronzes, are industrially important because of their high strength, excellent corrosion resistance, and good weldability. Generally, they contain from 1.5 to 4 percent silicon and 1.5 percent or less of zinc, tin, manganese or iron.

The addition of silicon to copper increases tensile strength, hardness, and work hardening rates. The ductility of silicon bronze decreases with increasing silicon content up to about 1 percent. Then ductility increases to a maximum value at 4 percent silicon. However, electrical and thermal conductivities decrease as the silicon content increases. Silicon bronzes should be stress relieved or annealed prior to welding, and heating to temperature should be slow. These bronzes are hot-short and require rapid cooling through the critical temperature range.

The addition of iron increases tensile strength and hardness. The addition of zinc or tin improves the fluidity of the molten bronze, which improves the quality of castings and weld metal deposited by oxyacetylene welding.

## COPPER-NICKEL ALLOYS

The commercial copper-nickel alloys have nickel contents ranging from 5 to 30 percent. The alloys most commonly used in welded fabrication contain 10 and 30 percent nickel. They may also contain minor alloying additions including iron, manganese, or zinc. Resistance to erosion type of corrosion requires that the iron be in solid solution. The thermal processing of the alloy must be done in a manner that does not cause precipitation of iron compounds.

The copper-nickel alloys have moderately high tensile strengths that increase with nickel content. They are ductile, relatively tough and their electrical and thermal conductivities are low.

Like nickel and some nickel alloys, the copper-nickel alloys are susceptible to lead or sulfur embrittlement. For welding, the phosphorus and sulfur levels in these alloys are normally held to a maximum of 0.02 percent. Contamination from sulfur-bearing marking crayons or cutting lubricants will cause cracking during welding.

## COPPER-NICKEL-ZINC ALLOYS (NICKEL SILVERS)

Nickel is added to copper-zinc alloys to make them silvery in appearance for decorative purposes and to increase their strength and corrosion resistance. The resulting alloys are called nickel silvers. These alloys are of two general types: (1) single phase alloys containing 65 percent copper plus nickel and zinc, and (2) two-phase (alpha-beta) alloys containing 55 to 60 percent copper plus nickel and zinc. The welding metallurgy of these alloys is similar to that of the brasses.

# JOINING PROCESS SELECTION

## ARC WELDING

Copper and most copper alloys can be joined by arc welding. The processes that use inert gas shielding are generally preferred although shielded metal arc welding works well for many noncritical applications.

Argon, helium, or mixtures of the two are used for shielding with gas tungsten arc (GTAW), plasma arc (PAW), and gas metal arc welding (GMAW). In general, argon is used when manually welding material less than 0.13-inch thick. Helium or a mixture of 75 percent helium–25 percent argon is recommended for machine welding of thin sections and for manual welding of sections 0.13-inch thickness and greater. However, helium or a helium-argon mixture may be needed for manual welding thin sections of high conductivity coppers to take advantage of the high arc energy.

Covered electrodes for shielded metal arc welding copper alloys are available in standard sizes ranging from 3/32 to 3/16 inch. Other sizes are available in certain electrode classifications. The process can be used to weld a range of thicknesses of various copper alloys with the appropriate electrode size and alloy.

Welding is best done in the flat position with all of the arc welding processes. Gas tungsten arc or shielded metal arc welding is preferred for welding in positions other than flat, particularly in the overhead position. Gas metal arc welding with pulsed power and small diameter electrodes may be suitable for the vertical and overhead positions with some copper alloys.

## OXYACETYLENE WELDING

Copper and many copper alloys can be oxyacetylene welded. The relatively low heat input of the oxyacetylene flame makes welding relatively slow compared to arc welding. Higher preheat temperatures or an auxiliary heat source may be required to counterbalance the low heat input, particularly with alloys with high thermal conductivities or with thick sections.

The oxyacetylene flame offers little protection of the weld zone from the atmosphere. Except for oxygen-free copper, a welding flux is generally required to exclude air from the weld metal at elevated temperatures.

This process should only be used for small, noncritical applications, including repair welding, when arc welding equipment is not available.

## RESISTANCE WELDING

Copper and high copper alloys are very difficult to join by resistance spot and seam welding because of their high electrical and thermal conductivities. Very high current densities would be required for spot and seam welding and this would cause electrode overheating, sticking to the work, and rapid deterioration. Therefore, joining of these materials by spot and seam welding is not recommended. However, they can be joined by flash welding.

Copper alloys with relatively low electrical and thermal conductivities may be spot, seam, and flash welded. The procedures are similar to those used for aluminum.

## BRAZING

Copper and its alloys are readily joined by brazing using an appropriate filler metal and flux

or protective atmosphere. Any of the common heating methods can be used. Certain precautions are required with specific base metals to avoid embrittlement, cracking, or excessive alloying with the filler metal. Silver alloy and copper-phosphorus filler metals are commonly used for brazing copper and its alloys.

## SOLDERING

Copper and most copper alloys are readily soldered with lead solders. Most copper alloys are easily fluxed, except for beryllium copper and aluminum and silicon bronzes. Special fluxes are required with these alloys to remove the refractory oxides that form on their surfaces.

Soldering is primarily used for electrical connections, plumbing, automotive radiators, and other room temperature applications. Joint strengths are much lower than those of brazed or welded joints.

## OTHER PROCESSES

Copper and many copper alloys can be joined by the following processes for special applications:

(1) Electron beam welding
(2) Friction welding
(3) Cold welding
(4) Diffusion welding
(5) Hot pressure welding
(6) Ultrasonic welding
(7) Laser beam welding
(8) High-frequency resistance welding

# WELDING

## FILLER METALS

Covered and bare welding electrodes and bare rods are available for welding copper and copper alloys to themselves and to other metals. Many of these filler metals meet the AWS Classifications listed in Table 7.3. The copper alloy base metals that may be welded with the filler metals in each classification are also given.

### Copper

Bare copper electrodes and rods (ERCu) are

generally produced with a minimum copper content of 98 percent for gas metal arc welding of deoxidized and electrolytic tough pitch copper. The weld metal electrical conductivity is about 25 to 40 percent of the conductivity of electrolytic copper. Covered electrodes (ECu) for shielded metal arc welding are generally manufactured with deoxidized copper core wire.

Most copper electrodes are designed for welding with direct current, electrode positive (dcep). The welding current for these electrodes should be 30 to 40 percent higher than that nor-

## Table 7.3
## Filler metals for fusion welding copper alloys

| AWS Classification | | Common name | Base metal applications |
|---|---|---|---|
| Covered electrode [a] | Bare wire [b] | | |
| ECu | ERCu | Copper | Coppers |
| ECuSi | ERCuSi-A | Silicon bronze | Silicon bronzes, brasses |
| ECuSn-A }<br>ECuSn-C } | ERCuSn-A | Phosphor bronze | Phosphor bronzes, brasses |
| ECuNi | ERCuNi | Copper-nickel | Copper-nickel alloys |
| ECuAl-A2 | ERCuAl-A2 | Aluminum bronze | Aluminum bronzes, brasses, silicon bronzes, manganese bronzes |
| ECuAl-B | ERCuAl-A3 | Aluminum bronze | Aluminum bronzes |
| ECuNiAl | ERCuNiAl | – | Nickel-aluminum bronzes |
| ECuMnNiAl | ERCuMnNiAl | – | Manganese-nickel-aluminum bronzes |
| | RBCuZn-A | Naval brass | Brasses, copper |
| | RCuZn-B | Low fuming brass | Brasses, manganese bronzes |
| | RCuZn-C | Low fuming brass | Brasses, manganese bronzes |

a. See AWS A5.6-76, *Specification for Copper and Copper Alloy Covered Electrodes*.
b. See AWS A5.7-77, *Specification for Copper and Copper Alloy Bare Welding Rods and Electrodes* or AWS A5.27-78, *Specification for Copper and Copper Alloy Gas Welding Rods*.

mally required for carbon steel electrodes of the same diameter.

Welding rods (ERCu) are used almost exclusively with gas tungsten arc or plasma arc welding, but sometimes with oxyacetylene welding.

## Copper-Zinc (Brass)

The copper-zinc welding rods are comprised of the following classifications: RBCuZn-A (naval brass), RCuZn-B (low-fuming brass), and RCuZn-C (low fuming brass). The RBCuZn-A welding rods contain 1 percent tin to improve corrosion resistance and strength. Electrical conductivity of these rods is about 25 percent that of electrolytic copper, and the thermal conductivity is about 30 percent that of copper. Due to the high zinc content, it is very difficult to use this type of welding rod with the gas tungsten arc process. The composition is primarily used for oxyacetylene welding of brass and also for braze welding copper, bronze, and nickel alloys.

The RCuZn-B welding rods are similar to naval brass but contain additions of manganese, iron, and nickel that increase hardness and strength. A small amount of silicon provides low fuming characteristics. The properties of this rod are quite similar to the properties of naval brass.

The RCuZn-C welding rods are similar to RCuZn-B rods in composition except that they do not contain nickel. The mechanical properties of as-deposited weld metal from both rods are about the same.

Copper-zinc filler metals cannot be used as electrodes for arc welding because of the low melting temperature and high vapor pressure of zinc. The zinc vapor would boil from the molten weld pool, and the weld metal would be porous.

## Copper-Tin (Phosphor Bronze)

Copper-tin welding electrodes and rods are usually referred to as phosphor bronze. The CuSn-A composition normally contains about 5 percent tin, and the CuSn-C composition usually has about 8 percent tin. Both electrodes are deoxidized with phosphorus. The electrodes can be used for welding bronze, brass, and also for copper if the presence of tin in the weld metal is not objectionable. The ECuSn-C electrodes provide weld metal with better strength and hardness than do ECuSi-A electrodes. The ECuSn-C electrodes are preferred for welding high strength bronzes. When welding with these electrodes, a preheat and interpass temperature of about 400°F is required, especially if the weldment is made up of heavy sections. ERCuSn-A rods can be used with the gas tungsten arc welding process for joining phosphor bronze.

## Copper-Silicon (Silicon Bronze)

Copper-silicon electrodes are often referred to as silicon bronze. Most of these electrodes are used in bare wire form (ERCuSi-A) for gas metal arc welding. They contain from 2.8 to 4.0 percent silicon with about 1.5 percent manganese, 1.5 percent tin, and 1.5 percent zinc. Copper-silicon (ERCuSi-A) welding rods are used with the gas tungsten arc process and sometimes with oxyacetylene welding. The tensile strength of copper-silicon weld metal is about twice that of copper weld metal. The electrical conductivity is about 6.5 percent of copper, and the thermal conductivity is about 8.4 percent of copper.

ECuSi covered electrodes are used primarily for welding copper-zinc alloys. However, they are occasionally used for welding silicon bronze and copper. The core wire of covered electrodes contains about 3 percent silicon and small amounts of tin and manganese. The mechanical properties of the weld metal are usually slightly higher than those of copper-silicon base metal. Direct current electrode positive is recommended for welding with all copper-silicon electrodes.

## Copper-Aluminum (Aluminum Bronze)

ECuAl-A2 covered electrodes and ERCuAl-A2 bare electrodes contain from 7 to 9 percent and 9 to 11 percent aluminum respectively. They are used for joining aluminum bronze, silicon bronze, copper-nickels, copper-zinc alloys, manganese bronze, and many combinations of dissimilar metals.

ERCuAl-A2 welding rods are used with the gas tungsten arc welding process, and the weld metal is characterized by its relatively high mechanical properties. These rods are used for the same applications as the wire electrodes.

ECuAl-B covered electrodes contain 8 to 10 percent aluminum, and produce a deposit with higher strength and hardness than the ECuAl-A2 electrodes. These electrodes are used for repair welding of aluminum bronze castings of similar compositions.

ERCuAl-A3 electrodes and rods also are used for repair welding aluminum bronze castings of similar composition with the gas metal arc and gas tungsten arc processes. Their high aluminum content produces welds with less tendency to crack in highly stressed sections.

Bare and covered electrodes are available for welding high manganese-nickel-aluminum and nickel-aluminum bronzes of similar compositions. These are the CuMnNiAl and CuNiAl types.

## Copper-Nickel

Copper-nickel electrodes and rods are nominally 70 percent copper and 30 percent nickel. Electrodes are available in both covered and wire forms. These filler metals are used for welding all the standard copper-nickel alloys. Welding with covered electrodes should be done in the flat position, although small sizes can be used in all positions. The electrodes are generally used with direct current, electrode positive.

## FIXTURING

The thermal coefficients of expansion of copper and its alloys are about 1.5 times that of steel (see Table 7.2). This means that distortion will be greater with copper alloys unless appropriate measures are used to control it.[1] Suitable clamping fixtures can be used to position and restrain thin components for welding. With thick sections, frequent tack welds may be used to align the joint for welding. The ends of the tack welds should be tapered to ensure good fusion with the first weld beads.

With multiple pass welds, the root bead should be rather large to avoid cracking. Fixturing and welding procedures must be designed to limit restraint with copper alloys that are hot-short and likely to crack when highly restrained.

Joint backing strips or rings may be helpful when welding copper and high copper alloys. Control of drop-thru can be a problem with these very fluid weld metals. The backing strips should be the same alloy as the base metal or a very similar one.

## SURFACE PREPARATION

The joint faces and adjacent surfaces should be clean. Oil, grease, dirt, paint, oxides, and mill scale should be removed by degreasing followed by suitable chemical or mechanical cleaning.[2] Wire brushing is not a suitable cleaning

---

1. Refer to *Welding Handbook*, Vol. 1, 7th Ed., 1976: 222-77, for a discussion of residual stresses and distortion and the means for control.
2. See the *Metals Handbook, Heat Treating, Cleaning, and Finishing*, Vol. 2, 8th Ed., Metals Park, OH: American Society for Metals, 1964; 635-41.

method for copper alloys that develop a tenacious surface oxide, such as the aluminum bronzes. These alloys should be cleaned by appropriate chemical or abrasive methods.

## PREHEATING

The relatively high thermal conductivities of copper and most copper alloys result in the rapid conduction of heat from the weld joint to the surrounding base metal. The heat conduction into a specific base metal depends upon its thickness or mass and its temperature. Heat loss from the weld zone will be greater with thick sections at low temperatures. It can be reduced by preheating the base metal to decrease the temperature differential.

The optimum preheat temperature depends upon the welding process, the alloy being welded, and the metal thickness or mass. Thin sections or high energy welding processes, such as gas metal arc and electron beam welding,

generally require less preheat than do thick sections or low energy welding processes, such as gas tungsten arc and oxyacetylene welding. This is shown in Fig. 7.1 for copper. When the welding conditions are similar, the coppers require higher preheat temperatures than many copper alloys because of their high thermal conductivities. Copper alloys that have low thermal conductivities, such as aluminum-bronze and copper-nickel alloys, should not be preheated. Suggested preheat temperatures are given in later tables for the particular alloy families and welding processes.

When preheat is used, the base metal adjacent to the joint must be heated uniformly to temperature. The temperature should be maintained until the joint is completed. When welding is interrupted, the joint area should be preheated again before welding is resumed.

Heat loss should be minimized, where appropriate, by insulating the parts with appropriate materials. High welding current should not be used to compensate for lack of preheat. It will

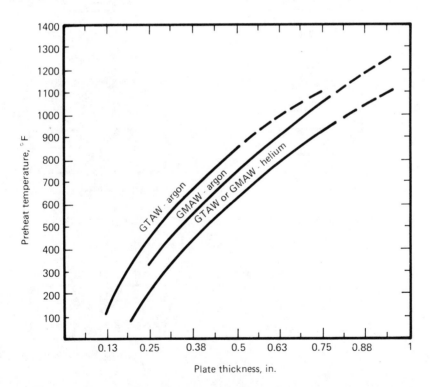

**Fig. 7.1–Effects of process, shielding gas, and metal thickness on preheat requirements for welding copper**

overheat the weld metal and cause a rough bead surface, undercut, porosity, and oxide inclusions.

## COPPERS

The coppers are normally welded with a filler metal of similar composition, although other compatible copper filler metals may be used to obtain desired properties. The high thermal conductivities of the coppers contribute to the rapid conduction of heat from the weld zone. Preheating should be used to reduce this heat loss. Otherwise, the welded joint may have incomplete fusion, inadequate joint penetration, or both.

## Gas Tungsten Arc Welding

Gas tungsten arc welding is best suited for joining sections up to 0.125 in. thick. It can be used advantageously for thicker sections when welding in a position other than flat. Typical joint designs for gas tungsten arc welding of copper are shown in Fig. 7.2.

*Shielding Gases.* Argon shielding is preferred for welding sections up to 0.06-in. thick. With thicker sections, low travel speeds and high preheat are required with argon shielding.

Helium is preferred for welding sections over 0.06-in. thick. The weld pool is more fluid and cleaner, and the risk of oxide entrapment is

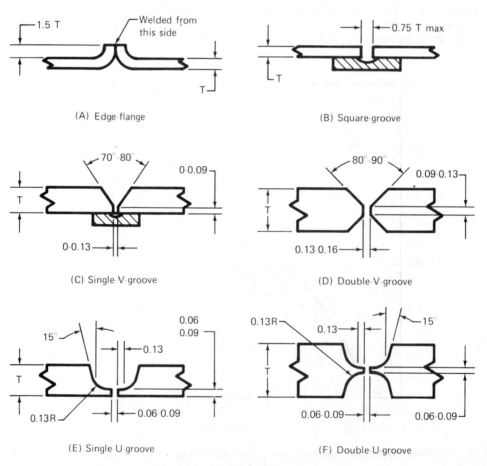

(A) Edge-flange

(B) Square-groove

(C) Single-V-groove

(D) Double-V-groove

(E) Single-U-groove

(F) Double-U-groove

**Fig. 7.2–Joint designs for gas tungsten arc welding of copper**

considerably reduced. Compared to argon, it permits deeper penetration or higher travel speeds at the same welding current. Figure 7.3 illustrates the the differences in penetration in copper with argon and helium as well as the effect of preheat temperature with argon.

Mixtures of these gases result in intermediate welding characteristics. For welding positions other than flat, a mixture of 65 to 75 percent helium-argon produces a good balance between the penetrating quality with helium and the ease of control with argon.

*Welding Technique.* Either forehand or backhand welding may be used. Forehand welding is preferred for all welding positions. It can provide a more uniform, smaller bead than with backhand welding. However, a larger number of beads may be required to fill the joint.

The joint should be filled with one or more stringer beads or narrow weave beads. Wide oscillation of the arc should be avoided because it intermittently exposes each edge of the bead to the atmosphere and consequent oxidation. The first bead should penetrate to the root of the joint and be fairly thick to provide time for deoxidation of the weld metal and to avoid cracking of the bead.

Typical preheating temperatures and welding conditions for gas tungsten arc welding of copper are shown in Table 7.4. These conditions should only be used as guides for establishing welding procedures for a particular application. The high thermal conductivity of copper and the rapid conduction of heat from the joint area into the components and the fixturing make it difficult to give welding conditions suitable for all applications, particularly the travel speed. With a suitable travel speed, the other welding conditions should be adjusted to produce the desired weld bead shape. With mechanized welding, the usual limitation on travel speed is the weld bead shape. At excessive speeds, the bead tends to be very convex in shape, causing underfill along the edges.

*Properties.* Representative mechanical and electrical properties of copper weld metal deposited by gas tungsten arc welding are presented in Table 7.5. The specimens were tested as-welded or after an annealing heat treatment.

## Gas Metal Arc Welding

*Shielding gases.* Argon and a mixture of 75 percent helium-25 percent argon are recommended for gas metal arc welding of copper. Argon is normally used for sections of 0.25-in. thickness and under. The helium-argon mixture is recommended for welding of thicker sections because the preheat requirements are lower, joint penetration is better, and filler metal deposition rates are higher.

*Welding Technique.* Type ERCu copper electrodes are recommended for welding copper. A copper alloy electrode may be used to obtain desired joint mechanical properties when good electrical or thermal conductivity is not a major requirement. The proper electrode size will depend upon the base metal thickness and the joint design.

The filler metal should be deposited in stringer beads or narrow weave beads using spray transfer. Wide weaving of the electrode may result in oxidation at the edges of the bead. Ap-

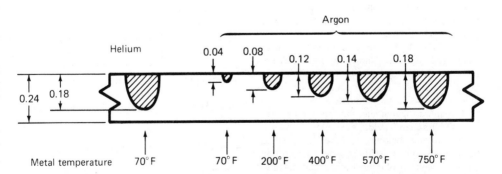

**Fig. 7.3–Effect of shielding gas and preheat temperature on weld bead penetration in copper when gas tungsten arc welding with 300A dc at a travel speed of 8 in./min**

## Table 7.4
### Typical conditions for manual gas tungsten arc welding of copper

| Metal thickness, in. | Joint design[a] | Shielding gas | Tungsten electrode diam, in. | Welding rod diam, in. | Preheat temp, °F | Welding current, A[b] | No. of passes |
|---|---|---|---|---|---|---|---|
| 0.01-0.03 | A | Ar | 0.02,0.04 | – | – | 15-60 | 1 |
| 0.04-0.07 | B | Ar | 0.04,0.062 | 0.062 | – | 40-170 | 1 |
| 0.09-0.19 | C | He | 0.094 | 0.094,0.125 | 100 | 100-300 | 1-2 |
| 0.25 | C | He | 0.125 | 0.125 | 200 | 250-375 | 2-3 |
| 0.38 | E | He | 0.125 | 0.125 | 450 | 300-375 | 2-3 |
| 0.5 | D | He | 0.125,0.156 | 0.125 | 650 | 350-420 | 4-6 |
| 0.62 and above | F | He | 0.188 | 0.125 | 750 min. | 400-475 | As req'd |

a. See Fig. 7.2
b. Direct current, electrode negative

## Table 7.5
### Typical properties of gas tungsten arc weld deposits of copper

| Test and conditions | Tensile strength, ksi | Yield strength, ksi | Elogation, percent | Impact strength, ft lb[a] | Electrical conductivity, percent IACS |
|---|---|---|---|---|---|
| All weld metal test | | | | | |
| As-welded | 27-32 | 15-20 | 20-40 | 20-40 | |
| Annealed at 1000°F | 27-32 | 12-18 | 20-40 | | |
| Transverse tension test | | | | | |
| As-welded | 29-32 | 10-13 | | | |
| Deposited metal conductivity[b] | | | | | |
| Oxygen-free copper | | | | | 95 |
| Phosphorous deoxidized copper | | | | | 83 |
| Phosphor bronze | | | | | 37 |
| Silicon bronze | | | | | 26 |

a. Charpy keyhole specimens
b. Copper base metal welded with the given filler metal

proximate minimum conditions for spray transfer with copper electrodes and argon shielding are given in Table 7.6.

Suggested joint designs for gas metal arc welding of copper are shown in Fig. 7.4. Typical preheat temperatures and welding conditions are given in Table 7.7. These should only be used as guides in establishing suitable welding conditions that are substantiated by appropriate tests.

The forehand welding technique should be used in the flat position. In the vertical position, the progression of welding should be up. Gas metal arc welding of copper is not recommended for the overhead position because the bead shape will be poor. Gas tungsten arc welding is the preferred method for this position.

### Plasma Arc Welding

Copper can be welded with the plasma arc process and Type ERCu welding rods. Argon, helium, or mixtures of the two are used for orifice and shielding gases depending upon the base metal thickness. As with gas tungsten arc welding, arc energy is higher with helium-rich mixtures. Hydrogen should not be added to either gas.

### Shielded Metal Arc Welding

Copper may be welded using ECu (copper) covered electrodes, but weld quality is not as good as that obtained with the gas shielded welding processes. Best results are obtained when

**Table 7.6**
**Approximate gas metal arc welding conditions for spray transfer with copper and copper alloy electrodes and argon shielding**

| Electrode | | Minimum welding current, A | Arc voltage, V | Filler wire feed, in./min | Minimum current density, kA/in.$^2$ |
|---|---|---|---|---|---|
| Type | Diam, in. | | | | |
| ERCu (copper) | 0.035 | 180 | 26 | 345 | 191 |
| | 0.045 | 210 | 25 | 250 | 134 |
| | 0.062 | 310 | 26 | 150 | 101 |
| ERCuAl-A2 (aluminum bronze) | 0.035 | 160 | 25 | 295 | 170 |
| | 0.045 | 210 | 25 | 260 | 134 |
| | 0.062 | 280 | 26 | 185 | 111 |
| ERCuSi-A (silicon bronze) | 0.035 | 165 | 24 | 420 | 176 |
| | 0.045 | 205 | 26-27 | 295 | 131 |
| | 0.062 | 270 | 27-28 | 190 | 88 |
| ERCuNi (copper-nickel) | 0.062 | 280 | 26 | 175 | 91 |

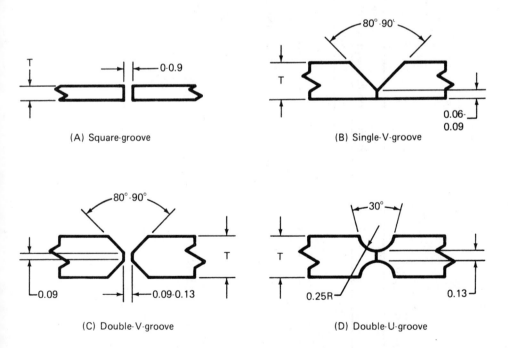

(A) Square-groove

(B) Single-V-groove

(C) Double-V-groove

(D) Double-U-groove

*Fig. 7.4–Joint designs for gas metal arc welding of copper*

## Table 7.7
## Typical conditions for gas metal arc welding of copper

| Metal thickness in. | Joint design[a] | Shielding gas | Electrode | | Preheat temp, °F | Welding current, A[b] | Travel speed in./min | No. of passes |
|---|---|---|---|---|---|---|---|---|
| | | | Diam, in. | Feed in./min | | | | |
| Up to 0.19 | A | Ar | 0.045 | 180-315 | 100-200 | 180-250 | 14-20 | 1-2 |
| 0.25 | B | 75% He-25% Ar | 0.062 | 150-210 | 200 | 250-325 | 10-18 | 1-2 |
| 0.38 | B | 75% He-25% Ar | 0.062 | 190-230 | 425 | 300-350 | 6-12 | 1-3 |
| 0.50 | C | 75% He-25% Ar | 0.062 | 210-270 | 600 | 330-400 | 8-14 | 2-4 |
| 0.63 and Above | D | 75% He-25% Ar | 0.062 0.094 | 210-270 150-190 | 800 800 | 330-400 500-600 | 6-12 8-14 | As req'd As req'd |

a. Refer to Fig. 7.4.
b. Constant potential power source; direct current, electrode positive using a short arc length that provides steady and quiet operation.

welding deoxidized copper. The electrodes may also be used to weld oxygen-free and tough-pitch coppers, but the welded joints will contain porosity and oxide inclusions associated with these coppers.

Copper may also be welded with one of the copper alloy covered electrodes, such as ECuSi or ECuSn-A electrodes. These electrodes are used for (1) minor repair of relatively thin sections, (2) fillet welded joints with limited access, and (3) welding copper to other metals.

Joint designs similar to those shown in Fig. 7.4 are suggested for this process. A grooved copper backing strip may be used to aid in control of root surface contour.

The electrode size should be as large as practical for the base metal thickness. Welding should be done using direct current, electrode positive of sufficient amperage to provide good weld metal fluidity. Weave or stringer beads may be used to fill the joint groove. Welding should be done in the flat position using a preheat of 500°F or higher for sections over 0.13-in. thick.

## Oxyacetylene Welding

Copper and most copper alloys can be welded with an oxyacetylene flame. However, the heat energy available is significantly less than that produced by a welding arc. Therefore, welding speeds are much slower with this process. Preheat and auxiliary heating are recommended with thicknesses over 0.13-in. to obtain good fusion.

Base welding rods of Type ERCu or ERCuSi can be used for copper, depending upon the desired joint properties. A commercial flux designed for welding copper alloys can be used. The welding rod and the joint surfaces should be coated with flux.

The flame should be adjusted neutral when flux is used and slightly oxidizing for welding without flux. The torch tip size should be one to two sizes larger than the one normally used for the same thickness of steel. Typical welding tip sizes and joint designs are given in Table 7.8.

Backhand welding is generally preferred for the flat position. This technique can give a thicker bead than forehand welding, and the danger of oxide entrapment is less. Control of the molten weld pool is greatly improved when the joint axis is tilted about 10 to 15 degrees and the direction of welding is uphill. Backhand or forehand welding can then be used.

Long seams should not be tack welded. The initial root opening should increase along the joint length with a taper that will close gradually as welding proceeds along the joint. A rule-of-thumb is to increase the root opening 0.015 in. per inch of joint length.

Completed weld beads should be peened to relieve welding stresses and increase the weld metal strength by cold working. Peening may be done while the weld metal is still warm or after it cools to room temperature.

## Friction Welding

Although limited in application, friction welding offers several advantages for joining copper and copper alloys. The heat-affected zone is very narrow, and the joint contains no cast metal. Joint properties are excellent.

The process can be used to join copper to itself, to copper alloys, and to other metals including aluminum, silver, carbon steel, stainless steel, and titanium.

## Electron Beam Welding

Copper can be readily joined by electron beam welding. Sound welds can be produced by this process, although multiple-pass welding is sometimes required to avoid discontinuities in the joint. Filler metal can be added to a weld with an auxiliary wire feed system.

## High-Frequency Welding

Copper and copper alloy tubing is frequently manufactured from strip in a tube mill using high-frequency resistance welding to join the seam. The edges are resistance heated to welding temperature utilizing the skin effect with high-frequency current. The heated edges are forged together continuously in the tube mill to consummate a weld.

## Solid-State Welding

Copper can readily be welded without melting using various combinations of temperatures, pressures, and deformations. Annealed copper can be cold welded at room temperature because of its excellent malleability. Copper tubing can be welded and pinched off using commercially available steel dies. The metal can be diffusion welded at elevated temperatures.

# HIGH COPPER ALLOYS

High copper alloys include the beryllium, cadmium, and chromium coppers. Cadmium

**Table 7.8**

**Suggested joint designs and torch tip sizes for oxyacetylene welding of copper**

| Metal thickness, in. | Joint design | Root opening, in. | Welding tip drill size no. | Remarks |
|---|---|---|---|---|
| 0.06 | Edge-flange | 0 | 55 to 58 | |
| 0.06 | Square-groove | 0.06–0.09 | 55 to 58 | |
| 0.13 | Square-groove | 0.09–0.13 | 51 to 54 | |
| 0.19 | 60° to 90° single-V-groove | 0.13–0.18 | 48 to 50 | Auxiliary heating required |
| 0.25 | 60° to 90° single-V-groove | 0.13–0.18 | 43 to 46 | Auxiliary heating required |
| 0.38 | 60° to 90° single-V-groove | 0.18 | 38 to 41 | Auxiliary heating required |
| 0.50–0.75 | 90° double-V-groove | 0.18 | 38 to 41 | Weld both sides simultaneously in vertical position |

copper has good electrical conductivity and is strengthened by cold working. Beryllium and chromium coppers can be strengthened by a precipitation-hardening heat treatment alone or in combination with cold working. Welding heat will soften the heat-affected zones in the high copper alloys by annealing or overaging. The characteristics of each alloy and its condition should be considered when establishing welding procedures and manufacturing sequence. For maximum strength properties, beryllium and chromium coppers should be precipitation hardened after welding.

### Cadmium and Chromium Coppers

Generally, the procedures recommended for arc and oxyacetylene welding of copper are good bases for developing welding procedures for cadmium and chromium coppers. The alloys have lower thermal and electrical conductivities than copper. They can be welded with lower preheats and heat inputs than those required for copper.

Cadmium copper can be joined by the gas shielded welding processes, oxyacetylene welding, and flash welding. Chromium copper can be welded with the gas shielded processes and flash welding, but oxyacetylene welding is not recommended because of problems caused by chromium oxide formation on the joint faces. When oxyacetylene welding cadmium copper, a flux containing sodium fluoride and fused borax or boric acid is recommended to dissolve cadmium oxides.

### Beryllium Coppers

Two types of beryllium copper are available. One type which contains about 0.5 percent beryllium and 1.5 or 2.5 percent cobalt, has relatively good electrical conductivity. The other type which contains about 2 percent beryllium, has good strength in the precipitation-hardened condition but low electrical conductivity.

A difficulty common to the beryllium coppers is the formation of surface oxide. Beryllium will form a tenacious oxide that inhibits wetting and fusion during welding. Cleanness of the joint faces and surrounding surfaces before and during welding is necessary for good results.

With gas tungsten arc welding, stabilized ac power is preferred for manual welding to take advantage of its surface cleaning action. Direct current, electrode negative can be used for mechanized gas tungsten arc welding provided adequate gas shielding is used to prevent oxidation.

Preheat is not usually required for welding sections of 0.13 in. and under in thickness. Preheat temperatures for the high conductivity alloys should be those recommended for coppers. For the high strength alloys, preheat temperatures of 300° to 400°F are sufficient.

Joint designs similar to those shown in Figs. 7.2 and 7.4 are suitable for gas tungsten and gas metal arc welding respectively. High beryllium copper is more readily welded than low beryllium copper. The addition of beryllium to copper lowers the melting point, increases the fluidity of the molten metal, and decreases thermal conductivity. All of these contribute to better weldability.

Sound welds can be made in the low beryllium alloy. However, cracking during welding or postweld heat treatment can be a problem. Typical conditions used for welding beryllium copper are given in Table 7.9. These data can be used as a guide in establishing suitable welding conditions for a particular application.

Optimum mechanical properties are obtained by solution heat treatment followed by artificial aging after welding, as the data in Table 7.10 indicate. When the heat-affected zone is not overaged during welding, direct aging of the weldment can produce joint strengths approaching those of a fully heat treated weldment. The characteristics of the weld metal must be considered when planning a postweld heat treatment when the filler metal composition is different from that of the base metal.

Components in the precipitation-hardened condition are not normally welded because of the danger of cracking the heat-affected zone. Thin sections should be welded in the solution heat treated condition. Where multiple-pass welding is required, the base metal should be in the overaged condition because it is more stable metallurgically in this condition than in the solution heat-treated condition.

Low beryllium copper can be joined more readily with a filler metal of higher beryllium content. Beryllium copper components can be repaired by shielded metal arc welding with aluminum bronze electrodes or by gas tungsten arc welding with silicon bronze welding rods when joints with high mechanical properties are not required.

## COPPER-ZINC ALLOYS (BRASSES)

The copper-zinc alloys, known as brasses, can be joined by conventional arc, oxyacetylene,

### Table 7.9
### Typical conditions for arc welding beryllium coppers

| Process | GTAW | | | | GMAW |
|---|---|---|---|---|---|
| Operation | Manual | | Automatic | | Manual |
| Alloy | C17200 | C17500 | C17200 | C17200 | C17000 |
| Thickness, in. | 0.125 | 0.25 | 0.020 | 0.090 | 1.0 |
| Joint design | Cᵃ | Eᵃ | Bᵃ | Bᵃ | Dᵇ |
| Preheat temp., °F | 70 | 300 | 70 | 70 | 300 |
| Filler metal diam., in. | 0.125 | 0.125 | – | 0.062 | 0.062 |
| Filler metal feed, in./min | | | – | | 190-200 |
| Shielding gas | Ar | Ar | Ar | 65% Ar-35% He | Ar |
| Welding power | achf | achf | dcen | dcen | dcep |
| Welding current, A | 180 | 225-245 | 43 | 150 | 325-350 |
| Arc voltage, V | | 22-24 | 12 | 11.5 | 29-30 |
| Travel speed, in./min | | | 27 | 20 | |

a. Refer to Fig. 7.2.
b. Refer to Fig. 7.4.

### Table 7.10
### Typical mechanical properties of welded joints in beryllium copper

| Alloy and condition | Tensile strength, ksi | Yield strength, ksi |
|---|---|---|
| C17200 (Cu-2Be)ᵃ | | |
| As-welded | 60-70 | 30-33 |
| Aged only | 130-155 | 125-150 |
| Solutioned and aged | 150-175 | 145-170 |
| C17500 (Cu-2.5 Co-0.5 Be)ᵃ | | |
| As-welded | 50-55 | 30-45 |
| Aged only | 80-95 | 65-85 |
| Solutioned and aged | 100-110 | 75-85 |

a. Welded in the solution heat-treated condition.

resistance spot, flash, and friction welding processes. The electrical and thermal conductivities of the brasses decrease with increasing zinc content. Therefore, high zinc brasses require less preheat and welding heat than low zinc brasses. However, zinc fuming is worse when arc welding the high zinc brasses.

## Gas Tungsten Arc Welding

The brasses are commonly joined by gas tungsten arc welding, particularly thin sections. The zinc tends to vaporize as fume from the molten brass. The welding procedures should be designed to minimize this fume.

Thin brass sheets can be welded together without filler metal addition. Addition of filler metal is recommended when welding sections over 0.062-in. thick. Phosphor bronze filler metal may give better color match with some brasses, but zinc fuming is less with silicon bronze filler metal.

With high zinc brasses, aluminum bronze (ERCuAl-A2) filler rod can be used to provide good joint strength. However, this filler rod is not so effective in controlling zinc fuming, and the welds tend to be porous.

Fuming can be minimized by directing the arc on the welding rod and molten weld pool rather than on the base metal. The base metal is heated to fusion temperature by conduction from the molten weld pool rather than by direct impingement of the arc.

When the metal thickness requires a V-groove weld, a groove angle of 75 to 90 degrees should be used to permit good joint penetration. A preheat in the 200° to 600°F range should be used. In general, the preheat temperature. can be lower with the high zinc brasses.

## Gas Metal Arc Welding

Gas metal arc welding is used primarily to join relatively thick sections of brass, but zinc fuming is more severe than with gas tungsten arc welding. Silicon bronze (ERCuSi-A), phosphor bronze (ERCuSn-A), or aluminum bronze (ERCuAl-A2) bare electrodes are recommended. The phosphor bronze electrode will produce weld metal of good color match with most brasses, but the silicon bronze electrode has better fluidity. The aluminum bronze electrode is best for welding high strength brasses to produce joints with equivalent strengths. A V-groove weld with a 60 to 70 degree groove angle or a U-groove weld is used, depending upon the metal thickness.

A preheat in the range of 200° to 600°F is recommended for the low zinc brasses because of their relatively high thermal conductivities. Preheat is not required with the high zinc brasses, but it can decrease the required welding current, and thus reduce zinc fuming.

During welding, the arc should be directed on the molten weld pool as much as possible to minimize zinc fuming. Argon shielding is normally used, but helium-argon mixtures can provide higher welding heat.

## Shielded Metal Arc Welding

Brasses can be welded with phosphor bronze (ECuSn-A or ECuSn-C), silicon bronze (ECuSi), or aluminum bronze (ECuAl-A2) covered electrodes. The applications are the same as those for equivalent bare electrodes used with gas metal arc welding.

Relatively large welding groove angles are needed for good joint penetration and avoidance of slag entrapment. Welding should be done in the flat position, for best results, using a backing strip of copper or brass.

The preheat and interpass temperature for the low brasses should be in the 400° to 500°F range, and in the 500° to 700°F range for the

high brasses. Low preheat temperature will provide better weld joint mechanical properties when using phosphor bronze electrodes. Fast welding speed and a welding current in the high end of the recommended range for the electrode should be used to deposit stringer beads in the joint. The arc should be directed primarily on the molten weld pool to minimize zinc fume.

## Oxyacetylene Welding

The low brasses are readily joined by oxyacetylene welding. This process is particularly suited for piping because it can be done in all welding positions. Silicon copper (ERCuSi-A) welding rod or one of the brass welding rods (RBCuZn-A, RCuZn-B, or RCuZn-C) may be used.

For oxyacetylene welding the high brasses, Type RCuZn-B or RCuZn-C welding rods are used. These low-fuming rods have compositions similar to many of the high brasses. A flux of AWS Type 3B or Type 5 is required and the torch flame should be adjusted to a slightly oxidizing flame to assist in controlling fuming. Preheating and an auxiliary heat source may also be necessary. The welding procedures for copper are also suitable for the brasses.

## COPPER-TIN ALLOYS (PHOSPHOR BRONZE)

The copper-tin alloys, called phosphor bronzes, have rather wide freezing ranges. They solidify with large, weak dendritic grain structures. Such structures in weld metal have a tendency to crack unless the welding procedures are designed to prevent it. Hot peening of each layer of multiple pass welds will reduce welding stresses and the likelihood of cracking.

## Joint Preparation

A single-V-groove weld should be used to join thicknesses of 0.15 through 0.50 inch. The groove angle should be 60 to 70 degrees for gas metal arc welding and 90 degrees for shielded metal arc welding. For greater thicknesses, a single- or double-U-groove weld having a 0.25-in. groove radius and a 70 degree groove angle is recommended for good access and fusion with the joint faces. A square-groove weld can be used for thicknesses under 0.15 inch.

## Preheat and Postheat

Phosphor bronze weld metal tends to flow

sluggishly because of its wide melting range. Preheating to about 400°F and maintaining this interpass temperature helps to improve metal fluidity when welding thick sections. However, the interpass temperature should not exceed this temperature to avoid weld hot cracking. Preheat is not essential when gas metal arc welding with spray transfer. A postweld heat treatment at about 900°F, followed by rapid cooling to room temperature is recommended for maximum weld ductility or stress corrosion resistance.

## Gas Metal Arc Welding

Gas metal arc welding is recommended for joining large phosphor bronze fabrications and thick sections. Direct current, electrode positive and argon shielding are normally used.

Table 7.11 gives suggested gas metal arc welding conditions that can be used when establishing welding procedures for the phosphor bronzes. The molten weld pool should be kept small and the travel speed rather high. Stringer beads should be used. Hot peening of each layer will reduce welding stresses and the likelihood of cracking.

## Gas Tungsten Arc Welding

Gas tungsten arc welding is used primarily for repair of castings and joining of phosphor bronze sheet. Type ERCuSn-A welding rod is recommended for filler metal. As with gas metal arc welding, hot peening of each layer of weld metal is beneficial. Either stabilized ac or direct current, electrode negative can be used with helium or argon shielding. The metal should be preheated to the 350° to 400°F range, and the travel speed should be as fast as practical.

## Shielded Metal Arc Welding

Phosphor bronze (ECuSn-A or ECuSn-C) covered electrodes are available for joining bronzes of similar compositions. These electrodes are designed for use with direct current, electrode positive. Filler metal should be deposited as stringer beads for best weld joint mechanical properties. Postweld annealing at 900°F is not always necessary, but it is desirable for maximum ductility, particularly if the weld metal is to be cold worked. Moisture, both on the work and in the electrode coverings, must be strictly avoided. Baking the electrodes at 250° to 300°F before use may be necessary to reduce moisture in the covering to an acceptable level.

## Oxyacetylene Welding

Normally, oxyacetylene welding is not recommended for joining the phosphor bronzes because of the resulting wide heat-affected zone and the slow cooling rate. These characteristics may result in cracking because of the hot-shortness of the phosphor bronzes. In an emergency, oxyacetylene welding with ERCuSn-A welding rods can be used if arc welding equipment is not available. If a color match is not essential, braze welding can be done with an oxyacetylene torch and RCuZn-C welding rod. A commercial brazing flux and a neutral or slightly oxidizing flame is used with either, depending upon the composition of the base metal.

## COPPER-ALUMINUM ALLOYS (ALUMINUM BRONZE)

### Weldability

Single-phase aluminum bronzes that con-

---

**Table 7.11**
**Suggested conditions for gas metal arc welding of phosphor bronze**

| Metal thickness, in. | Joint design | | Electrode diam. in.[a] | Arc voltage, V | Welding current, A[a] |
|---|---|---|---|---|---|
| | Groove type | Root opening, in. | | | |
| 0.06 | Square | 0.05 | 0.030 | 25-26 | 130-140 |
| 0.13 | Square | 0.09 | 0.035 | 26-27 | 140-160 |
| 0.25 | V-groove | 0.06 | 0.045 | 27-28 | 165-185 |
| 0.50 | V-groove | 0.09 | 0.062 | 29-30 | 315-335 |
| 0.75 | b | 0-0.09 | 0.078 | 31-32 | 365-385 |
| 1.00 | b | 0-0.09 | 0.094 | 33-34 | 440-460 |

a. ERCuSn-A phosphor bronze electrodes and argon shielding
b. Double V- or double-U-groove

tain less than 7 percent aluminum are hot-short, and weldments of these alloys are likely to crack in the heat-affected zone. These alloys are difficult to weld. Single-phase alloys with 7 percent aluminum or more and two-phase alloys are weldable using procedures designed to avoid cracking.

Alloys C61300 and C61400 (7 percent aluminum) are frequently used for welded fabrications including heat exchangers, piping, and vessels. Alloy C61300 is often preferred because of its good weldability, and it can be successfully arc welded. Alloys with a higher aluminum content are also joined by arc welding in both cast and wrought forms. Oxyacetylene welding of aluminum bronze is not recommended due to the problem with fluxing of aluminum oxide from the weld metal.

## Filler Metals

Welding electrodes and rods recommended for joining the weldable aluminum bronzes are given in Table 7.12. Typical mechanical properties of weld metal deposited by arc welding are shown in Table 7.13. In general, weld metal deposited with gas metal arc welding is slightly stronger and harder than that deposited with covered electrodes. This is attributed to the higher welding speeds and better shielding with gas metal arc welding.

## Joint Design

For section thicknesses up to and including 0.13 in., square-groove welds are used with a root opening of up to 75 percent of the thickness. For thicknesses of 0.15 through 0.75 in., a single-V-groove weld is recommended. The groove angle should be 60 to 70 degrees for gas tungsten arc welding and 90 degrees for gas metal arc and shielded metal arc welding. A double-V- or double-U-groove weld should be used for section thicknesses over 0.75 inch. A U-groove should have a 0.25-in. groove radius.

## Preheating

Preheating is often unnecessary when arc welding thin sections of aluminum bronzes. Generally, the preheat and interpass temperatures should not exceed 300°F for alloys with less than 10 percent aluminum, including the nickel aluminum bronzes. The weldments should be air cooled to room temperature.

When the aluminum content is from 10 to 13 percent, a preheat and interpass temperature

**Table 7.12**
**Suggested filler metals for arc welding aluminum bronzes**

| UNS no. | Shielded metal arc welding | Gas tungsten or Gas metal arc welding |
|---|---|---|
| C61300 C61400 C61800 C62300 | ECuAl-A2 | ERCuAl-A2 |
| C61900 C62400 | ECuAl-B | ERCuAl-A2 |
| C62200 C62500 | ECuAl-B | ERCuAl-A3 |
| C63000 C63200 | ECuNiAl | ERCuNiAl |

of about 500°F is recommended for thick sections. Cooling of the weldment should be rapid.

## Gas Metal Arc Welding

Gas metal arc welding is suitable for aluminum bronze sections of 0.18 in. and thicker. Proper gas shielding is essential for sound welds. Although argon shielding is recommended for most joining and surfacing applications, a 75 percent argon-25 percent helium mixture is helpful when welding thick sections of aluminum bronze where increased welding heat and penetration are required. To maintain proper gas coverage, the gun should be tilted to 25 to 40 degrees in the forehand direction of travel. Electrode extension of 0.38 to 0.50 in. is suggested. When welding in a position other than flat, a pulsed power source or globular type of metal transfer is recommended. Table 7.14 gives suggested welding operating conditions for various electrode sizes. Minimum conditions for spray transfer with ERCuAl-A2 electrodes are given in Table 7.6.

## Gas Tungsten Arc Welding

Gas tungsten arc welding is recommended for critical applications regardless of section thickness. Welding may be done with stabilized ac or dc power. Alternating current provides arc cleaning action during welding to remove oxides from the joint faces. Argon is used with ac for shielding. For better penetration or faster travel speed, dc power can be used with argon, helium,

**Table 7.13
Typical mechanical properties of aluminum bronze weld metal**

| Electrode[a] | Tensile strength, ksi | Yield strength, ksi[b] | Elongation in 2 in., percent | Brinell hardness number[c] |
|---|---|---|---|---|
| | Gas metal arc welding | | | |
| ERCuAl-A2 | 79 | 35 | 28 | 160 |
| ERCuAl-A3 | 116 | 45 | 18 | 207 |
| ERCuNiAl | 104 | 59 | 22 | 196 |
| ERCuMnNiAl | 110 | 67 | 27 | 217 |
| | Shielded metal arc welding | | | |
| ECuAl-A2 | 77 | 35 | 27 | 140 |
| ECuAl-B | 89 | 47 | 15 | 177 |
| ECuNiAl | 99 | 58 | 25 | 187 |
| ECuMnNiAl | 95 | 56 | 27 | 185 |

a. Refer to Specifications AWS A5.6-76 and AWS A5.7-77.
b. 0.5 percent offset.
c. 3000 kg load.

**Table 7.14
Typical operating conditions for arc welding aluminum bronze**

| Electrode size, in. | Gas metal arc welding | | Shielded metal arc welding current, A[a] |
|---|---|---|---|
| | Arc voltage, V | Welding current, A[a] | |
| 0.030 | 25-26 | 130-140 | |
| 0.035 | 26-27 | 140-160 | |
| 0.045 | 27-28 | 165-185 | |
| 1/16 | 29-30 | 315-335 | |
| 5/64 | 31-32 | 365-385 | |
| 3/32 | 33-34 | 440-460 | 50-70 |
| 1/8 | | | 60-80 |
| 5/32 | | | 100-120 |
| 3/16 | | | 130-150 |
| 1/4 | | | 170-190 |
| | | | 235-255 |

a. Direct current, electrode positive

or a mixture of the two. Preheat is used only for thick sections with this process.

## Shielded Metal Arc Welding

Shielded metal arc welding of aluminum bronzes is done with appropriate covered electrodes and direct current, electrode positive (see Tables 7.12 and 7.13). Use of a short arc length and stringer or weave beads are recommended. To avoid inclusions, each bead must be thoroughly cleaned of slag before the next bead is applied. Representative welding current ranges for covered electrodes are given in Table 7.14. This process should only be used where it is inconvenient or uneconomical to use gas metal arc welding because the welding speeds are significantly lower.

## COPPER-SILICON ALLOYS (SILICON BRONZE)

The silicon bronzes generally have good weldability. Characteristics of these bronzes that contribute to this are their low thermal conductivities, good deoxidation of the weld metal by the silicon, and the protection offered by the resulting slag. Silicon bronzes have a relatively narrow hot-short temperature range just below the solidus, and they must be rapidly cooled through this range to avoid weld cracking.

Since heat loss to the surrounding base metal is low, high welding speed can be used and preheating is unnecessary. Interpass temperature should not exceed 200°F. For butt joints, the groove angle of V-groove welds should be 60 degrees or larger. Square-groove welds can be used to join sections up to 0.13-in. thick with or without filler metal. Copper backing bars are often used to control melt-thru.

The weld metals have good fluidity but the molten slag that usually covers them is viscous. In some cases, it may be helpful to use a flux.

## Gas Tungsten Arc Welding

The silicon bronzes are readily gas tungsten arc welded in all positions using ERCuSi-A welding rods. Aluminum bronze welding rod ERCuAl-A2 is used sometimes. Welding is usually done with dc power and argon or helium shielding, but ac power with argon shielding may be used to take advantage of the arc cleaning action. Table 7.15 gives representative conditions for gas tungsten arc welding the silicon bronzes in thicknesses of from 0.062 to 0.50 inch.

## Gas Metal Arc Welding

Gas metal arc welding may be used for joining the silicon bronzes, particularly sections over 0.25-in. thick. ERCuSi-A electrodes, argon shielding, and relatively high welding speeds may be used with this process. With multiple-pass welds, the oxide should be removed by wire brushing between passes.

Representative conditions for gas metal arc welding of butt joints are given in Table 7.16. The conditions necessary for entry into spray transfer with ERCuSi-A electrodes are listed in Table 7.6.

## Shielded Metal Arc Welding

Silicon bronzes can be welded with covered electrodes when it is uneconomical or impractical to use gas metal arc welding. ECuAl-A2 or ECuSi electrodes are commonly used. Square-groove welds are suitable for thicknesses up to 0.156 in.; single- or double-V-groove welds are used for thicker sections.

Welding in the flat position is preferred, but ECuSi electrodes can be used in the vertical and

**Table 7.15**
**Typical welding rods and welding currents for gas tungsten arc welding of silicon bronze**

| Thickness, in. | Welding rod diam. in. | Welding current, A[a] |
|---|---|---|
| 0.06 | 0.062 | 100-130 |
| 0.13 | 0.094 | 130-160 |
| 0.19 | 0.125 | 150-225 |
| 0.25 | 0.125, 0.188 | 150-300 |
| 0.50 | 0.125, 0.188 | 250-325 |

a. Direct current, electrode negative with argon shielding in the flat position

overhead positions. Preheating is not needed, and the interpass temperature should not exceed 200°F. Stringer beads should be deposited using a welding current near the middle of the recommended range for the electrode. The arc length should be short and the travel speed should be adjusted to give a relatively small molten weld pool.

## Oxyacetylene Welding

Silicon bronze can be oxyacetylene welded using ERCuSi-A welding rod and a suitable flux. A slightly oxidizing flame should be used. Fixturing should not unduly restrict movement of the components during welding, and welding should be done rapidly. Either forehand or backhand welding can be used with the former preferred for thin sections. This process should only be used when arc welding equipment is not available.

## COPPER-NICKEL ALLOYS

The thermal and electrical conductivities of copper-nickel alloys are similar to those of carbon steel. These properties facilitate welding, preheating is not normally required, and the interpass temperature during welding should not exceed 150°F. For welding or brazing, the surfaces must be clean and free of lead, phosphorus, and sulfur. These materials cause intergranular cracking in the weld heat-affected zone.

Copper-nickel alloys can be joined by the gas shielded arc welding processes, shielded metal arc welding, and oxyacetylene welding. Filler metal of 70 percent copper-30 percent nickel composition (ECuNi or ERCuNi) is normally used for welding all of the copper-nickel

**Table 7.16**
**Typical conditions for gas metal arc welding of silicon bronze**

| Thickness, in. | Groove type | Root face, in. | Root opening, in. | Pass no. | Welding current, A | Arc voltage, V | Electrode feed, in./min [a] | Travel speed, in./min |
|---|---|---|---|---|---|---|---|---|
| 0.25 | Square | | 0.06 | 1 | 300 | 26 | 215 | 15 |
| 0.25 | Square | | 0.13 | 1 | 315 | 21 | 315 | 15 |
| 0.25 | 60° single-V | | 0 | 1 | 300 | 26 | 215 | 13 |
| 0.375 | 60° single-V | 0.06 | 0 | 1 | 300 | 26 | 215 | 10 |
| 0.375 | 60° single-V | 0.06 | 0 | 1 | 300 | 26 | 215 | 21 |
| | | | | 2 | 300 | 26 | 215 | 18 |
| 0.375 | 60° single-V | 0.13 | 0 | 1 | 300 | 26 | 215 | 15 |
| | | | | 2 | 300 | 26 | 215 | 16 |
| | | | | 3 | 300 | 26 | 215 | 36 |
| 0.375 | 60° single-V | 0.13 | 0 | 1 | 310 | 26 | 215 | 24 |
| | | | | 2 | 310 | 26 | 215 | 16 |
| 0.375 | 60° double-V | 0.06 | 0 | 1 | 310 | 26 | 215 | 18 |
| | | | | 2 | 310 | 26 | 215 | 21 |
| 0.50 | 60° single-V | 0.06 | 0 | 1 | 315 | 21 | 305 | 12 |
| | | | | 2 | 315 | 21 | 305 | 13 |
| | | | | 3 | 315 | 21 | 305 | 12 |
| 0.50 | 60° single-V | 0.06 | 0 | 1 | 320 | 21 | 305 | 13 |
| | | | | 2 | 320 | 21 | 305 | 7 |
| 0.50 | 60° single-V | 0.13 | 0 | 1 | 310 | 26 | 215 | 18 |
| | | | | 2 | 310 | 26 | 215 | 12 |
| | | | | 3 | 310 | 26 | 215 | 18 |
| 0.50 | 60° double-V | 0.06 | 0 | 1 | 310 | 26-28 | 215 | 12 |
| | | | | 2 | 310 | 26-28 | 215 | 13 |

a. 0.062-in. diam. ERCuSi-A electrode; direct current, electrode positive; argon shielding

alloys. Filler metals of matching composition may be used in special cases.

## Gas Tungsten Arc Welding

Copper-nickel alloys can be welded in all positions using gas tungsten arc welding. Direct current with the electrode negative is recommended. Alternating current can be used for automatic welding provided the arc length is accurately controlled. Manual welding is normally used for sheet and plate up to 0.25-in. thick and for pipe and tubing. Thicker sections can be welded with this process, but gas metal arc welding is more economical for such applications.

Argon or helium shielding may be used, but argon gives better arc control. Proper control of arc length and gas shielding is necessary to obtain sound welds. Arc length in the range of 0.02 to 0.03 in. is recommended.

Autogenous welds can sometimes be made in sheet up to 0.06-in. thick. However, weld porosity may be a problem because of the absence of deoxidation provided by a filler metal.

## Gas Metal Arc Welding

Copper-nickel alloys are welded with the gas metal arc process using either spray or short-circuiting transfer. Spray transfer is normally used for sections of 0.25 in. or greater in thickness. The approximate conditions for spray transfer with a 0.062-in. diameter ERCuNi electrode are given in Table 7.6. Spray transfer with pulsed power or short-circuiting transfer can be used to weld thin sections. These transfer methods also provide good control of the molten weld pool when welding in the vertical and overhead positions.

Argon is the preferred shielding gas but argon-helium mixtures give better penetration with thick sections. Direct current power with the electrode positive is recommended.

V-groove and U-groove joint designs are used with gas metal arc welding. The groove angle for V-grooves should be 70 to 80 degrees, and the root face and opening should be about 0.06-in. wide. This design is suitable for section thicknesses between 0.18 and 0.50 inch. Above 0.50-in. thickness, a U-groove with a 50 to 60 degree groove angle and a 0.12- to 0.16-in. groove radius is recommended. The root opening should be 0.06- to 0.10-in. wide. Backing bars are recommended with single groove welds to control root surface contour.

Representative conditions for gas metal arc welding of copper-nickel alloys are given in Table 7.17. Travel speeds should be relatively slow with a slight weaving of the electrode.

## Shielded Metal Arc Welding

Copper-nickel alloys may be welded with ECuNi covered electrodes. In general, the electrode size for a particular application should be one size smaller than the size of a steel electrode used for a similar steel application.

Copper-nickel weld metal is not as fluid as carbon steel weld metal. For this reason, careful manipulation of the electrode is required to produce a good bead contour. It is essential that a short arc be maintained. A weave bead is generally preferred, but weaving should not exceed three times the core wire diameter. A stringer bead with a minimum of weaving may be used for the first pass in a deep groove weld. The slag must be completely removed from each bead before the next bead is deposited.

Square-groove welds can be used to join

## Table 7.17
### Representative conditions for gas metal arc welding of copper-nickel alloy plate

| Thickness, in. | Electrode feed,[a] in./min | Arc voltage, V | Welding current,[b] A |
|---|---|---|---|
| 0.25 | 180-220 | 22-28 | 270-330 |
| 0.38 | 200-240 | 22-28 | 300-360 |
| 0.50 | 220-240 | 22-28 | 350-400 |
| 0.75 | 220-240 | 24-28 | 350-400 |
| 1.0 | 220-240 | 26-28 | 350-400 |
| over 1.0 | 240-260 | 26-28 | 370-420 |

a. ERCuNi electrode, 0.062-in. diam
b. Direct current, electrode positive and argon shielding

## Table 7.18
### Representative conditions for shielded metal arc welding of 0.25-in. thick 90 percent copper-10 percent nickel alloy plate

| Welding position | Joint design | | | Weld pass | Electrode size,[a] in. | Welding current,[b] A |
|---|---|---|---|---|---|---|
| | Groove type | Groove angle, degrees | Root opening, in. | | | |
| Flat | Square | – | 0.13 | 1, 2 | 1/8 | 115-120 |
| Vertical[c] | Double-V | 80-90 | 0.09-0.13 | 1, 2 | 3/32 | 85-90 |
| Vertical[c] | Fillet | 90 | 0 | 1 | 3/32 | 85 |
| Horizontal | Single-V | 90-100 | 0.06-0.13 | 1[d] | 3/32 | 100 |
| | | | | 2 | 1/8 | 100 |
| Flat and overhead | Single-V | 80-90 | 0.09-0.13 | 1[e] | 1/8 | 110-115 |
| | | | | 2 | 3/32 | 95-100 |

a.  ECuNi covered electrodes.
b.  Direct current, electrode positive.
c.  Direction of welding is up.
d.  Backing weld—Back gouge before welding the other side (Pass 2).
e.  Back gouge the root of the joint before completing the back weld (Pass 2).

section thicknesses of 0.13 in. and under. The root opening should be about one half of the section thickness. For thicker sections, V-groove or U-groove joint designs similar to those for gas metal arc welding are recommended. One exception is the U-groove angle and radius which should be about 30 degrees and 0.25 inch, respectively.

Representative joint designs and welding procedures for shielded metal arc welding of 0.25-in. thick 90 percent copper-10 percent nickel plate in various positions are given in Table 7.18. The appropriate joint design depends upon the welding position.

### Oxyacetylene Welding

Copper-nickel alloys can be oxyacetylene welded, but its use should be restricted to repairs when suitable arc welding equiment is not available. ERCuNi welding rods are used with a slightly reducing oxyacetylene flame. Preheat should not be used. Liberal use of a flux made especially for nickel or copper-nickel alloys is necessary to protect the welding rod and base metal from oxidation.

## COPPER-NICKEL-ZINC ALLOYS (NICKEL SILVERS)

The nickel silvers can be joined by gas tungsten arc and oxyacetylene welding, although welding of these alloys is not widely practiced. From a welding standpoint, nickel silvers are similar to brasses having comparable zinc contents. The nickel silvers are frequently used for decorative applications where color match is important. However, no zinc-free filler metals are available suitable for arc welding that give good color match. Therefore, gas tungsten arc welding is usually restricted to thicknesses of 0.094 in. or less without filler metal addition. Square groove butt, lap, or edge joints should be used. The joint faces must be in contact before and during welding.

A wide range of thicknesses may be oxyacetylene welded using RBCuZn-D welding rods and a slightly oxidizing flame. An AWS Type 5 brazing flux should be applied to both the joint area and the welding rod before and during welding.

# RESISTANCE WELDING

## SPOT WELDING

Resistance spot weldability of copper and copper alloys varies inversely with their electrical and thermal conductivities. Thin sheets of many copper alloys can be spot welded, but it is not practical with unalloyed copper. The high electrical conductivity of pure copper makes it impossible to resistance-heat the faying surfaces to welding temperature without also welding the copper alloy electrodes to the sheets. Spot welds can be made in copper alloys having electrical conductivities up to about 30 percent of that of copper, but they become less consistent as the electrical conductivity increases. When the electrical conductivity of the metal is over about 60 percent, it is almost impossible to obtain a good resistance spot weld with conventional methods.

Despite the restrictions, copper alloys that can be spot welded are as follows:

(1) Beryllium copper
(2) Jewelry bronze (87.5% Cu)
(3) Low brass (80% Cu)
(4) Cartridge brass (70% Cu)
(5) Yellow brass
(6) Muntz metal
(7) Inhibited admiralty brass
(8) Naval brass
(9) Phosphor bronze (5, 8, and 10% Sn)
(10) Aluminum bronze
(11) Silicon bronze
(12) Manganese bronze
(13) Aluminum brass
(14) Nickel silvers
(15) Copper-nickel

Electrode forces for both spot and seam welding are usually 50 to 70 percent of those used for welding the same thickness of steel. Also, the welding currents are generally higher and the weld times are shorter.

Spot welding of the leaded copper alloys is not advisable because of their cracking tendency.

## SEAM WELDING

It is difficult to seam weld copper alloys

because of excessive shunting of welding current, high thermal conductivity, and low electrode contact resistance. Seam welding is generally not practicable when the electrical conductivity exceeds 30 percent.

The copper alloys that can be seam welded are as follows:

(1) Beryllium copper
(2) Naval brass
(3) Phosphor bronze (5, 8, and 10% Sn)
(4) Aluminum bronze
(5) Silicon bronze
(6) Manganese bronze
(7) Aluminum brass
(8) Copper-nickels
(9) Nickel silvers

## FLASH WELDING

Flash welding techniques produce very good results in almost all applications and on all copper alloys. The design of the equipment must provide accurate control of all factors, including upset pressure, platen travel, flash-off rate, current density, and flashing time.

Almost all leaded copper alloys can be flash welded, but the integrity of the joint depends upon the alloy composition. Lead content of up to 1.0 percent is usually not detrimental.

Rapid upsetting at minimum pressure is necessary as soon as the abutting faces are molten because of the relatively low melting temperature and narrow plastic range of copper alloys. Low pressure is usually applied to the joint before the flashing current is initiated so that platen motion will begin immediately after flashing starts. Termination of flashing current is rather critical. Premature termination of current will result in lack of fusion at the weld interface. Excessive flashing will overheat the metal and result in improper upsetting.

# CASTINGS

Copper casting alloys are divided into fifteen families based on composition, as shown in Table 7.1. Some alloys contain sufficient lead to render them nonweldable. However, most nonleaded casting alloys and some of the leaded alloys can be welded. Welding is normally used for rebuilding, repairing, surfacing, and joining of castings. Filler metals recommended for copper casting alloys are given in Table 7.3.

## PREPARATION

Castings are more porous and have rougher surfaces than wrought mill products. Surfaces to be welded should be ground or machined smooth, unsound metal or porosity removed, and the parts cleaned of oil, grease, and scale.[3] It is of the utmost importance to adequately clean castings which have been in service, especially if they have been exposed to grease, oil, or chemicals. These contaminants will cause porosity when welding is attempted.

## REPAIRING

In repairing a casting, sufficient metal should be removed to eliminate all surface and subsurface defects. The defective area should be gouged to remove the defects and provide a weld groove with sloping sidewalls and rounded bottom, as indicated in Fig. 7.5. Gas tungsten arc, gas metal arc, shielded metal arc, or oxyacetylene welding can be used to fill the groove with weld metal.

The welding procedures for copper casting alloys are similar to those used for similar wrought copper alloys. They should be designed to minimize dilution, welding heat input, and welding stresses.

---

3. See the *Metals Handbook, Heat Treating, Cleaning, and Finishing*, Vol. 2, 8th Ed., Metals Park, OH: American Society for Metals, 1964; 635-41.

Manganese bronze castings are generally welded with aluminum bronze filler metal, but phosphor bronze covered electrodes can be used for shielded metal arc welding. Oxyacetylene welding is not recommended. Shielding with 75 percent argon-25 percent helium mixture is recommended for gas tungsten and gas metal arc welding. The weldment should be stress-relieved if it will be exposed to an environment that can cause stress corrosion.

When copper-nickel alloy castings are to be welded, the lead and phosphorus contents must be less than 0.01 and 0.02 percent respectively. Homogenization of the casting by heating it for two hours at 1800°F may help to avoid cracking during welding.

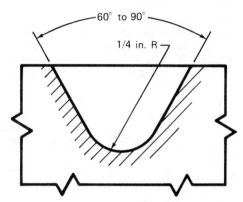

*Fig. 7.5–Excavation design for repair of copper alloy castings*

# POSTWELD HEAT TREATMENT

## REQUIREMENTS

The necessity for heat treating after welding depends upon the base metal composition and the application of the weldment. Postweld heat treatment is usually required if the base metal can be strengthened by a heat treatment or if the service environment can cause stress-corrosion cracking. Alloys that respond to precipitation hardening include some high coppers, some copper-aluminum alloys, and copper-nickel castings containing beryllium. If these alloys are not heat-treated, the hardness in the weld area will vary as a result of aging or overaging caused by the welding heat.

Alloys that are particularly susceptible to stress-corrosion cracking are high zinc brasses, manganese bronzes, nickel-manganese bronzes, aluminum bronzes, and nickel silvers. Stresses induced during forming or welding of these alloys can lead to premature failure in certain corrosive environments.

## STRESS RELIEVING

Stress relieving is accomplished by heating the weldment to some temperature well below the recrystallization temperature of the base metal for a minimum of one hour. The temper-

ature and time should be designed to reduce welding stresses to relatively low values. Heating time must be adequate for the entire weldment to reach temperature. Typical stress relieving temperatures for some copper alloys are given in Table 7.19.

## ANNEALING

Annealing and homogenization treatments are used only for weldments of hardenable copper alloys to produce a metallurgical structure

**Table 7.19**
**Typical stress relieving temperatures for weldments of copper alloys**

| Common name | UNS no. | Temperature, °F[a] |
|---|---|---|
| Red brass | C23000 | 550 |
| Admiralty brass | C44300-C44500 | 550 |
| Naval brass | C46400-C46700 | 500 |
| Aluminum bronze | C61400 | 650 |
| Silicon bronze | C65500 | 650 |
| Copper-nickel alloys | C70600-C71500 | 1000 |

a. Heat slowly to and hold at temperature for at least one hour.

that will respond to heat treatment satisfactorily. Annealing is carried out at temperatures considerably higher than those used for stress relieving, as shown in Table 7.20. Stress relaxation proceeds rapidly at the annealing temperature.

Extended annealing times or annealing at the top of the temperature range can cause undesirable grain growth. Excessive grain growth usually reduces tensile strength and can cause other undesirable metallurgical effects.

**Table 7.20**
**Annealing temperature ranges for copper and copper alloys**

| Common name | UNS no. | Temperature range, °F[a] |
|---|---|---|
| Phosphor deoxidized copper | C12200 | 700-1200 |
| Beryllium copper | C17000, C17200 | 1425-1475 |
| Beryllium copper | C17500 | 1675-1725 |
| Red brass | C23000 | 800-1350 |
| Yellow brass | C27000 | 800-1300 |
| Muntz metal | C28000 | 800-1100 |
| Admiralty | C44300-C44500 | 800-1100 |
| Naval brass | C46400-C46700 | 800-1100 |
| Phosphor bronze | C50500-C52400 | 900-1250 |
| Aluminum bronze | C61400 | 1125-1650 |
| Aluminum bronze | C62500 | 1100-1200 |
| Silicon bronze | C65100, C65500 | 900-1300 |
| Aluminum brass | C68700 | 800-1100 |
| Copper-nickel, 10 percent | C70600 | 1100-1500 |
| Copper-nickel, 30 percent | C71500 | 1200-1500 |
| Nickel silver | C74500 | 1100-1400 |

a. Time at temperature—15 to 30 min

# BRAZING

Brazing is an excellent process for joining copper and copper alloys.[4] Surface oxides are easily fluxed during brazing except for those on aluminum bronzes that contain more than 8 percent aluminum. Special techniques are required to braze those aluminum bronzes.

All of the common brazing processes can be used except for special cases, such as resistance or induction brazing of copper and copper alloys that have high electrical conductivities.

---

4. Additional information may be found in the *Welding Handbook,* Vol. 2, 7th Ed., 370-438, and the *Brazing Manual,* 3rd Ed., American Welding Society, 1976, 169-83.

## DESIGN

Both lap and butt joints may be used for brazements. The joint clearance must provide for capillary flow of the selected brazing filler metal throughout the joint at brazing temperature. When designing a brazed joint for a specific application, the properties and compatibility of the base metal-filler metal combination must be properly evaluated for the environment in which it will operate. Some important considerations are brazing temperature, type of loading, joint strength, galvanic corrosion, and interaction between the base and filler metals at the service temperature. In the case of electrical conductors, the electrical conductivity of the joint must be considered since the brazing filler metals have low conductivity compared to copper.

## FILLER METALS

All of the silver (BAg), copper-phosphorus (BCuP), gold (BAu), and copper-zinc (RBCuZn) brazing filler metals are suitable for brazing. Table 7.21 gives the commonly used brazing filler metals for copper and copper alloys.

The various BAg filler metals may be used with all copper alloys. The BAu filler metals are best suited for electronic applications where the vapor pressure of the brazing filler metal is important.

The BCuP filler metals can be used with most copper alloys, including some copper-nickel alloys. However, they should be evaluated by appropriate test for brazing a particular copper-nickel alloy. Copper can be brazed with these filler metals without flux. BCuP filler metals are not recommended for brazing beryllium copper because the joints will be porous and low in strength.

The RBCuZn filler metals may be used to join the coppers, copper-nickel, copper-silicon, and copper-tin alloys. Their liquidus temperatures are too high for brazing the brasses and nickel silvers. They are not recommended for torch brazing the alumuinum bronzes because of their high brazing temperatures.

## FLUXES AND ATMOSPHERES

Recommended brazing fluxes and furnace atmospheres for brazing the various coppers and copper alloys are also shown in Table 7.21. Types 3A and 3B fluxes are suitable for use with BAg and BCuP filler metals for brazing all copper alloys except the aluminum bronzes. A more reactive Type 4 flux is needed for these bronzes. Type 3B or 5 flux is required with RBCuZn filler metals because of their high brazing temperatures.

Combusted fuel gases are economical brazing atmospheres for copper and its alloys, except for the oxygen-bearing coppers. Atmospheres high in hydrogen cannot be used when brazing these coppers because hydrogen can diffuse into the copper, reduce the copper oxide, and form water vapor. This will rupture the copper. Inert gas may be used for the brazing atmosphere with copper and copper alloys.

Vacuum is a suitable atmosphere for brazing provided the base metal or filler metal does not contain elements that have high vapor pressures at brazing temperature. Zinc, phosphorus, and cadmium are examples of elements that vaporize when heated in vacuum.

## SURFACE PREPARATION

Good wetting and flow in brazed joints can be achieved when the joint surfaces are free of oxides, dirt, and other foreign substances. Standard solvent or alkaline degreasing procedures are suitable for cleaning copper base metals, and mechanical methods may be used to remove surface oxides. Chemical removal of surface oxides requires an appropriate pickling solution. Typical chemical cleaning procedures are as follows:

### Copper

Immerse in cold 5 to 15 percent by volume sulfuric acid.

### Beryllium Copper

(1) Immerse in 20 percent by volume sulfuric acid at 160° to 180°F until the dark scale is removed, then water rinse.

(2) Dip in cold 30 percent nitric acid solution for 15 to 30 seconds, then rinse in hot water and dry.

### Chromium Copper and Copper-Nickel Alloys

Immerse in hot 5 percent by volume sulfuric acid.

### Brass and Nickel-Silvers

Immerse in cold 5 percent by volume sulfuric acid.

### Silicon Bronzes

Immerse first in hot 5 percent by volume sulfuric acid, then in a cold mixture of 2 percent by volume hydrofluoric acid and 5 percent by volume sulfuric acid.

### Aluminum Bronzes

Successive immersions in two solutions are required. These are:

(1) Cold 2 percent hydrofluoric acid and 3 percent sulfuric acid mixture.

(2) A solution of 5 percent by volume sulfuric acid at 80° to 120°F.

### Copper Plate

It is often desirable to copper plate the sur-

## Table 7.21
## Guide to brazing copper and copper alloys

| Material | Commonly used brazing filler metals | AWS brazing atmospheres[a] | AWS brazing flux, No. | Remarks |
|---|---|---|---|---|
| Coppers | BCuP-2, BCuP-3, BCuP-5 RBCuZn, BAg-1a, BAg-1, BAg-2, BAg-5, BAg-6, BAg-18 | 1, 2, or 5 | 3 or 5 | Oxygen-bearing coppers should not be brazed in hydrogen-containing atmospheres. |
| High coppers | BAg-8, BAg-1 | | | |
| Red brasses | BAg-1a, BAg-1, BAg-2 BCuP-5, BCup-3, BAg-5, BAg-6, RBCuZn | 1, 2, or 5 | 3A 3, 5 for RBCuZn | |
| Yellow brasses | BCuP-4, BAg-1a, BAg-1, BAg-5, BAg-6 BCuP-5, BCuP-3 | 3, 4, or 5 | 3 | Keep brazing cycle short. |
| Leaded brasses | BAg-1a, BAg-1, BAg-2 BAg-7, BAg-18 BCuP-5 | 3, 4, or 5 | 3 | Keep brazing cycle short and stress relieve before brazing. |
| Tin brasses | BAg-1a, BAg-1, BAg-2 BAg-5, BAg-6 BCuP-5, BCuP-3 (RBCuZn for low tin) | 3, 4, or 5 | 3 | |
| Phosphor bronzes | BAg-1a, BAg-1, BAg-2 BCuP-5, BCuP-3 BAg-5, BAg-6 | 1, 2, or 5 | 3 | Stress relieve before brazing. |
| Silicon bronzes | BAg-1a, BAg-1, BAg-2 | 4 or 5 | 3 | Stress relieve before brazing. Abrasive cleaning may be helpful. |
| Aluminum bronzes | BAg-3, BAg-1a, BAg-1 BAg-2 | 4 or 5 | 4 | |
| Copper-nickel | BAg-1a, BAg-1, BAg-2 BAg-18, BAg-5 BCuP-5, BCuP-3 | 1, 2, or 5 | 3 | Stress relieve before brazing. |
| Nickel silvers | BAg-1a, BAg-1, BAg-2 BAg-5, BAg-6 BCuP-5, BCuP-3 | 3, 4, or 5 | 3 | Stress relieve before brazing and heat uniformly. |

a. Inert gas, hydrogen, or vacuum atmospheres are also usually acceptable (AWS Type 6, 9, or 10). Brazing atmospheres are listed below.

| AWS brazing atmosphere | Source | Maximum dew point of incoming gas | AWS brazing atmosphere | Source | Maximum dew point of incoming gas |
|---|---|---|---|---|---|
| 1 | Combusted fuel gas (low hydrogen) | Room temp. | 5 | Dissociated ammonia | −65°F |
| 2 | Combusted fuel gas (decarburizing) | Room temp. | 6 | Cylinder hydrogen | Room temp. |
| 3 | Combusted fuel gas, dried | −40°F | 9 | Purified inert gas | |
| 4 | Combusted fuel gas, dried (carburizing) | −40°F | 10 | Vacuum | − |

faces of copper alloys that contain strong oxide-forming elements to simplify brazing and fluxing requirements. Copper plate about 0.001-in. thick is used on chromium-copper while about 0.0005-in. thickness is sufficient on beryllium copper, aluminum bronze, and silicon bronze.

## COPPER

Oxygen-free, high-conductivity copper and deoxidized copper are readily brazed by furnace or torch methods. Boron-deoxidized copper is sometimes preferred when brazing at high tempertures because grain growth is less pronounced than with the other coppers.

Oxygen-bearing coppers are susceptible to oxide migration and hydrogen embrittlement at elevated temperatures. Therefore, these coppers should be furnace brazed in an inert atmosphere or torch brazed with a neutral or slightly oxidizing flame.

The copper-phosphorus and copper-silver-phosphorus filler metals (BCuP) are self-fluxing on copper. However, a flux is beneficial for massive assemblies where prolonged heating can result in excessive oxidation. Some phosphorus is lost from the filler metal, and this slightly increases the remelt temperature of the joint. Joints brazed with filler metal containing phosphorus should not be exposed to sulfurous atmospheres at elevated temperatures for long periods because of the danger of corrosive attack.

With the copper-zinc (RBCuZn) filler metals, the recommended brazing temperatures should not be exceeded to avoid volatilization of zinc and resulting voids in the joint. When torch brazing, an oxidizing flame reduces zinc fuming. The corrosion resistance of these filler metals is inferior to that of copper.

A lap joint will develop the full strength of annealed copper at room temperature when the overlap is at least three times the thickness of the thinner member. As the service temperature increases, the brazing filler metal strength decreases more rapidly than does the strength of the copper. Eventually, failure will occur through the joint. The tensile strengths at room, elevated, and sub-zero temperatures for single-lap brazed joints in copper are shown in Table 7.22. Typical creep properties for tough-pitch copper brazed with BAg-1A, BAg-6, and BCup-5 filler metals are shown in Fig. 7.6. At 77° and 260°F, the failures generally took place in the base metal.

### Table 7.22
### Tensile strength of single-lap brazed joints in deoxidized copper[a]

| Brazing filler metal | Tensile strength, ksi | | |
|---|---|---|---|
| | −321°F | 72°F | 400°F |
| BAg-1 | 30.1 | 19.0 | 9.7 |
| BAg-6 | 28.0 | 17.6 | – |
| BAg-8 | 24.7 | 17.6 | – |
| BCuP-2 | 17.9 | 18.6 | 10.7 |
| BCuP-4 | 21.4 | 19.1 | – |
| BCuP-5 | 21.9 | 17.9 | 10.8 |

a. Specimens were made from 0.25-in. thick sheet; joints had an overlap of 0.15-in. and no braze fillet.

## HIGH COPPER ALLOYS

### Beryllium Copper

The surfaces of beryllium copper components must be cleaned prior to brazing. The oxide scale can be removed by pickling.

The 2 percent beryllium copper can be brazed by two methods. The more common procedure involves simultaneous brazing and solution heat treatment at 1450°F in a furnace. The silver-copper eutectic filler metal, BAg-8, is generally used with an AWS Type 3A flux. The furnace temperature is quickly lowered to 1400°F to solidify the brazing filler metal. The brazement is then quenched in water, and finally age-hardened at 600° to 650°F.

A second method is used with thin sections that can be heated rapidly, preferably in one minute or less. Rapid heating permits brazing at a temperature below the solutioning temperature of the beryllium copper. The brazement can be aged without resolutioning. Sufficiently fast heating rates can usually be attained with an oxyacetylene torch or resistance heating. BAg-1 filler metal and Type 3A flux are commonly used. However, other silver brazing filler metals may also be suitable for special applications.

The 0.5 percent beryllium copper alloys are solution heat-treated at 1700°F and then aged between 850° and 900°F. They can be brazed rapidly with BAg-1 alloy at about 1200°F after the aging heat treatment. However, the base metal will be overaged during brazing with some loss in hardness and strength properties.

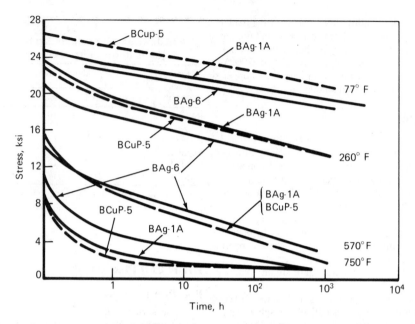

**Fig. 7.6 – Stress-rupture strength curves for copper brazed with three filler metals using a plug and ring creep specimen**

### Chromium and Zirconium Coppers

Chromium copper and zirconium copper are solution heat-treated at 1650° to 1850°F, quenched, and then age-hardened at about 900°F. Brazing with BAg-1 filler metal and a fluoride type of flux should be performed after a solution heat treatment and before aging. The brazement should then be aged. However, the mechanical properties of the base metal after brazing and aging will be lower than those of metal that is solution heat-treated and aged only.

Combined brazing and solution heat treatment of chromium copper can give near optimum mechanical properties after aging. However, distortion from rapid quenching from the solutioning temperture should be evaluated for each application.

### Cadmium Coppers

Cadmium coppers are brazed in the same manner as deoxidized copper.

## COPPER-ZINC ALLOYS (BRASSES)

All brasses can be brazed with BAg and BCuP filler metals. The high melting, low zinc brasses can also be brazed with RBCuZn filler metals. Type 3 flux should be used with BAg and BCuP filler metals, and Type 5 flux with RBCuZn filler metals.

Zinc fuming above 750°F can be reduced by fluxing the parts before furnace brazing, even when a protective atmosphere is used, and by torch brazing with an oxidizing flame. Brasses are subject to cracking when heated too rapidly and should, therefore, be heated uniformly and slowly. Sharp corners and other stress concentrations that can localize thermal strains during heating should be avoided. It is good practice to heat the parts to brazing temperature slowly.

Brasses containing aluminum or silicon require cleaning and brazing procedures similar to those used for aluminum or silicon bronze.

Lead in brasses can alloy with the brazing filler metal and cause brittleness when the lead content is over 3 percent. To maintain good flow and wetting during brazing, leaded brasses require complete flux coverage to prevent the formation of lead oxide or dross.

Rapid heating of high leaded brasses can

cause cracking. Stress-relieving before brazing and slow, uniform heating minimizes this cracking tendency.

## COPPER-TIN ALLOYS (PHOSPHOR BRONZES)

In a stressed condition, phosphor bronzes are subject to cracking during heating. It is good practice to (1) stress-relieve or anneal the parts before brazing, (2) support them in a stress-free condition during brazing, and (3) use a slow heating cycle. Adequate flux protection is necessary during brazing, particularly with the high tin alloys and the leaded phosphor bronzes. All phosphor bronzes can be brazed with BAg and BCuP filler metals. Alloys with low tin content can be brazed with the RBCuZn filler metals.

The phosphor bronzes are sometimes used in the form of powdered metal compacts. Before brazing, these compacts must be treated to seal the pores and restrict penetration by the brazing filler metal. This is done by painting the surface with a colloidal graphite suspension. The compact is then baked at a low temperature, followed by cleaning and degreasing.

## COPPER-ALUMINUM ALLOYS (ALUMINUM BRONZES)

Aluminum bronzes can be brazed with BAg brazing filler metals and Type 4 flux. Refractory aluminum oxide readily forms at brazing temperature on the surface of bronzes containing more than 8 percent aluminum. The brazing procedures must prevent oxide formation to obtain satisfactory flow and wetting in the joint. With furnace brazing, flux should be used with a protective atmosphere.

Copper plating on the surfaces to be brazed prevents the formation of aluminum oxide during brazing. Plated components can be brazed using procedures suitable for copper.

## COPPER-SILICON ALLOYS (SILICON BRONZES)

Copper-silicon alloys should be cleaned and then coated with flux or copper plated before brazing to prevent the formation of refractory silicon oxide. Mechanical cleaning is recommended. Light oxide contamination can be removed from the surfaces by chemical cleaning. Silver brazing filler metals and Type 3 flux are generally used.

When furnace brazing, a flux should be used in combination with a protective atmosphere. Silicon bronzes are subject to intergranular penetration by the filler metal and to hot-shortness under stress. Components should be stress-relieved before brazing, adequately supported during heating, and brazed below 1400°F.

## COPPER-NICKEL ALLOYS

Copper-nickel alloys can be brazed with BAg and BCuP filler metals. However, BCuP filler metals may form a brittle nickel phosphide with alloys containing more than 10 percent nickel. The structure and properties of joints brazed with BCuP filler metals should be thoroughly evaluated for the intended application. AWS Type 3 flux is suitable for most applications.

The base metal surfaces must be free of sulfur or lead because they may cause cracking during the brazing cycle. Standard solvent or alkaline degreasing procedures may be used to remove grease or oil. Surface oxides can be removed by mechanical cleaning or by pickling.

Copper-nickel alloys in the stressed conditions are susceptible to intergranular penetration by molten filler metal. To prevent cracking, they should be stress-relieved before brazing.

## COPPER-NICKEL-ZINC ALLOYS (NICKEL SILVERS)

Nickel silvers can be brazed readily with the same filler metals and procedures used for brazing the brasses. However, when RBCuZn filler metals are used, precautions must be taken to prevent problems resulting from the relatively high brazing temperatures. These alloys are subject to intergranular penetration by the filler metals when under stress. They should be stress-relieved before brazing. The poor thermal conductivity of these alloys can lead to non-uniform heating during brazing. They should be properly fluxed and heated slowly to brazing temperature.

## DISSIMILAR METALS

Dissimilar copper alloys can be brazed readily. Copper alloys can also be brazed to steel, stainless steel, and nickel alloys. Suggested brazing filler metals for dissimilar metal combinations are given in Table 7.23.

**Table 7.23**
**Recommended filler metals for brazing copper alloys to other metals**

| Carbon and low alloy steels | Cast iron | Stainless steels | Tool steels | Nickel alloys | Titanium alloys | Reactive metals | Refractory metals |
|---|---|---|---|---|---|---|---|
| BAg, BAu, RBCuZn | BAg, BAu, RBCuZn | BAg, BAu | BAg, BAu, RBCuZn BNi | BAg, BAu, RBCuZn | BAg | BAg | BAg |

# SOLDERING

Copper and copper alloys are among the most frequently soldered engineering materials.[5] Their solderability, as described in Table 7.24, ranges from excellent to difficult. There are no serious problems in soldering most of the copper alloys. However, those alloys that contain beryllium, silicon, or aluminum require special fluxes to remove surface oxides.

The high thermal conductivities of copper and some of its alloys require a high rate of heat input when localized heating is used.

## SOLDERS

The most widely used solders for joining copper and its alloys are the tin-lead solders. Tin readily alloys and diffuses with copper. Copper alloys will accept a certain amount of tin into solid solution. However, one or more intermetallic phases (probably $Cu_6Sn_5$) will form when the solid solubility limit is exceeded. The intermetallic phase will form at the interfaces of solder joints. Since intermetallic phases tend to be brittle, their thickness should be minimized by proper selection of process variables and service conditions. As the thickness of the intermetallic layer increases, the strength of the soldered joint will decrease. Service at elevated temperature will accelerate this change.

## FLUXES

Organic and rosin types of noncorrosive fluxes are excellent for soldering coppers. They may be used with some success on copper alloys containing tin and zinc, provided the surfaces are precleaned. These fluxes are recommended for soldering electrical connections. A light coat of flux should be applied to the precleaned surfaces to be joined.

Intermediate fluxes are used on copper, copper-tin, copper-zinc, copper-beryllium, and copper-chromium alloys. Some of the more active fluxes are adequate for the copper-nickels and silicon bronzes, but a generalization in this respect could be misleading.

The inorganic corrosive fluxes can be used on all the copper alloys, but they are required only on those alloys that develop refractory oxides, such as the silicon and aluminum bronzes. The aluminum bronzes are especially difficult to solder and require special fluxes or copper plating. Chloride fluxes are useful for soldering the silicon bronzes and copper-nickels.

Oxide films can reform quickly on cleaned copper alloys. Therefore, fluxing and soldering should be done within a short time after cleaning. The fluxes best suited for use on copper tube systems with 50 percent tin-50 percent lead and 95 percent tin-5 percent antimony solders are mildly corrosive liquid or petrolatum pastes containing zinc and ammonium chlorides. Many liquid fluxes for plumbing applications are self-cleaning, but there is a risk of corrosion in their use.

---

5. Additional information on soldering, solders, and fluxes is contained in the *Welding Handbook*, Vol. 2, 440-58 and the *Soldering Manual*, 2nd Ed., American Welding Society, 1977.

A highly corrosive flux can remove some oxides and dirty films. However, there is always an uncertainty as to whether uniform cleaning has been achieved and whether corrosive action continues after soldering. It is always best to start with clean surfaces and a minimum amount of flux.

## SURFACE PREPARATION

Solvent or alkaline degreasing and pickling are suitable for cleaning copper base metals. Mechanical methods may be used to remove oxides. Chemical removal of oxides requires proper choice of a pickling solution followed by thorough rinsing. Typical procedures used for chemical cleaning are the same as those described previously for brazing.

## COATED COPPER ALLOYS

The most commonly employed coatings for copper alloys are tin, lead, tin-lead, nickel, chromium, and silver. The soldering of a coated copper alloy component depends upon the characteristics of the coating. However, the thermal conductivity of the base metal must also be considered. Except for chromium plate, none of the coatings presents any serious soldering problems. For chromium-plated copper, the chromium should be removed from the joint faces before soldering.

## FLUX REMOVAL

After the joint is soldered, flux residues that may be corrosive or otherwise prove harmful to the serviceability of the joint must be removed. Removal of flux residues is especially important when the joints will be exposed to humid environments.

Inorganic flux residues containing salts and acids should be removed completely. Residues from the organic fluxes should also be removed.

Generally, rosin flux residues can remain on the joint unless appearance is important or the joint area is to be painted or otherwise coated. The activated rosin fluxes can generally be treated in the same manner as the organic fluxes for structural soldering, but they should be removed for critical electronic applications.

Zinc chloride fluxes leave a fused residue that absorbs water from the atmosphere. Removal of this residue is best accomplished by thorough washing in a hot solution of 2 percent concentrated hydrochloric acid per gallon of water, followed by a hot water rinse. The acid solution removes the white zinc oxychloride crust, which is insoluble in water. Complete removal sometimes requires subsequent washing in hot water that contains some washing soda (sodium carbonate), followed by a clear water rinse. Occasionally some mechanical scrubbing may also be required.

The residues from the organic fluxes are usually quite soluble in hot water. Double rinsing is always advisable. Oily or greasy paste flux residues are generally removed with an organic solvent. Soldering pastes are usually emulsions of petroleum jelly and a water solution of zinc-ammonium chloride. Because of the corrosive nature of the chlorides, residues must be removed to prevent subsequent corrosion of the soldered joints.

If rosin residues must be removed, alcohol or chlorinated hydrocarbons may be used. Certain rosin activators are soluble in water but not in organic solvents. These flux residues require removal by organic solvents, followed by a water rinse.

## MECHANICAL PROPERTIES

The mechanical properties of a soldered joint depend upon a number of process variables in addition to the solder composition. These variables include thickness of the solder in the joint, base metal composition, cleaning procedures, type of flux, soldering temperature, soldering time, and cooling rate.

Soldered joints are normally loaded in shear. Shear strength and creep strength in shear are the important mechanical properties. For specialized applications, such as auto radiators, peel strength and fracture initiation strength may be important. In a few cases, tensile strength is of interest.

### Shear Strength

Shear strength is determined using single- or double-lap flat specimens or sleeve type cylindrical specimens. Testing is done with a crosshead speed of 1.0 or 0.1 in./min. The shear strengths of copper joints soldered with lead-tin solders are shown in Fig. 7.7. The maximum joint shear strength is obtained with solder of eutectic composition (63 percent tin-37 percent lead). Strength may decrease up to 30 percent if the joints are aged at room temperature or moderately elevated temperature for several weeks

## Table 7.24
## Solderability of copper and copper alloys

| Base metal | Solderability |
| --- | --- |
| Copper (Includes tough-pitch, oxygen-free, phosphorized, arsenical, silver-bearing, leaded, tellurium, and selenium copper.) | Excellent. Rosin or other noncorrosive flux is suitable when properly cleaned. |
| Copper-tin alloys | Good. Easily soldered with activated rosin and intermediate fluxes. |
| Copper-zinc alloys | Good. Easily soldered with activated rosin and intermediate flux. |
| Copper-nickel alloys | Good. Easily soldered with intermediate and corrosive fluxes. |
| Copper-chromium and copper-beryllium | Good. Require intermediate and corrosive fluxes and precleaning. |
| Copper-silicon alloys | Fair. Silicon produces refractory oxides that require use of corrosive fluxes. Should be properly cleaned. |
| Copper-aluminum alloys | Difficult. High aluminum alloys are soldered with help of very corrosive fluxes. Precoating may be necessary. |
| High-tensile manganese bronze | Not recommended. Should be plated to ensure consistent solderability. |

prior to testing.

Shear strength of soldered joints decreases with increasing temperature, as shown in Fig. 7.8 for two solders commonly used for copper plumbing. Many solders remain ductile at cryogenic temperatures, and their strengths increase significantly as the temperature decreases below room temperature.

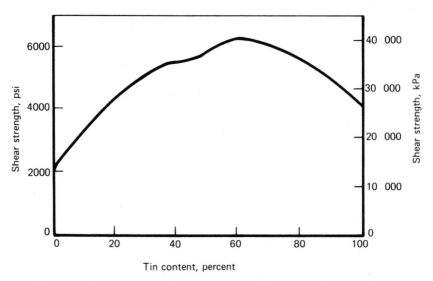

Fig. 7.7 – Shear strengths of copper joints soldered with tin-lead solders

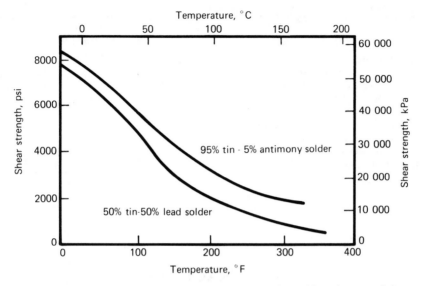

**Fig. 7.8–Shear strengths at elevated temperatures for soldered copper joints**

## Creep Strength

The creep strength in shear of a soldered joint is considerably less than its short-time shear strength, sometimes below 10 percent of it. The creep shear strengths of copper joints soldered with three solders are shown in Fig. 7.9. The 50 percent tin-50 percent lead and 95 percent tin-5 percent antimony solders have about the same short-time shear strengths, but the tin-antimony

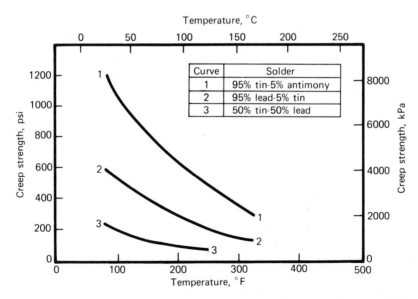

**Fig. 7.9–Creep strengths at elevated temperatures for soldered copper joints**

solder is much stronger in creep.

## Tensile Strength

Typical tensile strengths of soldered butt joints made with five compositions of tin-lead solder are presented in Table 7.25. Soldered joints are much stronger in tension than in shear. Tensile strength increases with increasing tin content up to the eutectic composition. Soldering is not normally recommended for butt joints in copper because any voids in the solder layer will cause premature failure through the solder when the joint is loaded.

**Table 7.25
Tensile strength of soldered copper butt joints**

| Solder composition, percent | | Tensile strength, ksi |
|---|---|---|
| Tin | Lead | |
| 20 | 80 | 11.3 |
| 30 | 70 | 13.8 |
| 40 | 60 | 16.8 |
| 50 | 50 | 18.2 |
| 63 | 37 | 19.6 |

# SAFE PRACTICES

Some brazing filler metals contain cadmium, and some copper alloys and solders contain beryllium, lead, or zinc. When these materials are used, adequate ventilation must be provided to protect personnel.

Copper and zinc fume and dust can cause irritation of the upper respiratory tract, nausea, and metal fume fever. They may also cause skin irritation and dermatitis as well as eye problems. Cadmium and beryllium fume are toxic when inhaled.

Fluxes used for welding, brazing, and soldering certain copper alloys may contain fluorides and chlorides. Fume from these fluxes can be very irritating to the eyes, nose, throat, and skin. Some fluorine compounds are very toxic.

When brazing, welding, soldering, or grinding copper or copper alloys, adequate ventilation is necessary to keep the fume and dust in the breathing zone within safe levels. Local exhaust ventilation should be used in confined spaces.

Good personal hygiene should be practiced, particularly before eating. Food and beverages should not be stored or consumed in the work area. Contaminated clothing should be changed.

Furnaces or retorts that use a flammable brazing atmosphere must be purged of air prior to heating. In addition, an atmosphere furnace must be purged with air before personnel enter it to avoid their suffocation.

## Metric Conversion Factors

$t_C = 0.56 (t_F-32)$
$X/°F = 1.9 (X/°C)$
1 in. = 25.4 mm
1 in./min = 0.423 mm/s
1 psi = 6.89 kPa
1 ksi = 6.89 MPa
1 ft·lb = 1.36 J
1 kA/in.$^2$ = 1.55 A/mm$^2$
1 Btu/(ft • h • °F) = 1.73 W/ (m•K)

# SUPPLEMENTARY READING LIST

Belkin, E., and Nagata, P. K., Hydrogen embrittlement of tough pitch copper by brazing, *Welding Journal,* 54(2): 54s-62s; 1975 Feb.

Brandon, E., The weldability of oxygen-free boron deoxidized and deoxidized low phosphorus copper, *Welding Journal,* 48(5): 187s-194s; 1969 May.

*Brazing Manual,* 3rd Ed., Miami: American Welding Society, 1976.

Dimbylow, C. S., and Dawson, R. J. C., Assessing the weldability of copper alloys, *Welding and Metal Fabrication:* 461-71; 1978 Sept.

Gaffoglio, C. J., Copper takes a lot of heat, *Welding Design and Fabrication:* 90-99; 1979 Oct.

Goldstein, S., TIG welding of beryllium copper, *Welding Engineer,* 58; 1979 Sept.

*Guide to the Welding of Copper-Nickel Alloys,* New York: International Nickel Company, 1979.

Hartsell, E. W., Joining copper and copper alloys, *Welding Journal,* 52(2): 89-100; 1973 Feb.

Howes, M. A. H., and Saperstein, Z. P., The reaction of lead-tin solders with copper alloys, *Welding Journal,* 48(2): 80s-85s; 1969 Feb.

Johnson, L. D., Some observations on the electron beam welding of copper, *Welding Journal,* 49(2): 55s-60s; 1970 Feb.

Lesuer, D. R., and Weil, R., Mechanism of solder fluxing on copper surfaces, *Welding Journal,* 51(6): 325s-328s; 1972 June.

Littleton, J., et al, Nitrogen porosity in gas shielded arc welding of copper, *Welding Journal,* 53(12): 561s-565s; 1974 Dec.

Miller, V. R., and Schwancke, A. E., Interface compositions of silver filler metals on copper, brass, and steel, *Welding Journal,* 57(10): 303s-310s; 1978 Oct.

Munse, W. H., and Alagia, J. A., Strength of brazed joints in copper alloys, *Welding Journal,* 36(4): 177s-184s; 1957 Apr.

Sapterstein, Z. P., and Howes, M. A. H., Mechanical properties of soldered joints in copper alloys, *Welding Journal,* 48(8): 317s-327s; 1969 Aug.

Savage, W. F., et al, Microsegregation in 70Cu-30Ni weld metal, *Welding Journal,* 55(6): 165s-173s; 1976 June.

Savage, W. F., et al, Microsegregation in partially heated regions of 70Cu-30Ni weldments, *Welding Journal,* 55(7): 181s-187s; 1976 July.

*Soldering Manual,* 2nd Ed., Miami: American Welding Society, 1977.

Wold, K., Welding copper and copper alloys, *Metal Progress,* 108(8): 43-47; 1975 Dec.

# 8
# Aluminum Alloys

## Chapter Committee

F.R. HOCH, *Chairman*
*Aluminum Company of America*
D.L. PEMBERTON
*Reynolds Metals Company*
I.B. ROBINSON
*Kaiser Aluminum and
Chemical Corporation*

G.R. ROTHSCHILD
*Welding Consultant*
H.L. SAUNDERS
*Aluminum Company of Canada*
G.N. SPENSER
*General Motors Corporation*

## Welding Handbook Committee Member

D.D. RAGER
*Reynolds Metals Company*

# 8
# Aluminum Alloys

## CHARACTERISTICS

Aluminum is known for its light weight and corrosion resistance. Its density is about 0.1 lb/in.$^3$, which is approximately one-third that of steel or copper. It has good corrosion resistance to air, water, oils, and many chemicals. This is attributed to the tenacious, refractory oxide film that reforms rapidly on a clean surface in air. This oxide is virtually insoluble in the metal and inhibits wetting by molten filler metals. Therefore, it must be removed or broken up during welding, brazing, or soldering to permit good wetting and fusion.

Aluminum has thermal and electrical conductivities approximately four times greater than steel. Consequently, higher heat inputs are required when fusion welding aluminum. Thick sections may require preheating to reduce the heat input needed for welding. For resistance spot welding, higher current density and shorter weld time are needed for aluminum than for steel of equivalent thickness.

This metal is highly reflective to radiant energy, including visible light and heat. Consequently, it does not assume a dark red color at 1200° F just prior to melting, as does steel. The lack of color change makes it difficult to judge when the metal is approaching the molten state during welding.

Aluminum is nonmagnetic and, therefore, arc blow is not encountered when arc welding with direct current. A welding arc or an electron beam will not be deflected from the intended path when aluminum components are located near it.

The linear coefficient of thermal expansion of aluminum is about twice that of steel. However, the melting range of aluminum alloys is roughly one-half that of steel. Therefore, the total expansion during welding for identical aluminum and steel components would be about equal since the differences in their coefficients of thermal expansion and melting ranges tend to balance each other.

Aluminum has a face-centered cubic crystal structure. The pure metal is very ductile, even at cryogenic (subzero) temperatures, and has very low strength in the annealed condition. The metal is strengthened by alloying, cold working, heat treatment, or a combination of these. Heating during welding, brazing, or soldering may alter these strengthening mechanisms and change the mechanical properties of the base metal. This must be considered in the component design, the selection of the joining process, and the manufacturing procedures.

## ALUMINUM ALLOYS

Aluminum is alloyed principally with copper, magnesium, manganese, silicon, and zinc. Small additions of chromium, iron, nickel, and titanium are sometimes added to specific alloy systems to obtain desired properties and grain refinement. Magnesium, manganese, silicon, and iron, singly or in various combinations, are used to strengthen aluminum by solid solution or by dispersed intermetallic compounds within the matrix. Silicon addition also lowers the melting point and increases the fluidity of aluminum.

Copper, magnesium, silicon, and zinc additions in appropriate amounts produce alloys that are heat treatable. These elements, singly or in various combinations, have increasing solid solubility in aluminum with increasing temperature. Therefore, such alloys can be strengthened by subjecting them to appropriate thermal treatments alone or in combination with cold working.

The effect of a strengthening heat treatment or cold working, or both, may be negated by the thermal cycle of a joining operation. Heat treating in conjunction with or after joining may provide optimum mechanical properties in a welded or brazed joint.

## FORMS

Aluminum products are produced by common casting and metalworking techniques. Alloys are available for sand, permanent mold, and die castings. Wrought alloys are produced in many forms including sheet, plate, tubing, rod, wire, extruded shapes, and forgings.

Substantially the same welding, brazing, or soldering practices are used for both cast and wrought products. Conventional die castings are not normally used for welded construction; however, such castings may be adhesive bonded and soldered to a limited extent. Vacuum die castings may be welded satisfactorily for some applications because they are less porous than conventional die castings.

Cast or extruded sections that are to be welded together may be designed to provide desired joint geometry, alignment, and reinforcement. Examples of such designs are shown in Fig. 8.1. An integral lip can be provided on one piece to facilitate joint alignment or serve as a weld backing, or both.

Some alloys are clad with high purity aluminum or a special aluminum alloy on one or both sides to improve corrosion resistance. A cladding layer is usually from 2.5 to 5 percent of

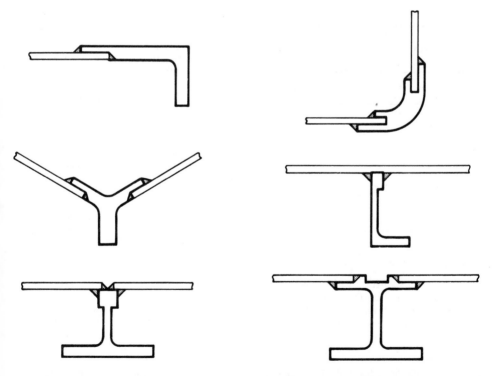

**Fig. 8.1—Typical extrusion designs incorporating desired joint geometry, alignment, and reinforcement**

the total thickness. Special composites are also produced for brazing or surface finishing purposes.

# WROUGHT ALLOYS

## Designations

A system of four-digit numerical designations is used to identify the various wrought aluminum alloys.[1] The first digit indicates the alloy group, as shown in Table 8.1. The second digit indicates a modification of the original alloy or the impurity limit in the case of unalloyed aluminum. The third and fourth digits identify the alloy or indicate the aluminum purity.

A temper designation system is used to indicate the condition of a product. It is based on the sequence of basic treatments used to produce the desired mechanical properties. This designation follows the alloy designation. The basic temper designations are shown in Table 8.2. Subsequent divisions of the basic letter tempers are indicated by one or more digits following the letter. These digits designate a specific sequence of basic treatments.

## Table 8.1
## Designations for wrought aluminum alloy groups

| Alloy group | Series designation |
|---|---|
| Aluminum, 99.00% min purity | 1XXX |
| Aluminum–copper | 2XXX |
| Aluminum–manganese | 3XXX |
| Aluminum–silicon | 4XXX |
| Aluminum–magnesium | 5XXX |
| Aluminum–magnesium–silicon | 6XXX |
| Aluminum–zinc | 7XXX |
| Other | 8XXX |

## Nonheat-Treatable Alloys

The initial strengths of the nonheat-treatable alloys depend primarily upon the hardening effect of alloying elements such as silicon, iron, manganese, and magnesium. These elements increase the strength of aluminum by formation of dispersed phases in the metal matrix or by solid solution. The nonheat-treatable alloys are mainly found in the 1XXX, 3XXX, 4XXX and 5XXX series depending upon their major alloying elements.[2] Iron and silicon are the major impurities in commercially pure aluminum but they do contribute to its strength. Magnesium is the most effective solution-strengthening addition. Aluminum-magnesium alloys of the 5XXX series have relatively high strength in the annealed condition. All of the nonheat-treatable alloys are work-hardenable.

The nonheat-treatable alloys may be annealed by heating to an elevated temperature to remove the effects of cold working and improve ductility. The proper annealing schedule to use will depend upon the alloy and its temper.[2] When fusion welding the nonheat-treatable alloys, the heat-affected zone may lose the strengthening effects of cold working. Thus, the strength in this zone may decrease to near that of annealed metal.

## Heat-Treatable Alloys

The heat-treatable alloys are found in the 2XXX, 4XXX, 6XXX and 7XXX series. The strength of any of these alloys depends only upon the alloy composition, in the annealed condition as do the nonheat-treatable alloys. However, copper, magnesium, zinc, and silicon, either singly or in various combinations, show a marked increase in solid solubility in aluminum with increasing temperature. Therefore, these alloys can be strengthened by appropriate thermal treatments.

Heat-treatable aluminum alloys develop their improved strength by solution heat treating followed by either natural or artificial aging. Cold working before or after aging may provide additional strength. Heat-treated alloys may be annealed to provide maximum ductility with a sacrifice in strength properties. Annealing is achieved by heating the component at an elevated temperature for a specified time, and then cooling it at a controlled rate.

During fusion welding, the heat-affected zone will be exposed to sufficiently high temperatures to overage heat-treated metal. As a result, this zone will be softened to some extent.

---

1. Refer to ANSI Publication H35.1-1978, *Alloy and Temper Designation Systems for Aluminum*, New York: American National Standards Institute.

2. Refer to *Aluminum Standards and Data*, ASD-1 or ASD-1M, published by The Aluminum Association, Washington, D.C., for further information.

**Table 8.2**
**Basic temper designations for aluminum alloys**

| Designation | Condition |
|---|---|
| F | As-fabricated |
| O | Annealed |
| H1 | Strain hardened only |
| H2 | Strain hardened and partially annealed |
| H3 | Strain hardened and thermally stabilized |
| W | Solution heat-treated |
| T1 | Cooled from an elevated temperature shaping process and naturally aged |
| T2 | Cooled from an elevated temperature shaping process, cold worked, and naturally aged |
| T3 | Solution heat-treated[a] cold worked, and naturally aged |
| T4 | Solution heat-treated and naturally aged |
| T5 | Cooled from an elevated temperature shaping process and then artificially aged |
| T6 | Solution heat-treated and then artificially aged |
| T7 | Solution heat-treated and stabilized |
| T8 | Solution heat-treated, cold worked, and then artificially aged |
| T9 | Solution heat-treated, artificially aged, and then cold worked |
| T10 | Cooled from an elevated temperature shaping process, cold worked, and then artificially aged |

a. Achieved by heating to and holding at a suitable temperature long enough to allow constituents to enter into solid solution, and then cooling rapidly to hold the constituents in solution.

## CASTING ALLOYS

A system of four-digit numerical designations is used to identify aluminum casting alloys. The first digit indicates the alloy group as shown in Table 8.3. The second two digits identify the aluminum alloy within the group. The last digit, which is separated from the others by a decimal point, indicates the product form, i.e., casting or ingot. A modification of the original alloy or impurity limits is indicated by a serial letter before the numerical designation. The serial letters are asigned in alphabetical sequence starting with A but omitting I, O, Q, and X. The temper designation system for castings is the same as that for wrought aluminum products (see Table 8.2.)

Casting alloys are either nonheat-treatable or heat-treatable. A second classification may be made according to the type of casting method for which the alloy is suitable, namely sand, permanent mold, or die casting.

**Table 8.3**
**Designations for cast aluminum alloy groups**

| Alloy group | Series designation |
|---|---|
| Aluminum, 99.00% min purity | 1XX.X |
| Aluminum–copper | 2XX.X |
| Aluminum–silicon–copper or aluminum–silicon–magnesium | 3XX.X |
| Aluminum–silicon | 4XX.X |
| Aluminum–magnesium | 5XX.X |
| Aluminum–zinc | 7XX.X |
| Aluminum–tin | 8XX.X |
| Other alloy systems | 9XX.X |

# SELECTION OF JOINING METHOD

Aluminum alloys can be joined by most fusion and solid-state welding processes as well as by brazing and soldering.[3] Fusion welding is commonly done by gas metal arc, gas tungsten arc, and resistance spot and seam welding. Plasma and electron beam welding are used in special applications. Shielded metal arc and oxyfuel gas welding may be used for applications where high strength and quality are not essential for the intended service. Aluminum has been joined by the vertical electrogas processs in the laboratory with limited success. Submerged arc welding is not applied commercially.

Solid-state welding processes suitable for joining most aluminum alloys are friction, diffusion, explosion, high frequency, and cold welding. These processes are also suitable for joining aluminum to certain other metals, particularly those combinations that are not metallurgically compatible.[4] Brazing, soldering, and adhesive bonding are also used for joining aluminum components. Adhesive bonding in combination with resistance spot welding, called weld bonding, produces joints with better strength properties than those produced by either process individually.

The relative ease with which most wrought and cast aluminum alloys can be joined by welding, brazing, and soldering is given in Tables 8.4 through 8.7, together with the alloy melting range. Table 8.8 gives the practical thickness or cross-sectional area that can be joined by various processes.

The applications and the service environment of an assembly often influence the selection of the joining method as well as the alloy and its temper. One should not be selected without consideration for the other.

Joint design is important in selecting the joining method. If butt joints are required, the choice is limited preferably to an inert gas or vacuum shielded fusion welding process or a solid-state welding process. In some applications, shielded metal arc or oxyfuel gas welding may be satisfactory, but the weld quality will be inferior to that obtainable with an inert gas

shielded process. Special processes adaptable to butt joints are flash, high-frequency, and electron beam welding.

Resistance spot and seam welding, ultrasonic welding, and adhesive bonding are applicable to lap joints. Lap, T, or line contact joints are widely used for brazed or soldered designs. Where joints are relatively inaccessible, furnace or dip brazing may be used to advantage. The choice of method may also depend upon whether the joint will be made in the shop or in the field, and whether the part can be moved to the joining equipment.

Service requirements of the assembly will also influence the process selection. Important considerations are static, impact, and fatigue strength requirements as well as corrosion resistance and service temperature. Joint quality requirements, applicable nondestructive testing techniques, and leak-tightness must be considered also.

Since most joining methods use heat to effect a metallic bond, softening of the base material is another consideration. Also, heating during the joining operation may affect the corrosion resistance of some alloys. Resistance, laser, electron beam, flash, cold, and ultrasonic welding can be done with minimum heat input to the parts. Adhesive bonding requires no heat, unless curing must be done at elevated temperature. Soldering can be done at relatively low temperatures and the metal properties are not changed significantly. A rapid welding process, such as gas metal arc, electron beam, or flash welding, is best when a narrow heat-affected zone is desired.

Appearance is often of considerable importance in selecting a joining method. Gas tungsten arc, gas metal arc, plasma arc, and electron beam welding can provide the best as-welded bead surface. Joints with excellent appearance and little or no finishing requirements can be made by brazing or adhesive bonding. Resistance spot and seam welding and ultrasonic welding may mar sheet surfaces. Cold welding requires large metal deformations to accomplish metallurgical bonding. The process is not suitable when a smooth surface is needed. The flash may be removed from flash or stud welds to improve appearance.

Welded parts may be given a chemical or anodic (electrochemical) surface treatment to provide corrosion resistance, coloring, or both.

---

3. Welding and other joining processes are covered in the *Welding Handbook*, Vols. 2 and 3, 7th Ed.
4. The welding of dissimilar metals is discussed in Chapter 12.

**Table 8.4**
**Relative ease of joining nonheat-treatable wrought aluminum alloys**

| Alloy | Process[1,2,3] | | | | Melting range, °F |
|---|---|---|---|---|---|
| | GMAW GTAW | RSW RSEW | B | S | |
| 1060 | A | B | A | A | 1195-1215 |
| 1100 | A | A | A | A | 1190-1215 |
| 1350 | A | B | A | A | 1195-1215 |
| 3003 | A | A | A | A | 1190-1210 |
| 3004 | A | A | B | B | 1165-1205 |
| 3105 | A | A | B | B | 1175-1210 |
| 5005 | A | A | B | B | 1170-1210 |
| 5050 | A | A | B | B | 1155-1205 |
| 5052 } 5652 | A | A | C | C | 1125-1200 |
| 5083 | A | A | X | X | 1095-1180 |
| 5086 | A | A | X | X | 1085-1185 |
| 5154 } 5254 | A | A | X | X | 1100-1190 |
| 5182 | A | B | X | X | 1070-1180 |
| 5252 | A | A | C | C | 1125-1200 |
| 5454 | A | A | X | X | 1115-1195 |
| 5456 | A | A | X | X | 1055-1180 |
| 5457 | A | A | B | B | 1165-1210 |
| 5557 | A | A | A | A | 1180-1215 |
| 5657 | A | A | B | B | 1175-1210 |
| 7072 | A | A | A | A | 1185-1215 |

1.  GMAW   – Gas metal arc welding
    GTAW   – Gas tungsten arc welding
    RSW     – Resistance spot welding
    RSEW   – Resistance seam welding
    B         – Brazing
    S         – Soldering

2.  Rating, based on the temper most readily joined:
    A   – readily joined by the process.
    B   – joinable by the process for most applications; may require special techniques or tests to establish suitable procedures.
    C   – joining by the process is difficult.
    X   – joining by the process is not recommended.

3.  All alloys can be joined by ultrasonic welding or adhesive bonding. Some alloys can be ultrasonic or abrasive soldered.

Resistance, ultrasonic, and cold welds are least noticeable after such a treatment. However, thorough rinsing is necessary to avoid color change and entrapment of chemical solutions in lap joints.

Filler metals that contain large amounts of silicon will darken during an anodic treatment. In addition, all flux must be completely removed from brazed, soldered, and welded joints prior to such a treatment.

Not all adhesives can be subjected to chemical or anodic treatment. However, adhesive bonding can be used successfully to join pretreated parts.

Economic considerations of process selection include costs of equipment, labor, and related operations, such as preparation and finishing. Production requirements, probability of repeat business, and availability of equipment also influence process selection. Automation is desirable when a number of assemblies of the same size and shape are to be produced.

## Table 8.5
## Relative ease of joining heat-treatable wrought aluminum alloys

| Alloy | Process[1,2,3] | | | | Melting range, °F |
|-------|------|------|---|---|------|
| | GMAW GTAW | RSW RSEW | B | S | |
| 2014 | C | A | X | C | 950-1180 |
| 2017 | C | A | X | C | 955-1185 |
| 2024 | C | A | X | C | 935-1180 |
| 2036 | C | A | X | | 1030-1200 |
| 2218 | C | A | X | C | 940-1175 |
| 2219 | A | A | X | C | 1010-1190 |
| 2618 | C | A | X | C | 1040-1185 |
| 6009 | A | A | A | B | 1040-1200 |
| 6010 | A | A | A | B | 1085-1200 |
| 6061 | A | A | A | B | 1080-1200 |
| 6063 | A | A | A | B | 1140-1210 |
| 6070 | A | A | C | B | 1050-1200 |
| 6101 | A | A | A | A | 1150-1210 |
| 6201 | A | A | A | B | 1125-1210 |
| 6951 | A | A | A | A | 1140-1210 |
| 7005 | A | A | B | B | 1125-1195 |
| 7039 | A | A | C | B | 1070-1180 |
| 7075 | C | A | X | C | 890-1175 |
| 7178 | C | A | X | C | 890-1165 |

1. GMAW –Gas metal arc welding
   GTAW –Gas tungsten arc welding
   RSW –Resistance spot welding
   RSEW –Resistance seam welding
   B –Brazing
   S –Soldering

2. Rating, based on the temper most readily joined:
   A –readily joined by the process.
   B –joinable by the process for most applications; may require special techniques or tests to establish suitable procedures.
   C –joining by the process is difficult.
   X –joining by the process is not recommended.

3. All alloys can be joined by ultrasonic welding or adhesive bonding. Some alloys can be ultrasonic or abrasive soldered.

**Table 8.6**
**Relative ease of joining nonheat-treatable wrought aluminum alloys**

| Alloy | Process[1, 2, 3] GMAW GTAW | RSW RSEW | B | S | Melting range, °F |
|---|---|---|---|---|---|
| | | Sand castings | | | |
| 208.0 | B | B | X | C | 970-1160 |
| B443.0 | A | A | A | C | 1065-1170 |
| 514.0 | A | B | X | X | 1110-1185 |
| B514.0 | B | B | X | X | 1090-1170 |
| A712.0 | B | C | B | B | 1105-1195 |
| D712.0 | B | C | A | B | 1135-1200 |
| | | Permanent mold castings | | | |
| 213.0 | B | C | X | C | 965-1160 |
| B443.0 | A | A | A | C | 1065-1170 |
| A514.0 | A | B | X | X | 1075-1180 |
| | | Die castings | | | |
| 360.0 | B | C | X | X | 1035-1105 |
| 380.0 | B | C | X | C | 1000-1100 |
| 413.0 | B | C | X | X | 1065-1080 |
| 518.0 | B | C | X | X | 995-1150 |

1. GMAW –Gas metal arc welding
   GTAW –Gas tungsten arc welding
   RSW –Resistance spot welding
   RSEW –Resistance seam welding
   B –Brazing
   S –Soldering

2. Rating, based on the temper most readily joined:
   A –readily joined by the process.
   B –joinable by the process for most applications; may require special techniques or tests to establish suitable procedures.
   C –joining by the process is difficult.
   X –joining by the process is not recommended.

3. All alloys can be joined by ultrasonic welding or adhesive bonding. Some alloys can be ultrasonic or abrasive soldered.

**Table 8.7**
**Relative ease of joining heat-treatable cast aluminum alloys**

| Alloy | Process[1,2,3] | | | | Melting range, °F |
|-------|------|------|---|---|---------|
| | GMAW GTAW | RSW RSEW | B | S | |
| *Sand castings* | | | | | |
| 222.0 | C | C | X | X | 965-1155 |
| 242.0 | C | C | X | X | 990-1175 |
| 295.0 | C | C | X | C | 970-1190 |
| 319.0 | B | B | X | X | 960-1120 |
| 355.0 | B | B | X | X | 1015-1150 |
| 356.0 | A | A | C | C | 1035-1135 |
| 520.0 | B | C | X | X | 840-1120 |
| *Permanent mold castings* | | | | | |
| 222.0 | C | C | X | X | 965-1155 |
| 242.0 | C | C | X | X | 990-1175 |
| A332.0 | B | B | X | X | 1000-1050 |
| F332.0 | B | C | X | X | 970-1080 |
| 330.0 | B | B | X | X | 960-1085 |
| 354.0 | B | B | X | X | 1000-1105 |
| 355.0 | B | B | X | X | 1015-1150 |
| C355.0 | B | B | X | X | 1015-1150 |
| 356.0 | A | A | C | C | 1035-1135 |
| A356.0 | A | A | C | C | 1035-1135 |
| A357.0 | A | B | B | C | 1035-1135 |
| 359.0 | B | B | B | C | 1045-1115 |

1. GMAW –Gas metal arc welding
   GTAW –Gas tungsten arc welding
   RSW –Resistance spot welding
   RSEW –Resistance seam welding
   B –Brazing
   S –Soldering

2. Rating, based on the temper most readily joined:
   A –readily joined by the process.
   B –joinable by the process for most applications; may require special techniques or tests to establish suitable procedures.
   C –joining by the process is difficult.
   X –joining by the process is not recommended.

3. All alloys can be joined by ultrasonic welding or adhesive bonding. Some alloys can be ultrasonic or abrasive soldered.

**Table 8.8**
**Practical aluminum thickness or area ranges for various joining processes**

| Joining process | Thickness, in. or (area, in.$^2$) | |
|---|---|---|
| | Min | Max |
| Gas metal arc welding | 0.12 | No limit |
| Gas tungsten arc welding | 0.02 | 1 |
| Resistance spot welding | Foil | 0.18 |
| Resistance seam welding | 0.01 | 0.18 |
| Flash welding | 0.05 | (12) |
| Stud welding | 0.02 | No limit |
| Cold welding–butt joint | (0.0005) | (0.2) |
| Cold welding–lap joint | Foil | 0.015 |
| Ultrasonic welding | Foil | 0.12 |
| Electron beam welding | 0.02 | 6 |
| Brazing | 0.006 | No limit |

# WELDING FILLER METALS

Aluminum welding rods and bare electrodes are generally used with the gas tungsten arc, gas metal arc, and oxyfuel gas welding processes. In special cases, these filler metals may be used with other fusion welding processes, such as electron beam or plasma arc welding. A limited number of aluminum covered electrodes are available for shielded metal arc welding.

## CLASSIFICATIONS

The filler metals are classified by the same four-digit systems used to designate wrought and cast aluminum alloys.[5] Filler metals for joining wrought aluminum alloys fall into the 1XXX (Al), 2XXX (Al-Cu), 3XXX (Al-Mn), 4XXX (Al-Si), or 5XXX (Al-Mg) series. All but the E3003 electrode can be used with the gas metal arc, gas tungsten arc, and plasma arc welding processes. The 1XXX and 4XXX series are the only two recommended for use with oxyfuel gas welding. The four-digit number is prefixed with the letter E to indicate suitability as only an

electrode, with R for a welding rod, or with ER for both applications.

## BARE RODS AND ELECTRODES

Aluminum filler metal ER1100 can be used to weld all of the 1XXX aluminum alloys, as well as the 3003 and 5005 alloys. It provides adequate tensile strength and ductility for butt joints, and very good electrical conductivity and corrosion resistance. A filler metal of equal or higher purity is recommended for high purity aluminum base metals for chemical applications.

ER2319 filler metal may be used for welding 2219 and 2014 aluminum alloys as well as the Al-Cu casting alloys. This filler metal is heat-treatable, and can provide high strength and good ductility with the Al-Cu casting alloys.

Filler metals ER4043 and ER4047 may be used to weld a wide variety of base metals. The former is available as a covered electrode for shielded metal arc welding. The latter is commonly used for brazing. Both filler metals may be used with all 1XXX, 3XXX, and 6XXX series alloys, as well as with 2014, 2219, 5005, 5050, 5052, 7005, and 7039 alloys. They are also used for welding Al-Si and Al-Si-Mg casting alloys, or any combinations of these cast and wrought alloys.

---

5. For additional information, refer to AWS A5.3-80, *Specification for Aluminum and Aluminum Alloy Covered Arc Welding Electrodes*, and AWS A5.10-80, *Specification for Aluminum and Aluminum Alloy Bare Welding Rods and Electrodes.*

ER4043 and ER4047 filler metals exhibit low sensitivity to cracking during welding, moderate strength, and good corrosion resistance. Their relatively high silicon content results in lower weld ductility than that obtained with the 1XXX, 2XXX, and 5XXX filler metals. Neither filler metal will respond to a precipitation-hardening heat treatment. If one of these filler metals is considered for welding a heat-treatable alloy, the application should be carefully evaluated prior to production.

ER4145 filler metal exhibits low sensitivity to weld cracking when used for welding the 2XXX wrought alloys and the Al-Cu or Al-Si-Cu casting alloys. It is excellent for repair welding these casting alloys. The welded joints should respond to heat treatment. This filler metal can be substituted for ER4043 and ER4047 filler metals for many applications, but the welded joints may have lower ductility.

Welds made with one of the 5XXX filler metals will possess higher as-welded strength than welds made with any of the other aluminum filler metals. They will also have higher ductility, except in the case of pure aluminum filler metal. The 5XXX filler metals can be used to weld the 5XXX and 6XXX alloys and 7005 alloy. Weld metal strength increases and sensitivity to cracking decreases with increasing magnesium content in the filler metal. These filler metals have excellent characteristics for cryogenic applications. Preheat and interpass temperatures should be held at 150° F or lower to avoid sensitization and subsequent stress-corrosion cracking.

Filler metals R242.0, R295.0, R355.0, and R356.0 are used for foundry repair of castings of similar alloy compositions.

## COVERED ELECTRODES

There are three classifications of covered electrodes for shielded metal arc welding of aluminum. Each classification is based on the chemical composition of the core wire and also the tensile strength and ductility of the weld metal. The core wires, unalloyed aluminum (1100), aluminum-1 percent manganese alloy (3003) and aluminum-5 percent silicon alloy (4043), are the bases for the classifications.

E1100 electrodes are designed for welding unalloyed aluminum, and E3003 electrodes for 3003 aluminum alloy. E4043 electrodes are preferred for general purpose welding because of their excellent fluidity. They can be used to weld a wide variety of base metals including 6XXX alloys, 5XXX alloys containing 2.5 percent magnesium or less, aluminum-silicon casting alloys, 3003 alloy, and unalloyed aluminum. All electrodes are used with direct current, electrode positive (DCEP).

The covering on these electrodes has three functions. It provides a gas to shield the arc, a flux to dissolve the aluminum oxide, and a protective slag to cover the cooling weld bead. It is important that all slag be removed from the completed weld to avoid corrosion in service.

Moisture in the covering of these electrodes is a major source of porosity in weld metal. To avoid this, electrodes should be stored in a heated cabinet until shortly before they are used. Any electrodes that have been exposed to moisture should be reconditioned (baked), as recommended by the manufacurer, before they are used.

One difficulty that may occur during welding is the fusing of slag over the end of the electrode when the arc is broken. This fused slag must be removed to restrike the arc.

## FILLER METAL SELECTION

Selection of an appropriate filler metal for a specific application depends upon a number of considerations including the following:

(1) Base metal composition
(2) Joint design
(3) Dilution
(4) Cracking tendencies
(5) Strength and ductility requirements
(6) Corrosion in service
(7) Appearance

Table 8.9 gives a suggested filler metal to satisfy one or more specific requirements of welded joints in commonly welded aluminum alloys. With most alloys, one filler metal meets the needs for two or more requirements.

### Base Metal Composition

Composition of the base metal as well as joint geometry determines if a filler metal is required for welding. Unalloyed aluminum and alloys with narrow melting ranges can generally be welded with or without filler metal. On the other hand, alloys with wide melting ranges usually require the addition of a suitable filler metal to avoid hot cracking, as described later.

### Weld Metal Dilution

When filler metal is added, the weld metal

**Table 8.9**
**Suggested filler metals for commonly welded aluminum alloys**
**to provide specific requirements**

| | Recommended filler metal for | | | | |
|---|---|---|---|---|---|
| Base metal | High strength | Good ductility | Color match after anodizing | Saltwater corrosion resistance | Least cracking tendency |
| 1100 | 4043 | 1100 | 1100 | 1100 | 4043 |
| 2219 | 2319 | 2319 | 2319 | 2319 | 2319 |
| 3003 | 4043 | 1100 | 1100 | 1100 | 4043 |
| 5052 | 5356 | 5654 | 5356 | 5554 | 5356 |
| 5083 | 5183 | 5356 | 5183 | 5183 | 5356 |
| 5086 | 5356 | 5356 | 5356 | 5356 | 5356 |
| 5454 | 5356 | 5554 | 5554 | 5554 | 5356 |
| 5456 | 5556 | 5356 | 5556 | 5556 | 5356 |
| 6061 | 5356 | 5356 | 5654 | 4043 | 4043 |
| 6063 | 5356 | 5356 | 5356 | 4043 | 4043 |
| 7005 | 5556 | 5356 | 5356 | 5356 | 5356 |
| 7039 | 5556 | 5356 | 5356 | 5356 | 5356 |

composition depends upon dilution of the filler metal by the base metal. The properties of the weld metal will have a direct effect on the behavior of the joint in service. Strength, ductility, resistance to cracking and corrosion, response to heat treatment, and other properties may depend upon the amount of dilution.

The extent of dilution will depend upon the joint design, welding process, and welding procedure. Weld cracking tendencies are generally minimized by keeping dilution to a minimum. This is particularly true with heat-treatable aluminum alloys. For example, dilution will be less with a V-groove butt joint than with a square-groove joint, as shown in Fig. 8.2. With the proper filler metal, the V-groove joint will ordinarily be less susceptible to cracking during welding. A U-groove joint will offer even less dilution than the V-groove joint, but edge preparation may be more costly.

## Weld Cracking

The choice of a filler metal for welding a specific aluminum alloy is an important factor in avoiding weld cracking. Cracking in aluminum welds is due to the low strength or ductility of the weld metal or the heat-affected zone at elevated temperatures. This behavior is known as hot-shortness.

A filler metal with a melting temperature similar to or below that of the base metal greatly reduces the tendency for intergranular cracking in the heat-affected zone. A filler metal with this characteristic minimizes the stresses imposed by the solidification shrinkage of the weld metal until any low melting phases in the heat-affected zone have solidified and developed sufficient strength to resist the stresses. Under highly restrained conditions, aluminum weld metal may crack if it does not possess sufficient strength during cooling to withstand the contraction stresses.

Weld metal cracking usually can be prevented by welding with a filler metal of higher alloy content than the base metal. For example, 6061 alloy, with a nominal silicon content of 0.6 percent, is extremely crack-sensitive when welded with 6061 filler metal. However, it is readily welded with 4043 filler metal, which contains 5 percent silicon. ER4043 filler metal melts at a temperature lower than that of the base metals for which it is commonly used. For this reason, it remains more plastic than the base metal and yields during cooling to relieve the contraction stresses that might cause cracking.

Under other conditions, an aluminum-magnesium alloy filler metal, such as ER5183, ER5356, or ER5556, provides good weld strength

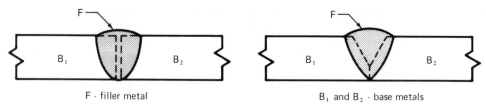

F - filler metal          B₁ and B₂ - base metals

***Fig. 8.2—Relative dilutions with square-groove and single-V-groove joint geometries***

and low crack sensitivity. ER4043 filler metal should not be used to weld the high magnesium alloys, such as 5083, 5086, or 5456. Excessive magnesium-silicide eutectics can form in the weld metal, and this will decrease ductility and increase crack sensitivity.

Weld metal compositions known to be sensitive to cracking should be avoided. Combining magnesium and copper in an aluminum weld metal is not desirable. Therefore, an ER5XXX series filler metal should not be used to weld a 2XXX series base metal. Conversely, ER2319 filler metal should not be used with a 5XXX series base metal.

In the case of Al-Si and Al-Mg alloy weld metals, sensitivity to cracking is greatest when the weld metal contains about 0.5 to 2 percent silicon or magnesium, respectively. Cracking tendencies decrease when the weld metal composition is below or above this range. For example, when 5052 alloy (2.5 percent Mg) is welded with ER5554 filler metal (2.7 percent Mg), the weld would exhibit greater susceptibility to cracking than if it were welded with ER5356 filler metal (5 percent Mg). Conversely, if ER4043 filler metal (5 percent Si) is diluted by 80 percent with 1100 aluminum base metal, the resulting weld metal composition would be in the crack-sensitive range (1 percent Si).[6] Such dilution could occur in a square-groove weld joint.

Aluminum filler metals are carefully balanced chemically to minimize weld cracking. Cracking is not usually a problem when an appropriate filler metal and welding technique are used to make a joint. Suitable procedures must be used to avoid crater cracks at termination of welding.

_____

6. See Ch. 12 for a discussion of dilution and alloying.

## Corrosion Resistance

Most aluminum base and filler metal combinations are satisfactory for service under normal atmospheric conditions. For best performance in specific corrosive environments, the filler metal composition should nearly match the chemistry of the base metal. For some corrosion applications, a special filler metal may be needed to prevent selective attack of the weld metal. The filler metal may have to be more pure or held within narrower composition limits than the standard filler metal.

Aluminum-magnesium filler metals are highly resistant to corrosion but tend to be anodic to many nonheat-treatable base metals, particularly when dilution is low. For this reason, these filler metals should be used with base metals that have a similar electrolytic solution potential if the weldment will be continuously or cyclically exposed to an electrolyte. Two examples are the use of ER5654 filler metal when welding 5254 aluminum alloy for hydrogen peroxide service, and ER1100 filler metal with 1100 aluminum base metal for some salt water or other corrosive applications.

## Anodic Coatings

Anodic coating requirements (anodizing) for a weldment should be considered in selecting a filler metal when appearance and color match are important. (See Table 8.9.) The weld zone may be darker or lighter than the base metal after anodizing unless an appropriate filler metal is used. The difference in color may be caused by one or more of the following conditions:

(1) The weld metal composition is significantly different from that of the base metal.

(2) Outlining of the cast structure of the weld metal by the anodizing process.

(3) Annealing of the base metal in the heat-affected zone.

Color differences can be minimized if a suitable filler metal is selected to produce the best color match with the particular anodizing process. However, service requirements must also be considered when selecting the appropriate filler metal. For example, a filler metal that would produce a crack-sensitive weld would be unacceptable in any case. More than one filler metal may be acceptable for a particular anodizing treatment. A choice can then be made on the basis of other service requirements.

If good strength and color match are desired, one of the magnesium alloy filler metals, such as ER5356 or ER5654, may be acceptable. Filler metals containing silicon, such as ER4043, turn dark gray with these treatments due to entrapment of elemental silicon within the anodic coating. They are therefore not recommended where anodizing is required unless a dark appearance is not objectionable.

The presence of an anodic coating on aluminum surfaces to be welded will likely cause discontinuities that are unacceptable for the application. These may include porosity, lack of fusion, and oxide inclusions in the weld metal. When welding must be done after anodizing, the coating should first be removed from the weld area. The heat of welding may discolor some types of dies and pigments on either side of the welded joint.

## Recommended Filler Metals

Selection of the correct filler metal greatly influences the service life of an aluminum weldment. Recommended filler metals for general purpose welding of various base metal combinations are given in Table 8.10. For welding castings, it is generally preferable to use a filler metal of the same composition as the casting. However, many casting alloys can be welded with the standard filler metals, as shown in Table 8.10.

When the welded joint must have specific design requirements, such as high strength or corrosion resistance, the filler metal should be selected to provide those requirements. (See Table 8.9.)

## Manufacture and Storage

Bare filler metals should be manufactured free of dissolved gases and nonmetallic impurities. They also should have a clean, smooth surface that is free of thick oxide, moisture, lubricant, and other contaminants. Care must be taken during storage and use to prevent contamination. Contaminants may be sources of hydrogen, which is the major cause of porosity in aluminum welds.

The quality of the filler metal is particularly important with the gas metal arc welding process. In this process, a relatively small diameter electrode is fed through the welding gun at a high rate of speed. The electrode can carry a relatively large amount of foreign material into the weld pool if its surface is not clean.

To avoid contamination, filler metal supplies should be covered and stored in a dry place at relatively uniform temperature. Electrode spools on welding machines should be covered. If the equipment will not be used for an extended period, the spool should be either returned to the original carton and tightly resealed or placed in a low humidity cabinet. Original electrode or rod containers should not be opened until the contents are needed for the job.

# PREPARATIONS FOR JOINING

Certain operations are required prior to fabrication to ensure that joint quality will be adequate for the intended service. These operations may include the preparation of the required joint geometry and cleaning, positioning, and fixturing the components. In some cases, preheating of the components may be needed.

## EDGE PREPARATION

Joint geometry requirements depend upon product design, section thickness, design requirements, and joining process. The joint geometry may be a simple lap, T, or corner joint with square edges or a butt joint with a groove.

## Table 8.10
### Recommended aluminum filler metals for general applications

| Base metal | 319.0, 333.0, 354.0, 355.0, C355.0 | 356.0, A356.0, A357.0, 359.0, 413.0, 433.0 | 514.0, A514.0, B514.0 | 7005, 7039, 7046, 7146, 710.0, 712.0 | 6070 | 6005, 6061, 6063, 6101, 6151, 6201, 6351, 6951 | 5456 | 5454 |
|---|---|---|---|---|---|---|---|---|
| 1060, 1350 | ER4145*c,i* | ER4043*i,f* | ER4043*e,i* | ER4043*i* | ER4043*i* | ER4043*i* | ER5356*c* | ER4043*e,i* |
| 1100, 3003, Alclad 3003 | ER4145*c,i* | ER4043*i,f* | ER4043*e,i* | ER4043*i* | ER4043*i* | ER4043*i* | ER5356*c* | ER4043*e,i* |
| 2014, 2024, 2036 | ER4145*g* | ER4145 | — | — | ER4145 | ER4145 | — | — |
| 2219 | ER4145*g,c,i* | ER4145*c,i* | ER4043*i* | ER4043*i* | ER4043*f,i* | ER4043*f,i* | ER4043 | ER4043*i* |
| 3004, Alclad 3004 | ER4043*i* | ER4043*i* | ER5654*b* | ER5356*e* | ER4043*e* | ER4043*b* | ER5356*e* | ER5654*b* |
| 5005, 5050 | ER4043*i* | ER4043*i* | ER5654*b* | ER5356*e* | ER4043*e* | ER4043*b* | ER5356*e* | ER5654*b* |
| 5052, 5652*a* | ER4043*i* | ER4043*b,i* | ER5654*b* | ER5356*e* | ER5356*b,c* | ER5356*b,c* | ER5356*b* | ER5356*e* |
| 5083 | — | ER5356*c,e,i* | ER5356*e* | ER5183*e* | ER5356*e* | ER5356*e* | ER5183*e* | ER5356*b* |
| 5086 | — | ER5356*c,e,i* | ER5356*e* | ER5356*e* | ER5356*e* | ER5356*e* | ER5356*e* | ER5356*b* |
| 5154, 5254*a* | ER4043*i* | ER4043*b,i* | ER5654*b* | ER5356*b* | ER5356*b,c* | ER5356*b,c* | ER5356*b* | ER5654*b* |
| 5454 | — | ER4043*b,i* | ER5654*b* | ER5356*b* | ER5356*b,c* | ER5356*b,c* | ER5356*b* | ER5554*c,e* |
| 5456 | — | ER4043*b,i* | ER5356*e* | ER5556*e* | ER5356*e* | ER5356*e* | ER5556*e* | |
| 6005, 6061, 6063, 6101, 6151, 6201, 6351, 6951 | ER4145*c,i* | ER5356*c,e,i* | ER5356*b,c* | ER5356*b,c,i* | ER4043*b,i* | ER4043*b,i* | | |
| 6070 | ER4145*c,i* | ER4043*e,i* | ER5356*c,e* | ER5356*c,e,i* | ER4043*e,i* | | | |
| 7005, 7039, 7046, 7146, 710.0, 712.0 | ER4043*i* | ER4043*b,i* | ER5356*b* | ER5039*e* | | | | |
| 514.0, A514.0, B514.0 | — | ER4043*b,i* | ER5654*b,d* | | | | | |
| 356.0, A356.0, A357.0, 359.0, 413.0, 443.0 | ER4145*c,i* | ER4043*d,i* | | | | | | |
| 319.0, 333.0, 354.0, 355.0, C355.0 | ER4145*c,d,i* | | | | | | | |

**Table 8.10 (cont.)**
**Recommended aluminum filler metals for general applications**

| Base metal | 5154, 5254a | 5086 | 5083 | 5052, 5652a | 5005, 5050 | 3004, Alc. 3004 | 2219 | 2014, 2024, 2036 | 1100, 3003, Alc. 3003 | 1060, 1350 |
|---|---|---|---|---|---|---|---|---|---|---|
| 1060, 1350 | ER4043e,i | ER5356c | ER5356c | ER4043i | ER1100c | ER4043 | ER4145 | ER4145 | ER1100c | ER1100c |
| 1100, 3003, Alclad 3003 | ER4043e,i | ER5356c | ER5356c | ER4043e,i | ER4043e | ER4043e | ER4145 | ER4145 | ER1100c | |
| 2014, 2024, 2036 | — | — | — | — | — | — | ER4145g | ER4145g | | |
| 2219 | ER4043i | ER4043 | ER4043 | ER4043i | ER4043 | ER4043 | ER2319c,f,i | | | |
| 3004, Alclad 3004 | ER5654b | ER5356e | ER5356e | ER4043e,i | ER4043e | ER4043e | | | | |
| 5005, 5050 | ER5654b | ER5356e | ER5356e | ER4043e,i ER5654a,b,c | ER4043d,e | | | | | |
| 5052, 5652a | ER5654b | ER5356e | ER5356e | | | | | | | |
| 5083 | ER5356e | ER5356e | ER5183e | | | | | | | |
| 5086 | ER5356b | ER5356e | | | | | | | | |
| 5154, 5254a | ER5654a,b | | | | | | | | | |

*a.* Base metal alloys 5254 and 5652 are sometimes used for hydrogen peroxide service. ER5654 filler metal is used for welding both alloys for service at 150° F and below.

*b.* ER5183, ER5356, ER5554, ER5556, and ER5654 may be used. In some cases, they provide: (1) improved color match after anodizing treatment, (2) highest weld ductility, and (3) higher weld strength. ER5554 is suitable for elevated temperature service.

*c.* ER4043 may be used for some applications.

*d.* Filler metal with the same analysis as the base metal is sometimes used.

*e.* ER5183, ER5356, or ER5556 may be used.

*f.* ER4145 may be used for some applications.

*g.* ER2319 may be used for some applications.

*i.* ER4047 may be used for some applications.

*j.* ER1100 may be used for some applications.

Notes:

1. Service conditions such as immersion in fresh or salt water, exposure to specific chemicals, or a sustained high temperature (over 150°F) may limit the choice of filler metals. Filler metals ER5356, ER5183, ER5556, and ER5654 are not recommended for sustained elevated temperature service.

2. Recommendations in this table apply to gas shielded arc welding processes. For oxyfuel gas welding, only ER1100, ER4043, ER4047, and ER4145 filler metals are ordinarily used.

3. Filler metals are listed in AWS Specification A5.10.-80.

4. Where no filler metal is listed, the base metal combination is not recommended for welding.

Methods of preparing the edges to be joined include shearing, sawing, machining, arc cutting, grinding, chipping, and filing.

Aluminum sheet and plate up to 0.5 in. thick is commonly sheared to size. The shear blades should be kept clean, sharp, and properly adjusted to provide a smooth edge that can be readily cleaned. Edges may be beveled with a portable beveling machine that operates with a progressive shearing action.

Sawing is another means of cutting and beveling aluminum. Special blades with adequate rake and chip clearance are required for free cutting. Cutting speeds must be at least 5000 ft/min.

Machining of edges may be done with several methods, most of which use high cutting speeds, such as milling and routing. These methods are best for producing more complex shapes, such as J- or U-grooves.

Edges may be prepared by plasma arc cutting and air carbon arc gouging. Plasma arc cutting can be done at high speeds with good accuracy and quality of cut. This is especially true with water injection plasma cutting.

Air carbon arc cutting is not commonly applied to aluminum, but it can be used for gouging. Its advantages are the simplicity, low cost, and portability of the equipment. The disadvantages include a lower quality surface than with machining and the presence of carbon residue. The residue must be removed by wire brushing.[7]

Filing has limited application because of the low rate of metal removed. However, there are occasions when filing is used to remove excessive roughness or for final cleaning. The correct file is a vixen or body type. Its wide-set, single-cut teeth are slow to load up.

Many woodworking tools can be used on aluminum. A good example is a standard power plane. It may be used for the same purposes as filing.

Many of the operations described are used with a lubricant. Any lubricant must be completely removed before proceeding with the joining operation.

## CLEANING

The surfaces to be joined must be clean to obtain good wetting between the filler metal and the base metal.[8] This means that they must be free of relatively thick oxide, moisture, greases, oils, paints, or any other substance. Many contaminants break down at elevated temperatures and produce hydrogen, which causes porosity in fusion welds. Degreasing may be done by wiping, spraying, or dipping in a suitable, safe solvent or by steam cleaning.

Surface oxides on aluminum can be removed by action of the welding arc as welding progresses. However, it is best to remove the oxide from the surfaces by appropriate chemical or mechanical methods. Chemical removal is accomplished with alkaline, acid, or proprietary cleaning solutions. Typical chemical treatments are given in Table 8.11.

Mechanical oxide removal is usally satisfactory if performed properly, but it is not generally as consistent as chemical methods. Wire brushing, scraping, filing, sanding, and scrubbing with steel wool are suitable methods. Wire brushing is by far the most widely used. Stainless steel brushes with 0.010 to 0.015 in. diameter bristles are recommended. To be effective, the brush must be kept clean and free of contaminants. It should be used with light pressure to avoid burnishing the aluminum surface.

Cleaning of components is best done prior to fitup and fixturing. Fitted joints may be very difficult to clean. Often, solvents or contaminants may be trapped in or near the joint and cause internal discontinuities.

## WELD BACKING

When fusion welding from one side only, some type of backing may be desirable to control the amount of root reinforcement and the shape of the root surface. Two types of backing may be used: permanent and temporary. The latter type is generally incorporated in special tooling, and inert gas may be fed into a groove in the backing material to protect the root surface from oxidation.

Strips or extruded sections of aluminum may be used for permanent backing when they are to be left in place after welding. Permanent backings may sometimes cause lack of fusion at the root of the weld if the backing material is not properly cleaned or if the fitup is poor. Permanent

---

7. Arc cutting is discussed later in this chapter.

8. The cleaning of aluminum is discussed in the *Metals Handbook*, Vol. 2, 8th Ed., Metals Park, OH: American Society for Metals, 1964: 611-20.

**Table 8.11**
**Typical chemical treatments for removal of oxide from aluminum surfaces**

| Solution | Concentration | Temperature, °F | Container material | Procedure | Purpose |
|---|---|---|---|---|---|
| Nitric acid | 50% water, 50% nitric acid, technical grade | 65-75 | Stainless steel | Immersion 15 min. Rinse in cold water, then in hot water. Dry. | Removal of thin oxide film for fusion welding. |
| Sodium hydroxide (caustic soda) followed by | 5% Sodium hydroxide in water | 160 | Mild steel | Immersion 10-60 seconds. Rinse in cold water. | |
| Nitric acid | Concentrated | 65-75 | Stainless steel | Immerse for 30 seconds. Rinse in cold water, then in hot water. Dry. | Removal of thick oxide film for all welding and brazing processes. |
| Sulfuric-chromic | $H_2SO_4$—1 gal. $CrO_3$—45 oz. Water—9 gal. | 160-180 | Antimonial lead-lined steel tank | Dip for 2-3 min. Rinse in cold water, then hot water. Dry. | Removal of films and stains from heat treating, and oxide coatings. |
| Phosphoric-chromic | $H_3PO_3$(75%)– 3.5 gal. $CrO_3$–1.75 lbs. Water–10 gal. | 200 | Stainless steel | Dip for 5-10 min. Rinse in cold water. Rinse in hot water. Dry. | Removal of anodic coatings. |

Note: There are many proprietary materials and methods available for removing aluminum oxides. Most of these are as efficient as the solutions listed.

backing should not be used in corrosive service unless all edges of the backing strip are completely sealed. Otherwise, crevice corrosion may take place at the root of the weld.

Temporary backing is generally in the form of a bar made of copper, carbon steel, or stainless steel. This bar is normally part of the fixturing. Care should be taken during welding to avoid melting the backing bar and contaminating the aluminum weld metal.

A groove may be machined into the backing bar to decrease the chilling effect at the root of the joint for good penetration. The groove should be designed to permit good fusion at the root of the weld and to provide the desired root reinforcement and contour. Groove dimensions may range from 0.13 to 0.5 in. wide and 0.01 to 0.09 in. deep. The actual dimensions will depend upon the thickness of the section being welded, the joint geometry, and the welding process.

Backing bars may be used to remove heat from welds in thin sections. This may permit welding with sufficient current for a stable arc. Water-cooled copper bars provide the greatest chilling effect; stainless steel bars, the least. However, excessive chilling may result in lack of fusion at the root of the weld.

## PREHEATING

Preheating is not normally necessary when fusion welding aluminum and should only be used to overcome or prevent problems. Its use may increase the width of the heat-affected zone and reduce the mechanical properties of welded joints in some alloys. One exception is thick sections where the heat of welding is conducted rapidly into the base metal. In this case, a moderate preheat can reduce the heat input needed for good fusion and penetration. For most alloys, the preheat temperature should not exceed 300° F, and the time of heat application should be held to a minimum. Aluminum alloys that contain magnesium in the range of about 3 to 5.5 percent should not be preheated above 250° F and the interpass temperature should not exceed about 300° F. Exposure of these alloys to temperatures in the range of 250° to 400° F for relatively short times can decrease their resistance to stress-corrosion cracking.

Preheating may be beneficial with large, complex weldments to minimize welding stresses, distortion, or cracking. The weldment should be slowly cooled to room temperature upon completion of welding.

## TACK WELDING

Tack welds are often used to hold compo-nent parts in proper relative positions or to attach starting and stopping tabs at the ends of the joint. Tack welds may have areas of porosity and incomplete fusion if they are improperly made. It is recommended that tack welds be removed as welding progresses. If this is not done, the ends of each tack weld should be tapered gradually to blend smoothly with the joint faces.

Tack welds should penetrate well and have sufficient length to provide the required strength. Craters at the ends should be filled, using techniques appropriate to the welding process. Crater cracks must be repaired prior to welding the joint. Joints that require welding from both sides should be back gouged to sound metal before welding the other side. This lessens the possibility of lack of fusion or porosity, or both, at the root of the weld.

# ARC WELDING

## JOINT GEOMETRY

In general, the recommended joint geometries for arc welding aluminum are similar to those for steel. However, smaller root openings and larger groove angles are generally used because aluminum is more fluid and welding gun nozzles are larger. Typical joint geometries for arc welding aluminum are shown in Fig. 8.3.

A special joint geometry used for aluminum is shown in Fig. 8.4. It is recommended for gas tungsten arc (GTAW) or gas metal arc welding (GMAW) when only one side is acccessible and a smooth root surface is required. The effectiveness of this design for complete joint penetration is dependent upon the surface tension of the weld metal. It can be used with section thicknesses over 0.12 in. and in all welding positions. The abutting lips are designed so that complete joint penetration is possible with the first welding pass. However, this design has a large groove area, and this requires a relatively large amount of filler metal to fill the joint. Distortion may be greater than with conventional joint designs. Its principal application is for circumferential joints in aluminum pipe.

V-groove joint designs are adequate for butt joints that are accessible from both sides. As a rule, a 60 degree groove angle is the minimum practical size for section thicknesses greater than 0.12 inch. Thick sections may require even larger groove angles, such as 75 or 90 degrees, depending upon the welding process.

With thicker plates, J-grooves are preferred to V-grooves to minimize the amount of deposited metal. Special joint geometries, such as those shown in Fig. 8.5, may be warranted when welding horizontally to minimize porosity caused by entrapment of hydrogen.

## GAS TUNGSTEN ARC WELDING

### Types of Welding Current

Aluminum can be gas tungsten arc welded using conventional ac (60 Hz), square-wave ac (SWAC), dc with the electrode either negative (DCEN) or positive (DCEP). Direct current can be pulsed to provide good control of the molten weld pool, particularly when welding in other than the flat position.

Surface cleaning of the aluminum takes place when the electrode is positive, but penetration is poor. Conversely, penetration is good with a negative electrode, but there is no cleaning

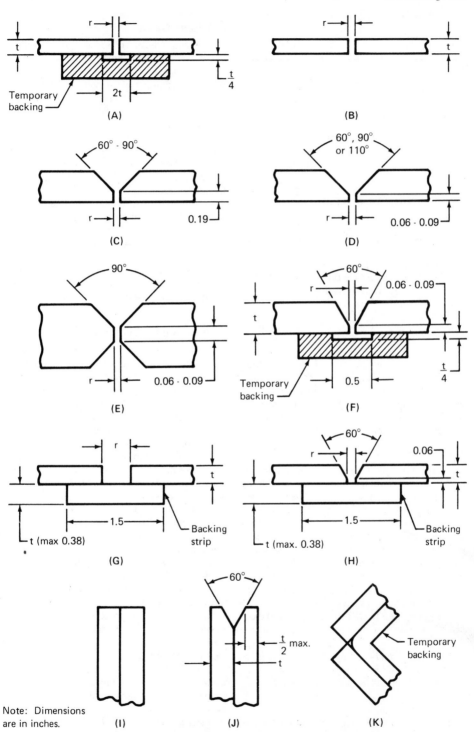

**Fig. 8.3—Typical joint geometries for arc welding of aluminum**

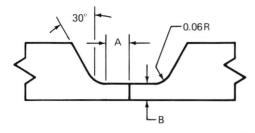

| Process | Dimensions, in. | |
|---------|------|------|
|         | A    | B    |
| GTAW    | 0.19 | 0.06 to 0.09 |
| GMAW    | 0.06 | 0.06 |

*Fig. 8.4—Special joint geometry for arc welding from one side only for complete joint penetration*

action. Alternating current can provide good cleaning action and acceptable penetration, particularly SWAC of variable frequency and pulse width. Direct current can provide good penetration (DCEN) or good cleaning (DCEP), but not both conditions simultaneously.

## Electrodes

The choice of tungsten electrode depends upon the type of welding current selected for the application. With conventional ac, better arc action is obtained when the electrode has a hemispherical-shaped tip, as shown in Fig. 8.6. AWS Classes EWP (pure tungsten) and EWZr (tungsten-zirconia) electrodes retain this tip shape well. Class EWTh-3 (tungsten-thoria) electrodes may also be used.[9] The electrode should be tapered as shown to facilitate melting the tip to form a hemispherical shape.

EWTh-1 or EWTh-2 (tungsten-thoria) electrodes are suitable with direct current or square-wave ac power. Both have higher electron emissivity, better current-carrying capacity, and longer life than do EWP electrodes. Consequently, arc starting is easier and the arc is more stable.

## Shielding Gases

Argon is the most commonly used shielding gas, particularly for manual welding, but helium is used in special cases. Arc voltage characteristics with argon permit greater arc length variations with minimal effect on arc power than does helium. Argon also provides better arc starting characteristics and improved cleaning action with alternating current.

Helium is used primarily for machine welding with DCEN power. It permits welding at higher travel speed or with greater penetration than does argon.

Helium-argon mixtures are sometimes used to take advantage of the higher heat inputs with helium while maintaining the favorable arc characteristics of argon. A mixture of 75 percent He-25 percent Ar will permit higher travel speeds with ac power. Cleaning action is still acceptable. A mixture of 90 percent He-10 percent Ar will provide better arc starting characteristics with dc power than does pure helium.

## Alternating Current Power

When ac is used in conjunction with shielding of argon or an argon-helium mixture, the surface oxide is removed by arc action. However, this cleaning action may not be satisfactory when the mixture contains 90 percent helium or more, and preweld cleaning is usually necessary. Pure helium shielding is seldom used with alternating current because the arc characteristics are poor.

The oxide removal action takes place only during the portion of the ac cycle when the electrode is positive. This action tends to rectify the ac power. To assure arc initiation during this half cycle, the power source must have either a high open-circuit voltage or an auxiliary circuit to superimpose high voltage on the welding circuit. The arc should be initiated by some means other than touching the electrode to the work to avoid tungsten contamination.

The magnitude of the current will be greater when the electrode is negative unless the power source contains appropriate electrical circuitry to balance the ac wave. For this reason, balanced-wave ac power sources are recommended for welding aluminum.

Tables 8.12, 8.13, and 8.14 give typical procedures for manual gas tungsten arc welding with ac power.[10] These tables are intended to serve

9. Refer to AWS1 A5.12-80, *Specification for Tungsten Arc Welding Electrodes*, Miami: American Welding Society.

10. Refer to AWS D10.7-60, *Recommended Practices for Gas Shielded Arc Welding of Aluminum and Aluminum Alloy Pipe* for pipe welding procedures.

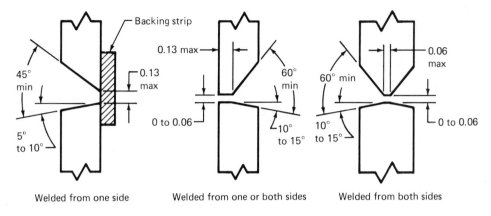

Welded from one side          Welded from one or both sides          Welded from both sides

Note: Dimensions are in inches.

**Fig. 8.5—Special weld joint geometries for arc welding aluminum in the horizontal position**

only as guides to establish welding procedures for a specific application. They are not considered to be accurate or substantiated by test results.

With aluminum alloys that are not crack sensitive, some joint designs may be fused without the addition of filler metal. Two examples are edge and corner welds in thin sheets. Close joint fitup is essential to obtain uniform melting and fusion of the sheets.

Proper gas shielding and arc cleaning action are indicated by a bright weld bead with silvery borders on each side. An oxidized weld bead may be a result of an unstable arc, low welding current, poor gas shielding, or excessive arc length.

### Direct Current, Electrode Negative Power

Gas tungsten arc welding with direct current, electrode negative (DCEN) has distinct advantages compared to ac power, particularly with machine welding where a consistent, short arc

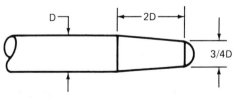

**Fig. 8.6—Approximate tungsten electrode tip shape for arc welding aluminum with ac power**

length can be easily maintained. The deep penetration possible with helium shielding is particularly useful for welding thick sections. Preheating is not normally required. With thin sections, DCEN permits much higher travel speed than does alternating current.

The surface appearance of a weld made with DCEN will be dull rather than bright because the cleaning action of the reverse polarity half cycle of ac is absent. A thin oxide film accounts for this appearance, but it is easily removed by wire brushing. Thorough preweld cleaning is essential, as is interpass cleaning with multiple-pass welds.

Argon shielding may be used with DCEN, but penetration will be less than with helium. Arc length control will not be so critical, and this may be beneficial when manually welding thin sections.

*Joint Geometry.* A square-groove joint geometry may be satisfactory for some section thicknesses that would usually require a V-groove joint with ac power. A square-groove weld in 0.75 in. thick aluminum plate is shown in Fig. 8.7. The weld was made with two passes, one from each side, without filler metal. Helium shielding and DCEN-SP power were used.

When a V-groove joint is necessary, the root face may have a greater width and the groove angle may be less than those needed with an ac arc. Examples of joint designs for thick sections are shown in Fig. 8.8.

*Manual Welding.* The technique for manual welding with DCEN power is somewhat different

**Table 8.12**
**Typical procedures for manual gas tungsten arc welding of butt joints in aluminum with ac and argon shielding**

| Section thickness, in. | Welding position[a] | Joint geometry[b] | Root opening, in. | No. of weld passes | Filler rod diam., in. | Electrode diam., in. | Welding current, A | Travel speed, in./min |
|---|---|---|---|---|---|---|---|---|
| 0.06 | F, V, H | B | 0-0.06 | 1 | 0.062,0.094 | 0.062 | 60-80 | 8-10 |
|  | O |  |  |  |  |  | 60-75 |  |
| 0.09 | F | B | 0-0.09 | 1 | 0.125 | 0.094 | 95-115 | 8-10 |
|  | V,H |  |  |  | 0.094,0.125 |  | 85-110 |  |
|  | O |  |  |  | 0.094,0.125 |  | 90-110 |  |
| 0.13 | F | B | 0-0.12 | 1-2 | 0.125,0.156 | 0.094 | 125-150 | 10-12 |
|  | V,H |  | 0-0.09 |  | 0.125 |  | 110-140 |  |
|  | O |  | 0-0.09 |  | 0.125,0.156 |  | 115-140 |  |
| 0.19 | F | D-60° | 0-0.12 | 2 | 0.156,0.188 | 0.125 | 170-190 | 10-12 |
|  | V | D-60° | 0-0.09 |  | 0.156 |  | 160-175 |  |
|  | H | D-90° | 0-0.09 |  | 0.156 |  | 155-170 |  |
|  | O | D-110° | 0-0.09 |  | 0.156 |  | 165-180 |  |
| 0.25 | F | D-60° | 0-0.12 | 2 | 0.188 | 0.156 | 220-275 | 8-10 |
|  | V | D-60° | 0-0.09 | 2 | 0.188 |  | 200-240 |  |
|  | H | D-90° | 0-0.09 | 2-3 | 0.156,0.188 |  | 190-225 |  |
|  | O | D-110° | 0-0.09 | 2 | 0.188 |  | 210-250 |  |
| 0.38[c] | F | D-60° | 0-0.12 | 2 | 0.188,0.25 | 0.188 | 315-375 | 8-10 |
|  | F | E | 0-0.09 | 2 | 0.188,0.25 |  | 340-380 |  |
|  | V | D-60° | 0-0.09 | 3 | 0.188 |  | 260-300 |  |
|  | V,H,O | E | 0-0.09 | 2 | 0.188 |  | 240-300 |  |
|  | H | D-90° | 0-0.09 | 3 | 0.188 |  | 240-300 |  |
|  | O | D-110° | 0-0.09 | 3 | 0.188 |  | 260-300 |  |

a. F–flat, V–vertical, H–horizontal, O–overhead.
b. See Fig. 8.3. Angle dimension is the appropriate groove angle.
c. May be preheated.

**Table 8.13**
**Typical procedures for manual gas tungsten arc welding of fillet welds in aluminum with ac and argon shielding**

| Section thickness, in. | Welding position[a] | No. of weld passes | Filler rod diam., in. | Electrode diam., in. | Welding current, A | Travel speed, in./min |
|---|---|---|---|---|---|---|
| 0.06 | F, H, V<br>O | 1 | 0.062, 0.094 | 0.062 | 70-110<br>65-90 | 8-10 |
| 0.09 | F<br>H, V<br>O | 1 | 0.094, 0.125<br>0.094<br>0.094 | 0.094 | 110-145<br>90-125<br>110-135 | 8-10 |
| 0.13 | F<br>H, V<br>O | 1 | 0.125 | 0.094 | 135-175<br>115-145<br>125-155 | 10-12<br>8-10<br>8-10 |
| 0.19 | F<br>H, V<br>O | 1 | 0.156 | 0.125 | 190-245<br>175-210<br>185-225 | 8-10 |
| 0.25 | F<br>H, V<br>O | 1 | 0.188 | 0.156 | 240-295<br>220-265<br>230-275 | 8-10 |
| 0.38[b] | F<br>V<br>H<br>O | 2<br>2<br>3<br>3 | 0.188 | 0.188 | 325-375<br>280-315<br>270-300<br>290-335 | 8-10 |

a. F–flat, H–horizontal, V–vertical, O–overhead.
b. May be preheated.

**Table 8.14**

**Typical procedures for manual gas tungsten arc welding of edge and corner joints in aluminum with ac and argon shielding**

| Section thickness, in. | Joint geometry[a] | No. of weld passes | Filler rod diam., in. | Electrode diam., in. | Welding current,[b,c] A | Travel speed,[c] in./min |
|---|---|---|---|---|---|---|
| 0.06 | I,K | 1 | 0.062, 0.094 | 0.062 | 60-85 | 10-16 |
| 0.09 | I,K | 1 | 0.125 | 0.062 | 90-120 | 10-16 |
| 0.13 | I,K | 1 | 0.125, 0.156 | 0.094 | 115-150 | 10-16 |
| 0.19 | J,K | 1 | 0.156 | 0.125 | 160-220 | 10-16 |
| 0.25 | J,K | 2 | 0.188 | 0.125 | 200-250 | 8-12 |

a.  See Fig. 8.3.
b.  Use current in low end of range for welding in the horizontal and vertical positions.
c.  Higher welding current and travel speed may be used for corner joints if temporary backing is employed.

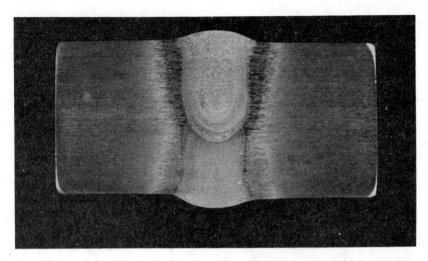

*Fig. 8.7–Two-pass square groove weld in 0.75 in. thick aluminum plate welded from both sides without filler metal with DCEN power ( ×2)*

from that used with ac power. Superimposed high voltage should be used to initiate the arc, and it should shut off automatically when the arc is established. Generally, it is not necessary to hesitate to form a molten weld pool because melting by the arc is rapid. However, a delay in travel may be necessary to achieve desired penetration at the start of the weld. A foot-operated control is recommended for adjusting the welding current as needed.

The welding torch should be moved steadily forward, and the filler rod should be fed evenly into the leading edge of the molten weld pool. Another approach is to lay the filler metal on the joint and melt it as the arc moves forward. In this case, the filler rod size must be selected to produce the desired weld bead size. In all cases, the crater should be completely filled to eliminate cracking. Typical procedures for manual welding with DCEN power are given in Tables 8.15 and 8.16. These are only guides and have not been substantiated by appropriate tests to a particular code or standard.

*Machine Welding.* Precision equipment is available for machine welding with DCEN power. It may provide upslope and downslope of welding current as well as automatic control of shielding gas flow, wire feed, and travel speed. In addition, a programmer may be included for automation of the welding torch movement, welding current, and arc voltage.

For welding square-groove butt joints, a very short arc is used, and the tip of the tungsten electrode may be positioned at or below the base metal surface. The position of the tungsten electrode tip affects weld bead penetration and width as well as undercutting and weld metal turbulence. The optimum position of the electrode tip will likely depend upon the base metal composition and thickness and the welding procedure. In most cases, weld backing is not needed when machine welding square-groove butt joints from one side.

Typical machine welding procedures that may be used as guides in establishing suitable procedures for a particular application are given in Tables 8.17 to 8.20.

## Direct Current, Electrode Positive Power

Welding with DCEP provides good surface cleaning action and permits welding of thin aluminum sections with sufficient current to maintain a stable arc. However, the weld bead tends to be wide and penetration is shallow. Application is limited to sections of about 0.05 in. thickness and under because of tungsten electrode heating. Argon shielding should be used. Helium or argon-helium mixtures would contribute to electrode overheating.

Edge or square-groove joint geometries with or without filler metal may be applicable. Typical conditions for manual welding with DCEP power

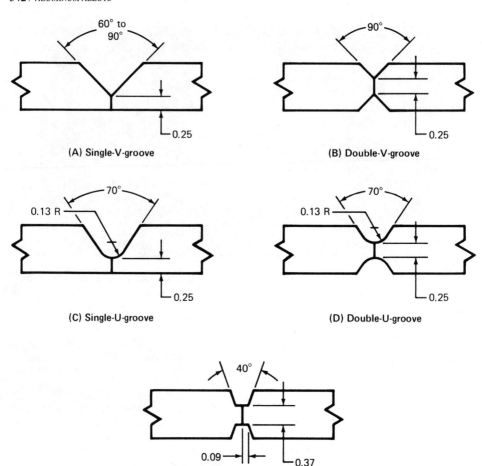

(A) Single-V-groove

(B) Double-V-groove

(C) Single-U-groove

(D) Double-U-groove

(E) Modified double-V-groove

Note: Dimensions are in inches.

**Fig. 8.8—Typical joint designs for gas tungsten arc welding thick aluminum sections with DCEN power**

are given in Table 8.21. A suitable weld backing is recommended.

### Square-Wave Alternating Current Power

Square-wave alternating current (SWAC) power differs from conventional ac balanced wave power with respect to the current wave form. The SWAC power source is designed to produce dc power and rapidly shift the polarity to produce a square alternating wave form of adjustable frequency. In addition, the relative per-centage of electrode negative time within one cycle of current is adjustable within limits.

This type of power combines the advantages of surface cleaning associated with conventional ac power and deep penetration obtainable with DCEN power. However, one is gained with some sacrifice in the other. If longer electrode positive time is needed for acceptable cleaning, penetration will decrease with a specific welding current and frequency.

The square-wave shape enhances arc re-

**Table 8.15**
**Typical procedures for manual gas tungsten arc welding butt joints in aluminum with DCEN and helium shielding in flat position**

| Section thickness, in. | Joint geometry[a] | No. of weld passes | Filler rod diam., in. | Electrode diam., in. | Arc voltage, V | Welding current, A | Travel speed, in./min |
|---|---|---|---|---|---|---|---|
| 0.03 | A | 1 | 0.045 | 0.040 | 21 | 20 | 17 |
| 0.04 | A | 1 | 0.062 | 0.040 | 20 | 26 | 16 |
| 0.06 | A | 1 | 0.062 | 0.040 | 20 | 44 | 20 |
| 0.09 | A | 1 | 0.094 | 0.062 | 17 | 80 | 11 |
| 0.13 | A | 1 | 0.125 | 0.062 | 15 | 118 | 16 |
| 0.25 | A | 1 | 0.156 | 0.125 | 14 | 250 | 7 |
| 0.50 | B | 2 | 0.156 | 0.125 | 14 | 310 | 5.5 |
| 0.75 | C | 2 | 0.156 | 0.125 | 17 | 300 | 4 |
| 1.00 | C | 5 | 0.25 | 0.125 | 19 | 360 | 1.5 |

a.  A–square-groove
    B–single-V-groove, view (A), Fig. 8.8
    C–double-V-groove, view (B), Fig. 8.8

**Table 8.16**
**Typical procedures for manual gas tungsten arc welding of fillet welds in aluminum with DCEN and helium shielding**

| Section thickness, in. | Welding position[a] | Fillet size, in. | Filler rod diam., in. | Electrode diam., in. | Arc voltage, V | Welding current, A | Travel speed, in./min |
|---|---|---|---|---|---|---|---|
| 0.09 | H | 0.13 | 0.094 | 0.094 | 14 | 130 | 21 |
| 0.13 | H | 0.13 | 0.094 | 0.094 | 14 | 180 | 18 |
| 0.25 | H | 0.19 | 0.156 | 0.125 | 14 | 255 | 15 |
| | V | | | | | 230 | 10 |
| 0.38 | H | 0.19 | 0.156 | 0.125 | 14 | 335 | 14 |
| | | 0.31 | 0.250 | | | 290 | 7 |
| 0.50 | H,V | 0.31 | 0.250 | 0.125 | 16 | 315 | 6-7 |

a.  H–horizontal, V–vertical

**Table 8.17**
**Typical procedures for machine gas tungsten arc welding of square-groove butt joints in aluminum with DCEN and helium shielding**

| Section thickness, in. | Electrode diam., in. | Filler wire feed rate,[a] in./min | Arc voltage, V | Welding current, A | Travel speed, in./min |
|---|---|---|---|---|---|
| 0.04 | 0.040 | 60 | 14 | 65 | 54 |
| 0.09 | 0.094 | 75 | 13 | 180 | 54 |
| 0.13 | 0.094 | 55 | 11 | 240 | 40 |
| 0.25 | 0.125 | 40 | 11 | 350 | 15 |
| 0.38 | 0.156 | 30 | 11 | 430 | 8 |

Note:  Single-pass weld in flat position
a.  Filler wire of 0.062 in. diam.

**Table 8.18**

**Typical procedures for machiné gas tungsten arc welding of square-groove butt joints in 2219 aluminum alloy with DCEN and helium shielding**

| Section thickness, in. | Welding position[a] | Electrode | | Filler wire feed rate,[c] in./min | Welding current, A | Travel speed, in./min |
|---|---|---|---|---|---|---|
| | | Diam., in. | Position,[b] in. | | | |
| 0.25 | F | 0.125 | 0.10 | 36 | 145 135 | 8 10 |
| 0.38 | F H | 0.125 | 0.10 | 32 | 220 180 | 8 10 |
| 0.50 | H,V | 0.125 | 0.10 | 10 | 250 | 8 |
| 0.63 | H,V | 0.125 | 0.10 | 5-7 | 300 | 7 |
| 0.75 | H,V | 0.125 | 0.10 | 5-7 | 340 | 6 |
| 0.88 | H,V | 0.156 | 0.12 | 4-6 | 385 | 5 |
| 1.00 | H,V | 0.188 | 0.15 | 3-5 | 425 | 4 |

Notes: 1. Weld with two passes, one from each side.
2. Arc voltage is adjusted at 11.5 to 12.5V.
a. F–flat, H–horizontal, V–vertical.
b. Distance of the electrode tip below the base metal surface.
c. ER 2319 filler metal, 0.062 in. diam.

**Table 8.19**

**Typical procedures for machine gas tungsten arc welding of square-groove butt joints in 7039 aluminum with DCEN and helium shielding**

| Section thickness, in. | Welding position[a] | Electrode diam., in. | Welding current, A | Arc voltage, V | Travel speed, in./min |
|---|---|---|---|---|---|
| 0.25 | F | 0.094 | 200 | 10 | 20 |
| | H | 0.125 | 280 | 10 | 20 |
| | V-up | 0.125 | 260 | 9.5 | 20 |
| | V-down | 0.125 | 285 | 9 | 26 |
| 0.38 | F | 0.125 | 300 | 12 | 16 |
| | H | 0.125 | 380 | 9 | 12 |
| | V-up | 0.125 | 300 | 11 | 16 |
| | V-down | 0.156 | 410 | 11 | 18 |
| 0.50 | F | 0.125 | 350 | 18 | 13 |
| | H | 0.156 | 400 | 9.5 | 9 |
| | V-up | 0.156 | 400 | 10 | 13 |
| | V-down | 0.156 | 400 | 10 | 10 |
| 0.75 | F | 0.125 | 375 | 22 | 6 |
| | H | 0.188 | 500 | 9 | 5 |
| | V-up | 0.156 | 470 | 11 | 5 |
| | V-down | 0.188 | 550 | 11 | 10 |

Notes: 1. No root opening.
2. Welded with two passes, one from each side, without filler metal.
a. F–flat; H–horizontal; V–vertical with direction of torch travel up or down.

**Table 8.20**

**Typical procedures for machine gas tungsten arc welding of single-V-groove joints in 7039 aluminum with DCEN and helium shielding**

| Section thickness, in. | Joint design[a] | Welding position | Pass no. | Welding current, A | Arc voltage, V | Travel speed, in./min. | Filler metal Diam., in. | Filler metal Feed, in./min. |
|---|---|---|---|---|---|---|---|---|
| 0.25 | 1 | Vertical-up | 1 | 175 | 10 | 12 | 0.063 | 70 |
|  |  |  | 2 | 185 | 9 | 9 | None | None |
| 0.38 | 2 | Vertical-up | 1, 2, 3 | 215-220 | 11-11.5 | 12 | 0.063 | 51-56 |
|  |  |  | 4 | 215-220 | 10 | 8 | None | None |
| 0.50 | 2 | Vertical-up | 1 | 250-270 | 11-13 | 10 | 0.063 | 60-69 |
|  |  |  | 2, 3 | 250-270 | 11-13 | 7.5 | 0.063 | 60-69 |
|  |  |  | 4 | 250-270 | 11.5 | 8.5 | None | None |
| 0.63 | 3 | Vertical-up | 1 | 320 | 12 | 12 | 0.093 | 32 |
|  |  |  | 2, 3, 4 | 320 | 12 | 12 | None | 28-44 |
|  |  |  | 5 | 320 | 11 | 12 | None | None |
| 0.75 | 3 | Vertical-up | 1 | 380 | 10.5 | 8 | 0.093 | 39 |
|  |  |  | 2, 3 | 370-380 | 12 | 8 | 0.093 | 39 |
|  |  |  | 4 | 370 | 11.5 | 8 | 0.093 | 33 |
|  |  |  | 5 | 380 | 10 | 6 | None | None |
| 0.25 | 1 | Horizontal | 1 | 225 | 9.5 | 12 | 0.093 | 25 |
|  |  |  | 2 | 225 | 9 | 9 | None | None |
| 0.38 | 2 | Horizontal | 1, 2 | 300 | 11.5 | 10 | 0.063 | 68 |
|  |  |  | 3 | 300 | 11 | 6.5 | 0.063 | 56 |
|  |  |  | 4 | 310 | 10 | 6.5 | None | None |
| 0.50 | 2 | Horizontal | 1 | 310 | 11 | 10 | 0.063 | 68 |
|  |  |  | 2, 3 | 310 | 10 | 8.5-9 | 0.063 | 56-68 |
|  |  |  | 4 | 320 | 9 | 6 | None | None |

**Table 8.20 (cont.)**

**Typical procedures for machine gas tungsten arc welding of single-V-groove joints in 7039 aluminum with DCEN and helium shielding**

| Section thick- ness, in. | Joint design[a] | Welding position | Pass no. | Welding current, A | Arc voltage, V | Travel speed, in./min. | Filler metal Diam., in. | Filler metal Feed, in./min. |
|---|---|---|---|---|---|---|---|---|
| 0.63 | 3 | Horizontal | 1 | 330 | 11 | 9 | 0.093 | 30 |
| | | | 2, 3 | 315-325 | 12-12.5 | 8-9 | 0.093 | 40-44 |
| | | | 4 | 310 | 13 | 8 | 0.093 | 34 |
| | | | 5 | 330 | 11 | 6 | None | None |
| 0.75 | 3 | Horizontal | 1 | 350 | 10.5 | 8 | 0.093 | 39 |
| | | | 2, 3 | 350 | 12.5 | 8 | 0.093 | 39 |
| | | | 4 | 350 | 12.5 | 8 | 0.093 | 33 |
| | | | 5 | 360 | 10 | 6 | None | None |

a. Joint designs.

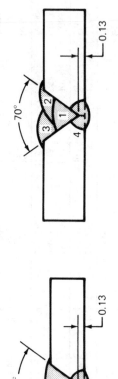

(1)

(2)

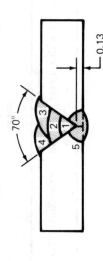

(3)

**Table 8.21**
**Typical conditions for manual gas tungsten arc welding of aluminum with DCEP and argon shielding**

| Section thickness, in. | Electrode diameter, in. | Welding current, A | Filler rod diam., in. |
|---|---|---|---|
| 0.010 | 0.094 | 15-25 | None |
| 0.020 | 0.125, 0.156 | 40-55 | 0.020 |
| 0.030 | 0.188 | 50-65 | 0.020-0.047 |
| 0.040 | 0.188 | 60-80 | 0.047 |
| 0.050 | 0.188 | 70-90 | 0.047-0.062 |

Note:    An edge or square-groove weld should be used.

ignition during polarity reversal. Often, superimposed high voltage is needed only to start the arc rather than continously during welding to stabilize the arc.

Welding techniques similar to those for conventional ac welding are suitable with SWAC welding, as is the electrode tip shape. Argon shielding is preferred, but argon-helium mixtures or helium alone can provide deeper penetration at some sacrifice in cleaning action.

## GAS METAL ARC WELDING

### Equipment

*Electrode Drive Systems.* In semiautomatic welding, there are three drive systems for delivering the electrode from the spool to the arc. These systems are designated *push, pull,* and *push-pull.*

In the push system, the drive rolls are located near the spool and the wire electrode is pushed to the gun through a conduit that is about 10 to 12 feet long. In the pull system, the rolls are located at the gun and the electrode is pulled through the conduit. The pull distance through a conduit is also limited to 10 to 12 feet for aluminum electrodes. In the push-pull system, the electrode is fed through the conduit by two or more sets of rolls, one near the spool and one in the gun. This permits conduit lengths of 25 feet or more. Some systems include conduit modules with a drive at one end of each. Several modules can be connected in series to feed over long distances.

System selection is based largely upon the diameter and tensile strength of the electrode and the distance between the electrode supply and the work. The push system is limited to high-strength electrodes of 0.047 in. diameter and larger that have adequate stiffness to overcome the drag in the conduit and welding gun.

Pull type feed systems are common with machine welding heads and with some semiautomatic welding guns. These systems work better with low-strength, small diameter electrodes. In automatic or machine welding, the electrode drive unit is normally located just above the welding gun. The push-pull systems are applicable to most electrodes and available sizes.

*Guns.* Semiautomatic welding guns may be either pistol-shaped or curved somewhat like an oxyacetylene torch. They may or may not have an electrode feeding mechanism. When they do, some designs have a mount for a small spool of wire. Both water- and air-cooled guns are available.

For machine welding, the guns are almost always straight and of circular cross section. Concentric gas shielding is necessary on all welding guns.

The choice of gun design depends upon a number of factors. Generally, the most important one is the ease with which the gun can be manipulated for a particular joint design and welding position.

### Shielding Gases

Argon is the most commonly used shielding gas for manual welding with the spray type of metal transfer. It provides excellent arc stability, bead shape, and penetration. This gas may be used in all welding positions. Helium is suitable for machine and automatic welding with high currents in the flat position. Helium-argon mixtures are sometimes used instead of helium to take advantage of the arc stability provided by argon. Such mixtures may contain from 25 to 90 percent helium. Arc voltage, penetration, and spatter increase with increasing helium content.

## Metal Transfer

With DCEP power, the filler metal will be transferred across the arc as a stream of fine, superheated droplets when the welding current and arc voltage are above certain threshold values. These values will depend upon the electrode alloy, size, and feed rate. This mode of transfer is known as *spray transfer.* It is the one normally used for GMAW of aluminum. The spray transfer may be continuous or intermittent. Intermittent transfer is called *pulsed spray welding.*

Pulsed spray welding may be used for welding in all positions. Metal transfer takes place during the periods of high welding current, but ceases during the intervening periods of low current. This action reduces the overall heat input to the base metal for good control of the molten weld pool and the penetration. The lower heat input makes it easier to gas metal arc weld thin aluminum sections.

The amount of spatter during welding may be more severe with electrodes that contain low vapor pressure elements. The aluminum-magnesium alloy electrodes (ER5XXX series) cause the most common spatter. The vapor pressure of the magnesium tends to cause disintegration of the droplets as they separate from the electrode tip. This produces small spatter balls that are often thrown clear of the arc.

When the arc voltage (length) is decreased to below a certain value for a specific electrode and feed rate, the size of the droplets will increase and the form of the arc will change. Electrode melting rate and penetration will increase. This has a distinct advantage when welding thick sections.

When the arc voltage is in the spray transfer range but the welding current is decreased to below some threshold value, metal transfer will change from spray to globular type. This latter type is not suitable for aluminum welding because fusion with the base metal will be incomplete. Conversely, when the arc voltage (length) is decreased significantly with adequate current, short circuiting will occur. This type of

transfer also is not recommended for aluminum for the same reason.

## Welding Procedures

Typical GMAW procedures using small and large diameter electrodes are given in Tables 8.22 through 8.25. Small diameter electrodes can be used for semiautomatic welding in all positions; large diameter electrode can be used only in the flat position with automatic welding.

Constant-voltage power and constant-speed electrode drive are normally used with small diameter electrodes. The electrode feed rate is adjusted to obtain the desired welding current for good fusion and penetration. The arc voltage is adjusted to give spray type transfer of filler metal.

A constant-current power source and variable speed electrode drive should be used with large diameter electrodes. The welding current is set at the desired value. The arc voltage is set to a value slightly below that for spray transfer. The electrode drive unit will adjust the feed rate to maintain the preset arc voltage. The voltage setting is critical with respect to good fusion with the groove faces. If the voltage is too high, lack of fusion may occur. If it is too low, short circuiting will take place between the electrode and the weld pool.[11]

Welding procedures for pulsed spray welding depend upon the design of the power source. The welding variables may include part or all of the following:

(1) Electrode diameter
(2) Electrode feed rate
(3) Shielding gas
(4) Pulse rate
(5) Pulse peak voltage
(6) Background voltage
(7) Average welding current
(8) Background current
(9) Travel speed

The recommendations of the welding machine (power source) manufacturer should be followed when developing procedures for pulse spray welding of aluminum.

## Semiautomatic Welding

The welding equipment should be set up according to the manufacturer's instructions. The

---

11. Refer to Liptak, J.A., Gas metal arc welding aluminum with large diameter filler, *Welding Journal*, 40 (9): 917-27; 1961 Sept.

**Table 8.22**

**Typical procedures for gas metal arc welding of groove welds in aluminum alloys with small diameter electrodes and argon shielding**

| Section thickness, in. | Welding position[a] | Joint geometry[b] | Root opening, in. | No. of weld passes | Electrode diameter, in. | Welding current,[c] A | Arc voltage, V | Travel speed, in./min |
|---|---|---|---|---|---|---|---|---|
| 0.06 | F | A | 0 | 1 | 0.030 | 70-110 | 15-20 | 25-45 |
| | F | G | 0.09 | 1 | | | | |
| 0.09 | F | A | 0 | 1 | 0.030-0.047 | 90-150 | 18-22 | 25-45 |
| | F,V,H,O | G | 0.12 | 1 | 0.030 | 110-130 | 18-23 | 23-30 |
| 0.13 | F,V,H | A | 0.09 | 1 | 0.030-0.047 | 120-150 | 20-24 | 24-30 |
| | F,V,H,O | G | 0.18 | 1 | 0.030-0.047 | 110-135 | 19-23 | 18-28 |
| 0.19 | F,V,H | B | 0.06 | 2 | 0.030-0.047 | 130-175 | 22-26 | 24-30 |
| | F,V,H | F | 0.06 | 1 | 0.047 | 140-180 | 23-27 | 24-30 |
| | O | F | 0.06 | 2 | 0.047 | 140-175 | 23-27 | 24-30 |
| | F,V | H | 0.09-0.18 | 2 | 0.047-0.062 | 140-185 | 23-27 | 24-30 |
| | H,O | H | 0.18 | 3 | 0.047 | 130-175 | 23-27 | 25-35 |
| 0.25 | F | B | 0.09 | 2 | 0.047-0.062 | 175-200 | 24-28 | 24-30 |
| | F | F | 0.09 | 2 | 0.047-0.062 | 185-225 | 24-29 | 24-30 |
| | V,H | F | 0.09 | 3F, 1R | 0.047 | 165-190 | 25-29 | 25-35 |
| | O | F | 0.09 | 3F, 1R | 0.047-0.062 | 180-200 | 25-29 | 25-35 |
| | F,V | H | 0.13-0.25 | 2-3 | 0.047-0.062 | 175-225 | 25-29 | 24-30 |
| | O,H | H | 0.25 | 4-6 | 0.047-0.062 | 170-200 | 25-29 | 25-40 |
| 0.38 | F | C-90° | 0.09 | 1F, 1R | 0.062 | 225-290 | 26-29 | 20-30 |
| | F | F | 0.09 | 2F, 1R | | 210-275 | | 24-35 |
| | V,H | F | 0.09 | 3F, 1R | | 190-220 | | 24-30 |
| | O | F | 0.09 | 5F, 1R | | 200-250 | | 25-40 |
| | F,V | H | 0.25-0.38 | 4 | | 210-290 | | 24-30 |
| | O,H | H | 0.38 | 8-10 | | 190-260 | | 25-40 |
| 0.75 | F | C-60° | 0.09 | 3F, 1R | 0.062-0.094 | 340-400 | 26-31 | 14-20 |
| | F | F | 0.12 | 4F, 1R | 0.094 | 325-375 | 26-31 | 16-20 |
| | V,H,O | F | 0.06 | 8F, 1R | 0.062 | 240-300 | 26-30 | 24-30 |
| | F | E | 0.06 | 3F, 3R | 0.062 | 270-330 | 26-30 | 16-24 |
| | V,H,O | E | 0.06 | 6F, 6R | 0.062 | 230-280 | 26-30 | 16-24 |

a. F—flat; V—vertical; H—horizontal; O—overhead.  b. Refer to Fig. 8.3.  c. Constant voltage power source and constant speed electrode feed unit.

## Table 8.23
**Typical procedures for gas metal arc welding of fillet welds in aluminum alloys with small diameter electrodes and argon shielding**

| Section thickness, in. | Welding position[a] | No. of passes | Electrode diameter, in. | Welding current,[b] A | Arc voltage, V | Travel speed, in./min |
|---|---|---|---|---|---|---|
| 0.09 | F,V,H,O | 1 | 0.030 | 100-130 | 18-22 | 24-30 |
| 0.13 | F | 1 | 0.030-0.047 | 125-150 | 20-24 | 24-30 |
|  | V,H | 1 | 0.030 | 110-130 | 19-23 | 24-30 |
|  | O | 1 | 0.030-0.047 | 115-140 | 20-24 | 24-30 |
| 0.19 | F | 1 | 0.047 | 180-210 | 22-26 | 24-30 |
|  | V,H | 1 | 0.030-0.047 | 130-175 | 21-25 | 24-30 |
|  | O | 1 | 0.030-0.047 | 130-190 | 22-26 | 24-30 |
| 0.25 | F | 1 | 0.047-0.062 | 170-240 | 24-28 | 24-30 |
|  | V,H | 1 | 0.047 | 170-210 | 23-27 | 24-30 |
|  | O | 1 | 0.047-0.062 | 190-220 | 24-28 | 24-30 |
| 0.38 | F | 1 | 0.062 | 240-300 | 26-29 | 18-25 |
|  | H,V | 3 | 0.062 | 190-240 | 24-27 | 24-30 |
|  | O | 3 | 0.062 | 200-240 | 25-28 | 24-30 |
| 0.75[c] | F | 4 | 0.094 | 360-380 | 26-30 | 18-25 |
|  | H,V | 4-6 | 0.062 | 260-310 | 25-29 | 24-30 |
|  | O | 10 | 0.062 | 275-310 | 25-29 | 24-30 |

a. F–flat; V–vertical, H–horizontal; O–overhead
b. Constant voltage power source and constant speed electrode feed unit.
c. For thicknesses of 0.75 in. and larger, a double-bevel joint with a 50 degree minimum groove angle and 0.09 to 0.13 in. root face is sometimes used.

## Table 8.24
### Typical conditions for gas metal arc welding of groove welds in aluminum alloys with large diameter electrodes

| Section thickness, in. | Joint geometry[a] | | | Electrode diam., in. | Shielding gas | Weld pass[b] | Arc voltage, V | Welding current,[c] A | Travel speed, in./min. |
|---|---|---|---|---|---|---|---|---|---|
| | Type | A, degrees | F, in. | | | | | | |
| 0.75 | A | 90 | 0.25 | 0.156 | Ar | 1 / 2 | 28 | 450 / 500 | 16 |
| 1.00 | A | 90 | 0.13 | 0.188 | Ar | 1,2 | 26.5 | 500 | 12 |
| 1.25 | A | 70 | 0.18 | 0.188 | Ar | 1,2 | 26.5 | 550 | 10 |
| 1.25 | B | 45 | 0.25 | 0.156 | Ar | 1 / 2 / Back | 25 / 27 / 26 | 500 | 10 / 10 / 12 |
| 1.50 | A | 70 | 0.18 | 0.188 | Ar | 1 / 2 / 3,4 | 26 / 27 / 29 | 550 / 575 / 600 | 10 |
| 1.50 | A | 70 | 0.18 | 0.219 | Ar | 1 / 2 | 27 / 27.5 | 650 / 675 | 8 |
| 1.75 | A | 70 | 0.13 | 0.219 | Ar | 1,2 / 3,4 | 26 / 27 | 650 / 600 | 10 |
| 1.75 | B | 45 | 0.25 | 0.188 | Ar | 1,2 / 3,4 / Back | 28 / 30 / 30 | 600 / 550 / 550 | 10 / 14 / 10 |
| 2.00 | A | 70 | 0.18 | 0.188 | He | 1,2,3,4 | 32 | 550 | 10 |
| 2.00 | B | 45 | 0.25 | 0.188 | Ar | 1,2 / 3-7 / Back | 28 / 26 / 28 | 600 / 500 / 550 | 10 / 14 / 10 |

## Table 8.24 (cont.)
### Typical conditions for gas metal arc welding of groove welds in aluminum alloys with large diameter electrodes.

| Section thickness, in. | Joint geometry[a] | | | | Shielding gas | Weld pass[b] | Arc voltage, V | Welding current,[c] A | Travel speed, in./min. |
| | Type | A, degrees | F, in. | Electrode diam., in. | | | | | |
|---|---|---|---|---|---|---|---|---|---|
| 3.00 | A | 70 | 0.18 | 0.219 | Ar-25%He | 1,2<br>3,4<br>5,6<br>7-10 | 25<br>23<br>26<br>27 | 650<br>500<br>650<br>625 | 9<br>10<br>9<br>9 |
| 3.00 | C | 30 | 0.50 | 0.219 | He | 1,2<br>3-6 | 29<br>31 | 650 | 10 |

a. Joint types

(A)  (B)  (C)

0.25 R

b. All passes are welded in the flat position, odd numbers from one side and even numbers from other side with joint designs (A) and (C). Joint is back gouged prior to depositing the back weld.

c. Constant dc power source and variable speed electrode drive unit (voltage sensitive).

**Table 8.25**
**Typical procedures for gas metal arc welding of fillet welds in aluminum alloys with large diameter electrodes and argon shielding**

| Fillet size, in. | Electrode diameter, in. | Weld pass[a] | Welding current,[b] A | Arc voltage, V | Travel speed, in./min |
|---|---|---|---|---|---|
| 0.5 | 0.156 | 1 | 525 | 22 | 12 |
| 0.5 | 0.188 | 1 | 550 | 25 | 12 |
| 0.63 | 0.156 | 1 | 525 | 22 | 10 |
| 0.75 | 0.156 | 1 | 600 | 25 | 10 |
| 0.75 | 0.188 | 1 | 625 | 27 | 8 |
| 1 | 0.156 | 1 | 600 | 25 | 12 |
|   |   | 2, 3 | 555 | 24 | 10 |
| 1 | 0.188 | 1 | 625 | 27 | 8 |
|   |   | 2, 3 | 550 | 28 | 12 |
| 1.25 | 0.156 | 1, 2, 3 | 600 | 25 | 10 |
| 1.25 | 0.188 | 1 | 625 | 27 | 8 |
|   |   | 2, 3 | 600 | 28 | 10 |

a. Welded in the flat position with one or three passes, using stringer beads.
b. Constant-current power source and variable speed electrode wire drive unit (voltage sensitive).

appropriate shielding gas should be connected and the desired flow rate set. The proper electrode and nozzle for the application should be installed on the equipment.

In multiple-pass welding, arc voltage for the first pass should be set on the low side of the range to ensure good root penetration. For subsequent passes, the voltage can be increased to widen the weld bead. A short arc length should be used for small fillet welds. Selection of correct arc voltage should be determined by appropriate tests. As a general rule, a harsh-sounding, low-voltage arc will cause excessive spatter and tend to increase porosity. On the other hand, a soft high-voltage arc may result in both incomplete root penetration and contamination of the weld metal. A compromise between a harsh, low-voltage arc and a soft, high-voltage arc will usually produce the best results.

The arc should be started at a location on the joint that will be melted into the weld metal. Alternatively, it may be started on a tab at the end of the joint groove. The arc should not be started on the base metal outside of the weld area. An arc strike might cause a surface discontinuity that would cause eventual failure of the weldment in service.

A good technique is to start the arc in the joint about one inch ahead of the weld starting point. As soon as the arc is initiated, it is quickly returned to the starting point. The direction of travel is reversed, and welding is started.

The end of the gas nozzle should be held approximately 0.75 in. above the work. This distance should be shorter for aluminum-magnesium alloy electrodes than for other alloy electrodes. The better shielding provided by this technique minimizes the loss of magnesium during transfer across the arc.

The forehand welding technique is recommended to take advantage of the arc cleaning action ahead of the molten weld pool. The electrode should be tilted approximately 7 to 12 degrees from the vertical and pointed in the direction of welding.

When welding in the horizontal position,

the electrode should also be directed slightly upward. For vertical welds, a vertical-up technique is recommended. When welding thick-to-thin material, the arc should be directed toward the heavier section.

The gun should be moved progressively along the joint without weaving to produce a stringer bead. With this type of bead, as much filler as can be controlled without sagging or undercutting should be deposited in each pass. If undercutting is encountered with vertical and horizontal butt joints, a slight reduction in welding current is recommended. For fillet welding, a slight circular motion will tend to agitate the molten pool. This will help to minimize porosity.

When breaking the arc, the travel should be accelerated to gradually reduce the size of the molten weld pool to the moment of arc outage. This will minimize the crater size and the incidence of cracking. An alternative is to avoid a crater at the end of the weld by reversing the direction of travel and breaking the arc on the weld bead. In some cases, welding is continued onto a runoff tab at the end of the joint.

Heat-treatable aluminum alloys should be welded using the stringer bead technique, and the joint should be cooled to below 150° F between passes. With this technique, the heat input will be minimized and the heat-affected zone will usually be narrower than when large passes are used. On the other hand, some of the nonheat-treatable alloys, such as the 5XXX series, may be welded with larger beads without adversely affecting the strength of the joint.

The largest electrode that is compatible with the section thickness, the joint design, and the welding position should be used. Larger diameter electrodes have a favorable surface-to-volume ratio, which minimizes porosity. They are also economical. However, there are some applications where small diameter electrodes are needed to control fillet weld size or weld reinforcement. Each application should be properly evaluated when selecting the electrode size.

## Automatic Welding

Automatic or machine welding can be done at higher travel speeds than are possible with semiautomatic welding. Longer joints can be made without interrupting welding; this reduces the number of weld craters and the possibility of crater cracks. Higher welding current can be used with the maximum amperage limited only by arc stability or adequate process control. With high amperages, welds can be made with little or no joint preparation and fewer weld passes. Square-groove butt joints in sections up to 0.5 in. thick can be welded with one pass. Similar joints in 1 in. thick sections can be welded with two passes, one from each side of the joint. Sections up to 3 in. thick can be welded in the same manner using high welding currents and large eletrodes. It may be necessary to bevel the plate edges to control the height of weld reinforcement.

Conversely, sections as thin as 0.02 in. can be welded automatically with adequate control of the welding variables and good fixturing.

## Spot Welding

Lap joints in aluminum sheet can be spot welded by gas metal arc welding. Basically, the arc melts through the front sheet and penetrates into the rear sheet. This action produces a round nugget of weld metal that joins the sheets together, similar to resistance spot welding. The filler metal forms a convex reinforcement on the face of the nugget. Penetration into the rear sheet may be partial or complete.

Equipment for spot welding is similar to that for semiautomatic welding. The gun must have an integral electrode drive system and an insulated shielding gas nozzle to bear against the front sheet. A series of spot welds is more consistent when the gun system exerts the same force against the sheet during each weld.

A control unit is needed to time the durations of welding current, electrode feed, and shielding gas flow. Welding current and electrode feed timers should have a range of 0 to 2 seconds with an accuracy of about 0.017 second.

Welding current should flow for a brief period after electrode feed stops to melt the electrode back a predetermined length. This prevents the end of the electrode from freezing in the weld metal.

Argon, helium, or mixtures of the two may be used for shielding. Argon is normally used, but helium is preferred for welding thin sheet because it produces a nugget with a larger cone angle than does argon. A rough weld surface and more spatter are disadvantages of spot welding with helium shielding.

Typical conditions for spot welding lap joints between two aluminum sheets with a 0.047 in. diameter electrode are given in Table 8.26. These conditions may vary somewhat with the compositions of the base metal and electrode, surface conditions, fitup, shielding gas, and

**Table 8.26**
**Typical settings for gas metal arc spot welding of lap joints between various aluminum sheet thicknesses**

| Sheet thickness, in. | | Partial penetration welds | | | Complete penetration welds | | |
|---|---|---|---|---|---|---|---|
| Top | Bottom | Open circuit voltage, V | Electrode feed,[a] in./min | Weld time, s | Open circuit voltage, V | Electrode feed,[a] in./min | Weld time, s |
| 0.020 | 0.020 | – | – | – | 27 | 250 | 0.3 |
| 0.020 | 0.030 | – | – | – | 28 | 300 | 0.3 |
| 0.030 | 0.030 | 25.5 | 285 | 0.3 | 28 | 330 | 0.3 |
| 0.030 | 0.050 | 25.5 | 330 | 0.3 | 31 | 430 | 0.3 |
| 0.030 | 0.064 | 30 | 360 | 0.3 | 31 | 450 | 0.3 |
| 0.050 | 0.050 | 31 | 385 | 0.4 | 32 | 450 | 0.4 |
| 0.050 | 0.064 | 32 | 400 | 0.4 | 32 | 500 | 0.4 |
| 0.064 | 0.064 | 32 | 420 | 0.4 | 32 | 550 | 0.5 |
| 0.064 | 0.125 | 32.5 | 650 | 0.5 | 34.5 | 675 | 0.5 |
| 0.064 | 0.187 | 35 | 700 | 0.5 | 39 | 700 | 0.5 |
| 0.064 | 0.250 | 39 | 775 | 0.5 | 41 | 800 | 0.5 |
| 0.125 | 0.125 | 39.5 | 800 | 0.5 | 41 | 850 | 0.6 |
| 0.125 | 0.187 | 41 | 850 | 0.75 | 41 | 900 | 0.75 |
| 0.125 | 0.250 | 41 | 900 | 1.0 | – | – | – |

a. Electrode is 0.047 in. diameter ER5554. The welding current in amperes is about 0.5 times the electrode feed rate in in./min.

equipment design. Welding conditions for each application should be established by appropriate destructive tests.

Penetration into the back sheet depends upon the arc voltage, electrode feed rate, and welding time for a particular electrode size and shielding gas. Increasing any of these variables will increase penetration. High electrode feed rates are needed to obtain good penetration with thick sheets. If a small adjustment in penetration is desired, only the electrode feed needs to be changed.

A welding time of approximately 0.5 second is usually satisfactory. Times less than 0.25 second may result in nonuniform welds because the arc starting time is appreciable compared to the total welding time. Long welding times may help to reduce porosity. Short welding times may be best for vertical and overhead welding, or when a flat weld nugget is desired. With short weld times, arc length must be adjusted by the open-circuit voltage setting on the welding machine.

Fitup between parts can cause variations in gas metal arc spot welds. If the welds are inconsistent despite careful control of the variables, better methods of fixturing should be tried.

## SHIELDED METAL ARC WELDING

Shielded metal arc welding (SMAW) of aluminum is primarily used in small shops for non-critical applications and repair work. It can be done with simple, low-cost equipment that is readily available and portable. However, welding speed is slower than that with gas metal arc welding. This process is not recommended for joining aluminum when good welding practices are required. One of the gas shielded arc welding processes should be used instead.

The electrode covering contains an active flux that combines with aluminum oxide to form a slag. This slag is a potential source of corrosion, and must be completely removed after each weld pass. Aluminum covered electrodes are prone to deterioration with time and exposure to moisture. Therefore, they must be stored in a dry atmosphere as described previously. Welding is done with DCEP power. Important factors to be considered in welding aluminum with the shielded metal arc welding process are (1) moisture content of the electrode covering, (2) cleanness of the electrode and base metal, (3) preheating of the base metal, and (4) proper slag removal between passes and after welding.

The minimum recommended base metal thickness for shielded metal arc welding of aluminum is 0.12 inch. For thicknesses less than 0.25 in., no edge preparation other than a relatively smooth, square cut is required. Sections over 0.25 in. thickness should be beveled to provide a 60 to 90 degree single-V groove. On very thick material, U-grooves may be used. Depending upon base metal thickness, root-face width should be between 0.06 and 0.25 inch. A root opening of 0.03 to 0.06 in. is desirable.

For welding applications involving plate or complicated welds, it is desirable to apply preheat. Preheating is nearly always necessary on thick sections to obtain good penetration. It also prevents porosity at the start of the weld due to rapid cooling and reduces distortion. Preheating may be done using an oxyfuel gas torch or a furnace.

Single-pass welding should be used whenever possible. Where multiple passes are needed, thorough removal of slag between passes is essential for optimum results. After the completion of welding, the bead and work should be thoroughly cleaned of slag. The major portion can be removed by mechanical means, such as a rotary wire brush or a slag hammer, and the rest by steam cleaning or a hot water rinse. To test for complete slag removal, weld area is swabbed with a solution of 5 percent silver nitrate. Foaming will occur when flux is present.

One difficulty encountered by interruption of the arc is the formation of a fused slag coating over the end of the electrode. It must be removed to restrike the arc.

## PLASMA ARC WELDING

Plasma arc welding (PAW) is similar to gas tungsten arc welding except that the arc is constricted in size by a water-cooled nozzle. Arc construction increases the energy density, directional stability, and focusing effect of the arc plasma. Welding with the plasma arc is done with two techniques: melt-in and keyhole. Welding with the melt-in technique is similar to gas tungsten arc welding. The keyhole technique involves welding with a small hole through the molten weld pool at the leading edge. This technique is normally used when welding relatively thick sections.

Plasma arc welding is normally done with DCEN and there is no arc cleaning action, as with gas tungsten arc welding. Surface cleaning prior to welding is necessary. Aluminum can be plasma arc welded also with conventional or square-wave ac or DCEP to take advantage of arc cleaning. Power sources for plasma arc welding are similar to those for gas tungsten arc welding.

Deeper penetration and higher welding speeds are advantages of plasma arc welding over gas tungsten arc welding. These conditions tend to increase the presence of porosity in aluminum welds because gases have less time to escape to the surface of the molten weld pool. Preweld cleaning of the base metal, clean filler wire, and adequate inert gas shielding of both sides of the weld are needed to minimize weld porosity.

## PLASMA-GMA WELDING

Plasma-GMA welding, sometimes called plasma MIG welding, is a combination of plasma arc and gas metal arc welding in a single gun. A gas metal arc welding electrode is fed through a concentric, transferred plasma arc that is surrounded by an inert gas shield. A typical gun design is shown in Fig. 8.9. The gun design assures laminar plasma gas flow and a high degree of plasma stability.

The gas metal arc is powered by a separate power source which permits ready control of electrode melting rate and joint penetration. The surrounding plasma arc provides cathodic cleaning of the aluminum surface around the gas metal arc as well as preheat to the workpieces. Heat loss to the base metal from the gas metal arc is less than with gas metal arc welding alone. Therefore, higher deposition rates are possible with this process.

## CRATER FILLING

Aluminum alloy weld metal has a tendency to crater crack when fusion welding is stopped abruptly. This problem can be avoided when the proper technique is used to fill the crater.

With manual welding, the forward travel should be stopped, and the crater filled with filler metal. When using gas tungsten arc or plasma arc welding with remote current control, the welding current should then be decreased gradually until the molten weld pool freezes. Without remote current control, the arc should be moved rapidly back on the weld bead as the welding torch is withdrawn to break the arc.

With semiautomatic gas metal arc or shielded metal arc welding, arc travel should be reversed and the crater filled. The arc is then moved back on the weld bead for a short distance before it is broken. With automatic welding, the electrode or filler wire feed, the welding current, and the orifice gas with plasma arc welding should be programmed to fill the crater when travel stops. These variables are gradually decreased as the crater fills.

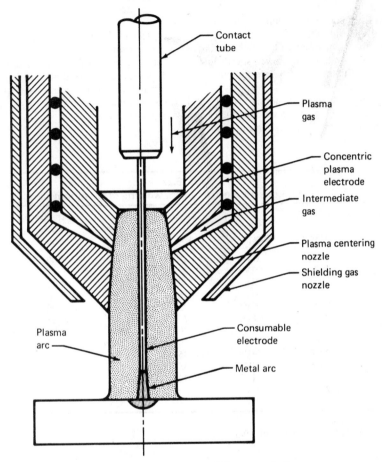

Contact tube

Plasma gas

Concentric plasma electrode

Intermediate gas

Plasma centering nozzle

Shielding gas nozzle

Consumable electrode

Metal arc

Plasma arc

**Fig. 8.9—Plasma-GMA welding gun design**

# ELECTRON BEAM WELDING

Electron beam welding is generally applicable to edge, butt, fillet, and lap joints. Filler metal is not used unless it is needed for metallurgical or cosmetic reasons. Welding can be done in a vacuum chamber or at atmospheric pressure using helium shielding. In the latter case, the electron beam is generated in vacuum and exited to atmospheric pressure through a series of ports.

Most aluminum alloys can be electron beam

welded, but cracking may be experienced with some of the heat-treatable alloys, such as 6061, 2024, and 7075. As with arc welding, the addition of filler metal may be helpful in reducing weld cracking.

Porosity may be a problem when welding some alloys at high travel speeds. This may be caused by the vaporization of low melting alloying elements. Welding at slower speeds with lower energy input should help overcome this problem.

Square-groove joints are normally used with electron beam welding. When welding thick sections with a single pass, it is desirable to maintain a constant weld bead width through the thickness to ensure good root fusion. In some instances, this may cause undercutting or a concave weld face. A second pass, with or without filler metal, may be used to produce a flat or convex weld face.

As with other metals, aluminum alloys can be welded with either low- or high-voltage equipment. The choice of equipment and procedures will depend upon the alloy, thickness, joint design, and service requirements. Typical conditions for welding several alloys and thicknesses are given in Table 8.27.

### Table 8.27
### Typical conditions for electron beam welding of aluminum alloys

| Thickness, in.[a] | Alloy | Welding atmosphere | Beam voltage, kV | Beam current, mA | Travel speed, in./min | Energy input, kJ/in. |
|---|---|---|---|---|---|---|
| 0.050 | 6061 | HV[b] | 18 | 33 | 100 | 0.36 |
| 0.050 | 2024 | HV | 27 | 21 | 71 | 0.48 |
| 0.120 | 2014 | HV | 29 | 54 | 75 | 1.3 |
| 0.125 | 6061 | HV | 26 | 52 | 80 | 1.00 |
| 0.125 | 7075 | HV | 25 | 80 | 90 | 1.3 |
| 0.38 | 2219 | Helium | 175 | 40 | 55 | 7.6 |
| 0.50 | 2219 | HV | 30 | 200 | 95 | 3.8 |
| 0.63 | 6061 | HV | 30 | 275 | 75 | 6.6 |
| 0.75 | 2219 | HV | 145 | 38 | 50 | 6.6 |
| 1.00 | 5086 | HV | 35 | 222 | 30 | 15 |
| 2.00 | 5086 | HV | 30 | 500 | 36 | 25 |
| 2.38 | 2219 | HV | 30 | 1000 | 43 | 42 |
| 6.00 | 5083 | HV | 58 | 525 | 10 | 182 |

a. Square-groove butt joint
b. High vacuum ($10^{-5}$ torr or $1.33 \times 10^{-3}$ Pa)

# LASER BEAM WELDING

Laser beam welding is done with the heat generated by a high power photon (light) beam impinging on the work surface. When the photon beam strikes the surface, the energy is either absorbed or reflected. The amount of energy reflected depends upon the reflective characteristics of the metal being welded, the condition of the metal surface, and the angle of incidence of the beam. Aluminum has high reflectivity (about 90 percent) for visible and infrared photon energy. Therefore, a laser beam of very high-power density is required for welding aluminum. Energy absorption is improved by a deep penetration cavity or keyhole in the molten weld metal.

Absorption levels greater than 50 percent have been observed with aluminum when welding with the keyhole technique.[12]

Aluminum is more difficult to weld than are most other common structural metals. A deep penetrating keyhole must be initiated with less than 10 percent of the incident power. Once the keyhole is formed, the power density is too high, and the molten weld metal overheats. This can cause serious weld metal discontinuities. Improvements in laser beam control are needed to overcome this problem. Despite these difficulties, some acceptable laser welded joints have been made in 5XXX alloys. Fillet welds in lap joints and T-joint welds have been made with promising results. The general application of laser welding to aluminum requires further development of equipment and procedures.

# OXYFUEL GAS WELDING

Aluminum can be welded by the oxyfuel gas welding process. However, it should only be used for noncritical applications or repair when suitable inert gas shielded arc welding equipment is not available. The advantage of the process is the simplicity, portability, and low cost of the equipment. However, the disadvantages when compared to arc welding are more numerous:

(1) An active welding flux is required to clean and protect the aluminum from oxidation.

(2) Welding speeds are slower.

(3) Heat-affected zones are wider.

(4) Weld metal solidification rates are slower, increasing the possibility of hot cracking.

(5) The gas flame offers no surface cleaning action.

(6) Welds are likely to contain a larger number of discontinuities.

(7) Distortion of the weldment is greater.

Standard oxyfuel gas welding torches are suitable for welding aluminum sections about 0.03 to 1 in. in thickness. Thicker sections are seldom welded because the limited heat available from the flame makes good fusion difficult.

## FUEL GASES

Acetylene is the best fuel for oxyfuel gas welding of aluminum because of its high combustion intensity and flame temperature. A neutral flame is best, but a slightly reducing flame (excess fuel gas) reduces the possibility of forming unwanted aluminum oxides. Hydrogen may be an adequate fuel for welding thin sheet.

## WELDING FLUX

Aluminum welding flux is designed to remove the aluminum oxide surface film and exclude oxygen from the molten weld pool. It is generally furnished in powder form, and mixed with water to form a thin free-flowing paste. The filler rod should be uniformly coated with flux either by dipping or painting. The joint faces and adjacent surfaces should also be coated with flux to prevent oxidation of these surfaces during the welding operation.

Flux residues are corrosive to aluminum when moisture is present. Thorough cleaning after welding is therefore of prime importance. Small weldments may be cleaned by immersion in one of the following baths:

(1) 10 percent sulfuric acid at room temperature for 20 to 30 minutes

(2) 5 percent sulfuric acid at 150° F for 5 to 10 minutes

(3) 40 to 50 percent nitric acid at room temperature for 10 to 20 minutes

In any case, the acid cleaning should be followed by a hot or cold water rinse. Steam cleaning also may be used to remove flux residue, particularly on parts that cannot be immersed. Brushing may be necessary to remove adhering flux particles.

## JOINT DESIGNS

Joint designs for oxyfuel gas welding are the same as those for GTAW, shown in Fig. 8.3. For

---

12. Refer to the *Welding Handbook*, Vol. 2, 7th Ed., 301-302 for a description of keyhole welding.

sections over 0.18 in. thick, penetration is best achieved by beveling the edges to be joined. Single V-groove joints are sometimes used on plate up to about 0.5 in. thick. Permanent backings are not recommended for gas welding due to possible entrapment of welding flux and the probability of subsequent corrosion. Fillet welded lap joints generally are not recommended for the same reason.

## PREWELD CLEANING

Grease and oil should be removed from the welding surfaces with a safe solvent. Fluxes will perform better if thick oxide layers are removed from the surfaces prior to welding.

## FILLER METALS

Filler rods ER1100, ER4043, ER4047, and ER4145 may be used for oxyfuel gas welding. Although the welding of the 5XXX series of alloys is not recommended due to poor wetting characteristics, thin sections may be welded with a single pass. The proper choice of filler metal depends upon the alloy being joined and the end-use requirements.

The size of the filler rod is related to the thickness of the sections being welded. A large rod may melt too slowly and tend to freeze the weld pool prematurely. A small rod tends to melt rapidly, and make addition of filler metal into the molten weld pool difficult.

## PREHEATING

Preheating is necessary when the mass of base metal is so great that the heat is conducted away from the joint too fast to accomplish welding. Preheat will also improve control of the molten weld pool.

## WELDING TECHNIQUE

Initially, the flame is moved in a circular motion to preheat both edges of the joint uniformly. The flame is then held where the weld is to begin until a small molten pool forms. The end of the filler rod is fed into the molten pool to deposit a drop of metal and then withdrawn. This is repeated as welding progresses using the forehand technique. The flame should be oscillated to melt both joint faces simultaneously. The inner flame cone should not touch the molten weld pool, but be kept from 0.062 to 0.25 in. away. The crater should be filled before removing the flame.

# WELDING OF CASTINGS

Castings are sometimes welded to correct foundry defects, repair castings damaged in use, or assemble castings into weldments. The relative weldability ratings of various casting alloys are given in Tables 8.6 and 8.7

Die castings tend to retain dissolved gases. Such castings are difficult to weld because these gases are released by the welding heat and erupt through the weld pool. The welds would likely be porous.

When repairing new castings, filler metal of the same composition is preferred. Gas tungsten arc welding is well adapted for this repair. The surfaces to be welded should be clean and dry. The defect must be completely removed to sound metal. Large castings may require preheating, but the temperatures and times must be carefully controlled.

Castings that have been in service must be thoroughly cleaned of oil, grease, or other contaminants before welding. It may not be practical to obtain a filler metal of exactly the same composition as the casting. In such circumstances, a filler metal may be selected from Table 8.10. When welding a casting to a wrought product, the selected filler metal should be properly evaluated for the intended service.

A welded casting may require heat treatment to fully restore its original properties. Solution heat treatment is essential for alloys high in magnesium to avoid the presence of the brittle beta phase, which is particularly susceptible to corrosion.

# STUD WELDING

There are two types of stud welding that employ an arc to obtain fusion. These are arc stud welding and capacitor discharge stud welding. Aluminum studs can be joined to aluminum components with both types of equipment.[13]

## ARC STUD WELDING

Arc stud welding equipment consists of a stud welding gun, a timing control device, a dc power source, and a regulated supply of shielding gas. The gun is equipped with a gas adapter foot that holds a ceramic ferrule around the stud and conducts shielding gas to the joint. The ferrule confines the weld metal and aids in forming a fillet at the base of the welded stud.

Studs are commonly made of aluminum-magnesium alloys, such as 5086 or 5356. Stud weld bases range from 0.125 through 0.5 in. diameter. Such studs may be readily welded to alloys of the 1XXX, 3XXX, and 5XXX series. Alloys of the 4XXX and 6XXX series are considered passable for stud welding applications, and the 2XXX and 7XXX series are considered poor. Typical conditions for arc stud welding of aluminum alloys are given in Table 8.28.

## CAPACITOR DISCHARGE STUD WELDING

Capacitor discharge stud welding uses an electrostatic storage system to power a dc arc. It is best suited for welding studs to thin sheet. Neither flux nor shielding gas is normally required to protect the weld metal because the welding time is very short. However, argon shielding should be used with the drawn arc method because the welding time is long enough for oxidation to take place.

Studs for capacitor discharge welding may have bases ranging from 0.062 through 0.375 in. diameter. Studs are commonly made from 1100, 5086, or 6063 alloy, and are readily welded to 1XXX, 3XXX, 5XXX, and 6XXX alloys.

Aluminum stud welding requires attention to the following points to assure good reliability:

(1) Correctly designed studs and proper matching of stud and component alloys

(2) Power source and welding equipment of sufficient capacity for the stud size

(3) Surfaces that are clean and free of lubricants, oxides, and other contaminants

(4) Proper positioning of the stud welding gun on the work surface, and correct stud 'lift' and 'plunge' settings

---

13. Stud welding is discussed in AWS C5.4-74, *Recommended Practices for Stud Welding*, and Vol. 2 of the *Welding Handbook*.

### Table 8.28
### Typical conditions for arc stud welding of aluminum alloys[a]

| Stud weld base diameter | | Weld time, cycles (60 Hz) | Welding current,[b] A | Shielding gas flow[c] | |
|---|---|---|---|---|---|
| in. | mm | | | ft³/h | liter/min |
| 1/4 | 6.4 | 20 | 250 | 15 | 7.1 |
| 5/16 | 7.9 | 30 | 325 | 15 | 7.1 |
| 3/8 | 9.5 | 40 | 400 | 20 | 9.4 |
| 7/16 | 11.1 | 50 | 430 | 20 | 9.4 |
| 1/2 | 12.7 | 55 | 475 | 20 | 9.4 |

a. Settings should be adjusted to suit job conditions.
b. The current shown is actual welding current and does not necessarily correspond to power source dial setting.
c. Shielding gas–argon.

# RESISTANCE WELDING

## SPOT WELDING

Spot welding is a practical joining method for fabricating aluminum sheet structures. It may be used with all wrought alloys as well as many permanent mold and sand casting alloys.

The procedures and equipment for spot welding of aluminum are similar to those used for steels. However, the higher thermal and electrical conductivities of aluminum alloys require some variations in the equipment and the welding schedules. For example, the welding current must be two to three times that required for a comparable joint between steel sections.

Quality criteria for spot welds should be established for the intended application. A range of quality standards, based upon service reliability requirements, is suggested when service conditions for spot welds vary in a particular product.

### Weldability

Some aluminum alloys are easier to spot weld than others. The relative weldabilities of various wrought and casting alloys are given in Tables 8.4 through 8.7. In general, the casting alloys considered weldable by other processes can also be spot welded. Permanent mold and sand casting alloys can be successfully spot welded, but die castings are considered difficult to join by this method. Casting alloys may be spot welded to themselves, to other casting alloys, and to wrought alloys.

The temper of an aluminum alloy affects its weldability. Electrodes used to spotweld the harder tempers have longer electrode life with improved spot weld consistency. Aluminum alloys in the annealed condition are more difficult to spot weld consistently than are the alloys in a wrought or age-hardened condition.

### Design

The best joint properties are obtained with equal sections having a thickness in the range of 0.028 to 0.125 inch. Acceptable spot welds can be made between unequal sections with thickness ratios ranging up to 3 to 1. As the thickness ratio increases, the welding conditions become more critical. They must be closely controlled to ensure acceptable weld quality. It is practical to spot weld through three thicknesses. However, spot welding of multiple thicknesses should be adequately verified by procedure qualification testing before the joint design is utilized in production.

Joints in aluminum require greater edge distances and joint overlaps than those used for steel. Suggested design dimensions are given in Table 9.29. When a flange is used on one or both components, it should provide the required overlap and be flat. Sometimes, spring-back will permit only the edge of the flange to contact the other component. If the faying surfaces cannot be brought together when an electrode force of 100 lbs maximum is applied, the flange should be reworked.

Table 8.29 also contains suggested spot weld spacings. Closer spacings than those indicated will require adjustment of the spot welding schedule to account for increased shunting of current through previous welds.

Spot welds should normally be used to carry shear loads, not tensile or peel loads. When tension or combined loadings are applied, special tests should be conducted to determine the expected strength of the joint under service conditions. The strength of spot welds in tension may vary from 20 to 90 percent of the shear strength.

The shear strength of individual spot welds will vary considerably with alloy composition, section thickness, welding schedule, weld spacing, edge distance, and overlap. Joint strength will depend upon the average individual spot strength, the number of welds, and the positioning of the welds in the joint.

The minimum strength per weld and the minimum average strength requirements of a military specification are given in Table 8.30. Somewhat lower strengths may be used for industrial applications where a lesser degree of quality is needed. The strengths listed in the table are based upon spot welds having the given minimum diameters. Weld nuggets of smaller diameter than those shown in the table are not recommended. Nugget diameter and penetration are directly affected by changes in the welding schedule and the electrode face geometry (wear).

### Equipment

Aluminum can be welded with both ac and dc power. High welding current is required be-

## Table 8.29
## Minimum design dimensions for spot welded joints in aluminum sheet

| Sheet thickness,[a] in. | Nugget diam., in. | Weld spacing,[b] in. | Edge distance,[c] in. | Joint overlap, in. |
|---|---|---|---|---|
| 0.032 | 0.14 | 0.50 | 0.25 | 0.50 |
| 0.040 | 0.16 | 0.50 | 0.25 | 0.56 |
| 0.050 | 0.18 | 0.63 | 0.31 | 0.63 |
| 0.063 | 0.20 | 0.63 | 0.38 | 0.75 |
| 0.071 | 0.21 | 0.75 | 0.38 | 0.81 |
| 0.080 | 0.23 | 0.75 | 0.38 | 0.88 |
| 0.090 | 0.24 | 0.88 | 0.44 | 0.94 |
| 0.100 | 0.25 | 1.00 | 0.44 | 1.00 |
| 0.125 | 0.28 | 1.25 | 0.50 | 1.13 |

a. Data for other thicknesses may be obtained by interpolation, or the data for the next smaller thickness may be used.
b. Distance between centers of adjacent spot welds.
c. Distance from the center of a spot weld to the edge of the sheet or flange.

## Table 8.30
## Minimum tension-shear strength for resistance spot welds in aluminum alloys
## (MIL-W-6858D)

| | | Base metal tensile strength, psi | | | | | | | |
|---|---|---|---|---|---|---|---|---|---|
| | | Below 19,500 | | 19,500 to 28,000 | | 28,000 to 56,000 | | 56,000 and above | |
| Thickness of thinner sheet, in. | Nugget diam., min, in. | Tension-shear strength, lbs/weld | | | | | | | |
| | | Min | Min avg.[a] | Min | Min avg.[a] | Min | Min avg.[a] | Min | Min avg.[a] |
| 0.016 | 0.085 | 50 | 65 | 70 | 90 | 100 | 125 | 110 | 140 |
| 0.020 | 0.10 | 80 | 100 | 100 | 125 | 135 | 170 | 140 | 175 |
| 0.025 | 0.12 | 110 | 140 | 145 | 185 | 175 | 200 | 185 | 235 |
| 0.032 | 0.14 | 165 | 210 | 210 | 265 | 235 | 295 | 260 | 325 |
| 0.040 | 0.16 | 225 | 285 | 300 | 375 | 310 | 390 | 345 | 435 |
| 0.050 | 0.18 | 295 | 370 | 400 | 500 | 430 | 540 | 465 | 585 |
| 0.063 | 0.20 | 395 | 495 | 570 | 715 | 610 | 765 | 670 | 840 |
| 0.071 | 0.21 | 450 | 565 | 645 | 810 | 720 | 900 | 825 | 1035 |
| 0.080 | 0.23 | 525 | 660 | 765 | 960 | 855 | 1070 | 1025 | 1285 |
| 0.090 | 0.24 | 595 | 745 | 870 | 1090 | 1000 | 1250 | 1255 | 1570 |
| 0.100 | 0.25 | 675 | 845 | 940 | 1175 | 1170 | 1465 | 1490 | 1865 |
| 0.125 | 0.28 | 785 | 985 | 1050 | 1315 | 1625 | 2035 | 2120 | 2650 |
| 0.140 | 0.30 | | | | | 1920 | 2400 | 2525 | 3160 |
| 0.160 | 0.32 | | | | | 2440 | 3050 | 3120 | 3900 |
| 0.180 | 0.34 | | | | | 3000 | 3750 | 3725 | 4660 |
| 0.190 | 0.35 | | | | | 3240 | 4050 | 4035 | 5045 |
| 0.250 | – | | | | | 6400 | 8000 | 7350 | 9200 |

a. Average of three or more tension-shear tests.

cause of the high electrical conductivity of aluminum. Consequently, the primary power demand is higher than that required when spot welding an equivalent thickness of steel. For the best quality, machines that produce continuous or pulsed dc power are preferred. However, acceptable welds for some applications can be made with single-phase ac equipment. In any case, the welding machine should be equipped with a low inertia force system to provide fast electrode follow-up as the weld nugget is formed. In addition, a forging force system may be necessary to consistently produce crack-free welds.

### Electrodes

Standard electrodes with radius faces are used on both sides of the joint for most welding operations. When surface marking is undesirable, a flat-faced electrode may be used on one side. Heat balance in the joint must be considered when this is done.

RWMA Group A, Class 1 copper alloy electrodes are recommended for spot welding aluminum. Class 1 alloy has the highest electrical conductivity of the Group A alloys. A Class 2 alloy is sometimes used when higher hardness is needed for wear resistance.

A copper-aluminum alloy tends to form on the face of each electrode during use. The condition is commonly called *tip pickup.* The alloy that forms has relatively low electrical conductivity, and it is brittle. As this alloy coating builds up, the contact resistance increases and the electrode tends to stick to the aluminum surface. Electrode sticking mars the surface and produces an unpleasing appearance. This sticking may also pull particles of the Cu-Al alloy from the electrode face. This action increases the surface roughness and the rate of electrode deterioration.

Excessive electrode pickup is generally the result of improper surface preparation of the parts prior to welding, insufficient electrode force, or excessive welding current. The coating on the electrode faces may be removed by dressing periodically with an appropriately shaped tool covered with a fine abrasive cloth. Care should be taken to maintain the original face contour.

### Welding Schedules

Suggested schedules for spot welding with three types of machines are given in Tables 8.31,

8.32, and 8.33. These tables can be used as guides for establishing production welding schedules.

## ROLL SPOT AND SEAM WELDING

Roll spot and seam welding are very similar to spot welding except that wheel type electrodes are used. Weld nuggets in thin sheets can be overlapped to form a seam welded joint that may be gas or liquid tight. Uniformly spaced spot welds can be produced by adjusting the *off* time. This procedure is called *roll spot welding.*

Welding may be done while the wheel electrodes and the work are in motion or while they are momentarily stopped. Surface appearance and weld quality will be better when the electrodes and work are stationary. The welding force is maintained on the nugget during solidification and cooling. With moving electrodes, the weld nugget moves from between them before it is adequately cooled.

Equipment used for roll spot or seam welding should have features similar to those of spot welding machines. Somewhat higher welding currents and electrode forces may be required for roll spot and seam welding because shunting of current through the previous nugget may be greater than with spot welding.

Excessive travel speed can contribute to pick up of aluminum on the wheel electrodes. This can be corrected by increasing the time between welds and decreasing the travel speed to give the desired number of welds per unit length.

Radius-faced wheel electrodes are normally used. Face radii generally range from 1 to 10 inches. Normally, the face radius should be about the same as the wheel radius to approach a spherical radius face in contact with the workpieces. The faces should be cleaned after each 3 to 5 revolutions with continuous welding, and after each 10 to 20 revolutions with roll spot welding. The electrodes may also be cleaned continuously with a medium-fine grade commutator stone bearing against each electrode face under 5 to 10 lbs. of force. An appropriate cutting tool may also be used.

Typical settings for seam welding 5052-H34 aluminum sheet with single-phase ac seam welding machines are given in Table 8.34. These data may be used as a guide when developing welding schedules for other alloys or tempers. Quality control for roll spot and seam welding is the same as for spot welding.

**Table 8.31**
**Suggested schedules for spot welding of aluminum alloys with single-phase ac machines**

| Thickness,[a] | Electrode face radius, in., top-bottom | Net electrode force, lbs. | Approximate welding current, kA | Welding time, cycles (60 Hz) |
|---|---|---|---|---|
| .032 | 2-2 or 2-Flat | 800 | 27 | 4 |
| .040 | 3-3 or 3-Flat | 880 | 28 | 5 |
| .050 | 3-3 or 3-Flat | 1000 | 29.5 | 6 |
| .062 | 3-3 or 3-Flat | 1150 | 33.2 | 8 |
| .070 | 4-4 | 1200 | 35.5 | 10 |
| .081 | 4-4 | 1430 | 38.5 | 10 |
| .090 | 6-6 | 1600 | 41.0 | 12 |
| .100 | 6-6 | 1800 | 44.0 | 15 |
| .110 | 6-6 | 2000 | 48.0 | 15 |
| .125 | 6-6 | 2400 | 53.0 | 15 |

a. Thickness of one sheet of a two-sheet combination.

**Table 8.32**
**Suggested schedules for spot welding of aluminum alloys with three-phase frequency converter machines**

| Sheet thickness,[a] in. | Electrode face radius, in. | Electrode force, lbs | | Current,[b] kA | | Time, cycles (60Hz) | |
|---|---|---|---|---|---|---|---|
| | | Weld | Forge | Weld | Post-heat | Weld | Post-heat |
| 0.020 | 3 | 500 | – | 26 | 0 | 1/2 | 0 |
| 0.025 | 3 | 500 | 1500 | 34 | 8.5 | 1 | 3 |
| 0.032 | 4 | 700 | 1800 | 36 | 9.0 | 1 | 4 |
| 0.040 | 4 | 800 | 2000 | 42 | 12.6 | 1 | 4 |
| 0.050 | 4 | 900 | 2300 | 46 | 13.8 | 1 | 5 |
| 0.063 | 6 | 1300 | 3000 | 54 | 18.9 | 2 | 5 |
| 0.071 | 6 | 1600 | 3600 | 61 | 21.4 | 2 | 6 |
| 0.080 | 6 | 2000 | 4300 | 65 | 22.8 | 3 | 6 |
| 0.090 | 6 | 2400 | 5300 | 75 | 30.0 | 3 | 8 |
| 0.100 | 8 | 2800 | 6800 | 85 | 34.0 | 3 | 8 |
| 0.125 | 8 | 4000 | 9000 | 100 | 45.0 | 4 | 10 |

a. Thickness of one sheet of a two-sheet combination.
b. Suitable for 2014-T3,4, and 6; 2024T-3 and 4; 7075-T6. Somewhat lower current may be used for softer alloys, such as 5052, 6009, and 6010.

**Table 8.33**
**Suggested schedules for spot welding of aluminum alloys with three-phase rectifier machines**

| Sheet thickness,[a] in. | Electrode face radius, in. | Electrode force, lbs | | Current,[b] kA | | Time, cycles (60Hz) | |
|---|---|---|---|---|---|---|---|
| | | Weld | Forge | Weld | Post-heat | Weld | Post-heat |
| 0.016 | 3 | 440 | 1000 | 19 | 0 | 1 | 0 |
| 0.020 | 3 | 520 | 1150 | 22 | 0 | 1 | 0 |
| 0.032 | 3 | 670 | 1540 | 28 | 0 | 2 | 0 |
| 0.040 | 3 | 730 | 1800 | 32 | 0 | 3 | 0 |
| 0.050 | 8 | 900 | 2250 | 37 | 30 | 4 | 4 |
| 0.063 | 8 | 1100 | 2900 | 43 | 36 | 5 | 5 |
| 0.071 | 8 | 1190 | 3240 | 48 | 38 | 6 | 7 |
| 0.080 | 8 | 1460 | 3800 | 52 | 42 | 7 | 9 |
| 0.090 | 8 | 1700 | 4300 | 56 | 45 | 8 | 11 |
| 0.100 | 8 | 1900 | 5000 | 61 | 49 | 9 | 14 |
| 0.125 | 8 | 2500 | 6500 | 69 | 54 | 10 | 22 |

a. Thickness of one sheet of a two-sheet combination.
b. Suitable for 2014-T3,4 and 6; 2024-T3 and 4; 7075-T6. Somewhat lower current may be used for softer alloys, such as 5052, 6009, and 6010.

## WELD QUALITY

The quality of spot and seam welds in aluminum alloys is more sensitive to process variations than are similar welds in steels. This is related to the high resistivity of aluminum oxides and the high electrical and thermal conductivities of the metal. The size of the weld nugget is very sensitive to the heat energy developed by the resistance to the current flowing through the work. The energy must be produced rapidly to overcome losses to the surrounding base metal and the electrodes.

The contact resistances between the faying surfaces and between the electrodes and the workpieces can be a significant part of the total resistance in the circuit. Significant variations in these contact resistances can cause large changes in the welding current density. Since the condition of the aluminum surfaces affects the contact resistances, uniform cleanness is essential for consistent weld quality.

The contact resistance between the electrodes and the workpieces increases with pick-up of aluminum onto the electrode faces. As this contact resistance increases, so does electrode heating and wear. As the contact area increases, welding current density decreases. This reduces nugget size and penetration. Weld strength decreases at the same time. Thus, electrode wear requires constant attention.

Other important factors that affect weld quality are surface appearance, internal discontinuities, sheet separations, metal expulsion, weld strength, and weld ductility. Uniform weld quality can be obtained only by the use of proper equipment and trained operators, rigid adherence to qualified welding procedures, and close control of machine settings. To ensure weld quality, a correct welding procedure and schedule should be developed and then maintained during production by a regular program of quality control.

Factors that tend to produce cracks or porosity in welds are excessive heating of the nugget, a high cooling rate, and improper application of the electrode force. Spot welds in some high strength alloys, such as 2024 and 7075, are subject to cracking if the welding current is too high or the electrode force is too low. Cooling rate can be controlled by application of current downslope or a postheat cycle. With dual force machines, proper adjustment of the forge delay time may also prevent cracking.

## QUALITY CONTROL

Quality of aluminum spot and seam welds depends upon the welding schedule, electrode

**Table 8.34**
**Typical schedules for gas-tight seam welds in 5052-H34 aluminum alloy with single-phase ac machines**

| Sheet thickness,[a] in. | Welds per in. | On + off time,[b] cycles (60 Hz) | Travel speed,[c] ft/min | On time, cycles, (60 Hz) Min | Max | Electrode force, lbs | Welding current, kA | Approx. weld width,[d] in. |
|---|---|---|---|---|---|---|---|---|
| 0.010 | 25 | 3½ | 3.4 | ½ | 1 | 420 | 19.5 | 0.08 |
| 0.016 | 21 | 3½ | 4.1 | ½ | 1 | 500 | 22.0 | 0.09 |
| 0.020 | 20 | 4½ | 3.3 | ½ | 1½ | 540 | 24.0 | 0.10 |
| 0.025 | 18 | 5½ | 3.0 | 1 | 1½ | 600 | 26.0 | 0.11 |
| 0.032 | 16 | 5½ | 3.4 | 1 | 1½ | 690 | 29.0 | 0.13 |
| 0.040 | 14 | 7½ | 2.9 | 1½ | 2½ | 760 | 32.0 | 0.14 |
| 0.050 | 12 | 9½ | 2.6 | 1½ | 3 | 860 | 36.0 | 0.16 |
| 0.063 | 10 | 11½ | 2.6 | 2 | 3½ | 960 | 38.5 | 0.19 |
| 0.080 | 9 | 15½ | 2.1 | 3 | 5 | 1090 | 41.0 | 0.22 |
| 0.100 | 8 | 20½ | 1.8 | 4 | 6½ | 1230 | 43.0 | 0.26 |
| 0.125 | 7 | 28½ | 1.5 | 5½ | 9½ | 1350 | 45.0 | 0.32 |

a. Thinner of a two-sheet combination.
b. Use next higher full cycle setting if timer is not equipped for synchronous initiation.
c. Should be adjusted to give the desired number of spots per inch.
d. Welding force, welding current, or both should be adjusted to produce the desired weld width. Use lower force for soft alloys or tempers and higher force for hard alloys or tempers.

condition, and surface preparation. All three must be controlled to maintain acceptable weld quality.

## Surface Preparation

Welds of uniform strength and good appearance depend upon a consistently low surface resistance. The surface condition of as-received material may be satisfactory for many commercial spot and seam welding applications. On the other hand, some applications, such as aircraft, require very consistent welds of high quality, particularly when failure of the weldment would result in loss of the equipment. Proper cleaning of parts and monitoring of the cleaning operation is one factor in maintaining uniform weld quality.

For most applications, some cleaning is necessary before spot or seam welding. The procedures are similar to those described earlier for arc welding. For critical applications, the cleanness of the components must be monitored, and time lapse between cleaning and welding must not exceed a specified period.

## Surface Contact Resistance

Measurement of surface contact resistance is an effective method of monitoring a cleaning operation. Figure 8.10 is a diagram of a surface resistance measuring device. Two cleaned coupons are overlapped and placed between two 3 in. radius-faced spot welding electrodes. A standardized current and electrode force are applied; a current of 50 mA and a force of 600 lbs. are frequently used. The voltage drop between the two coupons is measured with a millivolt meter or a Kelvin bridge. The resistance between the two coupons is then calculated from the current and voltage readings ($R = E/I$).

It is important that all tests are carried out under identical conditions. The results are sensitive to small changes in procedure. Average surface resistance is usually obtained from at least 10 readings on each set of coupons. The coupons must not move as the electrode force is applied. Movement may break the oxide coating and cause false readings.

With this test, the contact resistance between properly cleaned aluminum sheets will range from $10^{-5}$ to about $10^{-4}$ ohms. That of uncleaned stock may range up to $10^{-2}$ ohms or higher.

## FLASH WELDING

All aluminum alloys may be joined by the flash welding process. This process is particu-

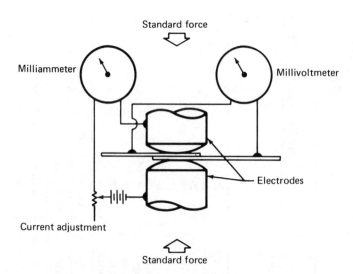

**Fig. 8.10—Arrangement for measuring surface contact resistance for monitoring cleanness**

larly adapted to making butt or miter joints between two parts of similar cross sections. It can be used to join aluminum to copper.

Good mechanical properties can be obtained in flash welded joints. Joint efficiencies of at least 80 percent are readily obtained. They are generally higher when the alloy is in a hard temper. Heat treatment after flash welding may provide a slight increase in joint efficiency.

## HIGH FREQUENCY RESISTANCE WELDING

Resistance welding with high frequency current is used primarily for high-speed production of tubing. In this application, squeeze rolls forge the edges of the sheet together after they are heated to welding temperture with high frequency current. Tubing with wall thicknesses of 0.03 to 0.125 in. can be welded at high travel speeds.

# SOLID-STATE WELDING

## COLD WELDING

Cold welding is done at room temperature with substantial deformation at the weld. Diffusion is minimal, and there is no cast structure or heat-affected zone in the weld area. Butt and lap joints can be made with this process. Since intimate contact between joint surfaces is essential, they must be clean and free of heavy oxide. Surface preparation includes degreasing followed by abrading or machining just prior to welding. Wire brushing of the faying surfaces is preferred just before welding lap joints.

Lap joints in soft nonheat-treatable alloys, such as 1100, 1350, and 3003, are easily cold welded. Greater deformation is required to weld harder alloys, especially with hard tempers. Lap joints in alloys containing more than 3 percent magnesium, those of the 2XXX and 7XXX series, and casting alloys are not readily cold welded.

Butt joints can be made in most aluminum alloy wire, rod, tubing, and simple extruded shapes. Welds in annealed material will require an upset distance equal to about 1 to 1.5 times the diameter or wall thickness. High-strength alloys may require an upset distance of 4 to 5 times the diameter or wall thickness. The weld is usually as strong as the base metal.

## ULTRASONIC WELDING

With this process, welding is accomplished by the local application of high-frequency vibratory energy to the workpieces while they are held together under low pressure. The process is used to join foil as well as sheet gages of aluminum alloys. While all aluminum alloys can be ultrasonically welded, the degree of weldability varies with the alloy and temper. Aluminum alloys can also be joined to other metals with this process.

Ultrasonic welding can be done (1) without extensive surface preparation, (2) with minimum deformation, and (3) with low compressive loads. Ultrasonic welds look much like resistance spot or seam welds, but are often characterized by a localized roughened surface.

For relatively low-strength aluminum alloys, ultrasonic spot welds exhibit approximately the same strength as resistance spot or seam welds. With the higher strength alloys, the strength of ultrasonic welds can exceed the strength of resistance welds. The primary reasons for this are that ultrasonic welding produces no fusion or heat-affected zones in the base metal, and the size of the weld is generally greater.

Ultrasonic welding generally requires less surface preparation than for resistance welding methods. Normally, degreasing of the aluminum is advisable. Heat-treated alloys and alloys containing high percentages of magnesium should be cleaned of surface oxides before welding to obtain uniform welds.

## EXPLOSION WELDING

Explosion welding uses energy from the detonation of an explosive to produce a solid-state weld. The force produced by the detonation of the explosive drives the two components together to create a high-strength weld with minimum diffusion and deformation at the interface. Explosion welding is limited to lap joints and to cladding of parts with a second metal having special properties. It is particularly useful for joining aluminum to other metals. Surface preparation for aluminum is similar to that for other metals. Although normal surface oxides are broken up and dispersed during welding, the faying surfaces should be cleaned within a specified time before welding.

## DIFFUSION WELDING

With this process, the principal mechanism for joint formation is solid-state diffusion. Coalescence is produced by the application of heat and pressure, but there is no macroscopic deformation or relative motion of the parts. A solid filler metal may be inserted between the faying surfaces to aid the diffusion process.

Aluminum alloys can be diffusion welded

provided some means is used to prevent, disrupt, or dissolve the surface oxides. Best weld strengths and ease of bonding are achieved with a thin intermediate layer of another metal such as silver, copper, or gold-copper alloy. A wide range of temperatures, times, and pressures may be used. The welding operation is normally done in vacuum or high-purity inert gas.

## FRICTION WELDING

Friction welding produces coalescence by the heat developed between two faying surfaces from mechanically induced sliding or rubbing motion. Pressure is applied during and after the heating cycle. The friction welding process is usually based on rotating one part at controlled high speed against a stationary member to which it is to be joined.

Aluminum can be welded to itself and several other metals with this process. Some combinations require precise control of the process variables. Equipment manufacturers should be consulted for specific applications.

# PROPERTIES AND PERFORMANCE OF WELDMENTS

## METALLURIGAL EFFECTS

The properties and performance of a welded joint are greatly influenced by the solidification behavior and microstructural changes that take place in the joint during and after the welding operation. An understanding of these changes is essential in predicting the properties and performance of a weldment.[14] When a fusion weld is made in aluminum, the metallurgical structures of both the weld metal and the base metal must be considered.

### Weld Metal

The microstructure of the weld metal is affected by its composition, the welding process, the cooling rate, and the metallurgical reactions

that take place. Chemical composition determines the temperature range over which the weld metal solidifies. The weld metal tends to hot crack if the solidification range is wide. Addition of a filler metal that narrows the solidification range of the weld metal can be beneficial, provided dilution is controlled. Increasing the weld metal cooling rate using external chills helps to reduce segregation during freezing.

The solubility of hydrogen in solid aluminum is much lower than in molten aluminum. Hydrogen is often introduced into the weld metal from hydrated oxides on the surface of the filler metal or the surfaces of the components. It is evolved during welding and causes porosity because of the rapid solidification. This can be a significant problem with welding processes that use relatively large amounts of bare filler wire. For this reason, the bare filler wire as well as the components should be properly cleaned. Then, they should be stored in a dry environment until needed in production.

---

14. Refer to the *Welding Handbook*, Vol. 1, 7th ed., 100-151, for a general discussion of welding metallurgy.

The final mechanical properties of the weld metal are generally established after solidification is complete. Relatively few commercial applications utilize a strengthening heat treatment of a completed weldment. In some cases, a stress relieving heat treatment may be used but it generally will not significantly improve mechanical properties. With some alloys and welding procedures, the weld metal and the fusion zone may be in the solution heat-treated condition. This metal may age at room temperature or be artificially aged to increase its strength.

### Heat-Affected Zone

During and immediately after welding, many metallurgical changes can take place in the heat-affected zone. These include recrystallization, resolutioning of precipitates, aging, and overaging.

With nonheat-treatable alloys, recrystallization of cold worked metal and some grain growth are likely to occur in this zone. The strength and ductility of the heat-affected zone will generally be similar to that of annealed metal. The behavior of this zone in service will depend somewhat on its width in relationship to the section thickness. Generally, the nonheat-treatable alloys are the easiest to weld because this ductile zone can yield to relieve welding stresses.

Heat-treatable alloys are strengthened by a precipitation-hardening treatment and, in some cases, in combination with cold working. These alloys respond to the heat of welding in a manner similar to the nonheat-treatable. However, the response is more complex because the effect depends upon the peak temperature to which the metal is exposed. These alloys are normally welded in the aged condition. Therefore, the metallurgical changes vary with distance from the weld interface, as shown in Fig. 8.11.

The weld metal and fusion zone will be in the solutioned condition, as will any adjacent base metal that is heated to a temperature where precipitates redissolve rapidly. These zones will be soft and ductile. Next to this, a narrow band of previously aged base metal will be heated to a temperature where the precipitates agglomerate into larger particles. This will overage and soften the metal somewhat. Subsequent aging will not alter the properties of this zone.

It is best to weld heat-treated alloys with low heat input to minimize the width of the overaged zone. The entire weldment must be solution heat treated and aged to restore the previous properties. Solution heat treatment requires a rapid quench, and this may cause unacceptable distortion of the weldment. Heat treatment of small weldments may be practical in some cases. Where this is not practical, aging alone after welding will improve the strength of zones that are in the solutioned condition. As an alternative, the base metal can be solution heat-treated before welding, and then the entire weldment aged.

## TENSILE STRENGTH AND DUCTILITY

### Nonheat-Treatable Alloys

A nonheat-treatable alloy loses the effects of strain hardening after a relatively short time at the annealing temperature, which is normally below 700 F. With arc welded joints, a portion of the heat-affected zone will reach the annealing temperature of the base metal. For this reason, the minimum annealed tensile strength of the base metal is generally considered as the design strength for butt joints in these alloys.

Welds in nonheat-treatable alloys are capable of extensive deformation prior to failure. The high-strength, solid solution alloys of the aluminum-magnesium or aluminum-magnesium-manganese types (5XXX series) are particularly good for welded construction because of the nearly uniform mechanical properties of the various zones in a welded joint.

### Heat-Treatable Alloys

The strengths of heat-treatable alloys are also decreased by the heat of welding. However, these alloys require a substantial time at temperature to fully soften them. As a result, the mechanical properties of the heat-affected zone in an as-welded joint are higher than those of fully annealed base metal. The properties may vary considerably with heat input and cooling rate, but they are generally lower than those of weld heat-affected zones in many of the 5XXX series of alloys. One major exception is the 7XXX series of alloys, which naturally age to high strengths after welding.

The speed of welding has a significant effect upon the mechanical properties of arc welds in heat-treatable alloys. High welding speed contributes to low heat input and a narrow heat-affected zone. This, in turn, reduces the possibility of eutectic melting, grain boundary precipitation, overaging, grain growth, or a combination of these. Mechanical properties of gas metal arc

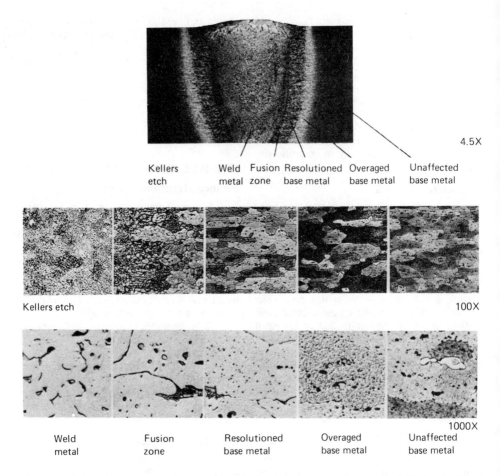

Kellers etch — Weld metal — Fusion zone — Resolutioned base metal — Overaged base metal — Unaffected base metal

4.5X

Kellers etch — 100X

Weld metal — Fusion zone — Resolutioned base metal — Overaged base metal — Unaffected base metal

1000X

**Fig. 8.11—Five microstructural zones in a single-pass gas tungsten arc weld in 0.5 in. thick 2219-T87 aluminum alloy plate (as-welded) (Reduced 71%)**

welded joints are generally higher than those of equivalent gas tungsten arc welded joints because of the higher travel speeds employed with the former process.

Section thickness also has an effect on the properties of arc welds. Welds in sheet have better strength properties than welds in plate of the same alloy because the former can be made at higher welding speeds and lower heat inputs. Other factors that affect weld strength are the size of the weld, the joint design, and the amount of reinforcement.

Welding of heat-treatable alloys in the solution heat-treated condition rather than in the aged

condition will often minimize cracking. Welds in solution heat-treated metal will have a more uniform microstructure, and the base metal will impose less restraint during weld metal solidification.

Typical as-welded tensile properties of gas shielded arc welds in aluminum alloys are given in Table 8.35. These data are useful for comparison only, since actual tensile properties of a welded joint depend upon many variables, as discussed previously.

Repair welding is usually more difficult with heat-treatable alloys because of microstructural changes that take place during the original

**Table 8.35**
**Typical as-welded tensile properties of gas metal arc or gas tungsten arc groove welds in aluminum alloys**

| Base metal | Filler metal | Avg. tensile strength, ksi | Avg. yield strength,[a] ksi | Avg. tensile elongation, percent in 2 in.[b] |
|---|---|---|---|---|
| | | Nonheat-treatable alloys | | |
| 1060 | ER1260 | 10 | 5 | 29 |
| 1100 | ER1100 | 13 | 5 | 29 |
| 1350 | ER1260 | 10 | 4 | 29 |
| 3003 | ER1100 | 16 | 7 | 24 |
| 5005 | ER5356 | 16 | 9 | 15 |
| 5050 | ER5356 | 23 | 12 | 18 |
| 5052 | ER5356 | 28 | 14 | 19 |
| 5083 | ER5183 | 43 | 22 | 16 |
| 5086 | ER5356 | 39 | 19 | 17 |
| 5154 | ER5654 | 33 | 18 | 17 |
| 5454 | ER5554 | 35 | 16 | 17 |
| 5456 | ER5456 | 49 | 23 | 14 |
| | | Heat-treatable alloys | | |
| 2014-T6 | ER4043 | 34 | 28 | 4 |
| 2219-T6,T8 | ER2319 | 35 | 26 | 3 |
| 6061-T4,T6 | ER4043 | 27 | 18 | 8 |
| 6061-T6 | ER5356 | 30 | 19 | 11 |
| 6063-T4 | ER4043 | 20 | 10 | 12 |
| 6063-T6 | ER4043 | 20 | 12 | 8 |
| 6063-T6 | ER5356 | 20 | 12 | 12 |
| 7005-T53 | ER5356 | 46 | 30 | 10 |

a. 0.2 percent offset in 2 in. gage length.
b. Elongation may not be uniform within the gage length.

welding. Restraint is usually greater when repair welding, and this may cause cracking in the heat-affected zone or in the previously deposited weld metal. Joint strength may also be lower after repairs are made.

## SHEAR STRENGTH

Fillet welds are designed on the basis of shear on a plane area through the weld. This area is the product of the effective throat and the length of the weld. It generally lies within the weld metal. Weld metal composition is nearly the same as that of the filler metal because dilution is normally low. The transverse and longitudinal shear strengths of fillet welds made with five aluminum filler metals are given in Table 8.36.

## CORROSION RESISTANCE

Many aluminum alloys can be welded with

no change in corrosion resistance, particularly the nonheat-treatable alloys. In general, the welding process does not affect corrosion resistance. An exception is shielded metal arc welding if the slag is not properly removed, as described previously.

Some heat-treatable alloys may have lower

**Table 8.36**
**Shear strengths of fillet welds made with several aluminum filler metals**

| Filler metal | Shear strength, psi[a] | |
|---|---|---|
| | Longitudinal | Transverse |
| ER1100 | 11,000 | 13,000 |
| ER4043 | 16,000 | 21,000 |
| ER5356 | 21,000 | 34,000 |
| ER5183 | 24,000 | 39,000 |
| ER5556 | 26,000 | 41,000 |

a. On a plane at 45 degrees to the legs of a right-angle joint.

resistance to corrosion after exposure to the heat of welding, particularly alloys containing relatively large amounts of copper (2XXX) or zinc (7XXX). Heat-affected zones in these alloys exhibit grain boundary precipitation of a second phase. The electrical potential of the metal near grain boundaries is normally anodic to the remainder of the microstructure. In the presence of an electrolyte, selective corrosion may take place at the grain boundaries. This corrosion can proceed faster when the metal is under welding or other stress. A postweld heat treatment can be used to produce a more homogeneous microstructure in these alloys for better corrosion resistance. If a completed weldment cannot be solution heat treated because of distortion, the components should be age hardened prior to welding for corrosion applications.

# BRAZING

## BRAZEABLE ALLOYS

Many aluminum alloys can be brazed with commercially available brazing alloys.[15] Some require special procedures to achieve acceptable joint quality. The commonly used alloys are rated for brazeability in Tables 8.4 through 8.7. The nonheat-treatable alloys most frequently brazed are the 1XXX and 3XXX series, 5005, and 5050. The high manganese alloys of the 5XXX series are difficult to braze because of poor wetting characteristics.

The 6XXX series are the easiest of the heat-treatable alloys to braze. The 2XXX series, 7001, 7075 and 7178 alloys are not brazeable because their melting temperatures are below those of commercially available filler metals.

Brazeable casting alloys include 356.0, 357.0, 359.0, 443.0 and 712.0. High-quality castings are as easy to braze as their equivalent wrought alloys. Problems arise when the casting quality is low and the metal is porous. Die castings are usually high in zinc and difficult to braze.

## FILLER METALS

Table 8.37 describes the commercial brazing filler metals used for aluminum. They are primarily aluminum-silicon alloys produced in the form of wire, rod, foil, or powder. Powdered filler metal is normally applied in paste form in combination with a brazing flux.

In some cases, the filler metal is applied to a core sheet as cladding on one or both sides. This product is known as brazing sheet which can be formed by conventional means. It is widely used for assemblies that can be furnace or dip brazed. In this case, the brazing filler metal is preplaced. Table 8.38 describes common brazing sheets. Those sheets clad with BAlSi-6,7,8,9 or 11 filler metal are designed for vacuum brazing applications.

Recommended filler metals for brazing several aluminum alloys are given in Table 8.39. Filler metal selection is based primarily on its melting range compared to that of the base metal to be brazed. The melting ranges of brazing filler metals lie within the overall temperature range in which aluminum alloys melt.

When selecting a filler metal, its liquidus must be below the solidus of the base metal. Generally, the difference in these temperatures will be small. This will require good control of the brazing temperature. For torch brazing, the filler metal should have a low melting temperature to minimize the possibility of melting the base metal. Also, it should have a wide melting range for good control of flow in the joint.

When furnace or dip brazing, a filler metal with a narrow melting range will minimize the necessary time at brazing temperature. This will permit a shorter cycle time and minimize diffusion. The filler metal should have good fluidity to fill narrow joint clearances.

---

15. Additional information is available in the *Brazing Manual*, 3rd Ed., Miami: American Welding Society, 1976.

**Table 8.37**
**Aluminum brazing filler metals**

| AWS class.[a] | Alloy designation | Nominal composition,[b] % | | Melting range, °F | Brazing range, °F | Standard forms[c] | Applicable brazing processes[d] |
|---|---|---|---|---|---|---|---|
| | | Si | Mg | | | | |
| BAlSi-2 | 4343 | 7.5 | | 1070-1142 | 1110-1150 | C,S | D,F |
| BAlSi-3[e] | 4145 | 10 | | 970-1085 | 1060-1120 | R,S | D,F,T |
| BAlSi-4 | 4047 | 12 | | 1070-1080 | 1080-1120 | P,R,S | D,F,T |
| BAlSi-5[f] | 4045 | 10 | | 1070-1110 | 1090-1120 | C,S | D,F |
| BAlSi-6[g] | | 7.5 | | 1038-1125 | 1110-1150 | C | VF |
| BAlSi-7[g] | 4004 | 10 | 1.5 | 1038-1105 | 1090-1120 | C | VF |
| BAlSi-8[g] | | 12 | 1.5 | 1038-1075 | 1080-1120 | C | VF |
| BAlSi-9[g] | | 12 | 0.3 | 1044-1080 | 1080-1120 | C | VF |
| BAlSi-10[g] | | 11 | 2.5 | 1038-1086 | 1080-1120 | R | VF |
| BAlSi-11[g,h] | | 10 | 1.5 | 1038-1105 | 1090-1120 | C | VF |
| — | 4044 | 8.5 | | 1070-1115 | 1100-1135 | C | D,F |

a. See AWS A5.8-81, *Specification for Brazing Filler Metal.*
b. Remainder Al
c. C—cladding on sheet, P—powder, R—rod or wire, S—sheet
d. D—dip, F—furnace, T—torch, VF—vacuum furnace
e. Also contains 4% Cu
f. Also contains 0.2% Ti
g. Melting range in air. Melting range in vacuum is different.
h. Also contains 0.1% Bi.

**Table 8.38**
**Aluminum brazing sheet**

| Designation number | Sides clad | Core alloy | Cladding alloy |
|---|---|---|---|
| 7 | 1 | 3003 | BA1Si-7 |
| 8 | 2 | | |
| 11 | 1 | 3003 | BA1Si-2 |
| 12 | 2 | | |
| 13 | 1 | 6951 | BA1Si-7 |
| 14 | 2 | | |
| 21 | 1 | 6951 | BA1Si-2 |
| 22 | 2 | | |
| 23 | 1 | 6951 | BA1Si-5 |
| 24 | 2 | | |
| 33 | 1 | 6951 | 4044 |
| 34 | 2 | | |
| 44[a] | 2 | 6951 | 4044 7072 |
| – | 1 or 2 | 3003 | BA1Si-6 BA1Si-8 BA1Si-9 |
| – | 1 or 2 | 3105 | BA1Si-11[b] |

a. One side is clad with 7072 alloy for corrosion resistance.
b. Max. brazing temperature is 1110° F.

The filler metals should be clean and free of oxide coating. If they are exposed to the atmosphere for a long time or handled prior to use, they should be cleaned before brazing using the same procedures applicable to the base metal.

## FLUXES

Chemical fluxes are required for all aluminum brazing operations except for furnace brazing in vacuum, inert gas, or nitrogen. Aluminum brazing fluxes consist of various combinations of fluorides and chlorides.[16] They are supplied as a dry powder. For torch and furnace brazing, the

16. These fluxes are toxic, and must be handled using appropriate safety measures to avoid ingestion and inhalation of flux and fumes. See ANSI Z49.1, *Safety in Welding and Cutting*, available from the American Welding Society.

flux is mixed with clean water or alcohol to make a paste. The paste is brushed, sprayed, dipped, or flowed onto the entire area of the joint and the filler metal as well. Most fluxes used with torch and furnace brazing may severely corrode thin aluminum. Therefore, excessive flux should not be used with these processes.

In dip brazing, the hot bath is molten flux. Less active fluxes can be used with this process, and thin components can be safely brazed. The molten flux must be dehydrated before use to remove chemically combined moisture. This is done by immersing coils of aluminum sheet in the flux. Hydrogen bubbles will evolve until all the moisture is gone.

## FLUX RESIDUE REMOVAL

After brazing, the corrosive flux residue must be removed. Flux residues form a hard, brittle layer, and can be removed by a hot water rinse. This is sometimes followed by chemical cleaning. Mechanical cleaning, such as wire brushing or grinding, is not adequate. It breaks up the residue into fine particles that may become embedded in the aluminum surface.

Several procedures are suitable for the removal of flux residue. Immersion in hot water before the brazement has cooled is effective in removing a major portion. For torch brazed

**Table 8.39**
**Recommended filler metals for commonly brazed aluminum alloys**

| Base metal | Brazing filler metal | |
|---|---|---|
| | AWS classification | Alloy designation |
| 1100 1350 3003 | BA1Si-2 BA1Si-4 | 4343 4047 |
| 3004 5005 5050 6063 6951 | BA1Si-2 BA1Si-4 BA1Si-5 | 4343 4047 4045 |
| 6061 7005 712.0 | BA1Si-3 BA1Si-4 | 4145 4047 |
| 356.0 443.0 | BA1Si-3 | 4145 |

joints, scrubbing with a fiber brush under running hot water is good practice.

Flux residues can be removed chemically. Acceptable solutions for this are given in Table 8.40. There are also a number of commercial proprietary cleaners available for this purpose. An acid dip is always required if the joints cannot be scrubbed during the water rinse. All methods require a thorough clean water rinse to remove the cleaning agent.

## DESIGN CONSIDERATIONS

Aluminum can be brazed by most methods of heating. Torch, furnace, and dip brazing are commonly used. A dip-brazed automobile radiator is shown in Fig. 8.12. Induction, resistance, and infrared brazing are more difficult because of the low electrical resistance and high reflectivity of aluminum.

The common brazed joint designs are suitable for aluminum. Lap and T-joints are normally used with sheet gages. Butt joints may be used to join large cross sections.

Joint clearance depends upon a number of factors including the component design, the brazing process, and the filler metal. Clearance may change during the heating cycle and this must be determined to establish the initial dimension. In general, a clearance of 0.002 to 0.004 in. is suitable for laps of less than 0.25 in. when dip brazing. For torch, furnace, or induction brazing, the clearance should be increased to 0.004 to 0.010 inch. For laps of 0.25 in. and greater, the clearance should be 0.010 to 0.025 in. because alloying between the molten filler metal and the base metal tends to inhibit flow. The optimum clearance for a given application should be determined by appropriate tests.

Joint clearance can generally be small with dip brazing because the capillary force is high in the molten flux bath. With brazing sheet or shim, the faying surfaces must be in contact and under slight pressure so that they will move together as the filler metal melts and flows in the joint.

Precleaning of the joint faces is essential for good wetting of the base metal and flow of the filler metal. The cleaning procedures described previously are acceptable for brazing.

In all cases, brazing temperature exceeds the annealing temperature of the base metals. Furnace or dip brazed assemblies of nonheat-treatable aluminum alloys have annealed mechanical properties.

## Table 8.40
### Chemical solutions for removing aluminum brazing flux residue

| Solution | Composition[a] | Bath temperature | Procedure[b] |
|---|---|---|---|
| Nitric acid | 1 gal $HNO_3$<br>7 gal water | Ambient | Immerse 10-20 min., rinse in hot or cold water.[c] |
| Nitric-hydro-fluoric acid | 1 gal $HNO_3$<br>0.06 gal HF<br>9 gal water | Ambient | Immerse for 5 min., rinse in cold and hot water.[d] |
| Hydrofluoric acid | 0.3 gal HF<br>10 gal water | Ambient | Immerse for 2-5 min., rinse in cold water, immerse in nitric acid solution, rinse in hot or cold water. |
| Phosphoric acid–chromium trioxide | 0.4 gal 85% $H_3PO_4$<br>1.8 lb $CrO_3$<br>10 gal water | 180° F | Immerse for 10-15 min., rinse in hot or cold water.[e] |

a. Acids are concentrated technical grades.
b. In all cases, remove the major portion of flux residue with a hot water rinse. To check for flux residue, place a few drops of distilled water on the area. After 1 min., collect the water and drop it into a 5 percent silver nitrate solution. A white precipitate indicates the presence of residue, and cleaning should be repeated.
c. Chloride concentration should not exceed 5 g/L. Not recommended for sections less then 0.020 in. thick.
d. Chloride concentration should not exceed 3 g/L. Not recommended for sections less than 0.020 in. thick.
e. Chloride concentration should not exceed 100 g/L. Suitable for thin sections.

*Fig. 8.12–A dip brazed aluminum automobile radiator (left) and a section through a fin-to-tube joint (right)*

Heat-treatable base metals are also annealed by brazing. However, the strength of a brazement can be substantially increased by a separate heat-treating operation or by quenching it from the brazing temperature. The latter procedure is not always feasible, depending upon the geometry of the part. Heat-treatable brazements can be quenched in an air blast, a water spray, or a tank of hot or cold water. A slight delay after removal from the heating source is necessary to permit the filler metal to solidify before quenching. Otherwise, dimensional changes during quenching may tear the joint.

Cracking of the joints may occur even though the filler metal has solidified. Complicated assemblies should be allowed to cool and then heat treated in a separate operation.

No useful purpose is served by using a heat-treatable aluminum alloy for a brazement unless it is heat treated after brazing. The nonheat-treatable alloys are often equal in strength to annealed heat-treatable alloys.

Resistance to corrosion of aluminum alloys is not lowered by brazing. Corrosion resistance of brazed joints is similar to that of welded aluminum joints. In certain cases, exposure tests under specific conditions are required for performance evaluation.

# SOLDERING

## SOLDERABILITY

The soldering of aluminum differs from the soldering of most other metals in several ways.[17] The most important concern is the oxide that forms rapidly on aluminum. In most cases, this requires the use of active soldering fluxes that are specifically designed for aluminum. Noncorrosive fluxes are not suitable. A second difference is that special techniques are required to obtain solder flow into certain types of joints. A third important difference is that the corrosion resistance of soldered aluminum joints is more dependent upon the composition of the solder than for similar joints in copper, brass, or steel.

The relative solderability of aluminum alloys is shown in Tables 8.4 through 8.7. Com-

---

17. Refer to the *Soldering Manual*, 2nd Ed., Miami: American Welding Society, 1978.

monly soldered aluminum alloys are 1060, 1100, 3003, 5005, 6061, and 7072.

Although aluminum and most aluminum alloys can be soldered, alloying elements influence the solderability. Readily soldered aluminum alloys contain no more than 1 percent magnesium or 5 percent silicon. Alloys containing greater amounts of these elements have poor flux wetting characteristics and may be rapidly attacked intergranularly by the solder. Aluminum castings generally have poor solderability because of their compositions. In addition, the surface condition of castings often makes them difficult to solder. The surface may be difficult to clean, and any surface porosity can hamper complete flux removal.

Residual stresses from cold working may accelerate intergranular penetration of solder into the aluminum and can cause cracking. Intergranular penetration can be minimized by stress relieving prior to soldering. Stress relieving will usually occur automatically when high-temperature soldering alloys are used.

## SOLDERS

The commercial solders for aluminum can be classified into low, intermediate, and high temperature groups according to their melting ranges. The general characteristics of these groups of solders are given in Table 8.41.

### Low Temperature

Low temperature solders are generally tin-zinc alloys in which tin is the predominant element. Lead-bismuth solders are also used to solder aluminum at low temperatures. One low temperature solder is the tin-zinc eutectic containing 91 percent tin and 9 percent zinc. Another solder contains 78.5 percent lead, 18.5 percent bismuth and 3 percent silver. These alloys melt at about 400° F, wet aluminum readily, flow easily, and have high resistance to corrosion.

Tin-lead solders, in general, form a highly anodic interface with aluminum, and they have poor corrosion resistance. Although the tin-lead solders are not recommended for aluminum the addition of a few percent zinc or cadmium to such solders improves both their soldering characteristics and their resistance to corrosion.

The lead-bismuth solders exhibit the best corrosion resistance of the low-temperature solders. In general, assemblies joined with low-temperature solders have poorer corrosion resistance than others made with high-temperature solders.

These solders should not be used in corrosive environments unless some protective coating is applied to the solder joints.

### Intermediate Temperature

Intermediate temperature solders are usually either tin-zinc or cadmium-zinc alloys containing from 30 to 90 percent zinc. They may also contain other metals, such as lead, bismuth, silver, nickel, copper, and aluminum. Among the more common of these solders are the 70% tin-30% zinc, 70% zinc-30% tin, and 60% zinc-40% cadmium solders. Because of their higher zinc content, these intermediate-temperature solders generally wet aluminum more readily, form larger fillets, and give stronger and more corrosion resistant joints than do the low-temperature solders.

### High Temperature

The high temperature solders contain 90 to 100 percent zinc and small amounts of other metals such as silver, aluminum, copper, and nickel. These additions are made to lower the soldering temperature, obtain a wider melting range, and improve wetting of the aluminum. The high-temperature solders are the strongest of the solders. They also produce joints that have corrosion resistance properties markedly superior to those of low and intermediate-temperature solders. To assure the best possible corrosion resistance, the high-temperature solders should be practically free of low melting metals.

## FLUXES

Fluxes for soldering aluminum may be divided into two general classes: chemical and reaction. Fluxes of both classes are corrosive to aluminum.

Chemical fluxes are composed of (1) a boron trifluoride-organic addition compound, such as boron trifluoride-monoethanolamine; (2) a flux vehicle such as methyl alcohol; (3) a heavy metal fluoborate, such as cadmium fluoborate; and (3) a plasticizer such as stearic acid. They may or may not contain other modifiers, such as zinc fluoride, zinc chloride, and ammonia compounds.

Chemical fluxes are most often used where the soldering temperature (actual temperature of the joint) is less than 525° F. For this reason the chemical fluxes are generally used with the low-temperature solders. In general, the flux residue

**Table 8.41**
**General characteristics of solders for aluminum**

| Group | Melting range, °F | Common constituents | Ease of application | Wettability on aluminum | Relative strength | Relative corrosion resistance |
|---|---|---|---|---|---|---|
| Low temperature | 300–500 | Tin, lead, bismuth, cadmium, zinc | Good | Poor to fair | Low | Low |
| Intermediate temperature | 500–700 | Cadmium-zinc or tin-zinc | Fair | Good to excellent | Medium | Medium |
| High temperature | 700–800 | Zinc, aluminum, copper, nickel, silver | Poor | Good to excellent | High | Good |

should be removed, especially if the assembly is used in electrical equipment.

The reaction fluxes usually contain zinc chloride, tin chloride, or both in combination with other halides. The metal halides are the primary fluxing agents. Other chemical compounds, such as ammonium chloride and metal fluorides, are added to improve fluidity, reduce the melting point, improve the wetting characteristics, and provide a flux cover that prevents reoxidation of the cleaned surface.

These fluxes penetrate the oxide film and contact the underlying aluminum. At some reaction temperature between 600° and 720° F, the metal chloride is reduced by the aluminum to the metal (tin or zinc) and gaseous aluminum chloride. The rapid formation of aluminum chloride breaks up the oxide film. The freshly deposited film of zinc or tin facilitates wetting by the solder. No significant fluxing action occurs below the reaction temperature of the flux. The solder will not likely flow below this temperature even though it may be above its liquidus temperature. Fluxes containing tin chloride generally react in the range of 600° to 640° F, and they are used primarily with tin-zinc solders of similar melting characteristics. The zinc chloride base fluxes, which react in the range of 640° to 720° F, are used with pure zinc and zinc base solders. Fluxes containing tin chloride should not be used with high-temperature solders because tin in the soldered joint can seriously reduce its corrosion resistance.

## JOINT DESIGN

The joint designs used for the soldering of aluminum are similar to those used with other metals. The most commonly used designs are lap, lock seam, and T-type joints. Joint clearance varies with the soldering method, base metal, solder composition, joint design, and flux composition. As a guide, a joint clearance in the 0.005 to 0.015 in. range is recommended when a chemical flux is used, and in the 0.002 to 0.010 in. range when a reaction flux is used.

## SURFACE PREPARATION

A prerequisite for soldering aluminum is careful surface preparation. Surface preparations designed to remove lubricant, dirt, and oxide from the surface of aluminum were described previously. When soldering fluxes formulated specifically for aluminum are used, no further surface preparation is necessary.

Other surface preparation techniques permit the soldering of aluminum without flux. These surface preparation methods can be divided conveniently into three groups: electroplating, solder coating, and cladding.

Aluminum can be electroplated with another metal, such as copper or nickel, to provide a surface that can be soldered. Deposition of copper is generally preceded by a zincate or stannate treatment in which aluminum oxide is removed from the surface, and zinc or tin is deposited by galvanic displacement.

## FLUXLESS SOLDERING

Solder coatings can be applied to aluminum by mechanically abrading the surfaces in the presence of molten solder. The solder wets and bonds with the aluminum as the oxide is mechanically removed. Among the best abrasion tools are fiberglass brushes, fine stainless steel wire brushes, and stainless steel wool. Ordinary carbon steel brushes should not be used because dislodged strands that are trapped in the solder will accelerate corrosion.

Some solders in rod form, called abrasion solders, have melting characteristics that enable them to perform as both the solder and the abrasion tool. The aluminum joint is heated by a torch or other method to soldering temperature. The solder stick is rubbed on the joint to break up the aluminum oxide and allow the molten solder to flow beneath it. The oxide is brushed aside with the solder stick to expose the surfaces wet with solder. Additional solder may now be applied to form a strong, stable joint between the surfaces. The process is not suitable where capillary flow is necessary to fill the joint.

Coating the joint surfaces with the aid of ultrasonic energy is better suited for production applications than is abrasion soldering. When ultrasonic vibrational energy is introduced into a molten solder, numerous voids form within it. This phenomenon is known as cavitation. The collapse of these voids creates an abrasive effect that removes the oxide film from the aluminum and permits the molten solder to wet it. This can be accomplished by transmitting the ultrasonic energy to a molten solder bath into which the assembly is dipped. Another approach is to use an ultrasonically vibrated soldering tip that is placed in contact with molten solder on the joint.

Fluxless soldering is generally done with 95% zinc-5% aluminum solder. It has excellent strength and corrosion resistance. An ultrasonic dip soldering pot is frequently replenished with commercially pure zinc since some aluminum from the component parts will be dissolved by the bath.

## SOLDERING PROCESSES

The high thermal conductivity of aluminum requires a relatively large heat source to rapidly bring the joint to soldering temperature. Uniform, well-controlled heating should be provided.

Aluminum can be joined with all of the common soldering processes, including torch, iron, furnace, and dip techniques. The procedures are similar to those used for other metals.

Line contact joints, such as fins on heat exchanger tubing, can be soldered with the zinc produced by reaction type flux. Zinc powder can be added to the flux to provide additional filler metal. Furnace or torch heating is suitable for this method of soldering. Fig. 8.13 shows an aluminum condenser fabricated by this technique.

## JOINT PROPERTIES

### Mechanical

While it is often of secondary importance, the mechanical strengths developed by soldered joints can range from the approximate shear strength of the solder to 40,000 psi or more. The intermediate and high temperature solders possess useful strengths at temperatures above the solidus of soft solders. High temperature solders can develop useful short time strengths at temperatures up to 600°F. However, excessive creep can take place at temperatures above 250°F.

### Corrosion Resistance

The corrosion resistance of soldered aluminum joints may be excellent to poor depending upon the choice of solder and the degree of removal of residual flux. When aluminum is soft soldered, a thin interfacial layer of either aluminum-tin or aluminum-lead alloy is formed between the solder and the aluminum. In most instances, the corrosion resistance of this interfacial layer is poorer than that of either the solder or the aluminum.

*Fig. 8.13 – Fin-to-tube joints in an aluminum condenser fabricated by reaction flux soldering in a furnace*

The rate of corrosion of soldered aluminum assemblies is greatest in the presence of a good electrolyte, such as a salt solution. It is considerably less in the presence of a poor electrolyte, such as condensed moisture, and insignificant in dry atmospheres. Painting or coating the joint area can minimize corrosion.

Generally, joints made with high zinc solders are highly resistant to corrosive attack. Assemblies prepared with pure zinc or zinc-aluminum solders are considered satisfactory for most applications requiring long outdoor service life. Corrosion resistance of assemblies joined with presently available intermediate or low temperature solders is usually satisfactory for interior or protected applications.

Flux residues are readily soluble in water. Warm water is satisfactory for cleaing zinc soldered joints. Alcohol is best for cleaning joints made with low temperature solders to avoid initiation of corrosion in the interfacial area.

Fluxless soldering methods should be evaluated for applications where residual flux products cannot be removed and good corrosion resistance is required. For less critical applications, a chemical flux may be used successfully.

# ADHESIVE BONDING

The oxide film on the surface of aluminum enhances a strong, durable bond with a suitable metal adhesive. Because of this, aluminum is easily joined with adhesives.[18] Ease of joining and high strength-to-weight ratio account for the widespread use of adhesive bonded aluminum assemblies.

Successful application requires that the assembly be properly designed for adhesive bonding and that the appropriate adhesive be used. A suitable chemical surface preparation is required to obtain the best and most consistent joint properties and performance.

# JOINING TO OTHER METALS

Special techniques are frequently required for joining aluminum to other metals. In many cases, a metal that is difficult to join directly to aluminum can be joined to a third metal applied to one of the other two by cladding, dip coating, electroplating, brazing, or buttering. Another technique is to employ a bimetal transition piece to which the aluminum and the other metal can be easily joined. These transition pieces are generally produced by a solid-state welding process, such as explosion or friction welding.

Bimetallic insert pieces of aluminum-carbon steel, aluminum-stainless steel, and aluminum-copper are commercially available as bars, rings, circles, sheet, and plate. These product forms make possible a wide range of welded aluminum-steel structures.

## ALUMINUM TO STEEL

An aluminum coating is the most common material applied to steel or stainless steel prior to brazing or welding it to an aluminum alloy. Aluminum coatings can be applied by dipping clean steel, with or without fluxing, into a molten aluminum bath held at 1275 to 1300° F. Small steel parts can be rub coated with aluminum or aluminum-silicon alloys using aluminum brazing flux.

The aluminum coated steel parts may be gas tungsten arc welded to aluminum with aluminum filler metal. The arc should be directed onto the aluminum member so that the molten weld pool will flow over the aluminum coating on the steel without penetrating it. Aluminum may be joined

---

18. Refer to the *Welding Handbook*, Vol. 3, 7th Ed., 338-65, for additional information.

directly to steel by soldering, explosion welding, friction welding, ultrasonic welding, and adhesive bonding.

## ALUMINUM TO COPPER

Aluminum may be joined to copper by flash welding, friction welding, soldering, brazing, explosion welding, ultrasonic welding, or adhesive bonding. It may also be done by gas welding, arc welding, or brazing after first precoating the copper with a solder or a silver-base filler metal.

## CORROSION

When aluminum is joined to certain other metals, the presence of moisture or an electrolyte may set up galvanic action between the two metals. This may cause preferential attack of one metal. After cleaning, the joint should be painted, coated, wrapped, or otherwise protected to exclude moisture or an electrolyte.

# ARC CUTTING

## PLASMA ARC

Plasma arc cutting can be used to sever aluminum alloys. The metal is melted and blown away by a high-velocity gas jet to form a kerf. Cutting can be done in any position. Nitrogen or Ar-$H_2$ is commonly used as the plasma gas; $CO_2$, nitrogen, or water vapor may be used as auxiliary shielding. Both manual and machine cutting equipment are available.

Aluminum can be cut with a rather wide range of operating conditions. The quality of the cut will be related to these conditions and the equipment used. Typical conditions for machine cutting are given in Table 8.42. Section thicknesses of from 0.125 to 5 in. can be cut with mechanized equipment. The maximum practical thickness for manual cutting is about 2 inches.

A plasma arc can also be used to gouge aluminum to produce J- and U-groove joint designs. Special torch orifice designs are needed to give the proper shape to the groove.[19]

The aluminum is melted during cutting and the heat produces a heat-affected zone on each side of the cut, similar to fusion welding. Consequently, the metallurgical behavior of aluminum alloys during cutting is similar to that during welding.

There are no significant problems when cutting nonheat-treatable alloys. However, shallow shrinkage cracks may develop in the cut surface with some heat-treatable alloys. An example of these cracks is shown in Fig. 8.14.

The heat-affected zone next to the cut surface in some high-strength, heat-treatable alloys, such as 2014, 2024, and 7075, may display reduced corrosion resistance. Cutting does not contaminate the metal as do machining methods that require a cutting fluid or a lubricant. Any oxide on the cut edges should be removed prior to welding.[20] Common methods of oxide removal may be used.

Plasma arc cutting apparatus is generally sold as a package including the torch and the power source. Since the fumes evolved with aluminum are voluminous, a water table and exhaust hood are recommended for operator safety.

## AIR CARBON ARC

This process uses a carbon arc and an air

---

19. Alban, J. F., Plasma arc gouging of aluminum, *Welding Journal*, 55 (11):954-59;1976 Nov.

---

20. Some welding codes require that the cut edges be machined prior to welding to remove any unsound or contaminated material.

### Table 8.42
### Typical conditions for plasma arc cutting of aluminum alloys

| Thickness | | Speed | | Orifice diam.[a] | | Current (DCEN), | Power, |
|---|---|---|---|---|---|---|---|
| in. | mm | in./min | mm/s | in. | mm | A | kW |
| 1/4 | 6 | 300 | 127 | 1/8 | 3.2 | 300 | 60 |
| 1/2 | 13 | 200 | 86 | 1/8 | 3.2 | 250 | 50 |
| 1 | 25 | 90 | 38 | 5/32 | 4.0 | 400 | 80 |
| 2 | 51 | 20 | 9 | 5/32 | 4.0 | 400 | 80 |
| 3 | 76 | 15 | 6 | 3/16 | 4.8 | 450 | 90 |
| 4 | 102 | 12 | 5 | 3/16 | 4.8 | 450 | 90 |
| 6 | 152 | 8 | 3 | 1/4 | 6.4 | 750 | 170 |

a. Plasma gas flow rates vary with orifice diameter and gas used from about 100 ft³/h (47 L/in.) for a 1/8 in. (3.2 mm ) orifice to about 250 ft³/h (120 L/in.) for a 1/4 in. (6.4 mm ) orifice. The gases used are nitrogen and argon with hydrogen additions from 0 to 35%. The equipment manufacturer should be consulted for each application.

blast to remove metal. It is more effective for gouging of grooves than for cutting. Grooves up to 1 in. deep can be made in a single pass, but depth increments of 0.25 in. provide better process control. The width of the groove is determined primarily by the size of the carbon electrode. The depth of the groove is affected by the torch angle and travel speed.

The arc is operated with DCEN power. Operating conditions must be closely controlled to ensure that all molten metal is blown from the work surface. The arc length must be great enough to permit the air stream to pass under the tip of the electrode. A layer of carbon particles left on the cut surface can be removed by wire brushing or machining prior to welding.

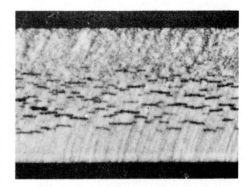

*Fig. 8.14(A) – A plasma arc cut surface in a heat treatable aluminum alloy showing cracking ( × 3)*

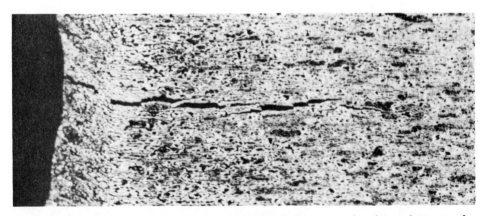

*Fig. 8.14(B) – A transverse section through the heat-affected zone showing an intergranular crack ( × 100)*

# SAFE PRACTICES

Aluminum and its oxides are not basically toxic, but good ventilation should be provided during welding, and inhalation of welding fumes should be avoided. Normal safe handling practices should be observed.

The high current density required for arc welding of aluminum alloys and the resultant high ultraviolet radiation requires dark viewing lenses to protect the eyes. All parts of limbs, hands, feet, and body should be completely covered with a suitable flame-retardant material to prevent radiation and spatter burns. Radiation from aluminum-magnesium alloys may be more intense than with the other alloys.

All brazing and soldering fluxes give off fumes or smoke when heat is applied. Some fluxes, such as the reaction soldering types, give off considerable smoke at high temperatures and long heating times. Other fluxes give off fumes that are harmful when inhaled. Prolonged inhalation of halides and some of the newer organic fluxes should be avoided. The aniline type fluxes and some of the amines also evolve fumes that are harmful and can cause dermatitis. Fluoride in a flux may create still another undesirable health condition.

Cadmium, lead, zinc, and other metals, are toxic when present as fumes or dusts in the atmosphere, as are their oxides. Since one or more of these may be present in a filler metal, the evolution of fumes should be minimized by not overheating the filler metal.

Several chemicals deserve special attention. Zinc chloride in flux may produce severe burns and dermatitis if allowed to remain on the skin for any length of time. Parts that have been degreased with organic solvents should be completely dry before welding, brazing, or soldering because their decomposition products may be toxic.

Some filler metals, fluxes, and cleaners contain materials that should not be ingested. Persons that handle these products should wash their hands thoroughly before eating.

# Metric Conversion Factors

1 lb·mass = 0.45 kg
1 lb·force = 4.45N
1 psi = 6.89 kPa
1 lb/in.$^3$ = 3.6 x 10$^{-5}$kg/m$^3$
1 in. = 25.4 mm
1 ft = 0.305 m
1 in.$^2$ = 645 mm$^2$
1 in./min = 0.42 mm/s
$t_C$ = 0.56 ($t_F$ − 32)
1 ft/min = 5.08 mm/s
1 ft$^3$/h = 0.47 L/min
1 gallon = 3.79 L

# SUPPLEMENTARY READING LIST

Anderson, W. A., Metallurgical studies on the vacuum brazing of aluminum, *Welding Journal*, 56 (10): 314s–18s; 1977 Oct.

Ashton, R. F., et al., The Effect of porosity on 5086–H116 aluminum alloy welds, *Welding Journal*, 54 (3): 95s–98s; 1975 Mar.

Ashton, R. F., and Wesley, R. P., One-side welding of aluminum plate, *Welding Journal*, 58 (4): 20-26; 1979 Apr.

Burk, J. D., and Lawrence, F. V., *Effects of Lack-of-Penetration and Lack-of-Fusion on the Fatigue Properties of 5083 Aluminum Alloy Welds*, New York: Welding Research Council, Bulletin 234, 1978 Jan.

Cooke, W. E., et al., Furnace brazing of aluminum with a noncorrosive flux, *Welding Journal;* 57 (12): 23–28; 1978 Dec.

Denslow, C. A., Ultrasonic soldering for aluminum heat exchangers, *Welding Journal*, 55 (2): 101-108; 1976 Feb.

Dockus, K. F., Fluxless bonding methods for aluminum components, *Welding Journal*, 57 (10): 36-42; 1978 Jan.

Enjo, T., et al., Diffusion welding of copper to aluminum, *Trans. of Japanese Welding Res. Inst.*, 8 (1): 77-84; 1979 June.

Essers, W. G., New process combines plasma with GMA welding, *Welding Journal*, 55 (5): 394-400; 1976 May.

Essers, W. G., et al., Plasma-MIG welding–a new torch and arc starting method, *Metal Construction*, 13 (1): 36-42; 1981 Jan.

Hay, R. A., and Adkins, H. E., Welding aluminum, *Welding Journal*, 53: 94-96, 155-57, 224-26, 309-11, 372-74, 441-43, 509-10, 709-11; 1974. 54: 28-30, 114-15, 182-84, 262-64, 362-63, 439-40; 1975.

Kijohara, M., et al., On the stabilization of GMA welding of aluminum, *Welding Journal*, 56 (3): 20-28; 1977 Mar.

Lawrence, F. V., et al., *Effects of Porosity on the Fatigue Properties of 5083 Aluminum Alloy Weldments*, New York: Welding Research Council, Bulletin 206, 1975 June.

Lippold, J.C., et al., An investigation of hot cracking in 5083-O aluminum alloy weldments, *Welding Journal*, 56 (6): 171s-78s; 1977 June.

MacArthur, I., Quality control in aluminum welding, *Welding Journal*, 54: 514-16, 595-97, 808-10, 888-90; 1975. 55: 44-45, 118-21, 200-201; 1976.

McNamara, P., and Singleton, O. R., Zinc distribution in alclad brazing sheet, *Welding Journal*, 58 (1): 7s-11s; 1979 Jan.

Montemarano, T. W., and Wells, M. E., Improving the fatigue performance of welded aluminum alloys, *Welding Journal*, 59 (6): 21-28; 1980 June.

Morley, R. A., and Caruso, J., The diffusion welding of 390 aluminum alloy hydraulic valve bodies, *Welding Journal*, 59 (8): 29-34; 1980 Aug.

Neiman, J. T., and Wille, G. W., Fluxless diffusion brazing of aluminum castings, *Welding Journal*, 57 (10): 285s-91s; 1978 Oct.

Papazoglou, V. J., and Masubuchi, K., Analysis and control of distortion in aluminum structures, *Welding Journal*, 57 (9): 251s-62s; 1978 Sept.

Patrick, E. P., Vacuum brazing of aluminum, *Welding Journal*, 54 (3): 159-63; 1975 Mar.

Rager, D. D., Direct current, straight polarity gas tungsten arc welding of aluminum, *Welding Journal*, 50 (5): 332-41; 1971 May.

Reichelt, W. R., Evancho, J. W., and Hoy, M. G., Effects of shielding gas on gas metal arc welding aluminum, *Welding Journal*, 59 (5): 147s-55s; 1980 May.

Reynolds, J., Adaptive spray–a new process for thin gauge aluminum, *Welding Journal*, 59 (7): 23-27; 1980 July.

Roest, C. A., and Rager, D. D., Resistance welding parameter profile for spot welding aluminum, *Welding Journal*, 53 (12): 529s-36s; 1974 Dec.

Sanders, W. W., Jr., and Gannon, S. M., *Fatigue Behavior of Aluminum Alloy 5083 Butt Welds*, New York: Welding Research Council, Bulletin 199, 1974 Oct.

Savage, W. F., et al., Hot cracking susceptibility of 3004 aluminum, *Welding Journal*, 58 (2): 45s-53s; 1979 Feb.

Sharpler, P., Aluminum brazing problems due to grain size, *Welding Journal*, 54 (3): 164-67; 1975 Mar.

Steenbergen, J. E., and Thorton, H. R., A quantitative determination of the conditions for hot cracking during welding of aluminum alloys, *Welding Journal,* 49 (2): 61-68; 1970 Feb.

Soldering aluminum, *Welding Journal,* 51: 571-73, 643-44, 714-16, 786-87; 1972.

Voigt, R. C., and Loper, C. R., Jr., GTA welding and heat treating of high-purity aluminum, *Welding Journal,* 58 (1): 20-26, 1979 Jan.

Warner, J. C., and Weltman, W. C., The fluxless brazing of aluminum radiators, *Welding Journal,* 58 (3): 25-32; 1979 Mar.

Woods, R. A., Metal transfer in aluminum alloys, *Welding Journal,* 59 (2): 59s-66s; 1980 Feb.

Woods, R. A., Porosity and hydrogen absorption in aluminum welds, *Welding Journal,* 53 (3): 97s-108s; 1974 Mar.

# 9

# Magnesium Alloys

## Chapter Committee

J.F. BROWN, JR., *Chairman*
*Kaiser Magnesium*

A.H. BRAUN
*Wellman Dynamics Corporation*

L.F. LOCKWOOD
*Dow Chemical, USA*

## Welding Handbook Committee Member

D.D. RAGER
*Reynolds Metals Company*

# 9

# Magnesium Alloys

## PROPERTIES

### CHEMICAL PROPERTIES

Magnesium and its alloys have a hexagonal close-packed crystal structure. The amount of deformation that they can sustain at room temperature is limited when compared to aluminum alloys. However, it increases rapidly with temperature, and the metal can be severely worked between 400° and 600° F. Forming operations, weld peening, and straightening are generally done at an elevated temperature.

When heated in air, magnesium will oxidize rapidly. This oxide will inhibit wetting and flow during welding, brazing, or soldering. For this reason, a protective shield of inert gas or flux must be used to prevent oxidation during exposure to elevated temperatures.

The oxide layer formed on magnesium surfaces will recrystalize at high temperatures and become flaky. It tends to break up more readily during welding than does the oxide layer on aluminum. Magnesium oxide (MgO) is highly refractory and insoluble in both liquid and solid magnesium. Magnesium nitride (MgN2) is relatively unstable and readily decomposes in the presence of moisture.

Under normal operating conditions, the corrosion resistance of many magnesium alloys in non-industrial atmospheres is better than that of ordinary iron and equal to some aluminum alloys. Formation of a gray oxide film on the surface is usually the extent of attack. For maximum corrosion resistance, chemical surface treatments and paint finishes as well as plating can be used. Galvanic type corrosion can be serious when magnesium is in direct contact with other metals in the presence of an electrolyte. This can be avoided by (1) proper design, (2) careful selection of metals in contact with the magnesium, or (3) insulation from dissimilar metals. Like other metals, some magnesium alloys are susceptible to stress-corrosion cracking if residual stresses from welding or fabrication are not reduced to a safe level by heat treatment.

### PHYSICAL PROPERTIES

Magnesium is well known for its extreme lightness, machinability, weldability, and the high strength-to-weight ratio of its alloys. It has a density of about 0.06 lb/in.[3] On an equal volume basis, it weighs about one-fourth as much as steel and two-thirds as much as aluminum.

Pure magnesium melts at 1200° F, about the same temperature as aluminum. However, magnesium boils at about 2025° F, which is low compared to other structural metals. The average coefficient of thermal expansion for magnesium alloys from 65° to 750° F is about $16 \times 10^{-6}$ per °F, about the same as that of aluminum and twice that of steel. The thermal conductivity is about 154 W/(m • °K) for magnesium and 70 to 110 W/(m • °K) for magnesium alloys. Its electrical resistivity is about 1.7 times that of aluminum.

Magnesium requires a relatively small amount of heat to melt it because of its comparatively low melting point, latent heat of fusion, and specific heat per unit volume. On an equal volume basis, the total heat of fusion is approximately two-thirds that for aluminum and one-fifth that for steel. The high coefficients of thermal expansion and conductivity tend to cause considerable distortion during welding. In this respect, the fixturing required for the welding of magnesium is very similar to that needed for aluminum. However, it must be more substantial than for the welding of steel.

## MECHANICAL PROPERTIES

Magnesium has a modulus of elasticity of about $6.5 \times 10^6$ psi compared to $10 \times 10^6$ psi for aluminum and $30 \times 10^6$ psi for steel. This means that magnesium is easier to distort under load than are aluminum and steel.

Cast magnesium has a yield strength of about 3 ksi and a tensile strength of about 13 ksi. Wrought magnesium products have tensile strengths in the range of 24 to 32 ksi. Their yield strengths in compression will be lower than those in tension. The reason for this is that it is easier for deformation to occur within a magnesium grain under compression. Alloying significantly increases the mechanical properties of magnesium.

Magnesium has low ductility when compared to aluminum. Tensile elongations range from 2 to 15 percent at room temperature. However, ductility increases rapidly at elevated temperatures.

Magnesium and its alloys are notch-sensitive, particularly in fatigue, because of the low ductility. Tensile properties increase and ductility decreases with decreasing testing temperature.

## APPLICATIONS

Magnesium alloys are used in a wide variety of structural applications where light weight is important. Structural applications include industrial, materials-handling, commercial, and aerospace equipment. In industrial machinery, such as textile and printing machines, magnesium alloys are used for parts that operate at high speeds and must be lightweight to minimize inertial forces. Materials-handling equipment includes dockboards, grain shovels, and gravity conveyors. Commercial applications include luggage and ladders. Good strength and stiffness at both room and elevated temperatures combined with light weight make magnesium alloys useful for some aerospace applications.

# ALLOYS

## ALLOY SYSTEMS

Most magnesium alloys are ternary types. They may be considered in four groups based on the major alloying element: aluminum, zinc, thorium, or rare earths.[1] There are also two binary systems employing manganese and zirconium.

Magnesium alloys may also be grouped according to service temperature. The magnesium-aluminum and magnesium-zinc alloy groups are suitable for only room temperature service. Their tensile and creep properties decrease rapidly when the service temperature is above about 300° F. The magnesium-thorium and magnesium-rare earth alloys are designed for elevated temperature service. They have good tensile and creep properties up to 700° F.

## DESIGNATION METHOD

Magnesium alloys are designated by a combination letter-number system composed of four parts. Part 1 indicates the two principal alloying elements by code letters arranged in order of decreasing percentage. The code letters are listed in Table 9.1. Part 2 indicates the percentages of the two principal alloying elements in the same order as the code letters. The percentages are rounded to the nearest whole number. Part 3 is an assigned letter to distinguish different alloys with the same percentages of the two principal alloying elements. Part 4 indicates the condition or temper of the product. It consists of a letter and number similar to those used for aluminum,

---

1. A group of 15 similar metals with atomic numbers 57 through 71.

**Table 9.1**
**Code letters for magnesium alloy designation system**

| Letter | Alloying element |
|--------|------------------|
| A | Aluminum |
| E | Rare earths |
| H | Thorium |
| K | Zirconium |
| L | Lithium |
| M | Manganese |
| Q | Silver |
| S | Silicon |
| T | Tin |
| Z | Zinc |

as shown in Table 9.2. They are separated from Part 3 by a hyphen.

An example is alloy AZ63A-T6. The *AZ* indicates that aluminum and zinc are the two principal alloying elements. The *63* indicates that the alloy contains nominally 6 percent aluminum and 3 percent zinc. The following *A* shows that this was the first standardized alloy of this composition. The fourth part, *T6*, shows that the product has been solution heat-treated and artificially aged.

# COMMERCIAL ALLOYS

Magnesium alloys are produced in the form of castings and wrought products including forgings, sheet, plate, and extrusions. A majority of the alloys produced in these forms can be welded. Commercial magnesium alloys designed for either room temperature or elevated temperature service are listed in Tables 9.3 and 9.4, respectively, with their nominal compositions.

## Wrought Alloys

AZ31B alloy is widely used for welded construction designed for room temperature service. It offers a good combination of strength, ductility, toughness, formability, and weldability in all wrought product forms. The alloy is strengthened by work hardening. AZ80A and ZK60A alloys can be artificially aged to develop good strength properties for room temperature applications.

AZ10A, M1A, and ZK21A alloy weldments are not sensitive to stress-corrosion cracking. Therefore, postweld stress relieving is not required for weldments made of these alloys. The alloys are strengthened by work hardening for room temperature service.

HK31A, HM21A, and HM31A alloys are designed for elevated temperature service. They are strengthened by a combination of work hardening followed by artificial aging.

## Cast Alloys

The most widely used casting alloys for room temperature service are AZ91C and AZ92A. These alloys are more crack-sensitive than the wrought Mg-Al-Zn alloys with lower aluminum content. Consequently, they generally require preheating prior to fusion welding.

EZ33A alloy has good strength stability for elevated temperature service and excellent pressure tightness. HK31A and HZ32A alloys are designed to operate at higher temperatures than EZ33A. QH21A alloy has excellent strength properties up to 500° F. All of these alloys require heat treatment to develop optimum properties. They have good welding characteristics.

## Mechanical Properties

Typical room temperature strength properties for magnesium alloys are given in Table 9.5.

**Table 9.2**
**Temper designations for magnesium alloys**

| | |
|---|---|
| F | As fabricated |
| O | Annealed, recrystallized (wrought products only) |
| H | Strain-hardened |

Subdivisions

| | |
|---|---|
| H1, | plus one or more digits...Strain-hardened only |
| H2, | plus one or more digits...Strain-hardened and then partially annealed |
| H3, | plus one or more digits...Strain-hardened and then stabilized |

| | |
|---|---|
| W | Solution heat-treated (unstable temper) |
| T | Thermally treated to produce stable tempers other than F, O, or H |

Subdivisions

| | |
|---|---|
| T1 | Cooled and naturally aged. |
| T2 | Annealed (cast products only) |
| T3 | Solution heat-treated and then cold-worked |
| T4 | Solution heat-treated |
| T5 | Cooled and artificially aged |
| T6 | Solution heat-treated and artificially aged |
| T7 | Solution heat-treated and stabilized |
| T8 | Solution heat-treated, cold-worked, and artificially aged |
| T9 | Solution heat-treated, artificially aged, and cold-worked |
| T10 | Cooled, artificially aged, and cold-worked |

## Table 9.3
### Commerical magnesium alloys for room temperature service

| ASTM Designation | Al | Zn | Mn | RE* | Zr | Th |
|---|---|---|---|---|---|---|
| | Nominal composition, % (remainder Mg) | | | | | |
| Sheet and plate | | | | | | |
| AZ31B | 3.0 | 1.0 | 0.5 | ... | ... | ... |
| M1A | ... | ... | 1.5 | ... | ... | ... |
| Extruded shapes and structural sections | | | | | | |
| AZ10A | 1.2 | 0.4 | 0.5 | ... | ... | ... |
| AZ31B | 3.0 | 1.0 | 0.5 | ... | ... | ... |
| AZ61A | 6.5 | 1.0 | 0.2 | ... | ... | ... |
| AZ80A | 8.5 | 0.5 | 0.2 | ... | ... | ... |
| M1A | ... | ... | 1.5 | ... | ... | ... |
| ZK21A | ... | 2.3 | ... | ... | 0.6 | ... |
| ZK60A | ... | 5.5 | ... | ... | 0.6 | ... |
| Sand, permanent mold or investment castings | | | | | | |
| AM100A | 10.0 | ... | 0.2 | ... | ... | ... |
| AZ63A | 6.0 | 3.0 | 0.2 | ... | ... | ... |
| AZ81A | 7.6 | 0.7 | 0.2 | ... | ... | ... |
| AZ91C | 8.7 | 0.7 | 0.2 | ... | ... | ... |
| AZ92A | 9.0 | 2.0 | 0.2 | ... | ... | ... |
| K1A | ... | ... | ... | ... | 0.6 | ... |
| ZE41A | ... | 4.2 | ... | 1.2 | 0.7 | ... |
| ZH62A | ... | 5.7 | ... | ... | 0.7 | 1.8 |
| ZK51A | ... | 4.6 | ... | ... | 0.7 | ... |
| ZK61A | ... | 6.0 | ... | ... | 0.8 | ... |

*As misch metal (approximately 52% Ce, 26% La, 19% Nd, 3% Pr)

For castings, the compressive yield strength is about the same as the tensile yield strength. However, the yield strength in compression for

## Table 9.4
### Commerical alloys for elevated temperature service

| ASTM Designation | Th | Zn | Zr | RE* | Mn | Ag |
|---|---|---|---|---|---|---|
| | Nominal composition, % (Remainder Mg) | | | | | |
| Sheet and plate | | | | | | |
| HK31A | 3.0 | ... | 0.7 | ... | ... | ... |
| HM21A | 2.0 | ... | ... | ... | 0.5 | ... |
| Extruded shapes and structural sections | | | | | | |
| HM31A | 3.0 | ... | ... | ... | 1.5 | ... |
| Sand, permanent mold or investment castings | | | | | | |
| EK41A | ... | ... | 0.6 | 4.0 | ... | ... |
| EZ33A | ... | 2.6 | 0.6 | 3.2 | ... | ... |
| HK31A | 3.2 | ... | 0.7 | ... | ... | ... |
| HZ32A | 3.2 | 2.1 | 0.7 | ... | ... | ... |
| QH21A | 1.1 | ... | 0.6 | 1.2 | ... | 2.5 |

*As misch metal (approximately 52% Ce, 26% La, 19% Nd, 3% Pr)

## Table 9.5
### Room temperature mechanical properties of magnesium alloys

| ASTM Designation | Tensile strength, ksi | Tensile yield strength,[a] ksi | Compressive yield strength,[a] ksi | Elongation, % in 2 In. |
|---|---|---|---|---|
| Sheet and Plate | | | | |
| AZ31B-O | 37 | 22 | 16 | 21 |
| AZ31B-H24 | 42 | 32 | 26 | 15 |
| HK31A-H24 | 33 | 30 | 22 | 9 |
| HM21A-T8 | 34 | 25 | 19 | 10 |
| M1A-O | 34 | 19 | .. | 18 |
| M1A-H24 | 39 | 29 | .. | 10 |
| Extruded shapes and structural sections | | | | |
| AZ10A-F | 35 | 22 | 11 | 10 |
| AZ31B-F | 38 | 29 | 14 | 15 |
| AZ61A-F | 45 | 33 | 19 | 16 |
| AZ80A-F | 49 | 36 | 22 | 11 |
| AZ80A-T5 | 55 | 40 | 35 | 7 |
| HM31A-T5 | 44 | 38 | 27 | 8 |
| M1A-F | 37 | 26 | 12 | 11 |
| ZK21A-F | 42 | 33 | 25 | 10 |
| ZK60A-F | 49 | 37 | 28 | 14 |
| ZK60A-T5 | 52 | 44 | 36 | 11 |
| Sand, permanent mold or investment castings | | | | |
| AM100A-T6 | 40 | 22 | 22 | 1 |
| AZ63A-F | 29 | 14 | | 6 |
| AZ63A-T4 | 40 | 13 | | 12 |
| AZ63A-T6 | 40 | 19 | 19 | 5 |
| AZ81A-T4 | 40 | 12 | 12 | 15 |
| AZ91C-F | 24 | 14 | | 2 |
| AZ91C-T4 | 40 | 12 | | 14 |
| AZ91C-T6 | 40 | 21 | 21 | 5 |
| AZ92A-F | 24 | 14 | | 2 |
| AZ92A-T4 | 40 | 14 | | 9 |
| AZ92A-T6 | 40 | 21 | 21 | 2 |
| EK41A-T5 | 25 | 13 | | 3 |
| EZ33A-T5 | 23 | 15 | 15 | 3 |
| HK31A-T6 | 32 | 15 | 15 | 8 |
| HZ32A-T5 | 27 | 14 | 14 | 4 |
| K1A-F | 25 | 7 | | 19 |
| QH21A-T6 | 40 | 30 | | 4 |
| ZE41A-T5 | 30 | 20 | 20 | 4 |
| ZH62A-T5 | 35 | 25 | 25 | 4 |
| ZK51A-T5 | 30 | 24 | 24 | 4 |
| ZK61A-T6 | 45 | 28 | 28 | 10 |

a. 0.2% offset yield strength.

wrought products is lower than in tension.

The tensile and creep properties of representative magnesium alloys at three elevated temperatures are given in Table 9.6. The alloys containing thorium (HK, HM, and HZ) have greater resistance to creep at 400° and 600° F than do the Mg-A1-Zn alloys.

**Table 9.6**

**Elevated temperature properties of some representative magnesium alloys**

| Alloy | 300° F | | | 400° F | | | 600° F | | |
|---|---|---|---|---|---|---|---|---|---|
| | Tensile strength, ksi | Tensile yield strength, ksi | Creep strength,[a] ksi | Tensile strength, ksi | Tensile yield strength, ksi | Creep strength,[a] ksi | Tensile strength, ksi | Tensile yield strength, ksi | Creep strength,[a] ksi |
| *Sheet and plate alloys* | | | | | | | | | |
| AZ31B-H24 | 22 | 13 | 1.0 | 13 | 8 | ... | 6 | 2 | ... |
| HK31A-H24 | 26 | 24 | ... | 24 | 21 | 6.0 | 12 | 7 | ... |
| HM21A-T8 | 23 | 21 | ... | 19 | 18 | 11.4 | 15 | 13 | 5.0 |
| *Extrusion alloys* | | | | | | | | | |
| AZ31B-F | 25 | 15 | 3.0 | 15 | 9 | ... | 6 | 2 | ... |
| AZ80A-T5 | 35 | 23 | 3.5 | 22 | 15 | ... | 9 | 3 | ... |
| HM31A-T5 | 28 | 25 | ... | 24 | 21 | 10.9 | 18 | 15 | 7.6 |
| ZK60A-T5 | 25 | 22 | 1.0 | 15 | 12 | ... | ... | ... | ... |
| *Casting alloys* | | | | | | | | | |
| AZ92A-T6 | 28 | 17 | 3.8 | 17 | 12 | ... | 8 | 5 | ... |
| AZ63A-T6 | 24 | 15 | 4.1 | 18 | 12 | ... | 8 | 6 | ... |
| EZ33A-T5 | 22 | 14 | ... | 21 | 12 | 8.0 | 12 | 8 | 1.2 |
| HK31A-T6 | 27 | 15 | ... | 24 | 14 | 9.5 | 20 | 12 | 2.9 |
| HZ32A-T5 | 22 | 12 | ... | 17 | 10 | 7.8 | 12 | 8 | 3.0 |
| QH21A-T6 | 33 | 29 | ... | 30 | 27 | 12 | 14 | 13 | ... |

a. Creep strength based on 0.2% total extension in 100 h.

## MAJOR ALLOYING ELEMENTS

With most magnesium alloy systems, the solidification range generally widens as the alloy addition increases. This contributes to a greater tendency for cracking during welding. At the same time, the melting temperature as well as the thermal and electrical conductivities decrease. Consequently, less heat input is required for fusion welding as the alloy content increases.

Aluminum and zinc show decreasing solubility in solid magnesium with decreasing temperature. These elements will form compounds with magnesium. Consequently, alloys containing sufficient amounts of these elements can be strengthened by a precipitation-hardening heat treatment. Other alloying elements also behave similarly in ternary alloy systems.

### Aluminum

When added to magnesium, aluminum gives the most favorable results of the major alloying elements. It increases both strength and hardness. Alloys containing more than about 6 percent aluminum are heat-treatable. The aluminum content of an alloy has no adverse effect on weldability. Weldments of alloys containing more than about 1.5 percent aluminum require a postweld stress-relief heat treatment to prevent susceptibility to stress-corrosion cracking.

### Beryllium

The tendency for magnesium alloys to burn during melting and casting is reduced by the addition of beryllium up to about 0.001 percent. Beryllium is also added to magnesium filler metals to reduce oxidation and the danger of ignition at elevated temperatures during the joining operations.

### Manganese

This element has little effect upon tensile strength, but increases yield strength slightly. Its most important function is to improve the salt water corrosion resistance of magnesium-aluminum and magnesium-aluminum-zinc alloys. Magnesium-manganese alloys have relatively high melting temperatures and thermal conductivities. Therefore, they require somewhat more welding heat input than do the magnesium-aluminum-zinc alloys. Joint efficiency is low in magnesium-manganese alloys because of grain growth in the heat-affected zone.

### Rare Earths

Rare earths are added either as misch metal or didymium. These additions are beneficial in reducing weld cracking and porosity in castings because they narrow the freezing range of the alloy.

### Silver

Silver improves the mechanical properties of magnesium alloys. The QH21A alloy in the T6 condition has the highest room temperature strength of the commercial magnesium casting alloys. This alloy has good weldability.

### Thorium

Thorium additions greatly increase the strengths of Mg alloys at temperatures up to 700° F. The most common alloys contain 2 to 3 percent thorium in combination with zinc, zirconium, or manganese. Thorium improves the weldability of alloys containing zinc.

### Zinc

Zinc is often used in combination with aluminum to improve the room temperature strength of magnesium. It increases hot shortness when added in amounts over 1 percent to magnesium alloys containing 7 to 10 percent aluminum. In amounts greater than 2 percent in these alloys, zinc is likely to cause weld cracking.

Zinc is also used in combination with zirconium, thorium, or rare earths to produce precipitation-hardenable magnesium alloys with good strength properties.

### Zirconium

Zirconium is a powerful grain-refining agent in magnesium alloys. It is usually added to alloys containing zinc, thorium, rare earth, or combinations of these. Zirconium is believed to confer a slightly beneficial effect on the weldability of Mg-Zn alloys by increasing the solidus temperature.

## HEAT TREATMENT

Magnesium alloys are usually heat-treated to improve mechanical properties.[2] The type of

---

2. For additional information, refer to *Heat Treating, Cleaning and Finishing, Metals Handbook*, Vol. 2, 8th Ed., Metals Park, OH: American Society for Metals, 1964: 292-97.

heat treatment depends upon the alloy composition, product form, and service requirements.

A solution heat treatment improves strength, toughness, and impact resistance. A precipitation heat treatment following solution heat treatment increases the yield strength and hardness at some sacrifice in toughness. A precipitation heat treatment alone will simultaneously increase the tensile properties and stress relieve as-cast components.

Combinations of solution heat treating, strain hardening, and precipitation hardening are often used with wrought products. This processing will give higher strength properties than when the strain-hardening operation is omitted.

## WELDABILITY

The relative weldability of magnesium alloys by gas shielded arc and resistance spot welding processes are shown in Table 9.7. Castings are not normally resistance welded. The Mg-Al-Zn alloys and alloys that contain rare earths or thorium as the major alloying element have the best weldability. Alloys with zinc as the major alloying element are more difficult to weld. They generally have a rather wide melting range, which makes them sensitive to hot cracking. With proper joint design and welding conditions, joint efficiencies will range from 60 to 100 percent, depending upon the alloy.

Most wrought alloys can be readily resistance spot welded. In this case, the metal is molten for a very short time and the cooling rate is very high. There is little time for metallurgical reactions to take place.

**Table 9.7**
**Releative weldability of magnesium alloys**

| Alloy | Gas shielded arc welding | Resistance spot welding |
|---|---|---|
| Wrought | | |
| AZ10A | A | A |
| AZ31B, AZ31C | A | A |
| AZ61A | B | A |
| AZ80A | B | A |
| HK31A | A | A |
| HM21A | A | B |
| HM31A | A | B |
| M1A | A | B |
| ZK21A | B | A |
| ZK60A | D | A |
| Cast | | |
| AM100A | B | |
| AZ63A | C | |
| AZ81A | B | |
| AZ91C | B | |
| AZ92A | C | |
| EK41A | B | |
| EZ33A | A | |
| HK31A | B | |
| HZ32A | B | |
| K1A | A | |
| QH21A | B | |
| ZE41A | B | |
| ZH62A | D | |
| ZK51A | D | |
| ZK61A | D | |

A – Excellent
B – Good
C – Fair
D – Poor

# SURFACE PREPARATION

As with other metals, the cleanliness of magnesium alloy components and filler metals is important for obtaining sound joints of acceptable quality. Any surface contamination will inhibit wetting and fusion. Magnesium alloys are usually supplied with an oil coating, an acid pickled surface, or a chromate conversion coating for protection during shipping and storage. The surfaces and edges to be joined must be cleaned just before joining to remove the surface protection as well as any dirt or oxide present.[3] Chemical cleaners commonly used for magnesium alloys are given in Table 9.8.

Oil, grease, and wax are best removed by

---

3. For additional information on cleaning refer to *Heat Treating, Cleaning and Finishing, Metals Handbook,* Vol. 2, 8th Ed., Metals Park, OH: American Society for Metals, 1964: 648-60.

**Table 9.8**

**Chemical cleaning of magnesium alloys**

| Type of treatment | Composition of solution | Method of application | Uses |
|---|---|---|---|
| Alkaline cleaner | Sodium carbonate 3 oz.<br>Sodium hydroxide 2 oz.<br>Water to make 1 gallon<br>Temperature 190-212° F<br>Solution pH 11 or greater | 3 to 10 min. immersion followed by cold water rinse and air dry. | Used to remove oil and grease films, as well as old chrome pickle or dichromate coatings. |
| Bright pickle | Chromic acid 1.5 lb.<br>Ferric nitrate 5.3 oz.<br>Potassium fluoride 0.5 oz.<br>Water to make 1 gallon<br>Temperature 60-100° F | 0.25 to 3 min. immersion followed by cold and hot water rinse and air dry. | Used after degreasing to remove oxide and prepare surfaces for welding and brazing. Gives bright clean surfaces, resistant to tarnish. |
| Spot weld cleaners | *No. 1 Bath*<br>Conc. sulfuric acid 1.3 fl. oz.<br>Water to make 1 gallon<br>Temperature 70-90° F<br>*No. 2 Bath*<br>Chromic acid 1.5 lb<br>Conc. sulfuric acid 0.07 fl. oz.<br>Water to make 1 gallon<br>Temperature 70-90° F<br>*No. 3 Bath*<br>Chromic acid 0.33 oz.<br>Water to make 1 gallon<br>Temperature 70-90° F | Immerse 0.25-1 min. in No. 1 bath. Rinse in cold water. Follow by immersing in either No. 2 or No. 3 bath. For No. 2 bath, immerse 3 min., and follow by cold water rinse and air dry. For No. 3 bath, immerse 0.5 min. followed by cold water rinse and air dry. | Used after degreasing to remove oxide layer and prepare surface for spot welding. Gives low consistent surface resistance. |
| Flux remover cleaner | Sodium dichromate 0.5 lb.<br>Water to make 1 gallon<br>Temperature 180-212° F | 2 hour immersion in boiling bath, followed by cold and hot water rinse and air dry. | Used after hot water cleaning and Chrome Pickling to remove or inhibit any flux particles remaining from welding or brazing. |
| Chrome pickle<br>MIL-M-3171 Type 1<br>AMS 2475 | Sodium dichromate 1.5 lb.<br>Conc. nitric acid 24 fl. oz.<br>Water to make 1 gallon<br>Temperature 70-90° F | 0.5 to 2 min. immersion, hold in air 5 sec., followed by cold and hot water rinse and air or forced dry, max. 250° F. When brushed on, allow 1 min. before rinse. | Used as paint base and for surface protection. Second step in flux removal. Applied with brush for touch-up of welds and treatment of large structures. |

either washing with organic solvents or vapor degreasing in a chlorinated hydrocarbon solvent.[4] Subsequent cleaning in alkaline or emulsion type cleaners is recommended to be sure that the surfaces are absolutely free of oil or grease. Alkaline (caustic) cleaner will also remove previously applied chemical surface treatments. Cleaners of this type that are suitable for steel are generally satisfactory for magnesium. Cleaning may be by either the immersion or the electrolytic method. Thorough water rinsing, preferably by spray, is necessary after alkaline cleaning to avoid degradation of subsequent acid chemical baths used to treat the parts.

After all oil or other organic material has been removed, the part or joint is ready for chemical or mechanical cleaning. A bright pickle will produce suitably clean surfaces for welding. A final mechanical cleaning is preferred for most critical production applications to ensure uniform surface cleanliness. Stainless steel wool or wire brush is recommended. Wire brushing should not gouge the surface. For resistance spot welding applications, chemical cleaning is usually preferred to provide a uniformly low surface resistance. (See Table 9.8.)

If neutralization is desired after chemical cleaning and prior to rinsing, a solution of 6.5 oz. of sodium metasilicate per gallon of water, operating at 180° F, may be used. After cleaning, special care must be taken to protect the components from contamination during all subsequent handling operations.

Any oxide film or smut deposited on the surface of weldments may be removed by wire brushing or by chemical treatment in a solution of 16 ounces of tetrasodium pyrophosphate and 12 ounces of sodium metaborate per gallon of water operating at 180° F.

# ARC WELDING

## APPLICABLE PROCESSES

The gas tungsten arc and gas metal arc welding processes are commonly used for joining magnesium alloy components. Inert gas shielding is required with these processes to avoid excessive oxidation and entrapment of oxide in the weld metal. Processes that use a flux covering do not provide adequate oxidation protection for the molten weld pool and the adjacent base metal.

## JOINT DESIGN

Joint designs suitable for gas shielded arc welding are shown in Fig. 9.1. The thickness limitations for welding these joint designs are given in Table 9.9. Because of the high deposition rate of the gas metal arc welding process, a root opening, a beveled joint, or both, should be used to provide space for the deposited metal. Increasing the travel speed to maintain a conventional bead size is not acceptable because undercutting,

incomplete fusion, or inadequate penetration may result.

A backing strip is usually employed when welding sheet metal components to help control joint penetration, root surface contour, and heat removal. Magnesium, aluminum, copper, mild steel, or stainless steel is usually employed as a backing material. When a temporary backing strip is used, the root side of the joint should be shielded with inert gas to prevent oxidation of the root surface. The gas is usually supplied through holes in the backing strip. In those instances where a backing strip cannot be used because of space limitations, a chemical flux of the type used in oxyfuel gas welding is sometimes painted on the root side of the joint to smooth the root surface and to control joint penetration.

## FILLER METALS

The weldability of most magnesium alloys is good when the proper filler metal is employed. A filler metal with a lower melting point and a wider freezing range than the base metal will provide good weldability and minimize weld

---

4. Vapors of these solvents are toxic. Refer to ANSI Publication Z49.1, *Safety in Welding and Cutting,* for additional information.

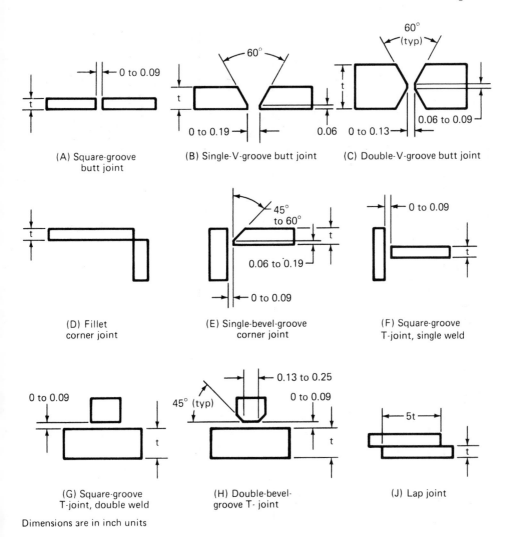

Dimensions are in inch units

**Fig. 9.1 – Typical arc welded joint designs for magnesium alloys**

cracking. The recommended filler metals for various magnesium alloys are given in Table 9.10.

ER AZ61A or ER AZ92A (Mg-Al-Zn) filler metal may be used to weld alloys of similar composition and also ZK21A (Mg-Zn-Zr) alloy. ER AZ61A filler metal is generally preferred for welding wrought products of those alloys because of its low cracking tendencies. On the other hand, ER AZ92A filler metal shows less crack sensitivity for welding the cast Mg-A1-Zn and AM 100A (Mg-A1) alloys. The deposited metal will respond to the precipitation heat treatments

normally applied to the repaired castings. ER AZ101A filler metal may also be used to weld those casting alloys.

ER EZ33A (RE-Zn-Zr) filler metal is used to weld wrought and cast alloys designed for high-temperature service, either to themselves or each other. The welded joints will have good mechanical properties at elevated temperatures.

ER AZ92A filler metal is recommended for welding the room temperature service wrought and cast alloys together or to one of the wrought or cast elevated service alloys. It will minimize

**Table 9.9**
**Thickness limitations for arc welded joints in magnesium alloys[a]**

| Joint design[b] | Gas tungsten arc welding with | | | | | | Gas metal arc welding with | | | | Remarks |
|---|---|---|---|---|---|---|---|---|---|---|---|
| | ac | | dcen (sp) | | dcep (rp) | | Short circuiting transfer | | Spray transfer | | |
| | t, in. | | t, in. | | t, in. | | t, in. | | t, in. | | |
| | Min. | Max. | Min. | Max. | Min. | Max. | Min. | Max. | Min. | Max. | |
| A | 0.025 | 1/4 | 0.025 | 1/2 | 0.025 | 3/16 | 0.025 | 3/16 | 3/16 | 3/8 | Single-pass, complete penetration weld. Used on lighter material thicknesses. |
| B | 1/4 | 3/8 | 1/4 | 3/8 | 3/16 | 3/8 | N.R.[c] | N.R.[c] | 1/4 | 1/2 | Multipass complete penetration weld. Used on thick material. On material thicker than suggested maximum, use the double-V-groove weld, butt joint C to minimize distortion. |
| C | 3/8 | d | 3/8 | d | 3/8 | d | N.R.[c] | N.R.[c] | 1/2 | d | Multipass complete penetration weld. Used on thick materials. Minimizes distortion by equalizing shrinkage stresses on both sides of joint. |
| D | 0.040 | 1/4 | 0.040 | 1/4 | 0.040 | 1/4 | 1/16 | 3/16 | 3/16 | 1/2 | Single-pass complete penetration weld. For material thicker than suggested maximum, use single-bevel-groove corner joint E. It requires less welding, especially if a square corner is required. |
| E | 3/16 | d | 3/16 | d | 3/16 | d | N.R.[c] | N.R.[c] | 1/4 | d | Single or multipass complete penetration weld. Used on heavy material thicknesses to minimize welding. Produces square joint corners. |

## Table 9.9 (cont.)

| Joint design[b] | Gas tungsten arc welding with | | | | | | | | | Gas metal arc welding with | | | | | Remarks |
|---|---|---|---|---|---|---|---|---|---|---|---|---|---|---|---|
| | ac | | dcen (sp) | | | dcep (rp) | | | | Short circuiting transfer | | | Spray transfer | | |
| | t, in. Min. | t, in. Max. | t, in. Min. | t, in. Max. | | t, in. Min. | t, in. Max. | | | t, in. Min. | t, in. Max. | | t, in. Min. | t, in. Max. | |
| F | 0.025 | 1/4 | 0.025 | 1/2 | | 0.025 | 5/32 | | | 1/16 | 5/32 | | 5/32 | 3/8 | Single welded T-joint. Suggested thickness limits based on 40% joint penetration. |
| G | 1/16 | 3/16 | 1/16 | 3/8 | | 1/16 | 1/8 | | | 1/16 | 3/32 | | 5/32 | 3/4 | Double welded T-joint. Suggested thickness limits based on 100% joint penetration. |
| H | 3/16 | d | 3/8 | d | | 1/8 | d | | | N.R.[c] | N.R.[c] | | 3/8 | d | Double welded T-joint. Used on heavy material requiring 100% joint penetration. |
| J | 0.040 | d | 0.040 | d | | 0.025 | d | | | 0.040 | 5/32 | | 5/32 | d | Single- or double-welded lap joint. Strength depends upon size of fillet welds. Maximum strength in tension on double-welded joints is obtained when lap equals five times thickness of thinner member. |

a. Based on good welding practices and the ability to gas tungsten arc weld with 300 A of ac or dcen(sp) power or with 125 A of dcep(rp) power, as well as gas metal arc weld with 400 A of dcep(rp) power.
b. Refer to Fig. 9.1 for the appropriate joint design.
c. N.R. – Not recommended.
d. Thickest material in commercial use may be welded this way.

**Table 9.10**
**Recommended filler metals for arc welding**
**magnesium alloys**

| Alloys | Recommended filler metal[a] | | | | |
|---|---|---|---|---|---|
| | ER AZ61A | ER AZ92A | ER EZ33A | ER AZ101A | Base Metal |
| Wrought | | | | | |
| AZ10A | X | X | | | |
| AZ31B | X | X | | | |
| AZ61A | X | X | | | |
| AZ80A | X | X | | | |
| ZK21A | X | X | | | |
| HK31A | | | X | | |
| HM21A | | | X | | |
| HM31A | | | X | | |
| M1A | | | | | X |
| Cast | | | | | |
| AM100A | | X | | X | X |
| AZ63A | | X | | X | X |
| AZ81A | | X | | X | X |
| AZ91C | | X | | X | X |
| AZ92A | | X | | X | X |
| EK41A | | | X | | X |
| EZ33A | | | X | | X |
| HK31A | | | X | | X |
| HZ32A | | | X | | X |
| K1A | | | X | | X |
| QH21A | | | | | X |
| ZE41A | | | X | | X |
| ZH62A | | | X | | X |
| ZK51A | | | X | | X |
| ZK61A | | | X | | X |

a. Refer to *AWS A5.19-80, Specification for Magnesium Alloy Welding Rods and Bare Electrodes*, for additional information.

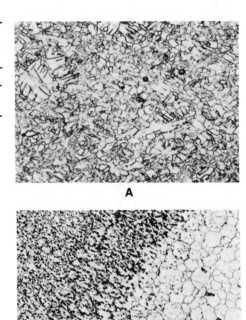

A

B

*Fig. 9.2 – (A) Microstructure of 1/8 in. thick AZ31B-H24 magnesium sheet (X250), (B) microstructure of weld interface between the AZ31B heat-affected zone and AZ61A weld metal (X100) (base metal to right, weld to left)*

weld cracking tendencies. ER EZ33A filler metal should not be used for welding aluminum-bearing magnesium alloys because of severe weld cracking problems.

Casting repairs should be made with a filler metal of the same composition as the base metal when good color match, minimum galvanic effects, or good response to heat treatment is required. For these and other unusual service requirements, the material supplier should be consulted for additional information.

Typical base metal and weld interface microstructures of AZ31B-H24, HK31A-H24, HM21A-T8 and HM31A-T5 alloys are shown in Figs. 9.2 to 9.5. Three of these alloys, excepting HM21A, which had been recrystallized prior to

welding, showed a significant amount of recrystallization and grain growth in the heat-affected zones. Radiographs of welds in alloys containing rare earths and thorium will often show segregation along the edges of the weld metal. This segregation is caused by incipient melting in the base metal. A white line will show along the weld interface because of the x-ray absorption characteristics of the rare earths and thorium segregated there. This type of microstructure is shown in Fig. 9.6.

## PREHEATING

The need to preheat the components prior to welding is largely determined by the product

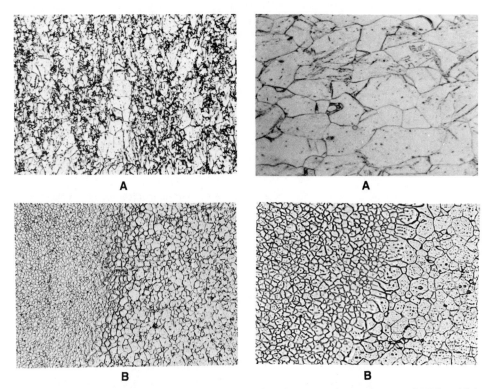

**Fig. 9.3 – (A) Microstructure of 1/8 in. thick HK31A-H24 magnesium sheet (X250), (B) microstructure of weld interface between the HK31A heat-affected zone and EZ33A weld metal (X100) (base metal to right, weld to left)**

**Fig. 9.4 – (A) Microstructure of 1/8 in. thick HM21A-T8 magnesium sheet (X250), (B) microstructure of weld interface between the HM21A heat-affected zone and EZ33A weld metal (X100) (base metal to right, weld to left)**

form, section thickness, and the degree of restraint on the joint. Thick sections may not require preheating unless the joint restraint is high. Thin sections and highly restrained joints generally require preheat to avoid weld cracking. This is particularly true of alloys high in zinc.

The recommended preheat temperature ranges for magnesium alloys are given in Table 9.11. The maximum preheat temperature generally should not exceed the solution heat treating temperature for the alloy. Otherwise, the mechanical properties of the weldment may be altered significantly.

The method of preheating will depend upon the component size. Furnace heating is preferred but large components may have to be preheated

locally. The welding fixture may also have to be heated when joining thin sections to maintain an acceptable interpass temperature.

An air circulating furnace with a temperature control of ± 10° F is recommended for preheating of castings. The furnace temperature should not cycle above the maximum temperature indicated in Table 9.11. A temperature limit control set at 10° F above the maximum acceptable temperature should be provided to override the automatic controls.

Solution heat-treated, or solution heat-treated and aged castings can be charged into a furnace operating at the preheat temperature without damage. They should remain in the furnace until they are uniformly heated throughout.

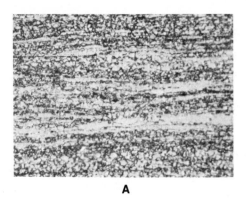

**A**

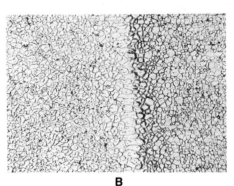

**B**

*Fig. 9.5 – (A) Microstructure of 1/8 in. thick HM31A-T5 magnesium extrusion (X500), (B) microstructure of weld interface between the HM31A heat-affected zone and EZ33A weld metal (X100) (base metal to right, weld to left)*

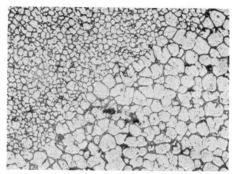

*Fig. 9.6 – Microstructure of the weld interface in an EZ33A magnesium alloy casting showing pools of a eutectic (X100) (base metal is on the right)*

Welding should proceed immediately after the castings are removed from the furnace. It should be discontinued if the temperature of the castings drops below the acceptable minimum preheat. In that case, the castings should be reheated in the furnace before proceeding with the welding operation.

Castings normally can be cooled in still, ambient air after welding without danger of cracking. However, castings of intricate design should be cooled more slowly to room temperature to avoid distortion from nonuniform cooling.

## GAS TUNGSTEN ARC WELDING

Gas tungsten arc welding is used for joining magnesium components and repair of magne-

sium castings. It is well suited for welding thin sections. Control of heat input and the molten weld pool is better than with gas metal arc welding.

### Welding Current

Magnesium alloys are welded by this process using techniques and equipment similar to those used for aluminum. They may be welded with alternating current or direct current. Alternating current is preferred because of the good arc cleaning action. Conventional ac power of 60 Hz with arc stabilization or square-wave alternating current (SWAC) may be used. With SWAC, the electrode positive and negative periods are adjustable within limits. This type of power can provide adequate cleaning action as well as good joint penetration and arc stability. A section through a weld made with 60 Hz ac power is shown in Fig. 9.7(A).

Direct current power with the electrode positive (DCEP-RP) provides an arc with excellent cleaning action. However, it can only be used to weld thin sections because the welding current is limited by heating of the tungsten electrode. Joint penetration tends to be wide and shallow, as shown in Fig. 9.7 (B). Welds in relatively thick sections are typified by low welding speeds, wide bead faces, and wide heat-affected zones with large grain size.

Direct current electrode negative (DCEN-SP) power is not commonly used for welding magnesium alloys because of the absence of arc

**Table 9.11**

**Recommended weld preheat and postweld heat treatments for magnesium alloys**

| Alloy | Metal temper before welding | Desired temper after welding | Weld preheat | Postweld heat treatment[a] |
|---|---|---|---|---|
| AZ63A | T4 | T4 | Heavy and unrestrained sections: none or local. Thin and restrained sections: 350°-720° F max | ½ hour at 730° F |
| | T4 or T6 | T6 | | ½ hour at 730° F + 5 hours at 425° F |
| | T5 | T5 | Heavy and unrestrained sections: none or local. Thin and restrained sections: None to 500° F (1½ hours max at 500° F) | 5 hours at 425° F |
| AZ81A | T4 | T4 | | ½ hour at 780° F |
| AZ91C | T4 | T4 | | ½ hour at 780° F |
| | T4 or T6 | T6 | Heavy and unrestrained sections: none or local. Thin and restrained sections: 350-750° F max | ½ hour at 780° F + either 4 hours at 420° F or 16 hours at 335° F |
| AZ92A | T4 | T4 | | ½ hour at 770° F |
| | T4 or T6 | T6 | | ½ hour at 770° F + either 4 hours at 500° F or 5 hours at 425° F |
| AM100A | T6 | T6 | | ½ hour at 780° F + 5 hours at 425° F |

## Table 9.11 (cont.)
### Recommended weld preheat and postweld heat treatments for magnesium alloys

| Alloy | Metal temper before welding | Desired temper after welding | Weld preheat | Postweld heat treatment[a] |
|---|---|---|---|---|
| EK41A | T4 or T6 | T6 | | 16 hours at 400° F |
| | T5 | T5 | None to 500° F (1½ hours max at 500° F) | 16 hours at 400° F |
| EZ33A | F or T5 | T5 | | 5 hours at 420° F; 2 hours at 650° F + 5 hours at 420° F |
| HK31A | T4 or T6 | T6 | None to 500° F | 16 hours at 400° F: 1 hour at 600° F + 16 hours at 400° F |
| HZ32A | F or T5 | T5 | | 16 hours at 600° F |
| K1A | F | F | None | None |
| ZE41A | F or T5 | T5 | | 2 hours at 625° F; 2 hours at 625° F + 16 hours at 350° F |
| ZH62A | F or T5 | T5 | | 16 hours at 480° F; 2 hours at 625° F + 16 hours at 350° F |
| ZK51A | F or T5 | T5 | None to 600° F | 16 hours at 350° F; 2 hours at 625° F + 16 hours at 350° F |
| ZK61A | F or T5 | T5 | | 48 hours at 300° F |
| | T4 or T6 | T6 | | 2-5 hours at 930° F + 48 hours at 265° F |

a. Temperatures shown are max. allowable; furnace controls should be set so temperature does not cycle above maximum.

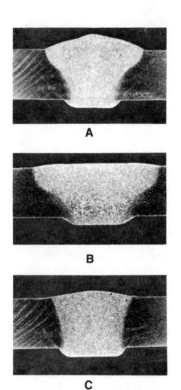

A

B

C

*Fig. 9.7 – Cross sections through gas tungsten arc welds in 0.188 in. AZ31B magnesium alloy with (A) AC (B) DCEP-RP, and (C) DCEN-SP power*

cleaning action. However, this type of power is sometimes used for mechanized welding of square-groove butt joints in sections up to 0.25 in. thickness. Careful preweld cleaning and good fit-up are needed to produce sound welds. DCEN-SP power with helium shielding can produce narrow, deep joint penetration, as shown in Fig. 9.7 (C).

## Shielding Gases

Argon, helium, and mixtures of these gases can be used. The factors governing the selection of the shielding gas for magnesium alloys are the same as those for other metals, particularly aluminum.

## Electrodes

Pure tungsten (EWP), tungsten-thoria (EWTh-1 or -2), and tungsten-zirconia (EWZr)

electrodes can be used with magnesium alloys. The selection depends primarily upon the type of welding power and the welding amperage to be used.

## Welding Conditions

Typical conditions for manual gas tungsten arc welding of butt joints in magnesium alloys are given in Table 9.12. Conditions for automatic gas tungsten arc welding butt joints in two thicknesses of AZ31B magnesium alloy are shown in Table 9.13. The welding machine should produce balanced ac power for good operating characteristics. These data may be used as guides for establishing joint welding procedures for a specific application.

## Welding of Castings

GTAW is generally recommended for the welding of magnesium alloy castings. Welding is usually limited to the repair of defects in clean metal including broken sections, sand or blow holes, cracks, and cold shuts. Repair welding is not recommended in areas containing gross porosity or inclusions of oxide or flux. Castings that have been organically impregnated for pressure tightness or that may contain oil in pores should not be welded. Many castings are parts of aircraft structures that are heat treated to meet strength requirements. These castings must be heat treated again if they are welded.

Several factors including the type of alloy, previous thermal treatment, size and intricacies of sections, and degree of restraint need to be considered when welding castings. Alloy composition can usually be identified by designation markings on the castings. If not, a chemical or spectrographic analysis should be made.

Castings can be welded in the as-cast, solution heat-treated, or solution heat-treated and aged condition. However, the welding of some alloys in the as-cast condition is not recommended because of the greater risk of cracking and the possibility of grain growth in the weld zone during the long solution heat treating times required. The heat-treated condition of the casting before welding may influence the preheat temperature selection.

Castings should be stripped of paint and degreased before welding. Conversion coatings should be removed from around the defective areas with stainless steel wool or wire brush. A rotary deburring tool is recommended for removing defects and preparing the area for weld-

**Table 9.12**
**Typical conditons for manual gas tungsten arc welding magnesium alloys**

| Thickness, in. | Joint design[1] | No. of passes | Welding current,[2] A | Electrode diam., in. | Welding rod diam., in. |
|---|---|---|---|---|---|
| 0.040 | A | 1 | 35 | 0.062 | 0.094 |
| 0.063 | A | 1 | 50 | 0.094 | 0.094 |
| 0.080 | A | 1 | 75 | 0.094 | 0.094 |
| 0.100 | A | 1 | 100 | 0.094 | 0.094 |
| 0.125 | A | 1 | 125 | 0.094 | 0.125 |
| 0.190 | A | 1 | 160 | 0.125 | 0.125 |
| 0.250 | B | 2 | 175 | 0.156 | 0.125 |
| 0.375 | B | 3 | 175 | 0.156 | 0.156 |
| 0.375 | C | 2 | 200 | 0.188 | 0.125 |
| 0.500 | B | 3 | 175 | 0.156 | 0.156 |
| 0.500 | C | 2 | 250 | 0.188 | 0.125 |

1. A - Square-groove butt joint, 0 root opening
   B - Single-V-groove butt joint, 0.06 in. root face, 0 root opening
   C - Double-V-groove butt joint, 0.09 in. root face, 0 root opening

2. With argon shielding. Helium shielding will reduce the welding current about 20 to 30 A. Thorium bearing alloys will require about 20% higher current.

ing. Broken pieces should be clamped in position for welding. The appropriate joint preparation should be determined from Table 9.12. Where large holes or defective areas are to be filled with weld metal, a backing strip can be used to prevent excessive melt-thru.

The casting should be preheated if the section to be repaired is relatively thick. Welding of broken pieces should commence at the center of the joint and progress toward the ends. Medium size weld beads are preferred. Low welding current may cause cold laps, oxide contamination, or porous welds. High welding current may cause weld cracking or incipient melting in the heat-affected zone. The filling of holes is usually the most critical type of repair from the standpoint of cracking. The arc should be struck at the bottom of the hole and welding should progress upward. The arc should not be held too long in one area to avoid the possibility of weld cracking or incipient melting in the heat-affected zone. The arc should be extinguished by gradually reducing the welding current to zero with appropriate current controls. This will permit the molten weld pool to solidify slowly and avoid crater cracking.

## GAS METAL ARC WELDING

The fundamental principles for gas metal arc welding of magnesium alloys are the same as for other metals. Welding can be done with this process at speeds that are 2 to 3 times faster than those with gas tungsten arc welding. Higher welding speeds reduce the heat input which, in turn, results in less distortion and some improvement in the tensile yield strength of the joint. The higher filler metal deposition rates reduce welding time and fabrication costs.

### Shielding Gases

Argon shielding is generally used for GMAW. Occasionally, mixtures of argon and helium are used to aid filler metal flow and alter the arc characteristics for deeper joint penetration. Pure helium is undesirable for shielding because it raises the current required for spray arc transfer and increases weld spatter.

### Metal Transfer

Typical melting rates for standard sizes of magnesium alloy electrodes using DCEP-RP power are given in Fig. 9.8. This figure shows the relationship between electrode feed rate and welding current for each size. It also illustrates the respective operating ranges for the three types of metal transfer used for GMAW. These are the short-circuiting, pulsed-spray, and spray transfer modes. The pulsed-spray operating region lies between the spray and short-circuiting transfer regions. Without pulsing, the welding amperages between the short-circuiting and

### Table 9.13
### Conditions for automatic gas tungsten arc welding of square-groove butt joints in AZ31B magnesium alloy

| Thickness, in. | Type of power | Welding speed, in./min.[a] | Welding current,[a] A | Filler metal[b] Diameter, in. | Filler metal[b] Feed Rate, in./min | Electrode diameter, in. | Arc length, in. |
|---|---|---|---|---|---|---|---|
| 0.063 | AC balanced wave | 12 | 55 | 0.063 | 35 | 0.094 | 0.025 |
| | | 24 | 60 | 0.063 | 50 | 0.094 | 0.025 |
| | | 36 | 70 | 0.063 | 54 | 0.125 | 0.025 |
| | | 45 | 95 | 0.063 | 96 | 0.125 | 0.025 |
| | | 70 | 170 | 0.063 | 160 | 0.188 | 0.025 |
| | | 80[c] | 195 | 0.063 | 190 | 0.188 | 0.025 |
| | | 95[c] | 200 | 0.063 | 203 | 0.188 | 0.025 |
| | DCEN-SP | 48 | 75 | 0.063 | 80 | 0.125 | 0.025 |
| | DCEP-RP | 80[c] | 120 | 0.063 | 184 | 0.250 | 0.020 |
| 0.190 | AC balanced wave | 34[c] | 300 | 0.063 | 159 | 0.250 | 0.020 |
| | DCEN-SP | 20[c] | 170 | 0.063 | 70 | 0.125 | 0.030 |
| | DCEP-RP | 7[c] | 120 | 0.094 | 10 | 0.250 | 0.020 |

a. With helium shielding
b. AZ61A or AZ92A filler metal
c. Maximum speed for arc stability and freedom from undercutting.

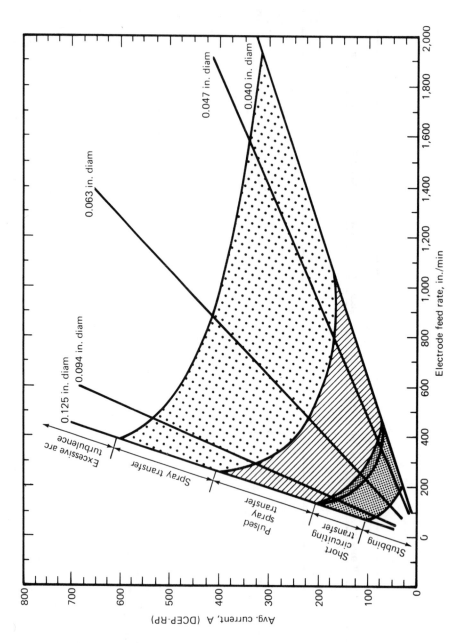

**Fig. 9.8 – Melting rates for bare magnesium alloy electrodes with argon shielding.**

spray transfer ranges would produce highly unstable globular transfer, which is not suitable for welding magnesium alloys.

Like short-circuiting transfer, spray transfer is only stable over a limited welding current range. Excessive welding current causes arc turbulence, which must be avoided. The approximate arc voltage ranges corresponding to each type of metal transfer are 13 to 16V for short-circuiting transfer, 17 to 23V for pulsed-spray transfer, and 24 to 28V for spray transfer.

### Equipment

The equipment used for gas metal arc welding magnesium alloys is similar to that used for other nonferrous alloys. An appropriate power source is used to produce the desired method of filler metal transfer. A constant current power source may be preferred for spray transfer at the lower end of the recommended current range for the applicable electrode size. It will minimize weld spatter. The power source for pulsed-spray welding must be designed to produce two current levels. Spray transfer takes place during the periods of high current and ceases between them when the current is low.

### Welding Conditions

Typical conditions for gas metal arc welding various thicknesses of magnesium alloys are given in Table 9.14. These may be used as a guide when establishing welding conditions for a specific application. The short-circuiting transfer mode is used for thin sections, and the spray transfer mode for thick sections. Pulsed-spray transfer is recommended for the intermediate thicknesses because there is less heat input than with continuous spray transfer.

Recommended electrode sizes for welding various thicknesses of magnesium alloys are given in Table 9.15. With both spray and pulsed-spray transfer, the lowest welding cost is achieved with the largest applicable electrode. With short-circuiting transfer, only one or two electrode sizes can be used to produce welds with good fusion and joint penetration.

### Spot Welding

Arc spot welding can be used to join magnesium sheet and extrusions in a variety of thicknesses. Welding schedules for suitable thickness combinations of AZ31B alloy are given in Table 9.16. These may be used as guides with other alloys.

Commercially available gas metal arc spot welding equipment is suitable for magnesium alloys. A constant potential power source, DCEP-RP power, and argon shielding are recommended. The strengths of gas metal arc spot welds may meet or exceed the strength requirements for resistance spot welds. Postweld stress relief of gas metal arc spot welds is required on all alloys that are sensitive to stress-corrosion cracking.

### STRESS RELIEVING

High residual stresses from welding or forming will promote stress-corrosion cracking in magnesium alloys that contain more than about 1.5 percent aluminum. Thermal treatments are used with these alloys to reduce residual stresses to safe levels to avoid this problem. Other magnesium alloys do not appear to be sensitive to this type of cracking.

Stress-corrosion cracking in welded structures usually occurs in the area adjacent to the weld bead. It is almost always a transcrystalline type of crack. Cracking may be delayed somewhat by painting. However, this will not ensure crack-free service for long periods, and should not be substituted for stress relieving of the weldment.

Stress relieving can be accomplished either in a furnace or with a torch. Furnace stress relieving is preferred. The time and temperature necessary to stress relieve weldments of the various alloys and product forms are shown in Table 9.17. When a furnace is used, a fixture should be used to support the weldment during heating to prevent distortion and correct any warpage. The temperature of large weldments should be monitored with thermocouples to make certain that all sections reach the proper temperature. In torch stress relieving, a temperature indicating device should be used to avoid overheating.

### POSTWELD HEAT TREATMENT

Welded castings are generally heat treated to obtain desired properties. The appropriate postweld heat treatment depends upon the temper of the casting before welding and the desired temper after welding, as shown in Table 9.11. Because of the fine grain size and extensive dispersion of the precipitates in the weld zone, aluminum bearing castings in the T4 or T6 condition may be solution heat treated for relatively short heating times after welding. In the case of

**Table 9.14**
**Typical conditions for gas metal arc welding of magnesium alloys[a,b,e]**

| Thickness, in. | Joint design[c] | No. of weld passes | Type of transfer | Electrode | | Voltage | | Welding current, A |
|---|---|---|---|---|---|---|---|---|
| | | | | Diam., in. | Feed rate, in./min | Pulse | Arc[d] | |
| 0.025 | A | 1 | Short-circuiting | 0.040 | 140 | — | 13 | 25 |
| 0.040 | A | 1 | Short-circuiting | 0.040 | 230 | — | 14 | 40 |
| 0.063 | A | 1 | Short-circuiting | 0.063 | 185 | — | 14 | 70 |
| 0.090 | A | 1 | Short-circuiting | 0.063 | 245 | — | 16 | 95 |
| 0.125 | B | 1 | Short-circuiting | 0.094 | 135 | — | 14 | 115 |
| 0.160 | B | 1 | Short-circuiting | 0.094 | 165 | — | 15 | 135 |
| 0.190 | B | 1 | Short-circuiting | 0.094 | 205 | — | 15 | 175 |
| 0.063 | A | 1 | Pulsed-spray | 0.040 | 360 | 55 | 21 | 50 |
| 0.125 | A | 1 | Pulsed-spray | 0.063 | 280 | 55 | 24 | 110 |
| 0.190 | A | 1 | Pulsed-spray | 0.063 | 475 | 52 | 25 | 175 |
| 0.250 | C | 1 | Pulsed-spray | 0.094 | 290 | 55 | 29 | 210 |
| 0.250 | C | 1 | Spray | 0.063 | 530 | — | 27 | 240 |
| 0.375 | C | 1 | Spray | 0.094 | 285-310 | — | 24-30 | 320-350 |
| 0.500 | C | 2 | Spray | 0.094 | 320-360 | — | 24-30 | 360-400 |
| 0.625 | D | 2 | Spray | 0.094 | 330-370 | — | 24-30 | 370-420 |
| 1.000 | D | 4 | Spray | 0.094 | 330-370 | — | 24-30 | 370-420 |

a. Argon shielding.
b. Arc travel speed of 24 to 36 in./min.
c. A - Square groove, no root opening.
   B - Square groove, 0.09 in. root opening.
   C - Single-V-groove, 0.06 in. root opening.
   D - Double-V-groove, 0.13 in. root opening.
d. Average voltage with pulsed-spray transfer.
e. These conditions may also be used for fillet welds in thicknesses of 0.25 to 1.0 inch.

**Table 9.15**
**Recommended electrode sizes for gas metal arc welding of magnesium alloys**

| Electrode diam., in. | Applicable base metal thickness range, in. | | |
| --- | --- | --- | --- |
| | Short-circuiting transfer | Pulsed-spray transfer[a] | Spray transfer[a] |
| 0.040 | 0.03-0.06 | 0.06-0.09 | 0.16-0.25 |
| 0.045 | 0.04-0.07 | 0.07-0.12 | 0.19-0.25 |
| 0.063 | 0.06-0.09 | 0.10-0.25 | 0.20-0.30 |
| 0.094 | 0.09-0.19 | 0.20-0.31 | 0.30 and over |

a. Pulsed-spray and spray transfer thickness schedules should provide good welding characteristics at minimum filler metal cost.

AM100A, AZ81A, AZ91C and AZ92A alloy castings, the solution heat-treating time must not exceed 30 minutes at temperature to avoid excessive grain growth in the weld zone. A protective atmosphere must be used when the solution treating temperature is above 750° F to prevent oxidation and active burning of the weldment.

The postweld heat treatments specified for the various alloys will produce the best weldment properties and also stress relieve castings to prevent cracking. If a postweld solution or temper heat treatment is not required, aluminum bearing castings should be stress relieved.

## WELD PROPERTIES

Typical tensile strength properties at room and elevated temperatures are given in Tables 9.18 and 9.19, respectively, for gas tungsten arc welds in wrought and cast magnesium alloys. Properties of joints made by gas metal arc welding are similar to or slightly higher than these strengths because of the reduced heat input. Table 9.20 gives the tensile properties of magnesium alloy weld metals produced by several filler metal-base metal combinations. The strengths of welds in most magnesium alloys are usually near to those of the base metals. This is shown by a comparison of the tensile strength data in Tables 9.18 and 9.20 for welded joints with similar data for the base metals in Table 9.5.

When the base metal is in the strain-hardened condition, recrystallization and some grain growth will take place in the heat-affected zone during welding. The heat-affected zone will then be weaker than the base metal, and sometimes have lower strength than the weld metal. The latter case is due to the fine grain size of the weld metal. The grain sizes of AZ61A, AZ92A, and EZ33A weld metals are shown in Fig. 9.9.

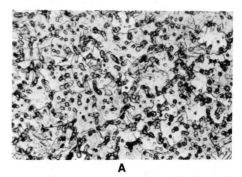

A

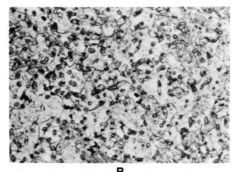

B

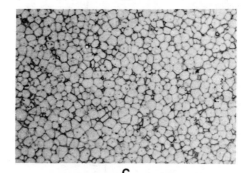

C

*Fig. 9.9 – Typical microstructures of (A) AZ61A, (B) AZ92A, and (C) EZ33A weld metal ( × 250, phospho-picral etch)*

Table 9.16
Typical conditions for gas metal arc spot welding of AZ31B magnesium alloy sheet

| Sheet thickness in. | | Electrode diam.,[a] in. | Welding current, A | Arc voltage, V | Weld time,[b] cycles | Shear strength, lb/spot |
|---|---|---|---|---|---|---|
| Front | Back | | | | | |
| 0.040 | 0.090 | 0.040 | 175 | 22-26 | 30-70 | 40-1085 |
| | 0.125 | | 200 | 24-26 | 25-50 | 760-1190 |
| | | 0.040 | 175 | 22-26 | 40-100 | 310-1370 |
| | 0.190 | | 200 | 22-28 | 20-90 | 460-1710 |
| | | 0.040 | 175 | 22-26 | 40-100 | 50-1265 |
| | | | 200 | 22-28 | 20-100 | 360-1725 |
| 0.063 | 0.063 | 0.040 | 200 | 26 | 50-60 | 348-889 |
| | 0.090 | 0.040 | 200 | 22-26 | 20-80 | 250-710 |
| | | | 225 | 24-26 | 25-45 | 360-850 |
| | 0.125 | 0.040 | 200 | 22-26 | 30-100 | 515-1165 |
| | | | 225 | 22-28 | 20-100 | 230-1340 |
| | 0.250 | 0.040 | 200 | 22-26 | 30-100 | 250-880 |
| | | | 225 | 22-28 | 20-100 | 320-2000 |
| 0.090 | 0.090 | 0.040 | 250 | 24-28 | 50-100 | 513-1129 |
| | | 0.063 | 275 | 25-28 | 40-90 | 314-1078 |
| | 0.125 | 0.063 | 275 | 22-26 | 30-100 | 520-1060 |
| | | | 300 | 24-28 | 25-80 | 230-960 |
| | 0.190 | 0.063 | 275 | 22-26 | 30-100 | 290-1145 |
| | | | 300 | 22-28 | 20-100 | 200-1700 |

## Table 9.16 (cont.)
### Typical conditions for gas metal arc spot welding of AZ31B magnesium alloy sheet

| Sheet thickness, in. | | Electrode diam.,[a] in. | Welding current, A | Arc voltage, V | Weld time,[b] cycles | Shear strength, lb/spot |
|---|---|---|---|---|---|---|
| Front | Back | | | | | |
| 0.125 | 0.125 | 0.094 | 325 | 24-27 | 40-100 | 583-1675 |
|  |  |  | 350 | 24-25 | 40-100 | 680-1337 |
|  | 0.156 | 0.094 | 350 | 22-26 | 30-100 | 530-1875 |
|  |  |  | 375 | 24-26 | 30-80 | 640-1600 |
|  | 0.190 | 0.094 | 350 | 22-24 | 30-100 | 190-1900 |
|  |  |  | 375 | 24-28 | 30-100 | 270-1640 |
| 0.156 | 0.156 | 0.094 | 375 | 24-26 | 80-150 | 517-1437 |
| 0.190 | 0.190 | 0.094 | 375 | 24-26 | 80-150 | 782-1323 |
|  |  |  | 400 | 24-26 | 60-110 | 853-1194 |

a. AZ61A electrode
b. 60Hz

## Table 9.17
### Recommended stress relieving heat treatments for magnesium alloys

| Sheet | | | Extrusions | | | Castings | | |
|---|---|---|---|---|---|---|---|---|
| Alloy | Temperature, °F | Time, min | Alloy | Temperature, °F | Time, min | Alloy | Temperature, °F | Time, min |
| AZ31B-O | 500 | 15 | AZ10A-F | 500 | 15 | AM100A | 500 | 60 |
| AZ31B-H24 | 300 | 60 | AZ31B-F | 500 | 15 | AZ63A | 500 | 60 |
| M1A-O | 500 | 15 | AZ61A-F | 500 | 15 | AZ81A | 500 | 60 |
| M1A-H24 | 400 | 60 | AZ80A-F | 500 | 15 | AZ91C | 500 | 60 |
| HK31A-H24 | 600 | 30 | AZ80A-T5 | 400 | 60 | AZ92A | 500 | 60 |
| HM21A-T81 | 750 | 30 | HM31A-T5 | 800 | 60 | | | |
| | | | M1A-F | 500 | 15 | | | |
| | | | ZK21A-F | 400 | 60 | | | |
| | | | ZK60A-F | 500 | 15 | | | |
| | | | ZK60A-T5 | 300 | 60 | | | |

**Table 9.18**

**Typical tensile properties of gas tungsten arc welds in magnesium alloys at room temperature**

| Alloy and temper | Filler metal | Tensile strength, ksi | Yield strength, ksi | Elongation, % in 2 in. | Joint efficiency, % |
|---|---|---|---|---|---|
| **Sheet** | | | | | |
| AZ31B-O | AZ61A, AZ92A | 35-36 | 17-19 | 10-11 | 95-97 |
| AZ31B-H24 | AZ61A, AZ92A | 36-37 | 19-22 | 5 | 86-88 |
| HK31A-H24 | EZ33A | 31-32 | 20-22 | 2-4 | 82-84 |
| HM21A-T8 | EZ33A | 28-31 | 19-20 | 2-4 | 80-89 |
| ZE41A-T5 | ZE41A | 30 | 20 | 4 | 100 |
| ZH62A-T5 | ZH62A | 38 | 25 | 5 | 95 |
| **Extrusions** | | | | | |
| AZ10A-F | AZ61A, AZ92A | 32-33 | 15-18 | 6-9 | 91-94 |
| AZ31B-F | AZ61A, AZ92A | 36-37 | 19-22 | 5-7 | 95-97 |
| AZ61A-F | AZ61A, AZ92A | 38-40 | 21-24 | 6-7 | 84-89 |
| AZ80A-F | AZ61A | 36-40 | 22-26 | 3-5 | 74-82 |
| AZ80A-T5 | AZ61A | 34-40 | 24-28 | 2 | 62-73 |
| HM31A-T5 | EZ33A | 28-31 | 19-24 | 1-2 | 64-70 |
| ZK21A-F | AZ61A, AZ92A | 32-34 | 17 | 4-5 | 76-81 |
| **Castings** | | | | | |
| AZ63A-T6 | AZ92A, AZ101A | 31 | – | 2 | 77 |
| AZ81A-T4 | AZ101A | 34 | 13 | 8 | 85 |
| AZ91C-T6 | AZ101A | 35 | 16 | 2 | 87 |
| AZ92A-T6 | AZ92A | 35 | 21 | 2 | 87 |
| EZ33A-T5 | EZ33A | 21 | 16 | 2 | 100 |
| HK31A-T6 | HK31A | 29 | 16 | 9 | 94 |
| HZ32A-T5 | HZ32A | 29 | 17 | 5 | 97 |
| K1A-F | EZ33A | 23 | 8 | 10 | 100 |

**Table 9.19**
**Typical elevated temperature tensile properties of gas tungsten arc welds in magnesium alloys[a]**

| Alloy[b] | Filler metal | Test temp., °F | Tensile strength, ksi | Yield strength, ksi | Elongation, % in 2 in. | Joint efficiency, % |
|---|---|---|---|---|---|---|
| | | | Sheet | | | |
| HK31A-H24 | EZ33A | 400 | 21 | 13 | 18 | 88 |
| | | 600 | 13 | 10 | 24 | 100 |
| HM21A-T8 | EZ33A | 400 | 18 | 12 | 16 | 100 |
| | | 600 | 14 | 10 | 14 | 93-100 |
| | | | Extrusions | | | |
| HM31A-T5 | EZ33A | 400 | 21 | 12 | 22 | 87 |
| | | 600 | 13 | 9 | 27 | 72 |
| | | | Castings | | | |
| EZ33A-T5 | EZ33A | 400 | 19 | 11 | 13 | 90 |
| | | 600 | 11 | 7 | 50 | 92 |
| HK31A-T6 | HK31A[c] | 400 | 18 | 11 | 33 | 82 |
| | | 600 | 15 | 9 | 25 | 79 |
| HZ32A-T5 | HZ32A[c] | 400 | 19 | 13 | 33 | 100 |
| | | 600 | 12 | 10 | 26 | 92 |

a. Weld reinforcement removed.
b. Alloys designed for elevated temperature service.
c. EZ33A filler metal will give equivalent joint strengths.

**Table 9.20**
**Tensile properties of magnesium alloy weld metals produced from various filler metal-base metal combinations**

| Filler metal | Base metal | Ultimate tensile strength, ksi | Tensile yield strength, ksi | Elongation, % in 2 in. |
|---|---|---|---|---|
| AZ61A | AZ31B | 34.3 | 14.5 | 10.0 |
| AZ92A | AZ31B | 36.8 | 18.9 | 8.0 |
| EZ33A | HK31A | 32.0 | 17.8 | 9.0 |
| EZ33A | HM31A | 26.8 | 19.8 | 3.5 |
| EZ33A | HM21A | 30.0 | 21.2 | 6.3 |
| HK31A | HM31A | 26.0 | 13.8 | 10.5 |
| HK31A | HM21A | 27.0 | 13.8 | 13.3 |

# RESISTANCE WELDING

## SPOT WELDING

Magnesium alloy sheet and extrusions can be joined by resistance spot welding in thicknesses ranging from about 0.02 to 0.13 inch. Alloys generally recommended for spot welding are M1A, AZ31B, AZ61A, HK31A, HM21A, HM31A, and ZK60A. Spot welding is normally used for low stress applications and occasionally for high stress applications where vibration is low or nonexistent. Magnesium alloys are spot welded using procedures similar to those for aluminum alloys.

## Preweld Cleaning

Careful preweld cleaning is essential for the production of spot welds of consistent size and soundness. A uniform electrical surface resistance of about 50 microhms or less is necessary to obtain consistency. Chemical cleaning procedures for spot welding are given in Table 9.8. Chemically cleaned parts will maintain a low, consistent surface resistance for about 100 hours when stored in a clean, dry environment. However, the time between cleaning and welding for critical applications should be limited to 24 hours. Mechanically cleaned surfaces will develop progressively higher, inconsistent surface resistance after 8 to 10 hours. For best results, mechanically cleaned material should be spot welded within this time.

## Equipment

Because of the relatively high thermal and electric conductivities of magnesium alloys, high welding currents and short weld times are required for spot welding. Spot welding machines designed for aluminum alloys are suitable for magnesium. As with aluminum, very rapid electrode follow-up is required to maintain pressure on the weld nugget as the metal softens and deforms rapidly. For this reason, low inertia welding machines should be used. A dual force system is not required for spot welding magnesium alloys. However, dual electrode force is sometimes used to reduce internal discontinuities by applying a higher forging force on the nugget during solidification. Timing of this forging force application is important for it to be beneficial.

## Electrodes

Spot welding electrodes for magnesium alloys should be made of RWMA Group A, Class 1 or Class 2 alloy. The faces of the electrodes must be kept clean and smooth to minimize the contact resistance between the electrode and the adjacent part. Cleaning should be done with an electrode dressing tool with the proper face contour covered with 280 grit abrasive cloth.

Electrode life between cleanings is limited by the transfer of copper to the adjacent part and subsequent sticking. The number of welds that can be produced between cleanings depends upon the electrode alloy and cooling efficiency, the method of base metal cleaning, the magnesium alloy composition, and the welding conditions. Table 9.21 shows the relative effectiveness of mechanical and chemical surface preparations on electrode life for three magnesium alloys. Chemical cleaning will give better electrode life than mechanical cleaning by wire brushing. The proper cleaning solution must be used for the magnesium alloy to be welded. In any case, the longest electrode life will be obtained when the welding conditions produce a weld nugget no larger than that necessary to meet design strength requirements.

Copper pickup on the spot weld surfaces increases the corrosion susceptibility of magnesium. Therefore, the copper should be completely removed from the surfaces by a suitable mechanical cleaning method. The presence of copper on spot welds can be determined by applying a 10 percent acetic acid solution. A dark spot will form if copper is present on the surface.

## Joint Design

The joint designs for spot welding magnesium alloys are much the same as those for aluminum alloys. Minimum recommended spot spacing and edge distance for the location of spot welds are given in Table 9.22.

Where two unequal thicknesses are to be spot welded, the thickness ratio should not exceed 2.5 to 1. With three thicknesses, the thickness variation should not exceed about 25 percent, and the thickest section should be in the center.

## Welding Schedules

The important factors that must be considered when developing a welding schedule are (1)

**Table 9.21**
**Effect of surface preparation on spot welding electrode life with magnesium alloys**

| Alloy | Electrode classification[a] | Wire brushing | No. of spot welds[b] Spot weld cleaner[c] No. 2 | No. 3 |
|---|---|---|---|---|
| AZ31B | Class 1 | 30 | over 200 | over 550 |
| AZ31B | Class 2 | 15 | 50 | 270 |
| HK31A | Class 2 | 40 | 400 | 195 |
| HM21A | Class 2 | 5 | 80 | 5 |

a. RWMA Group A
b. Between electrode cleanings
c. Refer to Table 9.8 for solution compositions

**Table 9.22**
**Suggested spot spacing and edge distance for spot welds in magnesium alloys**

| Thickness,[a] in. | Spot spacing, in. Minimum | Nominal | Edge distance, in. Minimum | Nominal |
|---|---|---|---|---|
| 0.020 | 0.25 | 0.50 | 0.22 | 0.31 |
| 0.025 | 0.25 | 0.50 | 0.22 | 0.31 |
| 0.032 | 0.31 | 0.62 | 0.25 | 0.36 |
| 0.040 | 0.38 | 0.75 | 0.28 | 0.38 |
| 0.050 | 0.44 | 0.80 | 0.31 | 0.41 |
| 0.063 | 0.50 | 1.00 | 0.38 | 0.48 |
| 0.080 | 0.63 | 1.25 | 0.44 | 0.54 |
| 0.100 | 0.88 | 1.50 | 0.47 | 0.56 |
| 0.125 | 0.94 | 1.75 | 0.56 | 0.67 |

a. Thinner section if thicknesses are unequal.

the dimensions, properties, and characteristics of the alloys to be welded, (2) the type of welding equipment to be employed, and (3) the joint design. A welding schedule can be established for any particular combination of these. Typical schedules for spot welding magnesium alloys with three types of equipment are given in Tables 9.23, 9.24, and 9.25. These data are intended only as guides in establishing schedules for specific applications.

The welding and postheat currents are approximate values. The magnitude of the welding current is usually adjusted by transformer taps, phase shift heat control, or both. To obtain the required current, simply start with a low value of weld heat and a corresponding percentage of postheat. The current is gradually increased until the desired shear strength, nugget diameter, and penetration are obtained. In some cases, it may be necessary to readjust the weld time to achieve the desired properties.

With single-phase ac equipment, welding current may be determined by primary or secondary measurement methods. The nugget diameter and the minimum indicated shear strength given in Table 9.24 should be obtained when the measured welding current is within 5 percent of the listed value.

When dual electrode force is used, timing of application of the forging force is very important. If the forging force is applied too late, the temperature of the nugget will be too low for this higher force to consolidate the nugget. If the forging force is applied too soon, the nugget size may be too small or the electrode indentation excessive. Insufficient electrode force may cause weld metal expulsion, internal discontinuities in the nugget, surface burning, or excessive electrode sticking. Excessive electrode force is evidenced by deep electrode indentation, large sheet separation and distortion, or unsymmetrical weld nuggets.

Weld nugget diameter and penetration can be determined by sectioning through the center of the nugget. The exposed edge is polished, and then etched with a 10 percent acetic or tartaric acid solution. Penetration should be uniform in equal sheet thicknesses. If not, subsequent welds may require the use of a smaller electrode radius against the side with the lesser penetration. It may also be necessary to clean the electrodes more frequently or the part surfaces more thoroughly.

When spot welding dissimilar alloys, differences in thermal and electrical conductivities can be compensated for by using an electrode with a smaller radius in contact with the alloy that requires the higher heat input. For example, to center the weld nugget in a joint between equal thicknesses of M1A and AZ31B sheets, a smaller radius face should be used against the M1A alloy.

## Joint Sealing

Spot welded assemblies can be given either a chrome pickle or a dichromate treatment, followed by painting and finishing as desired. Where sealed joints are required or the weldment is to be exposed to a corrosive atmosphere, a suitable sealing compound should be placed between the faying surfaces of the joint before welding. Several proprietary compounds are available for this purpose. Sealers should not be so viscous as to prevent metal-to-metal contact when the electrode force is applied. Welding should be done soon after applying the compounds, and frequent tests should be made to monitor weld quality.

## Joint Strength

Typical shear strengths for spot welds in several thicknesses of three magnesium alloys are given in Table 9.26. Although higher shear strengths are readily obtainable, these values represent the average strengths for welds of maximum soundness and consistency.

## SEAM WELDING

Seam welds can be made in magnesium alloys under conditions similar to those required for spot welding. Shear strengths of about 750 to 1500 lb/in. of seam can be obtained in M1A alloy in thicknesses of 0.040 to 0.128 inch. Strengths of seam welds in AZ31B alloy sheet material are approximately 50 percent higher.

## FLASH WELDING

Flash welding equipment and techniques similar to those used for aluminum alloys can be used for magnesium alloys. High current densities and extremely rapid flashing and upsetting rates are required. Upsetting current should continue for about 5 to 10 cycles (60 Hz) after upset. Special shielding atmospheres are not necessary. Flash welds in AZ31B, AZ61A, and HM31A

## Table 9.23
### Schedules for spot welding magnesium alloys with three-phase frequency converter machines

| Alloy | Thickness,[a] in. | Electrode[b] Diam., in. | Electrode[b] Face radius, in. | Electrode force, lb. Weld | Electrode force, lb. Forge | Forge delay time, cycles[c] | Weld heat or pulse time, cycles[c] | No. of pulses | Post-heat time, cycles[c] | Approx. current, A Weld | Approx. current, A Postheat | Nugget diam., in. | Min. aver. shear strength, lb. |
|---|---|---|---|---|---|---|---|---|---|---|---|---|---|
| AZ31B | 0.020 | 1/2 | 3 | 800 | --- | --- | 1 | 2 | --- | 25,400 | --- | 0.19 | 195 |
| | 0.025 | 1/2 | 3 | 800 | --- | --- | 1 | 1 | 2 | 20,200 | 4,000 | 0.14 | 200 |
| | 0.032 | 1/2 | 3 | 1,000 | --- | --- | 1 | 2 | --- | 26,400 | --- | 0.20 | 330 |
| | 0.040 | 5/8 | 3 | 1,200 | --- | --- | 1 | 2 | --- | 28,300 | --- | 0.21 | 425 |
| | 0.050 | 5/8 | 4 | 1,400 | 3,500 | 2 | 2 | 2 | 5 | 29,000 | 10,300 | 0.19 | 435 |
| | 0.050 | 5/8 | 4 | 1,600 | --- | --- | 2 | 1 | --- | 31,000 | --- | 0.19 | 440 |
| | 0.063 | 5/8 | 4 | 1,750 | --- | --- | 3 | 1 | --- | 35,200 | --- | 0.22 | 580 |
| | 0.063 | 7/8 | 4 | 1,200 | 3,900 | 3 | 3 | 1 | --- | 43,600 | --- | 0.25 | 690 |
| | 0.063 | 5/8 | 4 | 1,200 | 1,920 | 3 | 3 | 1 | 6 | 43,600 | 24,800 | 0.29 | 800 |
| | 0.090 | 7/8 | 4 | 2,000 | 4,300 | 2 | 3 | 1 | 5 | 42,700 | 15,000 | 0.26 | 910 |
| | 0.125 | 7/8 | 6 | 4,500 | --- | --- | 5 | 6 | --- | 66,900 | --- | 0.46 | 2,095 |
| HK31A | 0.040 | 1/2 | 3 | 1,000 | --- | --- | 1 | 1 | --- | 19,600 | --- | 0.17 | 310 |
| | 0.050 | 5/8 | 4 | 1,400 | --- | --- | 2 | 2 | --- | 31,600 | --- | 0.23 | 530 |
| | 0.063 | 3/4 | 4 | 2,400 | --- | --- | 3 | 1 | --- | 39,400 | --- | 0.25 | 660 |
| | 0.080 | 3/4 | 4 | 3,400 | --- | --- | 4 | 1 | --- | 50,500 | --- | 0.29 | 890 |
| | 0.125 | 7/8 | 6 | 5,000 | --- | --- | 5 | 6 | --- | 65,900 | --- | 0.33 | 1,300 |
| | 0.125 | 3/4 | 6 | 2,400 | 3,200 | 2 | 5 | 6 | --- | 50,900 | --- | 0.37 | 1,380 |
| HM21A | 0.040 | 1/2 | 3 | 800 | --- | --- | 1 | 2 | --- | 21,600 | --- | 0.18 | 355 |
| | 0.050 | 5/8 | 4 | 1,200 | --- | --- | 2 | 2 | --- | 30,700 | --- | 0.21 | 470 |
| | 0.063 | 5/8 | 4 | 1,600 | --- | --- | 3 | 2 | --- | 40,600 | --- | 0.23 | 560 |
| | 0.071 | 5/8 | 4 | 2,200 | --- | --- | 4 | 2 | --- | 47,400 | --- | 0.29 | 770 |
| | 0.090 | 3/4 | 4 | 3,000 | --- | --- | 4 | 2 | --- | 53,200 | --- | 0.26 | 950 |
| | 0.125 | 7/8 | 6 | 3,800 | --- | --- | 5 | 2 | --- | 66,700 | --- | 0.32 | 1,180 |
| | 0.125 | 7/8 | 6 | 2,000 | 3,600 | 5 | 5 | 6 | --- | 56,500 | --- | 0.37 | 1,405 |

a. Two equal thicknesses.
b. Spherical radius-faced electrodes on both sides.
c. Cycles of 60 Hz.

**Table 9.24**
**Schedules for spot welding magnesium alloys with single-phase ac machines**

| Alloy | Thickness,[a] in. | Electrode[b] Diam., in. | Electrode[b] Face radius, in. | Electrode force, lb. | Weld time, cycles[c] | Approx. welding current, A. | Nugget diam., in. | Min. aver. shear strength, lb. |
|---|---|---|---|---|---|---|---|---|
| AZ31B | 0.016 | 3/8 | 2 | 300 | 2 | 16,000 | 0.10 | 140 |
| | 0.020 | 3/8 | 3 | 350 | 3 | 18,000 | 0.14 | 175 |
| | 0.025 | 3/8 | 3 | 400 | 3 | 22,000 | 0.16 | 215 |
| | 0.032 | 3/8 | 3 | 450 | 4 | 24,000 | 0.18 | 270 |
| | 0.040 | 1/2 | 3 | 500 | 5 | 26,000 | 0.20 | 345 |
| | 0.050 | 1/2 | 4 | 550 | 5 | 29,000 | 0.23 | 430 |
| | 0.063 | 1/2 | 4 | 600 | 6 | 31,000 | 0.27 | 545 |
| | 0.071 | 1/2 | 4 | 650 | 7 | 32,000 | 0.29 | 610 |
| | 0.080 | 1/2 | 4 | 700 | 8 | 33,000 | 0.31 | 690 |
| | 0.090 | 1/2 | 4 | 750 | 9 | 34,000 | 0.32 | 770 |
| | 0.100 | 1/2 | 6 | 800 | 10 | 36,000 | 0.34 | 865 |
| | 0.125 | 1/2 | 6 | 1,000 | 12 | 42,000 | 0.38 | 1,080 |
| M1A | 0.016 | 3/8 | 2 | 300 | 3 | 17,000 | 0.08 | 70 |
| | 0.020 | 3/8 | 3 | 300 | 3 | 20,000 | 0.12 | 95 |
| | 0.025 | 3/8 | 3 | 350 | 4 | 24,000 | 0.14 | 130 |
| | 0.032 | 3/8 | 3 | 400 | 5 | 26,000 | 0.16 | 175 |
| | 0.040 | 3/8 | 3 | 450 | 6 | 28,000 | 0.18 | 225 |
| | 0.050 | 1/2 | 4 | 500 | 7 | 30,000 | 0.21 | 295 |
| | 0.060 | 1/2 | 4 | 550 | 8 | 32,000 | 0.24 | 385 |
| | 0.071 | 1/2 | 4 | 600 | 9 | 33,000 | 0.26 | 430 |
| | 0.080 | 1/2 | 4 | 650 | 10 | 35,000 | 0.28 | 495 |
| | 0.090 | 1/2 | 4 | 700 | 11 | 36,000 | 0.29 | 560 |
| | 0.100 | 1/2 | 6 | 750 | 12 | 38,000 | 0.31 | 630 |
| | 0.125 | 1/2 | 6 | 950 | 14 | 45,000 | 0.35 | 800 |

a. Two equal thicknesses.
b. Spherical radius-faced electrodes on both sides.
c. Cycles of 60 Hz.

**Table 9.25**
**Schedules for spot welding AZ31B magnesium alloy with dc rectifier machines**

| Thickness,[a] in. | Electrode[b] | | Electrode force, lb. | | Forge delay time, cycles[c] | Weld time, cycles[c] | Postheat time, cycles[c] | Approx. current, A | | Nugget diam., in. | Min. aver. shear strength, lb. |
| | Diam., in. | Face radius, in. | Weld | Forge | | | | Weld | Postheat | | |
|---|---|---|---|---|---|---|---|---|---|---|---|
| 0.020 | 5/8 | 3 | 300 | 600 | 0.6 | 1 | 1 | 21,000 | 14,700 | 0.14 | 145 |
| 0.032 | 5/8 | 3 | 400 | 880 | 1.0 | 2 | 1 | 24,000 | 16,900 | 0.18 | 245 |
| 0.040 | 5/8 | 3 | 480 | 1,000 | 1.2 | 2 | 2 | 26,000 | 18,000 | 0.20 | 336 |
| 0.051 | 5/8 | 3 | 580 | 1,270 | 1.5 | 3 | 2 | 28,500 | 20,000 | 0.22 | 435 |
| 0.064 | 5/8 | 4 | 700 | 1,540 | 1.8 | 3 | 3 | 29,300 | 20,500 | 0.27 | 560 |
| 0.081 | 7/8 | 4 | 860 | 1,890 | 2.4 | 4 | 4 | 35,750 | 25,000 | 0.31 | 740 |
| 0.093 | 7/8 | 6 | 970 | 2,150 | 3.9 | 6 | 4 | 38,750 | 27,100 | 0.32 | 855 |
| 0.102 | 7/8 | 6 | 1,050 | 2,320 | 4.5 | 7 | 4 | 41,300 | 28,800 | 0.34 | 985 |
| 0.125 | 7/8 | 6 | 1,270 | 2,780 | 7.7 | 10 | 6 | 48,000 | 33,400 | 0.38 | 1,208 |

a. Two equal thicknesses.
b. Spherical radius-faced electrodes on both sides.
c. Cycles of 60 Hz.

**Table 9.26**
**Typical shear strengths of single spot welds in wrought magnesium alloys**

| Thickness, in. | Avg. spot diam., in. | Spot shear strength, lb | | |
|---|---|---|---|---|
| | | AZ31B | HK31A | HM21A |
| 0.020 | 0.14 | 220 | .... | .... |
| 0.025 | 0.16 | 270 | .... | .... |
| 0.032 | 0.18 | 330 | 300 | .... |
| 0.040 | 0.20 | 410 | 375 | 360 |
| 0.050 | 0.23 | 530 | 550 | .... |
| 0.063 | 0.27 | 750 | 720 | 660 |
| 0.080 | 0.31 | 890 | .... | .... |
| 0.100 | 0.34 | 1180 | .... | .... |
| 0.125 | 0.38 | 1530 | 1490 | 1220 |

magnesium alloys have typical tensile strengths of 36, 42, and 38 ksi, respectively, with elongations of about 4 to 8 percent. The typical microstructures of various zones in a flash welded joint in an HM31A-T5 magnesium alloy extrusion are shown in Fig. 9.10.

# OXYFUEL GAS WELDING

Oxyfuel gas welding should only be used for emergency field repair work when suitable arc welding equipment is not available. Its use is restricted almost exclusively to simple groove welds where residual flux can be effectively removed. The repair welds should be considered only temporary until they can be replaced with arc welds or a new part can be put in service.

## FUEL GASES

The fuel gases most commonly used are acetylene or a mixture of about 80 percent hydrogen and 20 percent methane. The latter fuel gas is well-suited for welding sheets up to 0.064 in. thick because of its soft flame. For welding thicker gages, acetylene is desirable because of its higher heat of combustion. The oxyacetylene flame may cause slight pitting of the weld surface, but it is rarely serious enough to impair the strength of the weld.

## FLUXES

Fluxes specifically recommended for oxy-

fuel gas welding of magnesium should be used. These fluxes are prepared by mixing them with water or alcohol to form a heavy slurry or paste. They should be used soon after mixing. Prior to fluxing, the area to be welded should be cleaned to remove any dirt, oil, grease, oxide, or conversion coating.

One flux composition suitable for welding with various fuel gases is a mixture of 53% KCl, 29% $CaCl_2$, 12% NaCl, and 6% NaF, by weight. Another mixture suitable only for oxyacetylene welding consists of 45% KCl, 26% NaCl, 23% LiCl, and 6% NaF. The sodium compounds in these welding fluxes will give an intense yellow color to the flame. Welders should use suitable eye protection and ventilation when using these fluxes.

## WELDING TECHNIQUE

A liberal coating of flux should be applied to both sides of the joint and to the welding rod. If needed, the joint should be tack welded at 1 to 3 in. intervals depending upon the metal thick-

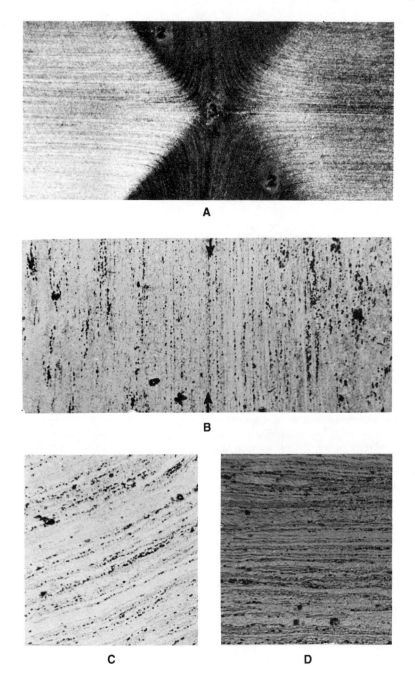

*Fig. 9.10 – (A) Flash weld in HM31A-T5 magnesium alloy rod, ×5, (B) weld interface, ×200, (C) upset metal, ×200, (D) unaffected base metal, ×200.*

ness. All tack welds and overlapping weld beads should be remelted to float out any flux inclusions. Parts of relatively large mass should be preheated.

All traces of flux must be removed from the weldment is washed in hot water. Then, it is given a chrome pickle followed by immersion for 2 hours in boiling flux remover. (See Table 9.8.)

# OTHER WELDING PROCESSES

## ELECTRON BEAM WELDING

In general, magnesium alloys that can be arc welded can also be electron beam welded. The same preweld and postweld operations apply to both processes.

Close control of electron beam operating variables is required to prevent overheating and porosity at the root of the weld. The high vapor pressures in vacuum of magnesium and zinc in alloys contribute to this problem. It is very difficult to produce sound welds in magnesium alloys containing more than 1 percent zinc. Beam manipulation may be helpful in overcoming porosity.

A photomicrograph of an electron beam weld in 0.25 in. thick HM31A-T5 magnesium alloy is shown in Fig. 9.11.

## STUD WELDING

The gas shielded arc stud welding process

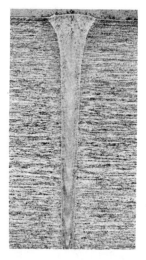

*Fig. 9.11 – Electron beam weld in 0.25 in. thick HM31A-T5 magnesium alloy extrusion (×10)*

Table 9.27
Typical welding conditions and breaking loads for AZ31B-F magnesium alloy studs joined to 0.25 in. AZ31B-O alloy plate

| Stud diam., in. | Welding current,[a] A | Weld time, cycles[b] | Lift, in. | Plunge, in. | Avg. breaking load, lb. |
|---|---|---|---|---|---|
| 0.25 | 125 | 45 | 0.125 | 0.25 | 1530 |
| 0.50 | 375 | 40 | 0.125 | 0.19 | 4100 |

a. DCEP-RP power and helium shielding
b. 60 Hz

used for aluminum is also applicable to magnesium alloys. A ceramic ferrule is not needed. Helium shielding and DCEP-RP power are used. The stud welding gun should be equipped with controlled plunge to avoid excessive spatter and undercutting of the base metal.

Typical conditions for welding 0.25 and 0.5 in. diameter AZ31B magnesium alloy studs to plate and the average breaking loads of the welded studs are given in Table 9.27. Figure 9.12 shows a cross section through a stud weld. The soundness of magnesium stud welds is, in general, very similar to that of aluminum stud welds.

*Fig. 9.12 – Typical arc stud weld in AZ31B magnesium alloy*

# BRAZING

Brazing techniques for magnesium alloys are similar to those used for aluminum alloys. However, the brazing of magnesium is not widely practiced. Furnace, torch, and dip brazing can be employed, but furnace and torch brazing experience is limited to M1A magnesium alloy.

## FILLER METAL

Only one magnesium brazing filler metal is covered by specifications.[5] It is BMg-1 filler metal having a nominal composition of 91 percent magnesium, 9 percent aluminum, and 2 percent zinc. Although it is similar to AZ92A magnesium alloy, the filler metal contains a small

amount of beryllium to prevent excessive oxidation while it is molten. The brazing temperature range for this filler metal is 1120° to 1160° F. It is suitable for brazing only AZ10A, K1A, and M1A magnesium alloys. These alloys will be annealed when exposed to brazing temperature.

## PREBRAZE CLEANING

As with other metals, all parts to be brazed should be thoroughly clean and free of burrs. All dirt, oil, or grease should be removed by vapor or solvent degreasing. Surface films, such as chromates or oxides, should be removed by mechanical or chemical cleaning. Abrasive cloth or steel wool is satisfactory for mechanical cleaning. Chemical cleaning should consist of immersion in hot alkaline cleaner and then in a suitable chemical cleaner. (See Table 9.8.)

---

5. Refer to AWS A5.8, *Specification for Brazing Filler Metals,* latest revision, for additional information.

**Table 9.28**
**Composition and melting point of magnesium brazing fluxes**

| Applicable brazing processes | Flux composition, % | | Approximate melting point, °F |
|---|---|---|---|
| Torch | KCl | 45 | 1000 |
| | NaCl | 26 | |
| | LiCl | 23 | |
| | NaF | 6 | |
| Torch, dip, furnace | KCl | 42.5 | 730 |
| | NaCl | 10 | |
| | LiCl | 37 | |
| | NaF | 10 | |
| | AlF$_3$-3NaF | 0.5 | |

## FLUXES

Chloride base fluxes similar to those used for oxyfuel gas welding are used for brazing. The composition and melting point of two suitable brazing fluxes are given in Table 9.28

## BRAZING PROCEDURES

### Furnace Brazing

Electric or gas furnaces with automatic temperature controls capable of holding the temperature within ± 5° F should be used for brazing. A special atmosphere is not required. Sulfur dioxide (SO$_2$) or products of combustion in gas-fired furnaces will inhibit brazing filler metal flow and must be avoided.

Parts to be brazed should be assembled with the filler metal preplaced in or around the joint. Joint clearances of 0.004 to 0.010-in. should be used for good capillary flow of the brazing filler metal. Best results are obtained when dry powdered flux is sprinkled along the joint. Flux pastes made with water or alcohol will retard the flow of brazing filler metal. Flux pastes made with benzol, toluene, or chlorbenzol may be used, but they are more difficult to apply because the pastes are not smooth. Flux pastes should be dried by heating the assembly at 350° to 400° F for 5 to 15 minutes in drying ovens or circulating air furnaces. Flame drying is not recommended because improper flame adjustment may cause a heavy soot deposit.

Brazing time will depend on the metal thickness at the joint and the amount of fixturing necessary to position the parts. The time should

be the minimum necessary to obtain complete filler metal flow with minimum diffusion between the filler and base metals. Normally, one to two minutes at the brazing temperature is sufficient.

### Torch Brazing

Torch brazing is done with a neutral oxyfuel gas or airfuel gas flame. Natural gas is well-suited for torch brazing because of its relatively low flame temperature. The brazing filler metal can be placed on the joint and fluxed before heating, or it may be face-fed. Flux pastes can be made with either water or alcohol. However, pastes made with alcohol give better results. Heat should be applied to the joint until the filler metal melts and flows into the joint. Overheating of the base metal must be avoided.

### Dip Brazing

Dip brazing is accomplished by immersing the assembly into a molten brazing flux held at brazing temperature. The flux serves the dual functions of both heating and fluxing. Temperature control should be accurate to within ± 5° F of the desired brazing temperature. Joint clearance should be from 0.004 to 0.010 inch.

After preplacing the filler metal, the parts should be assembled in a brazing fixture, preferably of stainless steel to resist the corrosive action of the flux. The fixtured assembly is preheated in a furnace to between 850° and 900° F. This is done to minimize distortion and the time in the flux bath. Immersion time in the flux bath should be relatively short because the parts are heated rapidly by the molten flux. For example, 0.063 in. thick sheet thickness can be heated in 30 to 45 seconds. Large assemblies with fixturing may require immersion for 1 to 3 minutes.

## POSTBRAZE CLEANING

Complete removal of all traces of flux from the brazement is required to avoid subsequent corrosion. Brazed parts should be rinsed thoroughly in flowing hot water to remove the flux from the surface of the part. A stiff-bristled brush may be used to scrub the surface and speed up flux removal. The brazement is then given a one to two minute immersion in chrome-pickle, followed by 2 hours in boiling flux remover cleaner. The compositions of these solutions are given in Table 9.8. The corrosion resistance of brazed joints depends primarily upon complete flux removal.

# SOLDERING

Bare magnesium alloys can be soldered only by the abrasion and ultrasonic methods. These methods can dislodge the oxide film on the surfaces to be soldered. No suitable flux is available to remove this film and permit the solder to wet the surfaces.

Conventional heating methods, including soldering irons and gas torches, may be used. Soldering is not recommended if the joint will be required to withstand moderately high stress. Soldered joints are low in strength and ductility. They are also unsatisfactory for service in the presence of an electrolyte. The marked difference in solution potential between a magnesium alloy and a solder can lead to severe galvanic attack. A suitable protective coating should be applied to soldered joints for good serviceability.

## SOLDERS

The solders listed in Table 9.29 are generally used for magnesium. Lead-containing solders, such as the 50% tin-50% lead alloy, can be used, but severe galvanic attack may take place in the presence of moisture. The tin-zinc solders generally have lower melting points and better wetting characteristics than the tin-zinc-cadmium solders, but they may form joints of low ductility. The high-cadmium solders generally produce the strongest and most ductile joints.

## SURFACE PREPARATION

Bare magnesium surfaces to be joined should be degreased with a suitable solvent and then mechanically cleaned immediately before soldering. A clean stainless steel wire brush,

stainless steel wool, or aluminum oxide abrasive cloth is a suitable cleaning tool.

Electroplated coatings on magnesium offer an excellent soldering base. A zinc-immersion (zincate) coating, the first step in plating magnesium, followed by a 0.0001 to 0.0002 in. thick copper plate over the zinc coating provides a solderable surface. Tin or silver plating may also be used for this purpose. Soldering of electroplated surfaces is carried out using the procedures normally used for the deposited metal. Fusing a tin coating improves its protective value by flowing the deposited tin and sealing the pores. This technique consists of electroplating a 0.0003 to 0.0005 in. tin coating over the copper electroplate. The part is then immersed in a hot oil bath to flow the tin coating and close the pores. The process is being used on a large number of magnesium electronic parts to permit easy soldering.

## JOINT TYPES

Fluxless soldering of bare magnesium alloys is limited to fillet joints and to the filling of surface defects in noncritical areas of wrought and cast products prior to painting. Conventional solder joints can be used with solderable electroplated surfaces.

## PROCEDURES

Bare magnesium surfaces must be precoated with a solder having good wetting characteristics. Solder coating with the friction method is done by rubbing the solder stick, soldering iron, or other tool on the magnesium un-

**Table 9.29**
**Solders for magnesium**

| Composition, % | Temperature, °F | | Use |
|---|---|---|---|
| | Solidus | Liquidus | |
| 60 Cd-30 Zn-10 Sn | 315 | 550 | Low temperature – below 300°F |
| 90 Cd-10 Zn | 509 | 570 | High temperature – above 300°F |
| 72 Sn-28 Cd | 350 | 470 | Medium temperature – below 300°F |
| 91 Sn-9 Zn | 390 | 390 | High temperature – above 300°F |
| 60 Sn-40 Zn | 390 | 645 | High temperature – above 300°F |
| 70 Sn-30 Zn | 390 | 592 | Precoating solder |
| 50 Sn-50 Pb | 361 | 421 | Filler solder on precoated surfaces |
| 80 Sn-20 Zn | 390 | 518 | Precoating solder |
| 40 Sn-33 Cd-27 Zn | ... | ... | Filler solder |

der the molten solder to break up the oxide film. The ultrasonic method of precoating utilizes a hot soldering bit vibrating at ultrasonic frequencies. When in contact with the molten solder on the magnesium, the vibration causes an abrasive

effect known as cavitation erosion. This action dislodges the surface oxides and permits wetting. After the surfaces are precoated with solder, the joint can be soldered using a soldering iron, torch, or hot plate.

# PLASMA ARC CUTTING

Magnesium alloys can be cut with a plasma arc cutting torch. An argon-hydrogen mixture is used for the orifice and shielding gases. A mixture of 80 percent Ar – 20 percent $H_2$ is recommended for manual operation, and 65 percent Ar – 35 percent $H_2$ for automatic cutting. Typical conditions for automatic cutting are given in Table 9.30.

**Table 9.30**
**Typical conditions for automatic plasma arc cutting of magnesium alloy plates[a]**

| Thickness, in. | Plasma | | Cutting speed, in./min | Shielding gas flow,[b] ft³/h | Remarks |
|---|---|---|---|---|---|
| | Current, A | Voltage, V | | | |
| 0.25 | 200 | 60 | 225 | 70 | Minimum Fume |
| 0.25 | 400 | 75 | 150 | 70 | Squarest cut |
| 0.25 | 400 | 80 | 300 | 70 | Maximum Speed |
| 0.5 | 240 | 60 | 150 | 70 | Minimum Fume |
| 0.5 | 420 | 75 | 100 | 100 | Squarest cut |
| 0.5 | 460 | 70 | 300 | 100 | Maximum Speed |
| 1 | 300 | 75 | 60 | 70 | Minimum Fume |
| 1 | 450 | 80 | 60 | 100 | Squarest cut |
| 1 | 660 | 80 | 75 | 100 | Maximum Speed |
| 2 | 350 | 100 | 25 | 100 | Minimum Fume |
| 2 | 520 | 100 | 25 | 100 | Squarest cut |
| 2 | 600 | 90 | 50 | 100 | Maximum Speed |
| 4 | 500 | 210 | 12 | 200 | Squarest cut |
| 6 | 750 | 225 | 12 | 300 | Squarest cut |

a. AZ31B magnesium alloy.
b. 65%Ar - 35% $H_2$

# SAFE PRACTICES

The welding fumes from all commercial magnesium alloys, except those containing thorium, are not harmful when the amount of fumes remains below the industrial hygiene standard for nuisance dusts. Welders should avoid inhalation of fumes from the thorium-containing alloys because of the presence of alpha radiation

in the airborne particles. However, the concentration of thorium in the fumes is sufficiently low so that good ventilation or local exhaust systems will provide adequate protection. No external radiation hazard is involved in the handling of the thorium-containing alloys.

The possibility of ignition when welding

magnesium alloys in thicknesses greater than 0.01 in. is extremely remote. Magnesium alloy product forms will not ignite in air until they are at fusion temperature. Then, sustained burning will occur only if the ignition temperature is maintained. Inert gas shielding during welding prevents ignition of the molten weld pool.

Magnesium fires usually occur with accumulations of grinding dust or machining chips. Accumulation of grinding dust on clothing should be avoided. Graphite-base (G-1) or proprietary salt-base powders, recommended for extinguishing magnesium fires, should be conveniently located in the work area. If large amounts of fine particles are produced, they should be collected in a waterwash type dust collector designed for use with magnesium. Special precautions pertaining to the handling of wet magnesium fines must be followed.

Some solvents, chemical baths, and fluxes used for cleaning, welding, brazing, or finishing of magnesium alloys contain chromates, chlorides, fluorides, acids, or alkalies. Adequate ventilation, protective clothing, and eye protection must be used when working with these materials to avoid toxic effects, burns, or other injuries that they may cause.

## Metric Conversion Factors

$$T_c = 0.56 \, (T_F - 32)$$
$$1 \, lb/in.^3 = 2.77 \times 10^4 \, kg/m^3$$
$$1 \, psi = 6.89 \, kPa$$
$$1 \, ksi = 6.89 \, MPa$$
$$1 \, oz = 28.3 \, g$$
$$1 \, gal = 3.79 \, L$$
$$1 \, lb = 0.45 \, kg$$
$$1 \, fl. \, oz. = 0.003 \, L$$
$$1 \, in. = 25.4 \, mm$$
$$1 \, in./min = 0.42 \, mm/s$$

# SUPPLEMENTARY READING LIST

Ayner, S.H., *Introduction to Physical Metallurgy,* 2nd ed., New York: McGraw-Hill Book Co., 1974: 498-507.

AWS C1.1-66, *Recommended Practices for Resistance Welding,* Miami, FL: American Welding Society, 1966: 45-76.

Brazing Manual, 3rd ed., Miami, FL: American Welding Society, 1976: 161-68.

Kenyon, D.M., Arc behavior and its effect on the tungsten arc welding of magnesium alloys, *Jnl of Inst of Metals,* 93: 85-89; 1964-65.

Koeplinger, R.D. and Lockwood, L.F., Gas metal arc spot welding of magnesium, *Welding Journal,* 43 (3): 195-201; 1964 Mar.

Lockwood, L.F., Pulse arc welding of magnesium, *Welding Journal,* 49 (6): 464-75; 1970 June.

Lockwood, L.F., Gas shielded stud welding of magnesium, *Welding Journal,* 46 (4): 168s-174s; 1967 Apr.

Lockwood, L.F., Repair welding of thin wall magnesium sand castings, *Trans. Amer Foundrymens Soc.,* 75: 530-40; 1967.

Lockwood, L.F., Automatic gas tungsten arc welding of magnesium, *Welding Journal,* 44 (5): 213s-220s; 1965 May.

Lockwood, L.F., Now you can dip braze magnesium, *Product Engineering,* 36: 113-16; 1965, Mar. 15.

Lockwood, L.F., Gas metal arc welding of AZ31B magnesium sheet, *Welding Journal,* 42 (10): 807-18; 1963 Oct.

*Metals Handbook,* Vol. 2, 9th ed., Metals Park, OH: American Society for Metals, 1979: 525-609.

Portz, A.G., and Rothgery, G.R., Flash welded magnesium rings meet space age needs, *Welding Design and Fabrication,* 1963 Jan: 44-45.

Sibley, C.R., *Arc Welding of Magnesium and Magnesium Alloys,* New York: Welding Research Council, 1962 Nov; Bulletin 83.

*Soldering Manual,* 2nd ed., Miami, FL: American Welding Society, 1978: 97-99.

# 10

# Titanium, Zirconium, Hafnium, Tantalum, and Columbium

## Chapter Committee

J. M. GERKIN, *Chairman*
*TRW, Incorporated*
D. W. BECKER
*Air Force Materials Laboratory*
S. GOLDSTEIN
*Kawecki Berylco Industries,*
*Incorporated*
M. H. HOROWITZ
*Walbar Metals, Incorporated*
H. NAGLER
*Rohr Industries*

J. R. ROPER
*Rockwell International*
H. ROSENBERG
*TIMET*
R. T. WEBSTER
*Teledyne Wah Chang*
K. C. WU
*Northrop Corporation*
H. C. ZIEGENFUSS
*Westinghouse Electric Corporation*

## Welding Handbook Committee Member

J.R. CONDRA
*E. I. DuPont de Nemours and Company*

# 10

# Titanium, Zirconium, Hafnium, Tantalum, and Columbium

## INTRODUCTION

The metals titanium, zirconium, hafnium, tantalum, and columbium have similar weldability because of two common characteristics. First, they readily react with oxygen and form very stable oxides. Second, they have high solubilities for oxygen, nitrogen, and hydrogen at elevated temperatures. These elements dissolve interstitially in the metals. Small amounts of dissolved oxygen and nitrogen significantly increase the hardness of the metal, while dissolved hydrogen reduces its toughness and increases its notch sensitivity. Therefore, these metals must be welded or brazed in a shield of high-purity inert gas or in a good vacuum to avoid embrittlement.

These metals also react with carbon at elevated temperatures to form carbides. Carbon is sometimes added intentionally as an alloying element to increase strength and hardness. However, excess carbides in these metals cause brittleness. The metals should be free of oil, grease, or other hydrocarbons before welding or brazing, and graphite should not be used for fixturing or for

dams to control the flow of molten filler metals.

The physical properties of these metals that are important in welding are listed in Table 10.1. Tantalum and columbium have melting points in excess of 4000°F, which places them in the family of refractory metals. Titanium has low density, and some titanium alloys are competitive with aluminum alloys or steels for aerospace application. Zirconium and hafnium are used primarily for nuclear applications because of their thermal neutron-absorption characteristics, corrosion resistance, and strength properties. Tantalum is known for its excellent corrosion resistance and finds wide use in the chemical industry. The electronics industry uses tantalum in electron tubes, capacitors, and special applications where its gas absorption and high temperature characteristics are important. Columbium is similar to tantalum in many respects, but weighs only half as much. It is resistant to several liquid metals at high temperatures.

## TITANIUM

### UNALLOYED TITANIUM

Pure titanium is a silver-colored metal that has a close-packed hexagonal crystal structure, known as alpha ($\alpha$) phase, up to 1625°F. Above

this temperature, the metal goes through an allotropic change to the body-centered cubic structure, which is called beta ($\beta$) phase. The transformation temperature, sometimes called the *beta transus*, is affected by the amount of im-

434

## Table 10.1
### Properties of titanium, zirconium, tantalum, columbium, and hafnium

| Property | Units | Titanium | Zirconium | Tantalum | Columbium | Hafnium |
|---|---|---|---|---|---|---|
| Density (20°C) | lb/in³ <br> Mg/m³ | 0.16 <br> 4.5 | 0.23 <br> 6.5 | 0.60 <br> 16.6 | 0.31 <br> 8.57 | 0.47 <br> 13.1 |
| Melting point | °F <br> °C | 3034 <br> 1668 | 3350 <br> 1852 | 5425 <br> 2996 | 4474 <br> 2468 | 4046 <br> 2230 |
| Coef. of thermal expansion (20°) | in.$^{-6}$ (in. °F) <br> m$^{-6}$/(m.°C) | 4.7 <br> 8.4 | 3.2 <br> 5.9 | 3.6 <br> 6.5 | 3.9 <br> 7.1 | 3.3 <br> 5.9 |
| Thermal conductivity | Btu/(ft²·h·in.·°F) <br> W/(m.·K) | 111 (68°F) <br> 16 (20°C) | 146 (77°F) <br> 21 (25°C) | 374 (68°F) <br> 54 (20°C) | 360 (32°F) <br> 52 (0°C) | 153 (122°F) <br> 22 (50°C) |
| Electrical resistivity (20°C) | (Ω·cm)$^{-6}$ | 42 | 45 | 13 | 16 | 35 |
| Specific heat | Btu/(lb.·°F) <br> J/(kg.·K) | 0.124 (68°F) <br> 522 (20°C) | 0.069 (68°F) <br> 290 (20°C) | 0.034 (77°F) <br> 142 (25°C) | 0.065 (32°F) <br> 272 (20°C) | 0.035 (68°F) <br> 147 (20°C) |
| Modulus of elasticity | 10⁶psi <br> GPa | 16.8 <br> 117 | 14.4 <br> 101 | 27 <br> 189 | 15 <br> 105 | 19.8 <br> 138 |

purities in the titanium or by deliberate alloying to improve properties.

Titanium has a strong chemical affinity for oxygen, and a stable, tenacious oxide layer forms rapidly on a clean surface, even at room temperature. This behavior leads to a natural passitivity that provides a high degree of corrosion resistance to salt or oxidizing acid solutions, and acceptable resistance to mineral acids.

The strong affinity of titanium for oxygen increases with temperature, and the surface oxide layer increases in thickness at elevated temperatures. At temperatures above about 1200°F, its oxidation resistance decreases rapidly, and the metal must be shielded from air to avoid contamination and embrittlement by oxygen and nitrogen.

Pure titanium is very ductile and has a relatively low tensile strength. Oxygen and nitrogen in solution markedly strengthen titanium, as do iron and carbon to a lesser degree. Hydrogen has a definite embrittling effect. One or more of these elements can be unintentionally added by contamination during processing or joining.

In the annealed condition, commercially pure titanium has tensile strengths ranging from 35 to over 80 ksi, yield strengths (0.2% offset) from 25 to over 70 ksi, and elongations in 2 inches from 15 percent to 25 percent, depending upon the purity.[1] Intentional additions of alloying elements increase the strength properties and simultaneously decrease the ductility. Titanium and some of its alloys have good strength, ductility, and toughness down to about −400°F. The low thermal expansion and thermal conductivity characteristics contribute to the good weldability of titanium.

## ALLOYING AND PHASE STABILIZATION

Alloying elements are added intentionally to titanium to improve properties. Table 10.2 lists these elements according to the phase that they tend to stabilize and the type of alloying. Oxygen, nitrogen, hydrogen, carbon, and silicon form compounds with titanium when present in amounts exceeding their solid solubility limits.

### Substitutional Alloying

An alloying element stabilizes the phase,

---

1. See ASTM Specifications B265 and B348 for additional information on the property requirements for titanium mill products.

**Table 10.2**
**Classification of alloying elements in titanium alloys**

| Alpha-phase stabilizers | Beta-phase stabilizers | Neutral |
|---|---|---|
| Substitutional alloying | | |
| Aluminum | Chromium | Tin |
| | Columbium | Zirconium |
| | Copper | |
| | Iron | |
| | Manganese | |
| | Molybdenum | |
| | Nickel | |
| | Palladium | |
| | Silicon | |
| | Tantalum | |
| | Tungsten | |
| | Vanadium | |
| Interstitial alloying | | |
| Oxygen | Hydrogen | |
| Nitrogen | | |
| Carbon | | |

alpha or beta, in which it has the greatest solid solubility. For example, aluminum additions stabilize the alpha phase and raise the allotropic or phase transformation temperature. On the other hand, chromium, molybdenum, and vanadium, as well as the other elements given in Table 10.3, stabilize the beta phase and decrease the allotropic transformation temperature. With large amounts of beta stabilizers, the beta phase is stable down to room temperature and below. In addition, alloying additions influence the rate of transformation from one phase to the other.

When both alpha- and beta-stabilizing elements are present, some beta phase is present in an alpha matrix at room temperature. The mechanical properties of the alloy depend partially on the phase ratio in the microstructure.

The alpha-to-beta phase ratio is important in several ways. The presence of beta phase has the following effects:

(1) Enhances grain refinement and hence strength

(2) Improves hot workability

(3) Provides response to heat treatment

(4) Enhances toughness except at very low temperatures

(5) Inhibits weldability and creep strength with some alloys

The type and amount of phase present at a

**Table 10.3**
**Classification of titanium alloys**

| Alloy | UNS No. |
|---|---|
| **Commercially pure** | |
| ASTM Grade 1 | R50250 |
| ASTM Grade 2 | R50400 |
| ASTM Grade 3 | R50550 |
| ASTM Grade 4 | R50700 |
| **Alpha alloys** | |
| Ti-0.2Pd (ASTM Grades 7 and 11) | R52400, R52250 |
| Ti-0.8Ni-0.3Mo (ASTM Grade 12) | – |
| Ti-5Al-2.5Sn (ASTM Grade 6) | R54520, R54521 |
| **Near-alpha alloys** | |
| Ti-8Al-1Mo-1V (AMS No. 4915) | R54810 |
| Ti-2Al-11Sn-5Zr-1Mo (AMS No. 4977) | R54790, R54560 |
| Ti-5Al-5Sn-2Mo-2Zr | R56210 |
| Ti-6Al-2Cb-1Ta-0.8Mo | R54620 |
| Ti-6Al-4Zr-2Mo-2Sn (AMS No. 4975) | |
| **Alpha-beta alloys** | |
| Ti-3Al-2.5V (AMS No. 4943, ASTM Grade 9) | R56320 |
| Ti-6Al-4V (ASTM Grade 5) | R56400, R56401 |
| Ti-7Al-4Mo (AMS 4970) | |
| Ti-6Al-6V-2Sn (AMS No. 4918) | R56740 |
| Ti-6Al-6Mo-4Zr-2Sn (AMS 4981) | R56620 |
| Ti-8Mn (AMS 4908) | R56260 |
| Ti-10V-2Fe-3Al | R56080 – |
| **Beta alloys** | |
| Ti-13V-11Cr-3Al (AMS 4917) | R58010 |
| Ti-11.5Mo-6Zr-4.5Sn (ASTM Grade 10) | R58030 |
| Ti-8Mo-8V-3Al-2Fe | R58820 |

given temperature are dependent on composition and thermal history. Titanium alloys are labeled as to type by the phase or phases present, i.e., alpha alloys, beta alloys, or alpha-beta alloys.

There are two common alloying elements that cannot be classified as either alpha or beta stabilizing. These are tin and zirconium that have extensive solid solubility in both the alpha and the beta phases. These elements do not strongly promote phase stability, but are used as strengthening elements in titanium alloys.

## Interstitial Elements

Carbon, hydrogen, nitrogen, and oxygen form interstitial solid solutions with titanium. Carbon, nitrogen, and oxygen are more soluble in the alpha phase and are thus classified as alpha stabilizers. Hydrogen is more soluble in the beta phase, and behaves as a beta-stabilizing element. All of these elements are found as impurities in titanium and its alloys. The difference in mechanical properties among the various grades of commercial unalloyed titanium is a result of differences in interstitial content, principally oxygen. These interstitial elements are the impurities that can contaminate titanium during welding or brazing. Insufficient protection of the hot titanium from the atmosphere usually leads to pick-up of oxygen because of preferential affinity of the metal for this element. Hydrogen and oxygen can be picked up if moisture is present in the protecting atmosphere. Residual oils, cleaning agents, and other contaminants on the pieces to be welded often lead to hydrogen or carbon contamination.

Nitrogen is the most effective strengthener for unalloyed titanium. Oxygen is next in effectiveness, followed by carbon. Carbon is only effective up to the limit of solubility in titanium.

In two-phase titanium alloys, the action of interstitials becomes more complicated. The interstitials partition to the alpha phase, raise the beta transus temperature, and promote instability of the beta phase. Because interstitials partition to the alpha phase, this phase is strengthened by their presence but the beta phase is only slightly affected. Thus, the strengthening effect of interstitials on alpha-beta alloys is greatly dependent on the relative amount and distribution of the alpha phase.

Hydrogen is only slightly soluble in alpha titanium at room temperature, but becomes soluble up to 8 percent above 570°F. When the titanium is cooled slowly, hydrogen in solution precipitates in plates throughout the alpha grains and at the grain boundaries and results in poor toughness. When the titanium is quenched, hydrogen precipitates in very fine particles throughout the alpha grains and has little effect on toughness. However, subsequent aging at room temperature results in agglomeration of the hydrides and decreased toughness.

Since beta titanium dissolves substantially more hydrogen than alpha titanium, alpha-beta alloys can dissolve considerable hydrogen without the formation of hydrides. However, alpha-beta alloys containing high percentages of hydrogen are subject to embrittlement at low strain rates.

## Solid-Solution Strengthening

All alloy additions to titanium contribute to

solid-solution strengthening. This mechanism is so important that the widely applied Ti-6Al-4V alloy is generally used in the annealed condition.

Both substitutional and interstitial alloying elements play significant roles. In general, substitutional alloying is preferred, and two or more alloying elements are generally used together. Multiple alloying generally exerts a greater overall effect than can be achieved with a binary alloy. Substitutional alloying usually leads to a better combination of mechanical properties including tensile strength, creep resistance, ductility, and toughness. The large number of suitable substitutional elements permits a wide range of alloying to obtain desirable properties. However, interstitial alloying is a very economical strengthening method.

The mechanical properties of alpha and near-alpha alloys are dependent entirely upon solid-solution strengthening. On the other hand, the strengths of the commercially pure grades of titanium vary with the impurity content. Oxygen, nitrogen, and iron are the principal impurities in these grades.

## TITANIUM ALLOYS

### General Alloy Classifications

Titanium and its alloys are generally classified into groups on the basis of their annealed microstructure. The groups and typical alloys in each group are shown in Table 10.3. There are other alloys, not listed, that are being evaluated for specific applications, or are seldom used. Some alloys are produced with extra low interstitial impurities (ELI) for applications where good ductility and toughness are needed at cryogenic temperatures. Titanium alloys can also be classified on the basis of their corrosion or heat resistant properties.

### Commercially Pure Titanium

Various grades of commercially pure titanium are classified on the basis of minimum mechanical properties and maximum interstitial impurities. In general, variations in strength are produced by differences in impurity content, primarily the interstitial elements (oxygen, nitrogen, and carbon) and iron. As the content of these elements increases, strength also increases. While strengthening by cold work is possible, it is seldom used.

### Alpha and Near-Alpha Alloys

The second classification is the alpha alloys.

Alpha alloys are not normally heat-treated to increase strength. They always contain a high percentage of alpha phase in their microstructure, and in some cases, they are totally alpha phase. They are commonly used where moderate elevated temperature strength or creep resistance is required. Near-alpha alloys are designed for outstanding creep strength and elevated temperature stability by the addition of small amounts of beta-stabilizing elements.

The Ti-0.2Pd and Ti-0.8Ni-0.3Mo alloys have better corrosion resistance to certain media than commercially pure titanium. The latter alloy is stronger than commercially pure titanium.

### Alpha-Beta Alloys

Alpha-beta alloys contain mixtures of alpha and beta phases in their microstructures, and can be strengthened by solution treating and aging heat treatments. Some alpha-beta alloys are also used in the annealed condition. They have excellent fracture toughness when annealed, and outstanding strength-to-density ratios in the heat-treated condition. The Ti-6Al-4V alloy is the most widely used alpha-beta titanium alloy.

### Beta Alloys

Commercial beta alloys contain a high percentage of beta phase stabilizing elements, but are not truly single phase. Their transformation to alpha is very sluggish, and during normal processing, the microstructure is nearly all beta phase at room temperature. They can be aged by a long-time, low-temperature heat treatment to greatly increase their strength. The hardening mechanism is the precipitation of alpha or a compound. Beta alloys in general are characterized by excellent formability in the single-phase condition, and are often used for this reason. When heat-treated, they exhibit high strength-to-weight ratios, but have relatively low ductility and fracture toughness. Beta alloys show exceptional work-hardening characteristics, and are used in fasteners and springs for this reason.

## THERMAL EFFECTS

Titanium and titanium alloys are heat-treated to relieve stresses or to improve properties, such as hardness, strength, and toughness. Annealing temperatures range from 1200°F for commercially pure titanium to 1650°F for alpha titanium alloys that contain moderate amounts of alloying elements. Heat treatment can result in increased grain size, and there is no heat treatment that can produce grain refinement in titanium. Grain refinement must be accomplished by cold work followed by a recrystallization anneal.

Transformation of the high-temperature beta phase of unalloyed titanium to alpha on cooling from temperatures above 1620°F cannot be suppressed by very high cooling rates, although the temperature at which the transformation takes place can be lowered slightly. Consequently, unalloyed titanium is not heat treatable in the general sense of the word. There are, however, differences in the microscopic form and arrangement of the alpha phase produced by different cooling rates from the high-temperature beta field.

Rapid cooling produces a microstructure referred to as acicular alpha. Slow cooling produces a structure known as equiaxed alpha, similar in appearance to very low-carbon annealed steel. Slight changes in properties can occur on reheating titanium to a temperature below the transformation temperature as a result of changes in the as-cooled microstructure.

Alpha-stabilized systems are generally nonheat treatable because only minor phase changes usually take place by subsequent heating in the low-temperature phase region.

The lowered temperature of transformation caused by beta-stabilizing elements and their slow diffusion rates retard the transformation of the beta phase on cooling, and equilibrium phases are not obtained. With low alloy content and rapid cooling, an alpha plus martensite microstructure can be developed.[2] As the alloy content is progressively increased to some limiting value, the transformed alloy exhibits an as-quenched microstructure containing more martensite until at some limiting value, it is essentially all martensite. Retained beta will appear with further increases in alloy content. Hardening can take place by heat treatment in those alloys in which the beta-to-alpha transformation has been sufficiently suppressed.

The increased strength of alpha-beta titanium alloys as a result of heat treatment is due to transformation, at a lower temperature, of an unstable high-temperature phase to a stronger and harder phase. This new phase may be a precipitate from the beta matrix, a eutectoidal

decomposition phase or phases, or transformation to a metastable transition phase. Generally, the high strength obtained by heat treatment is accompanied by a loss of ductility and toughness. The hardening heat treatment can consist of quenching the alloy from the beta phase, and then reheating it to approximately 300°F above the alpha transus temperature to cause precipitation. Heating to higher temperatures in the alpha-plus-beta field results in less hardening, but produces a structure of greater stability for use at elevated temperatures. In most cases, these alloys are used in the stabilized condition to produce a favorable combination of strengh and ductility, and to prevent further transformation upon the application of stress.

Alloys that can be hardened by heat treatment normally harden during a welding cycle. For this reason, such alloys might not be weldable, or they might require preheat and postweld heat treatments to restore desired properties.

Alloys with sufficient beta-stabilizing elements to retain the beta phase below room temperature, and also prevent compound or intermediate phase formation during short-time reheating, undergo no transformation during welding. Consequently, the thermal cycles associated with welding do not cause hardening, but only some grain coarsening.

## HEAT TREATMENT

### Annealing

Titanium and its alloys are annealed to improve ductility, dimensional or thermal stability, fracture toughness, and creep resistance.[3] Improvement in one or more properties is usually obtained at the expense of some other property. Therefore, the titanium producer should be consulted concerning a recommended heat treatment to provide desired properties.

The part should be held at the annealing temperature until it is uniformly heated and desired transformation reactions are completed. It should then be air- or furnace-cooled to room temperature. Suggested annealing treatments for several titanium alloys are given in Table 10.4.

---

2. In titanium, martensite is an alpha product produced by rapid cooling from the beta phase. It is supersaturated with beta stabilizer.

3. For additional information on heat treating titanium alloys, refer to the *Metals Handbook*, Vol. 4, *Heat Treating*, 9th Ed., American Society for Metals, 1981: 763-74.

### Table 10.4
### Annealing and stress-relieving heat treatments for titanium alloys

| Alloy | Annealing | | | Stress-relieving[a] | |
|---|---|---|---|---|---|
| | Temperature, °F | Time, h | Cooling medium | Temperature, °F | Time, h |
| Commercially pure grades | 1200-1400 | 0.1-2 | Air | 900-1100 | 0.5-1 |
| Alpha alloys | | | | | |
| Ti-5Al-2.5Sn | 1325-1550 | 0.2-4 | Air | 1000-1200 | 1-4 |
| Ti-8Al-1Mo-1V | 1450[b] 1450 | 1-8 0.25 | Furnace Air | 1075-1125 | 2 |
| Ti-5Al-5Sn-2Mo-2Zr | 1200-1450 | 0.5-2 | Air | 1100 | 2-8 |
| Ti-6Al-2Cb-1Ta-0.8Mo | 1450-1650 | 1-4 | Air | 1100-1200 | 0.25-2 |
| Ti-6Al-4Zr-2Mo-2Sn | 1650[b] 1450 | 0.5-1 0.25 | Air Air | 900-1200 | 1-4 |
| Alpha-beta alloys | | | | | |
| Ti-3Al-2.5V | 1200-1400 | 0.5-2 | Air | 1000-1100 | 0.5-2 |
| Ti-6Al-4V | 1300-1450 | 1-4 | Air or furnace | 1000-1100 | 2-4 |
| Ti-7Al-4Mo | 1300-1450 | 1-8 | Air | 900-1300 | 0.5-8 |
| Ti-6Al-6V-2Sn | 1300-1500 | 1-4 | Air or furnace | 900-1200 | 1-4 |
| Ti-6Al-6Mo-4Zr-2Sn | 1300-1350 | 2 | Air | 1100-1300 | 0.25-4 |
| Ti-8Mn | 1200-1400 | 0.5-1 | Furnace to 1000°F, then air | 900-1100 | 0.25-2 |
| Ti-10V-2Fe-3Al | 1450-1500[c] | 0.1-0.25 | Air | 1250-1300 | 0.5-2 |
| Beta Alloys | | | | | |
| Ti-13V-11Cr-3Al | 1300-1450 | 0.2-1 | Air or water | 900-1000 | 0.5-60 |
| Ti-11.5Mo-6Zr-4.5Sn | 1300-1400 | 0.2-1 | Air or water | 1325-1350 | 0.1-0.25 |
| Ti-8Mo-8V-3Al-2Fe | 1425-1450 | 0.1-0.25 | Air or water | 950-1100 | 1-4 |

a. Air cool
b. Duplex heat treatment
c. Not normally used in the annealed condition

## Solutioning and Aging

Alpha-beta and beta titanium alloys can be strengthened by heat treatment. A basic heat treatment consists of solutioning at a temperature high in the two-phase, alpha-beta field followed by quenching in air, water, or other suitable media. The beta phase present at the solutioning temperature may be completely retained during cooling or a portion may transform to alpha. The specific response depends upon the alloy composition, the solutioning temperature, the cooling rate, and the section size.

The solution-treated alloy is then aged at a suitable temperature, in the 900° to 1200°F range. During aging, fine alpha particles precipitate in the retained or transformed beta phase. This duplex structure is harder and stronger than an annealed structure. Solution treating and aging can increase the strength of alpha-beta alloys as much as 30 to 50 percent over that of annealed material. Typical solutioning and aging heat treatments for several titanium alloys are given in Table 10.5.

As with annealing, specific mechanical properties of titanium alloys are influenced by solutioning and aging heat treatments. The recommendations of the producers should be followed to obtain desired properties.

### Stress Relieving

Residual stresses in weldments are relieved during annealing or solution heat treatment. When these heat treatments are not required to produce desired properties in the weldment, a stress-relieving heat treatment might be applicable. Such a treatment can be beneficial in maintaining dimensions, reducing cracking tendencies, or avoiding stress-corrosion cracking with certain alloys. However, it can be detrimental in some cases.

Thermal stress-relieving might alter the mechanical properties of the weld zone by an aging reaction with heat-treatable alloys. This response might reduce the beneficial effects expected from stress-relieving because of reduced weld ductility as a result of aging. Suggested stress relieving temperatures for several titanium alloys are given in Table 10.4.

## MECHANICAL PROPERTIES

Typical tensile and impact strengths at room temperature of several titanium alloys are given in Table 10.6. Actual properties of the alloys vary with processing, product form, heat treatment, and service temperature.

## CLEANING

Prior to welding, brazing, or heat treating, titanium components must be cleaned of surface contaminants and dried. Oil, fingerprints, grease, paint, and other foreign matter should be removed using a suitable solvent cleaning method. Ordinary tap water should not be used to rinse titanium parts. Chlorides and other cleaning residues on titanium can lead to stress-corrosion cracking when the components are heated above about 550°F during welding, brazing, and heat-treating. Hydrocarbon residues can result in contamination and embrittlement of the titanium.

If the parts to be welded or brazed have a light oxide coating in the vicinity of the joint, it

**Table 10.5**
**Solutioning and aging heat treatments for titanium alloys**

| Alloy | Solutioning[a] | | | Aging[a] | |
| | Temperature, °F | Time, h | Cooling medium | Temperature, °F | Time, h |
|---|---|---|---|---|---|
| Alpha-beta alloys | | | | | |
| Ti-3Al-2.5V | 1600-1700 | 0.25-0.3 | Water | 900-950 | 2-8 |
| Ti-6Al-4V | 1650-1775 | 0.2-1 | Water | 900-950 | -12 |
| Ti-7Al-4Mo | 1675-1775 | 0.2-2 | Water | 950-1200 | 4-24 |
| Ti-6Al-6V-2Sn | 1550-1650 | 0.2-1 | Water | 875-1150 | 2-8 |
| Ti-6Al-6Mo-4Zr-2Sn | 1550-1700 | 0.2-1 | Air | 1050-1150 | 2-8 |
| Ti-10V-2Fe-3Al | 1450-1500 | 1 | Water | 925-975 | 8 |
| Beta alloys | | | | | |
| Ti-13V-11Cr-3Al | 1400-1500 | 0.2-1 | Air or water | 825-1000 | 2-60 |
| Ti-11.5Mo-6Zr-4.5Sn | 1350-1450 | 0.2-1 | Air or water | 900-1100 | 8 |
| Ti-8Mo-8V-3Al-2Fe | 1450-1475 | 0.1-1 | Air or water | 900-950 | 8 |

a. The time at temperature depends upon the temperature and section thickness. Longer times are used for lower temperatures and thicker sections.

## Table 10.6
## Typical room-temperature tensile and impact properties of titanium alloys

| Alloy | Form[a] | Con- dition[b] | Tensile strength, ksi | Yield strength, ksi | Elonga- tion, percent | Charpy V-notch impact energy, ft·lbs. |
|---|---|---|---|---|---|---|
| Commercially pure | | | | | | |
| ASTM Gr 1 | S | A | 38 | 27 | 30 | 100 |
| ASTM Gr 2 | S | A | 60 | 45 | 28 | 100 |
| ASTM Gr 3 | S | A | 75 | 60 | 25 | 55 |
| ASTM Gr 4 | S | A | 90 | 75 | 20 | 20 |
| Alpha alloys | | | | | | |
| Ti-0.2Pd | S | A | 62 | 46 | 27 | 25-40 |
| Ti-08Ni-0.3Mo | S | A | 74 | 60 | 33 | 11 |
| Ti-5Al-2.5Sn | S | A | 125 | 120 | 15 | 17 |
| Near-alpha alloys | | | | | | |
| Ti-8Al-1Mo-1V | S | A | 134 | 124 | 14 | 20-28 |
| Ti-2Al-11Sn-5Zr-1Mo | S | A | 148 | 133 | 12 | |
| Ti-5Al-5Sn-2Mo-2Zr | B | A | 130[c] | 120[c] | 10 | |
| Ti-6Al-2Cb-1Ta-0.8Mo | P | R | 110-115 | 95-100 | 10 | 27 |
| Ti-6Al-4Zr-2Mo-2Sn | B | A | 143 | 136 | 13 | |
| Alpha-beta alloys | | | | | | |
| Ti-3Al-2.5V | S | A | 104 | 85 | 22 | |
| Ti-6Al-4V | All | A | 135 | 130 | 15 | 10-20 |
| | B | STA | 162 | 150 | 12 | |
| Ti-7Al-4Mo | B | A | 148 | 138 | 10 | |
| | B | STA | 173 | 165 | 8 | |
| Ti-6Al-6V-2Sn | All | A | 165 | 150 | 15 | 14-18 |
| | B | STA | 185 | 170 | 11 | 10 |
| Ti-6Al-6Mo-4Zr-2Sn | B | A | 165 | 150 | 14 | |
| | B | STA | 185 | 170 | 10 | |
| Ti-8Mn | S | A | 140 | 125 | 15 | |
| Ti-10V-2Fe-3Al | F | STA | 185 | 175 | 7-11 | 22 |
| Beta Alloys | | | | | | |
| Ti-13V-11Cr-3Al | B | A | 132 | 126 | 10 | 8 |
| | B | STA | 175 | 165 | 4 | |
| Ti-11.5Mo-6Zr-4.5Sn | B | A | 122 | 115 | 15 | |
| | B | STA | 221 | 207 | 6 | |
| Ti-8Mo-8V-3Al-2Fe | B | A | 128 | 123 | 12 | |
| | B | STA | 180 | 175 | 8 | |

a. S—Sheet, B—bar, F—forging, P—plate
b. A—annealed, R—as-rolled, STA—solution-treated and aged
c. Minimum values

can be removed by pickling in an aqueous solution of 2 to 4 percent hydrofluoric acid and 30 to 40 percent nitric acid, followed by appropriate water rinsing and drying.[4] Hydrogen absorption

by titanium alloys is generally not a problem at bath temperatures up to 140°F. The part should be handled, at this point, with lint-free gloves during assembly in the welding or brazing fixture.

Scale formed at temperatures above 1100°F is difficult to remove chemically. Mechanical methods, such as vapor blasting and grit blasting,

4. Special safe practices are required when handling any acid, particularly hydrofluoric acid. Suitable protective clothing and equipment must be used.

are used for scale removal. Mechanical operations are usually followed by a pickling operation to insure complete removal of scale and any contaminated metal on the surfaces.

To control porosity in welding operations, the edges to be joined are often given special treatments. These treatments include draw filing, wire brushing, or abrading the joint edges and adjacent surfaces prior to fitup and final cleaning. Sheared joint edges frequently require such special treatments to remove entrapped dirt, metal slivers, and small cracks because these edge discontinuities promote weld porosity.

Where extended storage time is required before welding, it may be necessary to store the parts in sealed bags containing a desiccant or in a controlled humidity storage room. When this is not possible, thorough degreasing and light pickling of the parts just prior to welding or brazing is recommended. Mechanical abrasion of the faying surfaces followed by washing with a suitable solvent, is sometimes used in lieu of the pickling treatment. The fixturing itself should be thoroughly cleaned and degreased prior to loading.

## WELDING FILLER METALS

In general, the titanium filler metal and the base metal should have the same nominal composition. Normally, the filler metal is used in the form of bare rod or wire, depending upon the welding process and the type of operation (manual, semiautomatic, or automatic).[5] The compositions of standard titanium welding electrodes and rods are given in Table 10.7.

When welding commercially pure titanium, an unalloyed filler metal can tolerate some contamination from the welding atmosphere without significant loss in ductility. The ERTi-1,-2,-3 and -4 filler metal classifications are designed for this purpose, as are those in AMS Specification 4951.

Unalloyed filler metal may be used to weld titanium alloys when weld metal ductiliy is more important than joint strength. Joint efficiencies of less than 100 percent can be expected.

For cryogenic application where base metals with extra-low interstitial impurities are specified, the filler metals should also be low in those impurities. To be effective, the welding must be done with equipment and procedures that prevent contamination of the weld metal with carbon, oxygen, nitrogen, or hydrogen.

The quality and cleanliness of the filler metal are important considerations in the welding of titanium (and the other metals also). The filler metal can be a source of serious contamination of the weld metal from inclusions, dirt, oil, and drawing compounds on the surface. The relatively large surface area-to-volume ratios of commonly used wire or rod sizes make cleanliness very important. Physical defects in the wire, such as cracks, seams, or laps, can trap surface contaminants, and make their removal difficult or impossible. The filler rod or wire should be carefully inspected for mechanical defects, throughly cleaned, and then suitably handled, packaged, and stored to prevent contamination.

## WELDABILITY

### Metallurgical Effects

During fusion welding, solid grains of beta titanium exist immediately next to the molten metal. In the cooler regions farther from the weld metal, the solid titanium is all beta phase, mixtures of alpha and beta phases, or all single-phase alpha, depending on the alloy composition. The progressive solidification of the weld metal is accompanied by growth of solid beta grains in the direction of solidification. As cooling progresses, phase transformations occur in the weld metal and heat-affected zones, and are related to the temperature changes and cooling rates experienced in those zones. The greatest metallurgical changes occur in the weld metal and the adjacent beta phase in the heat-affected zone.

In commercially pure titanium, alpha alloys, and alpha-beta alloys, the large beta grains, formed at high temperatures, transform to an alpha or an alpha-beta structure when cooled below the beta transus. The microstructure within the prior beta grains will be in a pattern termed acicular and serrated for commercially pure and alpha alloys, and acicular and Widmanstätten for alpha-beta alloys, as shown in Fig. 10.1 and 10.2. The beta grains in beta and near-beta weld metal either remain stable and unchanged throughout the complete cooling cycle, or small amounts of a fine alpha form within the beta grains.

---

5. Refer to AWS A5.16-70, *Specification for Titanium and Titanium Alloy Bare Welding Rods and Electrodes,* and the appropriate Aerospace Material Specifications, published by the Society of Automotive Engineers, for additional information.

**Table 10.7**
**Titanium and titanium alloy bare welding electrodes and rods**

| AWS Classification | Chemical composition, percent [a] | | | | | | | | | | | | | |
|---|---|---|---|---|---|---|---|---|---|---|---|---|---|---|
| | Carbon | Oxygen | Hydrogen | Nitrogen | Aluminum | Vanadium | Tin | Chromium | Iron | Molybdenum | Columbium | Tantalum | Palladium | Titanium |
| ERTi-1[b] | 0.03 | 0.10 | 0.005 | 0.012 | – | – | – | – | 0.10 | – | – | – | – | Remainder |
| ERTi-2 | 0.05 | 0.10 | 0.008 | 0.020 | – | – | – | – | 0.20 | – | – | – | – | Remainder |
| ERTi-3 | 0.05 | 0.10-0.15 | 0.008 | 0.020 | – | – | – | – | 0.20 | – | – | – | – | Remainder |
| ERTi-4 | 0.05 | 0.15-0.25 | 0.008 | 0.020 | – | – | – | – | 0.30 | – | – | – | – | Remainder |
| ERTi-0.2Pd | 0.05 | 0.15 | 0.008 | 0.020 | – | – | – | – | 0.25 | – | – | – | 0.15-0.25 | Remainder |
| ERTi-3Al-2.5V | 0.05 | 0.12 | 0.008 | 0.020 | 2.5-3.5 | 2.0-3.0 | – | – | 0.25 | – | – | – | – | Remainder |
| ERTi-3Al-2.5V-1[b] | 0.04 | 0.10 | 0.005 | 0.012 | 2.5-3.5 | 2.0-3.0 | – | – | 0.25 | – | – | – | – | Remainder |
| ERTi-5Al-2.5Sn | 0.05 | 0.12 | 0.008 | 0.030 | 4.7-5.6 | – | 2.0-3.0 | – | 0.40 | – | – | – | – | Remainder |
| ERTi-5Al-2.5Sn-1[b] | 0.04 | 0.10 | 0.005 | 0.012 | 4.7-5.6 | – | 2.0-3.0 | – | 0.25 | – | – | – | – | Remainder |
| ERTi-6Al-2Cb-1Ta-1Mo | 0.04 | 0.10 | 0.005 | 0.012 | 5.5-6.5 | – | – | – | 0.15 | 0.5-1.5 | 1.5-2.5 | 0.5-1.5 | – | Remainder |
| ERTi-6Al-4V | 0.05 | 0.15 | 0.008 | 0.020 | 5.5-6.75 | 3.5-4.5 | – | – | 0.25 | – | – | – | – | Remainder |
| ERTi-6Al-4V-1[b] | 0.04 | 0.10 | 0.005 | 0.012 | 5.5-6.75 | 3.5-4.5 | – | – | 0.15 | – | – | – | – | Remainder |
| ERTi-8Al-1Mo-1V | 0.05 | 0.12 | 0.008 | 0.03 | 7.35-8.35 | .75-1.25 | – | – | 0.25 | .75-1.25 | – | – | – | Remainder |
| ERTi-13V-11Cr-3Al | 0.05 | 0.12 | 0.008 | 0.03 | 2.5-3.5 | 12.5-14.5 | – | 10.0-12.0 | 0.25 | – | – | – | – | Remainder |

a. Single values are maximum.
b. Extra-low interstitials for welding similar base metals.

**Fig. 10.1—Weld metal of ERTi-2 titanium composition showing serrated alpha phase with a small amount of fine, dispersed beta phase (× 250)**

**Fig. 10.2—Weld metal of ERTi-6Al-4V composition showing a acicular alpha and beta phases in Widmanstätten arrangement in prior beta grains (× 250)**

That portion of any titanium alloy weldment that is heated to less than about 1000°F during welding, generally is not affected metallurgically by welding. That portion of commercially-pure titanium or near-alpha titanium alloys heated to between 1000°F and the beta transus remains essentially unchanged metallurgically except for annealing or stress-relieving of the material. The heat-affected zone immediately adjacent to the weld metal in commercially pure titanium, alpha alloys, and alpha-beta alloys will have metallurgical and mechanical properties similar to those of weld metal of similar composition. The heat-affected zones of beta or near beta alloys can show a wide range of metallurgical and mechanical properties. The properties depend on the welding and post weld procedures, and on the original base metal conditions, which can have a wide range of properties depending on the preweld heat treatment.

The commercially pure alpha and alpha-rich alloys do not develop significant property variations from heat treatment, and therefore weldability is relatively unaffected by preweld or postweld heat treatments. Preweld and postweld heat treatments might be needed for the beta and beta-rich alloys to assure that the entire weldment has the intended properties, including the welded joints.

The toughness of commercially pure titanium weld metal, strengthened by oxygen, can be significantly decreased by a high hydrogen content as a result of the formation of brittle titanium hydride platelets. The presence of small amounts of iron and certain other beta stabilizers is beneficial, probably due to the formation of some beta phase that has high hydrogen solubility. Weld metal toughness of titanium alloys strengthened by the addition of alpha and beta stabilizing elements is not affected by hydrogen within the normal limits for commercial compositions.

## Weldability Rating

Unalloyed titanium and most titanium alloys are readily welded using procedures and equipment suitable for welding austenitic stainless steel or nickel alloys. Titanium alloys are sometimes rated according to their ability to produce tough, ductile welds in them. One weldability rating is shown in Table 10.8. Those alloys rated A or B are considered usable for most applications in the as-welded condition. Welded joints in all of the weldable alloys should have joint efficiencies near 100 percent if the base metals are initially in the annealed condition. Weldments of many alloys with limited weldability can be given an annealing heat treatment to im-

**Table 10.8**
**Relative weldability of titanium alloys**

| Alloy | Rating[a] |
|---|---|
| Commercially pure grades | A |
| **Alpha alloys** | |
| Ti-0.2 Pd | A |
| Ti-5Al-2.5Sn | B |
| Ti-5Al-2.5Sn ELI | A |
| **Near alpha-alloys** | |
| Ti-8Al-1Mo-1V | A |
| Ti-6Al-2Cb-1Ta-0.8Mo | A |
| Ti-6Al-4Zr-2Mo-2Sn | B |
| **Alpha-beta alloys** | |
| Ti-6Al-4V | B |
| Ti-6Al-4V ELI | A |
| Ti-7Al-4Mo | C |
| Ti-6Al-6V-2Sn | C |
| Ti-8Mn | D |
| **Beta alloy** | |
| Ti-13V-11Cr-3Al | B |

a. A—excellent
   B—fair to good
   C—limited to special applications
   D—welding not recommended

prove ductility. Two weldable alloys, Ti-6Al-4V and Ti-13V-11Cr-3Al, are heat-treated before welding to increase their strength properties.

An important criterion of weldability is the capacity to produce welds free from defects that would limit the effectiveness of the weldment in service. Especially important is the capacity to resist weld cracking, which is related to the restraint imposed by the fabrication and the strength and toughness of the alloy. Alloys with a tensile strength of more than about 100 ksi and a Charpy V-notch impact toughness of less than about 15 ft•lbs can develop weld cracking under adverse conditions. The high strength Ti-6Al-6V-2Sn and Ti-7Al-4Mo alloys are considered to have limited weldability because of a tendency for weld cracking under conditions of high restraint and the presence of minor discontinuities in the weld metal or heat-affected zone. The near-beta alloys, such as Ti-8Mn, are not recommended for welding because of weld cracking under conditions of moderate restraint or minor weld zone discontinuities. The resistance to weld cracking of the high-strength, low-toughness al-

loys can be improved by maintaining a preheat of 300° to 350°F during welding and by stress-relieving immediately after welding.

## Commercially Pure Titanium

Commercially pure titanium has moderate strength but good ductility. It is used primarily for its good corrosion resistance, formability, and weldability. If the iron content is above about 0.05 percent, preferential corrosive attack of weld metal can occur in nitric acid solutions. Welds are particularly vulnerable because of the acicular nature of any retained beta phase that is stabilized by the iron. Galvanic cells between the beta and the contiguous alpha phase initiate corrosion of the weld metal. The attack is continuous because the iron-rich beta platelets always intersect the surface. This is not true for the base metal where retained beta is finally divided and discontinuous, and corrosive attack is slight. Filler metal with low iron content should be used, and all sources of iron contamination during preparation and welding should be avoided. Obviously, steel wire brushes should not be used.

## Alpha Alloys

The alpha alloys and the near-alpha alloys have good weldability because of their good ductility. Welding or brazing operations have little effect on the mechanical properties of annealed material. However, the strength of cold-worked material in the weld heat-affected zone or in brazements is decreased as a result of heating. Therefore, these alloys are normally welded and brazed in the annealed condition.

The Ti-0.2 Pd alloy is essentially the same as commercially pure titanium from a weldability standpoint. The outstanding virtue of this alloy is its improved corrosion resistance in reducing environments. The crevice corrosion resistance of Ti-0.2 Pd weld metal is similarly high.

The Ti-0.3Mo-0.8Ni alloy exhibits some of these same characteristics. It has good resistance to crevice corrosion, and is stronger than ASTM Grade 2 commercially pure titanium and Ti-0.2 Pd alloy.

Precautions against iron contamination during preparation and welding are also good practice for these two alloys for the same reasons as for commercially pure titanium.

The Ti-5Al-2.5Sn alloy should be considered where good strength and toughness are needed with good weldability. Moreover the (ELI) extra-low-interstitial version provides high toughness down to −420°F, and is suitable for cryogenic applications. To retain this property, it must be joined using processes and procedures that avoid contamination by carbon, nitrogen, or oxygen. It has higher strength at elevated temperature than commercially pure titanium. Therefore, residual weld stresses can be proportionately higher. Stress relieving of weldments is commonly recommended.

## Near-Alpha Alloys

The near-alpha alloys exhibit excellent creep strength at elevated temperatures. They are weldable, but residual weld stresses can be high. Therefore, stress relieving of weldments is always recommended. In some cases, special welding procedures are needed to prevent weld cracking in heavy sections. As with other alloys, iron contamination degrades creep strength, and should be avoided.

## Alpha-Beta Alloys

The response of alpha-beta alloys to heat treatment depends upon the type and amount of alloying. Welding or brazing of these alloys may significantly change their strength, ductility, and toughness characteristics as a result of the thermal cycle to which the alloy is exposed.

The low ductility of most alpha-beta alloy welds is caused by phase transformations in the weld metal or the heat-affected zone, or both. Alpha-beta alloys can be welded with unalloyed titanium or alpha-titanium alloy filler metal to produce a weld metal that is low in beta phase. This improves weld ductility. However, this procedure does not overcome the low ductility of the heat-affected zone in alloys that contain large amounts of beta stabilizers.

The Ti-6Al-4V alloy has the best weldability of the alpha-beta alloys. It can be welded in either the annealed condition or the solution-treated and partially aged condition. Aging can be completed during a postweld stress-relieving heat treatment.

The Ti-10V-2Fe-3Al has good weldability because it is highly beta stabilized, and retains a fine grain size during welding.

Alpha-beta alloys that are highly beta stabilized have limited weldability. They tend to crack when welded under high restraint or when minor defects are present in the weld zone. As mentioned previously the Ti-7Al-4Mo and Ti-6Al-6V-2Sn alloys are examples of those types of alloys. The resistance to cracking may be improved by preheating in the range of 300° to 350°F, and then stress-relieving immediately

after welding. Welding of Ti-8Mn alloy is not recommended.

## Beta Alloys

Most beta alloys are weldable in either the annealed or the heat-treated condition. These include Ti-13V-11Cr-3Al, Ti-8Mo-8V-3Al-2Fe, and Ti-11.5Mo-6Zr-4.5Sn alloys. Welded joints have good ductility but relatively low strength in the as-welded condition. They are used most often in this condition because the welded joint can respond differently to heat treatment than the base metal, and have low ductility after aging. Aging can take place if the welds are exposed to elevated temperatures in service.

To obtain welded joints with 100 percent joint efficiency, the alloy can be welded in the annealed condition. The weld metal is cold-worked by peening or planishing, and then the entire weldment is solution-treated and aged. Adequate weld ductility for some applications can be obtained with this procedure.

## Dissimilar Alloys

Titianium alloys within the same group can generally be welded together. Commercially-pure titanium can be welded to alpha alloys and to some alpha-beta alloys. Commercially-pure titanium filler metal should be used. Sufficient weld reinforcement can provide good strength properties in the joints.

One precaution should be taken when welding alloys with unalloyed filler metal. Residual hydrogen can diffuse from the base metal to the weld metal, resulting in the formation of hydrides in the weld metal and delayed brittle fracture.

## Compatibility With Other Metals

Titanium can be successfully fusion welded to the other reactive metals, zirconium, columbium, tantalum, and hafnium. However, uniform melting is difficult when fusion welding it to all but zirconium. Titanium forms continuous solid solutions with these metals, but the welded joints are expected to have low ductility. On the other hand, titanium forms brittle intermetallic compounds with the more common structural alloys of iron, nickel, copper, and aluminum; thus extremely brittle welds are formed when the dissimilar metals are fused together. Titanium can be joined to some of these metals with certain solid-state welding processes where diffusion and alloying are very limited. Explosion, friction, and diffusion welding are examples. Titanium-clad steel is produced commercially by explosion welding.

## POSTWELD HEAT TREATMENT

Relief of welding stresses is generally recommended for complex weldments to avoid cracking or other undesirable effects of high residual stresses. Residual welding stresses can be sufficiently high to promote stress-corrosion in sensitive alloys. Also, such stresses can cause weldments to have low endurance limits in high- and low-cycle fatigue.

When residual welding stresses are expected to be a problem, the structure should be heated to the proper temperature and held for sufficient time to relieve the stresses (see Table 10.4). An annealing heat treatment automatically relieves residual stresses. Generally, the minimum stress-relieving temperature is in the 1000° to 1300°F range for practical short-time operations. It is important to determine the desired mechanical properties in the weldment before selecting the temperature and time because aging can take place during stress-relieving of heat-treatable alloys.

Before the weldment is heated, it should be cleaned to avoid contamination and stress-corrosion cracking. After the heat treatment, any surface discoloration can be removed by chemical cleaning, as described previously.

## PROTECTION DURING JOINING

Because of the sensitivity of titanium to embrittlement by oxygen, nitrogen, and hydrogen, the entire component or that portion to be heated above about 500°F must be protected from atmospheric contamination. Protection or shielding is commonly provided by a high-purity inert gas cover in the open or in a chamber, or by a vacuum of $10^{-4}$ torr or lower. (These requirements also pertain to the other metals discussed later in this chapter.)

During arc welding, the titanium must be protected from the atmosphere until it has cooled to below about 800°F. Adequate protection can be provided by an auxiliary inert gas shielding device when welding in the open. For critical applications, welding should be done in a gas-tight chamber that is thoroughly purged of air prior to filling with high-purity argon, helium, or mixtures of the two.

The purity of the shielding gas influences the mechanical properties of the welded joint. Both air and water vapor are particularly detrimental. The purity of commercial welding grade inert gases is normally satisfactory, but care must be taken to ensure that moisture and air are not added to the gas through a faulty system. The dew point of the gas should be measured at the welding location or as it is purged from a welding chamber. Figure 10.3 shows the effect of the dew point of argon shielding on the hardness and ductility of gas tungsten arc welds in Ti-6Al-4V alloy. The resultant hydrogen content is also shown. A dew point of −20°F at the point of weld is the approximate maximum moisture limit according to these data. Welding grade gases have a dew point of −65°F or lower.

The inert gas at the cylinder or other source must be sufficiently dry to allow a margin for some moisture pickup in the delivery system. One method of checking gas purity is to weld a piece of scrap titanium, prior to welding the work piece itself, and then to bend it. The surface appearance and the degree of bending are a good indication of the gas purity. A second sample should be welded and bent after the weldment is completed to assure that the shielding was satisfactory during welding.

The color of a weld bead on titanium is often used as a measure of the level of contamination or the shielding gas purity. A light bronze color indicates a small amount of surface contamination while a shiny blue color indicates a greater amount of surface contamination. Neither of these levels of surface contamination is desirable, but they might be acceptable on a single or final weld pass, provided the surface layer is removed before the weldment is placed in service. A white, flaky layer on the weld bead indicates excessive contamination, which is not an acceptable condition. If several passes are to be deposited in a groove weld, no contamination is acceptable before depositing additional passes. If a white or gray flaky oxide is present, the gas shielding system should be inspected, and the cause of contamination corrected. The contaminated weld metal should be removed because it is likely brittle.

When brazing or diffusion welding, the titanium must be protected by high-purity inert gas or processed in a vacuum. The time at temperature should be as short as practical because hot titanium absorbs (getters) oxygen, nitrogen, and hydrogen by diffusion when available in minute amounts in the system.

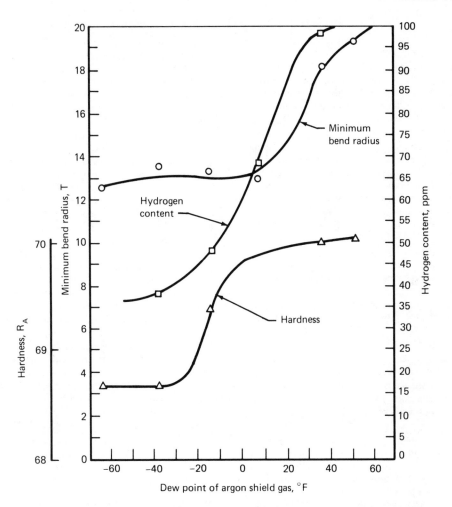

**Fig. 10.3—Effect of moisture content (dew point) of argon shielding on the bend ductility, hardness, and hydrogen content of gas tungsten arc welds in Ti-6Al-4V alloy**

## POROSITY IN WELDS

Porosity in titanium welds is a persistent problem. Most, if not all, porosity is apparently caused by gas bubbles formed during solidification of the weld metal, and welding procedures and techniques can affect this. There appears to be no single source of the porosity in welds.

Hydrogen in solution in the metal is not a prime cause of porosity when it is within specification limits. Most specifications limit the hydrogen content of titanium to 150 ppm or less,

and it is usually less than 100 ppm. Porosity-free welds can be made in titanium when the hydrogen content is within specification limits.

Filler metal addition can cause porosity if its hydrogen content is near the high end of the specification limit. Die lubricant worked into the wire during drawing is a major cause of porosity, and it must be removed prior to use. Filler metal cleanliness can be monitored by wiping with a solvent-moistened, lint-free cloth.

Another possible source of hydrogen is adsorbed moisture. Moisture can be evaporated by preheating the metal to about 250°F before welding. The welds from which the data in Fig. 10.3 were taken did not show an increase in porosity with increasing shielding gas dew point. Shielding gas that was bubbled through water to saturate it did not significantly affect the amount of porosity in the welds. Moisture in the shielding gas, however, seriously impairs weld properties, other than porosity.

Cleanness of the joint faces and surrounding area is a major factor in producing porosity-free welds. The number of foreign materials that can cause porosity is large. Plasticizers dissolved from rubber gloves by solvents, especially alcohol, are a cause of porosity, as are residues in cloths used for wiping joint areas. Grinding residue imbedded in the titanium is another source of porosity. Fingerprints, dirty rags, and lint-bearing gloves are contributory when placed in contact with the joint edges before welding. As-sheared edges of sheet metal can be contaminated by materials that cause porosity. Removal of all burrs eliminates or greatly reduces this source. Where porosity is a persistent problem, removal of metal along the joint edge by either mechanical or chemical methods a few hours before welding is recommended.

Porosity is also influenced by the welding procedure and technique. Many of the welding variables are interrelated and the important ones are not readily identified. However, improper tack welding techniques, wide root openings in welded joints, or the use of consumable inserts can cause porosity. Other variables that can play a part are heat input, cooling rate, welding speed, arc voltage, and shielding gas flow rates.

Although the data available appear to be contradictory, the temperature of the weld pool and its cooling rate might also be contributing factors that influence porosity. The likelihood of porous welds is less with slow cooling rates because there is more time for gases to escape to the surface.

In some cases, porosity increases as welding speed decreases when the heat input is also decreased to maintain a constant depth of joint penetration. This is attributed to a lower weld pool temperature. However, in other cases, a large decrease in welding speed results in a reduction in porosity. This is attributed to the slower cooling rate of the weld pool that provides more time for gas bubbles to escape.

Often, porosity can be reduced by the use of copper chills adjacent to the weld joint. The chilling effect requires high welding heat that, in turn, produces a more fluid weld pool from which dissolved gases are evolved at a faster rate.

Welds characteristically made with low heat input are porous. These include small tack welds, fillet welds, and partial-penetration groove welds. The only correction that can be employed is to increase the heat input. For example, tack welds should be several inches in length with properly controlled starts and stops. The weld pass should remelt the tack welds. Slight interruptions of the welding heat, that might be difficult to detect by visual examination of the completed weld, can also cause porosity. Conversely, a hot molten weld pool, such as that produced by the gas metal arc welding system operating in the spray transfer mode, seldom shows porosity.

## GAS SHIELDED ARC WELDING

The gas shielded arc welding processes are well suited for joining titanium and titanium alloys, provided the gas shielding arrangement adequately protects the weld area from the atmosphere. The three processes normally used are gas tungsten arc, gas metal arc, and plasma arc welding.[6]

Welding with all three processes can be done with manual, semiautomatic, or automatic equipment. Manual and automatic welding can be done in the open or in an inert gas-filled chamber. Semiautomatic welding is normally done in the open, but could conceivably be performed in a chamber.

### Welding in The Open

The primary concern with welding in the open is adequate inert gas shielding of (1) the molten weld pool and adjacent base metal (primary shielding), (2). the hot, solidified weld metal and heat-affected zone (secondary shielding), and (3) the back side of the weld joint (backing).

*Primary Gas Shielding.* Primary gas shielding is provided by the torch or gun nozzle. The nozzle size usually ranges from 0.5 to 0.75 inch. In general, the largest nozzle consistent with

---

6. Shielded metal arc, flux-cored arc, and submerged arc welding are unsuitable for joining titanium because of improper atmospheric shielding and undesirable reactions with the hot slag.

accessibility and visibility should be used. Nozzles that provide laminar flow of the shielding gas are desirable because they lessen the possibility of turbulent gas flow where air mixes into the gas stream at its periphery. Proper shielding of the molten weld pool is critical.

*Secondary Gas Shielding.* Secondary inert gas shielding is required for the solidified and cooling weld bead and the heat-affected zone. Normally, the primary shield has advanced with the welding and no longer protects the solidifying weld metal. The hot weld zone must be shielded from the atmosphere until it has cooled to a temperature where oxidation is not a problem. The low thermal conductivity, and consequent slow cooling, of titanium requires that a considerable length of the hot weld zone be shielded; more than is normally provided by gas flow from a torch nozzle.

The most common form of secondary shielding is a trailing shield; a typical design is shown in Fig. 10.4. It consists of a metal chamber fit to the torch nozzle and held by a clamp. The inert gas flows through a porous metal diffuser screen over the weld area. Its length is proportional to the heat input and welding speed. The shield must be wide enough to cover the heat-affected zone on each side of the weld bead.

A trailing shield is used more often for machine or automatic welding where travel speeds are higher. One important application of the trailing shield is making groove welds in pipe in the horizontal-rolled position. In this case, the trailing shield should be curved to match the pipe.

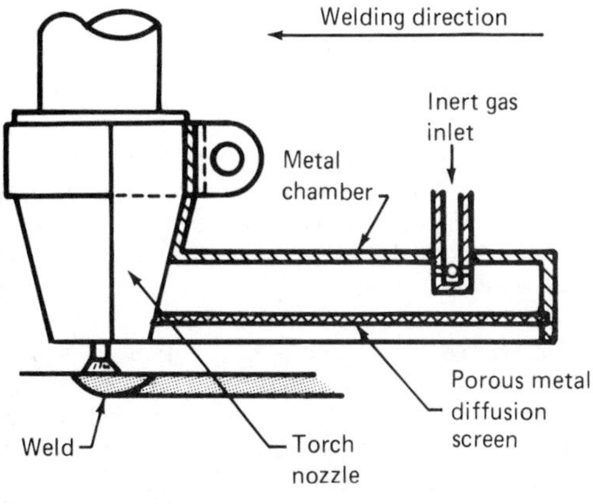

For manual welding, a large torch nozzle or an auxiliary annular nozzle can be used with slow welding speeds. A trailing shield can interfere with the visibility of the weld pool and the manipulation of the welding torch when welding manually.

Secondary shielding can be incorporated into the fixturing, as shown in Fig. 10.5. Inert

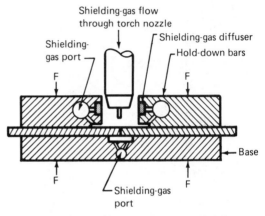

**Fig. 10.5—Secondary inert gas shielding incorporated in the welding fixture**

gas passages are provided in the hold-down bars on both sides of the weld seam. Shielding gas flows from the torch and hold-down bars into the channel formed by the bars where it displaces the air from above the weld.

*Backing Gas Shielding.* Inert gas shielding is also required to protect the root of the weld and the adjacent base metal from the atmosphere during welding. This is normally done through gas passages in a temporary backing bar (or ring) as shown in Fig. 10.6. The bar is usually incorporated into the fixturing (see Fig. 10.5). It contains a clearance groove under the joint that is purged of air prior to welding. Typical groove dimensions are given in Table 10.9. The pressure of the gas in the groove must be kept low to avoid a concave root contour on the welded joint. The titanium components must be in contact with the backing bar along its entire length to ensure a uniform weld.

The backing bar can also serve as a chill to remove heat from and accelerate cooling of the

**Fig. 10.4—Torch trailing shield for gas shielded arc welding of titanium and other reactive metals**

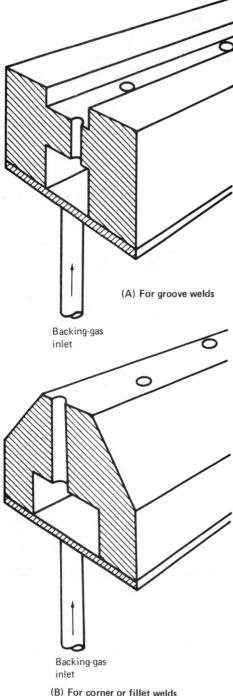

**(A) For groove welds**

Backing-gas
inlet

**(B) For corner or fillet welds**

*Fig. 10.6—Weld backing bars that provide
inert gas shielding*

Backing-gas
inlet

**Table 10.9
Typical groove dimensions for temporary
weld joint backing bars**

| Material thickness, in. | Groove width, in. | Groove depth, in. |
|---|---|---|
| 0.005-0.012 | 0.040 | 0.040 |
| 0.013-0.024 | 0.063 | 0.040 |
| 0.025-0.032 | 0.093 | 0.040 |
| 0.033-0.050 | 0.125 | 0.045 |
| 0.051-0.080 | 0.187 | 0.045 |
| 0.081-0.125 | 0.250 | 0.045 |
| 0.126-0.250 | 0.312 | 0.050 |

weld. This reduces gas shielding requirements. The bar is often made of copper, and can be water cooled. Stainless steel can be used when lower cooling rates are acceptable. The root opening of the joint must be near zero to prevent the arc from impinging on and fusing the titanium weld to the backing bar. Contamination of the titanium weld metal might embrittle it and result in a cracked weld.

When welding pipe or tubing, the interior of the pipe must be purged of air with inert gas. Usually a volume of inert gas that is at least six times the volume of the pipe is required to displace the air. In large systems, internal dams can be placed on both sides of the joint to confine the backing gas to the vicinity of the weld joint. Suitable dams are available commercially. They must have an inlet for the inert gas and an outlet for the displaced air and inert gas to escape. As with a backing bar, the internal gas pressure must be low, usually 2 or 3 in. of water.

## Welding in a Chamber

Many titanium weldment designs are not adaptable to welding in the open because adequate inert gas shielding of the weld joints would be very difficult to achieve. An alternate procedure is to weld the assembly in a gas-tight chamber filled with inert gas.

Two types of welding chambers are used, flow-purged and vacuum-purged. A typical flow-purge welding chamber is shown in Fig. 10.7. Other designs are available. The welding atmosphere in a flow-purged chamber is obtained by flowing inert gas through the chamber to flush out the air. The volume of inert gas needed to obtain a welding atmosphere of sufficent purity in the chamber is about 5 to 10 times the chamber volume.

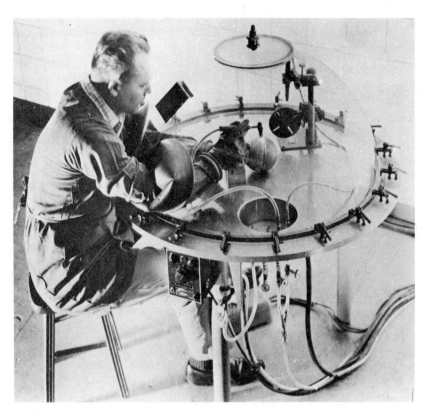

*Fig. 10.7—A flow-purged gas tungsten arc welding chamber*

The appropriate inert gas flow rate and purge time for a specific chamber should be established by welding tests. During welding, inert gas is flowed through the welding torch to insure adequate shielding of the molten weld pool. A low, positive gas pressure is always maintained in the chamber to prevent air from entering it. The welding atmosphere should be monitored during actual welding operations by running weld beads on coupons prior to, during, and after welding the actual assemblies. The coupons should be evaluated visually and mechanically to verify that the chamber atmosphere was satisfactory during the welding operation.

When contamination of the titanium during welding must be essentiallly avoided, welding is best done in a vacuum-purged welding chamber. One such chamber is shown in Fig. 10.8. With this system, the air is removed from the chamber by a vacuum pumping system to a pressure of usually $10^{-4}$ torr or lower. The chamber is then back-filled with inert gas of appropriate purity. The chamber should be equipped with an automatic gas delivery system that always maintains a positive pressure in the chamber and bleeds off excessive pressure during welding. Inert gas is not flowed through the welding torch or gun, which must be of water-cooled design.

Accessibility to the work is through gloved ports in the chamber. The gloves, welding torch or gun, fixturing, and other material installed or placed in the chamber must be impervious to air and water, and void of volatiles that can contaminate the titanium. As with a flow-purged chamber, the welding atmosphere should be monitored either by appropriate gas analyzing equipment or by welding and testing of coupons.

## Joint Design

The weld joint designs used for welding titanium are similar to those used for steels. Typical groove weld joint dimensions are given in

*Fig. 10.8—A vacuum-purged arc welding chamber*

Table 10.10, for use as a guide. Actual joint design depends upon several factors including the welding process, type of operation (manual or machine), joint accessibility, and inspection requirements.

Edge preparation should be done by a machining process that does not contaminate the titanium or leave embedded particles on the surface. As mentioned previously, root opening is important when welding with a temporary backing bar. Fixturing or tack welds should be used to maintain uniform root opening during welding.

The design of a weldment, the types of joints, and joint locations can be limited by shielding requirements during welding. When welding in a chamber, positioning of the weldment for welding each joint must be considered during the design phase.

### Preheat and Interpass Temperature

Preheat and interpass temperatures must be kept low for welding in the open to avoid surface oxidation that can be dissolved into the molten weld metal. A low preheat is generally employed

to drive off adsorbed surface moisture on the surfaces prior to welding. Preheat and interpass temperatures should not exceed 250°F. Prolonged exposure to air at temperatures above 250°F can cause an oxide film to form on the titanium weld and base metal. This oxide film must be removed with a stainless steel wire brush or rotary carbide burrs prior to welding to reduce weld metal contamination.

## GAS TUNGSTEN ARC WELDING

### Welding Procedures

Gas tungsten arc welding (GTAW) is commonly used to weld titanium and its alloys, particularly for sheet thicknesses up to 0.125 inch. Welding in the open is best done in the flat position to maintain adequate inert gas shielding with the welding torch and secondary shielding devices. Specially designed secondary shielding devices might be required when welding in positions other than flat. Welding position might also influence the amount of porosity in the weld metal.

**Table 10.10**
**Typical groove weld joint designs for arc welding titanium alloys**

| Joint type | Metal thickness, $t$, in. | Groove angle, degrees | Root opening | Root face width |
|---|---|---|---|---|
| Square-groove | 0.01-0.09<br>0.03-0.13 | 0<br>0 | 0<br>0-0.1t | t |
| Single-V-groove | 0.06-0.25<br>0.12-0.50 | 30-60<br>30-90 | | |
| Double-V-groove | 0.25-1.50 | 30-90 | 0-0.1t | 0.10-0.25t |
| Single-U-groove | 0.25-1.00 | 15-30 | | |
| Double-U-groove | 0.75-2.00 | 15-30 | | |

Direct current electrode negative is normally used with Type EWTh-2 tungsten electrodes of proper size. Contamination of the weld with tungsten should be avoided because it embrittles the titanium. Electrode extension from the gas nozzle should be limited to the amount required for good visibility of the weld pool. Excessive extension is likely to result in weld metal contamination.

Particular care must be exercised during manual welding operations when starting and stopping the welds. Scratch starting is not recommended and should be avoided. Preferably, a starting tab should be used or the arc should be initiated with high frequency power.

When welding in the open, welding should be terminated on a runoff tab or the torch should dwell over the weld with a postflow of shielding gas after shutting off the welding current. When filler metal is added, the heated end of the welding rod must be held under the gas nozzle at all times to avoid contamination. If the tip of the rod becomes contaminated, it must be cut off before continuing the weld.

Welding conditions for a specific application depend on joint thickness, joint design, the weld tooling design, and method of welding (manual or machine). For any given section thickness and joint design, various combinations of amperage, voltage, welding speed, and filler wire feed rate can be used to produce satisfactory welds. Table 10.11 gives typical welding conditions that can be used as a guide when setting up for machine GTAW of titanium. The welding conditions generally do not have to be adjusted radically to accommodate the various titanium alloys. However, certain adjustments are often made to control weld porosity.

**Table 10.11**
**Typical conditions for machine gas tungsten arc welding of titanium**

| Sheet thickness, in. | Filler wire Diam., in. | Feed, in./min | Shielding gas | Arc voltage, V | Welding current, A | Travel speed, in./min |
|---|---|---|---|---|---|---|
| 0.008 | none | | He | 14 | 10 | 16 |
| 0.030 | none | | Ar | 10 | 25-30 | 10 |
| 0.060 | none | | Ar | 10 | 90-100 | 10 |
| 0.060 | 0.062 | 22 | Ar | 10 | 120-130 | 12 |
| 0.090 | none | | Ar | 12 | 190-200 | 10 |
| 0.090 | 0.062 | 22 | Ar | 12 | 200-210 | 12 |
| 0.125 | 0.062 | 20 | Ar | 12 | 220-230 | 10 |

## Weld Mechanical Properties

The mechanical properties of gas tungsten arc welded joints as measured by smooth tension, notch tension, bend, and crack-susceptibility tests, compare very favorably with the base metal properties. Axial tension fatigue tests of welds generally show a significant decrease in strength compared to similar base metal specimens. Fracture-toughness tests on fusion welds are not easily interpreted, but are indicative of fair to good performance. The thermal stability of welded joints in Ti-8Al-1Mo-1V and Ti-6Al-4V alloys appears to parallel that of the base metals. The tensile, impact, and hardness properties of standard titanium and titanium alloy weld metals are given in Table 10.12. These data were obtained from multiple-pass gas tungsten arc welds in plates of 0.5-in. or greater thickness of the same composition as the filler metal.

The toughness of unalloyed titanium weld metal can be decreased significantly by high hydrogen content as a result of the formation of brittle titanium hydride platelets in the alpha titanium grains. Weld metal toughness of titanium alloys that are strengthened by substitutional alloying with beta stabilizing elements is not effected by hydrogen content within the normal commercial limits.

Hardness tests can provide useful information on the conditions of weld and heat-affected zones. The approximate relationship between tensile strength and hardness of titanium alloy weld metal is shown in Fig. 10.9. Hardness tests are useful for estimating the tensile strength of the weld metal and the heat-affected zone. From these data and a knowledge of the strength-toughness relationships for various alloys, many properties of the weld zone can be predicted without the need for extensive testing.

A common method of comparing the ductility of welds in titanium alloy sheet is a progressive-radius bend test.[7] The minimum bend radius is the one over which a weld can be bent without developing a crack, expressed in terms of the sheet thickness, $t$. Normally, the welded specimen is bent with the weld transverse to the bend axis until the permanent deformation equals 105 degrees. Typical minimum bend radii for welds in several alloys are given in Table 10.13.

---

7. Refer to *Welding Handbook*, Vol. 1, 7th Ed., 166-69, and AWS B4.0-77, *Standard Methods for Mechanical Testing of Welds*, 28-31.

## GAS METAL ARC WELDING

Gas metal arc welding (GMAW) can be used for joining titanium. It is more economical than gas tungsten arc welding because of the higher deposition rates, particularly with thick sections. Welding conditions should produce a smoothly contoured weld that blends smoothly with the base metal.

With GMAW, the droplets of filler metal being transferred across the arc are exposed to much higher temperatures than the filler metal fed into a GTAW molten weld pool. The combination of high temperature and fine particle size makes the filler metal highly susceptible to contamination by impurities in the arc atmosphere. Consequently, the welding gun and auxiliary gas shielding must be carefully designed to prevent contamination of the inert gas welding atmosphere.

## Equipment

Conventional GMAW power sources, welding guns, and control systems are sastisfactory for welding titanium. Conventional GMAW guns are modified to provide the necessary auxiliary gas shielding needed for titanium. Gas metal arc welding can be done in a closed chamber, but most weldments will probably be too large to be accommodated in such equipment. For welding in the open, supplemental shielding devices similar to those described for GTAW should be employed. Trailing shields for GMAW usually must be considerably longer than those used with GTAW because travel speeds are higher, and the weld beads are larger and cool slower. Trailing shields are sometimes water-cooled.

## Welding Consumables

Cleanliness and uniformity of the wire electrode are extremely important. It must be free of dirt, grease, oil, drawing compounds, and surface imperfections that may entrap foreign matter. Cleaned wire should be stored under clean conditions. Spools on the welding head should be covered when not in use, and moved to a clean storage area between jobs.

Shielding gases should be welding grade argon, helium, or mixtures of the two. Oxygen, nitrogen, hydrogen, or $CO_2$ should not be added to the inert gas because they embrittle titanium. The inert gas system must be leak-tight to avoid contamination by air or moisture.

**Table 10.12**
**Mechanical properties of titanium and titanium alloy weld metal**

| Filler metal | Impurity content [a] | | | Weld metal | | | | | | | | | |
| | H, ppm | O, percent | Fe, percent | Tensile properties | | | | Impact strength, Charpy V-notch, ft·lbs | | | Hardness | | |
| | | | | TS, ksi | 0.2%YS, ksi | Elong. in 1 in. percent | RA, percent | 68°F | 32°F | -80°F | BM | HAZ | Weld |
|---|---|---|---|---|---|---|---|---|---|---|---|---|---|
| ERTi-1[b] | 20 | 0.09 | <0.06 | 60 | 47 | 41 | 77 | 162 | 159 | 165 | 46RB | 47RB | 46RB |
| ERTi-2[b] | 60 | 0.07 | <0.06 | 58 | 43 | 42 | 76 | 40 | 42 | 35 | 43 | 45 | 44 |
| ERTi-3 | 50 | 0.16 | 0.13 | 82 | 64 | 24 | 48 | 16 | 14 | 13 | 53 | 55 | 56 |
| ERTi-4 | 40 | 0.20 | 0.28 | 88 | 69 | 22 | 46 | 27 | 22 | 25 | 55 | 57 | 57 |
| ERTi-0.2Pd | 45 | 0.13 | 0.16 | 72 | 58 | 28 | 59 | 41 | 38 | 33 | 52 | 53 | 53 |
| ERTi-3Al-2.5V | 75 | 0.08 | 0.08 | 100 | 79 | 20 | 59 | 46 | 42 | 35 | 60 | 60 | 60 |
| ERTi-3Al-2.5V-1 | 15 | 0.08 | 0.08 | 102 | 83 | 18 | 56 | 43 | 44 | 37 | 58 | 60 | 60 |
| ERTi-5Al-2.5Sn | 50 | 0.18 | 0.44 | 138 | 124 | 12 | 26 | 25 | 21 | 15 | 33RC | 34RC | 35RC |
| ERTi-5Al-2.5Sn-1 | 25 | 0.10 | 0.19 | 125 | 113 | 15 | 28 | 42 | 41 | 31 | 29 | 31 | 31 |
| ERTi-6Al-2Cb-1Ta-1Mo | 40 | 0.09 | 0.06 | 132 | 116 | 11 | 27 | 39 | 38 | 28 | 28 | 31 | 32 |
| ERTi-6Al-4V | 35 | 0.13 | 0.18 | 146 | 122 | 12 | 39 | 17 | 18 | 11 | 32 | 36 | 36 |
| ERTi-6Al-4V-1 | 45 | 0.10 | <0.06 | 139 | 120 | 10 | 25 | 18 | 16 | 13 | 30 | 34 | 34 |
| ERTi-8Al-1Mo-1V | 70 | 0.08 | 0.08 | 140 | 122 | 7 | 15 | 27 | 22 | 17 | 33 | 36 | 36 |
| ERTi-13V-11Cr-3Al | >75 | 0.10 | 0.13 | 116 | 111 | 22 | 39 | 4 | 4 | 4 | 29 | 30 | 30 |

a. Carbon—0.02 to 0.04 percent, nitrogen—0.005 to 0.012 percent
b. Differences in notch toughness of ERTi-1 and ERTi-2 filler metals are related to hydrogen content.

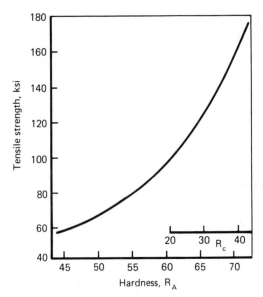

**Fig. 10.9—Approximate relationship between tensile strength and hardness of titanium alloy weld metal**

### Filler Metal Transfer

Titanium filler metal can be transferred by all three methods: short-circuiting, globular, and spray. Globular transfer is not recommended because of excessive spatter and incomplete fusion in the weld. Short-circuiting transfer can be used for welding thin sections in all positions and thicker sections in positions other than flat. However, incomplete fusion can be a problem with thick sections because of the inherent low heat input.

Spray transfer is preferred for welding thick sections in the flat and horizontal positions to take advantage of high heat input and high deposition rates. Pulsed spray welding provides spray transfer with lower heat inputs that is advantageous for thinner sections and welding in positions other than flat.

Regardless of the type of transfer, a trailing gas shield and gas backing are required. The length and width requirements for the trailing shield depend upon the heat input, joint design, and the bead width. In general, a relatively short, narrow shield can be used with short-circuiting transfer because of the relatively low heat input and a long, wide shield is needed for spray transfer because of the high heat input. The size needed with pulsed spray welding is between these two sizes.

### Welding Conditions

Suitable welded joints in a given thickness of titanium can be made by GMAW using several combinations of welding variables. The welding procedures should be qualified by appropriate tests prior to production welding. Table 10.14 gives typical conditions that have been used to GMAW titanium alloy plate. They are not necessarily optimum and should only be used as guides in setting up a GMAW operation.

For short-circuiting transfer, the groove angle and root opening of butt welds should be larger than with spray transfer. A 90-degree groove angle is recommended for V-groove welds. The larger dimensions provide better accessibility for the electrode to the root and faces of the joint to obtain good fusion.

Several laboratory studies have shown that a slight increase in the oxygen content of the weld metal can be expected when gas metal arc welding. However, the increase should not be excessive when proper inert gas shielding techniques and welding procedures are used.

### PLASMA ARC WELDING

Plasma arc welding (PAW) is an extension of gas tungsten arc welding in that the arc plasma is forced through a constricting nozzle. Inert gas shielding of the weld is provided by a shelding gas nozzle and an auxiliary trailing shield as with GTAW and GMAW. Welding is normally done with a transferred arc using direct current, electrode negative supplied by a constant current power source.

**Table 10.13**
**Typical minimum bend radii of as-welded joints in titanium alloy sheet**

| Alloy | Min bend radius, t[a] |
|---|---|
| Commercially pure Ti | 3-5 |
| Ti-0.15 Pd | 3-4 |
| Ti-5A1-2.5 Sn | 4-5 |
| Ti-5A1-2.5 Sn (ELI) | 4-5 |
| Ti-8A1-1Mo-1V | 4-5 |
| Ti-6A1-4V (ELI) | 8-12 |
| Ti-6A1-4V (ELI) | 8-10 |
| TI-7A1-4Mo | 16-18 |
| Ti-6A1-6V-2Sn | 16-18 |
| Ti-8Mn | 10-12 |
| Ti-13V-11Cr-3A1 | 3-5 |

a. t is the thickness of the base metal.

**Table 10.14**

**Typical conditions for gas metal arc welding of titanium alloy plate**

| Thickness, in. | Welding position[a] | Shielding gas | Electrode | | | Arc voltage, V | Welding current, A[c] | | Travel speed, in./min |
|---|---|---|---|---|---|---|---|---|---|
| | | | Type of transfer[b] | Diam., in. | Feed, in./min | | Back-ground | Mean pulse | |
| 0.125 | F | 75Ar-25He | S | 0.062 | 200-225 | 20 | 250-260 | — | — | 15 |
| 0.250 | F | 75Ar-25He | S | 0.062 | 300-320 | 30 | 300-320 | — | — | 15 |
| 0.375 | F | 85He-15Ar | S | 0.045 | 865 | 33 | 350-360 | — | — | 25 |
| 0.500 | F | 75Ar-25He | S | 0.062 | 375-400 | 40 | 340-360 | — | — | 15 |
| 0.625 | F | 75Ar-25He | S | 0.062 | 400-425 | 45 | 350-370 | — | — | 15 |
| 1.00 | F | Ar | S | 0.062 | 380 | 36-37 | 320-350 | — | — | 20-23 |
| 1.00 | F | Ar | S | 0.062 | 550 | 31-33 | 315-340 | — | — | 13 |
| 1.00 | F | 75Ar-25He | PS | 0.062 | 170 | 22-24 | — | 75 | 80 | 4 |
| 1.00 | V-up | 75Ar-25He | PS | 0.062 | 150 | 21-23 | — | 35 | 100 | 2.5 |
| 1.00 | F | 75Ar-25He | SC[d] | 0.030 | 650 | 17-21 | 140-185 | — | — | 7 |
| 1.00 | V-up | 75Ar-25He | SC[d] | 0.030 | 550 | 15-16 | 100-135 | — | — | 1.8 |

a. F—flat, V-up—vertical up.

b. S—Spray, electrode positive; PS—pulsed spray, electrode positive; SC—Short-circuiting, electrode negative.

c. Direct current

d. For groove welds, the groove angle and root opening should be larger than these used with spray transfer to provide good accessibility to the root and faces of the joint.

Welding grade argon is generally used as the orifice and shielding gas, but helium-argon mixtures are sometimes used particularly for shielding. Hydrogen must not be added to the inert gas because of it imbrittling effects on titanium.

Plasma arc welding can be done using two techniques, melt-in and keyhole. The melt-in technique is similar to gas tungsten arc welding. The keyhole technique provides deep joint penetration for welding square-groove joints in one or two passes, depending on the metal thickness. The two techniques can be combined for welding groove joints in the thick sections.

Square-groove joints in titanium alloys from about 0.062 to 0.50 inch thick can be welded with one pass or two passes, one from each side, with the keyhole technique. The welds tend to be undercut along the top edges and have very convex faces unless filler metal is added during welding or when a second "cosmetic" pass is made. Keyhole welds tend to be free of porosity.

An alternative design is a single-V- or single-U-groove with a root face height of 0.25 in. or more. The first pass is welded by the keyhole technique with without filler metal addition. The second pass can be welded with PAW or GTAW using filler rod or wire.

Typical conditions for plasma arc welding square-groove joints in titanium alloys are given in Table 10.15. The advantages of PAW are higher welding speeds and lower filler metal requirements than with gas tungsten arc welding.

## ELECTRON BEAM WELDING

Electron beam welding (EBW) in high vacuum is well suited for joining titanium with respect to atmospheric contamination of the weld can be tolerated in the application.[8] With nonvacuum EBW, the inert-gas shielding requirements for arc welding apply.

The process variables are accelerating voltage, beam current, beam diameter, and travel speed. Beam dispersion increases with atmospheric density, pressure, or both. Deep joint penetration in square-groove welds is obtained with high beam power density and a keyhole in the weld metal. Low power density melts the metal by conduction from the surface, as in arc welding, to produce a relatively wide, shallow weld bead.

Cleanliness and good joint fitup are important preparations for welding. Surface contaminants cause porosity in the weld. Poor fit-up causes underfill or discontinuous fusion unless filler metal is added during welding. The effects are greater with thin sheet than with plate thicknesses.

Typical EBW conditions that have been used to make square-groove joints in titanium alloys in high vacuum are given in Table 10.16. The variables that are not generally given are the electron beam diameter, the width of the bead face, and the depth-to-width ratio of the bead. Bead geometry is an important welding variable with respect to mechanical properties.

Electron beam welds in titanium are generally susceptible to the discontinuities experienced with other metals. These include undercut, underfill, porosity, shrinkage voids, incomplete penetration, incomplete fusion, and missed joints. The procedures for overcoming these problems are similar to those used with other metals.

Electron beam welding is a low heat input process when compared to gas tungsten arc welding. When evaluating the electron beam welding process for a specific titanium application, consideration must be given to the applicable mechanical properties, such as fatigue, fracture toughness, and tensile strength, as well as the permissible postweld heat treatment. Typical mechanical properties for three titanium alloys and electron beam welds in them are given in Table 10.17. In general, aged EB welds in alpha-beta titanium alloy are characterized by higher tensile strength and ductility than those of the unwelded base metal. Therefore, when examining test data, specimen configuration, intended application, welding conditions, and heat treatment must be considered. For example, the transverse EBW tensile properties for Ti-6Al-4V alloy are similar to those of the base metal. However, with longitudinal EBW tensile specimens, a moderate increase in tensile and yield strengths and a decrease in elongation would be expected. This effect is more pronounced with Ti-6Al-6V-2Sn alloy.

The as-welded microstructure and microsegregation of electron beam welds affect their mechanical properties both as-welded and after a postweld heat treatment. In the alpha-beta alloys, martensite (orthorhombic or hexagonal) predominates as the principle microstructural constituent in the weld. As previously stated, the weld metal and heat-affected zone usually exhibit

---

8. Refer to *Welding Handbook*, Vol. 3, 7th Ed., 170-215.

**Table 10.15**
**Typical conditions for plasma arc welding of square-groove joints in titanium alloys**

| Thickness, in. | Welding technique | Nozzle orifice | | Orifice and shielding gas[a] | Welding current, A[b] | Arc Voltage, V | Travel speed, in./min | Filler wire feed,[c] in./min |
| | | Diam., in. | Flow, ft³/h | | | | | |
| --- | --- | --- | --- | --- | --- | --- | --- | --- |
| 0.008 | Melt-in | 0.030 | 0.5 | Ar | 5 | – | 5 | – |
| 0.015 | Melt-in | 0.030 | 0.5 | Ar | 6 | – | 5 | – |
| 0.125 | Keyhole | 0.136 | 9 | Ar | 150 | 24 | 15 | 40 |
| 0.188 | Keyhole | 0.136 | 10-12 | Ar | 175 | 30 | 15 | 42 |
| 0.250 | Keyhole | 0.136 | 16 | Ar | 160 | 30 | 12 | 45 |
| 0.313 | Keyhole | 0.136 | 15 | Ar | 172 | 30 | 12 | 48 |
| 0.390 | Keyhole | 0.136 | 32 | 75He-25Ar | 225 | 38 | 10 | – |
| 0.500 | Keyhole | 0.136 | 27 | 50He-50Ar | 270 | 36 | 10 | – |

a. Backing with inert gas is necessary.
b. Direct current, electrode negative.
c. 0.062 in. diameter.

higher yield strengths and lower ductilities than the base metal, which is a direct result of aging. The starting microstructure and aging kinetics of the martensite constituent determine the mechanical properties after a postweld heat treatment. Both of these factors vary with alloy composition.

The weld metal and heat-affected zone in metastable beta alloys consists of single-phase beta, and possibly some omega phase. In these alloys, solidification segregation in the weld metal can cause poor properties at the weld centerline after postweld aging. Alloys of this type are typically welded in the solution-treated condition and aged directly after welding.

## LASER BEAM WELDING

Commercial laser beam welding (LBW) machines are capable of producing energy densities similar to those available with an electron beam operating in vacuum. The laser beam is diffused to a significantly lesser extent than an electron beam as it passes through the atmosphere. Consequently, higher energy densities are available with a laser beam for welding in the open.

The most practical lasers for welding are the Nd-YAG (solid-state) and the $CO_2$ continuous output types. As with electron beam welding, welds can be produced by the conventional melt-in technique or by the keyhole technique. With the keyhole technique, as much as 90 percent of the laser beam energy can be absorbed, depending upon the metal. Absorption efficiency is sig-

nificantly lower with the melt-in technique. At an energy level of 15 kW, the maximum thickness of Ti-6Al-4V alloy that can be welded in a single pass is about 0.60 inch.

When welding with a high power density, ionization of metal vapor above the molten weld pool diffuses the laser beam and interferes with welding. This can be prevented by blowing the metal ions from above the weld pool with inert gas, preferably helium. Helium-argon mixtures can also be used. At the same time, a titanium weld must be shielded from the atmosphere to prevent contamination and embrittlement, as described previously for arc welding.

The weld discontinuities in titanium normally encountered with electron beam welding are likely to be experienced with laser beam welding. Appropriate actions needed to prevent or remove discontinuities should be similar with both processes.

The microstructures and mechanical properties of laser beam welds do not vary significantly from those of electron beam welds. The weld and heat-affected zones have higher tensile strengths than the base metal, but exhibit lower values of fracture toughness. The reduction in fracture toughness in Ti-6Al-4V alloy is about the same as that with electron beam welds. Fatigue properties of laser welds in this alloy also are comparable to those of electron beam welds.

## DIFFUSION WELDING AND BRAZING

Titanium and its alloys are readily joined by

**Table 10.16
Typical conditions for electron beam
welding of titanium alloys**

| Thickness, in. | Accelerating voltage, kv | Beam current mA | Travel speed, in./min |
|---|---|---|---|
| 0.03 | 30 | 26 | 89 |
| 0.04 | 13 | 50 | 83 |
| 0.045 | 13 | 55 | 68 |
| 0.05 | 85 | 4[a] | 60 |
|  | 95 | 1.8 | 30 |
|  | 110 | 2[b] | 45 |
| 0.08 | 18.5 | 90 | 75 |
| 0.09 | 19 | 100 | 70 |
|  | 90 | 5 | 18 |
| 0.125 | 20 | 95 | 30 |
|  | 125-135 | 6-7 | 28-30 |
| 0.20 | 28 | 170 | 98 |
|  | 125 | 8[c] | 18 |
| 0.25 | 138 | 10 | 25 |
| 0.35 | 130 | 35 | 40 |
|  | 150 | 15 | 60 |
| 0.63 | 30 | 260 | 60 |
| 1.0 | 25 | 200 | 44 |
|  | 36.5 | 375 | 44 |
|  | 40 | 350 | 50 |
| 1.75 | 55 | 360[d] | 40 |
| 2.0 | 45 | 450 | 26 |

Beam diameter, in.
a. 0.006
b. 0.005
c. 0.01
d. 0.094

diffusion welding (DFW) and diffusion brazing (DFB).[9] For some applications, diffusion welded or brazed joints have better properties than do fusion welded joints.

A titanium diffusion weld should show no

9. Refer to the *Welding Handbook*, Vol. 3, 7th Ed., 312-35.

metallographic evidence of the original interface. With diffusion brazing, the metallographic structure should reflect the diffusion of the brazing filler metal into the base metal. There is generally no significant reduction in mechanical properties of the base metal.

Other inherent advantages of DFW and DFB, compared to fusion welding, are: minimization of atmospheric contamination, little or no distortion, potential weight savings, and corrosion resistance equivalent to the base metal. Joining can be done at temperatures where grain growth or phase transformation, that could adversely effect the mechanical properties of the weldment, do not take place.

Cleanliness of the surfaces to be joined has greater importance with these processes than with fusion welding because there is no melting or flow to flush entrapped impurities from the joint. Surface preparation for DFW and DFB is similar to that for fusion welding to remove oxides, organic matter, and any other contaminants. In addition, the surfaces to be joined must be flat and smooth so that intimate, uniform contact can be attained at the interface without the use of excessive pressure.

## Diffusion Welding

The yield strength of titanium diminishes rapidly above 1000°F, so that only a moderate applied stress is required to produce contact over the entire interface at diffusion welding temperatures. Also the diffusion of oxygen and nitrogen proceeds rapidly from the interface into the base metal at high temperatures.

With commercially pure titanium, about 96 percent of the joint area attains intimate contact within 10 minutes at 1600°F in vacuum and under a pressure of 1000 psi. This proceeds by creep of the metal to fill the voids at the interface. Figure 10.10 shows the weld microstructures during various stages of completion. In Fig. 10.10 (A), pressure was removed before completion of the first deformation stage, and significant joint void area is apparent. Figure 10.10 (B) shows the interface after the first deformation stage was essentially complete and pressure was removed. Only diffusion mechanisms were active during heating without pressure. The interfacial boundary has migrated to give an essentially homogeneous microstructure across the original joint interface. The joint mechanical properties are equivalent to base metal. Figure 10.10 (C) shows the joint after pressure was applied for the entire

**Table 10.17**

**Typical room-temperature mechanical properties of electron beam welds in titanium alloys**

| Alloy | Thickness, in. | Type of specimen[a] | Heat treatment[b] | Specimen design | Tensile test | | | | Fracture toughness[c] | | |
|---|---|---|---|---|---|---|---|---|---|---|---|
| | | | | | TS, ksi | YS, ksi | Elong., % | RA % | Specimen Type[d] | Orientation | $K_{Ic}$, ksi·in.$^{1/2}$ |
| Ti-6Al-6V-2Sn | 0.25 | BM | MA | Sheet | 161 | 153 | 13 | – | SEN | – | 44 |
| | | Weld | None | Sheet-long | 175 | 158 | 3.5 | – | SEN | – | 44 |
| | | Weld | 1400°F-4h | Sheet-long transv | 159 | 147 | 12 | – | SEN | – | 57 |
| Ti-6Al-4V | 1.0 | BM | MA | Round | 149 | 141 | 14 | 22 | CT | T-L | 100 |
| | | Weld | 1300°F-5h | Round-transv | 148[e] | 138 | 14 | 20 | CT | T-L | 57 |
| Ti-6Al-4V | 2.0 | BM | MA | Round | 136 | 126 | 9 | 10 | CT | T-L | 106 |
| | | Weld | 1300°F-5h | Round-transv | 133[e] | 126 | 10 | 18 | CT | T-L | 83 |
| Ti-11.5Mo-6Zr-4.5Sn | 1.0 | BM | ST, 1045°F-8h | Round | 172 | 166 | 10 | 15 | CT | T-L | 87 |
| | | Weld | ST, W, 1045°F-8h | Round-transv | 166 | 163 | 8 | 20 | CT | T-L | 42 |

a.  BM—base metal.
b.  MA—mill annealed, ST—solution treated, W—weld.
c.  Refer to ASTM E399-81 for additional information.
d.  SEN—single edge notch, CT—compact tension.
e.  Failed in base metal.

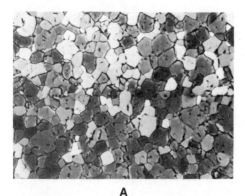

**A**

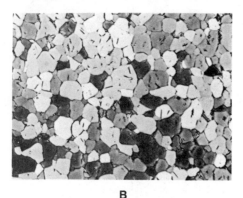

**B**

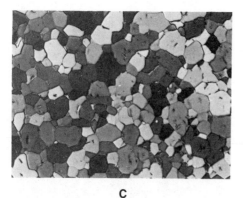

**C**

*Fig. 10.10—The interface of diffusion welds
in titanium made at 1600°F, (A) 1000 psi
for 5 min plus 25 min without pressure,
(B) 1000 psi for 10 min plus 20 min
without pressure, (C) 1000 psi for 30 min
(X250)*

period, and both deformation and diffusion stages were completed. All voids have been completely eliminated, and adequate boundary migration has taken place. The original joint interface has essentially disappeared.

A number of well-established diffusion welding methods are available for joining titanium alloys. Welding can be accomplished using pressures in the range of several hundred to several thousand psi. High pressures are used in conjunction with low welding temperatures and when the assembly is welded in a sealed jacket. Inserts may be used to hold the structure to dimensions. When welding at higher temperatures without an enclosure, maximum pressure is usually limited by the allowable deformation in the part, and this pressure must be determined empirically. Pressures of 300 to 500 psi work well in many cases. In some applications, total assembly deformation and deformation rate, rather than pressure per se, are controlled during welding for process control.

Welding temperature is probably the most influential variable in determining weld quality; it is set as high as possible without causing irreversible damage to the base metal. For the commonly used alpha-beta type titanium alloys, this temperature is about 75° to 100°F below the beta-transus temperature. For example, Ti-6Al-4V alloy with a beta-transus of approximately 1825°F is best diffusion welded between 1700° and 1750°F. The time required to achieve high weld strength can vary considerably with other factors, such as mating surface roughness, welding temperature, and pressure. Welding times of 30 to 60 minutes should be considered a practical minimum, with 2 to 4 hours being more desirable. Mating surface finish and preweld cleaning procedure are two other important considerations. Although the general rule that a smooth mating surface makes welding easier applies, parts with relatively rough, milled or lathe-turned mating surfaces can be successfully diffusion welded provided the welding temperature, time, and pressure are adjusted to accommodate the rough surfaces. Freshly machined mating surfaces only need to be degreased with a suitable solvent prior to welding. A preferred cleaning method is acid cleaning in $HNO_3$-HF acid solution. Any residue remaining from the cleaning operation must be removed by thorough rinsing.

A diffusion weld in Ti-6Al-4V alloy is shown in Fig. 10.11. The weld zone is virtually

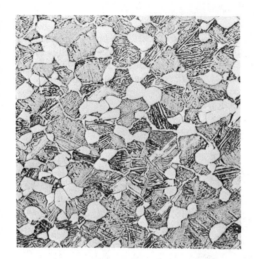

**Fig. 10.11—A diffusion weld in Ti-6Al-4V alloy made at 1700°F for 4 hours with an applied pressure of 500 psi (X250)**

indistinguishable from the base metal in both appearance and properties. This weld was made at 1700°F for 4 hours using an applied pressure of about 500 psi. The faying surfaces were machined on a lathe and acid cleaned prior to welding. Extensive testing of welds of this quality has demonstrated that the mechanical properties of the welds are equivalent to the base metal.

### Diffusion Brazing

Diffusion brazing techniques are also used for joining titanium alloys. Cycle times, temperatures, and preweld cleaning procedures are much the same as for diffusion welding. However, the applied pressure is just sufficient to hold the parts in contact. Mating surface finish requirements are not stringent.

The faying surfaces are electrolytically plated with a thin film of either pure copper or a series of elements, such as copper and nickel. When heated to the brazing temperature at 1650° to 1700°F, the copper layer reacts with the titanium alloy to form a molten eutectic at the joint interface. The assembly is held at temperature for at least 1.5 hours, or is given a subsequent heat treatment at this temperature for several hours, to reduce the composition gradient in the joint. A typical diffusion brazed joint between two Ti-8Al-1Mo-1V alloy sections is shown in Fig. 10.12. A Widmanstätten structure formed at

the joint as a result of copper diffusion into the titanium alloy.

Diffusion brazed joints made with a copper interlayer and a cycle of 1700°F for 4 hours exhibited tensile, shear, smooth fatigue, and stress-corrosion properties equal to those of the base metal. However, they had slightly lower notch fatigue and corrosion fatigue properties, and significantly low fracture toughness.

Diffusion brazing is being used to fabricate light-weight cylindrical cases of titanium alloys for jet engines. In this application, the titanium core is plated with a very thin layer of selected metals that react with the titanium to form a eutectic. During the brazing cycle in a vacuum of $10^{-5}$ torr, a eutectic liquid forms at 1650°F. This liquid performs the function of a brazing alloy between the core and face sheets. The eutectic quickly solidifies due to rapid diffusion at the joints. The assemblies are held at temperature for one to four hours to reduce the composition gradient at the joint.

## FRICTION WELDING

There are basically two variations of friction welding; continuous-drive and inertia. In the former, the parts to be joined are brought together while the rotational drive is still engaged. In the

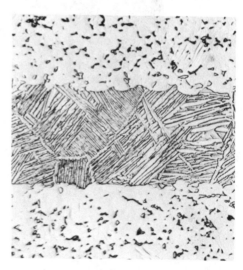

**Fig. 10.12—Diffusion brazed joint in Ti-8Al-1Mo-1V alloy made with a thin copper interlayer (500)**

latter, the rotational energy is stored in a fly wheel and then applied for welding. Both variations are suitable for welding titanium, and produce joints that are similar in many respects.

The basic requirements for obtaining a good friction weld are rapid heating of the joint surfaces to the forging temperature and immediate application of forging pressure to bring the interfaces into intimate contact and then extrude metal from the interface. The time to form a sound inertia friction weld, or the time from initial interface contact until rotation stops, ranges from less than one to four seconds. Inertia friction welds in titanium alloys are generally made using tangential speeds in the range of 1000 to 2500 ft/min and forging pressures in the range

of 8 to 15 ksi. The speeds are higher and the forging pressures are lower than those used for alloy steels and nickel-base, high-temperature alloys. Adequate upsetting of the joint is necessary to ensure that any contaminated metal is extruded from the weld interface. A typical cross section of a friction weld in Ti-8Al-1Mo-1V titanium alloy is shown in Fig. 10.13. It shows the small grain size normally found in an inertia weld. The small grain size and the absence of metallurgical inhomogeneity, normally associated with fusion welding, result in welded joints possessing mechanical properties that approach those of the base metal. Joint efficiencies in tension of 100 percent are readily obtained in titanium alloys.

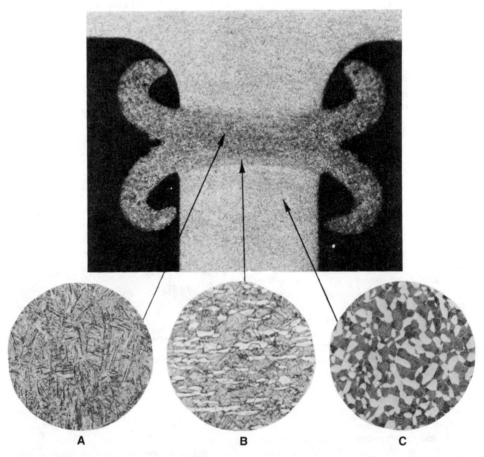

**Fig. 10.13—An inertia friction weld between a 0.19-in. thick ring and boss of Ti-8Al-1Mo-1V Alloy (X10); (A) weld interface, (B) upset heat-affected zone, (C) base metal (X250) (reduced to 77% for reproduction)**

Most friction welds in titanium alloys can be used in the as-welded condition. However, somewhat better tensile and fatigue properties can be expected if the welds are stress-relieved or solution treated and aged.

# RESISTANCE WELDING

## Surface Treatment

In resistance spot and seam welding, the welding heat is generated at the faying surfaces by resistance heating caused by the contact resistance between them. The contact resistance is normally not uniform in a joint when there are variations in surface roughness and cleanliness. Nonuniform contact resistance causes variation in weld quality.

When welding titanium, surface condition is important to provide consistency of weld strength and to avoid nugget embrittlement caused by dissolved surface oxides. Mechanical or chemical treatment can be used to reduce surface resistance. Cleaning with a stainless steel wire brush can produce an acceptable surface condition (about 100 microhms) for resistance welding. However, for large components, chemical cleaning is most efficient. A contact resistance of 50 microhms or lower can be obtained by pickling in HF-HNO$_3$ solution, as described previously. The efficiency of the pickling solution decreases with use. Therefore, the pickling time should be increased as the solution efficiency decreases or the efficiency of the pickling solution should be restored by adding hydrofluoric acid.

Because titanium has high affinity for oxygen, the contact resistance increases with storage time, and the rate of increase varies with the environment in which cleaned titanium is stored. Chemically cleaned titanium should be handled with clean, lint-free gloves and stored under low-humidity conditions for not more than 48 hours before welding.

## Electrodes

RWMA Class 2 copper alloy electrodes are generally used for spot and seam welding of titanium alloys. Spherical-faced spot welding electrodes with a 3-in. radius are generally recommended for titanium, but electrodes with other face contours can also be used. Radius-faced wheel electrodes with the same radius as the wheel are recommended for seam welding to produce a nearly round nugget.

## Gas Shielding

Inert gas shielding is not required when spot welding titanium sheet because (1) the faying surfaces are in contact under pressure prior to heating, (2) the weld times are generally less than 0.2 second, and (3) the cooling rates are very rapid. However, trailing shields are recommended when seam welding because the weld is still hot after it moves from between the wheel electrodes. Shields, similar to those used for arc welding, are placed on both sides of the joint, as shown in Fig. 10.14. The length of the shields depends upon the welding speed.

Inert gas shielding is also recommended for flash welding of titanium. The shield should conform to the geometry of the joint, and provide for easy removal of spatter from the flashing action.

## Welding Schedules

*Spot Welding.* Because of its relatively low electrical and thermal conductivity, titanium is considered to be easier to join by spot welding than aluminum and many carbon and low-alloy steels with respect to power requirements. It can be welded with most types of resistance welding machines, ac or dc, and it is less sensitive to electrode force variations and electrode geometry.

Metallurgically, titanium alloys can be considered as weldable by resistance welding processes because porosity, hot cracking, and cold cracking are rare. For commercially pure titanium, the welding schedules for stainless steel can be used because the thermal conductivity and electrical resistivity of both are about the same, as shown in Table 10.18. However, these properties of titanium alloys are different from those of stainless steels, and welding schedules must be modified for optimum results. Table 10.19 shows the recommended spotwelding schedules for two common titanium alloys.

The low electrical conductivity of titanium alloys results in less shunting of current through previous spot welds than with steel and aluminum alloys. Therefore, the spot spacing of multiple spot welds in titanium alloys can be closer than in the other metals.

Fig. 10.15 shows cross sections through spot welds in commercially pure titanium sheet and Ti-8Al-1Mo-1V alloy sheet. The nugget in the commercially pure titanium sheet has an equiaxed grain structure. The columnar grain structure in the alloy is attributed to the effect of alloying on solidification.

**Fig. 10.14—Inert gas trailing shields installed on a resistance seam welding machine**

*Seam Welding.* Titanium and titanium alloys are also easily seam welded. The wheel electrode diameter and face contour, electrode force, welding current, and heat time affect the size of the individual nuggets.

The electrode force and welding current for seam welding should be slightly higher than those used for spot welding the same sheet thickness.

**Table 10.18**
**Electrical resistivity and thermal conductivity of titanium alloys and steels**

| Material | Electrical resistivity, $10^{-6}\,\Omega\cdot$cm | Thermal conductivity, W/(M•K) |
|---|---|---|
| Commercially-pure Ti | 57 | 16 |
| Ti-5Al-2.5 Sn | 157 | 7.5 |
| Ti-6Al-4V | 171 | 6.6 |
| AISI 1025 steel | 17 | 51 |
| Type 410 sst | 57 | 25 |
| Type 304 sst | 72 | 16 |
| Type 6061-T6 Al | 4 | 167 |

## Flash Welding

The machine capacity required to weld titanium and titanium alloys does not differ greatly from that required for steel. This is especially true for transformer capacity. The upsetting force required for making titanium flash welds is not so high as that required for steel.

Joint designs for flash welds in titanium are similar to those used for other metals. Flat edges are satisfactory for welding sheet and plate up to about 0.25-in. thick. For thicker sections, the edges are sometimes beveled slightly. The metal allowance needed for flash welds, which include the metal lost during flashing and upsetting, is slightly less than for similar sections of steel.

External inert-gas shielding is not absolutely necessary, but is strongly recommended. For joints in tubing or assemblies with hollow cross sections, the air inside should be displaced with inert gas prior to welding.

The flash welding conditions that are of greatest importance are the flashing current, speed, and time, and also the upset pressure and distance. With proper control of these variables,

**Table 10.19**
**Typical spot welding schedules for two titanium alloys**

| Alloy | Thickness combination, in. | Welding current (ac), kA | Electrode face radius, in.[a] | Electrode force, lbs | Weld time, cycles[b] | Tension-shear strength, lbs | Cross-tension strength, lbs | Nugget diam. in. |
|---|---|---|---|---|---|---|---|---|
| Ti-5A1-2.5Sn | 0.062-0.062 | 10 | 3 | 1500 | 10 | 5750 | 1250 | – |
| Ti-6A2-4V | 0.035-0.035 | 5.5 | 3 | 600 | 7 | 1720 | 600 | 0.25 |
| | 0.062-0.062 | 10.6 | 3 | 1500 | 10 | 5000 | 1000 | 0.36 |
| | 0.093-0.093 | 12.5 | 3 | 2400 | 16 | 8400 | 2100 | 0.43 |
| | 0.020-0.040 | 6-8 | 4 | 400 | 8 | 1300 | – | – |
| | 0.020-0.062 | | | | | | | |
| | 0.020-0.040-0.020 | 5-6 | 4 | 600 | 8 | 1400 | – | – |
| | 0.040-0.063 | 8-9 | 4 | 600 | 10 | 2500 | – | – |
| | 0.040-0.063-0.020 | 8-9 | 4 | 800 | 10 | 3000 | – | – |

a. RWMA Class 2 copper alloy electrodes
b. 60 Hz

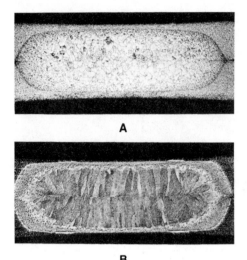

**A**

**B**

*Fig. 10.15—Cross sections through spot welds in (A) commercially-pure titanium sheet and (B) Ti-8Al-1Mo-1V alloy sheet*

the metal at the joint interface will be at the proper temperature for welding and all molten metal will be extruded from the joint.

Generally, fast flashing rates and short flashing times are used to weld titanium and titanium alloys. These conditions are desirable to minimize weld contamination. A parabolic flashing curve is more desirable than a linear flashing curve because maximum joint efficiency can be obtained with a minimum of metal loss. Low-to-intermediate upset pressures are used (7,000 to 20,000 psi).

Flash welds that have mechanical properties approaching those of the base metals are easily produced with conventional flash welding machines. The static and fatigue properties of flash welded joints are good. Most tension specimens fail at locations other than the weld interface with strengths that are about the same as those of the base metal. In fatigue tests, more failures occur in the weld than in tensions tests, but the incidence of weld failures is not high.

## CORROSION RESISTANCE OF WELDS

Welded joints in commercially-pure titanium, made with proper welding procedures, have essentially the same corrosion resistance to specific media as does the base metal. However, iron and possibly nickel or chromium exceeding 0.05 percent in the weld metal can cause selective accelerated corrosion of a weld in strong, oxidizing acid solutions.

There is no differential corrosion of weld metal or base metal in titanium alloy weldments that are made with matching filler metal and then properly heat treated. Discrete differences in the corrosion resistance of heat-treatable alpha-beta titanium alloys in halide acids or salts has been observed with different heat treatments of the same alloy. The corrosion rates of alpha and alpha-beta titanium alloys are definitely affected by minor changes in alloying elements present. This effect also applies to weld metal.

Loss of alloying elements from the filler metal during transfer across the arc, grain size, and varying microstructures in the heat-affected zone are potential causes of differences in the corrosion resistance of the weld metal, heat-affected zone, and base metal. The presence of residual stresses can accelerate corrosion and lead to stress corrosion cracking. Proper filler metal chemistry and appropriate welding procedures (including welding atmosphere) in combination with a suitable postweld heat treatment are important when fabricating a titanium alloy weldment for corrosion applications.

## BRAZING

### Base Metals

Brazing operations have very little effect on the properties of commercially-pure titanium and the alpha alloys. When used in the annealed condition, the beta alloys are virtually unaffected by a brazing thermal cycle. If they are to be heat treated, the brazing temperature can have an important effect on their properties. The best ductility is obtained when a beta titanium alloy is brazed at its solution treating temperature. As the brazing temperature is increased above this temperature, the ductility of the metal decreases.

Depending on composition, the mechanical properties of alpha-beta titanium alloys can be altered greatly by heat treatment and the resulting variations in microstructure. Wrought alpha-beta titanium alloys are generally processed to obtain a fine-grained, equiaxed, duplex microstructure that provides maximum ductility. When brazing an alpha-beta alloy, it is desirable to maintain this microstructure by brazing at a temperature that does not exceed the transformation temperature (beta transus), which usually is in the 1650° to 1900°F range. Alpha-beta alloys can be used in the annealed or the solution-treated and aged condition. If an annealed structure is desired after brazing, three alternatives are possible, namely:

(1) Anneal and then braze at or below the annealing temperatures.

(2) Braze above the annealing temperature and incorporate a step-cooling operation in the brazing cycle to obtain an annealed microstructure.

(3) Braze above the annealing temperature and then anneal after the brazing operation is completed.

Many commercial brazing filler metals flow at temperatures above titanium alloy annealing temperatures, and the latter two alternatives are most commonly used.

## Metallurgical Considerations

An understanding of the chemical and metallurgical characteristics of titanium is needed to ensure proper selection of a suitable brazing filler metal and brazing procedure. One characteristic is that titanium readily alloys with most molten brazing filler metals and, therefore, is easily brazed. However, there is a strong tendency for excessive alloying to occur in the brazed joint. This can result in undercutting, formation of brittle intermetallic compounds, low joint strength and ductility, and embrittlement of thin titanium sheet.

Two precautions can be taken to minimize alloying in brazed joints, namely: (1) Select a brazing filler metal that either alloys only slightly with titanium or does not form brittle intermetallic compounds. (2) Use a brazing cycle that keeps the filler metal molten for only a short time. Strong joints can be produced when these precautions are taken.

Some titanium alloys crack under tensile stress in contact with certain molten silver-base brazing filler metals, as well as molten cadmium and mercury. The behavior differs from stress-corrosion cracking but there are similarities. Liquid metal embrittlement appears to result from diffusion along grain boundaries and formation of brittle phases.

This problem should be evaluated when selecting a silver-base brazing filler metal for titanium. Titanium components should not come in contact with cadmium or mercury prior to welding or brazing, or during service at elevated temperatures.

## Processes

*Induction Brazing.* Induction heating is well suited for brazing titanium. It is especially recommended for use with filler metals that alloy readily with titanium to utilize short brazing cycles and minimize alloying in the joints. Inert gas protection is normally used for induction brazing operations although vacuum can be used with brazing alloys that have low vapor pressures when molten.

*Furnace Brazing.* Furnace heating is frequently used for brazing titanium. Furnace brazing is easily adapted to various joint designs and large assemblies. However, the filler metal is molten for a longer period of time than with induction brazing. Therefore, selection of a filler metal that does not alloy excessively with titanium is important.

Titanium can be brazed in a pure inert gas atmosphere or in a vacuum. The operation can be done in a cold-wall vacuum furnace that can achieve a vacuum of $10^{-4}$ torr or better, or in a sealed retort that is placed in an air furnace. The use of a retort generally requires longer heating and cooling cycles than when brazing is done directly in a furnace.

*Torch Brazing.* Oxyacetylene torch brazing can be used but a special flux and a skilled brazer are required. The flux is preplaced in the joint, and the filler metal is preplaced or fed into the joint during brazing. The use of preplaced filler metal is recommended with some fluxes. A slightly reducing flame is used. The brazing temperature is limited by the characteristics of the flux.

Torch brazing is more economical than induction and furnace brazing. However, the problems of (1) removing flux residue and surface contamination after brazing and (2) possible flux entrapment in the joint limit the use of torch brazing operations.

## Brazing Filler Metals

Pure silver and certain silver alloys, such as 95Ag-5Al and 92.5Ag-7.5Cu, can be used to braze titanium, and provide useful joint strengths up to about 800°F. The Ti-Ag intermetallic compound is reasonably ductile but the Ti-Cu intermetallic is not. Therefore, the copper content of silver brazing filler metals should be kept low. Lithium additions tend to accelerate the diffusion and alloying of silver base filler metals with titanium.

An Ag-5Al-0.5Mn brazing filler metal has good resistance to salt spray stress corrosion and excellent shear strengths in titanium joints. Its brazing temperature range is 1600° to 1650°F. The diffusion layer at the filler metal-titanium interface is a relatively ductile Ag-Al-Ti solid solution. Shear strengths in the 20 to 30 ksi range can be obtained when brazing Ti-6Al-4V alloy. Brazed joints show good fatigue life in shear and axial tension-tension modes.

Types 3003 and 4043 aluminum alloy and unalloyed aluminum as brazing filler metals for titanium can provide good shear strengths up to about 500°F and good corrosion resistance. The brazing temperature should be in the range of 1200° to 1275°F, depending upon the melting temperature of the filler metal.

For applications requiring a high degree of corrosion resistance or high strength up to about 1000°F, 48Ti-48Zr-4Be or 43Ti-43Zr-12Ni-2Be filler metal is recommended. The Ti-Zr-Be filler metal can also be used to braze titanium to carbon steel, austenitic stainless steel, refractory metals, and other reactive metals. A disadvantage of these filler metals is their high brazing temperatures (about 1600° to 2000°F).

The two brazing filler metals, 82Ag-9Pd-9Ga alloy and 81Pd-14.3Ag-4.6Si alloy, flow well on titanium, provide good strength, and have good corrosion and oxidation resistance. The Ag-Pd-Ga alloy has a brazing range of 1615° to 1690°F, and is available as foil, wire, and powder. Brazing must be done in an inert gas because of the high vapor pressure of silver and gallium at the brazing temperature. Lap-shear strengths with this filler metal average 26ksi.

Copper- and nickel-base filler metals are not recommended for conventional brazing of titanium. They form brittle intermetallic compounds and low-melting eutectic compositions. However, a thin layer of preplaced copper or other filler metal can be used with diffusion brazing to form an in-situ brazing filler metal, formed by diffusion and alloying at brazing temperature. The braze layer diffuses into the titanium during a diffusion heat treatment. For copper film, diffusion brazing can be done at 1425° to 1450°F for 15 min or longer, depending on the film thickness.

In the rather specialized field of electronic and high vacuum equipment, copper, ceramics, and magnetic alloys are commonly brazed to titanium. A binary titanium-beryllium filler metal is often used. A 49 Cu-49Ti-2Be alloy filler metal is also satisfactory.

In all applications, the time at brazing temperature should be as short as possible to limit diffusion and alloying in the joint.

## Surface Preparation

The preparation of clean, scale-free surfaces for brazing is as important as for welding. The cleaning procedures that are described previously for welding are satisfactory for brazing.

Special surface preparations are sometimes used when titanium is to be brazed to other metals. The most common of these is to deposit a thin layer of brazing filler metal on the titanium or the other metal in the joint area. The thin overlay is used primarily when the brazing flux is not satisfactory for both metals in torch-brazing operations, or when the filler metal does not readily wet one of the base metals. This procedure can be used to torch braze titanium to mild steel and stainless steel, and to induction or furnace braze titanium to stainless steel with silver-base filler metals in argon without a flux.

## Fluxes and Atmospheres

Argon, helium, and vacuum atmospheres are satisfactory for brazing titanium. Argon is preferred to helium because of its lower thermal conductivity. For torch brazing, special fluxes must be used. Fluxes for titanium are primarily mixtures of fluorides and chlorides of the alkali metals, sodium, potassium, and lithium. The possibility exists of stress-corrosion cracking by flux entrapment. All flux residue must be removed to avoid corrosion problems.

Vacuum, argon, or helium is used to protect titanium during furnace and induction brazing operations. Vacuums of $10^{-4}$ to $10^{-5}$ torr are required to prevent discoloration of the titanium. A dew point of $-70°F$ or lower is needed to prevent discoloration of titanium in inert gas at a brazing temperature in the range of 1400° to 1700°F.

## THERMAL CUTTING

The problems involved in the elevated temperature cutting of titanium and titanium alloys hinge on the tendency of titanium to absorb large quantities of oxygen, nitrogen, and hydrogen, and the deleterious effects of these elements upon the mechanical properties. Several methods have been tried with varying degrees of success. The optimum method is determined by the economic factors involved, as well as the quality of the cut.

Titanium can be severed by oxyfuel gas cutting. This method depends upon the chemical reaction and formation of the oxide. Consequently, the depth of cut and speed of travel are much better in titanium than in steel. Titanium can be cut approximately three times faster than an equivalent thickness of steel.

This type of cutting results in a contaminated and hardened surface, and some type of edge preparation is necessary after cutting and before welding. The main hardening depth is normally less than 0.010 in., but the overall hardened zone can extend up to 0.06-in. deep.

Chemical analysis of flame-cut edges indicates that most of the hardening is a result of oxygen contamination. Transformation products account for additional hardening. It is recommended that welds not be made between torch-cut edges. The contaminated titanium should be removed by machining to soft metal prior to joining the surfaces by welding or brazing. There does not appear to be a suitable method of thermal cutting titanium and titanium alloys without the additional prewelding step of machining or grinding. It should not be inferred, however, that oxyfuel gas cutting should not be done, since it is an economical method of cutting unusual shapes or heavy sections.

## SAFE PRACTICES

The possibility of spontaneous ignition of titanium and titanium alloys during welding is extremely remote. Like magnesium and aluminum, the occurrence of fires is usually encountered where an accumulation of grinding dust or machining chips exists. Even in extremely high surface-to-volume ratios, accumulations of clean titanium particles do not ignite at any temperature below incipient fusion temperature in the air.

However, spontaneous ignition of fine grinding dust or lathe chips saturated with oil under hot humid conditions has been reported. Water or water-based coolants should be used for all machining operations. Carbon dioxide is also a satisfactory agent. Large accumulations of chips, turnings, or other metal powders should be removed and stored in enclosed metal containers. Dry grinding should be done in a manner that will allow proper heat dissipation.

Dry compound extinguishing agents or dry sand are effective for titanium fires. Ordinary extinguishing agents such as water, carbon tetrachloride, and carbon dioxide foam are ineffective.

Violent oxidation reactions can occur between titanium and liquid oxygen or red-fuming nitric acid.

# ZIRCONIUM

## UNALLOYED ZIRCONIUM

Zirconium is very similar in characteristics to titanium except that its density is about 50 percent higher. Physical properties of importance in welding or brazing are given in Table 10.1. Zirconium has a hexagonal close-packed crystal structure (alpha phase) up to about 1600°F, and transforms to body-centered cubic (beta phase) above this temperature.

A visible oxide forms on zirconium in air at about 400°F, and it develops a loose white scale with long time exposure over 800°F. The metal reacts slowly with nitrogen at about 700°F and more rapidly with temperatures above 1500°F. At high temperatures, oxygen and nitrogen are readily soluble in the metal. Contami-

nation with these elements increases the tensile strength but reduces ductility. Commercially pure, annealed zirconium (ASTM Grade R60702) has a tensile strength of about 60 ksi, a yield strength (0.2 percent) of about 40 ksi, and an elongation of about 18 percent. These properties vary with the purity of the metal.

Zirconium is very resistant to corrosive attack by most organic and mineral acids, strong alkalies, and some molten salts. In addition it is corrosion resistant in water, steam, sea water, and liquid metals at elevated temperatures and pressures. The corrosion-resistant behavior is a result of the formation of a stable, dense, adherent, and self-healing zirconium oxide film on the metal surface.

Zirconium is used for corrosion application

in the petrochemical, chemical process, and food processing industries. The chemical composition of commercially pure zirconium is given in Table 10.20. Zirconium alloys are used in the nuclear industry for pressurized water reactor components because of their low thermal neutron absorption characteristic. Nuclear applications limit the hafnium content of zirconium alloys to 0.010 percent.

## ALLOYING AND PHASE STABILIZATION

Alloying and impurity elements in zirconium generally tend to stabilize the alpha or beta phase, as with titanium. Most alloying elements have only limited solid solubility in zirconium. Above this limit, they exist as intermetallic compounds in an alpha matrix.

### Alpha Stabilizers

Certain alloying elements stabilize the alpha phase and raise the alpha to beta transformation temperature of zirconium. This group includes aluminum, beryllium, cadmium, carbon, hafnium, oxygen, nitrogen, and tin. Tin is the only one used in commercial alloys. Small amounts of oxygen strongly affect the transformation temperature. With about 1000 ppm oxygen, the transformation to beta starts at about 1600°F; an increase to 1650 ppm raises the temperature to about 1690°F. Zirconium ores generally contain a substantial amount of hafnium that remains in the metal unless intentionally removed for nuclear applications.

### Beta Stabilizers

Alloying elements that stabilize the beta phase and lower the transformation temperature include cobalt, columbium, copper, hydrogen, iron, manganese, molybdenum, nickel, silver, tantalum, titanium, tungsten, and vanadium. Of these, columbium is the only one used commercially as an alloying element. Zirconium is embrittled by the absorption of hydrogen, which takes place rapidly between 600° and 1800°F.

### Low Solubility, Intermetallic Compounds

Carbon, silicon, and phosphorous have very low solubility in zirconium, even at temperatures in excess of 1830°F. They readily form stable intermetallic compounds, and are relatively insensitive to heat treatment.

As increasing amounts of most alloy elements are added to zirconium, strength and corrosion resistance are often increased. This is true up to the solubility limit of the alloying element in zirconium. Once this solubility limit is exceeded, intermetallic compounds or second phases are formed which tend to lower the ductility and degrade the corrosion resistance of the zirconium alloy.

## COMMERCIAL ALLOYS

Zirconium is alloyed commercially with columbium or tin and small additions of chromium, iron, and nickel. The commonly used commercial alloys are given in Table 10.20. The equiva-

### Table 10.20
### Chemical compositions of commercial zirconium and zirconium alloys

| ASTM Grade | Composition, percent[a] | | | | |
| --- | --- | --- | --- | --- | --- |
| | Zr + Hf, min[b] | Sn | Cb | Fe + Cr | 0,max |
| R60702[c] | 99.2 | – | – | 0.2 max | 0.16 |
| R60704[d] | 97.5 | 1.0-2.0 | – | 0.2-0.4 | 0.18 |
| R60705[e] | 95.5 | – | 2.0-3.0 | 0.2 max | 0.18 |
| R60706[e] | 95.5 | – | 2.0-3.0 | 0.2 max | 0.16 |

a. H–0.0005 percent max.
   N–0.025 percent max.
   C–0.05 percent max.
b. Hf–4.5 percent max.
c. Grade R60001 is nuclear grade unalloyed zirconium.
d. A similar nuclear Grade, R60802, is commonly called Zircaloy-2, and Grade R60804 is called Zircaloy-4.
e. The similar nuclear Grade is R60901.

lent nuclear grades, commonly known as Zircoloy-2, Zircoloy-4, and Zr-2.5Cb, are deliberately low in hafnium and impurity elements that form intermetallic compounds. The amount and distribution of the intermetallic phases are critical for corrosion resistance to high temperature steam and hot water.

The Zr-1.5Sn alloy has a tensile strength of about 70 ksi, a yield strength of about 55 ksi, and an elongation of about 20 percent at room temperature. Values for equivalent properties of Zr-2.5Cb alloy are 85 ksi, 65 ksi, and 20 percent respectively.

## HEAT TREATMENT

Zirconium is usually annealed at about 1450°F for 20 minutes at temperature in a neutral or slightly oxidizing atmosphere or in vacuum. Annealing of welds is not normally required because of the good ductility of zirconium alloys. However, weldments can be stress relieved at about 1250°F for 15 to 20 minutes at temperature, followed by air cooling. Commercial zirconium alloys are normally supplied and used in the annealed or cold-worked and stress-relieved condition.

## CLEANING

Zirconium and zirconium alloys are cleaned and stored for welding in the same manner as titanium. These procedures are described previously.

## WELDING FILLER METALS

Bare welding electrodes and rods are produced to weld the commercial grades of zirconium and zirconium alloys listed in Table 10.20.[10] Type ERZr2 is used to weld commercially pure zirconium (Grade R60702), Type ERZr3 to weld Zr-1.5Sn alloy (Grade R60704), and Type ERZr4 to weld Zr-2.5Cb alloy (Grades R60705 and R60706). These filler metals should not be used to weld nuclear grade zirconium and zirconium alloys because of the higher allowable hafnium and impurity contents. Sheared strip or specially produced wire should be used to weld nuclear grade material.

---

10. Zirconium welding filler metals are covered by AWS A5.24-79, *Specification for Zirconium and Zirconium Alloy Bare Welding Rods and Electrodes.*

## WELDABILITY

Zirconium and the aforementioned zirconium alloys are readily welded using the processes and procedures described previously for titanium. Zirconium has a low coefficient of thermal expansion, and this contributes to low distortion during welding. Residual welding stresses are low because of its low modulus of elasticity. Welded joints in unalloyed zirconium do not normally crack unless they are contaminated during welding. Inclusions are also not usually a problem in zirconium welds because the metal has a high solubility for its own oxide.

Some porosity might be found in zirconium welds. It can usually be prevented by applying one or more of the following corrections:

(1) Increasing the welding current

(2) Decreasing the welding speed

(3) Applying a low preheat to drive off volatiles

(4) Adjusting the inert gas coverage

(5) Using proper preweld cleaning procedures for base and filler metals

The welding technology is similar, if not identical, for both industrial and nuclear alloy grades with the exception that nuclear specifications generally require the use of high purity atmospheres. This can only be achieved by the use of vacuum-purged welding chambers that can be evacuated to a pressure of $10^{-4}$ torr and then back filled with high-purity inert gas. This precaution is required to preserve the corrosion resistance of zirconium alloys in high-temperature, pressurized water.

When the Zr-1.5Sn alloy is electron beam or laser beam welded in vacuum, tin can be lost by evaporation from the molten weld pool because of its high vapor pressure. One function of the tin in zirconium is to improve corrosion resistance. Therefore, excessive remelting of the weld metal in vacuum should be avoided for critical corrosion applications.

## RESISTANCE WELDING

Resistance welding of zirconium and zirconium alloys can be performed using approximately the same conditions described previously for unalloyed titanium. The same precautions, cleaning procedures, and shielding requirements as outlined for titanium should be observed. Clean zirconium sheet can be stored in a clean room under low humidity conditions for up to 14 days without an excessive increase in surface contact resistance.

Resistance welding processes that have been successfully employed to weld zirconium and zirconium alloys include spot, flash, upset, and high-frequency welding. Uniform resistance spot and seam welds can be made when the contact resistance between the faying surfaces is consistent and less than 100 microhms. Typical conditions for spot welding unalloyed zirconium sheet are given in Table 10.21. Zirconium sheet, 0.062 in. thick, can be seam welded using the conditions given in Table 10.22.

## BRAZING

Zirconium and its alloys are brazed using the same processes, equipment, cleaning methods, and procedures described previously for titanium. The development of brazing filler metals for zirconium has been limited to those with suitable corrosion resistance to high temperature water in a nuclear reactor environment. Suggested brazing filler metals for zirconium are given in Table 10.23.

The 95Zr-5Be filler metal, with a melting range of 1780 to 1815°F, can be used to braze zirconium to itself and other metals, such as stainless steel. This filler metal, which is usually applied in powder form, exhibits good wetting and flow properties. There is no noticeable erosion or attack of the base metal during brazing, and it forms a smooth braze fillet. Brazements show good corrosion resistance to water at 680°F.

Because of its ability to wet ceramic surfaces, the 95Zr-5Be filler metal can be used to braze zirconium to uranium oxide and beryllium oxide.

### Table 10.21
### Typical spot welding schedules and strengths for unalloyed zirconium sheet

| | Sheet thickness, in.[a] | |
|---|---|---|
| | 0.062 | 0.110 |
| Electrode force, lb[b] | 300 | 700 |
| Weld time, cycles (60 Hz) | 6 | 12 |
| Welding current, kA (ac) | 8-11 | 12-16 |
| Tensile-shear strength, lb. | 2900-3650 | 5900-6150 |
| Normal tensile strength, lb. | 450-550 | 1250-1400 |

a. Two equal thicknesses
b. RWMA Class 2 copper alloy electrodes with 4-in. spherical radius faces

### Table 10.22
### Typical conditions for seam welding two 0.062-in. zirconium sheets

| | |
|---|---|
| Electrode force, lb[a] | 500 |
| Heat time, cycles (60Hz) | 6 |
| Cool time, cycles (60Hz) | 9 |
| Welding current, kA (ac) | 8-11 |
| Welding Speed, in./min | 32 |
| Nuggets/in. | 8 |
| Nugget penetration, percent | 60-75 |
| Nugget overlap, percent | 12-30 |

a. Wheel electrodes of RWMA Class 2 copper alloy, 8-in. diameter, 4 in. radius faces

Like titanium, zirconium and zirconium alloys can be diffusion brazed using a thin film plating of copper or other metal to form an in situ liquid phase. Joints can be diffusion brazed using copper at 1900°F for 0.5 to 2 hours at temperature under a pressure of about 30 psi. Molten metal will form in a joint after about 0.25 hour at 1550°F. During subsequent exposure to a higher temperature, the copper diffuses into the zirconium and forms a solid joint. A distinct layer of filler metal does not exist.

## DISSIMILAR METALS

Zirconium can only be fusion welded successfully to the other reactive metals: titanium, columbium, tantalum and hafnium. When welded to other metals, such as iron or copper, extremely brittle intermetallics are formed that in turn, embrittle the weld metal.

Zr-1.5Sn alloy can be diffusion welded to stainless steels by hot rolling or with hot isostatic gas pressure. For hot rolling, the parts are enclosed in a welded steel jacket that is evacuated of air prior to rolling. The steel used for the

### Table 10.23
### Suggested brazing filler metals for joining zirconium

| Composition, percent | Brazing temperatures, °F |
|---|---|
| 95Zr-5Be | 1840 |
| 50Zr-50Ag | 2770 |
| 71Zr-29Mn | 2500 |
| 76Zr-24Sn | 3150 |

jacket should be pretreated to remove nitrogen. Titanium-killed mild steel can be used as a jacket for rolling temperatures up to about 1475°F. Rolling is normally carried out at temperatures between 1400° and 1600°F. Total reduction is generally about 67 percent with an average reduction of 20 percent per pass. The steel jacket is removed by chemical methods.

Hot isostatic gas pressure welding can be done in an autoclave at about 1500°F with a helium pressure of about 10 ksi. The specific conditions of pressure, temperature, and time at temperature must be developed for the specific application. The assembly itself may be evacuated and sealed by fusion welding. Otherwise, it must be encased in a thin, gas-tight steel envelope that is evacuated and sealed.

Zirconium is explosion welded to steel commercially in plate sizes up to 96 by 360 inches. A low-carbon zirconium plate is used that is more malleable than commercially pure zirconium. Bond strengths range from 20 to 35 ksi.

## SAFE PRACTICES

A strong pyrophoric reaction takes place between dry, red-fuming nitric acid and zirconium. Zirconium in powder or dust form is an extreme fire and explosion hazard in air or oxygen.

# HAFNIUM

## GENERAL CHARACTERISTICS

Hafnium is a bright, ductile metal having chemical properties similar to zirconium. It has a hexagonal close-packed crystal structure up to 3200°F, and a body-centered cubic structure above this temperature. Its density is approximately three times that of titanium. Hafnium has good corrosion resistance in water, steam, and molten alkali metals, superior to that of zirconium. It is resistant to dilute hydrochloric and sulfuric acids, various concentrations of nitric acid, and boiling or concentrated sodium hydroxide. Hafnium reacts slowly with oxygen in air at 750°F, with nitrogen at 1650°F, and with hydrogen at 1300°F. Typical mechanical properties of unalloyed, annealed nuclear grade hafnium are 60 ksi ultimate strength, 25 ksi yield strength, and 25 percent elongation in 2 inches.

Hafnium is used primarily for applications that make use of its unique neutron absorption and corrosion resistant properties. It is used for control rods in nuclear reactors and containers for corrosive media in spent nuclear fuel reprocessing plants.

## HEAT TREATMENT

Hafnium is annealed at about 1450°F for about 10 minutes. Recrystallization takes place in the range of 1475° to 1650°F, depending upon the amount of cold work. Weldments can be stress-relieved at about 1000°F.

## WELDABILITY

Hafnium is readily fabricated by welding using the process and procedures described for titanium. It has a low coefficient of thermal expansion, and this contributes to low distortion during welding. Because of its low modulus of elasticity, residual welding stresses are also low. The metal does not normally crack in welded joints up to 0.75-in. thick, unless it is grossly contaminated during welding.

Hafnium welds have a transformed beta structure with large grains. There is also grain coarsening in the heat-affected zone where the transformation temperature is exceeded. The transformed beta structure in a weld has higher strength and lower ductility than a fine-grained alpha structure in wrought and annealed material.

A weldment should not be given a recrystallization anneal unless it has been cold worked. It can be stress-relieved without altering the metallurgical structure of the weld or heat-affected zone.

The safest and most desirable equipment for a welding operation is in a vacuum-purge chamber capable of a vacuum of better than $10^{-4}$ torr.

Hafnium is subject to severe embrittlement by relatively small amounts of contamination by nitrogen, oxygen, carbon, or hydrogen.

One of the gas-shielded arc welding processes can be used in a chamber filled with inert gas. The parts should be cleaned with a suitable solvent immediately before placing them in the welding chamber. If practical, the joint faces should be abraded using stainless steel wool or a draw file prior to solvent cleaning.

High frequency arc starting is desirable with gas tungsten arc welding to reduce the possibility of contamination by the tungsten electrode.

Sample welds should be made prior to production welding to verify the procedures and welding atmosphere purity. The welds should be capable of being bent 90 degrees around a radius of three times the thickness of the weld sample.

## COMPATABILITY WITH OTHER METALS

Hafnium can be welded to titanium, zirconium, columbium, and tantalum. When welding hafnium to zirconium, the heat should be directed toward the hafnium because of its higher melting temperature. It cannot be welded to the more common metals, such as iron, nickel, cobalt, and copper because brittle intermetallic compounds are formed that embrittle the welds.

# TANTALUM

## GENERAL CHARACTERISTICS

Tantalum is an inherently soft, fabricable metal that has a high melting temperature, which categorizes it as a refractory metal. It has a body-centered cubic crystal structure that is stable to melting. It is one of the heavier metals with a density about twice that of steel (see Table 10.1). Unlike many body-centered cubic metals, tantalum retains good ductility to very low temperatures, and does not exhibit a ductile-to-brittle transition temperature.

Tantalum has excellent corrosion resistance to a wide variety of acids, alcohols, chlorides, sulfates, and other chemicals. For this reason, it is widely used for chemical equipment. The metal also is used in electrical capacitors and for high temperature furnace components.

The metal oxidizes in air above about 570°F and is attacked by hydrofluoric, phosphoric, and sulfuric acids, and by chlorine and fluorine gases above 300°F. Tantalum also reacts with carbon, hydrogen, and nitrogen at elevated temperatures. When dissolved interstitially, these elements and oxygen increase the strength properties and reduce the ductility of tantalum.

In the annealed condition, unalloyed tantalum has a tensile strength of 30 to 50 ksi, a yield strength of 24 to 32 ksi, and an elongation of 20 to 30 percent. Its Charpy V-notch impact strength at −320°F is over 200 ft•lbs.

Tantalum is available as powder metallurgy, vacuum arc melted, and electron beam melted products.[11] Welding of powder metallurgy material is not recommended because the weld would be very porous. Electron beam or vacuum arc melted material is recommended for welding applications.

## TANTALUM ALLOYS

Tantalum alloys are strengthened by solid solution, by dispersion or precipitation, and by combinations of the two. A high degree of mutual solubility exists among all pairs of the four refractory metals: molybdenum, tungsten, columbium, and tantalum. Vanadium is mutually soluble in all four. Zirconium and hafnium are soluble in tantalum to useful extent. Tungsten and molybdenum exert the greatest strengthening effect on tantalum; hafnium has a somewhat lesser effect. Vanadium lowers the melting point. Titanium has little effect as a solid solution strengthener, and it lowers the melting point.

Zirconium, titanium, and hafnium increase the strength of tantalum by formation of small, well-dispersed carbides, oxides, and nitrides with residual and added carbon, oxygen, and nitro-

---

11. See ASTM Specifications B364, B365, and B521 for additional information on tantalum and tantalum alloy mill products.

gen. Strengthening is achieved by pinning of dislocations (dispersion hardening).

Normal substitutional alloying elements tend to raise the ductile-to-brittle transition temperature of welds in tantalum. This increase is very pronounced in welds when the total metallic alloy content exceeds 10 atomic percent. Interstitials in solution also tend to raise the transition temperature, but their effects can be minimized by alloying elements that form stable carbides, oxides, or nitrides.

Tantalum alloys can contain intentional carbon additions that respond to thermal treatments during processing. They realize their strength properties in part from carbide dispersion and also from solution strengthening.

Typical tantalum alloys are listed in Table 10.24. They are basically solid solutions strengthened with tungsten (W) and smaller amounts of hafnium (Hf), columbium (Cb), rhenium (Re), and carbon (C). In wrought form, all of the alloys have good ductility at cryogenic temperatures.

## HEAT TREATMENT

Some complex tantalum alloys, such as T-222, can be age hardened as a result of precipitation of a second phase. Tantalum and its alloys are also given stress-relief and recrystallization heat treatments to lower stresses and improve ductility. Recommended stress-relief and recrystallization temperatures are given in Table 10.25.

## CLEANING

Tantalum and its alloys should be thoroughly cleaned prior to welding or brazing. Rough edges to be joined should be machined or filed smooth prior to cleaning. Components should be degreased with a detergent or suitable solvent and chemically etched in mixed acids. A

**Table 10.25**
**Recommended heat treating temperatures for tantalum and tantalum alloys**

| Composition | Temperature, °F | |
| | Stress-relief | Recrystallization |
|---|---|---|
| Unalloyed Ta | 1650 | 1800-2200 |
| Ta-2.5W-0.15Cb | 1800 | 2200-2400 |
| Ta-10W Ta-8W-2Hf Ta-10W-2.5Hf-0.01C | 2000 | 2400-3000 |

solution of 40 percent nitric, 10 to 20 percent hydrofluoric acid, up to 25 percent sulfuric acid, and remainder water is suitable for pickling, followed by hot and cold rinsing in deionized water and spot-free drying.

The cleaned components should be stored in a clean room of controlled low humidity until ready for use.

## WELDABILITY

Tantalum and tantalum alloys are readily welded using the processes and procedures described for titanium. Contamination by oxygen, nitrogen, hydrogen, and carbon should be avoided to prevent embrittlement of the weld. Preheating is not necessary.

The high melting temperature of tantalum can result in metallic contamination if fixturing contacts the tantalum too close to the weld joint. Copper, nickel alloys, or steel fixturing could melt and alloy with the tantalum. If fixturing is required close to the joint, a molybdenum insert should be used in contact with the tantalum. Graphite should not be used as fixturing because it will react with hot tantalum to form carbides.

Resistance spot welding of tantalum is feasible but adherence and alloying between copper alloy electrodes and tantalum sheet is a problem. Welding under water might be helpful because of the improved cooling of the copper electrodes. Weld time should not exceed 10 cycles (60 Hz).

Tantalum can be fusion welded to columbium, titanium, and zirconium because the combinations form solid solutions in all proportions. Reasonably ductile joints can be expected. However, fusion welds between tantalum and common structural metals are brittle because brittle intermetallics are formed in the weld metal.

**Table 10.24**
**Typical commercial tantalum alloys**

| Common designation | Nominal Composition, percent |
|---|---|
| KBI-10 | 97.5Ta-2.5W |
| Ta-10W[a] | 90Ta-10W |
| FS-63 | 97.4Ta-2.5W-o.15Cb |
| T-111 | 90Ta-8W-2Hf |
| T-222 | 87.5Ta-10W-2.5Hf-0.01C |
| Astar 811C | 90.3Ta-8W-1Re-0.7Hf-0.025C |

Tantalum can be joined to steel by explosion welding for cladding applications. Special techniques are required for the welding of tantalum clad steel to avoid contamination of the tantalum by the steel.[12]

## BRAZING

Successful brazing of tantalum and its alloys depends upon the application. For corrosion applications, the brazed joint must also be corrosion resistant. This presents problems in selecting a brazing filler metal because commercially available ones are not so corrosion resistant as tantalum. For high temperature applications, the brazed joint must have a high remelt temperature and possess adequate mechanical properties at the service temperature.

Tantalum must be brazed in a high purity inert atmosphere or in a high vacuum. Special equipment, such as a cold-wall vacuum furnace, is usually required when a high temperature brazing filler metal is used. For low temperature brazing, the tantalum can be plated with another metal, such as copper or nickel that is readily wet by the brazing filler metal. The brazing filler metal should alloy with and dissolve the plating during the brazing cycle.

Alloying with and erosion of the tantalum alloy by the filler metal can cause brittleness and low joint strength. Tantalum forms brittle intermetallic compounds with most commercial brazing filler metals. The composition of the filler metal, brazing temperature, and the heating cycle affect the degree of interaction between the two metals. In general, the brazing time should always be a minimum unless diffusion brazing techniques are used.

### Low Temperature Brazing

Tantalum can be brazed with nickel-base filler metals, such as the nickel-chromium-silicon alloys. However, tantalum and nickel form brittle intermetallic compounds. Such filler metals are satisfactory for service temperatures up to about 1800°F. Copper-gold filler metals with

---

12. Refer to Chapter 12 for information on joining clad steel.

---

less than 40 percent gold can be used. With higher gold content, brittle compounds are formed in the joint. Silver-copper filler metals can produce joints with useful strengths at room temperature.

### High Temperature Brazing

Brazing filler metals of the Ta-V-Ti and Ta-V-Cb alloy systems that might be suitable for certain high temperature applications are given in Table 10.26. Brazing should be done in vac-

**Table 10.26**
**Typical brazing filler metals for tantalum alloys for service up to 2500°**

| Filler metal composition, weight percent | Temperature, °F | |
|---|---|---|
| | Brazing | Remelt |
| 10Ta-40V-50Ti | 3200 | 4350 |
| 20Ta-50V-30Ti | 3200 | 4350 |
| 25Ta-55V-20Ti | 3350 | 4000 |
| 30Ta-65V-5Ti | 3350 | 4350 |
| 5Ta-65V-30Cb | 3300 | 4170 |
| 25Ta-50V-25Cb | 3400 | 4530 |
| 30Ta-65V-5Cb | 3400 | 4170 |
| 30Ta-40V-30Cb | 3500 | 3630 |

uum of $10^{-4}$ torr or better, but there is a tendency to vaporize titanium and vanadium from the molten filler metal during the brazing cycle because of their high vapor pressures. Reaction between the tantalum alloy and these filler metals should be low, and brazed joints in tantalum should have excellent stability at temperatures up to 2500°F. The filler metals have excellent room temperature ductility, and can be cold rolled to foil. As-brazed joints can be expected to be ductile at room temperature.

Brazing filler metals developed for brazing tantalum alloys for service temperatures in the 2500° to 3500°F range are given in Table 10.27. The strengths of three of the alloys are enhanced by a diffusion heat treatment, and they are all ductile at room temperature. Brazing with these alloys should also be done under high-vacuum conditions.

**Table 10.27**
**Typical brazing filler metals for tantalum alloys for service upt to 3500°F**

| Filler metal composition, weight percent | Brazing temperature, °F[a] | Diffusion treatment | | Remelt temperature, °F | Lap shear strength | |
|---|---|---|---|---|---|---|
| | | Temperature, °F | Time, h | | Temperature, °F | psi |
| 93Hf-7Mo | 3800 | 3700 | 0.5 | 4060 | 3000 | >3700 |
| | | | | | 3500 | 5250 |
| | | | | | 3800 | 1300 |
| 60Hf-40Ta | 3980 | – | – | 3800 | 3500 | 3700 |
| 66Ti-34Cr | 2700 | 2600 | 16 | 3780 | 2500 | 9100 |
| | | | | | 3000 | 2400 |
| 66Ti-30V-4Be | 2400 | 2050[b] | 4.5 | 3800 | 2500 | 5800 |
| | | 2400 | 16 | | 3000 | 1100 |
| | | | | | 3500 | 470 |

a. One minute at temperature
b. Duplex heat treatment

# COLUMBIUM

## GENERAL CHARACTERISTICS

Columbium, sometimes called niobium, is both a reactive metal and a refractory metal with characteristics similar to those of tantalum. Its density is only half that of tantalum and its melting temperature is lower (see Table 10.1). Like tantalum, this metal also has a body-centered cubic crystal structure, and does not undergo allotropic transformation.

Annealed high-purity columbium has a minimum tensile strength of 25 ksi, a minimum yield strength of 15 ksi, and a minimum elongation of 25 percent in 1 inch.[13] The mechanical properties of columbium are highly dependent on its purity, particularly with respect to interstitial impurities, oxygen, nitrogen, hydrogen, and carbon. For example, the Charpy-V-notch impact strength of electron-beam-melted columbium is 150 ft·lbs to below −200°F, while that of vacuum arc melted columbium of lower purity drops to about 10 ft·lbs at 0°F.

Columbium oxidizes rapidly at temperatures above about 750°F and absorbs oxygen interstitially at elevated temperatures, even in atmospheres containing small concentrations. It absorbs hydrogen between 500° and 1750°F. The metal also reacts with carbon, sulfur, and halogens at elevated temperatures. It forms an oxide coating in most acids that inhibits further chemical attack. Exceptions are diluted strong alkalis and hydrofluoric acid.

## COLUMBIUM ALLOYS

Columbium is alloyed with the other refractory metals (tantalum, tungsten, molybdenum, and hafnium), titanium, and zirconium to provide solid solution strengthening. Zirconium also provides age-hardening characteristics. Typical weldable columbium alloys and their compositions are listed in Table 10.28. These alloys can be processed into sheet, plate, and other forms having good strength and ductility at room temperature and below.

---

13. Refer to ASTM Specification B391-78a, B392-80, B393-78, and B394-78 for additional information.

### Table 10.28
### Typical columbium alloys

| Common designation | Nominal composition, percent |
|---|---|
| Cb-1Zr[a] | 99Cb-1Zr |
| B-66 | 89Cb-5Mo-5V-1Zr |
| C-103 | 89Cb-10Hf-1Ti |
| C-129Y | 80Cb-10W-1Hf-0.1Y |
| Cb-752 | 87.5Cb-10W-2.5Zr |
| FS-85 | 61Cb-28Ta-11W-1Zr |
| SCb-291 | 80Cb-10Ta-10W |

a. Commercial grade, UNS No. R04261; Reactor grade, UNS No. R04251

## HEAT TREATMENT

Typical annealing and stress-relieving temperatures for columbium alloys are given in Table 10.29. Heat treatment must be done in a high-purity inert gas or in high vacuum to avoid contamination by the atmosphere. Vacuum is generally the most practical.

Columbium alloys containing zirconium respond to aging after solutioning at high temperature. Fusion welds in such alloys are sensitive to aging during service in the range of 1500° to 2000°F, depending upon the alloy composition. For example, welds in Cb-1Zr alloy age in the 1500° to 1800°F range. Aging of welds in service can significantly decrease their ductility at room

### Table 10.29
### Typical heat treatments for columbium and columbium alloys

| Alloy | Temperature, °F | |
|---|---|---|
| | Annealing | Stress Releif[a] |
| Unalloyed Cb | 1650-2200 | 1400-1475 |
| Cb-1Zr | 1800-2200 | 1650-1800 |
| B-66 | 2200-2500 | 2000 |
| C-103 | 1900-2400 | 1600 |
| C-129Y | 1800-2400 | 1800 |
| Cb-752 | 2200-2400 | 1800-2000 |
| FS-85 | 2000-2500 | 1850 |
| SCb-291 | 2400-2550 | 2000 |

a. One hour at temperature

temperature or raise their ductile-to-brittle transition temperature. Annealing of the welded joint effectively overages the weld and heat-affected zones, and prevents aging in service. For the Cb-1Zr alloy, annealing at 2200°F for 1 hour is adequate to prevent aging of welded joints in service.

## CLEANING

Columbium and its alloys can be cleaned and pickled using the solvents and etchants described for tantalum. Cleaned components should be stored in a clean room under low humidity conditions.

## WELDABILITY

Most columbium alloys have good weldability, provided the tungsten content is less than 11 percent. With higher tungsten content in combination with other alloying elements, weld ductility at room temperature can be low.

The processes, equipment, procedures, and precautions generally used to weld titanium are also suitable for columbium. One exception is welding in the open with a trailing shield. Although columbium can be welded with this technique, contamination is likely to be excessive for columbium applications. Therefore, it is not recommended.

### Gas Tungsten Arc Welding

Gas tungsten arc welding of columbium

should be done in a vacuum-purged welding chamber backfilled with helium or argon. Helium is generally preferred because of the higher arc energy. Contamination by the tungsten electrodes should be avoided by either using a welding machine equipped with a suitable arc-initiating circuit or striking the arc on a run-on tab.

Typical conditions for gas tungsten arc welding columbium alloys are given in Table 10.30. They generally apply to B-66, C-129Y, Cb-752, FS-85, and SCb-291 alloys. To obtain uniform welds, automatic welding is preferred. Those alloys that are prone to aging should be welded at relatively high travel speeds with the minimum energy input needed to obtain desired penetration. Copper backing bars and hold downs can be used to extract heat from the completed weld. The purpose of these procedures is to minimize aging.

### Mechanical Properties

Typical room-temperature tensile strengths and bend transition temperatures of welds in columbium alloys are compared with those of the base metals in Tables 10.31.

### Resistance Welding

Resistance spot welding of columbium presents the same problems as those encountered with tantalum. Electrode sticking can be minimized by making solid-state pressure welds between two sheets rather than actual nuggets. However, weld consistency might be a problem. Projection welding might be an alternative to

---

**Table 10.30**
**Typical conditions for gas tungsten arc welding columbium alloy sheet**

| Sheet thickness, in. | Shielding gas | Welding current, A | Arc voltage, V | Travel Speed, in./min |
|---|---|---|---|---|
| 0.020 | 80He-20Ar | 40 | 12 | 7.5 |
| | | 55 | 12 | 15.0 |
| | | 70 | 11.5 | 30.0 |
| 0.030 | He | 80 | 17.5 | 7.5 |
| | 80He-20Ar | 90 | 11.5 | 15.0 |
| | 80He-20Ar | 130 | 11.5 | 30.0 |
| 0.060 | He | 130 | 13.5 | 7.5 |
| | | 160 | 15.0 | 15.0 |
| | | 200 | 15.0 | 30.0 |

**Table 10.31**
**Tensile and bend transition temperature properties of welded and unweld columbium alloy sheet**

| Alloy | Sheet thickness, t, in. | Type of specimen[a] | Tensile strength, ksi | Yield strength, ksi | Elong- ation, percent | Bend transition temp., °F[b] |
|-------|-------|-------|-------|-------|-------|-------|
| B-66 | 0.020 | GTAW | 103 | 81 | 10 | -120 |
|  |  | EBW | 107 | 82 | 18 | -120 |
|  |  | Unwelded | 109 | 87 | 25 | – |
|  | 0.060 | GTAM | 108 | 85 | 19 | 0 |
|  |  | EBW | 109 | 88 | 20 | 0 |
|  |  | Unwelded | 108 | 93 | 24 | -300 |
| C-129Y | 0.020 | GTAW | 102 | 78 | 10 | -250 |
|  |  | EBW | 89 | 70 | 9 | -250 |
|  |  | Unwelded | 97 | – | 24 | -300 |
| Cb-752 | 0.030 | GTAW | 82 | 66 | 20 | -150 |
|  |  | EBW | 84 | 64 | 20 | -150 |
|  |  | Unwelded | 85 | 70 | 24 | – |
|  | 0.060 | GTAW | 74 | 57 | 17 | -50 |
|  |  | EBW | 81 | 61 | 19 | -170 |
|  |  | Unwelded | 77 | 59 | 24 | -300 |
| FS-85 | 0.020 | GTAW | 79 | 70 | 2 | -250 |
|  |  | EBW | 90 | 77 | 3 | -270 |
|  |  | Unwelded | 110 | 91 | 13 | – |
|  | 0.060 | GTAW | 94 | 79 | 10 | -150 |
|  |  | EBW | 96 | 82 | 7 | -100 |
|  |  | Unwelded | 104 | 91 | 16 | -300 |

a.  GTAW – gas tungsten arc weld
    EBW – electron beam weld
b.  45 degree bend around a 1t radius.

spot welding, but some form of shielding might be required to avoid contamination by the atmosphere.

Cooling of the spot welding electrodes with liquid nitrogen can substantially decrease electrode sticking and deterioration. Techniques are available to minimize coolant consumption, but the cost of the liquid nitrogen must be considered when evaluating the process for a particular application.

## BRAZING

The brazing of columbium alloys for high-temperature service has been the subject of several studies. Experimental brazing filler metals based primarily on titanium and zirconium were evaluated, along with other alloys with high melting temperature. Those generally recommended for brazing columbium alloys, in order of increasing brazing temperature, are given in Table 10.32. They readily wetted and flowed on columbium alloys at brazing temperature in high vacuum. The strength of brazed joints was significantly improved with some filler metals by a postbraze diffusion heat treatment.

Prior to selecting a brazing filler metal for a columbium alloy fabrication, it should be thoroughly evaluated by appropriate brazing procedure tests, metallurgical examination, and mechanical property testing for the intended service. Erosion and alloying with the base metal should

**Table 10.32**
**Brazing filler metals for columbium alloys**

| Filler metal composition, weight percent | Brazing temperature °F |
|---|---|
| 48Ti-48Zr-4Be | 1920 |
| 75Zr-19Cb-6Be | 1920 |
| 66Ti-30V-4Be[a] | 2350-2400 |
| 91.5Ti-8.5Si | 2500 |
| 67Ti-33Cr[a] | 2650-2700 |
| 73Ti-13V-11Cr-3A1[b] | 2950 |
| 90Pt-10Ir | 3300 |
| 90Pt-10Rh | 3450 |

a. Diffusion heat treatment at about 300°F below the brazing temperature for several hours improves high-temperature joint strength.
b. Commercial titanium alloy, UNS No. R58010.

be carefully determined by appropriate tests.

Columbium alloy pipe has been brazed to Type 316 stainless steel pipe and Hastelloy X (UNS No. N06002) pipe using nickel-base brazing filler metals. For the columbium-to-stainless steel joint, the brazing filler metal composition was Co-21Cr-21Ni-5.5W-8Si-0.8B. Brazing was done at 2150°F in high vacuum. For the columbium to Hastelloy X joint, AWS Type BNi-4 brazing filler metal (Ni-3.5Si-2B) was used with a brazing temperature of 2025°F in high vacuum. A tongue-in-groove joint design was used with the tongue on either member. In any case, the joint should be designed so that the brazing layer is in compression after it has cooled to room temperature. The brazing filler metals are brittle and should not be stressed in tension.

# Metric Conversion Factors

$t_C = 0.56(t_F - 32)$
1 torr = 0.13 kPa
1 ft³/hr = 0.472 L/min
1 in. = 25.4 mm
1 in./min = 0.423 mm/s
1 ft·lb = 1.36 J
1 ksi = 6.89 MPa
1 MW/in.² = 1.55 X 10³ MW/m²
1 W/(m·K) = 418 cal/(cm·s·°C)

# SUPPLEMENTARY READING LIST

### GENERAL

*Metals Handbook,* Vol. 3, 9th Ed., Metals Park, OH: Amer. Soc. for Metals, 1980: 314-417, 755-64, 781-91.

Schwartz, M.M., *The Fabrication of Dissimilar Metal Joints Containing Reactive and Refractory Metals,* New York: Welding Research Council Bulletin 210, 1975 Oct.

### BRAZING

*Brazing Manual,* 3rd Ed., Miami: American Welding Society, 1976: 213-20.

Cole, N.C., *Corrosion Resistance of Brazed Joints,* New York: Welding Research Council Bulletin 247, 1979 Apr.

Pattee, H.E., *High Temperature Brazing,* New York: Welding Research Council Bulletin 187, 1973 Sept.

Schwartz, N.M., *Brazed Honeycomb Structures,* New York: Welding Research Council Bulletin 182, 1973 Apr.

### COLUMBIUM

Fox, C.W. Gilliland, R.G., and Slanghter G.M., Development of alloys for brazing columbium, *Welding Journal,* 42(12): 535s-40s; 1963 Dec.

Gerken, J.M., Welding characteristics of advanced columbium alloys, *Welding Journal,* 45(5): 201s-26s; 1966 May.

Kearns, W.H., Young, W.R., and Redden, T.K., Fabrication of a columbium alloy liquid metal loop, *Welding Journal,* 45(9): 730-39; 1966 Sept.

Korb, L.J., Beuyukian, C.S., and Rowe, J., Diffusion bonded columbium panels for the shuttle heat shield, SAMPE Quarterly, 3(4): 1-11; 1972 July.

Watson, G.K., and Moore, T.J., Brazed nickel-columbium dissimilar metal pipe joints for 720°C service, *Welding Journal,* 56(10): 306s-13s; 1977 Oct.

Yount, R.E. and Keller, D.L., Structural stability of welds in columbium base alloys, *Welding Journal,* 45(5): 227s-34s; 1966 May.

### TANTALUM

Kammer, P.A., Monroe, R.E., and Martin, D.C., Weldability of tantalum alloys, *Welding Journal,* 51(6): 304s-20s; 1972 June.

Stone, L.H., Freedman, A.H., Mikus, E.B., Brazing alloys and techniques for tantalum honeycomb structures, *Welding Journal,* 46(8): 343s-50s; 1967 Aug.

### TITANIUM

Becker, D.W. and Blaeslack, W.A. III, Property-microstructure relationships in metastable beta titanium alloy weldments, *Welding Journal,* 59(3): 85s-92s; 1980 March.

Blaeslack, W.A. III, and Banas, C.M., A comparative evaluation of laser and gas tungsten arc weldments in high-temperature titanium alloys, *Welding Journal,* 60(7): 121s-30s; 1981 July.

Elrod, S.D., Lovell, D.T., and Davis, R.A., Aluminum brazed titanium honeycomb sandwich structure, *Welding Journal,* 52(10): 425s-32s; 1973 Oct.

Greenfield, M.A., and Duvall, D.S., Welding of an advanced high strength titanium alloy, *Welding Journal,* 54(5): 73s-70s; 1975 May.

Kellerer, H.G. and Milacek, L.H., Determination of optimum diffusion welding temperatures for Ti-6A1-4V, *Welding Journal,* 49(5): 219s-24s; 1970 May.

Key, R.E., Burnett, L.I., and Inouye, S., Titanium structural brazing, *Welding Journal,* 53(10): 426s-31s; 1974 Oct.

Kimball, C.E., Aluminum brazed titanium acoustic structures, *Welding Journal,* 59(10): 26-30; 1980 Oct.

McHenry, H.E., and Key, R.E., Brazed titanium fail-safe structures, *Welding Journal,* 53(10): 432s-39s; 1974 Oct.

Messler, R.W. Jr., Electron beam weldability of advanced titanium alloys, *Welding Journal,* 60(5): 79s-84s; 1981 May.

Mitchell, D.R., and Feige, N.G., Welding of alpha-beta titanium alloys in one-inch plate,

*Welding Journal,* 46(5): 193s-202s; 1967 May.

Mullins, F.D., and Becker, D.W., Weldability study of advanced high-temperature titanium alloys, *Welding Journal,* 59(6): 177s-82s; 1980 June.

Nessler, C.S., Eng, R.D., and Vozzella, P.A., Friction welding of titanium alloys, *Welding Journal,* 50(9): 379s-85s; 1971 Sept.

Pease, C.C., Compatibility studies of capacitor discharge stud welding on titanium alloys, *Welding Journal,* 48(6): 253s-57s; 1969 June.

Titanium and Titanium Alloys, M1L-HDBK-697A, 1 June 1974.

Wells, R.R., Microstructural control of thin-film diffusion-brazed titanium, *Welding Journal,* 55(1): 20s-27s; 1976 Jan.

Wu, K.C., Correlation of properties and microstructure in welded Ti-6A1-6V-2Sn, *Welding Journal,* 60(11): 219s-26s; 1981 Nov.

Wu, K.C., and Krinke, T.A., Resistance spot welding of titanium alloy 8A1-1Mo-1V, *Welding Journal,* 44(8): 365s-71s; 1965 Aug.

ZIRCONIUM

Beal, R.E. and Saperstein, Z.P., Brazing filler metals for Zircaloy, *Welding Journal,* 50(7): 275s-91s; 1971 July.

Ferrill, D.A., Fatigue crack propagation in Zircaloy-2 weld metal, *Welding Journal,* 50(5): 206s, 230s, 234s; 1971 May.

Johnson, R.E., and Schaaf, B.W., A study of metallurgical effects in the multipass welding of Zircaloy, *Welding Journal,* 37(1): 1s-9s; 1958 Jan.

Nippes, E.F., Savage, W.F., and Wu, K.C., Spot and seam welding of Zircaloy 3, *Welding Journal,* 39(3): 97s-104s; 1960 March.

# 11
# Other Metals

## Chapter Committee

C. E. JACKSON, *Chairman*
    *Ohio State University*
J. BARTH
    *Ohio State University*
R. BURMAN
    *AMAX Specialty Metals Corporation*

W. D. KAY
    *Wall Colmonoy Corporation*
P. W. TURNER
    *EG&G, Incorporated*
C WALTER
    *TAPCO International*

J. F. SMITH
*Lead Industries Association*

## Welding Handbook Committee Member

I. G. BETZ
    *U.S. Army ARDC*

# 11

# Other Metals

## BERYLLIUM

### PROPERTIES

Beryllium has a hexagonal close-packed crystal structure which partly accounts for its limited ductility at room temperature. Some properties of beryllium that are important in welding are given in Table 11.1. The melting point and specific heat are about twice those of aluminum or magnesium. Its density is about 70 percent that of aluminum, but its modulus of elasticity is about 4 times greater. Therefore, beryllium is potentially useful for lightweight applications where good stiffness is needed. It is used in many nuclear energy applications because of its low neutron cross section.

Beryllium mill products are normally made by powder metallurgy techniques using several consolidation methods. Wrought products are produced from either cast or powder metallurgy billets. Cold-worked material may have good ductility in only one direction, and low ductility perpendicular to that direction (anisotropy). The tensile properties of beryllium may vary greatly, depending on the manner of processing.

An adherent refractory oxide film will rapidly form on beryllium, as with aluminum and magnesium. This oxide film will inhibit wetting, flow, and fusion during welding and brazing. Therefore, parts must be adequately cleaned prior to joining. The joining process and procedures must prevent oxidation during the operation by appropriate shielding media and techniques. Inert gas or vacuum is appropriate.

### SURFACE PREPARATION

Prior to welding or brazing, beryllium com-

ponents should be degreased, then pickled in either a 10 percent hydrofluoric acid solution or a 40 percent nitric acid-5 percent hydrofluoric acid solution and finally, ultrasonically rinsed with deionized water. Subsequent handling should be done in a manner that will not recontaminate the sufaces to be joined.

### FUSION WELDING

Two principal problems associated with the fusion welding of beryllium are (1) the control of grain size in the weld metal and heat-affected zone, and (2) the susceptibility to cracking because of the inherent low ductility of the metal. Studies have shown that intergranular microcracking in welds is caused by precipitates at the grain boundaries. These precipitates are binary or ternary compounds associated with impurities in the beryllium. These impurities include aluminum, iron, and silicon.

Electron beam welding appears to be the best fusion welding method. The low heat input with this process can provide a narrow heat-affected zone, minimal grain growth, and low distortion.

Limited success may be possible with the gas tungsten arc welding process. Welding with alternating current and argon shielding is recommended for manual operation. With automatic welding, direct current may be used. However, heat input is critical with both methods of welding. A low effective heat input and a high cooling rate are desirable. These conditions help to minimize the grain size in the weld metal and heat-affected zone.

Grain size has a significant effect on the

**Table 11.1**
**Physical properties of beryllium, uranium, molybdenum, and tungsten**

| Property | Units | Beryllium | Uranium | Molybdenum | Tungsten |
|---|---|---|---|---|---|
| Density (20°C) | Mg/m$^3$ | 1.85 | 19 | 10.2 | 19.3 |
| Melting point | °C (°F) | 1290 (2355) | 1130 (2065) | 2620 (4750) | 3390 (6130) |
| Coefficient of linear | | | | | |
| Thermal expansion (20°C) | μm/(m·K) | 11.5 | 12 | 4.8 | 4.5 |
| Thermal conductivity | W/(m·K) | 188 | 27.6 | 138 | 175 |
| Specific heat (20°C) | J/(kg·K) | 1886 | 117 | 276 | 138 |
| Electrical resistivity | nΩ·m | 44.5 | 300 | 51 | 54 |
| Modulus of elasticity | GPa (10$^6$ psi) | 300 (43.5) | 166 (24) | 324 (47) | 345 (50) |

mechanical properties of both the base metal and the weld metal. The strength and ductility of weldments are somewhat related to the grain sizes of the weld metal and the heat-affected zone. An average grain size of less than $30 \times 10^{-6}$m is desirable. Welded joints with fine grain structures exhibit better room temperature properties than those with coarse grain structures. However, coarse grain material has better mechanical properties above 1000°F.

Thermal stresses resulting from high cooling rates can cause weld cracking. Consequently, weld cooling rates must be adjusted to avoid cracking, whatever the grain size. In the majority of applications, weld metal grain sizes above 100 micrometers are deemed unacceptable.

Resistance spot welding can only be used for low strength applications. Cracking will likely occur around fine-grained weld nuggets of relatively high strength. When the weld heat input is adjusted to avoid cracking, a coarse-grained dendritic nugget of low strength is produced.

## DIFFUSION WELDING

Beryllium may be joined by diffusion welding. Lap joints in beryllium sheet and plate can be diffusion welded at 1400°F or above with a minimum applied pressure of 1500 psi. A minimum deformation, ranging from 5 percent for 0.062-in. thick sheet to 15 percent for 0.25-in. thick plate, is required to obtain a strong weld. The welding conditions and joint strengths appear to be related to the purity of the metal and the cooling rate from the welding temperature. Cracking, excessive grain growth, and other problems associated with fusion welding are not experienced with diffusion welding. A diffusion aid, such as a thin intermediate layer of silver, may be used to accelerate diffusion across the interface.

## BRAZING

Selection of a proper brazing technique and filler metal depends upon the service temperature, joint geometry, and required joint strength. Good capillary flow of filler metal is difficult wherever beryllium oxide is present on the surfaces to be brazed. Therefore, a brazing filler metal that has a low melting temperature and provides the desired mechanical properties is generally recommended.

Filler metals normally considered for brazing beryllium are the aluminum-silicon and silver-base alloys. Aluminum-silicon filler metals containing 7.5 or 12 percent silicon (BAlSi-2 or BAlSi-4) can provide high joint strengths up to about 300°F. However, they exhibit poor flow in capillary joints. Consequently, the filler metal must be preplaced in the joint.

Silver and silver-based filler metals are normally used for brazing beryllium for high temperature applications. Type BAg-19 filler metal (Ag-7%Cu-0.2%Li) is an example. Lithium is added to pure silver in small amounts to improve wettability, but even so, capillary flow is poor. Preplacement of the filler metal is recommended, but outside corners and edges of the joint sometimes show a lack of adequate filler metal flow. This leaves a notch at the joint. In this case, it is often necessary to machine the joint to a smooth surface to avoid the severe notch sensitivity of beryllium.

Alloying and grain boundary penetration into the beryllium can also be a problem. Brazing times should be short to minimize this problem.

Silver-copper filler metals that melt at relatively low temperatures may exhibit less grain boundary penetration. However, the copper will form brittle intermetallic compounds with beryllium, and this will cause low joint strength. Fast heating and short brazing times will minimize alloying in the joint. These metallurgical problems should be evaluated by appropriate tests before attempting to braze beryllium assemblies with a silver-copper filler metal.

Beryllium is most commonly brazed in argon, although helium or a vacuum of $10^{-3}$ torr or lower is also suitable. A mixture of 60 percent lithium fluoride-40 percent lithium chloride or tin-chloride may be suitable flux for furnace or induction brazing. However, flux should not be used when vacuum brazing.

Beryllium and beryllium compounds in flux residues are toxic. Only approved installations should be considered for brazing beryllium regardless of the methods used.

## BRAZE WELDING

Beryllium may be braze welded using a gas shielded arc welding process or electron beam welding. The filler metal is added in the form of wire or preplaced shim stock. The best results are obtained using aluminum-12 percent silicon (ER4047) filler metal. Joint strengths can equal that of the base or filler metal.

The joint must be completely filled with

filler metal. The face of the weld and the root surface must blend smoothly with the base metal because of the notch sensitivity of berryllium. Discontinuities at the fusion faces may cause the beryllium to crack at those locations. Excessive restraint during braze welding may also cause cracking.

## SOLDERING

On the basis of strength, wettability, and flow, high zinc solders appear best for soldering beryllium. A reaction type of flux suitable for aluminum is recommended. The flux residue should be removed in a safe manner to avoid any health hazards.

## SAFE PRACTICES

Beryllium and its compounds in the form of dusts, fume, or vapors are toxic and a serious health hazard. Inhalation or ingestion of beryllium in any form must be avoided. It can cause acute or chronic lung disease and other health problems.

Hazardous concentrations of beryllium vapor may build up in the atmosphere when metallic beryllium is heated above 1200°F or when beryllium oxide is heated above 2800°F. Therefore, all welding and brazing operations should be done in a welding chamber or brazing furnace that is exhausted through a system of high efficiency filters. Appropriate precautions must be taken when cleaning this equipment to avoid inhalation or ingestion of beryllium dust that may have collected on interior walls, components, or equipment during use. Users should comply with appropriate safety and health standards applicable to beryllium.

# URANIUM

## PROPERTIES

Uranium in its natural form is a controlled nuclear metal that emits weak alpha radiation. An enriched form is primarily used in nuclear reactors and explosives. Depleted uranium is used for counterweights in aircraft as well as x-ray shielding. The physical properties of uranium that are of concern in welding are given in Table 11.1. The metal transforms from an orthorhombic to a complex tetragonal crystal structure at 1250°F. Above 1520°F, the crystal structure changes to body-centered cubic.

Uranium is a very active metal chemically. It oxidizes rapidly in air and corrodes quickly in water. The oxide is very stable and difficult to reduce to the metal. For this reason, uranium cannot be successfully fluxed during welding or brazing. Uranium metal is always covered with an oxide film that varies in color from straw to dark brown and black, depending upon the time of exposure and the ambient temperature. Clean uranium has a silver-gray color. Welding must be done in an inert atmosphere or good vacuum to prevent oxidation.

## WELDING

Specific limitations on impurity levels must be maintained when welding uranium and its alloys. Proper cleaning and shielding practices must be followed. Certain impurities in uranium can cause welding problems. In particular, hydrogen can cause delayed cold cracking, iron can contribute to hot shortness, and carbon forms uranium carbides.

Alloys that contain molybdenum, columbium, titanium, or zirconium can be welded by careful selection of the welding process and conditions. Electron beam and gas tungsten arc welding are particularly suitable because they can provide excellent protection against oxidation. Friction and laser beam welding show promise in the laboratory.

### Surface Preparation

Uranium oxides on joint faces and adjacent surfaces may cause erratic arc action, incomplete fusion, or inclusions in the weld metal during fusion welding. The common method of removing oxides prior to welding is electropolishing in a solution of phosphoric acid saturated with chromium trioxide ($CrO_3$). An alternate method is pickling in a bath of equal parts of nitric acid and water.

Silver plating may be used to protect clean surfaces from reoxidation. During welding, the silver floats to the surface of the weld metal and offers some protection. However, adequate inert gas coverage is still required. Silver plating of the filler rod can help protect the heated section from oxidation during welding.

## Welding Position

Uranium is readily melted with a welding arc to form a fluid weld pool. The high fluidity of the weld metal makes control of the molten pool difficult, particularly when welding in other than the flat position. A grooved, nonconsumable backing of graphite or water-cooled copper can be used to control drop-thru and produce a smooth root surface. The metal can be electron beam welded in the horizontal position provided a narrow weld bead can be used.

## Gas Tungsten Arc Welding

Gas tungsten arc welding is well suited for joining uranium alloys. Welding should be done in a gas-tight chamber containing either argon or helium. Helium gives greater penetration and a smoother, narrower weld bead than argon. A mixture of 85 percent helium and 15 percent argon provides good arc stability.

Direct current electrode negative power is normally used, although some work indicates a slight oxide cleaning effect with ac power. As a guide in establishing suitable welding conditions for sheet gages, the welding amperage should be 1.3 to 1.7 x 10$^3$ times the thickness in inches. The welding speed should be about 18 in./min.

## Gas Metal Arc Welding

Uranium can be welded at higher speeds with gas metal arc welding than with gas tungsten arc welding. A silver-coated electrode has been particularly successful with this process. Arc spatter is frequently experienced. This can produce uranium oxide of aerosol fume dimensions which can remain suspended in the air for some time. Therefore, the welder must use a suitable face mask filter to prevent inhalation of the fume. Contamination of clothing and the surrounding area must also be controlled.

## Electron Beam Welding

This process is well suited for welding uranium alloys in a vacuum environment. The very intense heat of the beam permits welding at high travel speeds to produce very narrow, deep welds with correspondingly narrow heat-affected zones. Lack of fusion (spiking) may be observed when producing partial penetration welds. Improved

control of beam power may correct this problem.

Binary uranium alloys containing molybdenum, columbium, titanium, or zirconium have been electron beam welded. Uranium-7.5 percent columbium-2.5 percent zirconium alloy is considered weldable. A photomacrograph of a section through a typical multipass electron beam weld in a uranium-0.75 percent titanium alloy is shown in Fig. 11.1.

Most uranium alloys are hardenable. Therefore, a postweld heat treatment may be necessary to provide desirable mechanical properties.

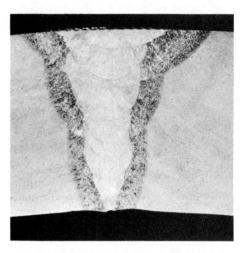

**Fig. 11.1–Section through a multiple pass electron beam weld in uranium–0.75 percent titanium alloy ( × 5)**

## SOLDERING

Soldering of uranium has been evaluated with a number of solders. Electroplating of the uranium with a solderable metal and the use of an ultrasonic soldering iron improve wetting by the solder. A 50 percent indium-50 percent tin solder may be suitable for uranium. However, its melting range is rather low (243°-260°F).

## SAFE PRACTICES

Adequate protection of personnel is of utmost importance with uranium. As mentioned previously, uranium emits alpha radiation which can be shielded with suitable protective clothing.

Uranium and uranium oxide can enter the body by abrasion, ingestion, and inhalation. Of these, inhalation is by far the most serious and must be controlled. Filtered exhaust facilities must be provided to control air contamination. When personnel are subjected to dust and fumes of a welding operation, respiratory protection utilizing particulate filters must be used. Air contamination should be monitored and in compliance with government standards.

The handling of uranium with bare hands is not recommended because it increases the possibility of introducing the metal or its oxide into the body through the mouth or skin abrasions. Shower and washing facilities should be provided for those handling the metal. Workers must change footwear and clothing when leaving areas where uranium fume or dust is present.

There are fire hazards with uranium under certain conditions. In massive form, spontaneous combustion is a remote possibility. However, chips, powder, or dust may ignite, especially in the presence of moisture. Some uranium alloys in coarser particle sizes can react at a sufficient rate to heat themselves to ignition. These materials should be stored in moisture-free oil, inert gas, or vacuum.

Uranium fires are extinguished by cooling the uranium and restricting its access to oxygen. Graphite powder or a dry powdered chemical extinguisher should be used to smother the fire. Water should never be used on uranium fires because explosive hydrogen is formed.

# MOLYBDENUM AND TUNGSTEN

## PROPERTIES

Molybdenum and tungsten are two metals that have very similar properties and weldability characteristics. The physical properties of importance in welding are given in Table 11.1.

Both metals have a body-centered cubic crystal structure and will show a transition from ductile to brittle behavior with decreasing temperature. A number of metallurgical factors can influence the temperature range where this transition will take place. These include the method of production, amount of strain hardening by warm working, recrystallization, grain size, metal purity and alloying, thermal effects of welding or brazing, types of loading, and stress concentrations.

The transition temperatures of recrystallized molybdenum alloys may vary from below to well above room temperature. With tungsten, the transition will be above room temperature. Consequently, fusion welds in these metals and their alloys will have little or no ductility at room temperature. In addition, preheating to near or above the transition temperature will be necessary to avoid cracking from thermal stresses.

These metals and their alloys are consolidated by powder metallurgy and sometimes by melting in vacuum. Wrought forms produced from vacuum melted billets generally have lower oxygen and nitrogen contents than do those produced from powder metallurgy billets.

Molybdenum and tungsten have low solubilities for oxygen, nitrogen, and carbon at room temperature. Upon cooling from the molten state or from temperatures near the melting point, these interstitial elements are rejected as oxides, nitrides, and carbides. If the impurity content is sufficiently high, a continuous, brittle grain boundary film is formed that severely limits plastic flow at moderate temperatures. Warm working below the recrystallization temperature breaks up the grain boundary films and produces a fibrous grain structure. This structure will have good ductility and strength parallel to the direction of working but not transverse to it (anisotropy).

When the warm-worked metal is reheated to melting or nearly to melting during welding, the interstitial compounds may be resolutioned. Upon cooling, the compounds may reform at the grain boundaries. At the same time, grain growth and an accompanying reduction in grain boundary surface area will take place. Then, the weld metal and heat-affected zones will be weaker and less ductile than the warm-worked base metal. The ductility of a welded joint is intimately related to the amount of interstitial impurities present and to the recrystallized grain size. Tungsten is inherently more sensitive to interstitial impurities than is molybdenum. Welds that were ductile at room temperature have been produced in molybdenum. However, tungsten welds are brittle at room temperature.

Oxygen and nitrogen may be present in the

metal, or they could be absorbed from the ambient atmosphere during welding. Therefore, welding should be done in a high-purity inert atmosphere or in high vacuum. Because grain size influences the distribution of the grain boundary films and the associated brittleness, welding should be done using procedures that produce fusion and heat-affected zones of minimum widths.

Molybdenum and tungsten are sensitive to the rate of loading and to stress concentration. The ductile-to-brittle transition range is shifted upward by increasing the strain rate. Welded joints are very notch sensitive and, where possible, the weld surface should be finished smooth and faired gradually into the base metal. Notches at the root of the weld should be avoided.

There is no evidence that the mechanical properties of welds in molybdenum and tungsten are improved by heat treatment. Heating to just below the recrystallization temperature of the base metal will relieve residual welding stresses. This could be beneficial by reducing the likelihood of cracking during subsequent handling.

## ALLOYS

Molybdenum is alloyed with small amounts of titanium, zirconium, and carbon to improve the high-temperature strength properties. An alloy designated TZM (0.5%Ti-0.087%Zr-0.015%C) is produced commercially.

An addition of about 20 atomic percent rhenium to molybdenum and tungsten greatly improves their ductility near room temperature. The improved ductility is accompanied by a considerable amount of coarse twinning, thereby reducing flow stresses. There is an indication that the addition of rhenium to molybdenum filler metal improves the ductility of welded joints at room temperature.

A tungsten-25 percent rhenium alloy, available commercially, has better weldability than unalloyed tungsten because of its improved ductility and lower melting temperature. However, it exhibits a tendency toward hot tearing because of segregation of a sigma phase at the grain boundaries upon freezing. This is more pronounced with arc welding than with electron beam welding. A low heat input and a high welding speed are beneficial.

## SURFACE PREPARATION

Prior to welding or brazing, the surfaces must be clean and free of dirt, grease, oil, oxides, and other foreign matter. These materials can inhibit wetting, flow, or fusion and also contaminate the metal.

The components should first be degreased with a suitable, safe solvent. They should then be cleaned by one of the methods given in Table 11.2. In any case, the cleaning media should be thoroughly removed by rinsing with clean water, and then the component dried with hot air.

## WELDING

Some general precautions should be taken when welding these metals. Fusion welding must be done in a pure inert atmosphere or in high vacuum to prevent contamination by oxygen. Fixturing should provide minimum restraint on a weldment, especially when welding a complex structure. The components should be preheated to above the transition temperature of the metal. Weldments should be stress-relieved promptly at a temperature below the recrystallization temperature of the base metal.

### Arc Welding

Molybdenum and tungsten may be joined by the gas tungsten arc welding process using direct current, electrode negative. This is best done in a gas-tight welding chamber capable of maintaining a high purity atmosphere. Argon or helium may be used for shielding. Welds should be made using procedures that give a narrow heat-affected zone with a minimum input of heat.

Run-off tabs are helpful when terminating a weld to avoiding crater cracks in the joint. Grinding of the weld surface to remove bead ripples, surface contamination, or both, may improve weld joint ductility.

### Electron Beam Welding

The electron beam welding process is well suited for joining molybdenum and tungsten because of its high energy density. Narrow, deep welds can be produced by this process using much less energy than with arc welding. Welding in a high vacuum will prevent contamination of the weld metal with oxygen or nitrogen. Figure 11.2 is a photomicrograph of an electron beam weld in TZM molybdenum alloy sheet.

### Friction Welding

Exploratory work suggests that friction welding may be a suitable method for joining

## Table 11.2
## Cleaning methods for molybdenum and tungsten

| Tungsten | Molybdenum |
|---|---|
| Method 1—Immerse in 20% potassium hydroxide solution (boiling) | Method 1—Immerse in a solution of 95% $H_2SO_4$, 4.5% $HNO_3$, 0.5% HF and $Cr_2O_3$ (equivalent to 18.8 g/l) |
| Method 2—Electrolytic etching in 20% potassium hydroxide solution | Method 2—(a) Immerse in an alkaline bath of 10% NaOH, 5% $KMnO_4$ and 85% $H_2O$ by weight operating at 150°-180°F (66-82°C), 5-10 min. |
| Method 3—Chemical etching in 50 vol. % $HNO_3$ − 50 vol. % HF solution | (b) Immerse 5-10 min. in second bath to remove smut formed during first treatment. Bath consists of 15% $H_2SO_4$ 15% HCl, 70% $H_2O$, and 6-10 wt. % chromic acid per unit volume |
| Method 4—Immerse in molten sodium hydroxide | Method 3—For Mo-0.5 Ti alloy |
| Method 5—Immerse in molten sodium hydride | (a) Degrease 10 min. in trichloroethylene |
|  | (b) Immerse in commercial alkaline cleaner for 2-3 min. |
|  | (c) Rinse in cold water |
|  | (d) Buff and vapor blast |
|  | (e) Repeat step (b) |
|  | (f) Rinse in cold water |
|  | (g) Electropolish in 80% $H_2SO_4$ at 120°F (54°C) |
|  | (h) Rinse in cold water |

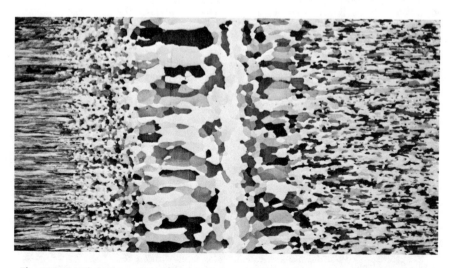

**Fig. 11.2–Section through an electron beam weld between stress-relieved (left) and recrystallized (right) TZM molybdenum alloy sheets ( × 40, reduced 27%)**

molybdenum. The short welding cycle and absence of a fusion zone are attributes of this process. A section through a friction weld in TZM molybdenum alloy is shown in Fig. 11.3.

## Diffusion Welding

Joints can be made in tungsten by diffusion welding in the 1800° to 2000°F temperature range using a thin layer of nickel or palladium as an activator for diffusion. The joint interface will exhibit recrystallization. The interface can be obliterated with an appropriate heat treatment.

Joints are inevitably accompanied by partial recrystallization of the base metal. As a result, they lack room temperature ductility, but there has been no indication of base metal cracking during the joining operation.

## BRAZING

### Filler Metals

A wide variety of brazing filler metals may be used to join molybdenum or tungsten. Typical ones are listed in Table 11.3. The brazing temperatures of these filler metals range from 1200 to 4500°F.

A brazing filler metal must be evaluated by suitable tests for a specific application. In many cases, the service temperature limits the available filler metals for consideration. The effects of brazing temperature, diffusion, and alloying on the base metal properties must also be determined. In any case, the brazing time should be kept as short as possible to minimize recrystallization and grain growth in the base metal.

### Table 11.3
### Typical brazing filler metals for molybdenum and tungsten

| Filler metal | Liquidus, °F |
|---|---|
| Ag | 1765 |
| Au | 1950 |
| Cb | 4475 |
| Cu | 1981 |
| Ni | 2647 |
| Pd | 2826 |
| Pt | 3217 |
| BAg Series[a] | 1145-1780 |
| BAu Series[a] | 1635-2130 |
| BNi Series[a] | 1610-2075 |
| BCo-1[a] | 2100 |
| Au-6Cu | 1815 |
| Au-50Cu | 1780 |
| Au-35Ni | 1970 |
| Au-8Pd | 2265 |
| Au-13Pd | 2380 |
| Au-25Pd | 2570 |
| Au-25Pt | 2570 |
| Cr-25V | 3185 |
| Pd-35Co | 2255 |
| Pd-40Ni | 2255 |
| Au-15.5Cu-3Ni | 1670 |
| Au-20Ag-20Cu | 1535 |

a. Refer to AWS A5.8-81, *Specification for Brazing Filler Metals.*

Silver and copper alloy filler metals are suitable for service at room or moderate temperatures. For somewhat higher temperatures, the stronger gold-nickel, gold-copper, or nickel-copper filler metals are recommended. Platinum,

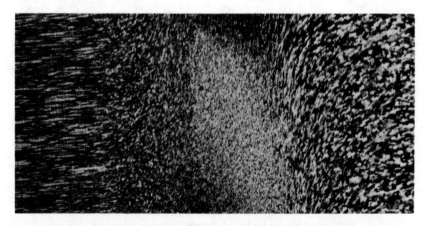

**Fig. 11.3–Section through a friction weld between stress-relieved (left) and recrystallized (right) TZM molybdenum alloy bars ( × 20)**

palladium, or a palladium alloy may be an appropriate brazing filler metal for high temperature service.

Nickel alloy filler metals have been used for brazing tungsten for aerospace applications. Tungsten can be brazed to molybdenum with a 75 percent chromium-25 percent vanadium alloy for high temperature service. Filler metals based on platinum and boron have been used for brazing tungsten assemblies. After a diffusion treatment at 2000°F, the brazements are capable of operating at temperatures up to 3900°F.

## Joint Clearance

Joint clearance requirements should be determined after a brazing filler metal has been selected. Usually, joint clearances in the range of 0.002-0.005 in. at the brazing temperature are suitable.

Molybdenum and tungsten have low coefficients of thermal expansion. This property should be considered in the design of brazed joints, particularly when these metals are joined to other metals. The selection of the fixturing material should also be based on this property.

## Processes and Equipment

Oxyfuel gas torches, controlled atmosphere furnaces, vacuum furnaces, and induction and resistance heating equipment can be used to braze molybdenum and tungsten. Oxyacetylene torch brazing can be done using a low melting silver or copper alloy filler metal and an appropriate flux.

## Fluxes and Atmospheres

When brazing with an oxyacetylene torch, fair protection may be obtained using a commercial silver brazing flux (AWS Type 3A or 3B) in combination with a high temperature flux containing calcium fluoride. The two fluxes should be active in the temperature range of 1050° to 2600°F. The molybdenum is coated first with the commercial silver brazing flux and then with the high temperature flux. The silver brazing flux is active at the lower end of the brazing temperaturea range, and the high temperature flux is active at the higher temperatures.

Pure dry hydrogen, inert gas (helium or argon), and vacuum atmospheres are suitable for brazing. For brazing unalloyed molybdenum or tungsten, the purity of the hydrogen atmosphere is not critical because it easily reduces the oxides at brazing temperature.

Two precautions should be observed when vacuum brazing. First, the vapor pressures of the filler metal components should be compatible with the brazing temperatures and pressures involved. Second, the effect of outgassing on the soundness of the base metal and the filler metal should be evaluated.

## Techniques

Tungsten parts must be handled and fixtured with care during assembly operations because of their inherent brittleness. Previously formed or welded tungsten parts should be stress relieved prior to brazing.

In most cases, filler metals and brazing cycles should be selected to provide minimum interaction between the filler metal and the base metal to maintain the integrity of the base metal. The recrystallization temperature of the base metal can be lowered by diffusion with the filler metal. A palladium-base filler metal in combination with a base metal having a high recrystallization temperature (Mo-0.5%Ti) has been used to minimize these problems. Molybdenum has been plated with a chromium barrier layer to prevent the formation of intermetallic compounds. The metals must not be exposed to carbon at brazing temperature because of possible formation of brittle carbides.

Vacuum tube applications cannot tolerate outgassing or volatilization of elements. Therefore, fluxes and brazing filler metals containing low-melting elements, such as cadmium and zinc, are not suitable. The filler metals for these applications are designed to have low vapor pressures at the service temperature of the component.

# GOLD

## PROPERTIES

Gold is a bright yellow, soft, and malleable metal. Its special properties include excellent oxidation and corrosion resistance, high reflectance, and good electrical and thermal conductivities. The properties of gold that are important in welding and brazing are given Table 11.4.

Gold is alloyed with copper and silver for many jewelry and dental applications. Zinc,

**Table 11.4**
**Physical properties of precious metals**

| Property | Units | Gold | Silver | Platinum | Palladium |
|---|---|---|---|---|---|
| Density (near 20°C) | Mg/m³ | 19.3 | 10.5 | 21.5 | 12.0 |
| Melting point | °C (°F) | 1064 (1950) | 962 (1765) | 1769 (3217) | 1552 (2826) |
| Coef. of linear thermal expansion (20°C) | μm/(m·K) | 14.2 | 19 | 9.1 | 11.8 |
| Thermal conductivity | W/(m·K) | 318 | 428 | 71 | 70 |
| Specific heat | J/(kg·K) | 131 (18°C) | 234 (0°C) | 132 (0°C) | 245 (0°C) |
| Electrical resistivity | nΩ·m | 23.5 | 14.7 | 106 | 108 |
| Modulus of elasticity | GPa (10⁶ psi) | 80 (11.6) | 75.8 (11) | 147 (21.3) | 112 (16.3) |

nickel, palladium, or platinum may be included in some alloys. Filler metals with good oxidation resistance for brazing iron, nickel, and cobalt alloys are alloys of gold with copper, nickel, palladium, or platinum. Brazing temperatures with these filler metals range from 1650° to 2300°F, and the brazed joints are ductile.

Oxidation during welding, brazing, or soldering is not a serious problem with pure gold. However, protection against oxidation during joining may be necessary with some gold alloys. In any case, the surfaces to be joined should be free of oil, grease, dirt, and other contaminants that would inhibit wetting and flow.

## FUSION WELDING

Gold and most gold alloys have melting temperatures below 2000°F. This property, together with good oxidation resistance, makes fusion welding of these metals rather easy.

Gold and its alloys can be welded using an oxyfuel gas torch. A small oxyacetylene torch is commonly used to weld fine gold and high karat gold alloys. A slightly reducing flame is recommended for gold alloys to avoid oxidation and porosity. The filler metal should have a composition similar to that of the base metal if color matching is important. Borax, boric acid, or mixtures of the two are suitable welding fluxes.

Gold alloys can also be joined by gas tungsten arc, plasma arc, and electron beam welding. These processes are particularly suited for alloys that may oxidize and discolor at elevated temperatures. Also, welding speeds are higher than with oxyacetylene welding. Contamination with tungsten could be a problem with gas tungsten arc welding.

The welding filler metal composition should be similar to that of the base metal, particularly where joint properties, color match, or gold content are important considerations.

## RESISTANCE WELDING

Gold-copper alloys and many gold ternary alloys containing copper with silver or nickel may be resistance welded. Small bench type resistance welding machines with synchronous initiation are generally used. This type of welding is suitable for joining small components of jewelry, optical aids, and electrical contacts.

## SOLID STATE WELDING

Gold and many alloys can be cold welded or hot pressure welded because of their excellent malleability. They can also be friction welded, but applications for this process are limited.

With cold welding, treatment prior to welding may be required to remove strongly adsorbed films of moisture or gas. Such films can be broken up with deformations exceeding 20 percent after degreasing and wire brushing the surfaces to be joined. The films may be desorbed by heating at an elevated temperature for a sufficient time.

## BRAZING

### Filler Metals

Brazing is commonly used for joining gold jewelry and dental fabrications. A number of brazing filler metals are needed for color and karat matching with the many gold alloys used for these fabrications. Further, successive joints in an assembly may require filler metals of the same color but with progressively lower melting ranges.

Brazing filler metals are primarily gold alloyed with silver, copper, and zinc. Other metals may be added in small amounts to obtain desired color or melting range. Typical filler metal compositions and their melting ranges are given in Table 11.5.

High silver filler metals flow freely with minimum alloying with a gold alloy. High copper filler metals tend to wet and alloy with the base metal before they are completely melted. As the temperature is increased, alloying increases with a high copper filler metal as it flows onto the base metal. Therefore, brazing with high copper filler metals should be done rapidly to minimize this action.

## Flux

A flux should be used when brazing gold alloys. A mixture of 50 percent fused borax, 43 percent boric acid, and 7 percent sodium silicate is a suitable flux. With torch brazing, a reducing flame should be used. In general, the brazing procedures are similar to those used with silver brazing filler metals.

## Procedures

Most brazing in jewelry and dental applications is done with a torch. A suitable fuel gas, such as natural gas, MPS, or acetylene, and air or oxygen may be used with an appropriately designed heating torch. The gas selection depends upon the size of the parts to be brazed. A neutral or reducing flame is used for brazing gold alloys to avoid oxidation.

The procedure employed in small-scale torch brazing operations consists of first applying a small amount of flux to the fitted joint. Bits of wire, foil, or filings of filler metal are then placed on the fluxed joint. The joint is heated slowly with a soft flame until the flux melts without bubbling. The intensity of the gas flame is then increased and directed on the joint until the filler metal melts and flows into the joint. Continuity of grain structure across the interface and the absence of dendritic structure in etched cross sections are the criteria of good brazed joints.

Small parts may be brazed using resistance heating. The positioned joint is placed between suitable electrical contacts. High current is passed across the joint to heat it. Filler metal in wire form is fed into the joint when it reaches brazing temperature. Staining can be prevented with a suitable flux.

## SOLDERING

Gold is sometimes used as a coating on other metals, ceramics, and glass where solder connections are required. In thin film circuitry, the gold coating may be the electrical conductor.

Solder selection and soldering conditions

## Table 11.5
### Typical filler metals for brazing gold alloys

| Type | Composition, percent | | | | | Solidus | | Liquidus | |
| | Au | Ag | Cu | Zn | Other | °F | °C | °F | °C |
|------|----|----|----|----|-------|----|----|----|----|
| | | | | | Yellow | | | | |
| 10 K easy | 42 | 24 | 16 | 9 | 9 Cd | 1170 | 630 | 1295 | 700 |
| 10 K hard | 42 | 35 | 22 | 1 | | 1340 | 730 | 1375 | 745 |
| 14 K easy | 58 | 18 | 12 | 12 | | 1330 | 720 | 1390 | 755 |
| 14 K hard | 58 | 21 | 15 | 6 | | 1425 | 775 | 1470 | 800 |
| | | | | | White | | | | |
| 10 K easy | 42 | 30 | 8 | 15 | 5 Ni | 1295 | 700 | 1350 | 730 |
| | | | | | Dental | | | | |
| | 47 | 15 | 35 | | 3 Sn | 1215 | 657 | 1425 | 775 |
| | 62 | 17 | 15 | 4 | 2 Sn | 1420 | 770 | 1490 | 810 |
| | 65 | 16 | 13 | 4 | 2 Sn | 1365 | 740 | 1470 | 800 |
| | 82 | 8 | | | 8 Pd, 2 Sn | 1995 | 1090 | 2020 | 1105 |

for an application must be established by appropriate tests. Some molten solders may diffuse rapidly and alloy with a thin gold film. This is the case with 60 percent tin-40 percent lead solder. Two suitable solder compositions are 95 percent indium-5 percent bismuth and 53 percent tin-29 percent lead-17 percent indium-0.5 percent zinc.

Rosin type fluxes are used with a suitable heating method. After soldering, the flux residue is removed by washing in a suitable solvent such as alcohol or a chlorinated hydrocarbon.

# SILVER

## PROPERTIES

Silver is a very ductile, malleable metal with excellent electrical and thermal conductivities. It is rather passive chemically and is, therefore, used for many chemical applications. The metal is attacked by most sulfur compounds, which tarnish the surface. The properties of silver that are of concern in welding, brazing, and soldering are given in Table 11.4.

The high thermal conductivity of silver requires a high rate of heat input for welding or brazing. Molten silver is very fluid, and requires skill in controlling a molten weld pool. In addition, molten silver has high solubility for oxygen. Dissolved oxygen is rejected during solidification, resulting in spatter and porosity. Therefore, the molten weld metal must be adequately shielded with inert gas or flux to minimize exposure to the atmosphere.

Copper is added to silver to improve mechanical properties and hardness. At the same time it decreases the melting point and lowers the thermal and electrical conductivities.

## FUSION WELDING

Silver and its alloys can be joined by gas tungsten arc, gas metal arc, electron beam, and oxyfuel gas welding. Argon or helium is effective in preventing oxygen contamination of the molten weld pool during arc welding.

A suitable flux must be used with oxyfuel gas welding. Equal parts of borax and boric acid mixed with equal parts of water and alcohol is a satisfactory paste flux. Acetylene is the best fuel, although hydrogen may be used with very small components. A neutral or reducing flame is required to prevent exposure of the molten weld metal to oxygen.

The filler metal composition should be similar to that of the base metal, particularly where color matching is an important consideration.

## RESISTANCE WELDING

It is difficult to resistance weld silver because of its low electrical resistance. Projection welding techniques can be used to join silver contacts to relay or switch parts that have higher resistance. A projection is coined on the base of the contact to produce a high current density in the silver and melt it. The contacts may be readily welded to arms of phosphor bronze, copper-nickel alloy, beryllium copper, brass, and other metals.

## COLD WELDING

Silver can be cold welded because of its excellent malleability. Annealed silver bars can be cold welded end-to-end at room temperature by upsetting to an area of 150 to 200 percent of the original. Silver sheets can be cold welded together provided the surface oxide is removed and not permitted to reform before welding. Deformation in the range of 65 to 80 percent is required to obtain good weld strength.

## BRAZING

Silver and silver-copper alloys are easily joined by brazing with low-melting silver (BAg) filler metals.[1] AWS BAg-9 or -10 filler metals are recommended for brazing sterling silver (93%Ag-7%Cu). Filler metal selection depends upon the base metal composition and the difference in the melting temperatures of the two alloys. This dif-

---

1. Refer to AWS A5.8-81, *Specification for Brazing Filler Metals,* for information on these filler metals.

ference should be as large as possible to minimize alloying between them at brazing temperature. Alloying can be minimized by controlling the amount of filler metal used, avoiding excessive heating, and limiting the time at brazing temperature. The common methods of brazing may be used with silver and its alloys.

Commercial fluxes designed for use with silver filler metals are satisfactory for brazing silver alloys. These are usually paste mixtures of boric acid, borates, and sometimes fluorides. Flux residue should be removed by appropriate means. Furnace brazing can be done in a reducing gas atmosphere free of sulfur compounds.

Sulfur will react with and tarnish silver alloys.

Brazing of silver is not recommended for corrosion applications.

## SOLDERING

Silver may be joined with a solder composed of 62 percent Sn, 36 percent Pb, and 2 percent Ag. This solder contains sufficient silver to saturate itself at soldering temperature. This minimizes diffusion with the silver base metal. A rosin type of flux is used. The flux residue should be washed off with a suitable solvent after soldering.

# PLATINUM

## PROPERTIES

Platinum is a white, ductile metal that resists oxidation in air at all temperatures up to its melting point. Properties of importance when joining platinum are given in Table 11.4. The metal can be strengthened by alloying it with other platinum-group metals including iridium, palladium, rhodium, and ruthenium. Platinum and platinum-group alloys can be readily welded or brazed because of their excellent oxidation resistance at high temperatures.

Platinum is also alloyed with nickel or tungsten for electrical applications. These alloys are not oxidation resistant at high temperatures. Selective oxidation of the nickel or tungsten takes place.

## WELDING

Platinum may be joined by oxyfuel gas, inert

gas shielded arc, electron beam, and resistance welding processes. Hydrogen is a better fuel than acetylene for oxyfuel gas welding. If acetylene is used, the flame should be adjusted with excess oxygen to avoid possible carburization and embrittlement of the platinum. Pure platinum can be forge (hammer) welded in air when heated in the range of 1800° to 2200°F.

## BRAZING

Platinum may be brazed with pure gold filler metal if color match is not important. Gold-platinum filler metals may be used if color match is desirable. These filler metals may contain 20 to 30 percent platinum, which raises the melting temperature and increases the strength and hardness.

# PALLADIUM

## PROPERTIES

Palladium is a white, ductile metal with properties similar to those of platinum. It is less dense than platinum and gold, as shown in Table 11.4. When heated in air in the range of 750° to 1475°F, a thin oxide film forms on the surface. Above this range, the film decomposes and leaves the surface bright when cooled rapidly to room temperature. Palladium absorbs hydrogen rapidly at elevated temperatures.

The metal is alloyed with silver, copper, gold, and ruthenium for various electrical and jewelry applications. Alloys containing silver and copper are used for electrical contacts.

## WELDING AND BRAZING

Palladium can be welded or brazed using procedures similar to those for gold or platinum, except that hydrogen should not be used as a fuel for oxyfuel gas welding. Palladium absorbs hydrogen rapidly at elevated temperatures. Acetylene is the preferred fuel, and the flame should be adjusted neutral or slightly oxidizing. Several gold-palladium filler metals are available for brazing operations. Palladium content ranges from 8 to 35 percent, and the melting temperatures vary between 2260° and 2625°F.

# OTHER PRECIOUS METALS

Iridium, osmium, rhodium, and ruthenium have a few industrial uses as pure metals. The metals or their alloys are best joined by gas tungsten arc or electron beam welding. Osmium and ruthenium should be welded in a controlled atmosphere chamber to avoid oxide fume because their oxides are toxic.

# LEAD

## PROPERTIES

Lead is a very soft, malleable metal with low strength properties. It is comparatively heavy, having a density of 11.34 Mg/m³. The metal is easy to weld because of its low melting point (621°F) and low thermal conductivity.

Lead has good resistance to corrosion when exposed to a number of acids, sea water, and other chemicals. When cut, the surface has a bright silvery luster. A gray oxide film soon forms on this surface and protects the metal from further oxidation.

Lead is alloyed with antimony, calcium, and tin to increase its strength and hardness for many commercial applications. Lead alloys are used as solders for other metals and as coatings on steel sheet for corrosion applications.

## WELDING

### Process

Lead is welded primarily by oxyfuel gas welding, but it can be joined by other welding processes. Acetylene, natural gas, or hydrogen may be used as fuel gas. Oxyhydrogen or oxyacetylene mixtures can be used for welding in all positions. The hydrogen flame gives better control of the molten weld pool because of the lower heat input. Oxynatural gas welding is recommended for the flat position only.

A neutral flame should be used with acetylene or natural gas fuel. A reducing flame tends to deposit soot on the joint. Excess oxygen in the flame will oxidize the lead and inhibit wetting.

Manipulation of the welding torch and the flame intensity depend upon the type of joint being welded as well as the position of the welding. In general, the torch is moved in a semicircular or V-shaped pattern. The molten lead is controlled and directed with the flame to produce a circular or herringbone appearance.

When welding in the horizontal position, a soft, diffused flame is desirable. A more pointed, forceful flame is generally used in the vertical and overhead positions.

### Filler Metal

Filler metal is generally added in rod form of convenient size. Its composition should be similar to that of the base metal. Strips can be sheared from sheet, or the metal can be cast in a suitable mold.

### Surface Preparation

The joint area should be cleaned with a suitable, safe solvent to remove dirt, oil, and grease. The surfaces to be joined, the adjacent surfaces, and the filler metal should be cleaned of surface oxide by mechanical shaving or wire brushing. This includes the interior of pipe.

### Joint Design

Joint designs commonly used in welding lead sheet or plate are shown in Fig. 11.4. These joints can be used interchangeably for sheet, de-

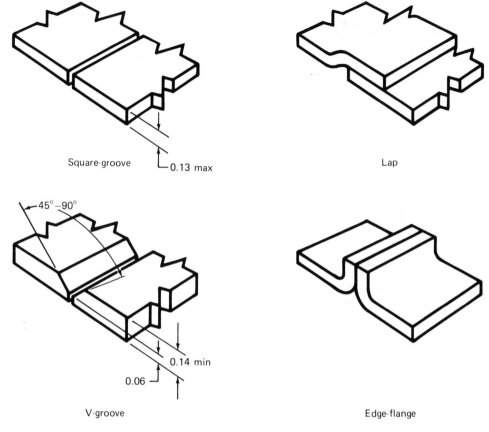

Square-groove    0.13 max

Lap

45°–90°

0.14 min

0.06

V-groove

Edge-flange

Note: Dimensions are in inches.

**Fig. 11.4–Joint designs for welding lead sheet**

pending upon service requirements. For vertical and overhead welding, the lap joint is almost always used. An edge-flange joint is used only under special conditions.

A lap joint is generally designed with one part overlapping the other about 0.5 to 2 in., depending upon the thickness of the lead. The thicker the lead, the greater is the overlap required. The backside, edge, and face side of the top part, as well as the face side of the bottom part should be scraped clean for welding.

For butt joints in sheet of 0.13-in. thickness or less, a square-groove joint design is used. With pipe, the ends should be flared slightly to compensate for thermal contraction. A V-groove

design should be used for sections over 0.13-in. thick.

An edge-flange joint, for use with thin sheet, is prepared by flanging the edge of each sheet to be joined to a right angle. The flange width should be about 1.5 times the sheet thickness. The flange faces should be scraped clean before fitting the joint.

A butt joint is usually preferred for pipe, but a lap or cup joint may be appropriate with wall thicknesses of 0.16 in. and under. The cup joint design is shown in Fig. 11.5. A sleeve is often used on large diameter pipe joints. It is made either by casting or by forming sheet to shape and welding the seam. The sleeve is

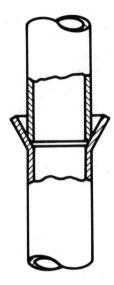

**Fig. 11.5 – Cup joint design for welding lead pipe in the vertical position**

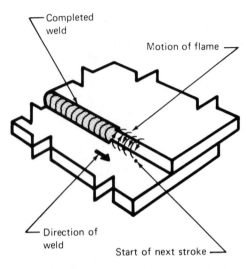

**Fig. 11.6 – Technique for welding a lap joint in lead sheet in the flat position**

slipped over the joint and joined to both pipes with fillet welds.

### Sheet Welding

*Flat Position.* A butt joint design is commonly used for welding sheet in the flat position, but a flange joint is preferred for thin sheet. A lap joint is sometimes used for simplicity of fitting. This joint is welded by oscillating the welding torch in a semicircular path along one edge of the lap to produce a fillet weld, as shown in Fig. 11.6. Filler metal is generally not added to the first pass.

*Vertical Position.* Lap joints are used almost exclusively when welding in the vertical position. Welding should begin at the bottom of the joint using a backing to support the initial weld metal. Figure 11.7 shows the welding technique.

The flame is first directed at the root of the joint. A circular motion of the flame is used to first melt the surface of the back sheet, and then to melt a small amount of the front sheet into the molten pool. The flame is then moved higher to repeat the process as the first pool solidifies. The weld bead is thus carried to the top of the joint.

Butt joints are used in special cases. For this type of joint, a mold must be used to contain the weld metal. The mold is held in place while the filler rod is melted and fused with the base metal by directing the flame into the cavity. The mold

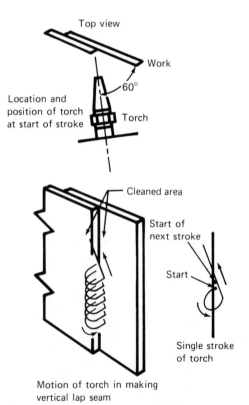

**Fig. 11.7 – Technique for welding a lap joint in lead sheet in the vertical position**

is repositioned to weld succeeding sections of the joint.

*Overhead Position.* This position is the most difficult for welding lead and should be avoided wherever possible. If it is necessary to weld in this position, a lap joint is the most suitable type. The weld beads should be small, and the welding speed should be as fast as possible. Filler metal is not normally added.

*Horizontal Joints.* There are two methods for welding horizontal joints in vertical sheets, overhand and underhand. A lap joint is used for both. With the overhand method, the lower sheet is formed slightly to overlap the upper sheet and form a trough, as shown in Fig. 11.8(A). The edge of the upper sheet may be beveled to fit. The flame is directed into the trough, and a fillet weld is made along the edge. Filler rod is added as required.

The underhand method is used if the upper sheet must lap over the lower sheet for the application. The overlapping section is forced tightly against the lower sheet, and the edge of this section is melted and fused to the lower sheet, as shown in Fig. 11.8(B). No filler rod is added. This joint design is difficult to weld in this manner and should be avoided if possible.

### Pipe Welding

When welding pipe joints, the job should be

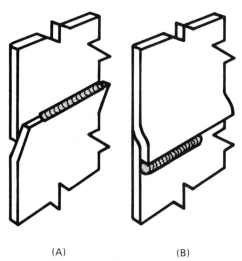

(A)              (B)

**Fig. 11.8 – Joint designs for welding horizontal joints in vertical lead sheet using (A) overhand and (B) underhand techniques**

planned so that a minimum number of joints will have to be made in a fixed position. Where the pipe can be rotated with its axis horizontal, welding can be performed in the flat (1G) position. In the horizontal fixed (5G) position, a butt joint is generally used. To avoid overhead welding, a V-shaped section is usually cut from the upper half of one pipe end. This permits welding of the lower half of the joint from the inside of the pipe. The V-section is then repositioned and welded to the pipe from the outside, as is the upper half of the joint. This design is shown in Fig. 11.9.

An alternative method is to cut T-slots in the upper halves of both pipe ends. The wall sections are bent outward radially to expose the lower half of the joint. After the lower half of the joint is welded from the inside, the wall sections are bent back to original shape. The slots and the upper half of the joint are then welded from the outside.

For pipe joints in the vertical (2G) position, the cup joint shown in Fig. 11.5 should be used. The welding procedure is similar to that used for overhand welding of sheet in the horizontal position, as described previously.

### Weld Size

One or more layers of weld metal may be required, depending upon the section thickness. In general, a weld reinforcement of about 0.13 in. is used for butt joints. With lap joints, the weld metal should be built up to the thickness of the lap, and then one more layer applied for reinforcement. For vertical and overhead lap joints, multiple passes are seldom practical. For cup joints welded in the vertical pipe position, weld metal should be built up even with the lip of the cup using multiple passes.

## SOLDERING

Lead and certain lead alloys are easily soldered when proper care is taken to avoid melting the base metal. Soldered joints in lead are generally confined to the plumbing field, some ar-

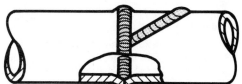

**Fig. 11.9 – Joint design for welding lead pipe in the horizontal fixed position**

chitectural uses, and joining lead-sheathed cables. The use of soldered joints in lead for containment of corrosive chemicals is not generally recommended. The joints for such applications should be welded.

## Solders

The solder for joining a lead alloy should be chosen so that it can be worked without melting the base metal. Careful preparation of the areas to be soldered and close joint tolerances can alleviate most of the problems connected with the soldering of lead.

Wiping is a technique unique to lead soldering, and it requires the proper type of solder to obtain a smooth, gas-tight joint. Wiping solders for lead contain between 30 and 40 percent tin, up to 2 percent antimony, and the balance lead. These solders start to melt at approximately 360°F and are completely liquid at approximately 480°F, providing a wide pasty or working range. Lead solder containing 34.5 percent tin, 1.25 percent antimony, and 0.11 percent arsenic is widely used in cable joining. A 50 percent tin-50 percent lead solder is widely used for joining lead sheet.

## Flux

Activated rosin flux, organic (stearic acid) flux, or tallow may be used for soldering lead.

## Surface Preparation

The areas to be joined should be thoroughly cleaned by wire brushing or shaving. Flux should then be applied promptly to prevent reoxidation of the cleaned surfaces. A very thin film of flux is sufficient, and it should not spread beyond the joint area upon application of heat.

Excessive cleaning with tools should be avoided. A rough surface or thinning of the lead near a critical section of the joint may cause fatigue failure. Gummed paper strips or plumber's soil are often used to limit the solder flow to the joint area and help form a solder bead at the joint.

## Heating Methods

The low melting point of lead alloys limits the choice of heating methods. Soldering irons are commonly used to heat joints in sheet lead. The heat for a joint made by the wiping technique is supplied by the molten solder. The solder is poured over the joint until it wets and flows on the base metal.

## Joint Designs

*Lap Joint*. A lap joint should have a minimum overlap of 0.38 inch for metal thicknesses up to and including 0.13 inch. The joint and surrounding areas of the two parts are first cleaned and then fluxed with tallow. The joint is fixtured by joining at intervals with solder. Soldering is usually done with 50 percent tin-50 percent lead solder.

The application of additional flux is often advisable. It may be rosin or stearine in cored wire solder. When bar solder is used, stearine or powdered rosin may be applied to the joint, as required.

*Lock Joint*. This type of joint provides considerable strength, and is preferred for supporting tensile loads. It is made in much the same manner as a lap joint, using a seam width of 0.5 in. or more. The solder should flow between the surfaces of the lock seam.

*Butt Joint*. This joint is made in a manner similar to braze welding. It is the least desirable design because the joint strength is limited to the strength of the solder. The edges to be joined are beveled 45 degrees or more. Gummed paper strips are placed parallel to each edge of the joint, 0.25 to 0.38 in. from it, to confine the solder to the joint. The joint is aligned, fluxed, and spot soldered at 4- to 6-in. intervals.

The joint is filled progressively with solder while heating with an iron drawn along the joint. Flux should be added as needed, and solder must wet the groove faces. Sufficient solder should be added to produce a slightly convex face on the joint.

*Pipe Joint*. The joint design for soldering lead pipe is similar to that shown in Fig. 11.5 for welding. The inlet or spigot end is beveled to fit snugly into the flared end.

The joint area is lightly shaved clean. A thin coat of tallow is then applied. Plumber's soil or gummed paper is applied next to the joint to restrict solder flow. The joint is assembled, and the flared section is swaged against the inner pipe. The pipes are braced so that they do not move during soldering.

With the pipes in the horizontal position, the joint is made by slowly pouring solder at the proper temperature on top of the joint while the operator manually directs the solder in place. For this purpose, the operator uses a tallow-coated cloth to wipe or form the joint while the solder is in a pasty stage. When completed, the joint should be cooled rapidly.

In the vertical position, joints are prepared and wiped in much the same manner. The solder is applied around the joint from a ladle. The wiping cloth is held directly under the ladle at the bottom of the joint. It is then wiped across the joint to form the solder.·

Branch joints in lead pipe are made by cutting a small oval-shaped hole in the main line. The pipe wall is flared outward around the hole to form a collar or hub into which a beveled branch line is fitted snugly. Preparation and wiping of the joint are essentially the same as previously described.

A cup joint is similar to the other joints except that the flared end is not swaged against the inside pipe (see Fig. 11.5). A soldering iron is used to melt and flow the solder into the joint. It is not wiped. These joints can only be made in the vertical position. Plumber's soil or gummed paper should be applied to restrict solder flow. The pipes are fitted together and tacked with solder. Solder is flowed around the joint with a hot, sharp-pointed iron until the cup is filled about half way. The remainder of the cup is then filled with solder using a hot blunted iron.

## SAFE PRACTICES

Lead and its compounds are toxic. Exposure to these materials can be a serious health hazard if proper precautions are not taken to keep fume to a safe level. Exposure to lead can occur in a variety of situations where cleaning and welding are performed on lead or lead containing products. Specifically, lead exposure results when workers must handle and weld lead sheets or pipes, or when lead surfaces are cleaned with an abrasive.

The control of fume concentrations within the breathing zone of a welder or cutting operator can be accomplished by either of two ventilation methods. The fume can be dispersed by diluting fume-laden air with uncontaminated air, or the fume can be captured by a hood collector connected to an exhaust system.

Dilution ventilation can be provided either naturally or mechanically. Natural ventilation relies on wind currents or vertical temperature gradients to move the air. General mechanical ventilation is done with fans that exhaust contaminated air and provide clean make-up air to dilute the concentration of contaminants in the atmosphere.

The use of local exhaust ventilation is a practical means of controlling the exposure of welders and cutting operators to fumes produced during their work. Compared with general ventilation, local exhaust ventilation can control fumes more effectively. It is the preferred means of ventilation, provided the exhaust hood can be positioned close enough to capture air contaminants.

Under some circumstances, approved respirators may be necessary to adequately protect the workers. Concentrations of lead within a worker's breathing zone may reach an unsafe level with poor ventilation or very high fume level. In that case, the worker should be required to wear an approved air-supplied respirator or mask. Additional information may be obtained in ANSI Z49.1, *Safety in Welding and Cutting,*[2] or from the Lead Industries Association.[3]

---

2. Published by the American Welding Society.
3. Lead Industries Association, 292 Madison Ave., New York, NY 10017

# Metric Conversion Factors

1 psi = 6.89 kPa
$t_C = 0.56 (t_F - 32)$
1 in. = 25.4 mm
1 torr = 0.13 kPa
1 in./min = 0.42 mm/s
1 lb/in.$^3$ = 27.7 Mg/m$^3$
(1 Btu • ft)/(h • ft$^2$ • °F) = 1.73 W/(m • K)
(1 in. x 10$^{-6}$)/(in. • °F) = 0.56 μm/(m • K)
1 Btu/(1b • °F) = 4.19 kJ/(kg • K)

# SUPPLEMENTARY READING LIST

## General

*Brazing Manual*, 3rd Ed., Miami: American Welding Society, 1976.

*Metals Handbook*, Vol. 2., 9th Ed., Metals Park, OH: American Society for Metals, 1979.

*Safety in Welding and Cutting*, ANSI Z49.1, Miami; American Welding Society, 1973.

Schwartz, M. M., *The Fabrication of Dissimilar Metal Joints Containing Reactive and Refractory Metals*, New York: Welding Research Council, Bulletin 210; 1975 Oct.

## Beryllium

Bosworth, T. J., Diffusion welding of beryllium, Part I, *Welding Journal*, 51(12): 579s-90s; 1972 Dec. Part II, *Welding Journal*, 52(1): 38s-48s; 1973 Jan.

Hauser, D., et al., Electron beam welding of beryllium, Part I, *Welding Journal*, 46(12): 525s-40s; 1967 Dec. Part II, *Welding Journal*, 47(11): 497s-514s; 1968, Nov.

Hauser, H. H., *Beryllium*, Univ. of California Press, 1965.

Keil, R. W., et al., Brazing and soldering of beryllium, *Welding Journal* 39(9): 406s-10s; 1960 Sept.

Knowles, J. L., and Hazlett, T. H., High-strength low-temperature bonding of beryllium and other metals, *Welding Journal*, 49(7): 301s-10s; 1970 July.

Maloof, S. R., and Cohen, J. B., Brazing of beryllium for high temperature service, *Welding Journal*, 40(3): 118s-22s; 1961 Mar.

Passmore, E. M., Fusion welding of beryllium, *Welding Journal*, 43(3): 116s-25s; 1964 Mar.

Passmore, E. M., Solid state welding of beryllium, *Welding Journal* 42(4): 186-89s; 1963 Apr.

White, D. W. Jr., and Burke, J. E., *The Metal Beryllium*, Metals Park, OH: American Society for Metals, 1955: 283-94.

## Uranium

Bradburn, E. H., et al., Multipass electron beam welding for controlled penetration, *Welding Journal* 50(4): 190s-93s; 1971 Apr.

Turner, P. W., and Johnson, L. D., *Joining of Uranium Alloys*, Proceedings of the Third Army Materials Technology Conference, 1974: 145-85.

Turner, P. W., and Lundin, C. D., Effect of iron on the hot cracking of uranium weld metal, *Welding Journal* 49(12): 579s-87s; 1970 Dec.

Wood, D. H., and Mara, G. L., Eliminating cold-shut defects in deep single-pass electron beam welds in uranium, *Welding Journal* 56(3): 88s-92s; 1977 Mar.

## Molybdenum and Tungsten

Bryant, W. A., and Gold, R. E., Weldability of three forms of chemically vapor deposited tungsten, *Welding Journal*, 54(11): 405s-408s; 1975 Nov.

Cole, N. C., et al., Weldability of tungsten and its alloys, *Welding Journal*, 50(9): 419s-26s; 1971 Sept.

Cole, N. C., et al., Development of corrosion resistant filler metals for brazing molybdenum, *Welding Journal*, 52(10): 466s-73s; 1973 Oct.

Farrell, K., et al., Hot cracking in fusion welds in tungsten, *Welding Journal*, 49(3): 132s-37s; 1970 Mar.

Lessman, G. G., and Gold, R. E., The weldability of tungsten base alloys, *Welding Journal*, 48(12): 528s-42s; 1969 Dec.

Lundberg, L. B., et al., Brazing molybdenum and tungsten for high temperature service, *Welding Journal* 57(10); 311s-18s; 1978 Oct.

Moorhead, A. J., Welding studies on arc cast molybdenum, *Welding Journal*, 53(5): 185s-91s; 1974 May.

Stone, L. H., et al., Recrystallization behavior and brazing of the TZM molybdenum alloy, *Welding Journal*, 46(7): 299s-308s; 1967 July.

## Precious Metals

Atkinson, R. H., Fabricating techniques for jewelry, *Metal Progress*, 94(6); 1957 Dec.

Butts, A, and Coxe, C. D., ed., *Silver – Economics, Metallurgy and Use*, Princeton, N.J.: D. Van Nostrand, 1967.

Wise, E. M., *Gold – Recovery, Properties and Applications*, Princeton, N.J.: D. Van Nostrand, 1964.

Wise, E. M., *Palladium – Recovery, Properties, and Uses,*, N.Y.: Academic Press, 1968.

**Lead**

Bowser, L. S., Lead welding practices, *Welding Journal*, 26(9): 777-81; 1947 Sept.

Canary, W. W., and Lundin, C. D., Properties of weld metal in a lead-calcium alloy, *Welding Journal*, 52(8): 347s-54s; 1973 Aug.

# 12

# Dissimilar Metals

## Chapter Committee

S. J. MATTHEWS, *Chairman*
*Cabot Corporation*

N. F. BRATKOVICH
*Consultant*

D. G. HOWDEN
*Ohio State University*

J.F. KING
*Oak Ridge National Laboratories*

D. L. OLSEN
*Colorado School of Mines*

## Welding Handbook Committee Member

J. R. CONDRA
*E. I. duPont de Nemours & Company*

# 12

# Dissimilar Metals

## DEFINITION AND SCOPE

In this chapter, dissimilar metals are considered to be either those that are chemically different (aluminum, copper, nickel) or alloys of a particular metal that are significantly different from a metallurgical standpoint (carbon steel vs stainless steel). They can be base metal, filler metal, or weld metal.

Most combinations of dissimilar metals can be joined by solid state welding,[1] brazing, or soldering where alloying between the metals is normally insignificant. In these cases, only the differences in the physical and mechanical properties of the base metals and their influence on the serviceability of the joint should be considered. When dissimilar metals are joined by a fusion welding process, alloying between the base metals and a filler metal, when used, becomes a major consideration. The resulting weld metal can behave significantly different from one or both base metals during subsequent processing or in service.

A combination of metals with significantly different chemical, mechanical, and physical properties can easily present problems during and after welding. The combination can be two different base metals or three different metals, one of which is a filler metal. The resulting weld metal composition will differ from that of any of the components, and can vary with the joint design, the welding process, the filler metal, and the welding procedure. Consequently, these factors and also any thermal treatment of the weldment must be established and properly evaluated prior to production. The major goal of dissimilar metal welding should be to produce a weldment that meets the intended service requirements.

The information presented here primarily addresses the fusion welding of dissimilar metals where the effects of significant dilution and alloying must be considered in the design of the joint as well as in the selection of the welding process and procedures.

## FUNDAMENTALS

### WELD METAL

In the fusion welding of dissimilar metal joints, the most important consideration is the weld metal composition and its properties. Its

composition depends upon the compositions of the base metals, the filler metal, if used, and the relative dilutions of these. The weld metal composition is usually not uniform, particularly with multiple pass welds, and a composition gradient is likely to exist in the weld metal adjacent to each base metal.

These solidification characteristics of the weld metal are also influenced by the relative

---

1. Cold, explosion, friction, and ultrasonic welding.

514

dilutions and the composition gradients near each base metal. These characteristics are important with respect to hot cracking of the weld metal during solidification.

The basic concepts of alloying, the metallurgical characteristics of the resultant alloy, and its mechanical and physical properties must be considered when designing a dissimilar metal joint. If the two base metals will form a continuous series of solid solutions when melted together, such as copper and nickel, production of a fusion weld between them is readily accomplished. On the other hand, if complex phases or intermetallic compounds are formed when the two base metals are melted together, successful fusion welding likely depends on the availability of a filler metal and a welding procedure that will avoid such compounds or phases and will produce sound weld metal with acceptable properties for the intended service.

## DILUTION AND ALLOYING

During fusion welding, metal from each member and the filler metal, if used, are melted together into the weld pool. Upon solidification, the weld metal will be either a single phase or a mixture of two or more phases. A phase can be a solid solution (Cu-Ni), an intermetallic compound ($CuAl_2$), or an interstitial compound ($Fe_3C$, TiC). The number, type, amount, and metallurgical arrangement of the phases present largely determine the properties and soundness of the weld metal. Solidification and cooling rates also have a significant effect on the phases present and the metallurgical structure of the metal.

Some basic metallurgy associated with the alloying of metals and the metallurgy of welding are discussed in Volume 1 of this *Handbook*. A review of this material will provide background information on the reactions that take place during welding and an insight into the mechanics of dilution and alloying.

In dissimilar metal welding, the filler metal must alloy readily with the base metals to produce a weld metal that has a continuous, ductile matrix phase. Specifically, the filler metal must be able to accept dilution (alloying) by the base metals without producing a crack-sensitive microstructure. The weld metal microstructure must also be stable under the expected service conditions. The weld metal strength should be equivalent to or better than that of the weaker base metal.

Significant agitation occurs in the molten weld pool with most fusion welding processes. This tends to produce weld metal with a substantially uniform composition, except for narrow bands next to each unmelted base metal. The band of melted base metal is usually wider when the filler metal has a higher melting point than the base metal.[2]

In multiple pass welding, the composition of each weld bead should be relatively uniform. However, definite compositional differences are likely in succeeding weld beads, especially between a root bead, the beads adjacent to the base metals, and the remaining fill beads. The average composition of the whole weld metal can be calculated when two things are known: (1) the ratio of the volumes of base metals melted to the entire weld metal volume, and (2) the compositions of the base and filler metals. The dilution can be based on area measurements on a transverse cross section through a test weld. Fig. 12.1 illustrates how to determine the dilution[3] by two base metals, A and B, when welding with filler metal F, and examples of total dilution.

The average percentage of a specific alloying element in the diluted weld metal can be calculated using the following equation:

$$X_W = (D_A)(X_A) + (D_B)(X_B) + (1 - D_T)(X_F)$$

where

$X_W$ is the average percentage of element X in the weld metal.

$X_A$ is the percentage of element X in base metal A.

$X_B$ is the percentage of element X in base metal B.

$X_F$ is the percentage of element X in the filler metal F.

$D_A$ is the percent dilution by base metal A, expressed as a decimal.

$D_B$ is the percent dilution by base metal B, expressed as a decimal.

$D_T$ is the percent total dilution by base metals A and B, expressed as a decimal.

To illustrate the calculation of weld metal composition, assume that Type 316 stainless steel is welded to a 2-1/4Cr-1Mo alloy steel with a

---

2. Baeslack, W.A., Lippold, J.C., and Savage, W.J., Unmixed zone formation in austenitic stainless steel weldments, *Welding Journal*, 58 (6): 168s-76s; 1979 June.

3. Dilution is the change in composition of the deposited (filler) metal by mixture with the base metal(s) or the previously deposited weld bead(s), expressed as a percentage.

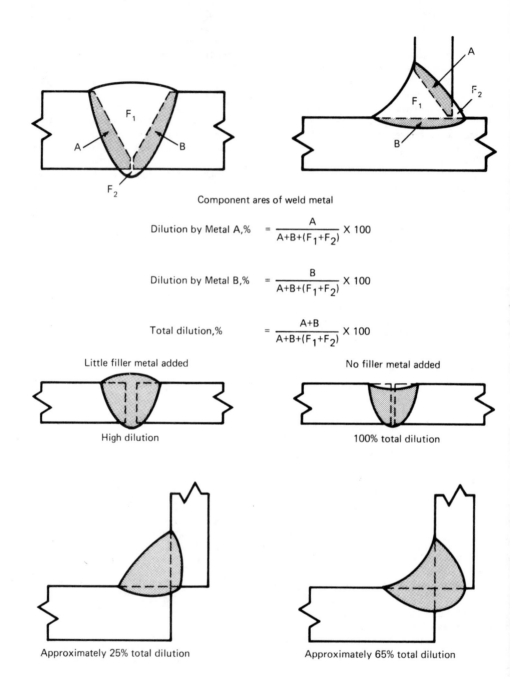

Component ares of weld metal

$$\text{Dilution by Metal A,}\% = \frac{A}{A+B+(F_1+F_2)} \times 100$$

$$\text{Dilution by Metal B,}\% = \frac{B}{A+B+(F_1+F_2)} \times 100$$

$$\text{Total dilution,}\% = \frac{A+B}{A+B+(F_1+F_2)} \times 100$$

Little filler metal added

High dilution

No filler metal added

100% total dilution

Approximately 25% total dilution

Approximately 65% total dilution

*Fig. 12.1—Dilution in a dissimilar metal welded joint*

nickel-chromium alloy filler metal (Type ERNiCr-3). The nominal chemical compositions of the three alloys are given in Table 12.1. Assuming that the total dilution is 35 percent, 15 percent by the Cr-Mo alloy steel and 20 percent from the Type 316 stainless steel, the average percentages of Cr, Ni, and Mo in the weld metal are calculated as follows:

Cr, $\% = 0.15(2.5) + 0.20(17) + 0.65(20) = 16.8$
Ni, $\% = 0.20(12) + 0.65(72) = 49.2$
Mo, $\% = 0.15(1) + 0.20(2.5) = 0.65$

**Table 12.1**
**Compositions of base and filler metals
chosen as an example**

| Element | Nominal Composition, percent | | |
|---------|-----------|-------------|-------------|
| | Cr-Mo steel | Type 316 sst | Filler metal |
| Cr | 2.5 | 17 | 20 |
| Ni | — | 12 | 72 |
| Mo | 1.0 | 2.5 | — |
| Fe | 95.5 | 63 | 3 |

## MELTING TEMPERATURES

Joining of dissimilar metals by fusion welding requires melting of both base metals. If the melting temperatures are close, say within 200°F, normal welding techniques and procedures are satisfactory. When there is a wide difference in melting temperatures, the problems of welding are quite complex. In fact, it might be necessary to use brazing, braze welding, or solid-state welding techniques for joining those dissimilar metals.

Table 12.2 compares the melting temperatures and other important physical properties of some metals to those of carbon steel. It is obviously difficult to weld aluminum to steel or to nickel alloys, according to these data.

Significant difference in melting temperatures of the two base metals or of the weld metal and a base metal can result in rupture of the metal having the lower melting temperature. Solidification and contraction of the metal with the higher melting temperature will induce stresses in the other metal while it is in a weak, partially solidified condition. This problem may be solved by depositing one or more layers of a filler metal of intermediate melting temperature on the face of the higher melting base metal. This procedure is known as *buttering*. The weld is then made between the buttered face and the other base metal. The buttering layer should serve to reduce the melting temperature differential.

## THERMAL CONDUCTIVITY

Most metals and alloys are relatively good conductors of heat, but some are much better than others. Rapid conduction of heat from the molten weld pool by an adjacent base metal may affect the energy input required to locally melt that base metal. When two dissimilar metals of significantly different thermal conductivities are to be welded together, the welding procedure must provide for this difference. Often the welding heat source must be directed at the metal having the higher thermal conductivity to obtain proper heat balance.

The thermal conductivity of a metal is a function of its temperature. When welding dis-

**Table 12.2**
**Relationship of physical properties of various metals with those of carbon steel**

| Relative property | Ratio of properties | | | | | |
|-------------------|-----------------|--------|----------|------------------------------|-----------|----------------|
| | Carbon steel | Copper | Aluminum | Austenitic stainless steel | 70Ni-30Cu | 76Ni-16Cr-8Fe |
| Mean coefficient of thermal expansion | 1.0 | 1.5 | 2.1 | 1.4 | 1.2 | 1.0 |
| Thermal conductivity | 1.0 | 5.9 | 3.1 | 0.7 | 0.4 | 0.2 |
| Heat capacity | 1.0 | 0.8 | 1.9 | 1.0 | 1.1 | 0.9 |
| Density | 1.0 | 1.1 | 0.3 | 1.0 | 1.1 | 1.1 |
| Melting temperature | 1.0 | 0.7 | 0.4 | 0.9 | 0.9 | 0.9 |

similar metals, heat loss to the base metals can be balanced somewhat by selectively preheating the metal having the higher thermal conductivity. Dilution is more uniform with balanced heating.

Preheating the base metal of higher thermal conductivity also reduces the cooling rate of the weld metal and the heat-affected zone.[4] The net effect of preheating is to reduce the heat needed to melt that base metal.

## THERMAL EXPANSION

The thermal expansion characteristics of the two dissimilar base metals and the weld metal are important considerations. Large differences in thermal expansion coefficients of adjacent metals during cooling will induce tensile stress in one metal and compressive stress in the other. The metal subject to tensile stress may hot crack during welding, or it may cold crack in service unless the stresses are relieved thermally or mechanically.

The linear thermal expansion coefficient, $\alpha$, can be defined as the change in strain, $\Delta\varepsilon$, with respect to a change in temperature, $\Delta T$, or

$$\alpha = \frac{\Delta\varepsilon}{\Delta T}$$

This is a characteristic property of each alloy. It is more commonly represented as:

$$\alpha = \frac{\Delta l}{l\Delta T}$$

where the change in length over the original length, $\Delta l/l$, represents the change in strain, $\Delta\varepsilon$.

The stress, $\sigma$, in the heat-affected zone of one of the metals associated with a dissimilar metal interface can be estimated using the following equation:

$$\sigma = E\Delta\alpha\Delta T$$

where:
E is the elastic modulus of that metal.
$\Delta\alpha$ is the difference in linear thermal expansion coefficients between the two metals.

The difference or mismatch in coefficients of thermal expansion between the metals in a dissimilar metal joint produces stresses in the

---

4. For additional information, refer to the *Welding Handbook,* Vol. 1, 7th Ed.: 84-88.

joint. This factor is particularly important in joints that will operate at elevated temperatures in a cyclic temperature mode. A common example of this is austenitic stainless steel-to-ferritic steel transition pipe joints used in fossil-fired power plants.

Ideally, the linear thermal expansion coefficient of the weld metal should be intermediate between those of the base metals, especially if the difference between those of the two base metals is large. If the difference is small, the weld metal may have an expansion coefficient equivalent to that of one of the base metals.

Dilution of the deposited metal can affect its expansion coefficient. For example, dilution of pure nickel with copper tends to increase the thermal expansion coefficient, but dilution with certain amounts of iron, chromium, or molybdenum tends to reduce it.

## PREHEAT AND POSTHEAT TREATMENTS

Selection of an appropriate preheat or postweld heat treatment for a welded joint can present a problem with some dissimilar metal combinations. Welding an alloy that requires preheat to another alloy that does not can be done if the preheat can be independently applied to the proper side of the joint.

Postweld heat treatment requirements for dissimilar metal joints may pose a challenge to the designer. The appropriate heat treatment for one component of the weldment may be deleterious to the other component for the intended service conditions. For example, if an age-hardenable nickel-chromium alloy is welded to a non-stabilized austenitic stainless steel, exposure of the weldment to the aging treatment for the nickel-chromium alloy would sensitize the stainless steel and decrease its resistance to intergranular corrosion.

One solution is to use a stabilized austenitic stainless steel, if this is acceptable. Another solution might be to butter the face of the age-hardenable, nickel-chromium alloy component with a similar alloy that is not age hardenable. This component is then heat treated to obtain the desired properties. Finally, the buttered surface is welded to the stainless steel component.

## OTHER WELDING CONSIDERATIONS
### Magnetic Effects

Magnetic fields, either induced or permanent, can interact with a dc arc or an electron beam to produce force fields that cause deflection of the welding arc or electron beam.[5] Arc blow, beam deflection, or metal transfer may be affected by the force of a magnetic field. Magnetic flux may be generated by the flow of welding current in one or both components, residual magnetism, or a nearby external source.

When only one of the metals being welded is ferromagnetic, a dc arc or electron beam can be deflected toward that side of the joint. An excessive amount of that metal can be melted, resulting in excessive dilution. Also, there may be incomplete fusion at the root of the weld. Such behavior can take place when welding carbon steel to a nickel base alloy. The arc or electron beam is deflected to the steel side of the joint unless special precautions are taken to offset magnetic effects. Magnetic deflection is not a problem with an ac arc. Beam deflection can be used with electron beam welding.

## Weld Metal-Base Metal Interaction

Weld metal penetration into the grain boundaries in the heat-affected zone of the base metal can occur in certain alloy systems.[6] The result is likely to be intergranular cracking in the heat-affected zone. Usually, the tendency for this is governed by the laws of liquid metal embrittlement. For instance, molten copper-rich weld metal can penetrate carbon steel grain boundaries during welding. The degree of penetration can be greater when the base metal is preheated or under tensile stress, or both.

## Joint Design

When designing butt joints between dissimilar metals, consideration must be given to the melting characteristics of each base metal and the filler metal, as well as to dilution effects. Large grooves decrease dilution, permit better control of viscous weld metal, and provide room for better manipulation of the arc for good fusion.

The joint design should particularly provide for appropriate dilution in the first few passes placed in the joint when welding from one side. Improper dilution could result in a layer of weld metal possessing mechanical properties that are inappropriate for the intended service, particularly when the joint will be exposed to cyclic stresses. When welding from both sides, back gouging of the first weld can provide better control of dilution in the first few passes of the second weld.

# SERVICE CONSIDERATIONS

## MECHANICAL AND PHYSICAL PROPERTIES

A dissimilar metal joint normally contains weld metal having a composition different from one or both base metals. The properties of the weld metal depend on the filler metal composition, the welding procedures, and the relative dilution with each base metal. There are also two different heat-affected zones, one in each base metal adjacent to the weld metal. The mechanical and physical properties of the weld metal as well as those of the two heat-affected zones must be considered for the intended service.

Special considerations are generally given to dissimilar metal joints intended for elevated temperature service. A favorable situation exists when the joint will operate at constant temperature. During elevated temperature service, internal stresses can decrease by relaxation and reach an equilibrium.

However, it is best to reduce the effects of large differences in coefficients of thermal expansion when large temperature fluctuations cannot be avoided in service. The problem can be avoided by selecting base metals with similar thermal expansion characteristics. If this is not feasible, an alternative is to place a third metal between the two base metals. This third metal should have thermal expansion characteristics intermediate to those of the other two.[7] Similar

---

5. See *Welding Handbook,* 7th Ed., Vol. 1: 56-59, and Vol. 3: 210-11.

6. Matthews, S.J. and Savage, W.J., Heat-affected zone infiltration by dissimilar liquid weld metal, *Welding Journal,* 50(4): 174s-82s: 1971 Apr.

---

7. In the case of pipe fabrication, the intermediate pipe section is know as a *safe end.*

reasoning can be applied in the selection of the filler metal. If possible, dissimilar metal welds should be located in low-stress areas because of the possible additive effect of thermally induced stresses to those produced by external loading.

The properties of the various metallurgically different regions in a welded joint are particularly important when the joint will see cyclic temperature service. Such properties include coefficient of thermal expansion, elastic modulus, yield strength, and crack propagation characteristics. When subjected to cyclic temperature service, differences in the properties of the two base metals and the weld metal may induce fluctuating stresses in the heat-affected zones and adjacent weld metal. Cracks may develop as a result of metal fatigue, and cause early failure of the joint. Service life under cyclic temperature conditions will depend on the ability of the metal to resist cracking as well as crack propagation.

Since various metal properties change with temperature, it is difficult to mathematically predict the behavior of a dissimilar metal weld joint in service. Thermal cycling of sample weldments followed by mechanical and metallurgical evaluations can provide information on expected service performance.

## MICROSTRUCTURAL STABILITY

There is likelihood of significant chemical composition gradients in the weld metal, particularly in the regions adjacent to the base metals. In addition, operation at elevated temperatures can cause interdiffusion between the weld metal and the base metal that, in turn, results in microstructural changes. A joint between two dissimilar metals has variations in atom concentration and activity across the weld. These activity gradients can be interpreted in terms of chemical potential gradients that can result in atom diffusion with or against the concentration gradient in multiple-component alloys. Thus, it is difficult to predict with absolute certainty from concentration data alone the migration of specific atoms across a dissimilar metal joint during high temperature service. Movement of atoms over a period of time alters local alloy compositions and produces changes in the mechanical and physical properties of the metal in the region.

A dissimilar metal weld made between a low alloy steel and an austenitic stainless steel with an austenitic stainless steel filler metal illustrates this problem. The carbon content of low alloy steel is generally higher than that of aus-tenitic stainless steel weld metal. The relatively large amount of carbide forming elements, such as chromium, in stainless steel tends to lower the chemical activity of the carbon. This produces a large chemical potential gradient for carbon diffusion from a low alloy steel to stainless steel weld metal, which can occur during postweld heat treatment or during service at elevated temperatures. As a result, decarburization and sometimes grain growth take place in the heat-affected zone of the low alloy steel that lower its mechanical properties. At the same time, the adjacent stainless steel weld metal is carburized, and complex carbides form. The buildup of carbides will substantially increase the hardness of the weld metal in this zone and increase the likelihood of cracking.

## CORROSION AND OXIDATION RESISTANCE

The weld metal and both base metals have specific corrosion behaviors that must be considered by the designer in the initial selection of materials. For example, the fact that sensitization of certain austenitic stainless steels promotes corrosion in specific environments is an important concern in welds involving those steels, especially when the filler metal is also austenitic stainless steel.

With dissimilar metal weldments, the formation of galvanic cells can cause corrosion of the most anodic metal or phase in the joint. Also, the weld metal is usually composed of several microstructural phases, and very localized cells between phases can result in galvanic corrosion at the microstructural level. To minimize galvanic corrosion, the composition of the weld metal can be adjusted to provide cathodic protection to the base metal that is most susceptible to galvanic attack. However, other design requirements should not be seriously compromised to do this. Instead, some other form of protection should be used.

A galvanic cell associated with a high-strength steel may promote hydrogen embrittlement in the heat-affected zone of that steel if it is the cathode of the cell. Hydrogen embrittlement must be considered if the service temperature of the weldment will be in the range of −40° to 200°F, and the weld will be in a highly stressed area of the assembly. Residual stresses developed in the weld zone are often sufficient to promote hydrogen embrittlement and stress corrosion.

Chemical compositional differences in a dissimilar metal weld can also cause high temperature corrosion problems. Compositional variations at the interfaces between the different metals can result in selective oxidation when operating at high temperature in air and formation of notches at these locations. Such notches are potential stress raisers in the joint and can cause stress-oxidation failure along the weld interface under cyclic thermal conditions.

If the resistance of a dissimilar metal weld joint to a specific corrosive environment is not certain, accelerated corrosion tests should be used to predict the estimated life of the proposed joint. Accepted corrosion test procedures are specified in various ASTM and NACE[8] publications.

# FILLER METAL SELECTION

## REQUIREMENTS

Selection of a suitable filler metal is an important factor in producing a dissimilar metal joint that will perform well in service. One objective of dissimilar metal welding is to minimize undesirable metallurgical interactions between the metals. The filler metal should be compatible with both base metals, and also be capable of being deposited with a minimum of dilution. Ideally, the filler metal must provide a welded joint that has the following characteristics.

*Soundness.* The filler metal must be capable of some dilution with the base metals without forming a crack-sensitive weld metal. It must not cause flaws, such as porosity or inclusions, in the weld metal.

*Structural Stability.* The resultant weld metal must remain structurally stable under the design service conditions.

*Physical Properties.* The physical properties of the weld metal should be compatible with those of both base metals. Thermal expansion coefficients are particularly important with respect to internal stresses during cyclic temperature service. The thermal expansion coefficient of the weld metal should be intermediate between those of the base metals. Equal consideration should be given to the thermal and electrical conductivities when these are important design requirements.

*Mechanical Properties.* The weld metal should be at least as strong and ductile as the weaker of the base metals under service conditions.

*Corrosion Properties.* The corrosion resistance of the weld metal should equal or exceed the resistance of both base metals to avoid preferential attack in the welded joint. This applies to corrosion, high temperature oxidation, and embrittlement by sulfur, phosphorus, lead, or other low melting metals.

## SELECTION CRITERIA

Two important criteria that should govern the selection of a proper filler metal for welding two dissimilar metals are as follows:

(1) The candidate filler metal must provide the joint design requirements, such as mechanical properties or corrosion resistance.

(2) The candidate filler metal must fulfill the weldability criteria with respect to dilution, melting temperature, and other physical property requirements of the weldment.

In addition to the above, the following suggestions should be considered:

(1) The filler metal should be one normally recommended for welding the base metal having the lower melting temperature when the difference between the melting temperatures of the base metals is wide. During welding a suitable filler metal will fuse satisfactorily with that base metal. The molten filler metal should also wet and bond to the face of the other part in a manner similar to braze welding. If a high-melting filler metal was used inadvertently, dilution from the low melting base metal would likely be excessive.

(2) The filler metal should have sufficient ductility to tolerate stresses induced by temperature changes because of differences in thermal expansion characteristics of the dissimilar metals. For instance, some nickel alloy filler metals are very ductile, and can tolerate dilution by

---

8. National Association of Corrosion Engineers

certain base metals without cracking or suffering a significant decrease in mechanical properties.

(3) The filler metal should be low in the interstitial elements carbon, oxygen, hydrogen, and nitrogen. Such a filler will generally be more ductile, more stable, and less prone to hot cracking than a similar filler metal that has high interstitial element content.

Commercially available electrodes and welding rods should be evaluated first for a dissimilar metal welding application because of economy and availability. If a standard filler metal is not suitable, the feasibility of using special metal-cored electrodes, modified covered electrodes, or stranded-wire welding rods should be investigated.

## WELDING PROCESS SELECTION

Selecting the welding process to make a given dissimilar metal joint is almost as important as selecting the proper filler metal. The depth of fusion into the base metals and the resulting dilution may vary with different welding processes and techniques.[9]

It is not uncommon with shielded metal arc welding for the filler metal to be diluted up to 30 percent with base metal. The amount of dilution can be modified somewhat by adjusting the welding technique. For example, the electrode can be manipulated so that the arc impinges primarily on the previously deposited weld metal. The dilution rate can be kept below 25 percent with this technique. However, the weld bead reinforcement will be high, but not necessarily objectionable. If dilution from one base metal is less detrimental than from the other, the arc should be directed toward that metal. This technique is also applicable to gas tungsten arc welding.

Dilution rates with gas metal arc welding can range from 10 to 50 percent depending upon the type of metal transfer and the welding gun manipulation. Spray transfer gives the greatest dilution, and short-circuiting transfer the least

dilution. Penetration with submerged arc welding can be greater, depending on polarity, and can result in more dilution.

Narrow welds (high depth-to-width ratio) can be produced by electron beam and laser beam welding. A high density beam can provide very narrow welded joints with minimum melting of the base metal. If filler metal is required, an interlayer of appropriate thickness can be pre-placed in tightly-fitted, square-groove joints, or filler metal may be added with an auxilary wire feed device.

Regardless of the process, dilution is also affected by other factors, including joint design and fit-up. It is always best to have a minimum uniform dilution along the joint. Variations in dilution may produce inconsistent joint properties.

With resistance spot and seam welding of lap joints, an interlayer of appropriate thickness can be placed between the overlapped sheet metal parts. This interlayer can melt and alloy with the molten base metals to form a nugget. An alternative is to join the interlayer to each base metal with separate nuggets, produced simultaneously.

## SPECIFIC DISSIMILAR METAL COMBINATIONS

### STAINLESS STEEL TO CARBON OR LOW ALLOY STEEL

Austenitic, ferritic, or martensitic stainless

---

9. Dilution control with arc welding is discussed in the *Welding Handbook*, Vol. 2, 7th Ed.: 524-27.

steel can be readily fusion welded to carbon or low alloy steel using a filler metal that can tolerate dilution by both base metals without formation of flaws in the joint. An austenitic stainless steel or nickel alloy filler metal is commonly used. The choice depends on the application and the service conditions.

## Carbon Migration

Chromium in steel has a greater affinity for carbon than does iron. When a carbon or low alloy steel is welded with a steel filler metal containing a significant amount of chromium, carbon will diffuse from the base metal into the weld metal at temperatures above about 800°F. The diffusion rate is a function of temperature and exposure time, and increases rather rapidly at 1100°F and above. Carbon migration can take place during postweld heat treatment or service at elevated temperature.

Austenitic steel has a greater solubility for carbon than does ferritic steel. Therefore, carbon depletion in a carbon or low alloy steel is greater when an austenitic stainless steel filler metal is used in preference to a ferritic steel filler metal. On the other hand, carbon migration is not a problem when a nickel-chromium-iron filler metal, such as Type ENiCrFe-2 electrodes, is used.

If carbon migration is extensive, it will be indicated by a lightly etching, low-carbon band in the carbon steel heat-affected zone and a dark, high-carbon zone in the stainless steel weld metal as seen in a transverse metallographic section. The extent of carbon migration during postweld heat treatment or elevated temperature service should be determined during welding procedure qualification. It is known to influence long-time stress-rupture strength during elevated temperature service.

During cyclic temperature service, the heat-affected zone will be subjected to varying shear stresses because of the differences in coefficients of thermal expansion of the base and weld metals. These stresses may produce fatigue failure in the decarburized band next to the weld interface.

## Austenitic Stainless Steel Filler Metal

A number of austenitic stainless steel filler metals are available commercially.[10] The phases in the microstructure of the deposited metal from several of these filler metals are indicated on the Schaeffler diagram shown in Fig. 12.2. In some cases, the microstructure is fully austenitic. In other cases, the microstructure will contain some delta ferrite, the amount depending on the alloy composition.

Type ER16-8-2 filler metal is recommended for welding stainless steel piping systems for high-temperature service, but its corrosion resistance is not as good as Type 316 stainless steel. If the weldment will be exposed to strong corrodants, the exposed weld passes should be made using a more corrosion-resistant filler metal.

## Nickel Alloy Filler Metal

A nickel alloy filler metal may be used to weld a stainless steel to carbon or low alloy steel.[11] Those filler metals generally recommended for this application are Type ENiCrFe-2 or -4 covered electrodes and Type ERNiCr-3 bare electrodes and welding rods, but other nickel alloy filler metals are also suitable for this application. For example, Fig. 12.3 shows two transverse bend specimens of Type 304 stainless steel welded to mild steel, one with Type ERNiMo-3 filler metal and the other with a 67Ni-16Cr-15Mo-2Co (Hastelloy S) filler metal. They indicate the good ductility of nickel alloy filler metals.

## Service Considerations

Figure 12.4 gives the mean linear thermal expansion coefficients as a function of temperature for several alloys commonly associated with transition joints for steam plant application.[12] The thermal expansion coefficient of 2-1/4Cr-1Mo steel is about 25 percent less than that of Types 304 and 316 austenitic stainless steel. In certain applications, transition joints between austenitic stainless steel and low alloy steel may experience numerous temperature changes during operation. For a given change in temperature, the stress imposed at the weld joint is proportional to the difference in their coefficients of thermal expansion. Stress analyses of welded joints between these two types of steel indicate that the stresses introduced by thermal change are nearly an order of magnitude greater than those produced by the operating pressures and any thermal gradients across the joint.

10. See AWS Specifications A5.4-81, A5.9-81, and A5.22-80 for information on stainless steel electrodes and welding rods.

11. Refer to AWS Specification A5.11-76 and A5.14-76 for information on nickel and nickel alloy filler metals.
12. King, J.F., Sullivan, M.D., and Slaughter, G.M., Development of an improved stainless steel to ferritic steel transition joint, *Welding Journal*, 56(11): 354s-58s; 1977 Nov.

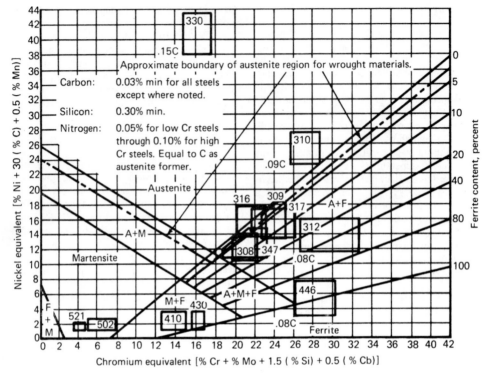

*Fig. 12.2—Positions of stainless steel weld metals on the Schaeffler diagram*

Experience with dissimilar metal transition joints has shown a significant number of failures in less than the expected service life. The majority of the transition joint failures in austenitic to ferritic steel joints occur in the ferritic steel heat-affected zone, adjacent to the weld interface, as shown in Fig. 12.5. These failures are generally attributed to one or more of the following causes:

(1) High stresses and the resulting creep at the interface as a result of the differences between the coefficients of thermal expansion of the weld and base metals.

(2) Carbon migration from the ferritic steel into the stainless steel, which weakens the heat-affected zone in the ferritic steel.

(3) Preferential oxidation at the interface, accelerated by the presence of stress.

## Austenitic Stainless Steel to Steel

Selection of the filler metal is an important step for reasonable service life when designing a fusion weld between a carbon or low alloy steel and an austenitic stainless steel. Proper choice depends upon the expected service conditions of the welded joint and the effect of dilution on the composition of the weld metal.

*Moderate Temperature Service.* For service temperatures below 800°F, an austenitic stainless steel filler metal is normally used. An Ni-Cr-Fe filler metal is also suitable, but its higher cost should be justified by the service requirements.

The selection of an austenitic stainless steel filler metal for joining dissimilar steels requires a prediction of the resulting weld metal composition and microstructure after dilution with the base metals. In multiple pass welds, the dilution may vary with the joint buildup sequence. More than one filler metal may be used to accommodate variations in dilution as succeeding weld passes are made.

The modified Schaeffler diagram in Fig. 12.6 indicates steel weld metal compositions,

**Fig. 12.3—Typical 2T bend specimens of Type 304 stainless steel welded to mild steel with 67Ni-16Cr-15Mo-2Co (Hastelloy S) (left) and Type ERNiMo-3 (right) filler metals**

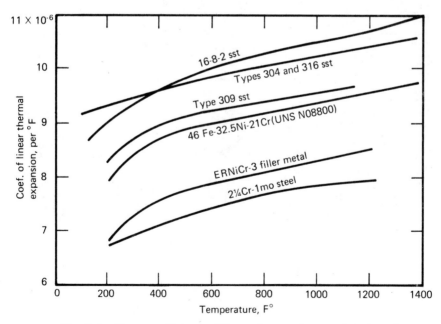

**Fig. 12.4—Mean coefficients of thermal expansion as a function of temperature for transition joint alloys**

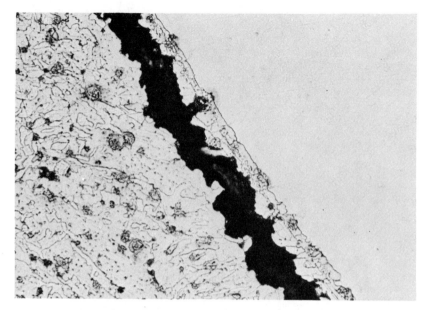

*Fig. 12.5—A typical crack along the weld interface between austenitic stainless steel weld metal (right) and a ferritic steel base metal (left) ×400*

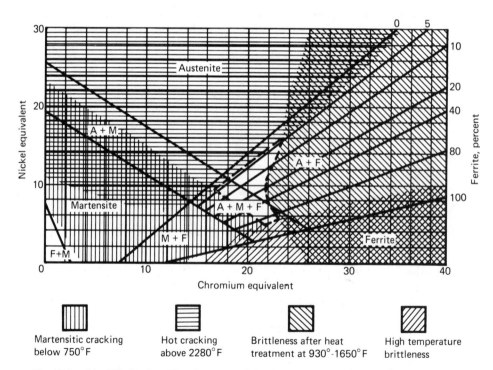

*Fig. 12.6—Modified Schaeffler diagram for stainless steel weld metal indicating those compositions prone to brittleness or cracking*

based on nickel and chromium equivalents, that may develop undesirable brittleness or cracking.[13]

The small, unshaded central region of the diagram indicates weld metal compositions that should be suitable for most service conditions. The microstructure of an austenitic stainless steel weld metal that falls in this region will consist primarily of an austenitic matrix with 3 to 8 percent delta ferrite. The delta ferrite assists in accommodating sulfur as a result of dilution and prevents hot cracking of the austenitic matrix. However, this phase transforms to a brittle sigma phase during high temperature service (930° to 1650°F), and lacks ductility at service temperatures below its ductile-to-brittle transition temperature. Therefore, excessive ferrite in austenitic weld metal is undesirable. Selection of both a suitable filler metal and a welding procedure should provide a weld metal composition that falls within the unshaded region of Fig. 12.6.

The Schaeffler diagram may be used to estimate the weld metal microstructure when joining a stainless steel to a carbon or low alloy steel. Figure 12.7 illustrates the procedure with an example of a single-pass weld joining mild steel to Type 304 stainless steel with Type ER 309 stainless steel filler metal. First, a connecting line is drawn between the two points representing the base metal compositions, based on their chromium and nickel equivalents. Point X representing the relative dilutions contributed by each base metal is then located on this line. If the relative dilutions are equal, point X is at the midpoint of the line. A second line is drawn between point X and the point representing the ER309 filler metal composition. The composition of the weld pass lies somewhere on this line, the exact location depending upon the total dilution. With 30 percent dilution, the composition would be at point Y. This composition is located in the unshaded area of Fig. 12.6, and would be considered acceptable. If a succeeding pass joins the first pass to mild steel, the dilution with the mild steel should be kept to a minimum to avoid martensite formation in the weld metal.

If dilution of austenitic stainless steel filler metal is a problem, it may be controlled by first buttering the joint face of the carbon or low alloy steel with one or two layers of Type 309 or 310 stainless steel filler metal, as shown in Fig. 12.8 (A) and (B). After machining and inspecting the buttered layer [Fig. 12.8 (C)], the joint between the stainless steel component and the buttered steel part can be made using conventional welding procedures and the appropriate filler metal for welding the stainless steel base metal [Figs. 12.8 (D) and (E)]. A low-alloy steel component can be heat treated after the buttering operation, and then joined to the stainless steel part. This avoids a postweld heat treatment that might sensitize the austenitic stainless steel to intergranular corrosion.

*High Temperature Service.* For applications above 700°F, it is general practice to use one of the nickel alloy filler metals. These filler metals offer a number of advantages for transition joints that are exposed to cyclic temperature service. During welding, they can tolerate dilution from a variety of base metals without becoming crack sensitive. Their solubility for carbon is low, which minimizes carbon migration from the low alloy steel to the weld metal.

Nickel alloy filler metals have coefficients of thermal expansion closer to those of low alloy steel (see Fig. 12.4). During thermal cycling, the stresses at the weld interface between one of these filler metals and the low alloy steel base metal are much lower than when an austenitic stainless steel filler metal is used. Internal stresses are introduced at the interface between the stainless steel and the nickel alloy filler metal because of the difference in their coefficients of thermal expansion. However, the adequate oxidation resistance and the high creep-rupture strengths of the metals at this interface can maintain suitable mechanical integrity in service.

*Design Considerations.* When carbon or low alloy steel to austenitic stainless steel transition joints are required, certain considerations should be given during the system design phase to provide long-time service. As mentioned previously, the stresses imposed at the weld interface in a dissimilar metal joint are high as a result of the differences in the coefficients of thermal expansion. Therefore, other system stresses at the joint should be kept low. External loads on the joint should be minimized by proper system design and placement of the joints.

An alternative joint design is to interpose a third base metal (safe end) between the austenitic

13. Bystram, M.C.T., Discussion of Paper 19, Payne, B.E., Nickelbase welding consumables for dissimilar metal welding applications, *Metal Construction and British Welding Journal*, 1 (12s): 150-51; 1969 Dec.

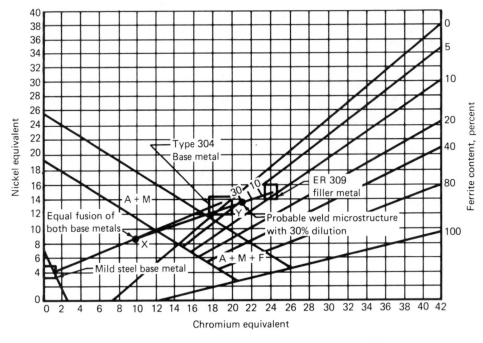

**Fig. 12.7—Estimating weld metal composition from the Schaeffler diagram**

stainless steel and the carbon or low alloy steel.[14] This third base metal should have a coefficient of thermal expansion intermediate to those of the other two base metals, and possess acceptable mechanical properties for the application. Iron-nickel-chromium alloys, such as 46Fe-32.5Ni-21Cr (UNS N08800) or 43Fe-36Ni-19Cr (UNS N08330) are likely candidates (see Fig. 12.4). An example of this design concept is shown in Fig. 12.9. Its main disadvantage is that two welded joints are required rather than one. When an Fe-Ni-Cr alloy is used as an intermediate piece, Type ERNiCr-3 filler metal is recommended for joining it to a Cr-Mo alloy steel. For the weld to the austenitic stainless steel piece, an austenitic stainless steel filler metal is recommended.

## Chromium Stainless Steel to Steel

The selection of a filler metal for joining a chromium stainless steel (4XX Series) to a car-

bon or low alloy steel can be selected using the following general rules:

(1) For welding one hardenable chromium steel to another with a higher chromium content, filler metal containing chromium equal to that of either steel may be used. Furthermore, any filler metal whose chromium content lies between these limits is equally satisfactory provided the weldment is properly heat treated.

(2) A general rule for welding any chromium steel to any low alloy steel is to use a filler metal that has the same composition as the low alloy steel, provided that it meets the service requirements of the application. With any low-alloy steel filler metal, the chromium that is picked up by dilution with the chromium steel base metal must be considered.

(3) For welding any chromium steel to a carbon steel, carbon steel filler metal can be used. A chromium steel filler metal can alternatively be used, but it is preferable to use a less hardenable filler metal.

(4) An austenitic stainless steel filler metal can be used for welding one chromium steel to another, or to any other steel, provided one of the steels is not a pressure vessel.

When determining the proper heat treatment

14. King, J.F., Sullwain, M.D., and Slaughter, G.M. Development of an improved stainless steel-to-ferritic steel transition joint, *Welding Journal,* 56(11): 354s-58s; 1977 Nov.

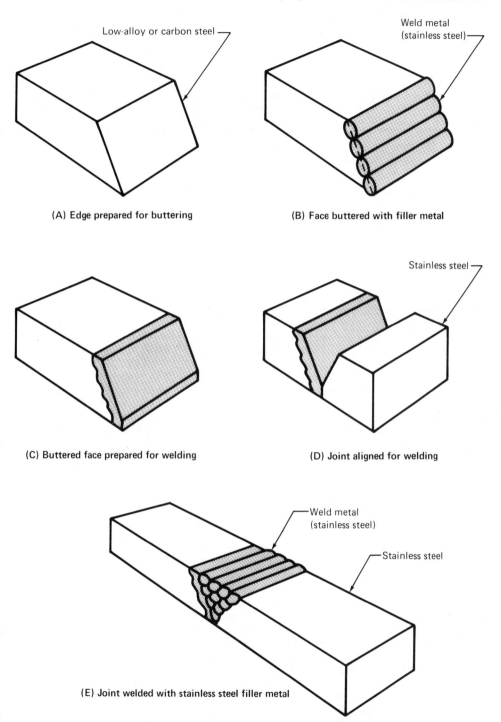

(A) Edge prepared for buttering

(B) Face buttered with filler metal

(C) Buttered face prepared for welding

(D) Joint aligned for welding

(E) Joint welded with stainless steel filler metal

*Fig. 12.8—Technique for welding stainless steel to carbon or low alloy steel*

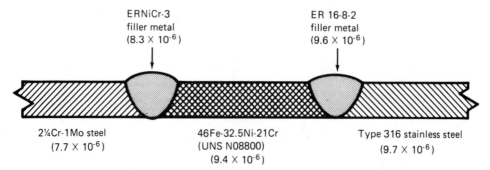

ERNiCr-3
filler metal
(8.3 × 10⁻⁶)

ER 16-8-2
filler metal
(9.6 × 10⁻⁶)

2¼Cr-1Mo steel
(7.7 × 10⁻⁶)

46Fe-32.5Ni-21Cr
(UNS N08800)
(9.4 × 10⁻⁶)

Type 316 stainless steel
(9.7 × 10⁻⁶)

Note: Mean coefficient of thermal expansion (in./in./°F) from 70° to 1000° F is noted below each material.

*Fig. 12.9—Example of the design concept for a transition piece between low alloy steel and austenitic stainless steel for elevated temperature service*

for a weld that joins a chromium steel to any other steel, it is inadvisable to apply general rules. The advice of a competent metallurgist should be sought because of the infinite number of possible metal combinations.

In a limited way, a general rule can be applied for the heat treatment of such welded joints where mechanical properties are the only concern. In such cases, a heat treatment that will soften the heat-affected zones of both base metals is usually suitable for the weld metal, provided its composition conforms to the recommendations previously given. The heat treatment that requires the highest temperature or longest soaking time, or both, is usually suitable for the other metal if lower mechanical properties are acceptable.

It is generally agreed that a chromium steel can be welded to any other alloy steel, provided the right procedures are used. Joints so produced should have satisfactory properties for operation at temperatures up to about 800°F.

When the properties of the heat-affected zone in the chromium stainless steel are important, both the stainless steel and other steel can be buttered up with Type 309 or 310 austenitic stainless steel weld metal. An appropriate preheat or postweld heat treatment can be used to obtain desired properties in the buttered components. The stainless steel surfaces can then be welded together without preheat using a suitable austenitic stainless steel filler metal, such as Type 308. (See Fig. 12.8.)

## NICKEL AND COBALT ALLOYS TO STEELS

### Nickel Alloys

Nickel alloys can easily be welded to steels using a suitable filler metal and proper control of dilution. Nickel base filler metals are generally used because of their good ductility and tolerance of dilution by iron.

Sulfur and phosphorus in nickel and nickel alloys cause hot cracking. The melting techniques used to produce nickel and its alloys are designed to keep the content of these elements to low levels. By contrast, the sulfur and phosphorus contents in some steels are typically higher. Consequently, dilution should be carefully controlled when joining a steel to a nickel alloy with a nickel alloy filler metal to avoid hot cracking in the weld metal.

Nickel alloys should be welded to steels using nickel alloy filler metals. These filler metals generally have good ductility and toughness, and some can tolerate considerable dilution by iron, depending upon the welding process.

*Iron Dilution.* Most nickel alloy weld metals can accept a substantial amount of iron dilution, but the dilution limit generally varies with the welding process, and sometimes with heat treatment. Figure 12.10 shows the limits of iron dilution in four types of weld metal deposited by the commonly used arc welding processes. These limits are based on practical experience rather than metallurgy.

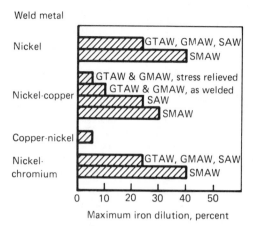

Weld metal

**Fig. 12.10—Limits for iron dilution of nickel and nickel-alloy weld metals**

Weld metal deposited with nickel or nickel-chromium covered electrodes can tolerate up to about 40 percent iron dilution. On the other hand, the dilution should be limited to about 25 percent when using nickel or nickel-chromium bare filler metal.

Acceptable limits of iron dilution for nickel-copper weld metal vary greatly, depending on the welding process. With shielded metal arc welding, iron dilution of up to about 30 percent normally can be tolerated. Submerged arc weld metal should not be diluted by more than 25 percent.

With the gas shielded welding processes, nickel-copper weld metal is less tolerant of iron dilution, especially if the weld is to be thermally stress-relieved. The maximum limits are not closely defined, but conservative guidelines for iron dilution in a welded joint are 10 percent when it will be used as-welded and 5 percent when it will be thermally stress-relieved. To avoid exceeding these limits, a buttering layer of nickel or nickel-copper weld metal should be applied to the steel face by shielded metal arc or submerged arc welding prior to welding the joint with a gas shielded process.

*Chromium Dilution.* As shown in Fig. 12.11, chromium dilution must be controlled with all nickel alloy weld metals. Dilution of nickel weld metal should be limited to 30 percent. Nickel-chromium filler metals are commonly used for joints involving dilution by chromium. The total chromium content of nickel-chromium weld

metal should not exceed about 30 percent. Most nickel-chromium alloys, including filler metals, contain less than 30 percent chromium, and dilution is not a problem.

Nickel-copper weld metal has a maximum dilution tolerance for chromium of 8 percent. Consequently, nickel-copper filler metal should not be used to join nickel-copper alloys to stainless steel.

*Silicon Dilution.* Dilution of nickel-chromium weld metal by silicon should also be considered, especially if one or both components are castings. Total silicon content in the weld metal should not exceed about 0.75 percent.

*Filler Metals.* Suggested nickel alloy filler metals for welding nickel alloys to steel or stainless steel are given in Table 12.3. Where two or three filler metals are given for a particular dissimilar metal combination, the choice should depend upon the specific type of steel or stainless steel to be joined to the nickel alloy. In some cases, the amount of dilution with the base metal is the controlling factor, and this may vary with the welding process.

*Mechanical Properties.* Typical tensile strengths of welded joints between several nickel alloys and steels are given in Table 12.4.

## Cobalt Alloys

Most cobalt alloys contain 10 to 20 percent nickel, 20 to 30 percent chromium, and 2 to 15 percent total of tungsten or molybdenum, or both. Metallurgically, they behave similar to the high-temperature nickel-chromium alloys with respect to welding. When joining a cobalt alloy

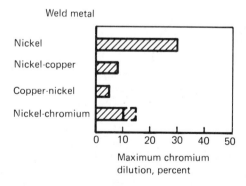

**Fig. 12.11—Limits for chromium dilution of nickel and nickel-alloy weld metals**

## Table 12.3
## Suggested filler metals for welding nickel alloys to steels

| Nickel alloy | | | Filler metal for welding to | |
|---|---|---|---|---|
| UNS No. | Common designation[a] | Filler metal form | Carbon or low alloy steel | Stainless steel |
| N02200 | Commercially pure nickel | Covered electrode<br>Bare wire | ENi-1,<br>ENiCrFe-2<br>ERNi-1,<br>ERNiCr-3 | ENi-1, ENiCrFe-2,<br>ENiCrFe-3<br>ERNi-1, ERNiCr-3,<br>ERNiCrFe-6 |
| N04400<br>N05500<br>N05502 | Monel 400<br>Monel K-500<br>Monel 502 | Covered electrode<br><br>Bare wire | ENiCu-7,<br>ENi-1<br><br>ERNi-1 | ENiCrFe-2,<br>ENiCrFe-3<br><br>ERNiCr-3,<br>ERNiCrFe-6 |
| N06600<br>N08800 | Inconel 600<br>Incoloy 800 | Covered electrode<br><br>Bare wire | ENiCrFe-2,<br>ENiCrFe-3<br><br>ERNiCr-3<br>ERNiCrFe-6 | ENiCrFe-2,<br>ENiCrFe-3<br><br>ERNiCr-3<br>ERNiCrFe-6 |
| N08825 | Incoloy 825 | Covered electrode<br><br>Bare wire | ENiCrMo-3<br><br>ERNiCrMo-3 | ENiCrMo-3<br><br>ERNiCrMo-3 |
| N10665 | Hastelloy B-2 | Covered electrode<br><br>Bare wire | ENiMo-7<br><br>ERNiMo-7 | ENiMo-7<br><br>ERNiMo-7 |
| N10276 | Hastelloy C-276 | Covered electrode<br><br>Bare wire | ENiCrMo-4<br><br>ERNiCrMo-4 | ENiCrMo-4<br><br>ERNiCrMo-4 |
| N06455 | Hastelloy C-4 | Covered electrode<br><br>Bare wire | ENiCrMo-4<br><br>ERNiCrMo-7 | ENiCrMo-4<br><br>ERNiCrMo-7 |
| N06007 | Hastelloy G | Covered electrode<br><br>Bare wire | ENiCrMo-9<br><br>ERNiCrMo-1 | ENiCrMo-9<br><br>ERNiCrMo-1 |

a. Some of these designations are tradenames. There may be other similar alloys having different designations.

**Table 12.4**
**Typical mechanical properties of welded joints between nickel alloys and steels**

| UNS No. | Nickel alloy Common designation[a] | Steel | Filler metal[b] | Tensile strength, ksi | Elongation, percent | Failure location |
|---------|-----------|-------|-------------|----------|-------------|----------|
| N04400 | Monel 400 | 410sst | ENiCrFe-2 | 81.8 | 34 | Monel |
| N04400 | Monel 400 | 304sst | ENiCrFe-2 | 83.4 | 45 | Monel |
| N06600 | Inconel 600 | 347sst | ENiCrFe-2 | 95.1 | 29 | Inconel |
| N06600 | Inconel 600 | 405sst | ERNiCrFe-6 | 90.0 | 35 | Stainless steel |
| N06625 | Inconel 625 | 304sst | ENiCrMo-3 | 91.2 | — | Stainless steel |
| N06625 | Inconel 625 | 410sst | ERNiCrMo-3 | 67.6 | — | Stainless steel |
| N08800 | Incoloy 800 | 347sst | ERNiCrFe-6 | 90.6 | 33 | Incoloy |
| N10001 | HASTELLOY B | Mild steel | ENiMo-1 | 60.0 | — | Mild steel |
| N10002 | HASTELLOY C | 316sst | ENiCrFe-2 | 90.5 | 33 | Stainless steel |
| N10002 | HASTELLOY C | Mild steel | ENiCrMo-5 | 61.0 | — | Mild steel |

a. Several of these may be registered tradenames. Alloys of similar composition may be known by other designations.
b. Refer to AWS Specifications A5.11-76 and A5.14-76 for information on nickel alloy filler metals.

to a stainless steel, a filler metal with a composition similar to that of the cobalt alloy is recommended. A nickel alloy filler metal may also be suitable for the application. In any case, the filler metal selection, welding process, and welding procedure for the application should be established by suitable tests. Figure 12.12 shows two bend-tested welds between a cobalt base alloy and an iron-base, high temperature alloy made with two different nickel alloy filler metals. In both cases, the weld metal and heat-affected zones exhibited good ductility.

## WELDING COPPER ALLOYS TO STEELS

Copper and iron mix completely in the liquid state but have limited mutual solubility in the solid state. Most copper-iron alloys produce two-phase solid solutions. The absence of brittle intermetallic compounds is advantageous from a weldability point of view. However, a two-phase alloy weld metal can lead to corrosion problems in certain applications. Iron dilution can be minimized by the use of appropriate welding procedures or placement of a buttering layer of nickel on the steel. Gas tungsten arc and shielded metal arc welding are preferred over gas metal arc welding because of better control of penetration and depth of fusion.

In the copper-iron alloy system, a large

number of compositions have wide freezing ranges. Therefore, hot cracking is likely in copper-iron alloy weld metal.

The surface activity of copper on iron is high. Molten copper will often attack iron along its grain boundaries, and produce hot cracking or fissuring in the heat-affected zone of steel. The phenomenon is known as *infiltration*.

## Copper

Copper will alloy with nickel in all proportions to produce a single-phase alloy. A copper component can be buttered with nickel filler metal to provide a high nickel joint face. Then, the nickel face can be welded to carbon or stainless steel using techniques and filler metals suitable for the particular combination of metals. The copper component should be preheated in the range of 400° to 1000°F, depending on its thickness, for heat balance across the joint during welding.

Copper can be welded directly to steel with silicon bronze (CuSi-A) or aluminum bronze (CuAl-AX) filler metal using one of the common arc welding processes. Preheating of the copper might be necessary. In this case, control of dilution with the steel is important.

### Copper-Nickel Alloys

Figure 12.13 shows the weld metal compo-

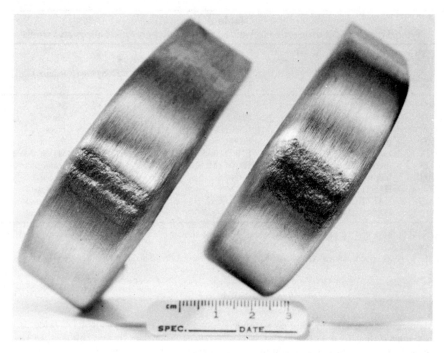

**Fig. 12.12—Typical 2T bend-test specimens of Haynes alloy 188 (41Co-22Cr-22Ni-15W) welded to Multimet alloy (34Fe-20Cr-20Ni-20Co-3Mo-3W) with Hastelloy S (left) and Hastelloy W (right) filler metals**

sitions in the copper-nickel-iron alloy system that are and are not susceptible to hot cracking.[15] The dilution of copper-nickel filler metal by iron or chromium, or the total of the two, should be limited to less than 5 percent. (See also Figs. 12.10 and 12.11.) This limitation is generally applicable to other copper filler metals.

Copper-nickel alloys are sometimes joined to carbon steel or stainless steel for marine applications. One method is to butter the steel face with nickel or nickel-copper filler metal to limit dilution of the weld metal by iron and, in the case of stainless steel, chromium. The joint can then be completed with copper-nickel or nickel-copper filler metal, the choice depending upon the service requirements. Preheat should not exceed 150°F.

Another method is to butter the copper-

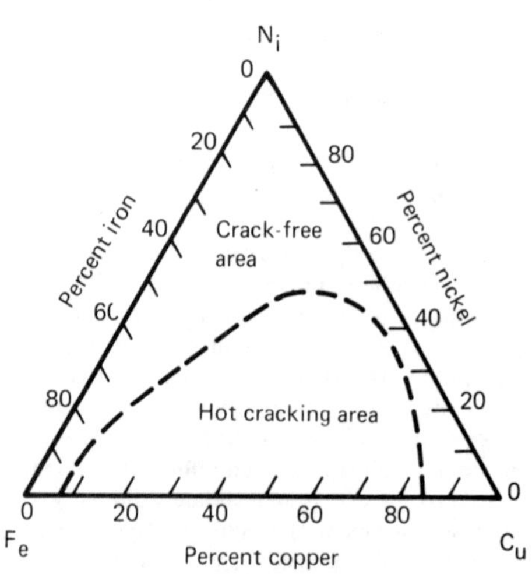

15. Thorneycroft, D.R., Constructional steels in the dissimilar metal joint, *British Welding Journal,* 12(3): 102-116; 1965 Mar.

**Fig. 12.13—Diagram showing the region of hot crack susceptibility of iron diluted nickel-copper weld deposits**

nickel alloy with nickel filler metal, and then weld the joint with the same filler metal.

## Aluminum Bronze

Aluminum bronze can be joined to carbon and stainless steel with aluminum bronze filler metal. Preheat and interpass temperature requirements depend upon the type of steel being welded. For carbon and low alloy steels, the preheat temperature should be in the range of 300° to 500°F, depending on the hardenability of the steel. For stainless steels, the temperature should not exceed 150°F.

The steel should be clean and free of oxide. With multiple pass welds, a joint buildup sequence using stringer beads helps to control dilution by the steel. Interpass cleaning with a stainless steel wire brush is recommended to remove oxides from the deposited metal. With gas tungsten arc welding, alternating current provides good cleaning action.

## Brass

Low-zinc brasses can best be welded to steel with the gas tungsten arc welding process. The zinc content of the brass should be 20 percent or less to minimize fuming and porosity in the weld. Welding procedures similar to those used for aluminum bronze are satisfactory. The steel should be buttered first with a copper-tin filler metal using direct current, electrode negative. The weld is then made using the same filler metal and alternating current to promote cleaning. Preheat is·not normally used with brasses. The arc should not impinge directly on the brass to avoid overheating and fuming. Instead, the arc should be directed onto the filler metal to limit the depth of fusion into the buttered layer on the steel and thus, dilution by iron.

## COPPER ALLOYS TO NICKEL ALLOYS

There are some applications for the welding of copper and copper-nickel alloys to nickel and nickel-alloys, but very few for brasses and bronzes. Copper and nickel are mutually soluble in each other. Therefore, welding together of these two metals and their alloys does not present serious problems. Copper-nickel, nickel-copper, and nickel filler metals are available. Nickel-copper weld metal has a minimum strength about 40 percent greater than that of copper-nickel or nickel weld metal (70, 50, and 55 ksi respectively).

Copper and copper-nickel alloys can be welded to nickel or nickel-copper alloys with either copper-nickel or nickel-copper filler metal, or without filler metal. The nickel may be buttered with nickel-copper filler metal, and the joint welded with copper-nickel. If a stronger joint is required, the copper can be buttered with copper-nickel filler metal and the joint welded with nickel-copper filler metal.

Nickel filler metal is recommended for joining copper or copper-nickel alloys to nickel alloys containing either chromium or iron, or both. The copper or copper-nickel joint face can be buttered with nickel filler metal prior to welding to control dilution by copper. As an alternative, the nickel buildup on the copper alloy can be welded to the nickel alloy component with a nickel alloy filler metal of similar composition.

## WELDING ALUMINUM ALLOYS TO STEEL

With respect to fusion welding, iron and aluminum are not compatible metals. Their melting temperatures differ greatly: 1220°F for aluminum vs 2800°F for iron. Both metals have almost no solubility for the other in the solid state, especially iron in aluminum, and several brittle intermetallic phases can form ($FeAl_2$, $Fe_2Al_5$, or $FeAl_3$). Consequently, fusion welds joining iron and aluminum would be brittle. In addition, high welding stresses would be expected because of the significant differences in their thermal expansion coefficients, thermal conductivities, and specific heats.

Aluminum can be joined to carbon or stainless steel by brazing or welding if the steel is first coated with a metal that is compatible with a suitable filler metal. Aluminum, silver, tin, and zinc coatings can be used, but aluminum is the most common one. The coating can be applied to clean steel by dipping the steel into a molten aluminum bath operating at 1275° to 1300°F, with or without fluxing. Steel can also be aluminum coated by electrodeposition or by vapor deposition. Small steel parts can be rub-coated with aluminum or aluminum-silicon alloy using an aluminum brazing flux. Alternatively, the steel can be coated with a silver brazing filler metal, and commercial zinc- or tin-coated steel might be satisfactory for some applications.

After cleaning, the coated steel part can be joined to the aluminum part using gas tungsten arc welding and an aluminum alloy filler metal.

The arc should be concentrated on the aluminum member while flowing the molten weld metal over the coating on the steel. The steel must not be melted.

The strength of such a joint is related to:
(1) The metal used to coat the steel
(2) The thickness of the coating
(3) The bond strength between the coating and the steel surfaces

Joint design also affects the strength of the weld because it determines the area of loading and the presence of stress concentrations. Under carefully controlled conditions, joint strengths in the range of 15 to 30 ksi can be obtained. Experimental evidence indicates that optimum strength is achieved when the weld is void of brittle intermetallic compounds and the adherence of the coating to the steel is optimum.

The addition of 3 to 5 percent of silicon, copper, or zinc to the aluminum filler metal helps to limit the thickness of the intermetallic layer between the fused zone and the zinc coating on galvanized steel. A copper flash on the steel prior to zinc coating improves adherence of the zinc and thus the strength of the joint.

A technique used with electron beam welding to minimize the amount of intermetallic compounds in the weld metal and also control their distribution is to vapor deposit a thin layer of aluminum on the steel. A heat treatment can be used to modify the composition and structure of the deposited layer. During electron beam welding of that layer to the aluminum, the amount of intermetallics is greatly reduced, and they are favorably distributed within the weld metal.

When a coating technique is used to avoid the mixing of aluminum with iron in a weld, the service temperature of the joint must be limited to avoid diffusion during operation. Otherwise, the weld would eventually be embrittled. A maximum service temperature of 500°F is considered safe for most applications.

Aluminum and steel can best be joined together by one of the solid-state welding processes: friction or explosion welding.[16] A transition piece can be fabricated by one of these processes. Then, the respective ends of that piece can be fusion welded to the aluminum and steel parts.

After tack welding the transition piece in place, the aluminum joint is welded first by gas metal arc welding. The steel side is then welded, taking care not to exceed an interpass temperature of 400°F. Gas metal arc, flux cored arc, or shielded metal arc welding can be used to make the steel joint.

When aluminum and other metals are coupled together, the presence of moisture or an electrolyte sets up galvanic action between the two metals and causes preferential attack. The joints should be painted, coated, wrapped, or protected by any convenient method to avoid this problem.

## ALUMINUM ALLOYS TO NONFERROUS ALLOYS

Aluminum readily mixes with most nonferrous metals in liquid state, but brittle intermetallic compounds commonly are formed in the solid state. These brittle compounds greatly limit the use of fusion welding to join aluminum to other metals. This is true in the alloy systems of greatest commercial interest, including aluminum-copper, aluminum-nickel, and aluminum-magnesium. Only limited success has been achieved with barrier layers of a third, more compatible metal, but they may be useful for certain applications.

To join aluminum to copper, a layer of silver or silver alloy may be applied to the copper surface before welding with an aluminum or aluminum-silicon filler metal.[17] The weld is then made without penetrating through the silver layer on the copper. Tin, aluminum, and zinc have also been used to coat the copper surface prior to welding.

Gas metal arc spot welding may be used to join aluminum to copper for electrical connections.[18] As shown in Fig. 12.14, one member is sandwiched between two layers of the other metal. A hole is drilled through two of the three members. An arc spot weld is made through this hole with penetration into the third member as the hole is filled. A copper or aluminum filler metal is selected to match the top and bottom members. Brittle Al-Cu compounds are formed

16. Refer to the *Welding Handbook*, Vol. 3, 7th Ed., 239-78.

17. Cook, L. A., and Stavish, M. F., Welding aluminum to copper using inert gas metal arc process, *Welding Journal*, 35(4): 348-55; 1956 Apr.

18. Stoehr, R. A., and Collins, F. R., Gas metal arc spot welding joins aluminum to other metals, *Welding Journal*, 42(40): 302-308; 1963 Apr.

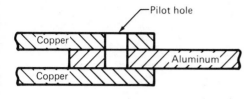

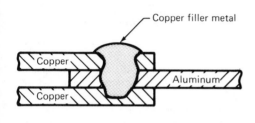

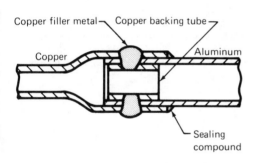

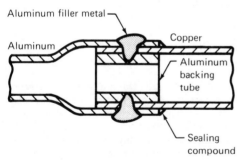

**Fig. 12.14—Gas metal arc spot welding of aluminum to copper**

at the periphery of the weld metal, but they remain there.

## TITANIUM ALLOYS TO OTHER METALS

Titanium has limited solid solubility for

aluminum, copper, iron, nickel, and chromium. When the solubility limit is exceeded, as in fusion welding, brittle intermetallic compounds are formed. Weld metal containing such compounds has inadequate ductility for structural applications.

Titanium forms a continuous series of solid solutions with columbium, molybdenum, tantalum, vanadium, and zirconium. The first four elements stabilize the beta phase. Although it is not of commercial significance, fusion welding of titanium alloys to columbium, tantalum, vanadium, or zirconium alloys is feasible. Because of the inherent brittleness of recrystallized molybdenum, special techniques are required to weld titanium to that metal.

Since vanadium is compatible with both titanium and iron, it has potential as an interlayer or filler metal for welding titanium to steel. Resistance spot welds between titanium and low carbon steel or Type 302 stainless steel sheets using a vanadium interlayer might have suitable shear strengths for some applications, provided fusion does not extend through the vanadium layer.[19]

Titanium is metallurgically compatible with columbium, and most alloys of these metals can be welded by fusion techniques to produce strong, ductile joints. For this reason, columbium can be used as an intermediate metal for joining titanium to other non-ferrous metals. For example, titanium has been joined to nickel alloys using a transition piece of columbium and copper alloy, similiar to the aluminum-steel type.[20] The titanium was welded to the columbium and the nickel alloy to the copper alloy. Similar joints made with the electron beam welding process also proved successful in sheet thicknesses up to 0.080 inch.

Copper and titanium are difficult to weld together. Two general approaches have been tried. Useful properties can be obtained when the two metals are joined with an intermediate layer of columbium by the gas tungsten arc welding process. Another approach is to take advantage of the solubility of beta titanium for copper. Gas tungsten arc welds between two beta-tita-

19. Mitchell, D. R., and Kessler, H. D., The welding of titanium to steel, *Welding Journal,* 40(12):546s-52s; 1961 Dec.
20. Gorin, I. G., Welding of titanium alloys to nickel base alloys, *Welding Production,* 11(12): 46-53; 1964 Dec.

nium alloys (Ti-30Cb and Ti-3Al-6.5Mo-11Cr) and copper showed good strength and ductility.

The fusion welding of titanium to aluminum poses several problems related to their greatly different melting temperatures and the brittle intermetallic compounds that are formed. Braze welding aluminum to titanium using aluminum filler metal can produce a minimal layer of brittle intermetallic compound at the aluminum-titanium interface. An example of this technique with a lap joint configuration is shown in Fig. 12.15.[21] A weld bead is produced in the titanium sheet with partial penetration using a gas tungsten arc torch. The underlying aluminum sheet is melted by heat conducted from the titanium. Molten aluminum will wet clean titanium and form a fillet if the underside of the joint is well protected with an inert gas.

## DISSIMILAR REFRACTORY METALS

The refractory metals of interest are columbium, tantalum, molybdenum, and tungsten. A high degree of mutual solubility exists among all pairs of the four metals.

Columbium and tantalum have ductile-to-brittle transition temperatures below room tem-

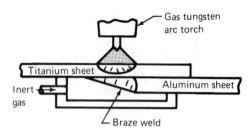

*Fig. 12.15—Indirect braze welding of aluminum to titanium*

perature. Therefore, these metals can be fusion welded together when adequately protected by inert gas shielding or vacuum to avoid contamination by oxygen and nitrogen.

Molybdenum and tungsten have relatively high transition temperatures, and fusion welds between these metals are essentially brittle at room temperature. These metals can be fusion welded to columbium and tantalum, but the welded joints will lack ductility at room temperature. They very likely would crack unless special precautions are taken to cool them to room temperature very slowly.

# WELDING CLAD STEEL

## CLAD STEEL

Plate or tube of carbon or low alloy steel can be clad with another metal or alloy to take advantage of its corrosion or abrasion resistance. Steel is less expensive than the cladding metal and, in some cases, it is stronger.

The cladding may be chromium stainless steel, austenitic stainless steel, copper or copper alloy, nickel or nickel alloy, silver, or titanium. The cladding may be applied to the steel by hot rolling, explosion welding, surfacing (welding), or brazing. Cladding thickness may vary from 5 to 50 percent of the total thickness, but it is

generally 10 to 20 percent for most applications. A minimum practical thickness is about 0.06 inch.

A joint between two clad sections becomes an integral part of the structure. It must not only support the applied load but also maintain uniformly the characteristics of the cladding. Most clad sections are readily welded together, but proper joint design and welding procedures must be used to ensure successful performance in service.

## JOINT DESIGNS

Suggested joint designs for arc welding clad steel from both sides are shown in Figs. 12.16 and 12.17. A square-groove weld can be used with thin sections [Figs. 12.17, (A) and (B)]. Single U-groove, double-V-groove, or combina-

21. Osokin, A. A., Technological characteristics of the fusion welding of aluminum alloys to titanium, *Welding Production*, 23(2): 14-15; 1976 Feb.

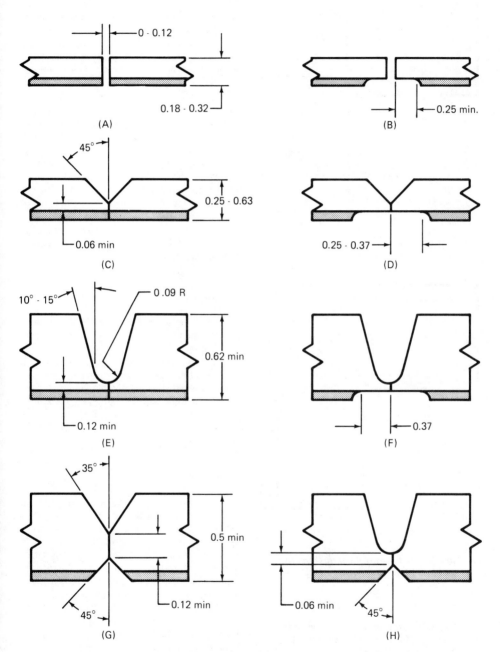

*Fig. 12.16—Butt joint designs for welding clad steel from both sides*

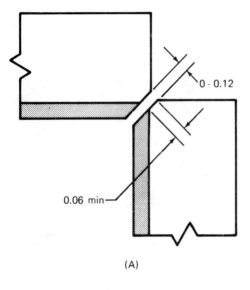

0 - 0.12

0.06 min

(A)

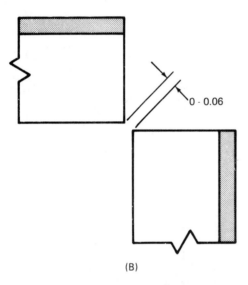

0 - 0.06

(B)

***Fig. 12.17—Corner joint designs for welding clad steel from both sides***

tion U- and V-groove welds can be used with thick sections [Figs. 12.17 (E) through (H)].

The cladding may be machined back from the joint edge for a short distance to ensure that the first pass with steel filler metal is not diluted with clad metal [Figs. 12.16 (B), (D), (F), (G), and (H)]. Excessive dilution by some clad metals,

such as copper or titanium, embrittles steel weld metal.

With a corner joint, the cladding can be on the inside or outside. In either case, the steel is welded first. When the cladding is on the inside, as shown in Fig. 12.17 (A), the steel weld is back-gouged from the inside corner prior to joining the cladding with a fillet weld. When the cladding is on the outside, Fig. 12.17 (B), the steel is joined on the inside corner with a fillet weld. The root of that weld is gouged smooth prior to placing the first weld pass in the outside corner. The first pass may be steel or a buttering filler metal, depending upon the section thickness.

When the melting temperature of the clad metal is higher than that of steel and the metals are metallurgically incompatible, the integrity of the cladding must be maintained with a backing strip of clad metal. The strip is fillet welded to the cladding after the steel is welded. This is done in the case of titanium cladding, as described later.

## FILLER METALS

Filler metals that are suitable with various cladding metals are given in Table 12.5. In most cases, an intermediate filler metal is used to butter the steel and thus control the amount of iron in the final passes of cladding metal. The steel should be welded with a suitable filler metal that provides a joint with the required mechanical properties for the intended service.[22]

## WELDING PROCEDURES

### Composite Welds

The base steel is normally welded first with a steel filler metal. The first pass of carbon steel weld metal must not penetrate into the clad metal. Dilution by most cladding metals can embrittle or crack a steel weld metal, and this is unacceptable for most applications. The appropriate joint design must be established for the particular clad steel during welding procedure qualification.

Dilution of the steel weld metal is not a problem when the cladding is stripped back from the joint, but the operation adds to costs and increases clad filler metal requirements. If strip-

---

22. Refer to Chapter 1 for information on welding carbon and low alloy steels.

### Table 12.5
### Filler metals for welding the cladding layer of clad steel

| Cladding metal | Buttering passes | | Fill passes | |
|---|---|---|---|---|
| | Covered electrode | Bare electrode or rod | Covered electrode | Bare electrode or rod |
| **Austenitic Cr-Ni stainless steels** | | | | |
| 304 | E309, E309L | ER309, ER309L | E308, E308L | ER308, ER308L |
| 304L | E309L | ER309L | E308L | ER308L |
| 309 | E309, E309L | ER309, ER309L | E309, E309L | ER309, ER309L |
| 309S | E309L | ER309L | E309L | ER309L |
| 310, 310S | E310, E310Cb | ER310 | E310, E310Cb | ER310 |
| 316 | E309Mo | ER309 | E316, E316L, E318 | ER316, ER316L, ER318 |
| 316L | E309L, E309Mo | ER309L | E316L, E318 | ER316L, ER318 |
| 317 | E309Mo | ER309 | E317, E317L | ER317, ER317L |
| 317L | E309L, E309Mo | ER309L | E317L | ER317L |
| 321 | E309Cb | ER309L | E347 | ER321 |
| 347 | E309Cb | ER309L | E347 | ER347 |
| 348 | — | ER348, ER309L | — | ER348 |
| **Chromium stainless steel** | | | | |
| 405, 429, and 430 | ENiCrFe-2 or -3[a] E309[a] E310[a] E430[b] | ERNiCrFe-5 or -6[a,c] ER309[a] ER310[a] ER430[b] | ENiCrFe-2 or -3[a] E309[a] E310[a] E430[b] | ERNiCrFe-5 or -6[a,c] ER309[a] ER310[a] ER430[b] |
| 410 and 410S | ENiCrFe-2 or -3[a] E309[a] E310[a] E430[b] | ERNiCrFe-5 or -6[a,c] ER309[a] ER310[a] ER430[b] | ERNiCrFe-2 or -3[a] E309[a] E310[a] E410[b], E410NiMo[b], E430[b], | ERNiCrFe-5 or -6[a,c] ER309[a] ER310[a] ER410[b], ER410NiMo[b], ER430[b], |
| **Nickel alloys** | | | | |
| Nickel | ENi-1 | ERNi-1 | ENi-1 | ERNi-1 |
| Nickel-copper | ENiCu-7 | ERNiCu-7 | ENiCu-7 | ERNiCu-7 |
| Nickel-chromium-iron | ENiCrFe-1, 2, or 3 | ERNiCrFe-5 | ENiCrFe-1 or 3 | ERNiCrFe-5 |
| **Copper alloys** | | | | |
| Copper | ENiCu-7 ECuAl-A2 ENi-1 | ERNiCu-7 ERCuAl-A2 ERNi-1 | | ERCu |
| Copper-nickel | ENiCu-7 | ERNiCu-7 ERNi-1 | ECuNi | ERCuNi |
| Copper-aluminum | ECuAl-A2 | ERCuAl-A2 | ECuAl-A2 | ERCuAl-A2 |
| Copper-silicon | ECuSi | ERCuSi-A | ECuSi | ERCuSi-A |
| Copper-zinc | ECuAl-A2 | ERCuAl-A2 RBCuZn-C[d] | ECuAl-A2 | ERCuAl-A2 RBCuZn-C[d] |
| Copper-tin-zinc | ECuSn-A | ERCuSn-A | ECuSn-A | ERCuSn-A |

a. Welding on material colder than 50° F is not recommended.
b. Preheat of 300° F min is recommended, particularly with plate over 0.5 in. thick.
c. ERNiCrFe-6 weld metal is age-hardenable. Consult the supplier.
d. Deposited by oxyacetylene welding.

ping is not done, the root face and root opening must be designed to limit penetration by the steel weld metal. In any case, the root of the steel weld must be gouged or machined to sound metal.

The effects of dilution must be considered when welding the clad side of the joint. Some cladding metals have low tolerances for dilution by steel. Therefore, one or more buttering layers of a selected filler metal should be applied before depositing any of the cladding weld metal over the buttering layer. The buttering filler metal must be tolerant of some dilution by the base steel.

In practice, special equipment for weld cladding operations is not always available when making joints in clad steel. Without this equipment, special procedures must be used to minimize dilution. The following measures are recommended:

(1) Use small diameter electrodes and deposit stringer beads.

(2) Use more highly alloyed electrodes than the cladding to allow for dilution.

(3) Allow for several layers of weld metal, and remove part of the first layer if necessary.

(4) With automatic welding, oscillate the welding head, as in surfacing.

(5) Where possible, use direct current, electrode negative with the arc on the molten weld pool as it is advanced, and directed against the previously deposited bead.

When the cladding is not stripped back prior to welding, the steel weld should be back-gouged to sound metal and thus, produce a groove in the steel. This procedure permits the deposition of several layers of cladding weld metal to control dilution in the final layer.

One condition that should be avoided with submerged arc welding, or other deeply penetrating welding processes, is excessive penetration and melt-thru of the steel weld. Proper control of penetration is necessary to avoid dilution of steel weld metal by the cladding metal. An appropriate flux for submerged arc welding is one that permits welding with direct current, electrode negative, or alternating current for minimum dilution.

When welding the first pass on the steel side of a joint where the cladding forms part of the root face, the best practice is to use low-hydrogen welding procedures. Low-hydrogen steel weld metal is less likely to crack if some cladding metal is inadvertently melted and alloyed with it. Although not recommended, some dilution of carbon steel by a stainless steel cladding can be tolerated, but limited dilution by other metals, such as copper or its alloys, must be avoided. In some structures, partial penetration welds in clad plate are adequate. Such welds simplify control of dilution.

The root of the steel weld can be back-gouged with an air-carbon arc torch, by chipping, or by grinding. The gouged groove must be cleaned of any residue before buttering is done.

## Thin Clad Plate

When welding clad steel plates of 0.38 in. thickness or less, it might be more economical to weld the entire joint with a filler metal that is similar to the cladding. However, the welded joint must have the required mechanical properties and corrosion resistance. A square-groove or single-V-groove weld can be used, depending upon joint thickness. Buttering of the steel faces should be considered.

## Austenitic Stainless Steel Cladding

Some austenitic stainless steels, such as Types 304 and 316, contain sufficient carbon to form stable chromium carbides when cooled slowly through the 1500° to 800°F range. When this takes place, the stainless steel is susceptible to intergranular corrosion in some environments. Low-carbon stainless steels, such as 304L or 316L, and stainless steels stabilized with Cb or Ti are normally immune to this behavior.

When a stainless steel that is prone to carbide precipitation is slowly cooled, it must be reheated to above 1800°F, to redissolve the carbides, and then rapidly cooled to restore corrosion resistance. Such treatment may be detrimental to the base steel.

When joining a steel that is clad with a corrosion-sensitive austenitic stainless steel, some steps may help to maintain corrosion resistance. The cladding should be stripped back from the joint edge. Preheat and interpass temperatures should be kept to a suitable low level while welding the steel base plate to avoid overheating the cladding. The clad layer should then be restored by welding.

When depositing austenitic stainless steel filler metal on the steel, welding heat input should be low and the joint allowed to cool between passes. With thin plate, some means of removing heat during cladding may be appropriate.

## Chromium Stainless Steel Cladding

Chromium stainless steel cladding may be welded with an austenitic stainless steel or an Ni-Cr-Fe filler metal to avoid the low ductility of chromium stainless steel weld metal (see Table 12.5). Where this is not suitable, a matching chromium stainless steel filler metal should be used together with a preheat of about 300°F. In this case, the welded joint should be stress-relieved at a temperature that is both compatible with the steel base plate and ensures good corrosion resistance and ductility in the cladding.

Alternatively, Type 430 clad steel may also be heat treated at 1600° to 1650°F followed by air cooling. This treatment transforms any grain boundary martensite, increases the ductility of the stainless steel cladding, and normalizes the carbon steel. For optimum corrosion resistance, the 1600° to 1650°F heat treatment should be followed by a carbide stabilization treatment at 1250°F. The clad plate should not be given the 1450° to 1500°F annealing heat treatment recommended for Type 430 stainless steel. This temperature range will partially transform the carbon steel and will alter its mechanical properties, particularly toughness.

## Copper and Copper Alloy Cladding

Steel plate is usually clad with either deoxidized or oxygen free copper. The best method of welding copper and copper-nickel alloy cladding is with one of the gas shielded arc welding processes. Suitable joint designs are those employed for welding other types of clad steel (see Figs. 12.16 and 12.17). For copper cladding thicknesses over 0.125 in., a preheat of about 300°F or higher is recommended when surfacing with consumable copper electrodes smaller than 0.062 inch in diameter. When preheat is used with relatively thin cladding, part of the surfacing layer should be removed prior to deposition of additional layers of filler metal to control low iron dilution and maintain corrosion resistance.

Where the cladding is less than 0.09 in. thick, copper weld metal may be deposited directly on the steel with care. Semiautomatic gas metal arc welding with the backhand technique can provide first-layer iron contents of less than 5 percent, when the arc is directed onto the molten weld pool rather than on the steel.

It might be advantageous to cover the steel with a buttering layer of nickel-copper or nickel filler metal prior to surfacing with a copper or copper alloy filler metal. This procedure is particularly recommended with copper-nickel cladding. Nickel-copper and nickel filler metals are more tolerant of iron dilution than is deoxidized copper or copper-nickel filler metal. Furthermore, buttering avoids possible copper penetration of the steel grain boundaries, which can cause cracking.

## Nickel and Nickel Alloy Cladding

Nickel, nickel-copper, and nickel-chromium-iron filler metals can tolerate some dilution by iron. Therefore, they can be applied directly on steel using welding techniques that minimize dilution. Two or more layers of cladding may be required to reduce dilution by iron to an acceptable level.

## Silver Cladding

Steel may be clad with silver to take advantage of its corrosion resistance. For welded fabrications, the silver cladding should be at least 0.06 inch thick. One method of cladding is to vacuum furnace braze a silver sheet of desired thickness to the base plate. Lithium-deoxidized silver sheet and Type BAg-8A brazing filler metal[23] are recommended. Silver-clad plates also can be produced using roll bonding techniques.

When welding silver-clad steel plates together, contamination of the silver must be avoided. Iron and silver have very limited mutual solubility, and the melting temperature of iron is several hundred degrees above that of silver. These conditions help to maintain the integrity of the silver cladding.

The sequence recommended for welding other clad metals is generally followed with silver clad plates. Gas tungsten arc welding is recommended for depositing the silver filler metal. A low temperature preheat can be used to reduce the required welding heat. Welding should be performed in the flat position for best control of the molten silver.

## Titanium Cladding

Titanium and steel are not metallurgically

---

23. Alloy of 71.5 percent silver, 28 percent copper, and 0.5 percent lithium

compatible metals. Therefore, the steel and the titanium cladding must be welded independently. Joint designs shown in Fig. 12.16 (B) (D), and (F) are suitable. The titanium cladding must be stripped back to a point where the heat from welding the steel will not overheat it. Titanium absorbs oxygen when heated above 1200°F in air with a resulting decrease in ductility.

Figure 12.18 shows the arrangement for making a butt joint. The steel is welded first from that side of the joint. The root of the steel weld is back-gouged to sound metal, and a back weld made from the clad side. This back weld is then finished flush with the steel surface.

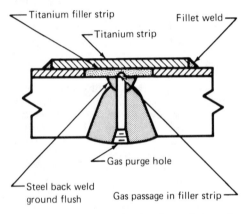

**Fig. 12.18—Design of a titanium-clad steel welded joint**

A titanium filler strip is placed in the shallow groove formed by the cladding and steel base. The strip is tack welded to the cladding at 4- to 6-inch intervals by gas tungsten arc welding. The tack welds must not penetrate to and melt the steel. Welding procedures suitable for titanium must be used.[24]

A titanium cover strip is placed over the filler strip, and positioned to overlap the cladding on both sides of the joint. Air in the cavities underneath the strip must be displaced with argon prior to welding the cover strip to the cladding. This can be done through holes in the steel weld, as shown in Fig. 12.18. The filler strip should be scored along its length before the cover strip is installed to provide for gas passage along the joint.

The titanium cover strip is fillet welded to the titanium cladding with the gas tungsten arc welding process and a suitable titanium filler metal. The welding torch gas nozzle must be large enough to adequately shield the titanium weld metal and the tip of the titanium welding rod from air contamination. A gas trailing shield might be needed on the welding torch. The titanium weld must not extend to the steel. Two or more small weld passes should be used to make the fillet weld.

It is good practice to isolate each weld joint in a structure from the voids underneath adjoining titanium cover strips. For example, the voids behind the cover strips on each circumferential joint and each longitudinal joint of a vessel should be isolated from each other. Each isolated void should be provided with a minimum of two 0.25-inch diameter gas purge holes through the steel weld. The holes should be located as closely as possible to the high and low points of longitudinal joints and to the ends of each cover strip segment of a circumferential joint. These holes are also used for leak testing the titanium fillet welds prior to and during service.

A suitable design for entrance nozzles for titanium clad vessels is shown in Fig. 12.19. Long weld-neck nozzles are usually preferred for pressure vessels to facilitate the installation of titanium liners. The inner end of the nozzle is prepared with a smooth transition radius, and is installed flush with the vessel cladding. The void behind each nozzle liner is provided with two 0.25-inch diameter gas purge holes extending through the base metal to the liner. The flange face and nozzle attachment areas are protected from corrosive fluids by titanium components. The titanium flange facing is braze welded to the steel flange face. A titanium sheet having the same diameter as the flange facing should be braze welded to the steel cover plate for the nozzle.

Titanium can be braze welded to steel with pure silver (BVAg-O) or silver-copper-lithium (BAg-19) brazing filler metal. The alloy filler metal may be easier to control because it melts over a range of 1400° to 1635°F, while silver melts instantly at 1760°F. Heat for braze welding is applied with a gas tungsten arc welding torch using argon shielding. The shielding gas nozzle and gas flow rate must be large enough to prevent oxidation of the titanium during the operation. The steel flange face should be precoated with the brazing filler metal to improve wetting during the braze welding operation. Alternatively, the

24. Refer to Chapter 10.

titanium component can be furnace brazed to the steel component in high purity argon or vacuum.

Designs for lap and T-joints in titanium clad steel are shown in Figs. 12.20 and 12.21 respectively. The principles of these designs are the same as those for a butt joint (Fig. 12.18).

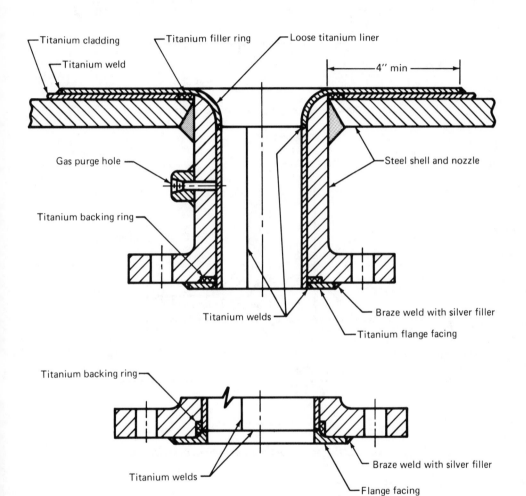

**Fig. 12.19—Typical nozzle design for titanium-clad steel vessels**

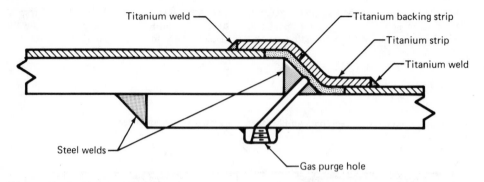

**Fig. 12.20—Design of a lap joint in titanium-clad steel**

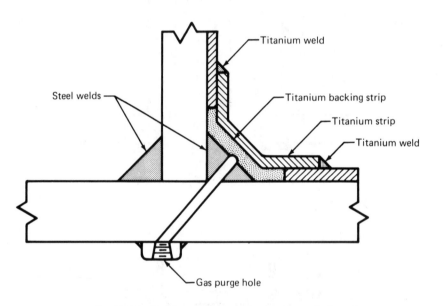

**Fig. 12.21—Design of a T-joint in titanium-clad steel**

# Metric Conversion Factors

°C = 0.56°F
$t_F$ = 1.8 $t_C$ + 32
$t_C$ = 0.56($t_F$ − 32)
in./in./°F = 1.8 mm/mm/°C
1 psi = 6.89 kPa
1 ksi = 6.89 MPa
1 in. = 25.4 mm
1 deg = 0.0175 rad

# SUPPLEMENTARY READING LIST

Castro, R., and deCadenet, J. J., *Welding Metallurgy of Stainless and Heat Resisting Steels,* Cambridge University Press, 1974.

Dalcher, A. W., Yang, T. M., and Chu, C. L., High temperature thermalelastic analysis of dissimilar metal transition joints, *Journal of Engineering Materials and Technology,* 99: 65-69; 1977 Jan.

Eckel, J. F., Diffusion across dissimilar metal joints, *Welding Journal,* 43(4): 170s-178s; 1964 Apr.

Funk, W. H., *Interpretive Report on Welding of Nickel Clad and Stainless Clad Steel Plate,* New York: Welding Research Council, Bulletin No. 61, 1960 June.

Lang, F. H. and Kenyon, N., *Welding of Maraging Steels,* New York: Welding Research Council, Bulletin No. 159, 1971 Feb.

Pattee, H. E., Evans, R. M., and Monroe, R. E., *The Joining of Dissimilar Metals,* Columbus, Ohio: Battelle Memorial Institute, DMIC Report S-16, 1968 Jan.

Payne, B. E., Nickel-base welding consumables for dissimilar metal welding applications, *Metal Construction and British Welding Journal,* 1(12s): 79-87; 1969 Dec.

Pease, G. R. and Bott, H. B., Welding high nickel alloys to other metals, *Welding Journal,* 29(1): 19-26; 1950 Jan. Discussion by N. C. Jessen, ibid, 29(5): 241s-42s; 1950 May.

Rutherford, J. J. B., Welding stainless steel to carbon or low alloy steel, *Welding Journal,* 38(1): 19s-26s; 1959 Jan.

# Welding Handbook
# Index of Major Subjects

|  | Seventh Edition Volume | Sixth Edition Section | Chapter |
|---|---|---|---|
| **A** | | | |
| Adhesive bonding of metals | 3 | | 11 |
| Air carbon arc cutting | 2 | | 13 |
| Aircraft | | 5 | 91 |
| Alternating current power sources | 2 | | 1 |
| Aluminum and aluminum alloys, welding of | 4 | | 8 |
| Applied liners | | 5 | 93 |
| Arc characteristics | 1 | | 2 |
| Arc cutting | 2 | | 13 |
| Arc stud welding | 2 | | 8 |
| Arc welding power sources | 2 | | 1 |
| Atmosphere, brazing | 2 | | 11 |
| Atomic hydrogen welding | 3 | | 13 |
| Austenitic manganese steel, welding of | 4 | | 4 |
| Austenitic (Cr-Ni) stainless steels, welding of | 4 | | 2 |
| Automotive products | | 5 | 90 |
| **B** | | | |
| Beryllium, welding of | 4 | | 11 |
| Boilers | | 5 | 84 |
| Bonding, adhesive | 3 | | 11 |
| Brass | 4 | | 7 |

| | Seventh Edition Volume | Sixth Edition Section | Chapter |
|---|---|---|---|

# INDEX